W9-AFZ-718

TEXTBOOK OF

Anatomy and Physiology

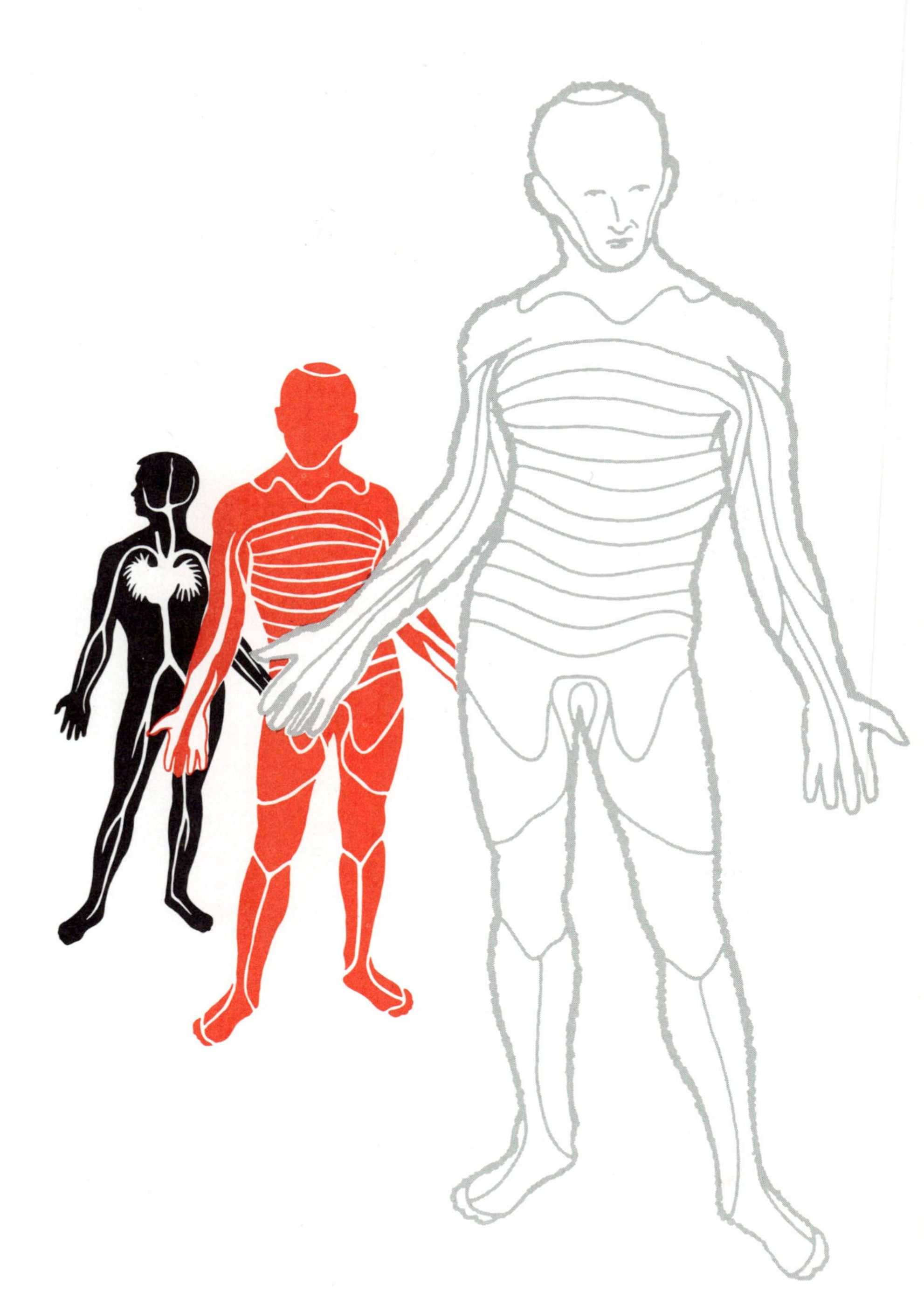

TEXTBOOK OF

Anatomy and Physiology

NINTH EDITION

CATHERINE PARKER ANTHONY, R.N., B.A., M.S.

Formerly Assistant Professor of Nursing, Science Department, and Assistant Instructor of Anatomy and Physiology, Frances Payne Bolton School of Nursing, Case Western Reserve University, Cleveland, Ohio; formerly Instructor of Anatomy and Physiology, Lutheran Hospital and St. Luke's Hospital, Cleveland, Ohio

with the collaboration of

NORMA JANE KOLTHOFF, R.N., B.S., Ph.D.

Professor of Nursing and Project Director for the School of Nursing Institutional Nursing Research Development Program, University of Wisconsin, Madison, Wisconsin

With 336 figures (145 in color), including 239 by Ernest W. Beck, and an insert on human anatomy containing 15 full-color, full-page color plates, with six in transparent Trans-Vision® (by Ernest W. Beck)

The C. V. Mosby Company

Saint Louis 1975

Ninth edition

Printed in the United States of America
Distributed in Great Britain by Henry Kimpton, London

Library of Congress Cataloging in Publication Data

Anthony, Catherine Parker, 1907-
Textbook of anatomy and physiology.

Includes bibliographies and index.
1. Human physiology. 2. Anatomy, Human.
I. Kolthoff, Norma Jane, 1921- joint author.
II. Title. [DNLM: 1. Anatomy. 2. Physiology.
QS4 A628t]
QP34.5.A59 1975 612 74-20991
ISBN 0-8016-0254-8

TS/K/B 9 8 7 6 5 4 3 2

PREFACE

This ninth edition of *Textbook of Anatomy and Physiology* has the same dominant aim as previous editions—to present accurate, up-to-date, significant information about the human body in ways that seem most likely to lead to understanding, learning, and enjoyment. In short, we hope you will both learn from the book and enjoy it. The latter aids (perhaps almost guarantees) the former.

This edition resembles previous ones in certain features of its organization and in its literary style. Outline surveys introduce each chapter. Outline summaries and review questions conclude each chapter. Summarizing diagrams and tables appear in nearly all chapters. Suggested supplementary readings, a list of commonly used abbreviations and prefixes, and a glossary are again included. The style of this book, as in its predecessors, is concise. We have tried to be more generous with ideas than with words and to transform complexity to clarity. Often we have chosen to pose questions rather than impose answers. This may, we hope, help students see science for what it is—a continual asking of questions and searching for answers, not merely a collection of facts and final answers.

You will find several major changes in the organization of this edition. For example, instead of one long chapter on the nervous system, this edition has three shorter chapters: cells of the nervous system, the somatic nervous system, and the autonomic nervous system. Metabolism is now discussed in a separate chapter that includes the biochemistry introduced in previous editions in the chapter on cells. The unit on reproduction in this ninth edition consists of separate chapters on cell reproduction and on the male and the female reproductive systems. The unit on stress now devotes one chapter to Selye's concepts and another chapter to current concepts about stress.

Substantial changes in subject matter appear in this edition. Look, for example, at the new material on "brain waves," altered states of consciousness, and the "emotional brain." You might find particularly relevant for our world of today and tomorrow the new information about biofeedback training and about physiological changes that occur during meditation (yoga). Other changes include a fuller discussion of liver functions, an extensive revision of the physiology of circulation, a brief discussion of the countercurrent hypothesis of urine concentration, and information about cyclic AMP, electrocardiograms, and central venous pressure.

Mr. Ernest W. Beck, one of the United States' most illustrious medical artists, has enriched this edition with a number of new drawings. A few of them are Figs. 5-45, 5-46, 5-47, 7-4, 8-4, and 12-8.

One more change you may find helpful is the many new references included in the supplementary reading list. They can add greatly to a knowledge and appreciation of physiological research and may even stimulate the student's desire for continued learning. Students may choose topics for independent study from them.

Literally a multitude of people have contributed to this ninth edition of *Textbook of Anatomy and Physiology.* Students and instructors who have used earlier editions of the book have made hundreds of thoughtful and thought-provoking comments orally and by letter. The value of their contributions cannot be measured nor can appreciation for them be adequately expressed. To the artist, Mr. Beck, and to my coauthor, Dr. Kolthoff, I owe a special indebtedness and hereby express heartfelt appreciation.

CATHERINE PARKER ANTHONY

CONTENTS

TEXTBOOK OF

Anatomy and Physiology

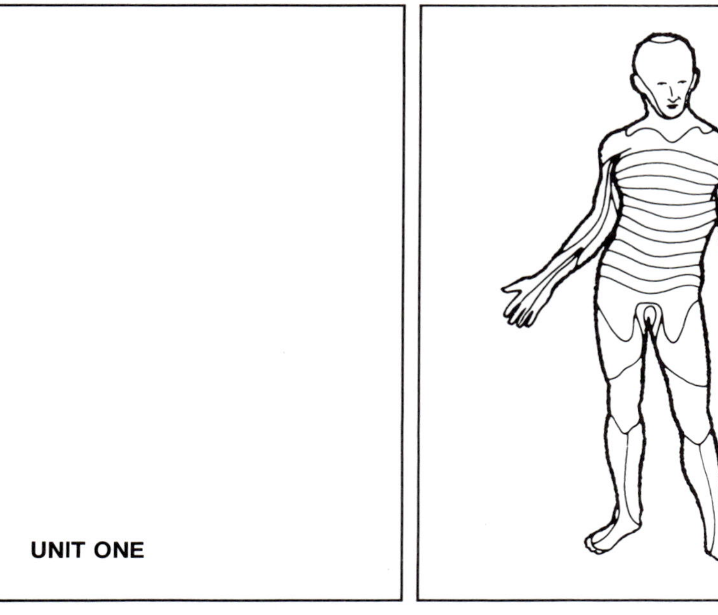

THE BODY AS A WHOLE

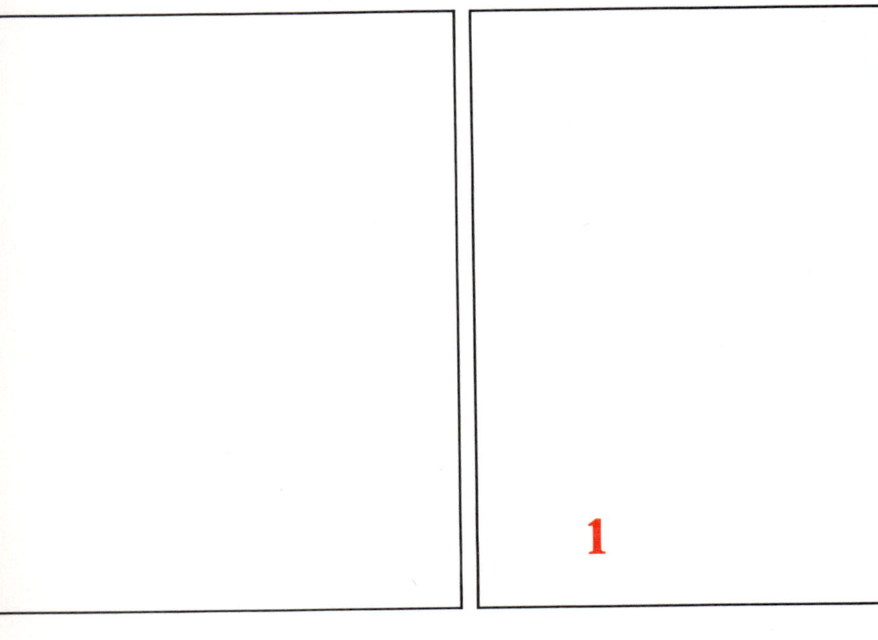

1

Organization of the body

What a piece of work is man! how noble in reason! how infinite in faculty! in form and moving how express and admirable! in action how like an angel! in apprehension how like a god! the beauty of the world! the paragon of animals!

Hamlet, in *Hamlet, Prince of Denmark*, Act II, Scene 2

Anatomy and physiology are biological sciences; that is, they are branches of knowledge about living things. *Human anatomy* is the science of the structure and *human physiology* is the science of the function of the most wondrous of all structures—the human body.

Before attempting to explore details about the body, it seems reasonable to examine some of its general characteristics, much as it is prudent to study the map of a strange city before embarking upon an excursion through it. This chapter, therefore, will present some generalizations about the body's structure, define some terms used in describing its structure, and conclude with some generalizations about body function.

Generalizations about body structure

1 *The body is a large structural unit made up of many millions of smaller parts.* The body's many parts consist of four kinds of smaller units: cells, tissues, organs, and systems. Smallest and most numerous of these are the cells. How many cells are in the body? Far too many for our imaginations to visualize. One average-sized adult body, according to one author,* consists of 100,000,000,000,000 cells. In case you cannot translate this number—one with fourteen zeros after it—it is 100 trillions! or 100,000 billions! or 100 million millions!

Cells have long been recognized as the simplest units of living matter that can maintain life and reproduce themselves. In truth, however, they are far from simple. How extremely complex cells are can be discovered by reading Chapter 2.

Tissues are somewhat more complex units than cells. By definition, a *tissue* is an organization of a great many similar cells with varying amounts and kinds of nonliving, intercellular substance between them.

Organs are more complex units than tissues. An *organ* is an organization of several different kinds of tissues so arranged that together they can perform a special function. For example, the stomach is an organization of muscle, connective, epithelial, and nervous tissues. Muscle and connective tissues form its wall, epithelial and connective tissues form its lining, and nervous tissue extends throughout both its wall and its lining.

Systems are the most complex of the component units of the body. A *system* is an organization of varying numbers and kinds of organs so arranged that together they can perform complex functions for the body. Nine major systems compose the human body: skeletal, muscular, nervous, endocrine, circulatory, respiratory, digestive, urinary, and reproductive. (For names and descriptions of the organs of a particular system, see the chapter on that system.)

2 *A ventral and a dorsal cavity, each with subdivisions, house the numerous internal organs of the body.*

The *ventral cavity* consists of the *thoracic* or *chest cavity* and the *abdominopelvic cavity*. The thoracic cavity consists of a right and a left pleural cavity and a midportion called the mediastinum. Fibrous tissue forms a wall around the mediastinum, completely separating it from the right pleural sac, in which the right lung lies, and from the left pleural sac, which contains the left lung. Thus the only organs in the thoracic cavity that are not located in the mediastinum are the lungs. Organs located in the mediastinum are the fol-

*Swanson, C. P.: The cell, ed. 3, Englewood Cliffs, N. J., 1969, Prentice-Hall, Inc.

lowing: the heart (enclosed in its pericardial sac), the trachea and right and left bronchi, the esophagus, the thymus, various blood vessels (for example, thoracic aorta, superior vena cava), the thoracic duct and other lymphatic vessels, various lymph nodes, and nerves (such as the phrenic and vagus). The abdominopelvic cavity consists of an upper portion, the *abdominal cavity*, and a lower portion, the *pelvic cavity*. The abdominal cavity contains the liver, gallbladder, stomach, pancreas, intestines, spleen, kidneys, and ureters. The bladder, certain reproductive organs (uterus, uterine tubes, and ovaries in the female; prostate gland, seminal vesicles, and part of the seminal ducts in the male), and part of the large intestine (namely, the sigmoid colon and rectum) lie in the pelvic cavity.

The *dorsal cavity* consists of the cranial cavity and the spinal cavity. The *cranial cavity* lies in the skull and houses the brain. The *spinal cavity* lies in the spinal column and houses the spinal cord.

3 *Organization is an outstanding and, in fact, a vital characteristic of body structure.* Every one of its component units—every system, every organ, every tissue, and even every cell—consists of smaller parts arranged in an orderly and precise manner. Organization is surely one of the most important characteristics of life since, as one scientist wrote, it "stands in direct opposition to the behavior of lifeless matter. The latter moves toward ever greater randomness. . . . A living organism, on the contrary, draws out from its chaotic environment particular substances and builds them into a system of ever greater and more organized complexity, thus steadily decreasing the randomness of matter. An organism is not an aggregate, but an integrate. . . . Life *is* organization."*

*From Sinnott, E. W.: Two roads to truth, New York, 1953, The Viking Press, Inc., p. 131.

4 *Body structures change gradually over the years.* They develop and grow during the years before maturity (young adulthood). After maturity, they age and, in general, atrophy. We shall mention many specific examples of aging in the chapters that follow.

Terms used in describing body structure

Directional terms

superior or *cranial*—toward the head end of the body; upper (example, the hand is part of the superior extremity).

inferior or *caudal*—away from the head; lower (example, the foot is part of the inferior extremity).

anterior or *ventral*—front (example, the kneecap is located on the anterior side of the leg).

posterior or *dorsal*—back (example, the shoulder blades are located on the posterior side of the body).

medial or *mesial*—toward the midline of the body (example, the great toe is located at the medial side of the foot).

lateral—away from the midline of the body (example, the little toe is located at the lateral side of the foot).

proximal—toward or nearest the trunk or the point of origin of a part (example, the elbow is located at the proximal end of the forearm).

distal—away from or farthest from the trunk or the point of origin of a part (example, the hand is located at the distal end of the forearm).

Planes of body

sagittal—a lengthwise plane running from front to back; divides the body or any of its parts into right and left sides (Fig. 1-1).

median—sagittal plane through midline; divides the body or any of its parts into right and left halves.

coronal or *frontal*—a lengthwise plane running from side to side; divides the body or

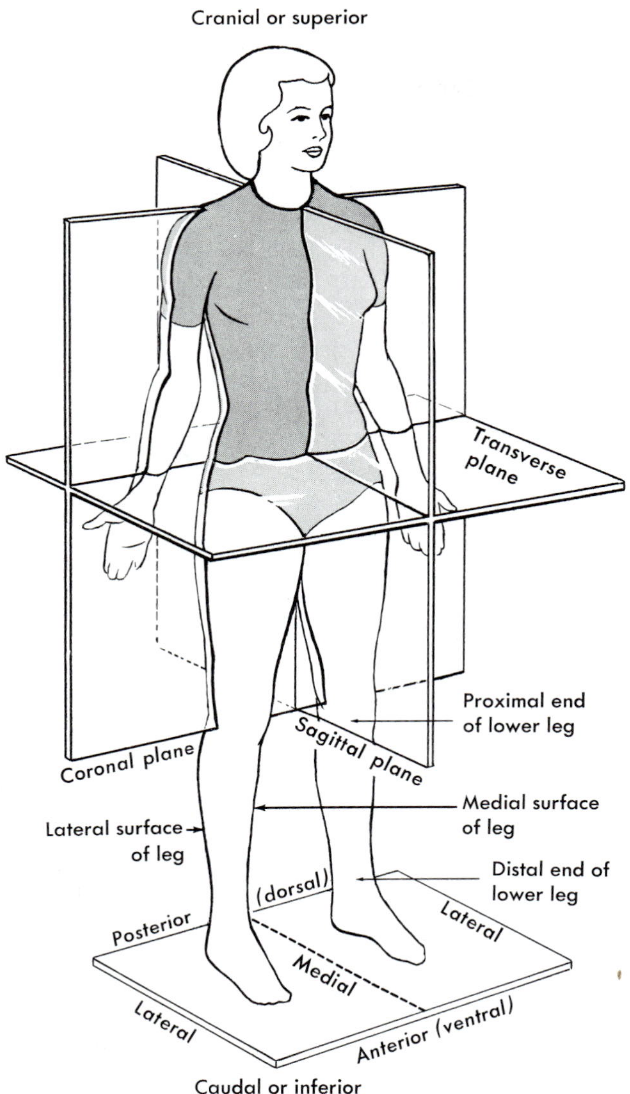

Fig. 1-1

Anterior view of the human figure demonstrating meanings of terms used in describing the body.

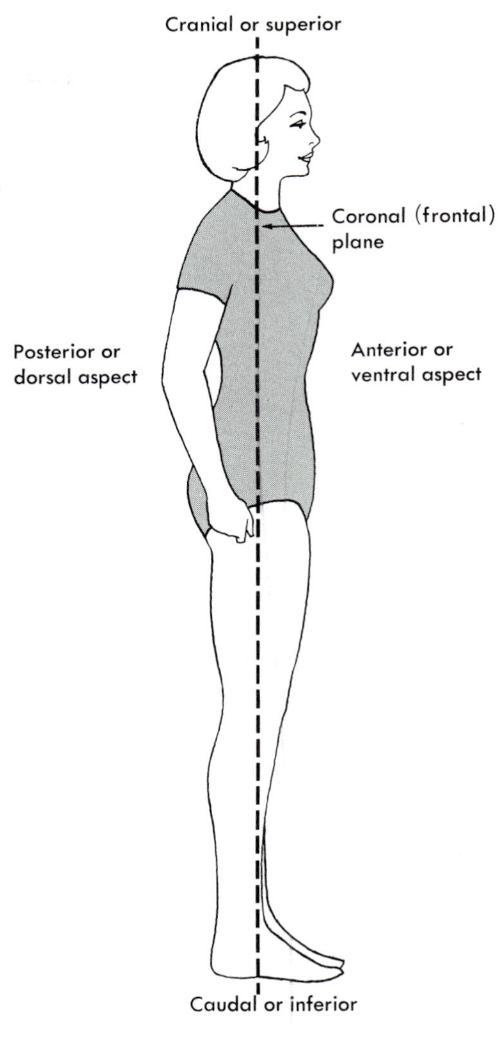

Fig. 1-2

Lateral view of the human figure demonstrating several terms used in describing the body.

Fig. 1-3

The nine regions of the abdomen. The top horizontal line crosses the abdomen at the level of the ninth rib cartilages. The lower horizontal line crosses it at the level of the iliac crests. The vertical lines pass through the midpoints of the right and left inguinal (Poupart's) ligaments.

Epigastric region
Right hypo-chondriac region
Left hypo-chondriac region
Umbilical region
Right lumbar region
Left lumbar region
Right iliac region
Hypogastric region
Left iliac region
Anterior superior iliac spine
E.W.B.

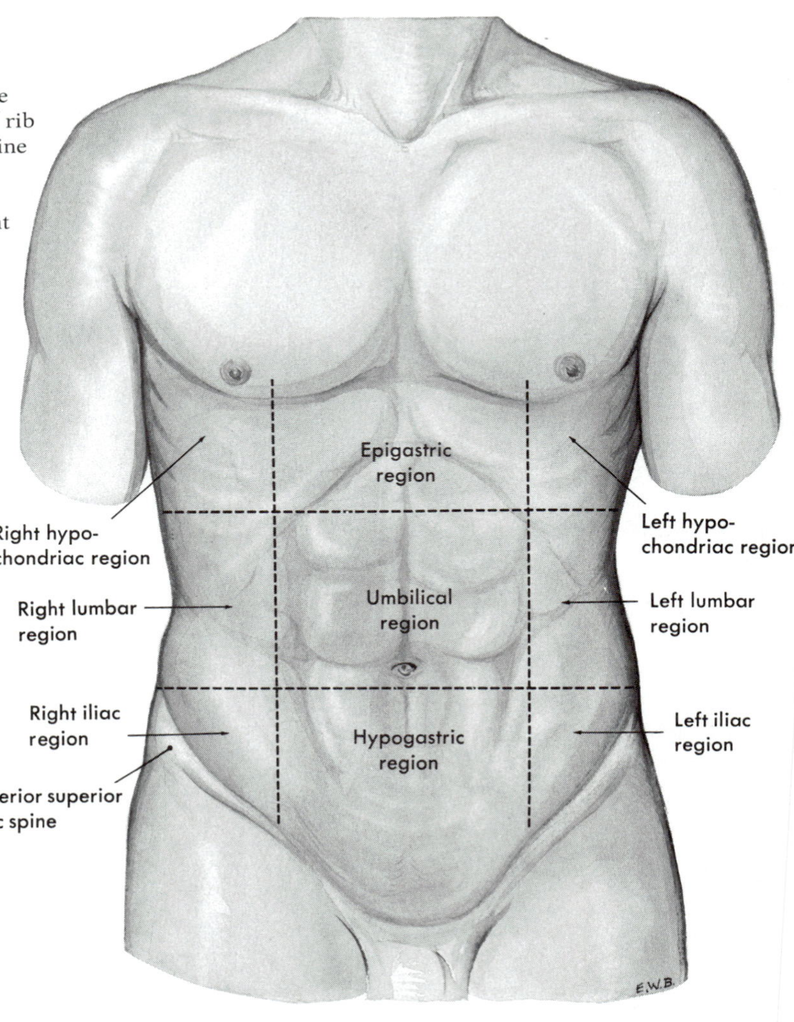

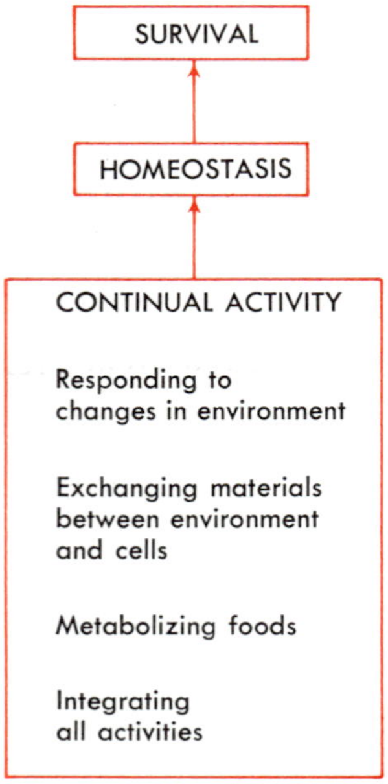

SURVIVAL
HOMEOSTASIS
CONTINUAL ACTIVITY
Responding to changes in environment
Exchanging materials between environment and cells
Metabolizing foods
Integrating all activities

Fig. 1-4

Scheme to illustrate some generalizations about body function. See text for explanation.

any of its parts into anterior and posterior portions (Figs. 1-1 and 1-2).

transverse or *horizontal*—a crosswise plane; divides the body or any of its parts into upper and lower parts (Fig. 1-1).

Abdominal regions

For convenience in locating abdominal organs, anatomists divide the abdomen into nine imaginary regions. See Fig. 1-3 for their names and boundaries.

Anatomical position

The term *anatomical position* means an erect position of the body, with arms at sides, palms turned forward (supinated).

Generalizations about body function (Fig. 1-4)

1 Survival is the body's most important business—survival of itself and survival of the human species.

2 Survival depends upon the body's maintaining or restoring homeostasis of its internal environment.

3 Homeostasis depends upon the body's ceaselessly carrying on many activities. Its major activities or functions are responding to changes in its environment, exchanging materials between its environment and its cells, metabolizing foods, and integrating all of its diverse activities.

4 The body's functions are ultimately its cells' functions.

5 The body's ability to perform many of its functions changes gradually over the years.

Homeostasis

Homeostasis is a key word in physiology. It comes from two Greek words—*homoios,* meaning the same, and *stasis,* meaning standing. "Standing or staying the same," then, is the literal meaning of the word. Walter B. Cannon, a noted physiologist, suggested homeostasis as the name for the steady states maintained by the body. He emphasized, however, that a steady state or homeostasis does not mean something set and immobile that stays exactly the same all the time. In his words, homeostasis "means a condition which may vary, but which is relatively constant."*

Human cells must have a relatively constant favorable environment if they are to survive in a healthy condition. This requirement applies to many aspects of the cellular environment—its chemical composition, its osmotic pressure, its hydrogen-ion concentration, its temperature, etc. Even a small shift away from homeostasis of any of these factors makes cells unable to function normally. A larger deviation may cause death of cells and of the body.

Cellular environment, you must remember, is not synonymous with body environment. The body lives in the atmosphere surrounding it, the gaseous atmosphere of the external world. Body cells live in the atmosphere surrounding them, but this is the liquid atmosphere of an internal world.

The liquid environment around cells is called *extracellular fluid* for the obvious reason that it lies exterior to cells. Extracellular fluid occupies two main locations. It fills in the microscopic spaces between body cells, where it is called appropriately *intercellular fluid* (or *interstitial fluid*), and it flows through blood vessels, where it is called *blood plasma.*

More than a century ago, a great French physiologist, Claude Bernard, recognized that the extracellular fluid (or *milieu interne,* as he called it) constituted the internal environment, or, in other words, the cellular environment. He also noted that its physical and chemical composition must be maintained relatively uniform if cells are to main-

*From Cannon, W. B.: The wisdom of the body, rev. ed., New York, 1939, W. W. Norton & Co., Inc., p. 24.

tain their life and health. Many years later a noted American physiologist, Walter B. Cannon, named the relative constancy of the internal environment *homeostasis*.

To maintain homeostasis, numerous complex processes called homeostatic mechanisms must operate. They involve the functioning of virtually all of the body's organs and systems. Hence homeostatic mechanisms constitute the major theme of physiology and of this book.

To illustrate what we mean by relative constancy or homeostasis and by homeostatic mechanisms, let us cite some examples. Homeostasis of blood glucose concentration means that the amount of glucose in blood normally varies (depending upon the method of measuring) between a low of 70 mg of glucose per 100 ml of blood to a high of 120 mg of glucose per 100 ml of blood. Glucose molecules are continually leaving the blood to enter body cells. This, of course, tends to decrease blood glucose concentration. Obviously, if other glucose molecules do not enter the blood to replace those leaving it, the blood glucose concentration would soon fall below the lower level of normal. In other words, homeostasis of blood glucose would not be maintained. Normally, however, this does not happen. Why? Because as soon as the amount of glucose in the blood decreases to the midpoint of its normal or homeostatic range, a homeostatic mechanism starts to operate. It acts to reverse the change in blood

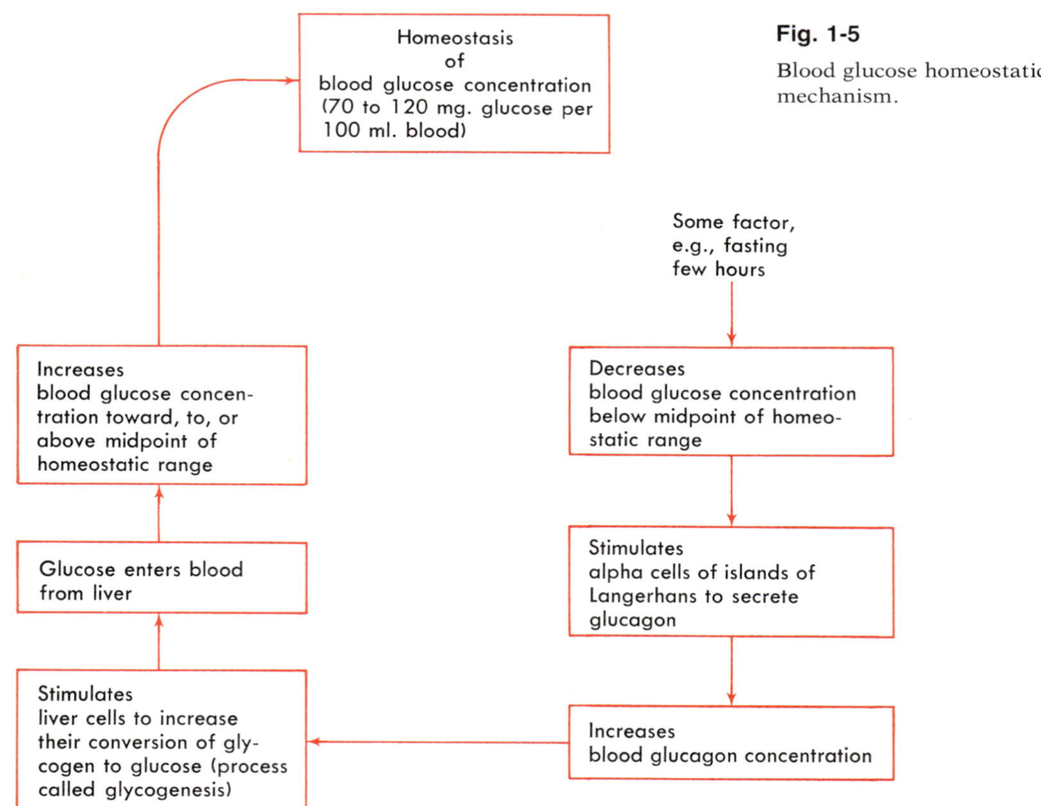

Fig. 1-5
Blood glucose homeostatic mechanism.

glucose concentration that has occurred. It causes new molecules of glucose to enter the blood to compensate for those that have left it. To discover how this particular homeostatic mechanism operates, study Fig. 1-4. It illustrates the following basic facts about homeostatic mechanisms.

1 A change in the internal environment acts as a stimulus that initiates a complex response, a homeostatic mechanism. In this example, note the change that serves as this stimulus—a decrease in blood glucose below the midpoint of its homeostatic range.

2 A homeostatic mechanism consists of responses that reverse the initial change in the internal environment that "turned on" the homeostatic mechanism. Hence, a homeostatic mechanism tends to maintain or to restore homeostasis. In Fig. 1-1, increased secretion of glucagon and the increased glycogenesis it causes are the responses that reverse the initial change and tend to maintain or restore homeostasis of blood glucose concentration.

3 The responses that make up a homeostatic mechanism are called appropriately *adaptive responses.* They enable the body to adapt to changes in its environment in ways that tend to maintain homeostasis and to promote healthy survival. Adaptive responses make it possible for the body to thrive as well as survive. Collectively, adaptive responses bring about adaptation. *Adaptation* consists of two kinds of changes: changes in the body induced by changes in the environment and changes in the environment produced by changes in the body. Adaptation, in other words, is a great complex of interactions between the body and its environment. Together these interactions work toward the well-being of the body. In short, adaptation is the successful coping of an organism with its environment. Successful adaptation means healthy survival. Unsuccessful adaptation means disease or death.

Body functions and age

Body functions, like body structures, change gradually over the years. In general, the body performs its functions least well at both ends of life—in infancy and in old age. During childhood, body functions gradually become more and more efficient and effective. During late maturity and old age the opposite is true. They gradually become less and less efficient and effective. During young adulthood, they normally operate with maximum efficiency and effectiveness. The changes in functions that occur during the early years are called *developmental processes.* Those that occur during the late years are called *aging processes.* Developmental processes improve the functional capacities of the body. Aging processes, in contrast, diminish many functions and have no effect on others. Functional age changes will be noted in almost every chapter of this book.

■ ■ ■

Chapter 2 will describe cell structure and explain some of the intricacies of cell function.

Outline summary

Generalizations about body structure

1 The body is a large structural unit made up of countless smaller parts of four kinds
- **a** Cells—smallest units that can maintain life and reproduce
- **b** Tissues—organizations of many similar cells with nonliving intercellular substance between them
- **c** Organs—organizations of several different kinds of tissues so arranged that they can perform a special function
- **d** Systems—organizations of different kinds of organs so arranged that they can perform complex functions

2 Ventral and dorsal cavities house the numerous internal organs (viscera) of the body
- **a** Ventral cavity
  - **(1)** Thoracic (or chest) cavity
    - **(a)** Pleural cavities—right pleural cavity contains right lung; left pleural cavity contains left lung
    - **(b)** Mediastinum—midportion of the thoracic cavity, completely separated from pleural cavities by fibrous tissue wall; contains heart (enclosed in its pericardial sac), trachea, bronchi, esophagus, thymus, and various blood vessels, lymphatic vessels, and nerves

(2) Abdominopelvic cavity
(a) Abdominal portion—contains liver, gallbladder, stomach, pancreas, intestines, spleen, kidneys, and ureters
(b) Pelvic portion—contains bladder, certain reproductive organs, and part of large intestine
b Dorsal cavity
(1) Cranial cavity—contains brain
(2) Spinal cavity—contains spinal cord
3 Organization is a vital structural characteristic of body and all its parts
4 Body structures change gradually over years; in general, body structures develop and grow during the years before young adulthood; after the period of young adulthood, they age and, typically, atrophy

Terms used in describing body structure

1 Directional terms
a superior—upper
b inferior—lower
c anterior—front
d posterior—back
e medial—toward midline of body
f lateral—away from midline of body
g proximal—toward or nearest trunk or point of origin of part
h distal—away from or farthest from trunk or point of origin of part
2 Planes of body
a sagittal—lengthwise, dividing body or any of its parts into right and left portions
b median—sagittal plane through midline
c frontal or coronal—lengthwise, dividing body or any of its parts into anterior and posterior portions
d transverse or horizontal—crosswise, dividing body or any of its parts into upper and lower sections
3 Abdominal regions—see Fig. 1-3
4 Anatomical position—erect, arms at sides, palms forward

Generalizations about body function

1 Survival is the body's most important business—survival of individual and of species.
2 Survival depends upon the body's maintaining or restoring homeostasis.
3 Homeostasis depends upon the body's continually carrying on many activities, notably, responding to changes in its environment, exchanging materials between its environment and its cells, metabolizing foods, and integrating all of its diverse activities.
4 Body functions are ultimately cell functions.
5 Body functions change gradually over the years. They are least efficient and effective during infancy and old age. Changes in functions that occur during the years before young adulthood improve the body's ability to maintain homeostasis and to survive in a healthy condition and are called developmental processes. After young adulthood changes in function lessen the body's ability to maintain homeostasis and healthy survival and are called aging processes.

Review questions

1 Name and briefly define the four kinds of structural units that compose the body.
2 Name the two cavities on the dorsal surface of the body. What organs does each cavity contain?
3 Identify the divisions of the thoracic cavity and of the abdominopelvic cavity. Name the organs in each.
4 Define briefly each of the following terms: anatomical position, anatomy, anterior, distal, frontal, hypochondriac region, hypogastric region, iliac region, inferior, lateral, median plane, organ, physiology, posterior, proximal, sagittal plane, tissue, transverse plane.
5 Make a generalization about the changes in body structures over the years.
6 What is the body's one all-inclusive function?
7 Define the term *homeostasis*. What, specifically, does homeostasis of blood glucose concentration mean?
8 What is a homeostatic mechanism? Explain three of its basic characteristics.
9 Make a generalization about changes in body functions over the years.

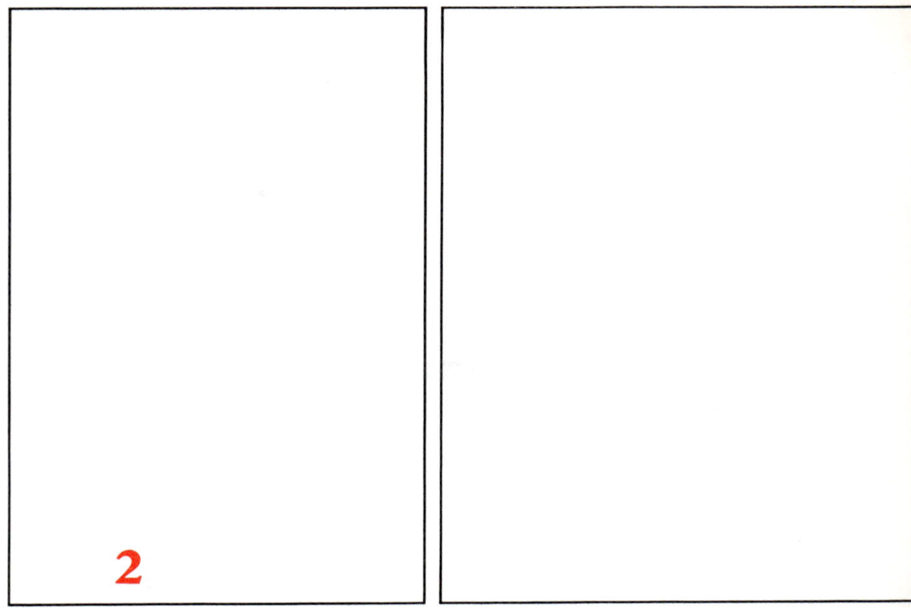

2

Cells

For the past 20 years or more, research into the mysteries of the body's smallest units—its cells—has moved at a rapid pace. The study of cells is a branch of biology called *cytology*. This chapter relates many facts and accepted ideas about this fascinating, expanding science. It starts with a discussion of cell structure, mentioning briefly the functions served by each cell part, and then considers in some detail the mechanisms by which substances move through cell membranes.

Cell structure

Protoplasm

The term *protoplasm* means living substance. Protoplasm consists of a complex of chemicals, including twenty-four elements* and hundreds of compounds. None of the elements is unique. All of them also occur in nonliving matter. The most abundant elements in protoplasm are hydrogen, oxygen, carbon, and nitrogen. Atoms of these four elements constitute about 99.4% of all of protoplasm's atoms.† Although living matter contains no unique elements, it does contain some unique compounds. These compounds —proteins (including enzymes), carbohydrates, lipids, and nucleic acids—occur naturally only in living matter or matter that once was living. Protoplasm's most abundant compound by far is water. It also contains appreciable amounts of salt compounds, mainly in the form of ions. Potassium and phosphate are the most plentiful ions in human protoplasm.

A prominent feature of protoplasm is its organization. It never exists as mere shapeless blobs of disorganized, mixed-up chemicals. On the contrary, its molecules are highly organized and occur in well-defined units called cells. Since the discovery of cells by Robert Hooke in 1665, ideas about their structure have undergone many changes. Hooke visualized them as "hollow rooms." Later biologists thought of them as closed sacs containing a semiliquid material with little objects floating in it. They named the semiliquid material *cytoplasm* (meaning cell protoplasm) and called the little objects *organelles* (meaning little organs). Today's biologists tend to think of a cell as a congested complex of membranes composed of molecules arranged in definite patterns. This concept has grown largely from what investigators have actually seen under the electron microscope. Its power to magnify objects several hundred thousand times has even made large molecules visible—for instance, those of DNA and proteins.* Other sophisticated tools and techniques have also contributed to our knowledge of cells.

One of the remarkable techniques that has uncovered many secrets of both cell structure and function is radioautography. It combines the use of radioactive atoms with a photographic technique.† And yet, however thoroughly we may manage to expose protoplasm's intricacy—millions of atoms in a single cell, organized into thousands of compounds, participating in numberless reactions—we shall probably never learn all of

*Before 1970 scientists believed that only twenty elements composed protoplasm, namely, hydrogen, oxygen, carbon, nitrogen, calcium, phosphorus, chlorine, potassium, sulfur, sodium, magnesium, iron, iodine, copper, manganese, zinc, cobalt, chromium, selenium, and molybdenum. Since 1970 various animal experiments have established that traces of four other elements—fluorine, silicone, tin, and vanadium—are also essential constituents of living matter. See Frieden, E.: The chemical elements of life, Sci. Am. **227**:52-60, July, 1972.

†Frieden, E.: The chemical elements of life, Sci. Am. **277**:54, July, 1972.

*Howland, John L.: Cell physiology, New York, 1973, The Macmillan Co., p. 6.

†See Baserga, R., and Kisieleski, W. E.: Autobiographies of cells, Sci. Am. **209**:103-110, Aug., 1963.

the cell's secrets. From studies on the microscopic structure of human cells, we now know that they consist of the following major parts:

Plasma membrane
Cytoplasm
- Endoplasmic reticulum
- Golgi apparatus
- Mitochondria
- Lysosomes
- Ribosomes
- Centrosome with centrioles

Nucleus

Examine Plate XII of the color insert (p. 54) to see what the various cell structures look like under a light microscope. Then note in Fig. 2-1 how different they appear under the much greater magnification of an electron microscope.

Both light and electron microscopes give us only flat views of cells. In Plate XII and in Fig. 2-1 the artist has given us his interpretation of the way a cell would look if we could see it in three dimensions with these microscopes. Now a newer kind of microscope, called the scanning electron microscope, has given us three-dimensional pictures of cells.*

Cell membranes

There are two kinds of cell membranes—external and internal—classified according to their location. The external membrane of a cell forms its outer boundary and is called the *plasma membrane.* A few tiny holes or pores penetrate it. It also folds inward in an intricate manner to form many of the cell's internal membranes. Fig. 2-1 shows several infoldings of the plasma membrane. Note, for example, the structure in this figure labelled "pinocytic vesicle." Internal membranes form the boundaries of many organelles.

In addition to their infoldings, some plasma membranes also have hundreds of finger-shaped projections called *microvilli.* Note in Fig. 2-1 how greatly the microvilli increase the surface area of the plasma membrane. Obviously, the larger its surface area, the more material can move through the membrane per minute. Hence microvilli tend to accelerate absorption. Not surprisingly, therefore, the plasma membranes of cells specializing in absorption have numerous microvilli.

Both facts and speculations characterize today's information about the molecular structure of plasma membranes. Examples of facts are that these membranes are composed of protein and phospholipid molecules and that they are incredibly thin. From its outer to inner surface, a typical plasma membrane measures only about 75 angstroms (Å),†or roughly 3/10,000,000 of an inch! Con-

*For example, see Everhart, T. E., and Hayes, T. L.: The scanning electron microscope, Sci. Am. **226**:65, 67, Jan., 1972. Almost surely you will find the three-dimensional pictures of human red blood cells and of the nerve cell of a sea hare shown on those pages dramatic and beautiful.

†Since 1 Å, as stated below, equals about 1,250,000,000 of an inch, 75 Å equals about 3/10,000,000 of an inch. The diameter of a hydrogen atom is 1 Å.

1 meter = 100 centimeters (cm)
1,000 millimeters (mm)
39.37 inches

1 micron = 0.001 mm
1,000 mμ
10,000 Å
about 1/25,000 inch

1 angstrom = 0.1 mμ
about 1/250,000,000 inch

1 millimeter = 0.001 meter
1,000 microns (μ)
1,000,000 millimicrons (mμ)
10,000,000 angstroms (Å)
about 1/25 inch

1 millimicron = 0.001 μ
10 Å
about 1/25,000,000 inch

1 inch = approximately
2.5 cm
25 mm
25,000 μ
25,000,000 mμ
250,000,000 Å

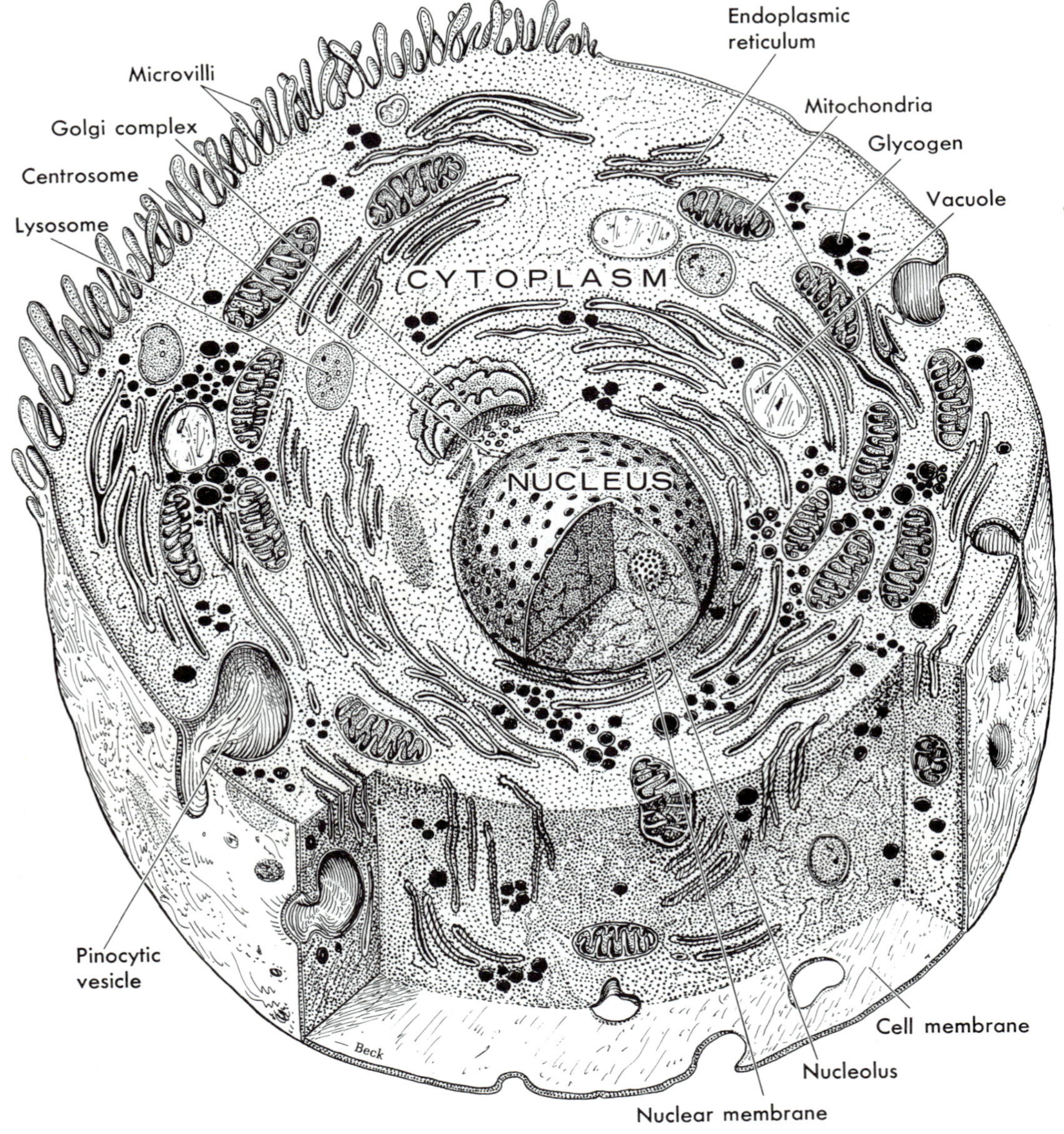

Fig. 2-1

A typical cell as seen under an electron microscope. Note the many mitochondria, popularly known as the "power plants of the cell." Note, too, the innumerable dots bordering the endoplasmic reticulum. These are ribosomes, the cell's "protein factories."

cepts about the arrangement of a plasma membrane's protein molecules in relation to its phospholipid molecules are still somewhat speculative. Less uncertain, however, is the relation of its phospholipid molecules to each other. X-ray diffraction, a sophisticated procedure, has recently indicated that they are arranged in the highly orderly manner shown in Fig. 2-2. Each phospholipid molecule has a "head" and a "tail." Its head consists mainly of phosphates and its tail consists of two fatty acids. Notice how the phospholipid molecules diagrammed in Fig. 2-2 are arranged in the plasma membrane. They lie parallel to each other but at right angles to the surface of the membrane. There are two rows of them, a bilayer. The tails of one row of phospholipid molecules face the tails of the other row. Consequently, the heads of the outer row of molecules face the outer surface of the plasma membrane and the heads of the inner row face the membrane's inner surface.

The arrangement of protein molecules in

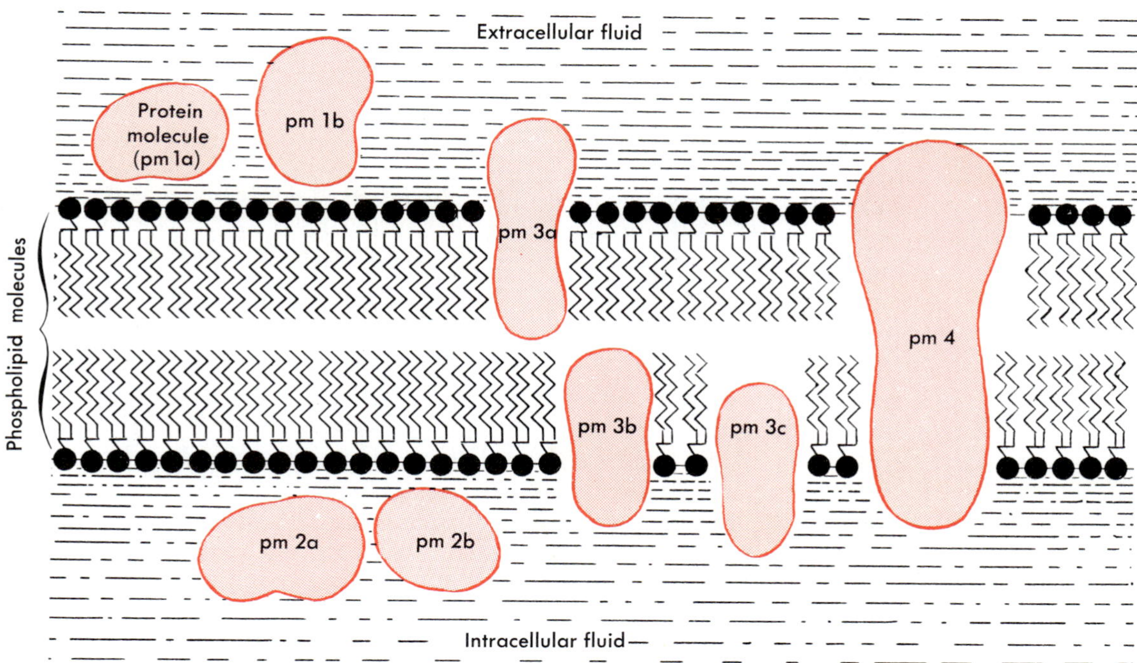

Fig. 2-2

Postulated structure of the plasma membrane. Note the following features of its molecular structure: the bilayer of phospholipid molecules arranged with their "tails" pointing toward each other; the location of protein molecules (pm's) in all possible positions with relation to the phospholipid bilayer—entirely outside (pm 1a and 1b), in the extracellular fluid; entirely inside (pm 2a and 2b), in the intracellular fluid; partially penetrating (pm 3a, 3b, and 3c) the bilayer; and extending all the way through the membrane (pm 4).

a plasma membrane, according to recent persuasive evidence, is less definite and orderly than that of the phospholipid molecules. It now seems that some protein molecules lie external and some internal to the bilayer of phospholipids, some penetrate the membrane's outer surface and others its inner surface, and still other protein molecules extend all the way through the membrane.* Find examples of these several locations of protein molecules in Fig. 2-2.

By virtue of its structure, the plasma membrane performs two general functions: it serves as a boundary or barrier that maintains the cell's integrity, and it helps determine what substances can penetrate this barrier to enter or leave the cell. Despite its incredible delicacy, the plasma membrane has sufficient firmness to maintain the arrangement of cell structures necessary for their carrying on the activities that maintain the cell's life. This sounds like a fantastic claim for a material only about 3/10,000,000 of an inch thick, yet it is true. If a cell's plasma membrane becomes torn, the cell loses its integrity, its organized wholeness. Its inner structure becomes disorganized as the cell's contents leak out. Result? The cell dies.

Summarizing, the plasma membrane serves both as a barrier and at the same time as a gateway. As a barrier, it separates the living substance of the cell from its fluid environment. As a barrier, it keeps some things inside the cell and in their proper places and other things outside the cell. But while it is acting as a barrier, the cell membrane is also serving as a gateway—not an ordinary gateway, however, through which any and all substances may pass, but a highly selective one. It entirely excludes some substances. It allows some substances to move through it easily, accelerates the movement of others, and actively pumps through itself still other materials.

*Fox, C. F.: The structure of cell membranes, Sci. Am. **226**:31-38, Feb., 1972.

Cytoplasm

Cytoplasm is the part of a cell between its membrane and its nucleus. In other words, it is all of a cell's protoplasm except its nucleus. Far from being the homogeneous substance once thought, it contains several different kinds of small structures. Not only that, in many cells, some of these number in the thousands. Collectively, they are known as *organelles*—that is, the "little organs" of cells. Each organelle consists of molecules arranged or organized in such a way that they can perform some function essential for maintaining the cell's life or for its reproduction. Membranes form the walls of most kinds of organelles. We shall discuss these membranous organelles—the endoplasmic reticulum (or ER), Golgi apparatus, mitochondria, and lysosomes—and then consider the nonmembranous ribosomes and the centrosome.

Endoplasmic reticulum

Endoplasm means the cytoplasm located toward the center of a cell. Reticulum means network. Therefore, the name endoplasmic reticulum (ER) means literally a network located deep inside the cytoplasm. And when first seen, it appeared to be just that. Later on, however, more highly magnified electron photomicrographs showed quite clearly that the endoplasmic reticulum has a much wider distribution than in the endoplasm alone. As you can observe in Fig. 2-1, it is scattered throughout the cytoplasm.

There are two types of endoplasmic reticulum: rough and smooth. Innumerable small granules—ribosomes, by name—dot the outer surface of the membranous wall of the rough type and give it its "rough" appearance. Ribosomes are themselves organelles. The rough endoplasmic reticulum seems to consist, as shown in Fig. 2-1, of flat, curving sacs arranged in parallel rows. Actually, it is a system of connected sacs and canals. The canals wind tortuously through the cytoplasm, extending all the way from the plasma membrane to the nucleus. The membrane forming the walls of the endoplasmic reticulum resembles the plasma membrane. The two membranes look alike under the microscope. Both measure only about 75 Å. Both consist of proteins and phospholipids, as do all cell membranes of higher organisms. Different membranes, however, contain different specific compounds of proteins and phospholipids.

The structural fact that the endoplasmic reticulum is an interconnected system of canals suggests that it might function as a miniature circulatory system for the cell. And in fact, proteins do move through the canals. The ribosomes attached to the rough endoplasmic reticulum synthesize proteins. These proteins enter the canals and move through them to the Golgi apparatus. Thus the rough endoplasmic reticulum functions in both protein synthesis and intracellular transportation.

No ribosomes border the membranous wall of the smooth endoplasmic reticulum—hence its smooth appearance and its name. Its functions are less well established and probably more varied than those of the rough type. In liver cells, for instance, it is thought that the smooth endoplasmic reticulum functions in lipid and cholesterol metabolism. On the other hand, in cells of the testes and adrenals, it apparently takes part in steroid hormone synthesis.

Golgi apparatus

The Golgi apparatus consists of tiny sacs stacked one upon the other and located near the nucleus. Note in Fig. 2-3 that the sacs look more and more distended in successive layers of the pile—as if some material were filling them up to the bulging point. And this actually is the case. Recently, several research teams have presented convincing evidence that the Golgi sacs synthesize large carbohydrate molecules and then combine them with proteins (brought to them through the canals of the endoplasmic reticulum) to form compounds called glycoproteins. As the amount of glycoproteins in the sacs increases, their shape changes from flat to "fat." They turn into perfect little spheres or globules. Then, one by one, they pinch away from the top of the stack. The Golgi apparatus, in other words, not only synthesizes carbohydrate and combines it with protein, but it even packages the product! Neat little globules of glycoprotein migrate outward away from the Golgi apparatus to and through the cell membrane. Once outside the cell, the globules break open, releasing their contents. The cell has secreted its product.

One might think that only in the secreting cells of glands would the Golgi apparatus function as just described. However, various other kinds of cells also make products for "export," products that move out of the cells that make them. In other words, various nonglandular cells secrete substances. Some examples—liver cells secrete blood proteins, plasma cells secrete antibodies, and connective tissue cells of certain types secrete substances used in bone and cartilage formation. The Golgi apparatus in all of these cells is believed to synthesize carbohydrate, to combine it with protein, and to package the product in globules for secretion. The Golgi apparatus no longer seems mysterious or insignificant. Two scientists who have contributed greatly to our knowledge of this organelle make clear their high regard for it in the following words: "All in all, it looks as if the Golgi apparatus is a creative mechanism in the cell ranking in importance with the ribosome. Just as the ribosomes are responsible for the construction of proteins, so the Golgi apparatus seems to be the main agency

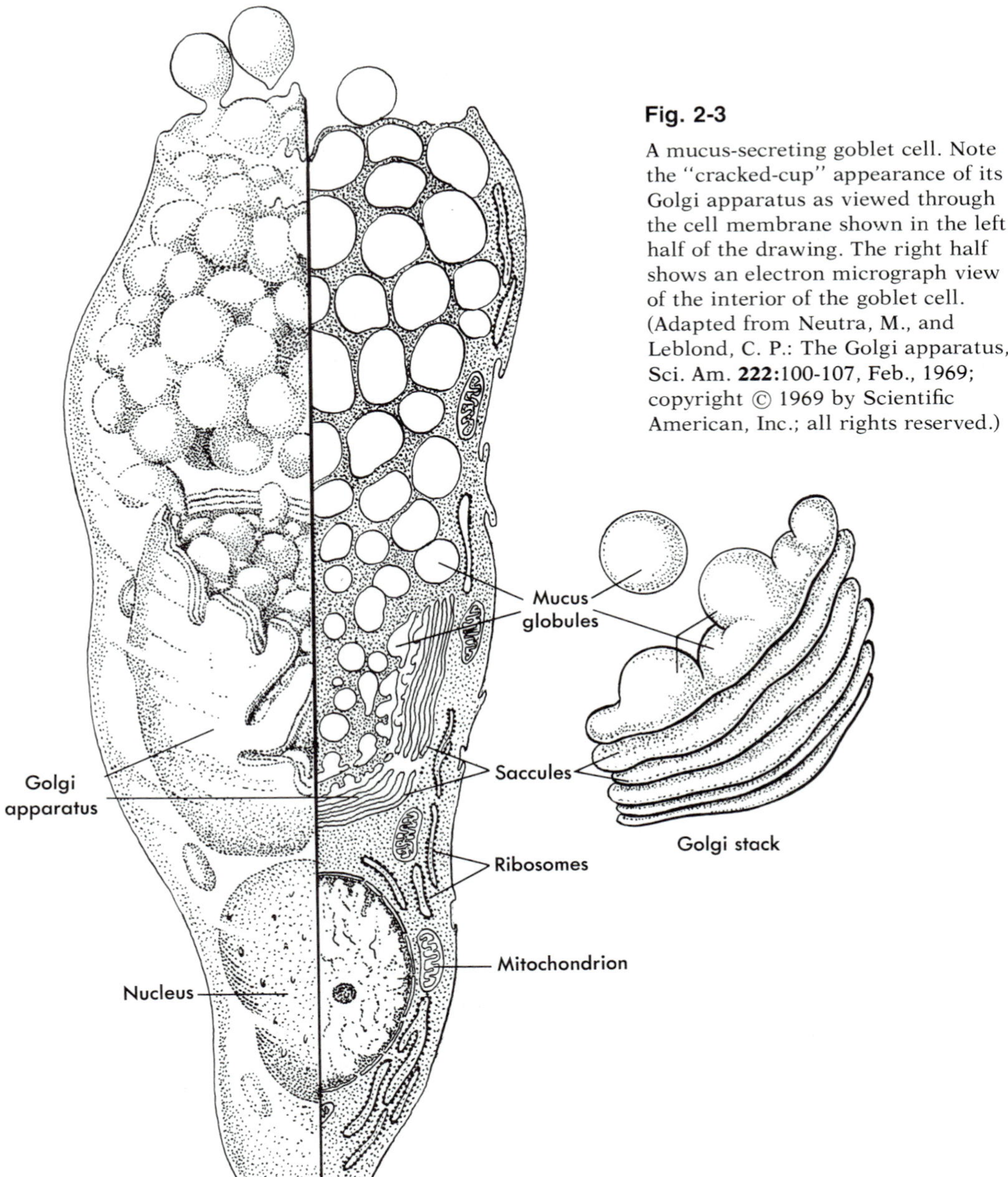

Fig. 2-3

A mucus-secreting goblet cell. Note the "cracked-cup" appearance of its Golgi apparatus as viewed through the cell membrane shown in the left half of the drawing. The right half shows an electron micrograph view of the interior of the goblet cell. (Adapted from Neutra, M., and Leblond, C. P.: The Golgi apparatus, Sci. Am. **222:**100-107, Feb., 1969; copyright © 1969 by Scientific American, Inc.; all rights reserved.)

for building a variety of large carbohydrates that serve many vital purposes."*

Mitochondria

Note the mitochondria shown in Fig. 2-1. Magnified thousands of times, as they are there, they look like small, partitioned sausages—if you can imagine sausages only 15,000 Å long and a third as wide. (In case you can visualize inches better than angstroms, 15,000 Å equals about 3/50,000 of an inch.) Yet, like all organelles, and tiny as they are, mitochondria have a highly organized structure. The membranous wall of a mitochondrion consists not of one but of two delicate membranes. They form a sac within a sac. The mitochondrion's inner membrane folds into a number of extensions called *cristae.* Note in Fig. 2-4 how they jut into the interior of the mitochondrion like so many little partitions. With the very powerful magnification of an electron microscope, one can see small round knobs attached to the cristae by short stalks and projecting inward from them. A single mitochondrion may contain thousands of these tiny knobs. Each knob, in turn, contains enzymes essential for the making of one of the most important chemicals in the world. Without this compound, life cannot exist. Its long name, adenosine triphosphate, its abbreviation, ATP, and its function will all become very familiar to you in later chapters of this book.

Both inner and outer membranes of the mitochondrion resemble the plasma membrane in their molecular structure. Many of the proteins in the membranes of the cristae are known to be enzymes. Also, all evidence so far indicates that they are arranged most precisely in the order of their functioning. This is another example, but surely an impressive one, of the principle that organization is a foundation stone and a vital characteristic of life.

*From Neutra, M., and Leblond, C. P.: The Golgi apparatus, Sci. Am. **220:**100-107, Feb., 1969; copyright © 1969 by Scientific American, Inc.; all rights reserved.

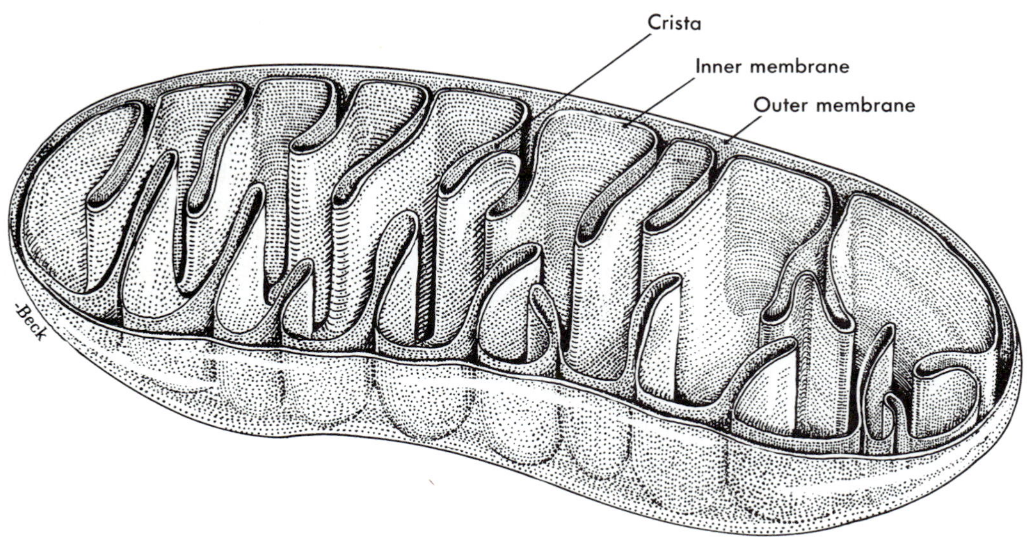

Fig. 2-4

Mitochondrion. Cutaway diagram shows the two-layered structure of its delicate membranous wall.

Enzymes in the mitochondrial inner membrane catalyze oxidation reactions. These are the chemical reactions that provide cells with most of the energy that does all of the many kinds of work that keep them and the body alive. Thus do mitochondria earn their now familiar title, the "power plants" of cells (discussed in Chapter 15).

From the knowledge that mitochondria generate most of the power for cellular work, one might deduce that the number of mitochondria in a cell would be directly related to its amount of activity. This principle does seem to hold true. In general, the more work a cell does, the more mitochondria its cytoplasm contains. Liver cells, for example, do more work and have more mitochondria than sperm cells. Swanson* estimates that a single liver cell may contain as many as 1,000 mitochondria, whereas a single sperm cell has only about 25 mitochondria.

Lysosomes

Lysosomes are membranous-walled organelles whose size and shape change with the stage of their activity. In their earliest, inactive stage, they look like mere granules. Later, as they become active, they take on the appearance of small vesicles or sacs and often contain tiny particles such as fragments of membranes or pigment granules (Fig. 2-1). The interior of the lysosome contains various kinds of enzymes capable of breaking down all the main components of cells. These enzymes can, and under some circumstances actually do, destroy cells by digesting them. The graphic nickname "suicide bags," therefore, seems appropriate for lysosomes. Little wonder that these powerful and dangerous substances are usually kept sealed up in lysosomes. However, lysosomal enzymes more often protect than destroy cells. Large molecules and larger particles (for example, bacteria) that find their way into cells enter lysosomes and their enzymes dispose of them by digesting them. Therefore, "digestive bags" and even "cellular garbage disposals" are other nicknames for lysosomes. White blood cells contain a great many lysosomes. They serve as scavenger cells for the body, engulfing bacteria and destroying them in their lysosomes.

*Swanson, C. P.: The cell, ed. 3, Englewood Cliffs, N. J., 1969, Prentice-Hall, Inc.

Ribosomes

Every cell contains thousands of ribosomes. The small spherical organelles you can see attached to the endoplasmic reticulum and scattered through the cytoplasm in Fig. 2-1 are ribosomes. Because ribosomes are too small to be seen with a light microscope, no one knew they existed until the electron microscope revealed them. Now scientists even know a good deal about their molecular structure. For instance, they know that they consist of approximately two thirds ribonucleic acid (RNA) and one third protein. They also know that ribosomes consist of two subunits of different sizes.*

The function of ribosomes is protein synthesis. Ribosomes are the molecular machines that make proteins. Or, to use a popular term, they are the cell's "protein factories." Ribosomes attached to the endoplasmic reticulum, as already mentioned, synthesize proteins for "export," whereas the ribosomes free in the cytoplasm make proteins for the cell's own domestic use. They make its structural proteins, in other words, and its enzymes. Working ribosomes, those that are actually making proteins, appear to function in groups, called polyribosomes. Under the electron microscope polyribosomes look like short strings of beads. In Chapter 15 we shall relate some current ideas about the complicated and still incompletely understood process of protein synthesis.

*Nomura, R.: Ribosomes, Sci. Am. **221**:28-35, Oct., 1969.

Centrosome

The name of this organelle suggests its location near the center of the cell (Fig. 2-1). Actually, this means that the centrosome is located near the nucleus since the nucleus takes up the center space of most cells. With the light microscope and suitably prepared slides, one can see two dots (called *centrioles*) in the centrosome. The electron microscope, however, reveals them not as mere dots but as tiny cylinders made up of nine extremely fine double tubules. How centrioles function no one yet knows for sure. They seem, however, to play some part in the formation of the spindle that appears during mitosis (cell division).

Nucleus

The nucleus occupies the central portion of the cell. Both the shape of the nucleus and the number of nuclei present in a cell vary. One spherical nucleus per cell, however, is common. As seen with a microscope, a nucleus consists of a nuclear membrane, nuclear fluid (or nucleoplasm), numerous chromatin granules, and one or more nucleoli. A nucleolus is a small, dense, usually spherical body. Today much is known about the molecular structure of both chromatin granules and nucleoli. Chromatin granules consist partly of molecules of proteins, but chiefly of molecules of nucleic acids. The specific nucleic acids in chromatin bear a name almost as tongue-twisting as these acids are important—deoxyribonucleic acid. Almost always, for an obvious reason, deoxyribonucleic acid is referred to by its abbreviated name, DNA. In Chapter 17, we shall have a good deal to say about both the structure and function of DNA, "the most golden molecule of all," Watson called it.*

The nuclear membrane is a two-layered membrane with essentially the same molecular structure as the plasma membrane. It differs, however, from the plasma membrane in certain respects; notably, it contains many more pores and it disappears for a short time when a cell is in the process of dividing.

Nucleoli have no membrane enclosing them. They consist, like chromatin, chiefly of nucleic acid and protein molecules. The specific nucleic acid in nucleoli, however, is RNA, not DNA, as it is in chromatin. RNA, like DNA, is a substance of vital importance. The composition of RNA molecules and their functions we shall consider in Chapter 17. For the present, let us say only that nucleoli play an essential role in the formation of ribosomes, the protein synthesizers of cells. You might guess, therefore, and correctly so, that the more protein a cell makes, the larger its nucleoli appear. Cells of the pancreas, to cite just one example, make large amounts of protein and have large nucleoli.

Table 2-1

Some major cell structures and their functions

Cell structures	Functions
Plasma membrane	Serves as the boundary of the cell, maintaining its integrity; transports some substances through itself into or out of cells; prevents some substances from moving into or out of a cell's interior
Endoplasmic reticulum (ER)	Serves as a cell's own circulatory system
Golgi apparatus	Synthesizes carbohydrate, combines it with protein, and packages the product as globules of glycoprotein
Mitochondria	Catabolism; ATP synthesis; a cell's "powerhouse"
Lysosomes	A cell's "digestive system"
Ribosomes	Synthesize proteins; a cell's "protein factories"
Nucleus	Controls cell's activities; the "executive" of a cell
Nucleoli	Play essential role in the formation of ribosomes

*From Watson, J. D.: The double helix, New York, 1968, The New American Library, Inc., p. 21.

What functions does the nucleus perform? Its general function is to serve as the executive of the cell. It acts in complex ways to control or direct a cell's metabolism and reproduction and, in fact, all of its activities.

Cell physiology

Every cell carries on a number of functions that maintain its own life—transportation and metabolism, for example. If a cell is to survive, it must continually move substances through its membranes and must metabolize foods (use them for energy and for building complex compounds). Also, from time to time, all but a few kinds of cells perform another function, that of reproducing themselves. In addition to self-serving activities, every cell in the body also performs some special function that serves the body as a whole. Muscle cells provide the function of movement, nerve cells contribute communication services, red blood cells transport oxygen, etc. In return, the body performs vital functions for all of its cells. It brings food and oxygen to them and removes waste from them, to mention only two examples. In short, a relationship of mutual interdependence exists between the body as a whole and its various parts. Optimum health of the body depends upon optimum health of each of its parts, down even to the smallest cell. Conversely, optimum health of each individual part depends upon optimum health of the body as a whole. In equation form, this physiological principle of mutual interdependence might be expressed as follows:

body health ⇌ system health ⇌ organ health ⇌ tissue health ⇌ cellular health

Some parts of the body, of course, are more important for healthy survival than others. Obviously, the heart is far more important than the appendix. And the nerve cells that control respiration are infinitely more important for survival than muscle cells that move the little finger.

Cell physiology deals with all the different kinds of functions cells perform. It is, as you might guess, an enormous subject. Therefore we have chosen to discuss only one aspect of it at this time—a cell's transportation activities. Later chapters will deal with the following cell functions: contractility, conductivity, metabolism, and reproduction. The rest of this chapter contains some known and some postulated answers to the question: "How do substances move through cell membranes?"

Movement of substances through cell membranes

Heavy traffic moves continuously in both directions through cell membranes. Streaming in and out of all cells, in endless procession, go molecules of water, foods, gases, wastes, and many kinds of ions. Several processes carry on this mass transportation. They are classified under two general headings as physical (or passive) and physiological (or active) processes.

The main distinction between these two kinds of processes lies in the source of the energy that does the work of moving a substance through a membrane. If the energy comes from chemical reactions taking place in a living cell, the transport mechanism is classified as an *active or physiological process*. If the energy for moving a substance stems from some other source—not from a living cell's chemical reactions—then the transport mechanism is classified as a *passive or physical process*. The two preceding sentences, you may have noticed, implied another distinction between active and passive transport mechanisms. Active mechanisms can move substances only through living cell membranes. Passive mechanisms, on the other hand, can move materials through either living or dead cell membranes and even through artificial membranes. The two major passive or physical transport processes are diffusion and osmosis. There are several kinds of active transport processes: those called "pumps," for example, and the mechanisms of phagocytosis and pinocytosis.

Diffusion

Diffusion means scattering or spreading. It occurs because small particles such as molecules and ions are forever on the go. They move continuously, rapidly, and in all directions. Perhaps the easiest way for you to learn the essential facts and principles about diffusion is to discover them for yourself in the following example, illustrated in Fig. 2-5. Suppose a 10% sodium chloride (NaCl) solution is separated from a 20% sodium chloride solution by a membrane. Suppose further that the membrane is permeable to both NaCl and to water. This means that both of these substances can and do pass through the membrane in both directions. Sodium chloride particles and water molecules racing in all directions through each solution collide with each other and with the membrane. Some inevitably hit the membrane pores from the 20% side and some from the 10% side. Just as inevitably, some bound through the pores in both directions. For a while more NaCl particles enter the pores from the 20% side simply because they are more numerous there than on the 10% side. More of these particles therefore move through the membrane from the 20% solution into the 10% than diffuse through it in the opposite direction. Using different words for the same thought, *net diffusion* of NaCl takes places from the solution where its concentration is greater into the one where its concentration is lesser. Thus, net diffusion of NaCl occurs "down" the *sodium chloride concentration gradient*—that is, from

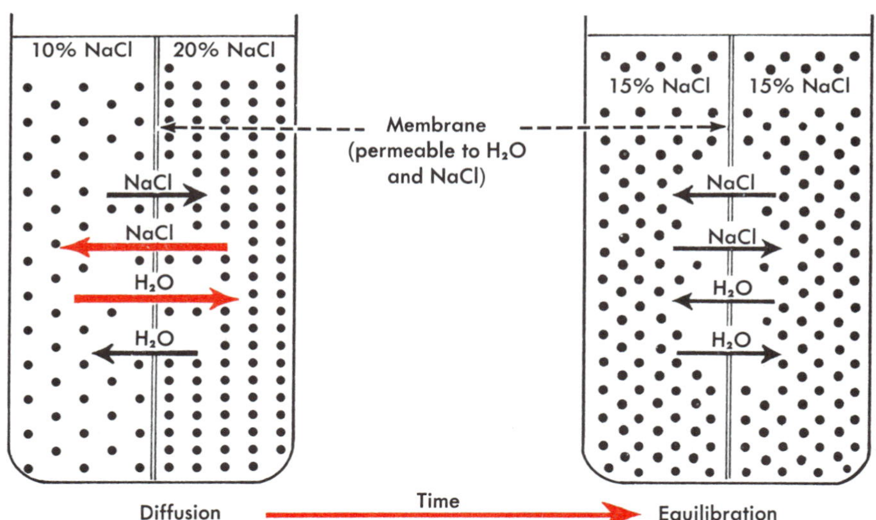

Fig. 2-5

Diffusion. Because the membrane separating the 10% NaCl from the 20% NaCl is freely permeable to both NaCl and H_2O, both substances diffuse rapidly through the membrane in both directions. But, as the red arrows indicate, more sodium and chloride ions move out of the 20% solution, where there are more of them, into the 10% solution, where there are fewer of them, than in the opposite direction. Simultaneously, more water molecules move from the 10% solution, where there are more of them, into the 20% solution, where there are fewer of them. Result: equilibration of the concentrations of the two solutions after an elapse of time. From then on, equal numbers of Na ions and Cl ions diffuse in both directions, as do equal numbers of H_2O molecules.

the higher concentration down to the lower concentration.

During the time that net diffusion of NaCl is taking place between the 20% and 10% solutions, net diffusion of water is also going on. The direction of net diffusion of any substance is always down that substance's concentration gradient. Applying this principle, net NaCl diffusion occurs down the NaCl concentration gradient and net water diffusion occurs down the water concentration gradient. Since the greater concentration of water molecules lies in the more dilute 10% solution (where fewer water molecules have been displaced by NaCl molecules), more water molecules diffuse out of the 10% solution into the 20% solution than diffuse in the opposite direction. Thus the net diffusion of water removes water from the more dilute solution and adds it to the more concentrated one. How then does the net diffusion of water affect the concentrations of the two solutions? Does it tend to make them more equal or more different? How does the net diffusion of NaCl affect their concentrations? The answers are quite obvious. Both the net diffusion of water and the net diffusion of NaCl tend to equalize (equilibrate) the concentrations of the two solutions. Note that whereas net diffusion of NaCl and of water both go on at the same time, they go on in opposite directions.

Diffusion of NaCl and water continues even after equilibration has been achieved. But from that moment on, it is equal diffusion in both directions through the membrane and not net diffusion of either substance in either direction.

Dialysis is diffusion under certain conditions. It takes place when a solution that contains both crystalloids and colloids is separated from plain water by a membrane that is permeable to crystalloids but impermeable to colloids (Fig. 2-6). (Crystalloids or true solutes are solute particles with diameters less than 1 mμ; for example, ions, glucose, oxygen, etc. Colloids are solute particles whose diameters range from

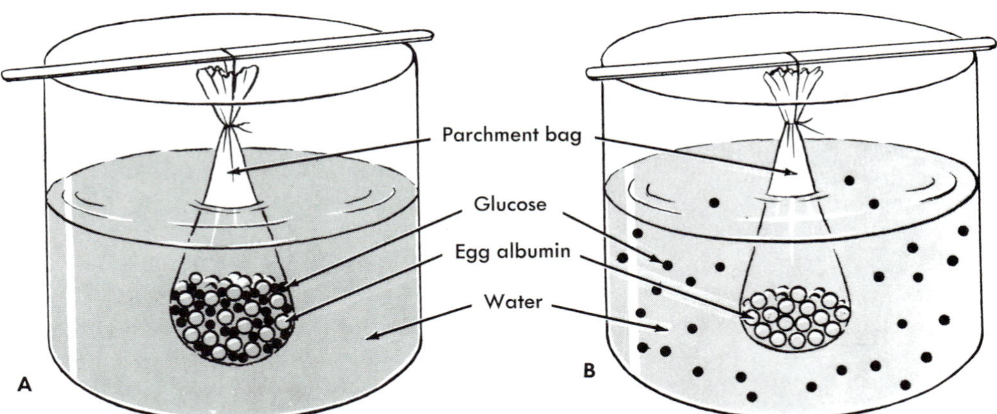

Fig. 2-6

Dialysis, the separation of crystalloids from colloids by means of a membrane permeable to crystalloids and impermeable to colloids. The parchment bag in **A,** which contains a solution of glucose and raw egg white, is permeable to glucose but not to albumin. Therefore, glucose moves out of the bag into the surrounding water while albumin stays inside the bag. Time elapses between **A** and **B. B** shows the result of dialysis—separation of the crystalloid, glucose, from the colloid, albumin.

about 1 to 100 mμ. Enzymes and all other proteins are colloids.) When the membrane that separates a solution of crystalloids from water is permeable to crystalloids and impermeable to colloids, the crystalloids, as you would expect, diffuse through the membrane but the colloids do not. Net diffusion of crystalloids occurs down their concentration gradient, that is, out of the solution into the water. The colloids remain behind. Hence dialysis may be described as diffusion that separates crystalloids from colloids.

The following paragraphs summarize some facts and principles worth remembering about diffusion.

1 *Diffusion* is the movement of solute and solvent particles in all directions through a solution or in both directions through a membrane.

2 *Net diffusion* is the movement of more particles of a substance in one direction than in the opposite direction.

3 Net diffusion of any substance occurs down its own concentration gradient, which means from the higher to the lower concentration of that substance. Following are two applications of this principle: (a) net diffusion of solute particles occurs from the more concentrated to the less concentrated solution; (b) net diffusion of water molecules, in contrast, occurs from the more dilute to the less dilute solution.

4 Net diffusion of the solute in one direction through a membrane and of water in the opposite direction eventually makes the concentrations of the two solutions equal. In short, it *equilibrates* them. We can also state this principle in another way. Net diffusion of both solute and water cause the solute and water concentration gradients to gradually decrease. Eventually—at the point of equilibration—they disappear entirely. There is no difference in concentration (which is what the term concentration gradient means) between equilibrated solutions. Equilibrated solutions have the same, not different, concentrations.

5 Equal diffusion means that the number of solute and water particles diffusing in one direction equals the number diffusing in the opposite direction. Net diffusion means more particles diffusing in one direction than in the other.

6 Diffusion is a passive transport mechanism because cells are passive, not active and working in this process. Cellular chemical reactions do not supply the energy that moves diffusing particles. The continual random movements characteristic of all molecules and ions (molecular kinetic theory) furnish the energy for diffusion.

Think about these diffusion principles. Make sure that you understand them, for they have many applications in physiology. Our very lives, in fact, depend upon diffusion. Evidence? Oxygen, the "breath of life," enters cells by diffusion through their membranes.

FACILITATED DIFFUSION

Facilitated diffusion resembles both ordinary diffusion and active transport. Like ordinary diffusion, facilitated diffusion is a passive process that moves a substance down its own concentration gradient. Like active transport, facilitated diffusion is a "carrier-mediated" process. By this we mean that in or near the outer surface of a cell membrane, a specific compound—the carrier—combines with the substance to be moved. The complex thus formed then rapidly diffuses through the membrane to its inner surface. There, the carrier compound dissociates from the substance. It has fulfilled its function of facilitating (accelerating) the substance's diffusion through the membrane. Glucose and amino acids are believed to move through a cell's plasma membrane by the mechanism of facilitated diffusion.

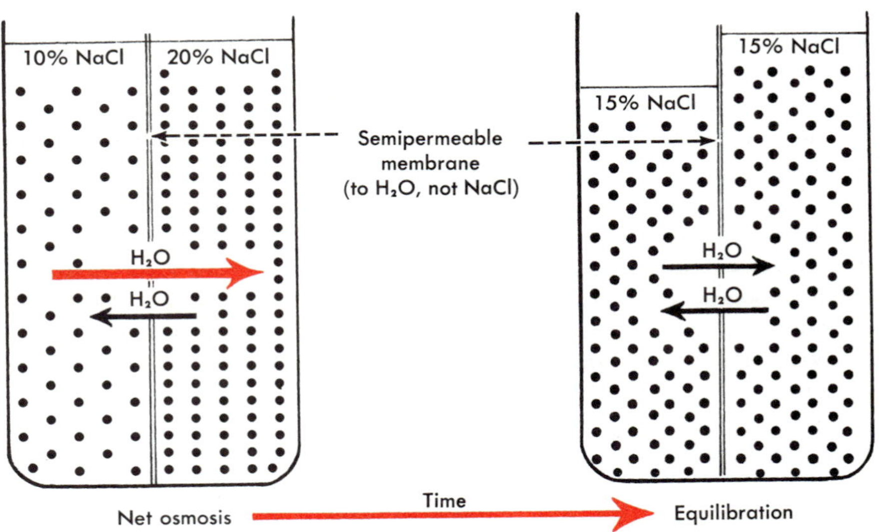

Fig. 2-7

Osmosis. Osmosis is the diffusion of water through a selectively permeable membrane. The membrane shown in this diagram is permeable to water but not to NaCl. Because there are relatively more water molecules in 10% NaCl than in 20% NaCl, more water molecules osmose from the more dilute into the more concentrated solution (as indicated by the red arrow in the left-hand diagram) than osmose in the opposite direction. The *net* direction of osmosis, in other words, is toward the more concentrated solution. Net osmosis produces the following changes in these solutions: their concentrations equilibrate, both the volume and the pressure of the originally more concentrated solution increase, and the volume and the pressure of the other solution decrease proportionately.

Osmosis

Osmosis is the diffusion of water through a selectively permeable membrane (Fig. 2-7). As its name suggests, a selectively permeable membrane is one that is not equally permeable to all solute particles present. It permits some solutes to diffuse through it freely but hinders or prevents entirely the diffusion of others. Those solutes allowed to diffuse freely through the membrane obey the law of diffusion. Hence they eventually equilibrate across it. Their concentrations on both sides of the membrane become equal. But can particles not permitted to diffuse freely through a membrane also obey the law of diffusion? Can they, too, equilibrate across the membrane? The answer to both questions, as you can readily deduce, is "No." When a membrane hinders or prevents a substance from moving through it, the concentration of that substance necessarily remains higher on one side of the membrane than on the other. That solute cannot equilibrate across the membrane. In short, the selectively permeable membrane maintains a concentration gradient of the not freely diffusible solute.

Summarizing the preceding paragraph, a *selectively permeable membrane* may be defined either as one that does not permit free, unhampered diffusion of all the solutes present or as one that maintains at least one solute concentration gradient across itself. *Osmosis* is the diffusion of water through a selectively permeable membrane. Or, osmosis is the diffusion of water through a membrane that maintains at least one concentration gradient across itself.

Because normal living cell membranes are

selectively permeable membranes that maintain various solute concentration gradients, water moves through them by osmosis. Here is an example that may help you understand this passive transport process. Imagine that you have a 20% NaCl solution separated from a 10% NaCl solution by a membrane. Assume that the membrane is impermeable to sodium and chloride ions but that it is freely permeable to water particles (Fig. 2-7). Obviously then, sodium and chloride ions will not diffuse through this membrane. It maintains a sodium chloride concentration gradient across itself, in other words. Therefore, water will move through this membrane by the process of osmosis. It osmoses through the membrane in both directions, but not in equal amounts.

To deduce the direction in which the greater volume of water will osmose (the direction, that is, of net osmosis), we must first decide which solution contains the greater concentration of water molecules. Pretty clearly, the more dilute of two solutions contains the greater concentration of water molecules. In this case, then, water concentration is greater on the 10% side of the membrane than it is on the 20% side. Next, we need to apply the principle already stated that "net diffusion of any substance occurs down the concentration gradient of that substance." Therefore, net osmosis (net water diffusion through the selectively permeable membrane) occurs down the water concentration gradient. In our example, net osmosis takes place from the more dilute 10% salt solution into the more concentrated 20% solution. Thus, the solution that was at first more concentrated gains water by net osmosis and becomes more dilute. And the solution that was at first more dilute loses water and becomes more concentrated. Net osmosis, in other words, tends to make the concentrations of the two solutions equal. In briefest form, the principle is this: net osmosis occurs down a water concentration gradient and tends to equilibrate solutions separated by a selectively permeable membrane.

Net osmosis into the originally more concentrated of our two salt solutions increases the volume of this solution. Further, the increase in its volume causes an increase in its pressure, called osmotic pressure—a logical term since it is pressure caused by net osmosis. By definition, then, *osmotic pressure* is the pressure that develops in a solution as a result of net osmosis into that solution. An important principle stems from this definition—osmotic pressure develops in the solution that originally contains the higher concentration of the solute that does not diffuse freely through the membrane. For instance, in the example previously given, osmotic pressure would develop in the solution that originally had the 20% salt concentration.

Potential osmotic pressure is the maximum osmotic pressure that could develop in a solution if it were separated from distilled water by a selectively permeable membrane. (Actual osmotic pressure, on the other hand, is a pressure that already has developed, not just one that could develop.) What determines a solution's potential osmotic pressure? The answer is this: the number of solute particles in a unit volume of solution directly determines its potential osmotic pressure—the more solute particles per unit volume, the greater the potential osmotic pressure. The number of solute particles per unit volume (for example, per liter) of solution, in turn, is determined by the molar concentration of the solution and also, if the solute is an electrolyte, by the number of ions formed from each molecule of solutes. This concept can be expressed by a diagram (Fig. 2-8) or by formula (see footnote to Fig. 2-8).

Since it is the number of solute particles per unit volume that directly determines a solution's potential osmotic pressure, one might at first thought jump to the conclusion that all solutions having the same percent concentration also have the same potential osmotic pressures. Obviously, it is true that all solutions containing the same percent concentration of the same solute do also have the same potential osmotic pressures. All 5% glucose solutions, for example, have a potential osmotic pressure at body temperature of somewhat more than 5,300 mm Hg pressure. But all solutions with the same percent concentrations of different solutes do not have the same molar concentrations. Hence they do not have the same potential osmotic pressures. By applying the equation in the footnote, you will discover that 5% NaCl at body temperature has a potential osmotic pressure of approximately 33,000 mm Hg—quite different from 5% glucose's potential osmotic pressure of about 5,300 mm Hg.

Two solutions that have the same potential osmotic pressures are said to be *isosmotic* to each other. Because they have the same potential osmotic pressures, the same amount of water osmoses in both directions between them when they are separated by a selectively permeable membrane. No net osmosis, however, occurs in either direction between isosmotic solutions. Hence, no actual osmotic pressure develops in either solution. Their pressures remain the same. And because they do, isosmotic solutions are also called isotonic (Gr. *isos,* the same; *tonos,* tension or pressure).

By definition, *isotonic solutions* are those whose volumes and pressures will stay the same if the two solutions are separated by a membrane. For example, 0.85% NaCl is referred to as "isotonic saline," meaning that it is isotonic to the fluid inside human cells. Translated more fully, it means that if 0.85% NaCl is injected into human tissues or blood, no net osmosis occurs into or out of cells. Therefore, no change in intracellular volume or pressure takes place. Cells in contact with isotonic solutions neither lose nor gain water. They become neither dehydrated nor hydrated, to use more technical terms. Physicians make use of this information frequently. For intravenous and intramuscular injections, they usually give isotonic solutions.

When two solutions have unequal potential osmotic pressures, their volumes and pressures will change if they are separated by a selectively permeable membrane. Net osmosis will occur into the solution that has the higher potential osmotic pressure. Net osmosis, as we have noted, always occurs down the water concentration gradient. Since the solution having the lower potential osmotic

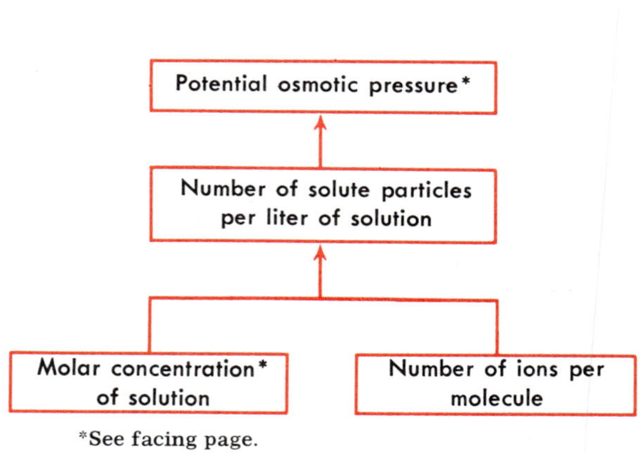

Fig. 2-8

Factors that determine potential osmotic pressure. Interpreted, the diagram shows that the number of solute particles (ions and molecules) present in a liter of solution determines its potential osmotic pressure. Note, however, that the number of solute particles per liter is itself determined by two factors—the number of molecules in the solution (indicated by its molar concentration) and also, if the solute is an electrolyte, by the number of ions formed from each molecule.

pressure has the lower concentration of solute particles, it necessarily has the higher concentration of water molecules. Therefore, net osmosis occurs out of this solution into the solution with the higher potential osmotic pressure. This increases the volume of the solution with the higher potential osmotic pressure. And the increase in its volume produces an increase in its pressure. This increase in pressure is osmotic pressure—but actual osmotic pressure and not just a potential osmotic pressure.

When two solutions have unequal potential osmotic pressures, one of them is described as hypertonic and the other as hypotonic. Here are several definitions of the term *hypertonic solution*. A hypertonic solution is one that has a higher potential osmotic pressure than another. A hypertonic solution is one into which net osmosis occurs when a selectively permeable membrane separates it from another solution. A hypertonic solution is one whose volume and pressure increase when a selectively permeable membrane separates it from another solution. A hypertonic solution is one in which osmotic pressure develops when a selectively permeable membrane separates it from another solution (Fig. 2-9).

A *hypotonic solution* has characteristics opposite to those of a hypertonic solution. Example: The fluid inside human cells is hypertonic to distilled water and, conversely,

---

**Formulas:*

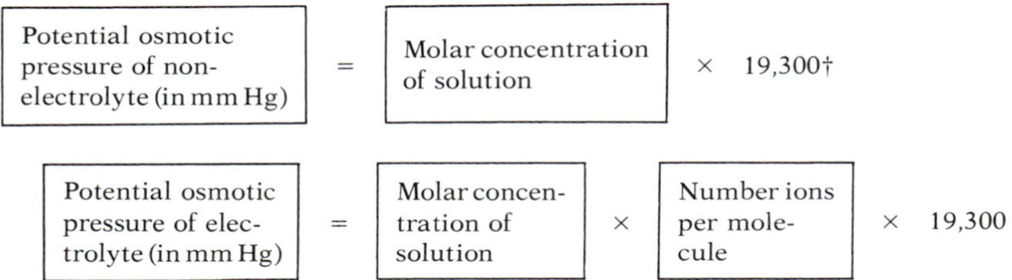

†Experimentation has shown that a solution concentration 1.0 molar of any nonelectrolyte has a potential osmotic pressure of 19,300 mm Hg pressure (at body temperature, 37° C).

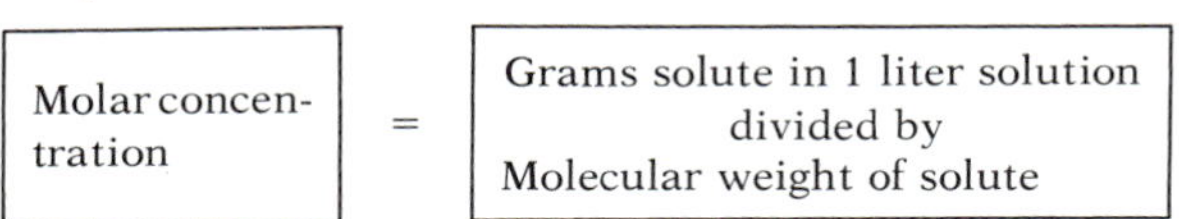

Example: Two solutions commonly used in hospitals are 0.85% NaCl and 5% glucose. What is the potential osmotic pressure of 0.85% NaCl at body temperature? (0.85% NaCl = 8.5 gm NaCl in 1 liter solution.)

Molecular weight of NaCl = 58 (NaCl yields 2 ions per molecule in solution)

Using the formula given for computing potential osmotic pressure of an electrolyte:

$$\boxed{\text{Potential osmotic pressure of 0.85\% NaCl}} = \frac{8.5}{58} \times 2 \times 19{,}300 = \textit{5,658.6 mm Hg pressure}$$

Problem: What is potential osmotic pressure of 5% glucose at body temperature? Molecular weight glucose = 180. Glucose does not ionize. It is a nonelectrolyte.*
*A 5% glucose solution has potential osmotic pressure of 5,359.6 mm Hg pressure.

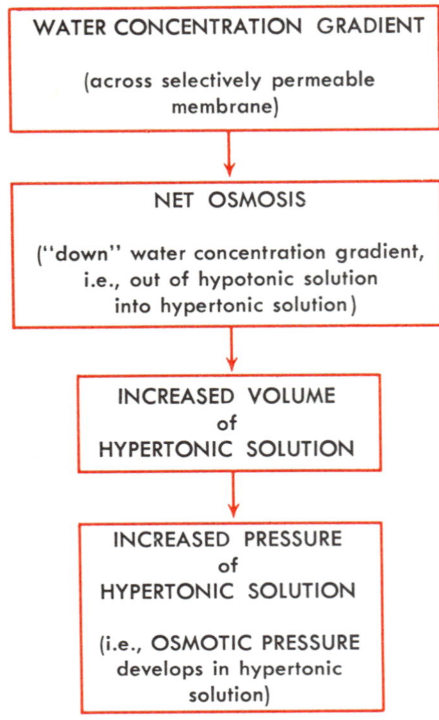

Fig. 2-9

Effects produced by existence of a water concentration gradient across a selectively permeable membrane.

distilled water is hypotonic to intracellular fluid. If, therefore, distilled water were injected into a vein, net osmosis would occur into blood cells. And eventually, if intracellular volume and pressure increased beyond a certain limit, blood cell membranes would rupture and the cells would die. When this happens to red blood cells, their hemoglobin leaks out and they are said to be hemolyzed. (*Hemolysis* means the destruction of red blood cells with the escape of hemoglobin from them into the surrounding medium.)

Concentrate on remembering the following facts and principles about osmosis:

1 Osmosis is the diffusion of water through a selectively permeable membrane, a membrane that maintains at least one concentration gradient across itself.

2 Net osmosis means the osmosis of more water in one direction through a membrane than in the opposite direction. The direction of net osmosis may be described in several ways. Net diffusion occurs out of the solution that contains the lower concentration of solute particles into the one that contains the higher solute concentration. Net diffusion occurs out of the solution that has the lower potential osmotic pressure into the solution with the higher potential osmotic pressure. Net diffusion occurs from a hypotonic solution into a hypertonic solution.

3 Net osmosis produces these results: it decreases the volume of the less concentrated or hypotonic solution, thereby increasing the volume and pressure of the more concentrated or hypertonic solution.

Active transport mechanisms

The active transport of substances through cell membranes is a major and a vital kind of cellular work. The energy that does this work comes from chemical reactions that take place within a living cell. Based on this criterion, we can define an active transport mechanism as a device that uses energy from cellular chemical reactions to move a substance through a cell membrane. One type of active transport mechanism is called a pump—a physiological or biological pump. By definition, a *physiological pump* is an active transport mechanism that moves molecules or ions through cell membranes in an uphill direction, meaning up their concentration gradients or against their natural tendency. Their natural tendency, as you know, is to diffuse down their concentration gradients. The law of diffusion requires that the net diffusion of any substance takes place from the area of its higher concentration to that of its lower concentration.

Every cell employs various physiological

pumps. Of these, one of the most important for healthy human cell survival is the sodium pump. Fluid called interstitial fluid bathes our human cells. It contains a much higher concentration of sodium ions than does the fluid inside cells (intracellular fluid). Sodium ions, therefore, diffuse down their concentration gradient from interstitial fluid into intracellular fluid. But about as fast as they diffuse inward through the plasma membrane, the sodium pump transports them back out into the interstitial fluid again. With energy supplied by cellular chemical reactions, the sodium pump thus moves sodium ions uphill against the sodium concentration gradient. Normally, therefore, sodium does not equilibrate across living cell membranes. The sodium pump maintains a steep sodium concentration gradient* across the plasma membranes of healthy human cells. Of all the many processes that keep us alive, few are more important than active transport mechanisms. Sodium transport alone plays an essential part in several vital functions—in nerve impulse conduction, in the maintenance of water balance, and in the maintenance of acid-base balance, for example.

Another important physiological pump is the potassium pump. In contrast to the sodium pump, which actively transports sodium ions out of cells, the potassium pump actively transports potassium into cells. A calcium pump actively transports calcium ions into mitochondria through their membranes. Almost surely, there are also many other physiological pumps operating in human cells.

A few facts and many conjectures characterize current concepts about active transport mechanisms. Accepted as facts are the following features that distinguish active from passive transport mechanisms. Active transport mechanisms are powered by energy from cellular chemical reactions. Passive transport mechanisms, in contrast, are powered by the kinetic energy of ever-moving molecules and ions. Many active transport mechanisms make use of a carrier substance; most passive transport mechanisms do not. Active transport mechanisms increase the concentration gradient of the transported substance across the membrane. In other words, they establish a higher concentration of the transported substance on one side of the membrane than on the other. (This last fact has given rise to the name *concentrative transport* as a synonym for active transport.) Passive transport mechanisms decrease the concentration gradient of the transported substance across the membrane—eventually to zero. That is, passive transport mechanisms equilibrate the concentrations of the transported substance on both sides of the membrane.

Conjectures about active transport mechanisms deal mainly with the nature of the carrier compound and its mode of functioning. So far all the carrier compounds identified have turned out to be proteins.* Several theories postulate the following events as the basic parts of the type of active transport mechanisms called physiological pumps (see Fig. 2-10).

1 On one surface of a cell membrane, a molecule of the substance to be transported (A) combines with a molecule of carrier compound (B) to form a new compound (AB).

$$A + B \rightarrow AB$$

2 Molecule AB passes through the cell

*One way of designating sodium concentration is as milligrams per liter. Expressed this way, interstitial fluid has a sodium concentration of about 3,200 mg per liter, whereas intracellular fluid has a sodium concentration of only about 345 mg per liter. Expressed in terms of milliequivalents per liter, interstitial fluid has a sodium concentration of about 139 mEq per liter and intracellular fluid, about 15 mEq per liter.

*Fox, C. F.: The structure of cell membranes, Sci. Am. **226**:36-37, Sept., 1972.

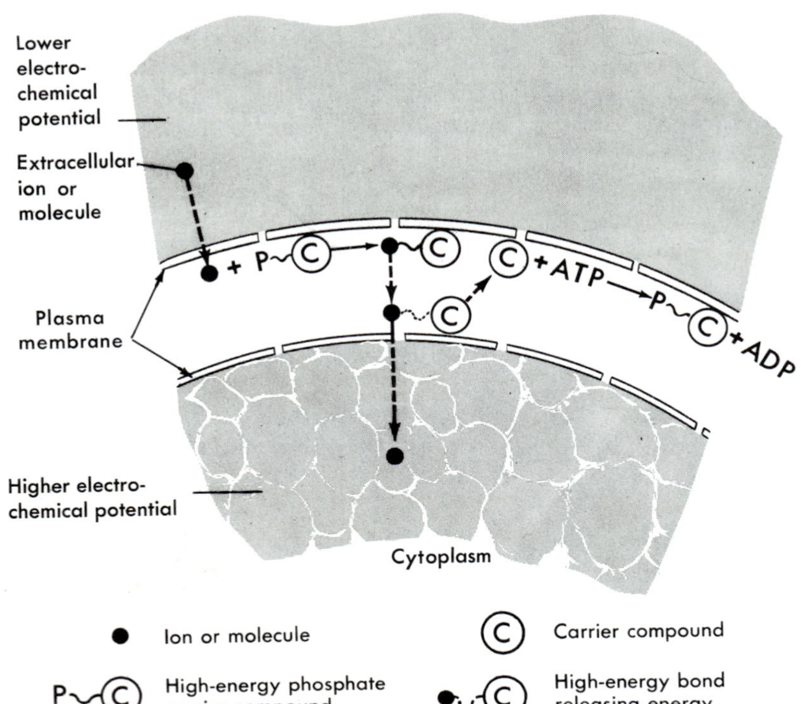

Fig. 2-10

Scheme to illustrate one hypothesis about active transport mechanisms. Diagram shows them consisting of the following steps: at the outer surface of cell's plasma membrane, an ion or a molecule combines with a high-energy phosphate carrier compound; the compound thus formed moves through the membrane to its inner surface; its high-energy bond breaks down, releasing energy which propels the transported particle, but not the carrier compound, into the cytoplasm of the cell; the carrier compound moves back toward the outer part of the plasma membrane and combines with ATP to again become a high-energy phosphate compound.

membrane into (or out of) the cell. According to recent proposals, the carrier protein molecule may change its shape and move in a revolving-door manner as it transports a substance from one surface to the other of a cell membrane.*

3 At the other surface of the cell membrane, AB dissociates, releasing the transported substance.

$$AB \rightarrow A + B$$

4 Carrier compound molecule B moves back to the surface form from which it came, ready to shuttle another molecule of A through the membrane.

*Fox, C. F.: The structure of cell membranes, Sci. Am. **226**:36-37, Sept., 1972.

Phagocytosis and pinocytosis

Phagocytosis and pinocytosis, like the physiological pumps, are active transport mechanisms for moving substances through cell membranes—but only in one direction, namely inward. About three quarters of a century ago, Elie Metchnikoff of the Pasteur Institute saw white blood cells engulf bacteria. It reminded him of eating. Therefore, from the Greek words for eating, cell, and action, he coined the word phagocytosis. Some 30 years later, in 1931, W. H. Lewis of Johns Hopkins University saw something similar in time-lapse photographs of some tissue culture cells. These cells, however, were engulfing droplets of fluid instead of solid particles. They seemed to be drinking rather than eating, so he named the process pinocytosis (from the Greek word for drinking).

Both phagocytosis and pinocytosis consist

of the same essential steps. A segment of a cell's plasma membrane forms a small pocket around a bit of solid or liquid material outside the cell, pinches off from the rest of the membrane, and migrates inward as a closed vacuole or vesicle. Later, it releases its contents into the cell's cytoplasm.

Summary

This chapter has presented information about the structure and some of the self-serving functions of cells. The next one will relate facts and principles about the structure and body-serving functions of tissues.

Outline summary

CELL STRUCTURE

See Fig. 2-1

Protoplasm

1 Definition—living matter; all cells composed of *protoplasm*
2 Composition
  a Main elements—carbon, hydrogen, oxygen, and nitrogen
  b Main compounds—water, inorganic salts, and compounds unique to living matter, i.e., proteins (including enzymes), carbohydrates, lipids, and nucleic acids

Plasma membrane

1 Structure
  a About 3/10,000,000 of an inch thick, according to present estimates
  b Composed of protein and phospholipid molecules; Fig. 2-2, p. 15, shows the arrangement of phospholipid and protein molecules
  c Has tiny opening or pores; few in number; spaced far apart
2 Function
  a Serves as the cell's boundary; maintains its integrity
  b Transports some substances through itself into or out of the cell's cytoplasm; prevents other substances from moving into or out of the cytoplasm

Cytoplasm

1 Definition—protoplasm located between cell membrane and nucleus
2 Contains thousands of organelles ("little organs")
  a Membranous organelles—endoplasmic reticulum, Golgi apparatus, mitochondria, and lysosomes
  b Nonmembranous organelles—ribosomes and centrosome
3 Endoplasmic reticulum
  a Structure—complicated network of canals and sacs extending through cytoplasm and opening at surface of cell; many ribosomes attached to membranes of rough endoplasmic reticulum but not to smooth
  b Functions—ribosomes attached to rough endoplasmic reticulum synthesize proteins; canals of reticulum serve cell as its inner circulatory system—e.g., proteins move through canals on way to Golgi apparatus
4 Golgi apparatus
  a Structure—membranous vesicles near nucleus
  b Function—synthesizes large carbohydrate molecules, combines them with proteins, and secretes product (glycoproteins)
5 Mitochondria
  a Structure—microscopic sacs; walls composed of inner and outer membranes separated by fluid; thousands of particles made up of enzyme molecules attach to both membranes
  b Function—"power plants" of cells; mitochondrial enzymes catalyze citric acid cycle, series of reactions that provide about 95% of cell's energy supply
6 Lysosomes
  a Structure—microscopic membranous sacs
  b Function—a cell's own digestive system; enzymes in lysosomes digest particles or large molecules that enter them; under some conditions, digest and thereby destroy cells
7 Ribosomes
  a Structure—microscopic spheres, large numbers of which attached to endoplasmic reticulum
  b Function—"protein factories"; ribosomes attached to endoplasmic reticulum synthesize proteins to be secreted by cell, and those lying free in cytoplasm make proteins for cell's own use, i.e., its structural proteins and enzymes; groups of ribosomes, called polysomes, synthesize proteins
8 Centrosome
  a Structure
    (1) Centrosome is spherical body near center of cell—i.e., near nucleus
    (2) Centrioles, located in centrosome, are tiny cylinders, the walls of which are composed of fine tubules, nine groups of two or three tubules each
  b Function—centrioles thought to play some part in formation of the mitotic spindle

Nucleus

1 Definition—spherical body in center of cell; enclosed by pore-containing membrane
2 Structure—consists of
  a Nuclear membrane (two-layered, with essentially the same molecular structure as the plasma membrane)
  b Numerous chromatin granules—consist chiefly of DNA and molecules of proteins
  c Nucleoli—consist chiefly of RNA and protein molecules; no membrane encloses nucleoli
3 Functions—nucleus serves as the executive of a cell; acts in complex ways to control all cellular activities

CELL PHYSIOLOGY

Movement of substances through cell membranes

1 By physical (or passive) processes
  a Energy that moves substances comes from random, never-ceasing movements of atoms, ions, and molecules, not from chemical reactions in cell
  b Diffusion and osmosis are physical processes that move substances through cell membrane and, also, through nonliving membranes
2 By physiological (or active) processes
  a Energy that moves substances comes from chemical reactions in living cell

b Active transport mechanisms include various physiological pumps, phagocytosis, and pinocytosis

Diffusion

1 Movement of solute and solvent particles in all directions through solution or in both directions through membrane
  a Net diffusion of solute particles—down solute concentration gradient, i.e., from more to less concentrated solution
  b Net diffusion of water—down water concentration gradient, i.e., from less to more concentrated solution
2 Diffusion tends to produce equilibration of solutions on opposite sides of membrane, but many exceptions to this rule when membrane living
3 Dialysis—separation of crystalloids from colloids by diffusion of crystalloids through membrane permeable to them but impermeable to colloids
4 Facilitated diffusion—diffusion in which carrier substance combines with particle being moved

Osmosis

1 In living systems, movement of water in both directions through membrane that maintains at least one solute concentration gradient across it
2 Net osmosis—more water osmoses in one direction through membrane than in opposite; net osmosis occurs down water concentration gradient (which is up solute concentration gradient and up potential osmotic pressure gradient); net osmosis tends to equilibrate two solutions separated by selectively permeable membrane
3 Osmotic pressure—pressure that develops in solution as result of net osmosis into it
4 Isotonic solution—one that has same potential osmotic pressure as solution to which it is isotonic; no net osmosis between isotonic solutions
5 Hypertonic solution—has greater potential osmotic pressure and higher solute concentration but lower water concentration than solution to which it is hypertonic; net osmosis into hypertonic solution from hypotonic solution
6 Hypotonic solution—has lower potential osmotic pressure and lower solute concentration but higher water concentration than solution to which it is hypotonic; net osmosis out of hypotonic solution into hypertonic solution

Physiological pumps

Active transport mechanisms that move ions or molecules through cell membranes against their concentration gradient, i.e., direction opposite from net diffusion or net osmosis; energy supplied by cellular chemical reactions

Phagocytosis and pinocytosis

1 Phagocytosis—physiological process that moves solid particles into cell; segment of cell's plasma membrane forms pocket around particle outside cell, then pinches off from rest of membrane and migrates inward
2 Pinocytosis—physiological process that moves fluid into cell; process similar to phagocytosis

Review questions

1 Define briefly or give a synonym for each of the following terms: cytology, protoplasm, cytoplasm, plasma membrane.
2 What is the most abundant compound in protoplasm?
3 What four kinds of compounds occur naturally only in protoplasm?
4 Describe a current concept about the molecular structure of the plasma membrane.
5 The diameter of one hydrogen atom is said to be 1 Å. Approximately what fraction of an inch is 1 Å?
6 Approximately how many angstroms thick is the plasma membrane believed to be? What fraction of an inch is this?
7 The sodium concentration of interstitial fluid is how many times as high as that of the intracellular fluid? (See footnote, p. 31.)
8 What is the sodium concentration gradient across the plasma membranes of human cells? Use the figures given in the footnote on p. 31 to complete the following equation:

Na^+ concentration gradient across plasma membrane	=	ICF Na^+ concentration (mg Na^+/l ICF)	−	IF Na^+ concentration gradient (mg Na^+/l IF)

9 Identify the following organelles with a brief statement about the structure and functions of each: endoplasmic reticulum, Golgi apparatus, mitochondria, lysosomes, ribosomes, centrosome.
10 One inch equals approximately how many angstroms? microns? millimeters? millimicrons?
11 One millimeter equals approximately what fractional part of an inch?
12 Explain briefly what each of the following terms mean: active transport, dialysis,, diffusion, net diffusion, facilitated diffusion, osmosis, phagocytosis, pinocytosis, physiological pump.
13 In what essential ways do active processes for moving substances through cell membranes differ from passive processes?
14 State the principle about the direction in which net osmosis occurs.
15 Differentiate between actual osmotic pressure and potential osmotic pressure.
16 What factor directly determines the potential osmotic pressure of a solution?
17 Explain the terms isotonic, hypotonic, and hypertonic.
18 State the principle about which solution develops an osmotic pressure, given appropriate conditions.

3

Tissues

Epithelial tissue
Simple squamous epithelium
Stratified squamous epithelium
Simple columnar epithelium

Muscle tissue

Connective tissue
Reticular tissue
Loose, ordinary connective tissue (areolar)
Adipose tissue
Dense fibrous tissue
Bone and cartilage
Hemopoietic tissue
Blood
Reticuloendothelial system

Nerve tissue

Tissues are organizations of cells. Nonliving intercellular substances fill in any spaces between cells. Using appearance and functions as criteria for classification, there are four basic types of tissues—epithelial, muscle, connective, and nerve—and many subtypes (Table 3-1, pp. 40-41). Tissues differ in structure, and because structure is a primary determinant of function, they also differ in function. Not only do cell size, shape, and arrangement vary in different kinds of tissues, but so too does the amount and kind of intercellular substance present. Some tissues contain almost no intercellular material. Others consist predominantly of it. Some intercellular substance contains many fibers, some is unformed gel, and some is fluid, the interstitial fluid that bathes most living human cells.

Epithelial tissue

Epithelial tissue is widely distributed throughout the body. For example, it composes the surface layer of the skin and of blood vessels (where it is called endothelium). Different types of epithelial tissue perform different functions. Some sturdy types serve to protect underlying tissues. Other, more delicate kinds of epithelial tissue allow substances to move through them, notably by diffusion or active transport, and they thereby serve as absorbing and secreting tissues. All types of epithelial tissue are composed largely or entirely of cells. In other words, they contain little or no intercellular substance. They also contain no blood vessels. These are located in the connective tissue that always undergirds epithelial tissue. Between the epithelial and connective tissues lies a very thin structure known as the *basement membrane*. Another common characteristic of all types of epithelial cells is that they undergo mitosis—a fact of practical importance since it means that old or destroyed epithelial cells can be replaced by new ones. There are several types of epithelial tissue. Three are described in the following paragraphs.

Simple squamous epithelium

Simple squamous epithelium consists of only one layer of flat, scalelike cells. Consequently, substances can readily diffuse or filter through this type of tissue. The microscopic air sacs of the lungs, for example, are composed of this kind of tissue, as are the linings of blood and lymphatic vessels and the surfaces of the pleura, pericardium, and peritoneum. (Blood and lymphatic vessel linings are called *endothelium*, and the surfaces of the pleura, pericardium, and peritoneum are called *mesothelium*. Some histologists classify these as connective tissue.)

Stratified squamous epithelium

Stratified squamous epithelium such as shown in Fig. 3-1 lines the mouth and esophagus. Its several layers of cells serve a protective function. The surface of the skin is composed of a special kind of stratified squamous epithelium (pp. 49-50).

Simple columnar epithelium

Simple columnar epithelium lines the stomach and intestines and parts of the respiratory tract. A single layer of cells composes this tissue, but two types of cells—goblet and columnar—may be present (Fig. 3-2). Goblet cells are the mucus-secreting specialists of the body. Columnar cells are its absorption specialists.

Muscle tissue

The main specialty of muscle tissue is contraction. Because not all muscle tissue is alike—in location, microscopic appearance, and nervous control—these criteria are used to classify its types. Thus, using location as

Tissues

Epithelial tissue
Simple squamous epithelium
Stratified squamous epithelium
Simple columnar epithelium

Muscle tissue

Connective tissue
Reticular tissue
Loose, ordinary connective tissue (areolar)
Adipose tissue
Dense fibrous tissue
Bone and cartilage
Hemopoietic tissue
Blood
Reticuloendothelial system

Nerve tissue

Tissues are organizations of cells. Nonliving intercellular substances fill in any spaces between cells. Using appearance and functions as criteria for classification, there are four basic types of tissues—epithelial, muscle, connective, and nerve—and many subtypes (Table 3-1, pp. 40-41). Tissues differ in structure, and because structure is a primary determinant of function, they also differ in function. Not only do cell size, shape, and arrangement vary in different kinds of tissues, but so too does the amount and kind of intercellular substance present. Some tissues contain almost no intercellular material. Others consist predominantly of it. Some intercellular substance contains many fibers, some is unformed gel, and some is fluid, the interstitial fluid that bathes most living human cells.

Epithelial tissue

Epithelial tissue is widely distributed throughout the body. For example, it composes the surface layer of the skin and of blood vessels (where it is called endothelium). Different types of epithelial tissue perform different functions. Some sturdy types serve to protect underlying tissues. Other, more delicate kinds of epithelial tissue allow substances to move through them, notably by diffusion or active transport, and they thereby serve as absorbing and secreting tissues. All types of epithelial tissue are composed largely or entirely of cells. In other words, they contain little or no intercellular substance. They also contain no blood vessels. These are located in the connective tissue that always undergirds epithelial tissue. Between the epithelial and connective tissues lies a very thin structure known as the *basement membrane*. Another common characteristic of all types of epithelial cells is that they undergo mitosis—a fact of practical importance since it means that old or destroyed epithelial cells can be replaced by new ones. There are several types of epithelial tissue. Three are described in the following paragraphs.

Simple squamous epithelium

Simple squamous epithelium consists of only one layer of flat, scalelike cells. Consequently, substances can readily diffuse or filter through this type of tissue. The microscopic air sacs of the lungs, for example, are composed of this kind of tissue, as are the linings of blood and lymphatic vessels and the surfaces of the pleura, pericardium, and peritoneum. (Blood and lymphatic vessel linings are called *endothelium*, and the surfaces of the pleura, pericardium, and peritoneum are called *mesothelium*. Some histologists classify these as connective tissue.)

Stratified squamous epithelium

Stratified squamous epithelium such as shown in Fig. 3-1 lines the mouth and esophagus. Its several layers of cells serve a protective function. The surface of the skin is composed of a special kind of stratified squamous epithelium (pp. 49-50).

Simple columnar epithelium

Simple columnar epithelium lines the stomach and intestines and parts of the respiratory tract. A single layer of cells composes this tissue, but two types of cells—goblet and columnar—may be present (Fig. 3-2). Goblet cells are the mucus-secreting specialists of the body. Columnar cells are its absorption specialists.

Muscle tissue

The main specialty of muscle tissue is contraction. Because not all muscle tissue is alike—in location, microscopic appearance, and nervous control—these criteria are used to classify its types. Thus, using location as the criterion, there are three kinds of muscle tissue:

1 *Skeletal muscle*—attached to bones
2 *Visceral muscle*—in the walls of hollow internal structures such as blood vessels, intestines, uterus, and many others
3 *Cardiac muscle*—composes the wall of the heart

With microscopic appearance as the basis of classification, there are only two types of muscle tissue: striated (named for cross striations seen in these cells) and nonstriated or smooth (no cross striations in cells).

On the basis of nervous control, there are also two kinds of muscle tissue: voluntary and involuntary.

Voluntary muscle receives nerve fibers from the somatic nervous system. Therefore, its contraction can be voluntarily controlled. Involuntary muscle, on the other hand, receives nerve fibers from the autonomic nervous system so that its contraction cannot be voluntarily controlled (except in a few rare individuals). Skeletal muscle is voluntary muscle. Visceral and cardiac muscles are involuntary muscles. Visceral and cardiac muscles are also automatic, meaning that even without nervous stimulation they continue to contract. Skeletal muscle, in contrast, cannot contract automatically. Anything that cuts off its nerve impulses paralyzes it—that is, puts it immediately out of working order. Poliomyelitis acts this way, for example. It damages nerve cells that conduct impulses to skeletal muscles so that they no longer conduct, and, deprived of stimulation, the muscles are paralyzed.

Finally, combining these classifications, we have the following:

1 *Skeletal* or striated voluntary muscle
2 *Cardiac* or striated involuntary muscle
3 *Visceral* or nonstriated (smooth) involuntary muscle

Look at Fig. 3-4, and you can observe the

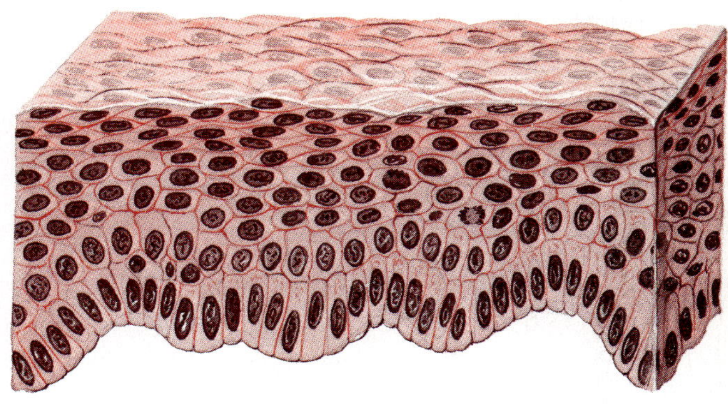

Fig. 3-1

Stratified squamous epithelium such as lines the mouth.

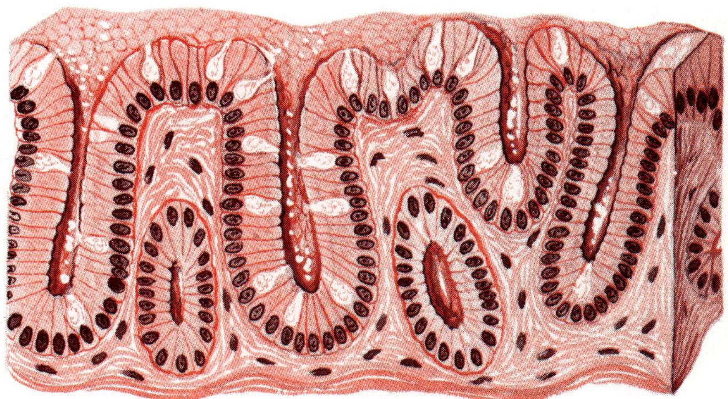

Fig. 3-2

Simple columnar epithelium with goblet cells such as lines intestines.

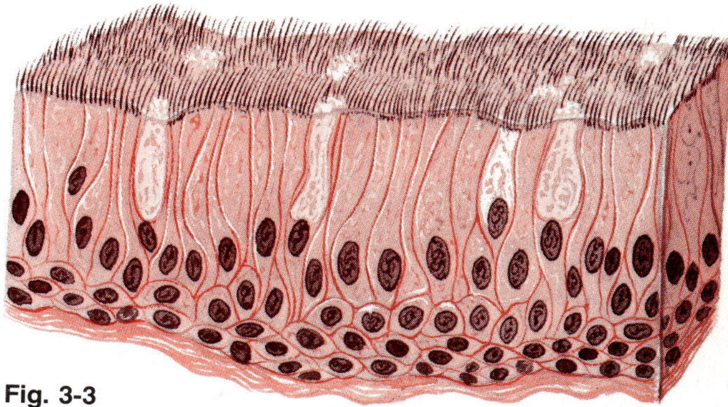

Fig. 3-3

Pseudostratified ciliated columnar epithelium with goblet cells. Stratified tissue consists of two or more layers of cells. Pseudostratified tissue appears, in certain sections, to meet this requirement but actually does not.

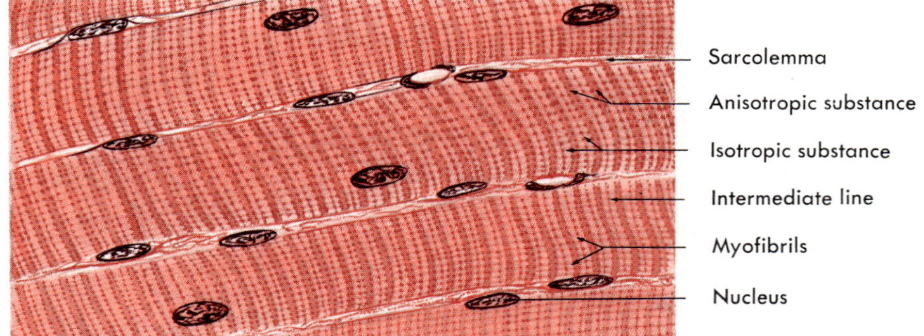

Fig. 3-4

Skeletal or striated voluntary muscle tissue.

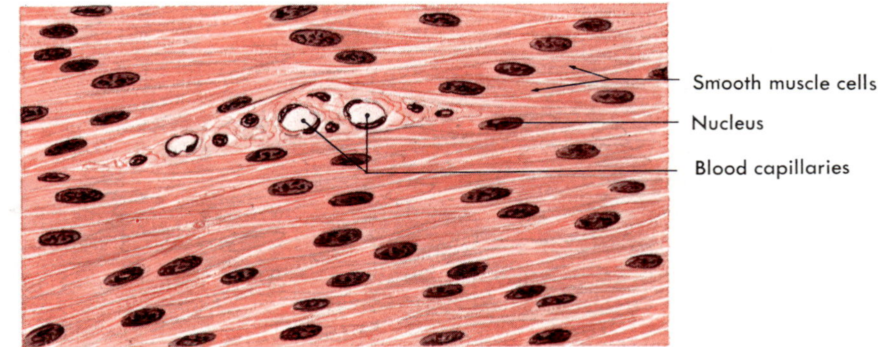

Fig. 3-5

Visceral or nonstriated (smooth) involuntary muscle tissue.

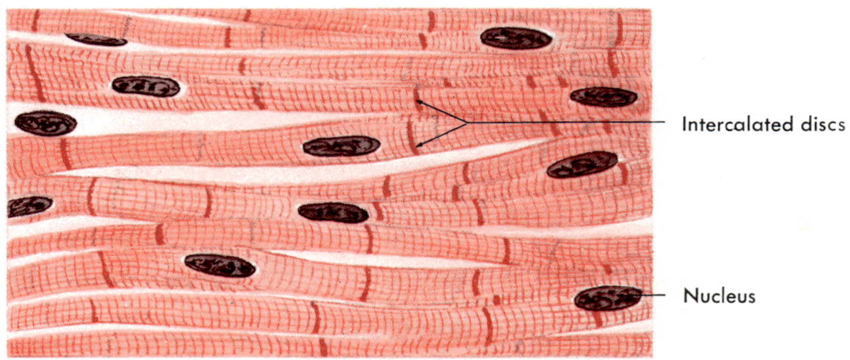

Fig. 3-6

Cardiac or striated involuntary muscle tissue.

following structural characteristics of skeletal muscle cells: They have many cross striations; each cell has many nuclei; the cells are long and narrow in shape—for example, often $1^1/_2$ or more inches in length but only between 0.01 and 0.1 mm (about 1/2,500 to 1/250 of an inch) in diameter. Because such measurements give muscle cells a threadlike appearance, they are probably more often called muscle fibers than cells. The discussion of details about skeletal muscle structure appears in Chapter 6.

Smooth muscle cells are also long narrow fibers but not nearly so long as striated fibers. One can see the full length of a smooth muscle fiber in a microscopic field but only part of a striated fiber. (According to one estimate, the longest smooth muscle fibers measure about 500 μ and the longest striated fibers about 40,000 μ. Can you translate these measurements into millimeters and inches? If not, and if you are curious, see footnote on p. 13.) As Fig. 3-5 shows, smooth muscle fibers have only one nucleus per fiber and are nonstriated or smooth in appearance.

Under the light microscope, cardiac muscle fibers (Fig. 3-6) have cross striations and unique dark bands (intercalated disks). They also seem to be incomplete cells that branch into each other to form a big continuous mass of protoplasm. The electron microscope, however, has revealed that the intercalated disks are actually places where the plasma membranes of two cardiac fibers abut (Fig. 3-7). Cardiac fibers do branch and anastomose but a complete plasma membrane encloses each cardiac fiber—around its ends (at intercalated disks) as well as its sides.

Connective tissue

Connective tissue is the most widespread and abundant tissue in the body. It connects and supports—connects tissues to each other, for example, and muscles to bones and bones to other bones. It forms a supporting framework for the body as a whole and for its organs individually. Connective tissue exists in more varied forms than the other three basic tissues. Delicate tissue paper webs, strong, tough cords, rigid bones—all are made of connective tissue.

One scheme of classification lists the following main types of connective tissue:

1 Reticular
2 Loose, ordinary (areolar)
3 Adipose
4 Dense fibrous
5 Bone
6 Cartilage
7 Hemopoietic
8 Blood

Connective tissue consists predominantly of intercellular material and relatively few cells. Hence, the qualities of the intercellular material largely determine the qualities of each type of connective tissue. One connective tissue, for example, is a fluid because its intercellular material is a fluid. Some connective tissues have the consistency of a soft gel, some are firm but flexible, some are hard and rigid, some are tough, others are delicate—and in each case it is their intercellular substance that makes them so.

Intercellular substance may contain one or more of the following kinds of fibers—collagenous (or white), reticular, and elastic. They differ considerably in their properties. Collagenous fibers are tough and strong, reticular fibers are delicate, and elastic fibers are extendable and elastic. Collagenous or white fibers often occur in bundles—an arrangement that provides great tensile strength. Reticular fibers, in contrast, occur in networks and, although delicate, support small structures such as capillaries and nerve fibers. The protein collagen composes collagenous fibers. You know the hydrated form of collagen as gelatin. Of all the hundreds of different protein compounds in the body,

Table 3-1
Tissues

Tissue	Location	Function
Epithelial		
Simple squamous	Alveoli of lungs	Diffusion of respiratory gases between alveolar air and blood
	Lining blood and lymphatic vessels (called endothelium; classed as connective tissue by some histologists)	Diffusion; filtration; osmosis
	Surface layer of pleura, pericardium, peritoneum (called mesothelium; classed as connective tissue by some histologists)	Diffusion; osmosis
Stratified squamous	Surface of lining of mouth and esophagus	Protection
	Surface of skin (epidermis)	Protection
Simple columnar	Surface layer of lining of stomach, intestines, and part of respiratory tract	Protection; secretion; absorption; moving of mucus (by ciliated columnar)
Muscle		
Skeletal (striated voluntary)	Muscles that attach to bones	Movement of bones
	Extrinsic eyeball muscles	Eye movements
	Upper third of esophagus	First part of swallowing
Visceral (nonstriated involuntary or smooth)	In walls of tubular viscera of digestive, respiratory, and genitourinary tracts	Movement of substances along respective tracts
	In walls of blood vessels and large lymphatics	Change diameter of blood vessels, thereby aiding in regulation of blood pressure
	In ducts of glands	Movement of substances along ducts
	Intrinsic eye muscles (iris and ciliary body)	Change diameter of pupils and shape of lens
	Arrector muscles of hairs	Erection of hairs (gooseflesh)
Cardiac (striated involuntary)	Wall of heart	Contraction of heart
Connective (most widely distributed of all tissues)		
Reticular tissue	Spleen, lymph nodes, bone marrow	Filtering of injurious substances from blood and lymph by reticular network; and phagocytosis by reticular cells; synthesis of reticular fibers by reticular cells
Loose, ordinary (areolar)	Between other tissues and organs	Connection
	Superficial fascia	Connection
Adipose (fat)	Under skin	Protection
	Padding at various points	Insulation
		Support
		Reserve food
Dense fibrous	Tendons	Flexible but strong connection
	Ligaments	
	Aponeuroses	
	Deep fascia	
	Dermis	
	Scars	
	Capsule of kidney, etc.	

Tissue	Location	Function
Connective continued		
Bone	Skeleton	Support Protection
Cartilage		
Hyaline	Part of nasal septum Covering articular surfaces of bones Larynx Rings in trachea and bronchi	Firm but flexible support
Fibrous	Disks between vertebrae Symphysis pubis	
Elastic	External ear Eustachian tube	
Hemopoietic		
Myeloid (bone marrow)	Marrow spaces of bones	Formation of red blood cells, granular leukocytes, platelets; also reticuloendothelial cells and some other connective tissue cells
Lymphatic	Lymph nodes Spleen Tonsils and adenoids Thymus gland	Formation of lymphocytes and monocytes; also plasma cells and some other connective tissue cells
Blood	In blood vessels	Transportation Protection
Nerve	Brain Spinal cord Nerves	Irritability and conduction

collagen is the most abundant. Biologists now estimate that it constitutes somewhat over one fourth of all the protein in the body. And interestingly, one of the most basic factors in the aging process, according to some researchers,* is the change in the molecular structure of collagen that occurs gradually with the passage of the years.

Like intercellular fibers, intercellular ground substance also varies in properties. The ground substance (matrix) in loose, ordinary connective tissue, for example, is a soft, viscous gel, whereas the matrix of bone is a very hard substance—as "hard as bone," in fact. A compound called hyaluronic acid gives the viscous quality to intercellular gels, but it can be converted to a watery consistency by the enzyme hyaluronidase. Physicians have made use of this latter fact for some time now. They frequently give a commercial preparation of hyaluronidase intramuscularly or subcutaneously with drugs or fluids. By decreasing the viscosity of intercellular material, the enzyme hastens diffusion and absorption and thereby lessens tissue tension and pain.

*Kohn, R. R.: Principles of mammalian aging, Englewood Cliffs, N. J., 1971, Prentice-Hall, Inc., pp. 22-24.

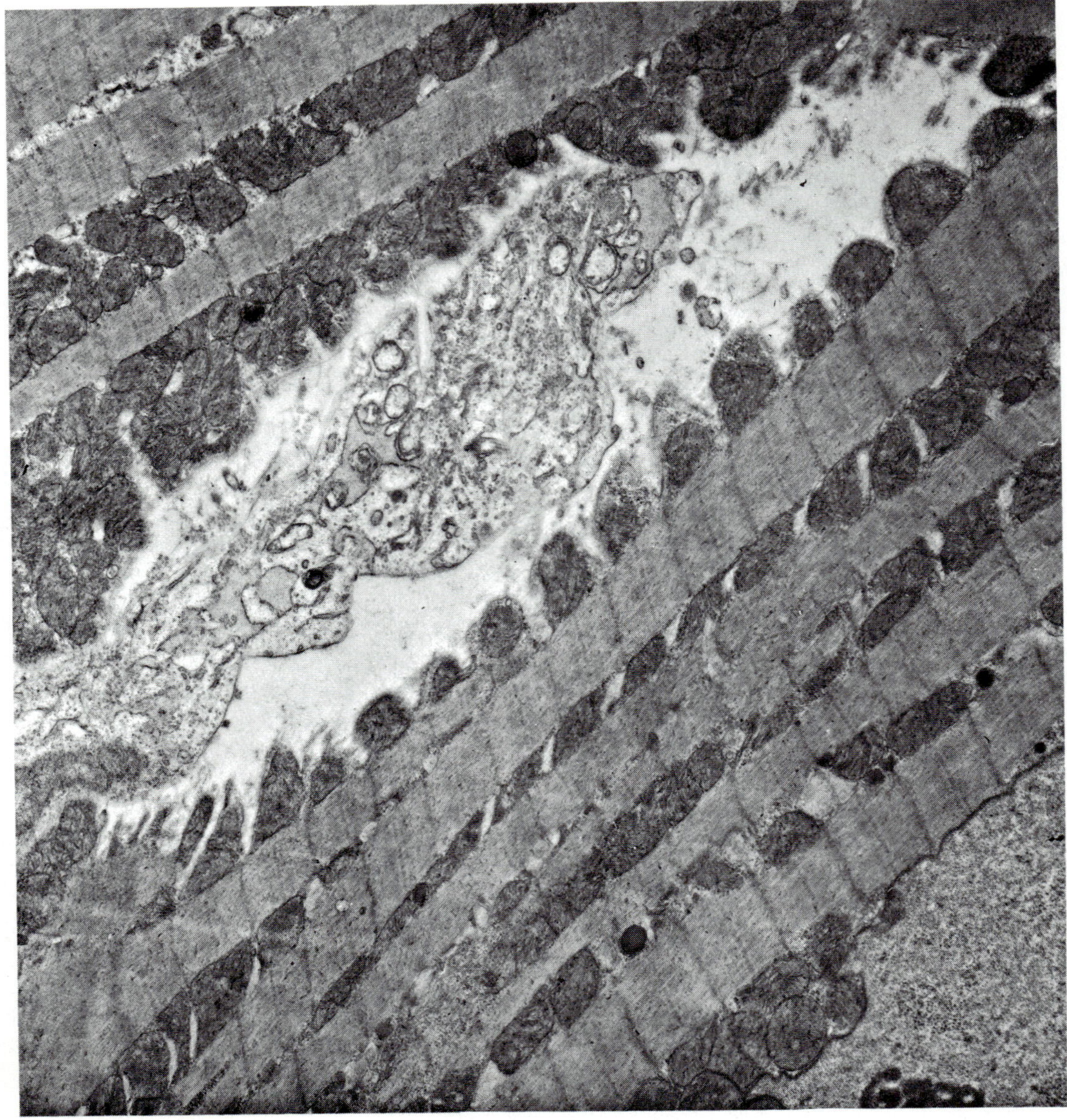

Fig. 3-7

Electron photomicrograph of rabbit heart muscle (×12,500). (Courtesy Department of Pathology, Case Western Reserve University, Cleveland, Ohio.)

Reticular tissue

A three-dimensional web, or reticular network, identifies reticular tissue. Slender, branching reticular fibers with reticular cells overlying them compose the reticular meshwork. Branches of the cytoplasm of reticular cells follow the branching reticular fibers.

Reticular tissue forms the framework of the spleen, lymph nodes, and bone marrow. It functions as part of the body's complex immune mechanism in that the reticular meshwork filters injurious substances out of the blood and lymph, and reticular cells and certain other kinds of cells phagocytose (engulf and destroy) them. Another and probably the major function of reticular cells is to make reticular fibers.

Loose, ordinary connective tissue (areolar)

First, a few words of explanation about the names of this kind of tissue. It is called loose because it is stretchable and ordinary because it is one of the most widely distributed of all tissues. It is common and ordinary, not special like some kinds of connective tissue (bone and cartilage, for example) that help form comparatively few structures. Areolar was the early name for the loose, ordinary connective tissue that connects many adjacent structures of the body. It acts like a glue spread between them, but an elastic glue that permits movement. The word areolar means "like a small space" and refers to the bubbles that appear as areolar tissue is pulled apart during dissection.

Although intercellular substance is prominent in loose, ordinary connective tissue, cells also are numerous and varied (Fig. 3-8). Collagenous and elastic fibers are interwoven loosely and embedded in a soft viscous ground substance. Of the half dozen or so kinds of cells present, *fibroblasts* are the most

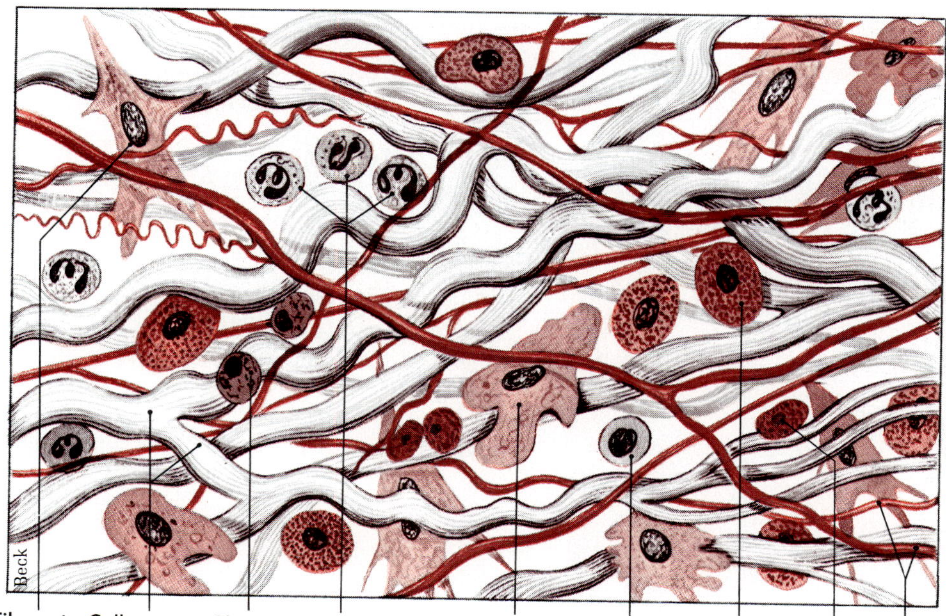

Fig. 3-8

Areolar connective tissue. The large white fibers are collagenous fibers. Each of the red strands consists of a bundle of elastic fibers. Several fibroblasts are shown between the fibers. Also shown are macrophages, a plasma cell, a mast cell, and three types of white blood cells: polymorphonuclear leukocytes, eosinophils, and a monocyte.

common and *macrophages* are second. Fibroblasts synthesize intercellular substances of both types—that is, both fibers and gels. Macrophages carry on phagocytosis and so are classified as phagocytes. Phagocytosis is part of the body's vital complex of defense mechanisms. Other cells besides macrophages also engage in phagocytosis. *Plasma cells, fat cells, mast cells,* and some *white blood cells* (leukocytes) are also found in loose ordinary connective tissue but in smaller numbers than fibroblasts and macrophages. Plasma cells function as antibody producers.

Adipose tissue

Adipose tissue differs from loose, ordinary connective tissue mainly in that it contains predominantly fat cells and many fewer fibroblasts, macrophages, and mast cells (Fig. 3-9). Adipose tissue forms protective pads around the kidneys and various other structures. It also serves two other functions—it constitutes a storage depot for excess food and it acts as an insulating material to conserve body heat.

Dense fibrous tissue

Dense fibrous tissue consists mainly of bundles of fibers arranged in parallel rows in a fluid matrix. It contains relatively few fibroblast cells. It composes tendons and ligaments. Bundles of collagenous fibers endow tendons with great tensile strength and nonstretchability—desirable characteristics for these structures that anchor our muscles to bones. In ligaments, on the other hand, bundles of elastic fibers predominate. Hence, ligaments exhibit some degree of elasticity.

Bone and cartilage

Bone and cartilage are discussed in Chapter 5.

Hemopoietic tissue

Hemopoietic tissue is discussed in Chapter 10.

Blood

Blood consists of billions of cells afloat in a fluid intercellular substance called plasma. We shall discuss both the cells and the inter-

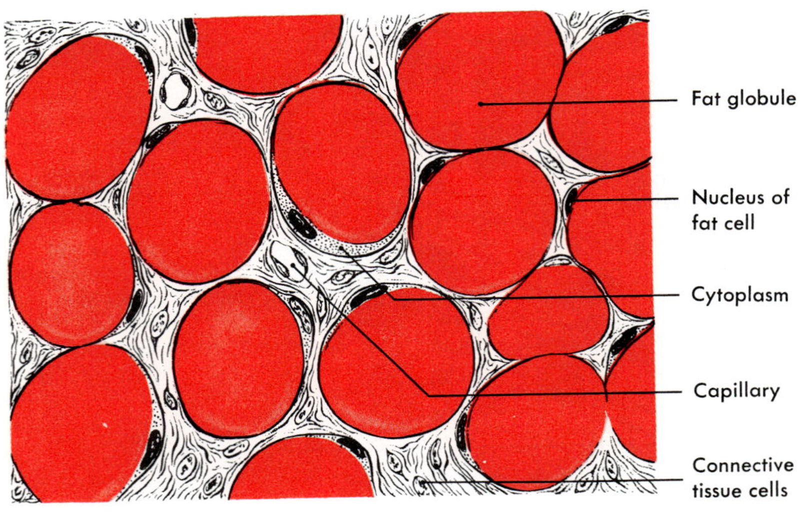

Fig. 3-9

Adipose tissue. Fat droplet occupies nearly the entire area of the cell. Cytoplasm and nucleus are forced to the periphery of the cell.

cellular substance of this vital connective tissue in considerable detail in Chapter 10.

Reticuloendothelial system

Various types of connective tissue cells that carry on phagocytosis, although widely scattered throughout the body, are sometimes spoken of as reticuloendothelial tissue or more often as the *reticuloendothelial system.* Phagocytic cells (that is, phagocytes) constitute an important part of the body's defense mechanism. Opinions differ somewhat as to which kinds of cells compose reticuloendothelial tissue. There does seem to be agreement, however, that among them are the following types of phagocytes:

1. *Macrophages*—two types of these cells identified, namely, fixed macrophages and free macrophages. Reticular cells is another name for the fixed or nonmotile macrophages. Free macrophages, in contrast, are motile cells. For example, in the presence of infectious organisms, free macrophages become activated and wander out into the inflamed area—hence the picturesque names "wandering cells" and "resting wandering cells" also used to designate free macrophages.
2. *Reticuloendothelial cells*—located in the lining of blood sinusoids* in the liver, spleen, and bone marrow and in the lining of lymph channels in lymph nodes. Another name for the reticuloendothelial cells of liver sinusoids is stellate cells of von Kuppfer, for the man who first described them.

Nerve tissue

Nerve tissue is discussed in Chapter 7.

*Sinusoids are tiny blood vessels. They are analogous to capillaries in that they connect the arterial side of circulation to the venous side but differ in minor ways from capillaries.

Outline summary

1 Definition—organization of cells with nonliving intercellular substances
2 Basic types
  a Epithelial
  b Muscle
  c Connective
  d Nerve

Epithelial tissue

1 General functions—protection, secretion, diffusion, filtration, and absorption
2 Main types
  a Simple squamous
  b Stratified squamous
  c Simple columnar

Simple squamous epithelium

1 Single layers of flat cells
2 Function—diffusion and filtration

Stratified squamous epithelium

1 Several layers of cells
2 Function—protection

Simple columnar epithelium

1 Single layer of columnar and goblet-shaped cells and, in some places, ciliated cells
2 Functions—absorption, secretion, and moving mucus

Muscle tissue

1 General functions—contraction and therefore movement
2 Types
  a Skeletal; also called striated voluntary
  b Visceral; also called nonstriated or smooth involuntary
  c Cardiac; also called striated involuntary

Connective tissue

1 General functions—connection, support, and protection
2 Main types
  a Reticular
  b Loose, ordinary connective (areolar)
  c Adipose
  d Dense fibrous
  e Cartilage
  f Bone
  g Hemopoietic
  h Blood
3 General characteristics—intercellular material predominates in most connective tissues and determines their physical characteristics; consists of fluid, gel, or solid matrix, with or without fibers (collagenous, reticular, and elastic)

Reticular tissue

Forms a three-dimensional network consisting of branching reticular fibers and reticular cells overlying them; reticular tissue makes the framework of the spleen, lymph nodes, and bone marrow; it functions as part of the body's complex immune mechanism, filtering out of the blood and lymph injurious substances and phagocytosing them; reticular cells synthesize reticular fibers

Loose, ordinary connective tissue (areolar)

1 One of most widely distributed of all tissues; intercellular substance is prominent and consists of collagenous and elastic fibers loosely interwoven and embedded in soft viscous ground substance; several kinds of cells present, notably fibroblasts and macrophages, also mast cells, plasma cells, fat cells, and some white blood cells
2 Function—connection

Adipose tissue

1 Similar to loose, ordinary connective tissue but contains mainly fat cells
2 Functions—protection, insulation, support, and reserve food

Dense fibrous tissue

1 Fibrous intercellular substance (collagenous and elastic fibers); few fibroblast cells
2 Function—furnishes flexible but strong connection

Reticuloendothelial system

1 Macrophages
   a Fixed or nonmotile—also called reticular cells
   b Free or motile—when activated by infections, they wander out into inflamed area; other names, wandering cells, resting wandering cells, histiocytes, and clasmatocytes
2 Reticuloendothelial cells that line small lymph and blood channels in lymph nodes, liver, spleen, and bone marrow; function as phagocytes

Review questions

1 Name the four basic types of tissue.
2 What are the main functions of each basic type of tissue?
3 What are the names of the subtypes of connective tissue?
4 Describe intercellular substance.
5 Explain the scientific basis for giving hyaluronidase with fluids or certain injected drugs.
6 Name several kinds of connective tissue cells.
7 What kind of cells produce intercellular substances?
8 Name three subtypes of muscle tissue. Give more than one name for each.
9 What special function do reticuloendothelial cells perform?
10 What are the main locations of reticuloendothelial cells?
11 Make a list of terms you have encountered for the first time in this chapter. Define each in your own words.
12 Recent evidence suggests that one of the most basic factors in the aging process is a molecular change in what intercellular protein?

4

Membranes and glands

Membranes

Membranes constitute a special class of organs in that they are merely thin sheets of tissues covering or lining various parts of the body. Of the numerous membranes in the body, four kinds are particularly important: mucous, serous, synovial, and cutaneous (skin). Other miscellaneous membranes (periosteum, fascia, dura mater, sclera, etc.) will be described from time to time.

Mucous membrane

Mucous membrane lines cavities or passageways of the body that open to the exterior, such as the lining of the mouth and entire digestive tract, the respiratory passages, and the genitourinary tract. It consists, as do serous, synovial, and cutaneous membranes, of a surface layer of epithelial tissue over a deeper layer of connective tissue. Mucous membrane performs the functions of protection, secretion, and absorption—protection, for example, against bacterial invasion, secretion of mucus, and absorption of water, salts, and other solutes.

Serous and synovial membranes

Serous and synovial membranes line the closed cavities of the body—that is, those that do not open to the exterior. Serous membrane that lines the thoracic cavity is called *pleura,* that which lines the abdominal cavity is called *peritoneum,* and that which lines the sac in which the heart lies is called *pericardium.*

Not only does serous membrane line the thoracic and abdominal cavities and the pericardial sac, but it also covers the organs lying in these spaces. The term *visceral layer* is applied to the part of the membrane that covers the organs, while that which lines the cavity is called *parietal layer* (Fig. 4-1). Between the two layers there is a potential space kept moist by a small amount of serous fluid. When an organ moves against the body wall (as the lungs do in respiration, or when the heart beats in its serous sac), friction between the moving parts is prevented by the presence of the smooth moist serous sheets lining the wall surface of the cavity and covering the organ surfaces. The mechanical principle that moving parts must have lubricated surfaces is thereby carried out in the body.

Synovial membrane lines joint cavities, tendon sheaths, and bursae. Its smooth moist surfaces protect against friction.

Cutaneous membrane

Vital, diverse, complex, extensive—these adjectives describe in part the body's largest and one of its most important organs, the skin. In terms of surface area, the skin is as large as the body itself—probably 2,500 to 3,000 square inches in most adults. Skin functions are crucial to survival. They are also diverse, including such different functions as protection, excretion, and sensation. The skin also plays a part in maintaining fluid and electrolyte balance and normal body temperature. It protects us against entry of unconquerable hordes of microorganisms. It minimizes mechanical injury of underlying structures. It bars entry of excess sunlight and of most chemicals. Even water does not penetrate it under most circumstances.

Millions of microscopic nerve endings are distributed throughout the skin. These serve as antennas or receivers for the body, keeping it informed of changes in its environment—information essential for health and at times even vital for survival.

Epidermis and dermis

Two main layers compose the skin: an outer and thinner layer, the *epidermis,* and an inner, thicker layer, the *dermis* (Fig. 4-2). The epidermis consists of stratified squamous

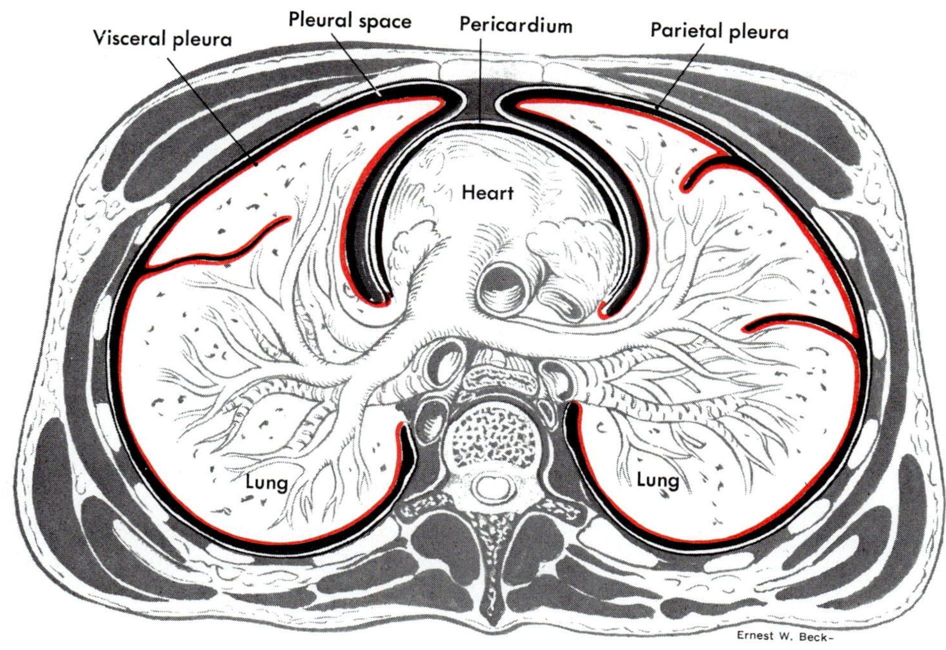

Fig. 4-1

Transverse section through the chest showing the parietal pleura in white (lining walls of chest cavity), the visceral pleura in red (covering lungs), and the pleural space in black (between parietal and visceral pleuras).

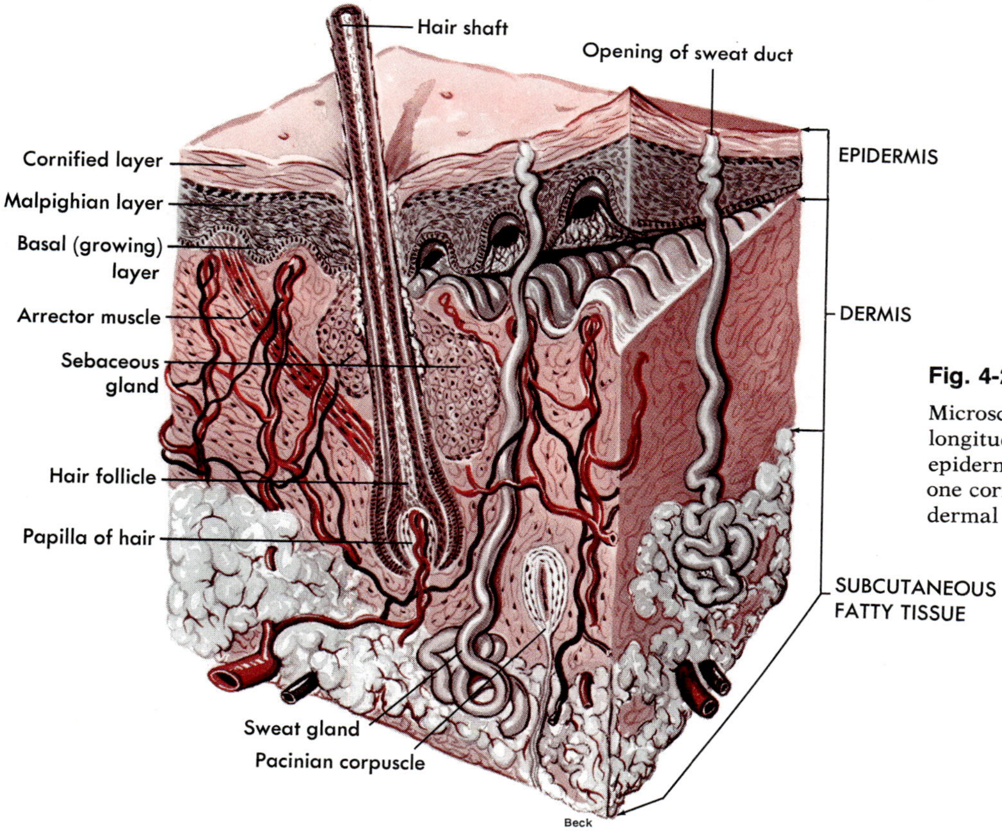

Fig. 4-2

Microscopic view of the skin in longitudinal section. The epidermis is shown raised at one corner to reveal the dermal papillae.

epithelial tissue and the dermis of fibrous connective tissue. Underlying the dermis is subcutaneous tissue or *superficial fascia* made of areolar and, in many areas, adipose tissue. The epidermis in all parts of the body except the palms of the hands and soles of the feet has four layers. In the skin of the palms and soles, there are five layers of epidermis. From the outside in, they are as follows:

1 *Stratum corneum (horny layer)*—dead cells converted to a water-repellant protein called keratin that continually flakes off (desquamates)
2 *Stratum lucidum*—so named because of the presence of a translucent compound (eleidin) from which keratin forms; present only in thick skin of palms and soles
3 *Stratum granulosum*—so named because of granules visible in cytoplasm of cells (cells die in this layer)
4 *Stratum spinosum (prickle cell layer)*—several layers of irregularly shaped cells
5 *Stratum germinativum* (or *basal layer*)—columnar-shaped cells, the only cells in the epidermis that undergo mitosis; new cells produced in this deepest stratum at the rate old keratinized cells lost from the stratum corneum; new cells continually push surfaceward from the stratum germinativum into each successive layer, only to die, become keratinized, and eventually flake off as did their predecessors*

An interesting characteristic of the dermis or deep layer of the skin is its parallel ridges, suggestive on a miniature scale of the ridges of a contour plowed field. Epidermal ridges, the ones made famous by the art of fingerprinting, exist because the epidermis conforms to the underlying dermal ridges.

*Incidentally, this fact illustrates nicely the physiological principle that while life continues the body's work is never done; even at rest it is producing new cells to replace millions of old ones.

As everyone knows, human skin comes in a wide assortment of colors. Explanation for this fact lies in a constellation of factors that act and interact in complex ways (Fig. 4-3). The basic determinant, however, of skin color is the quantity of *melanin*—the main skin pigment—deposited in the epidermis. Heredity, first of all, determines this. Geneticists now tell us that four to six pairs of genes exert the primary control over the amount of melanin formed by special cells called *melanocytes*. Thus, heredity determines how dark or how light one's basic skin color will be. But other factors can modify the genetic effect. Sunlight is the obvious example. Prolonged exposure of the skin to sunlight causes melanocytes to increase melanin production and darken skin color. So, too, does an excess of adrenocorticotropic hormone (ACTH) or an excess of melanocyte-stimulating hormone (MSH), two of the hormones secreted by the anterior pituitary gland.

An individual's basic skin color changes, as we have just observed, whenever its melanin content changes appreciably. But skin color can also change without any change in melanin. In this case, the change is usually temporary and most often stems from a change in the volume of blood flowing through skin capillaries. A marked increase in skin blood volume may cause the skin to take on a pink or red hue, and an appreciable decrease in skin blood volume may turn it frighteningly pale. In general, the sparser the pigments in the epidermis, the more transparent the skin is and, therefore, the more vivid the change in skin color will be with a change in skin blood volume. Conversely, the richer the pigmentation, the more opaque the skin is and the less skin color will change with a change in skin blood volume.

In some abnormal conditions, skin color changes because of an excess amount of unoxygenated hemoglobin in the skin capillary blood. If skin contains relatively little mela-

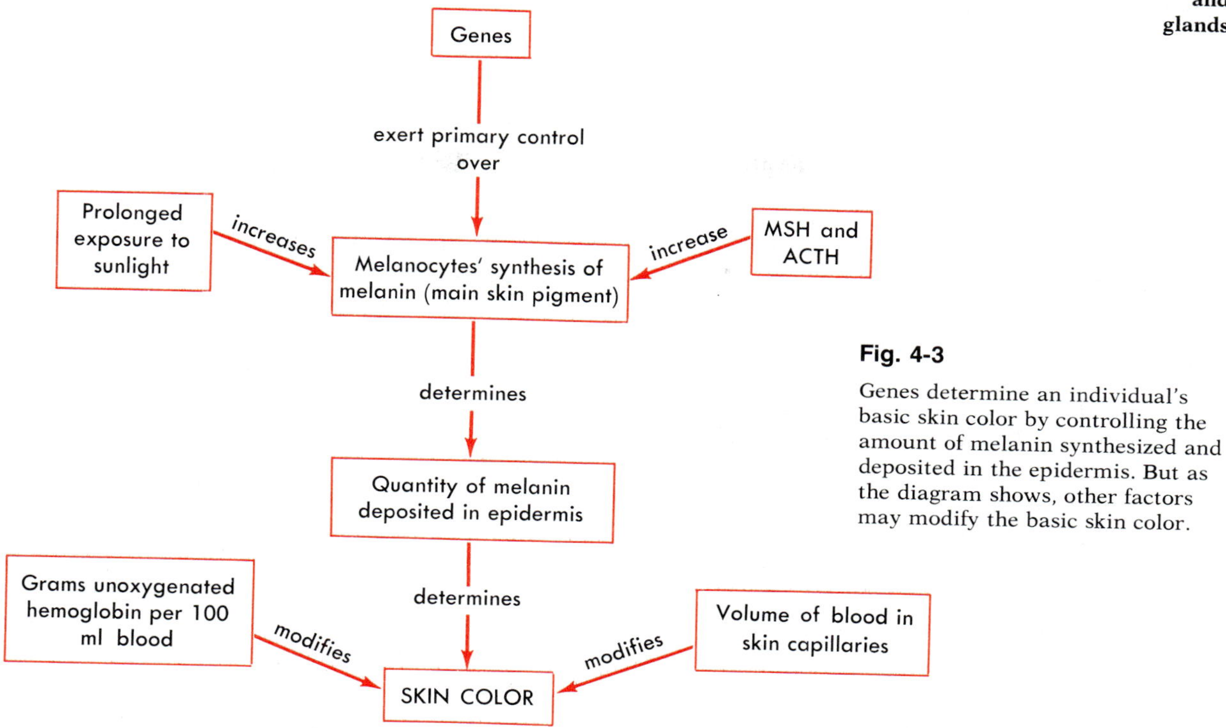

Fig. 4-3

Genes determine an individual's basic skin color by controlling the amount of melanin synthesized and deposited in the epidermis. But as the diagram shows, other factors may modify the basic skin color.

nin, it will develop a bluish cast known as *cyanosis* when 100 ml of blood contains about 5 gm of unoxygenated hemoglobin. In general, the darker the skin pigmentation, the more unoxygenated hemoglobin must be present before cyanosis becomes visible.

Accessory organs of skin

The accessory organs of the skin consist of hair, nails, and microscopic glands.

Hair. Hair is distributed over the entire body except the palms and soles. The structure of a hair has several points of similarity to that of the epidermis. Just as the epidermis is formed by the cells of its deepest layer, reproducing and forcing the daughter cells, which become horny in character, upward, so a hair is formed by a group of cells at its base multiplying and pushing upward and in so doing becoming keratinized. The part of the hair that is visible is the *shaft*, whereas that which is embedded in the dermis is the *root*. The root, together with its coverings (an outer connective tissue sheath and an inner epithelial coating that is a continuation of the stratum germinativum), forms the hair *follicle*. At the bottom of the follicle is a loop of capillaries enclosed in a connective tissue covering called the hair *papilla*. The clusters of epithelial cells lying over the papilla are the ones that reproduce and eventually form the hair shaft. As long as these cells remain alive, hair will regenerate even though it be cut or plucked or otherwise removed.

Each hair is kept soft and pliable by two or more *sebaceous glands*, which secrete varying amounts of oily *sebum* into the follicle near the surface of the skin. Attached to the follicle, too, are small bundles of involuntary muscle known as the *arrector pili muscles*. These muscles are of interest because when they contract, the hair "stands on end," as it does in extreme fright or cold, for example. This mechanism is responsible also for gooseflesh. As the hair is pulled into an upright position, it raises the skin around it into the familiar little goose pimples.

Hair color results from different amounts of melanin pigments in the outer layer (cortex) of the hair. White hair contains little or no melanin.

Some hair, notably that around the eyes and in the nose and ears, performs a protective function in that it keeps out some dust and insects. For the hair on the bulk of the skin, however, no function seems apparent.

Nails. The nails are epidermal cells that have been converted to keratin. They grow from epithelial cells lying under the white crescent (lunula) at the proximal end of each nail.

Skin glands. The skin glands include three kinds of microscopic glands—namely, sebaceous, sweat, and ceruminous.

Sebaceous glands secrete oil for the hair. Wherever hairs grow from the skin, there are sebaceous glands, at least two for each hair. The oil, or *sebum*, secreted by these tiny glands has value not only because it keeps the hair supple but also because it keeps the skin soft and pliant. Moreover, it prevents excessive water evaporation from the skin and water absorption through the skin. And because fat is a poor conductor of heat, the sebum secreted onto the skin lessens the amount of heat lost from this large surface.

Sweat glands, though very small structures, are very important and very numerous—especially on the palms, soles, forehead, and axillae (armpits). Histologists estimate, for example, that a single square inch of skin on the palms of the hands contains about 3,000 sweat glands.

Sweat secretion helps maintain homeostasis of fluid and electrolytes and of body temperature. For example, if too much heat is being produced, as in strenuous exercise, or if the environmental temperature is high, these glands secrete more sweat, which, in evaporating, cools the body surface. Inasmuch as sweat contains some nitrogenous wastes, the sweat glands also function as excretory organs.

Ceruminous glands are thought to be modified sweat glands. They are located in the external ear canal. Instead of watery sweat, they secrete a waxy, pigmented substance, the *cerumen.*

Terms used in connection with skin

hypodermic—beneath or under the skin
subcutaneous—same as hypodermic
intracutaneous—within the layers of the skin
diaphoresis—profuse perspiration
pores—minute openings of the sweat gland ducts on the surface of the skin; do not "open" or "close" since no muscle tissue enters into their formation; however, any agent that has an astringent action on the skin causes them to shrink or, in effect, to close
furuncle—a boil, an infection of a hair follicle

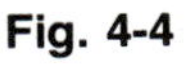

Fig. 4-4

Simple tubular gland, an exocrine gland with an unbranching duct.

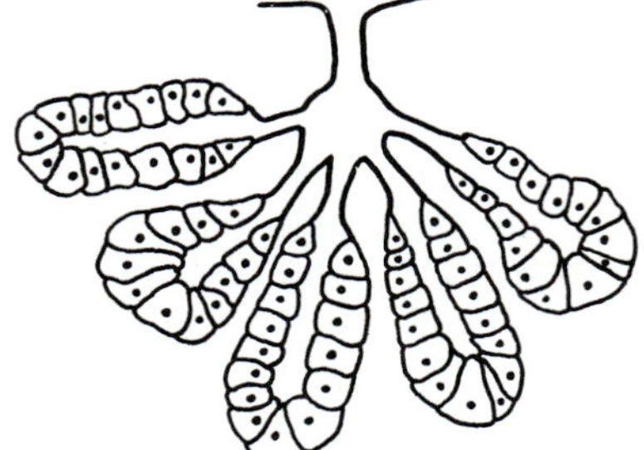

Fig. 4-5

Compound tubulo-alveolar gland, an exocrine gland with a branching duct.

Glands

Glands consist of epithelial cells specialized for synthesizing compounds that they secrete either into ducts or blood. Relatively simple compounds enter a gland's cells from the blood, and chemical reactions within the cells build them up into the more complex compounds that make up the gland's secretion. *Exocrine glands* secrete into ducts and *endocrine glands* into blood capillaries. Exocrine glands are further classified in several ways. Sweat glands, for example, are *simple tubular exocrine glands* (Fig. 4-4). Interpreted, simple means that each gland has a single nonbranching duct, and tubular means that its secretory unit is tubular shaped. A salivary gland, in contrast, is a *compound tubuloalveolar gland* (Fig. 4-5) because it has a branching duct and some tubular and some flask-shaped secretory units.

Outline summary

MEMBRANES

1 Definition—thin sheet of tissues that either covers or lines a part of body or divides an organ
2 Types—mucous, serous, synovial, cutaneous, and miscellaneous

Mucous membrane

1 Location—lines cavities and passages that open to exterior
2 Structure—surface layer of epithelial tissue over connective tissue
3 Functions—protection, secretion, and absorption

Serous membrane

1 Location—lines cavities that do not open to exterior
2 Functions—protection and secretion

Synovial membrane

1 Location—lines joint cavities, tendon sheaths, and bursae
2 Functions—protection and secretion

Cutaneous membrane

1 Functions
  a Protection against various factors—e.g., microorganisms, sunlight, and chemicals
  b Excretion of sweat
  c Sensations
  d Fluid and electrolyte balance—skin helps maintain this by secreting varying amounts of sweat
  e Normal body temperature—skin helps maintain this by varying amounts of blood flow through it and also by varying amounts of sweat secretion
2 Structure—two main layers: epidermis, outer, thinner layer of stratified squamous epithelium, and dermis, inner, thicker layer of connective tissue

Epidermis

1 Outer layer of stratified squamous epithelial cells
2 Surface cells dead, keratinized, and practically waterproof
3 Only deepest layer of cells undergoes mitosis to replace surface cells that continually desquamate
4 Melanin pigments mainly in deepest layer

Dermis

1 Dense fibrous connective tissue layer underlying epidermis
2 Dermis of palms and soles has numerous parallel ridges
3 Subcutaneous tissue (also called superficial fascia), underlying dermis, composed of areolar tissue or areolar and adipose tissues

Accessory organs of skin

1 Hair
  a Distribution—over entire body except palms and soles
  b Shaft—visible part of hair
  c Root—part of hair embedded in dermis
  d Follicle—root with coverings
  e Papilla—loop of capillaries enclosed in connective tissue covering
  f Germinal matrix—cluster of epithelial cells lying over papilla; these cells undergo mitosis to form hair; must be intact in order for hair to regenerate
  g Sebaceous glands and arrector pili muscles—attach to follicle; contraction of latter produces gooseflesh
  h Color—results from different amounts of melanin pigments in cortex of hair
2 Nails
  a Are epidermal cells converted to hard keratin
  b Grow from epithelial cells under lunula ("moons")
3 Skin glands
  a Sebaceous—secrete oil (sebum) that keeps hair and skin soft; helps prevent excess evaporation and absorption of water and excess heat loss
  b Sweat—numerous throughout skin, especially on palms, soles, forehead, and axillae; important in heat regulation
  c Ceruminous—thought to be modified sweat glands; located in external ear canal; secrete ear wax or cerumen

Terms used in connection with skin

See p. 52.

GLANDS

1 Composed of epithelial cells specialized for synthesizing compounds that they secrete either into ducts or blood capillaries
2 Types
   a Exocrine glands—secrete into ducts
      1 Simple glands—have nonbranching ducts
      2 Tubular glands—tubular-shaped secretory unit
      3 Compound glands—have branching ducts
      4 Alveolar glands—flask-shaped secretory units
   b Endocrine glands—secrete into blood

Review questions

1 What membranes line closed cavities? Cavities that open to the exterior?
2 What general functions do membranes serve?
3 What functions does the skin perform?
4 Describe the epidermis.
5 Describe the dermis.
6 What is keratin?
7 What is melanin and where is it found?
8 What is superficial fascia?
9 Name and describe the skin glands.
10 Define the following terms: alveolar gland, exocrine gland, endocrine gland, compound gland.
11 What organelle(s) would you expect to be prominent in secreting cells?

HUMAN ANATOMY

FULL-COLOR PLATES WITH SIX IN TRANSPARENT

"TRANS-VISION"® SHOWING STRUCTURES OF THE HUMAN TORSO

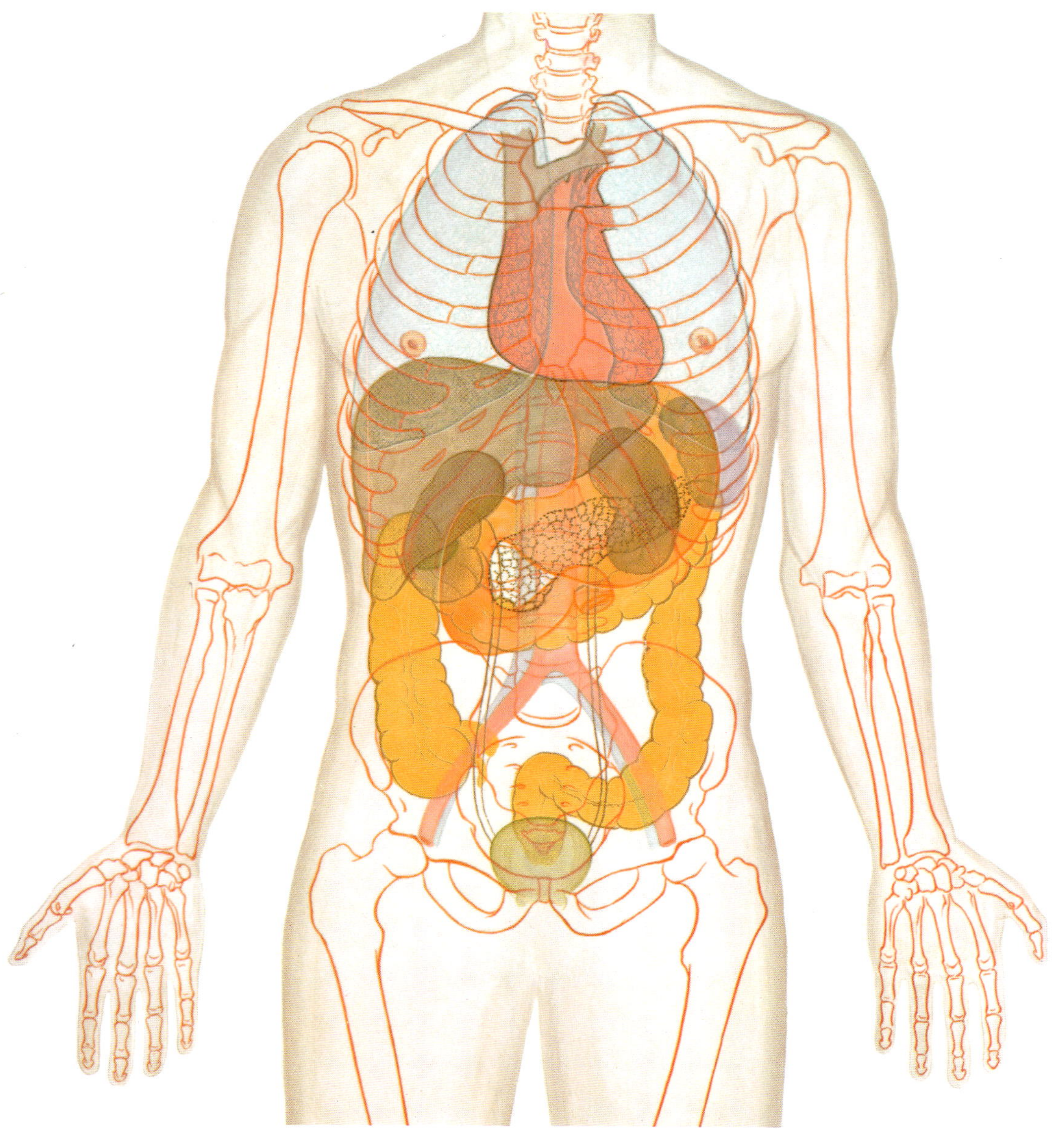

Plate I

ERNEST W. BECK, medical illustrator
in collaboration with
HARRY MONSEN, Ph.D.
Professor of Anatomy, College of Medicine, University of Illinois

Plate VIII

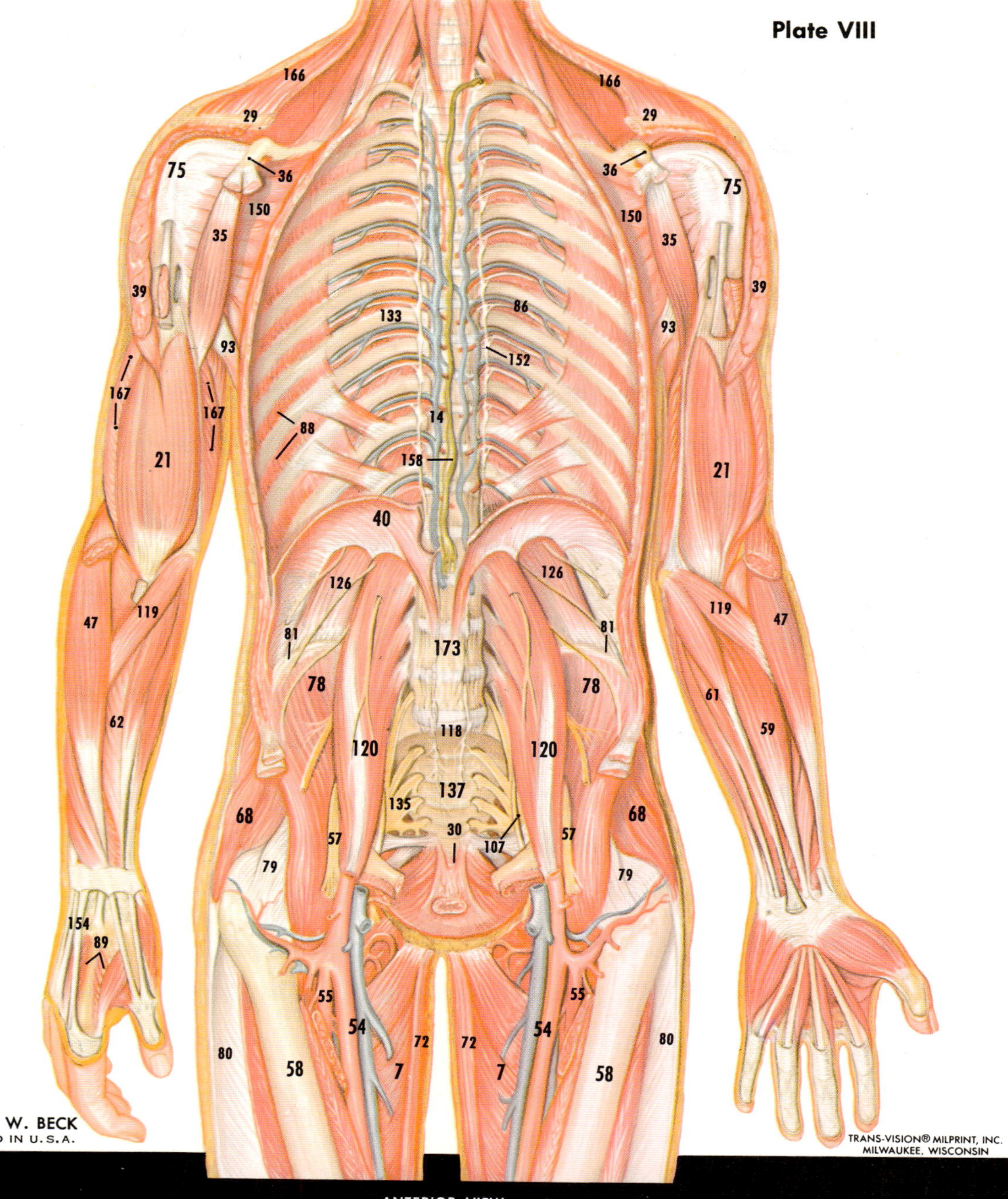

ANTERIOR VIEW

7. Adductor magnus muscle
14. Azygos veins
21. Brachialis muscle
29. Clavicle
30. Coccyx
35. Coracobrachialis muscle
36. Coracoid process of the scapula
39. Deltoid muscle
40. Diaphragm
47. Extensor carpi radialis longus muscle
54. Femoral artery and vein
55. Femoral artery, deep
57. Femoral nerve
58. Femur
59. Flexor carpi radialis muscle
61. Flexor digitorum profundus muscle
62. Flexor digitorum superficialis muscle
68. Gluteus medius muscle
75. Humerus
78. Iliacus muscle
79. Iliofemoral ligament
80. Iliotibial tract
81. Ilium
86. Intercostal artery, vein and nerve
88. Intercostal muscle, internal
89. Interosseous muscles, dorsal
93. Latissimus dorsi muscle
107. Obturator nerve
118. Promontory
119. Pronator teres muscle
120. Psoas muscles (major and minor)
126. Quadratus lumborum muscle
133. Rib
135. Sacral nerves
137. Sacrum
150. Subscapularis muscle
152. Sympathetic (autonomic) nerve chain
154. Tendons of extensor muscles of hand
158. Thoracic duct
166. Trapezius muscle
167. Triceps brachii muscle
173. Vertebral column

Plate IX

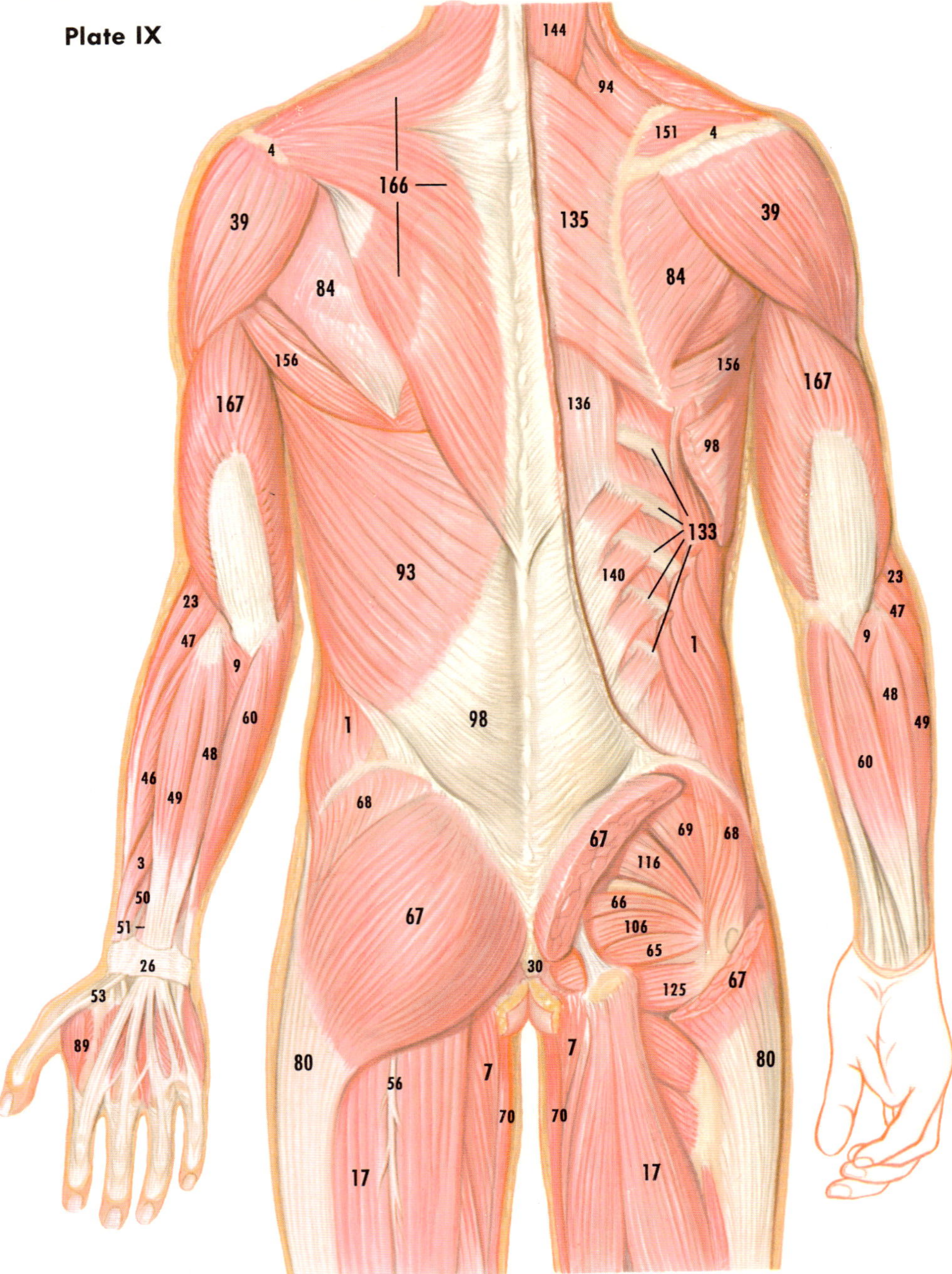

POSTERIOR VIEW

1. Abdominal oblique muscle, external
3. Abductor pollicis longus muscle
4. Acromion process of the scapula
7. Adductor magnus muscle
9. Anconeus muscle
17. Biceps femoris muscle
23. Brachioradialis muscle
26. Carpal ligament, dorsal
30. Coccyx
39. Deltoid muscle
46. Extensor carpi radialis brevis muscle
47. Extensor carpi radialis longus muscle
48. Extensor carpi ulnaris muscle
49. Extensor digitorum communis muscle
50. Extensor pollicis brevis muscle
51. Extensor pollicis longus muscle
56. Femoral cutaneous nerve, posterior
60. Flexor carpi ulnaris muscle
65. Gemellus inferior muscle
66. Gemellus superior muscle
67. Gluteus maximus muscle
68. Gluteus medius muscle
69. Gluteus minimus muscle
70. Gracilis muscle
80. Iliotibial tract
84. Infraspinatus muscle
89. Interosseous muscle, dorsal
93. Latissimus dorsi muscle
94. Levator scapulae muscle
98. Lumbodorsal fascia
106. Obturator internus muscle
116. Piriformis muscle
125. Quadratus femoris muscle
133. Ribs (VII-XII)
135. Rhomboideus muscle
136. Erector spinae muscl
140. Serratus posterior inferior muscle
144. Splenius capitis muscl
151. Supraspinatus muscle
156. Teres major muscle
166. Trapezius muscle
167. Triceps brachii muscle

BONES AND SINUSES OF THE SKULL

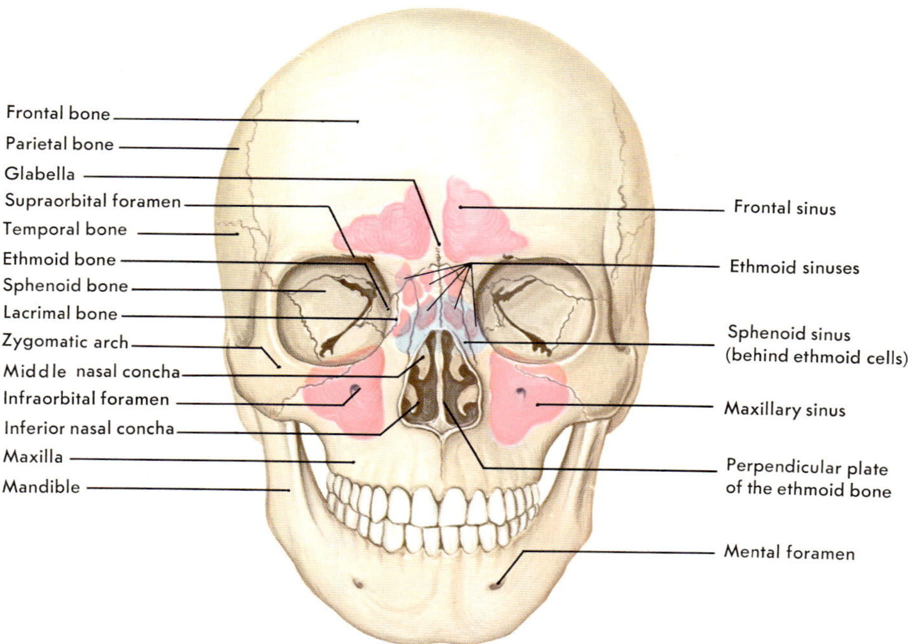

HEMISECTION OF THE HEAD AND NECK

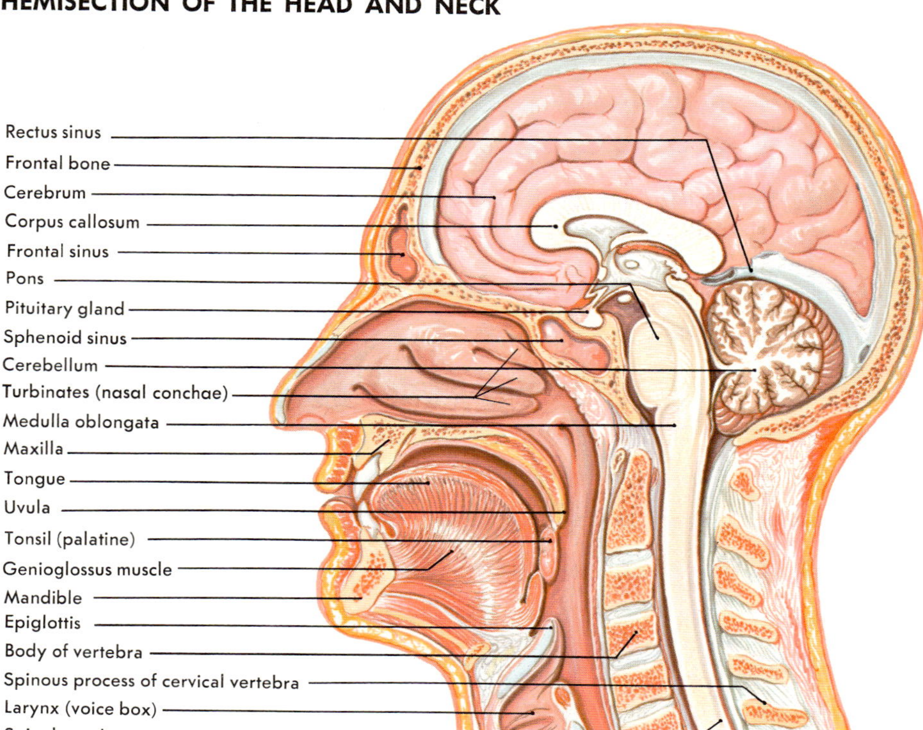

Plate XI

ANATOMY OF THE EAR

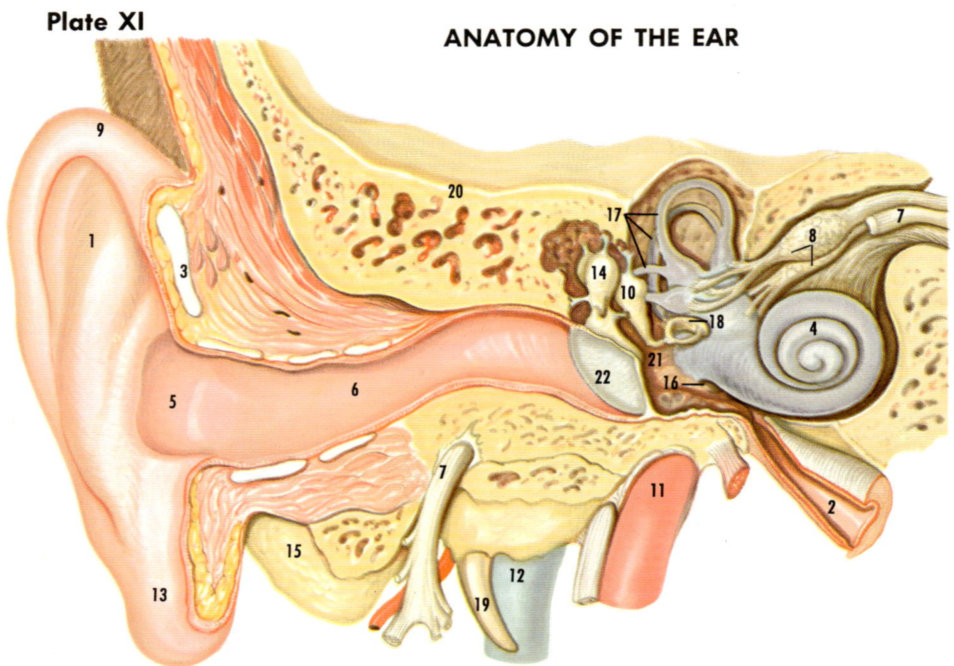

1. Anthelix
2. Auditory tube
3. Cartilage
4. Cochlea
5. Concha
6. External acoustic meatus
7. Facial nerve
8. Ganglia of the vestibular nerve
9. Helix
10. Incus (anvil)
11. Internal carotid artery
12. Internal jugular vein
13. Lobe
14. Malleus (hammer)
15. Mastoid process
16. Round window
17. Semicircular canals
18. Stapes (stirrup)
19. Styloid process
20. Temporal bone
21. Tympanic cavity
22. Tympanic membrane (eardrum)

ANATOMY OF THE EYE

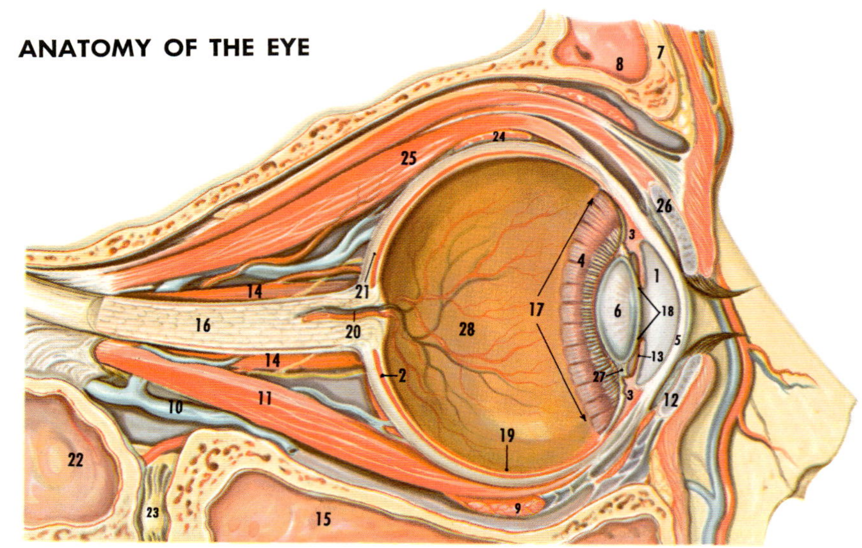

1. Aqueous chamber
2. Choroid
3. Ciliary muscle
4. Ciliary processes
5. Cornea
6. Crystalline lens
7. Frontal bone
8. Frontal sinus
9. Inferior oblique muscle
10. Inferior ophthalmic vein
11. Inferior rectus muscle
12. Inferior tarsus
13. Iris
14. Lateral rectus muscle
15. Maxillary sinus
16. Optic nerve
17. Ora serrata
18. Pupil of the iris
19. Retina
20. Retinal artery and vein
21. Sclera
22. Sphenoid sinus
23. Pterygopalatine ganglion
24. Superior oblique muscle
25. Superior rectus muscle
26. Superior tarsus
27. Suspensory ligament
28. Vitreous chamber

SCHEMATIC BODY CELL

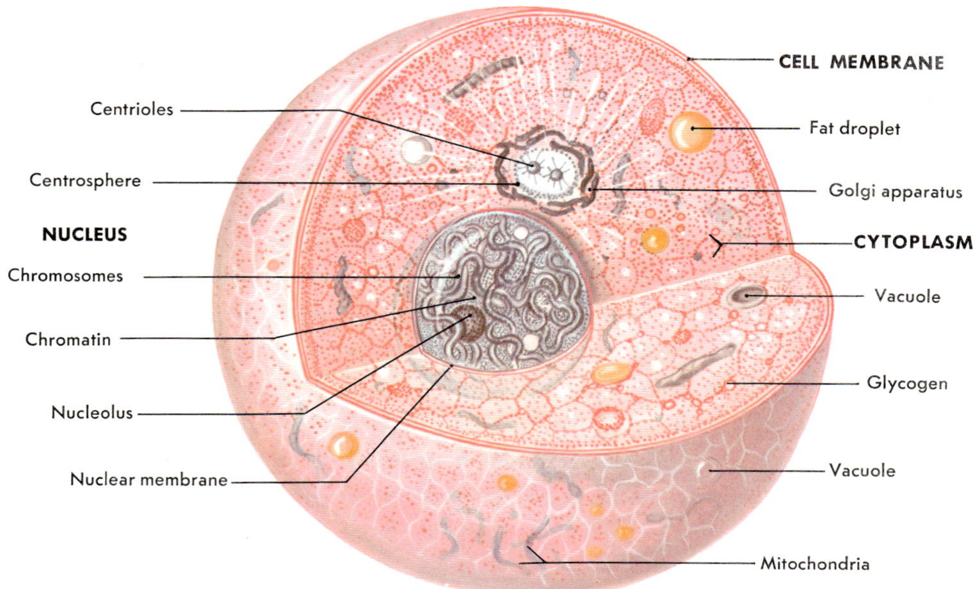

Every living cell, regardless of its shape or size, has three main parts: the cell membrane, cytoplasm, and nucleus. Together they constitute protoplasm. Billions of such cells as shown above make up the tissues of our bodies.

TYPES OF CELLS

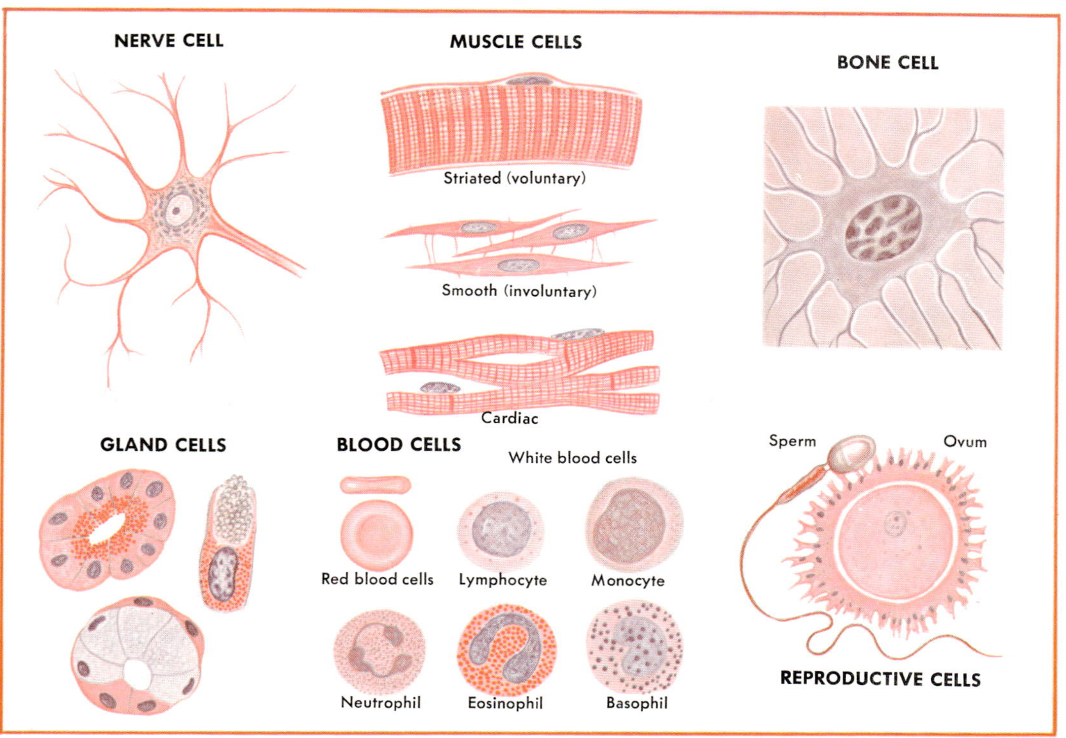

Plate XIII

SKELETON

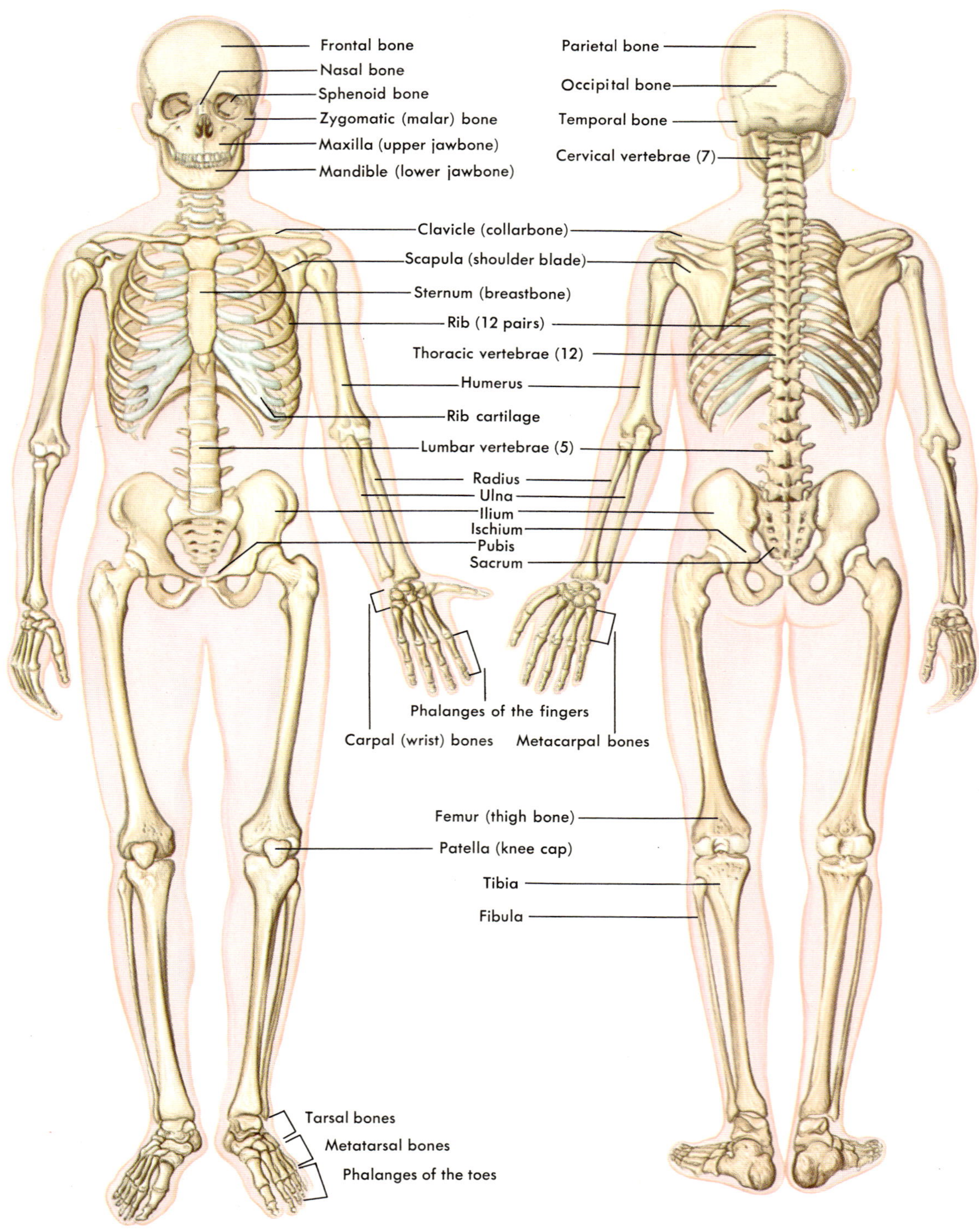

FEMALE PELVIC ORGANS

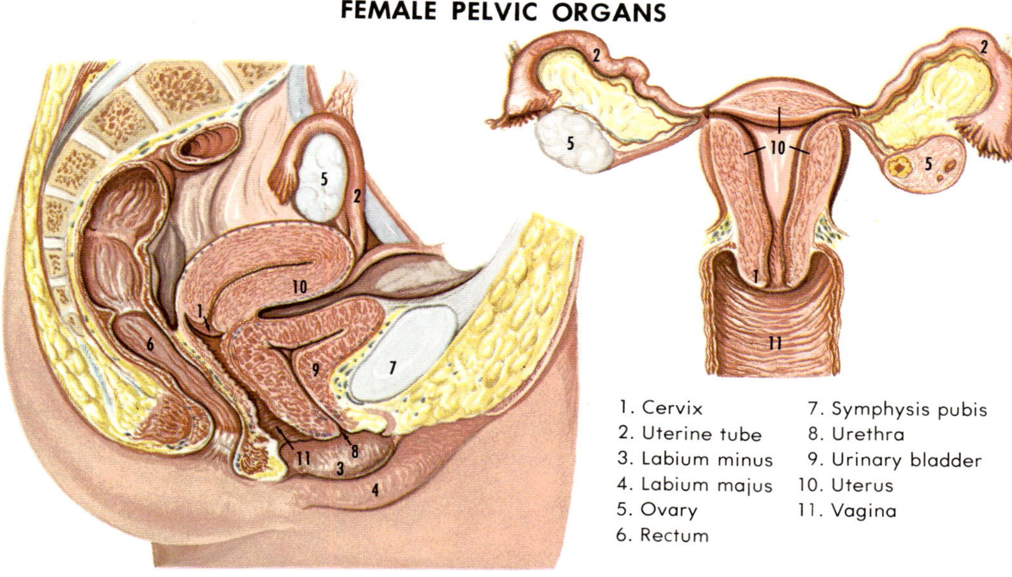

1. Cervix
2. Uterine tube
3. Labium minus
4. Labium majus
5. Ovary
6. Rectum
7. Symphysis pubis
8. Urethra
9. Urinary bladder
10. Uterus
11. Vagina

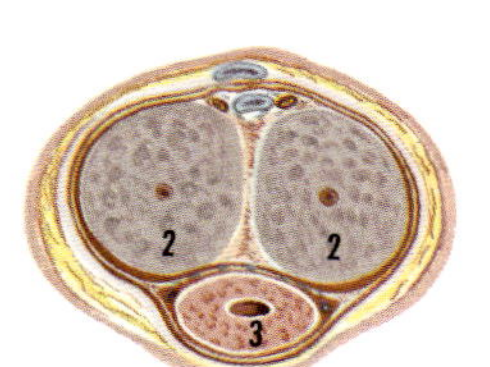

SECTION THROUGH PENIS

MALE PELVIC ORGANS

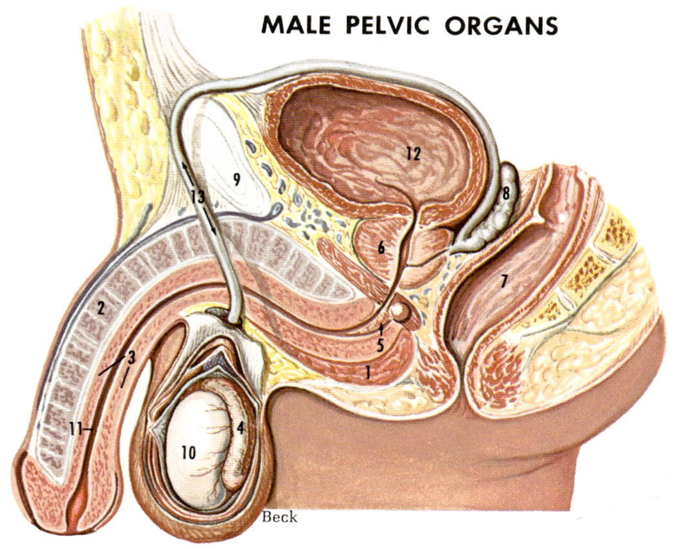

1. Bulb of urethra
2. Corpus cavernosum
3. Corpus spongiosum
4. Epididymis
5. Duct of bulbourethral gland
6. Prostate gland
7. Rectum
8. Seminal vesicle
9. Symphysis pubis
10. Testis
11. Urethra
12. Urinary bladder
13. Ductus deferens

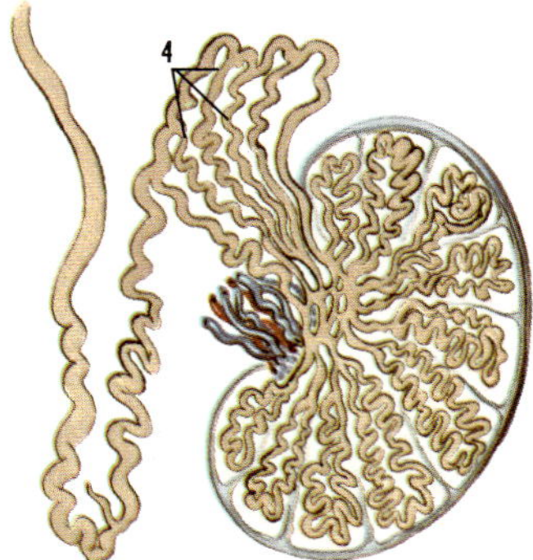

SCHEME OF DUCT ARRANGEMENT IN THE TESTIS AND EPIDIDYMIS

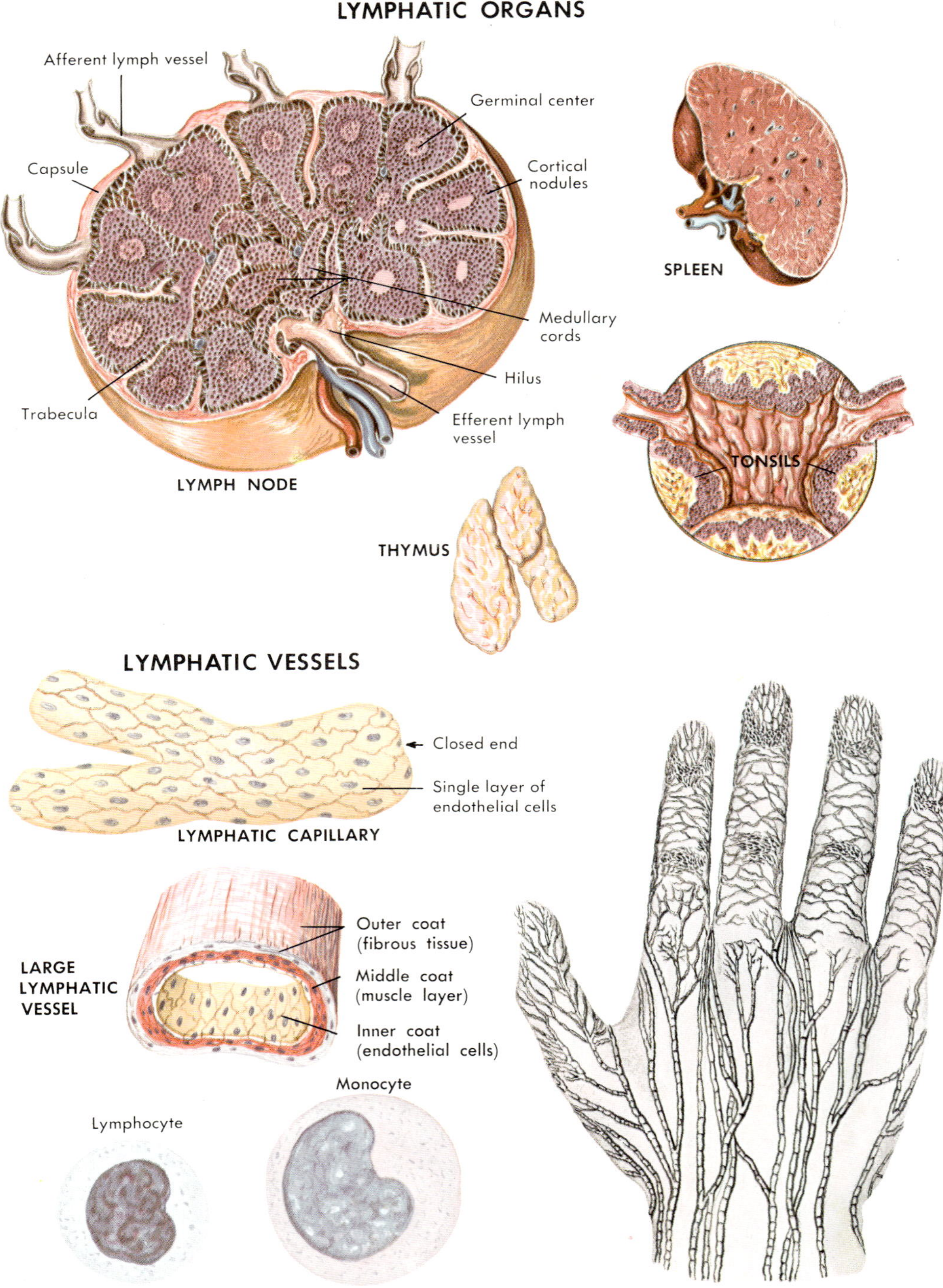

FREE CELLS OF THE LYMPHATIC SYSTEM

LYMPHATICS OF THE HAND

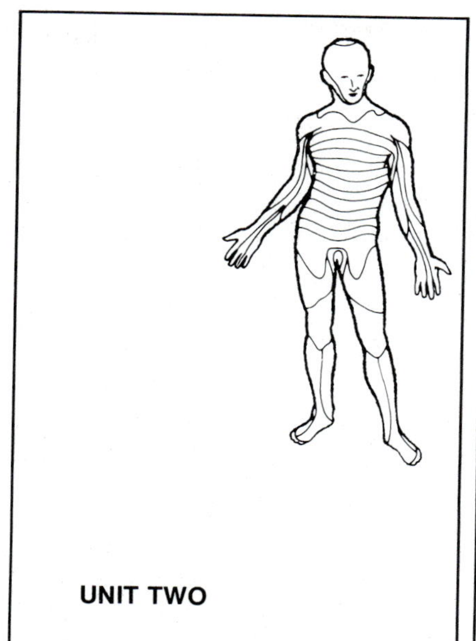

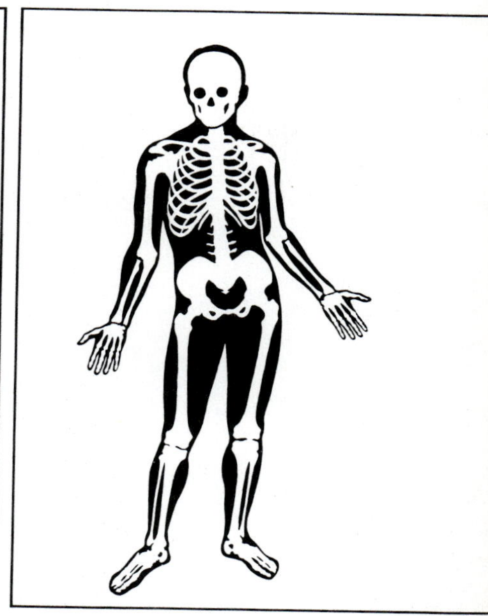

UNIT TWO

THE ERECT AND MOVING BODY

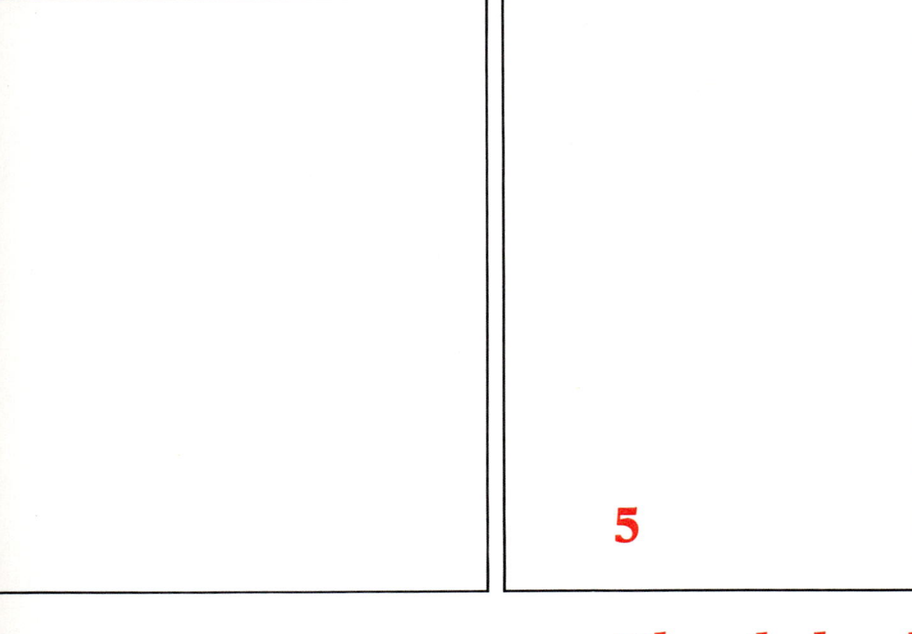

5

The skeletal system

Like all body structures, the skeletal and muscular systems play a part in the body's achievement of its overall goal of survival. These two systems work together to move the body and its parts. This is a function of tremendous importance not only for the enjoyment of life, but also for life itself, since without movement a favorable cellular environment cannot possibly be maintained. The body must adjust to many changes in its external environment in order to maintain homeostasis. Sometimes it adjusts the external environment. Sometimes it adjusts itself. In either case, movements play a part. Suppose, for instance, that environmental temperature drops below the comfort zone. The body then needs to make some kind of adjustment in order to maintain homeostasis of its internal temperature. It may change the environmental temperature back to a comfortable level (by building a fire or setting up the room thermostat, for example), or it may change itself in some way to counteract the environmental change (for instance, shivering to produce more heat and surface blood vessel constriction to lose less heat).

Whichever methods the body uses, movements are necessary. Movements require the coordinated activities of nearly all of the body's systems. In order to gain an understanding of how movements are accomplished, we shall start by investigating the two systems whose primary business is movement: the skeletal and the muscular systems.

Meaning

The term *skeletal system* means all the bones of the body plus the joints formed by their attachments to each other. Predominant tissues of the system are three types of connective tissue: bone, cartilage, and hemopoietic tissue.

Functions

The skeletal system serves the body by performing these functions: support, protection, movement, and hemopoiesis.

Support. Bones support the body much as steel girders support our modern buildings.

Protection. Hard, bony "boxes" protect delicate structures enclosed by them. For example, the skull protects the brain and the rib cage protects the lungs and heart.

Movement. Bones with their joints constitute levers. Muscles are anchored firmly to bones. As muscles contract and shorten, force is applied to the bony levers and movement results.

Hemopoiesis. Hemopoiesis means the process of forming blood cells. Tissues that carry on this process are specialized connective tissues of two kinds, namely, myeloid tissue and lymphatic tissue. Red bone marrow is another name for myeloid tissue. In the adult, it is found only in a few bones—in the sternum and ribs, in the bodies of the vertebrae, in the diploë of the cranial bones, and in the proximal epiphyses of femur and humerus. Red marrow occurs in many more bones in newborn infants and children. Because it forms blood cells, the red bone marrow is one of the most important tissues of the body. Its very location, hidden in the bones, like valuables in a safety deposit box, suggests its vital importance.

Bone and cartilage

Microscopic structure

BONE

Bone, like other tissues, consists of living cells and nonliving intercellular substance. And in bone, like other connective tissues, intercellular substance predominates over cells. But in bone the intercellular substance (matrix) is calcified. Calcium salts impregnate the cement substance of the matrix, a fact that explains the rigidity of bones and the familiar expression "as hard as bone." Embedded in the calcified matrix are many fibers of the body's most abundant protein, collagen.

Another unique feature of bone structure is the arrangement of its intercellular substance. As Fig. 5-1 shows, concentric cylindrical layers of calcified matrix (usually fewer than six of them) enclose a central longitudinal canal that contains a blood vessel. Each layer of bone matrix is called a *lamella*, the central canal is an *haversian canal*, and the entire unit of canal and surrounding lamellae is an *haversian system*. Many haversian canals contain a single large capillary, but some contain a small arteriole and venule and lymphatics. Bone cells *(osteocytes)* occupy minute spaces called *lacunae* between the lamellae. Microscopic canals (canaliculi), great numbers of them, radiate in all directions from the lacunae to connect them with haversian canals and provide

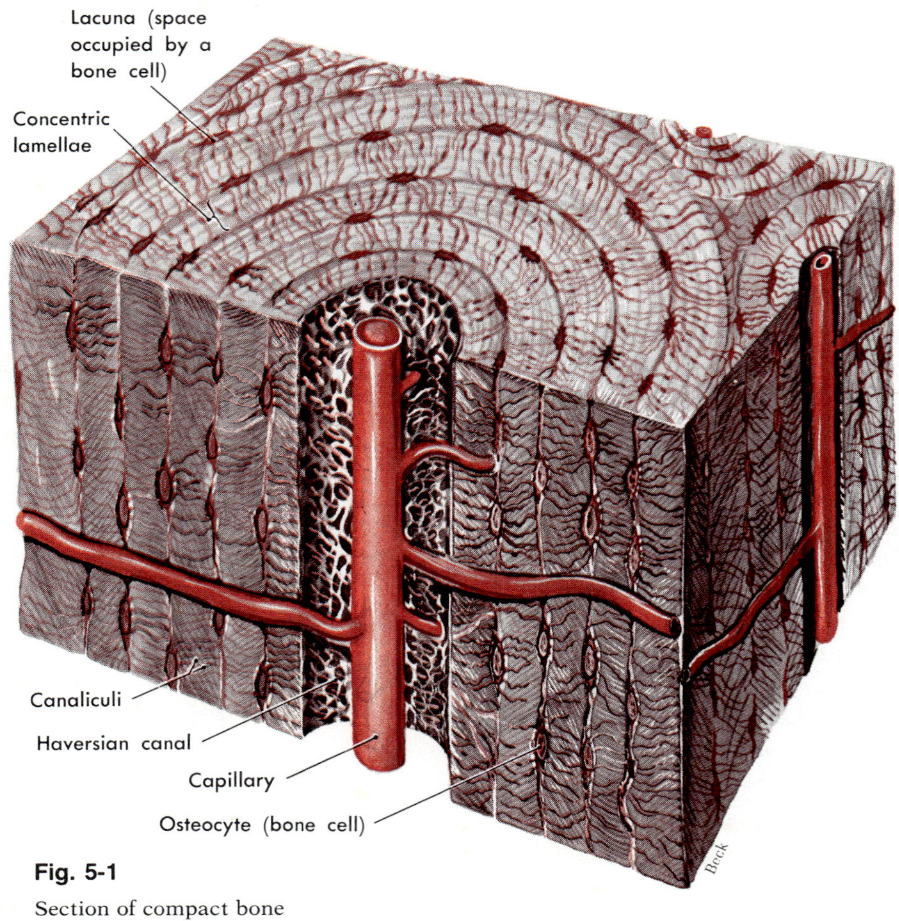

Fig. 5-1

Section of compact bone showing details of an haversian system.

routes for tissue fluid to reach bone cells. Each bone cell is said to lie not farther than 0.1 mm from an haversian canal.

There are two types of bone based on the arrangement of lamellae—compact or dense and cancellous or spongy. In *compact bone,* adjacent haversian units fit closely together with the spaces between them filled in with interstitial lamellae. In *cancellous bone,* on the other hand, there are many open spaces between thin processes of bone *(trabeculae),* which are joined together somewhat like the beams of wood in a scaffold. Arrangement of trabeculae in different ways in different bones gives structural strength along the lines of strain on individual bones.

Bones are not the lifeless structures they seem to be. We tend to think of them as lifeless, perhaps because what we see when we look at a bone is its nonliving intercellular substance. But within this hard, lifeless material lie many living bone cells that must continually receive food and oxygen and excrete their wastes. So blood supply to bone is important and abundant. For example, numerous blood vessels from the periosteum (bone covering) penetrate bone by way of *Volkmann's canals* to connect with blood vessels of an haversian canal. Also, one or more arteries supply the bone marrow in the internal medullary cavity of long bones.

CARTILAGE

Cartilage both resembles and differs from bone. Like bone, cartilage consists more of intercellular substance than of cells. Innumerable collagenous fibers reinforce the matrix of both tissues. But in cartilage the fibers are embedded in a firm gel instead of in a calcified cement substance as they are in bone. Hence, cartilage has the flexibility of a firm plastic material rather than the rigidity of bone. Another difference is this—no canal system and no blood vessels penetrate cartilage matrix. Cartilage is avascular and bone is abundantly vascular. Cartilage cells, like bone cells, lie in lacunae. However, because no canals and blood vessels interlace cartilage matrix, nutrients and oxygen can reach the scattered, isolated chondrocytes (cartilage cells) only by diffusion. They diffuse through the matrix gel, from capillaries in the fibrous covering of cartilage (perichondrium) or from synovial fluid in the case of articular cartilage.

Three types of cartilage are hyaline, fibrous, and elastic. They differ structurally mainly as to matrix fibers. Collagenous fibers are present in all three types but are most numerous in fibrocartilage. Hence it has the greatest tensile strength. Elastic cartilage matrix contains elastic fibers as well as collagenous fibers so has elasticity as well as firmness. Hyaline is the most common type of cartilage. It resembles milk glass in appearance. In fact, its name is derived from the Greek word meaning glassy. A thin layer of hyaline cartilage covers articular surfaces of bones, where it helps to cushion jolts. Fibrocartilage disks between the vertebrae also serve this purpose. For other locations of cartilage, see Table 3-1, p. 40.

Bones

Types

There are four types of bones, which are classified according to their shapes as follows:

1 *Long bones*—femur, tibia, fibula, humerus, radius, ulna, and phalanges
2 *Short bones*—carpals and tarsals (wrist and ankle bones)
3 *Flat bones*—several cranial bones, such as frontal and parietal; also ribs and scapulae
4 *Irregular bones*—vertebrae, sphenoid, ethmoid, sacrum, coccyx, and mandible

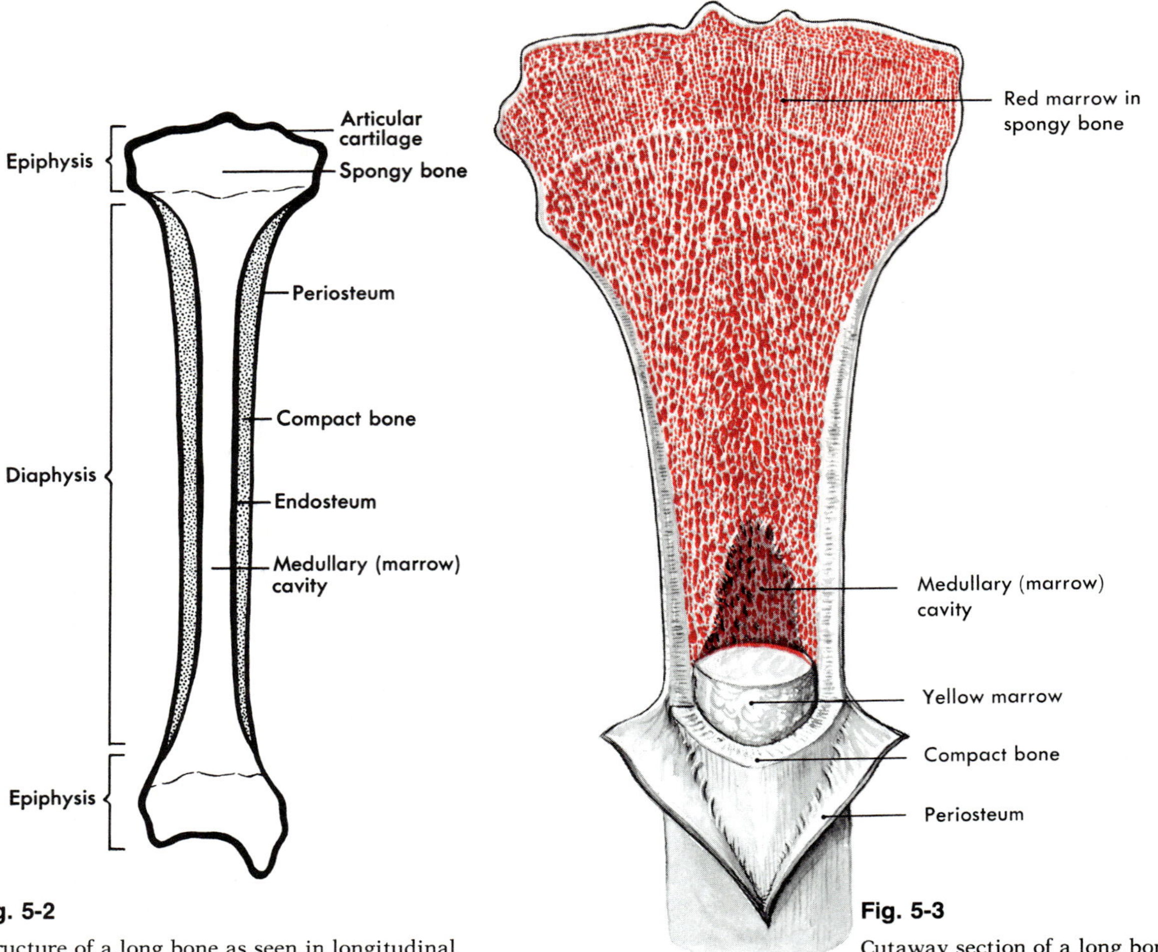

Fig. 5-2

Structure of a long bone as seen in longitudinal section (also see Fig. 5-3).

Fig. 5-3

Cutaway section of a long bone.

Structure

Identify the following parts of a long bone in Figs. 5-2 and 5-3: (1) diaphysis, (2) epiphyses, (3) articular cartilage, (4) periosteum, (5) medullary (or marrow) cavity, and (6) endosteum.

The *diaphysis* is the main shaftlike portion of a long bone. Several structural features accommodate it to its function of providing strong support without cumbersome weight—the thick compact bone used as construction material, for example, and the hollow cylindrical shape that offers the dual advantages of greater strength with less weight compared with a solid cylinder of the same size.

The *epiphyses* are the extremities of long bones. Their somewhat bulbous shape provides generous space for muscle attachment near joints and makes for greater stability of the joint. Lightness despite size is achieved by construction of porous cancellous bone with only an outer layer of dense compact bone. Arrangement of the lamellae corresponding to the lines of stress gives added strength to the epiphyses. Marrow fills the cancellous spaces—red marrow in the proximal epiphyses of the humerus and femur and yellow marrow in other epiphyses in the adult.

The *articular cartilage* is the thin layer of hyaline cartilage covering the articular surface of each epiphysis. Resiliency of this material cushions jars and blows.

The *periosteum* is a dense white fibrous membrane that covers bone except at joint surfaces, where articular cartilage forms the covering. Many of the fibers of the periosteum penetrate the underlying bone to weld these two structures to each other. (The penetrating fibers are called Sharpey's fibers.) Muscle tendon fibers interlace with periosteal fibers to anchor muscles firmly to bone.

The inner layer of the periosteum of growing bones contains osteoblasts (bone-forming cells). Because of its bone-forming cells and blood vessels, the periosteum is necessary for bone growth and repair and for its nutrition and, therefore, for survival of its cells. In addition, it serves as the means for attaching muscle tendons and ligaments to the bone.

The *medullary* (or *marrow*) *cavity*, running the length of the diaphysis, contains yellow or fatty bone marrow in the adult.

The *endosteum* is the membrane that lines the medullary cavity and haversian canals. It is composed of cells that become active osteoblasts as needed.

Short bones consist of a core of cancellous bone encased in a thin layer of compact bone.

Flat bones consist of a layer of cancellous bone between two plates of compact bone. Cancellous bone of the skull bones (diploë), ribs, and sternum contains red marrow.

Irregular bones are similar in structure to short bones—that is, a thin layer of compact bone forms a casing over cancellous bone.

Formation and growth

The embryo skeleton, when first formed, consists of "bones" that are not really bones at all but hyaline cartilage or fibrous membrane structures shaped like bones. Gradually the process of *ossification* (or *osteogenesis*) replaces these prebone structures with bone. Ossification of any structure begins with the appearance in it of special bone-forming cells called *osteoblasts*. The Golgi apparatus in an osteoblast specializes in synthesizing and secreting carbohydrate compounds of the type named mucopolysaccharides and its endoplasmic reticulum makes and secretes collagen, a protein. In time, relatively large amounts of the mucopolysaccharide substance (or cement substance, as it is more commonly called) accumulates around each individual osteoblast and numerous bundles of collagenous fibers become embedded in it. Together, they constitute the organic intercellular substance of bone called the *bone matrix*. If you like nicknames, you might call bone matrix the reinforced concrete of the body. Bundles of collagenous fibers in bone increase its strength much as iron rods in reinforced concrete strengthen it.

About as fast as the organic bone matrix forms, inorganic compounds—complex calcium salts, not yet positively identified—begin depositing in it. This calcification of the matrix is the process that makes bone "hard as bone." To summarize briefly, ossification is a complex, still not completely understood mechanism consisting of two processes—synthesis of the organic bone matrix by osteoblasts, immediately followed by calcification of the matrix.

In long bones, endochondral ossification starts in the diaphysis and in both epiphyses and proceeds toward each other. A layer of cartilage known as *epiphyseal cartilage* remains between the diaphyseal and epiphyseal centers of ossification. As long as bone growth continues, proliferation of epiphyseal cartilage cells brings about a thickening of the layer of cartilage from time to time. Ossification of this additional cartilage then follows—that is, osteoblasts synthesize organic bone matrix and the matrix undergoes calcification. As a result, the bone becomes longer. When epiphyseal cartilage cells

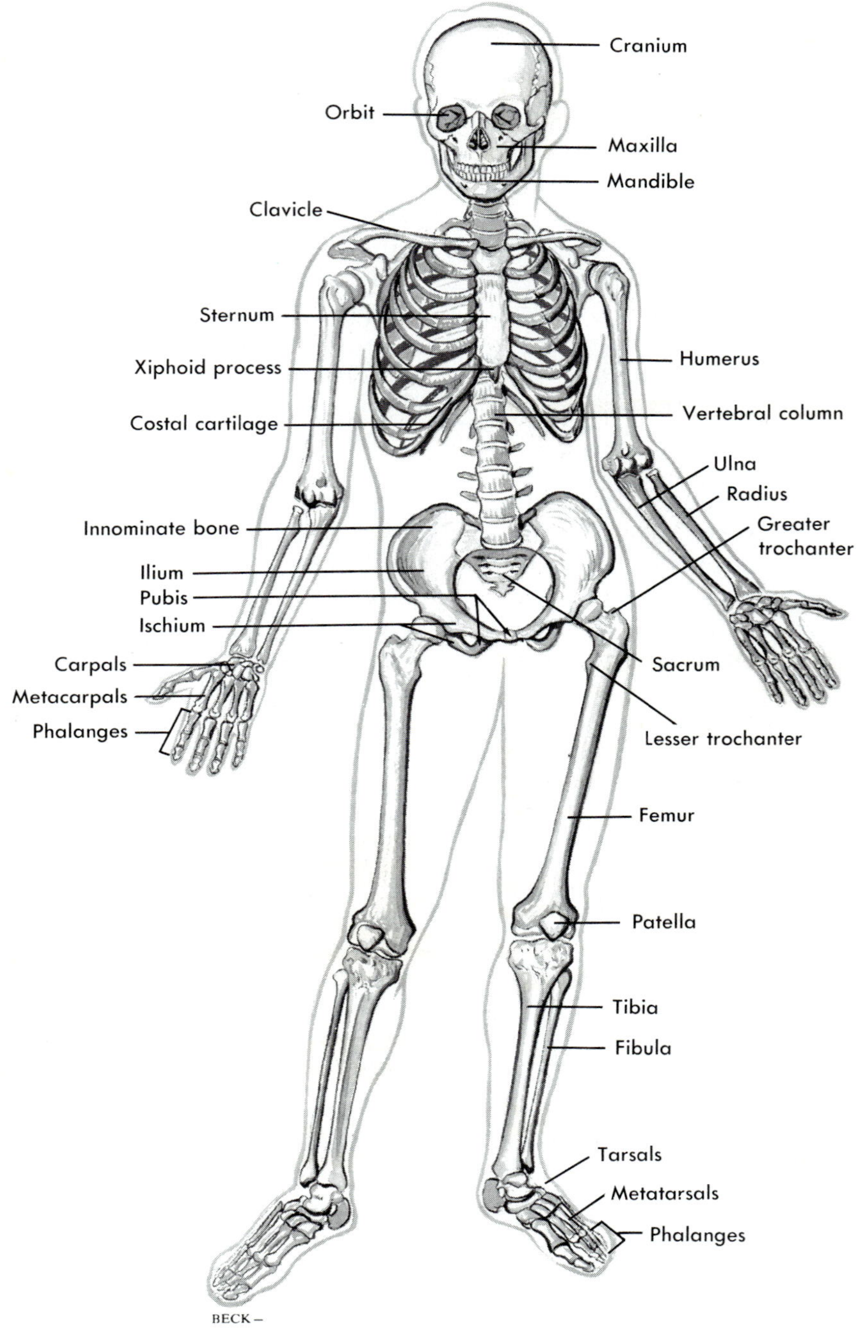

Fig. 5-4
Skeleton, anterior view.

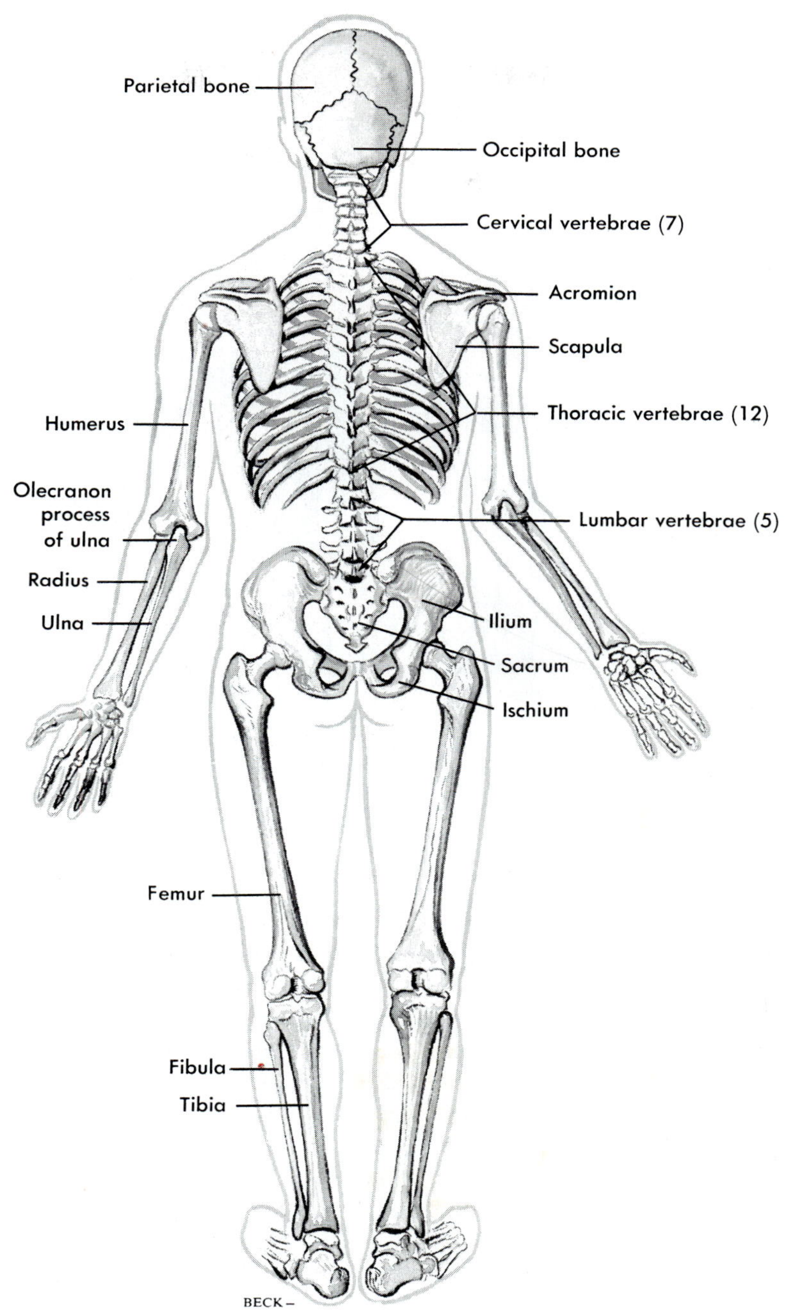

Fig. 5-5

Skeleton, posterior view.

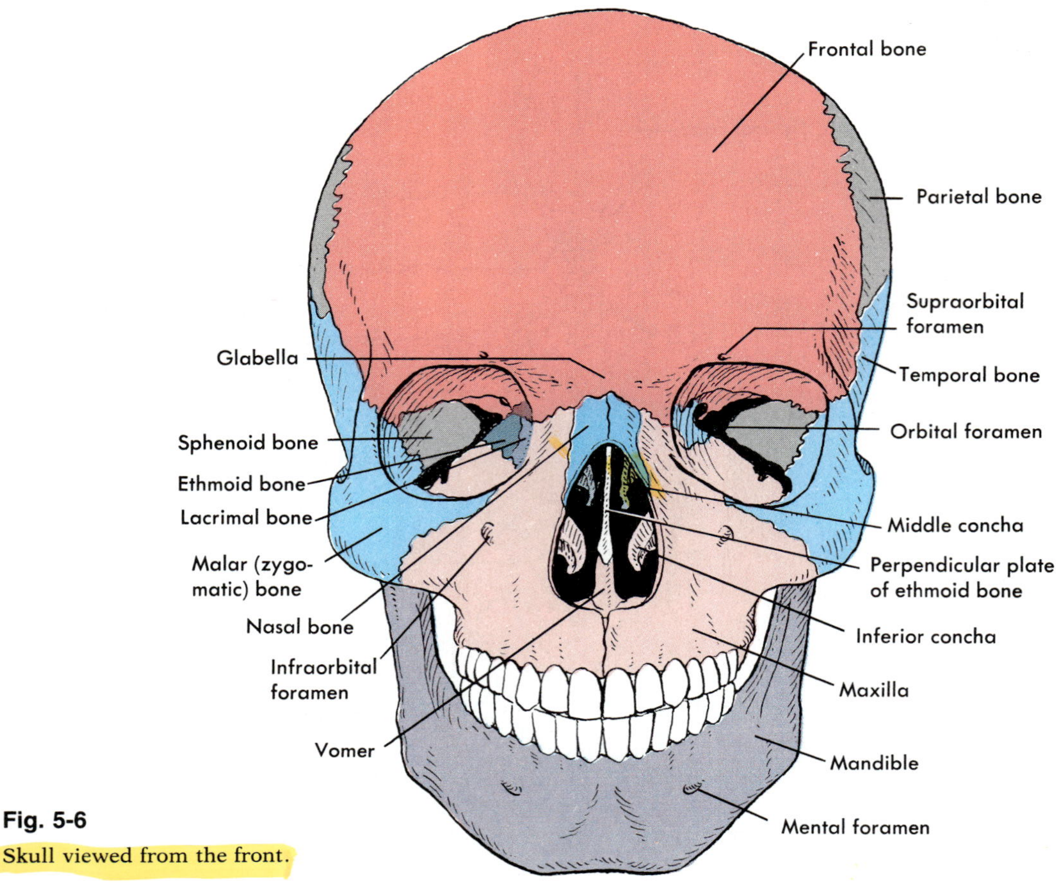

Fig. 5-6

Skull viewed from the front.

stop multiplying and the cartilage has become completely ossified, bone growth has ended.

Bones grow in diameter by the combined action of two special kinds of cells: osteoblasts and osteoclasts. Osteoclasts enlarge the diameter of the medullary cavity by eating away the bone of its walls. At the same time, osteoblasts from the periosteum build new bone around the outside of the bone. By this dual process a bone with a larger diameter and larger medullary cavity has been produced from a smaller bone with a smaller medullary cavity.

Bone is not the lifeless, inert substance it appears to be but a living tissue that is continually being formed and destroyed. During childhood, bones grow in size because bone formation (ossification) goes on at a faster rate than does bone destruction. During the early and middle years of adulthood, the opposing processes balance each other and bones neither grow nor shrink (atrophy). In old age, bone often degenerates and is reabsorbed faster than it is formed. As a result, it becomes smaller and also porous and fragile, a condition known as *senile osteoporosis.* Well known is the fact that this disease may have serious consequences, such as fracturing a hip or suffering severe back pains. Not known, however, is what causes osteoporosis. Evidence now at hand indicates that not a single factor but a complex of at least several factors produces it. Among the factors postu-

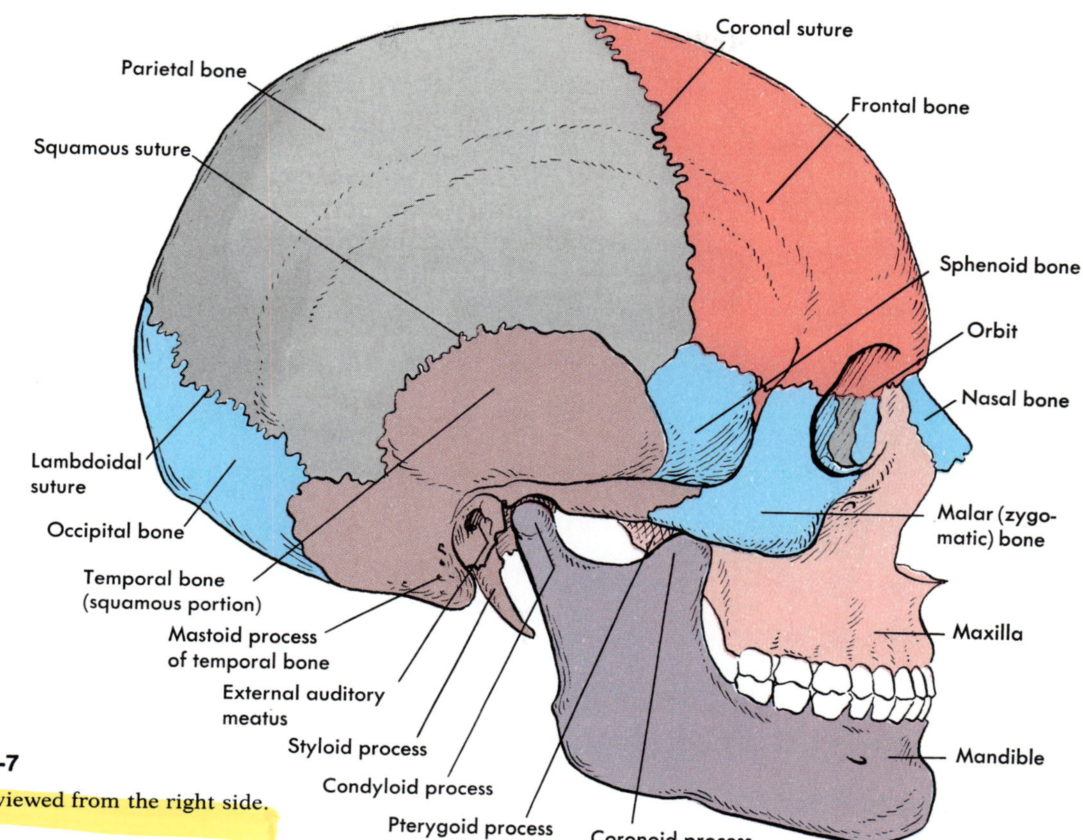

Fig. 5-7

Skull viewed from the right side.

lated but not firmly established as contributing to the development of osteoporosis are too little exercise, too low a concentration of estrogens in the blood, and too high a ratio of calcium to phosphorus in the blood.* Parathyroid hormones and calcitonin (discussed in Chapter 11) also influence bone calcification and decalcification.

Names and numbers

The human skeleton consists of two main parts: the *axial skeleton*, composed of the bones forming the upright part or axis of the body (the skull, hyoid bone, vertebral column, sternum, and ribs), and the *appendicular skeleton*, made up of the bones attached to the axial skeleton as appendages (that is, the upper and lower extremities). The names and numbers of the bones in each division of the skeleton, with an identifying remark about each, are given in Table 5-1 (pp. 94-95). When you are trying to learn these names, locate each bone on your own body. Feel its outline whenever possible. Locate each bone on a skeleton if one is available. Study Figs. 5-4 to 5-7.

AXIAL SKELETON

Skull. Twenty-eight irregularly shaped bones form the skull. Twenty-two of these consist of eleven pairs of bones; the remain-

*When bones wither; ailment of aging, Sci. News **104:** 247, Oct., 1973.

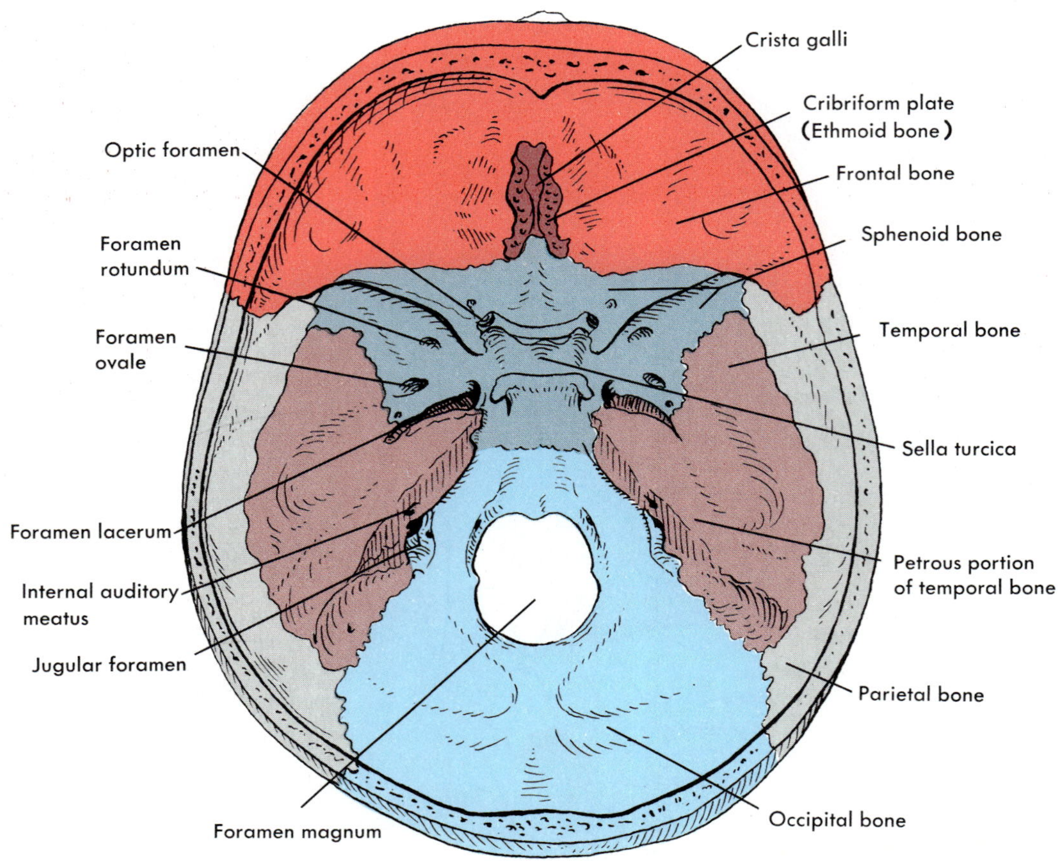

Fig. 5-8

Floor of the cranial cavity.

ing six are single. All but one of the skull bones are so joined to each other as to be immovable. Only the lower jawbone (mandible) is movable. The skull consists of two major divisions: the cranium or brain case and the face.

Cranium. The frontal, parietal, and occipital bones form the top of the cranium, whereas the temporal bones and the great wings of the sphenoid form its sides. These same bones, plus the small cribriform plate of the ethmoid bone, make up the lower part of the cranium called the *cranial floor* or *base* (Fig. 5-8). Its midportion is formed by the sphenoid bone, which serves as a keystone anchoring the frontal, parietal, occipital, and ethmoid bones.

The *frontal bone* constitutes the skeletal framework for the forehead. It contains mucosa-lined air-filled spaces, the *frontal sinuses,* and it forms the upper part of the orbits. It unites with the two parietal bones posteriorly in an immovable joint, the *coronal suture*. Several of the more prominent frontal bone markings are described in Table 5-2, p. 96.

The two *parietal bones* give shape to the bulging topsides of the cranium. They form immovable joints with several bones: the *lambdoidal suture* with the occipital bone,

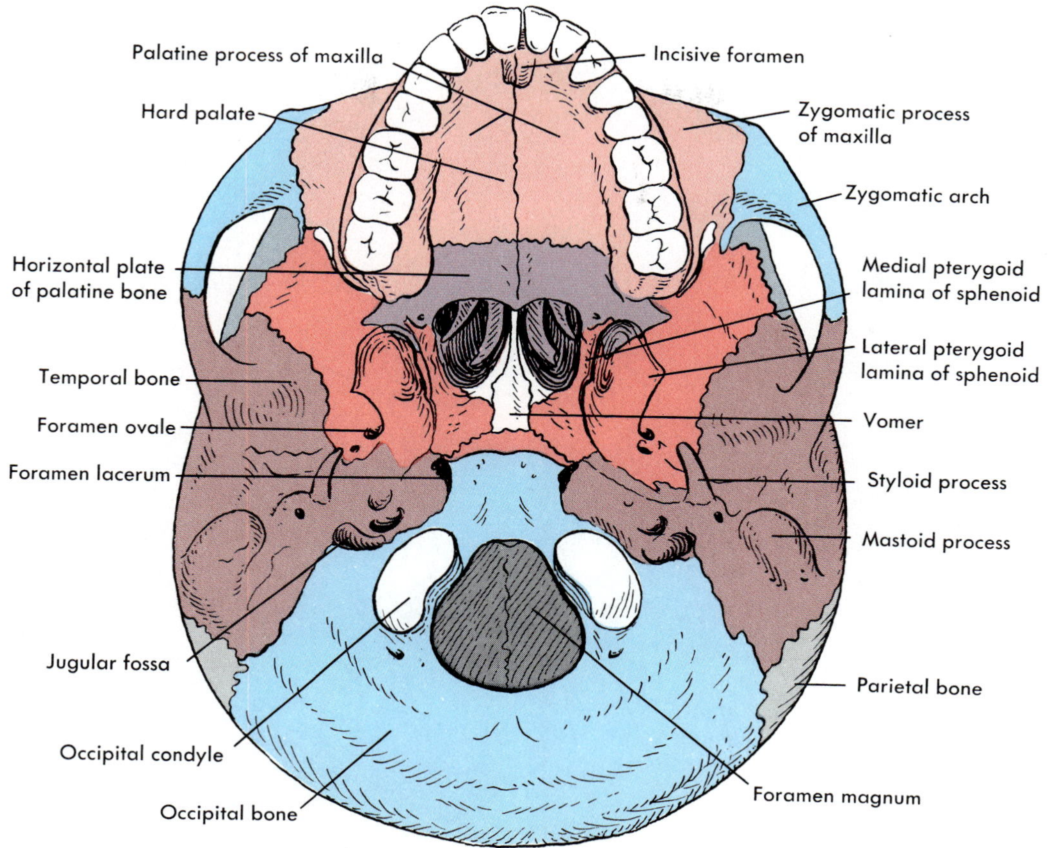

Fig. 5-9

Skull viewed from below.

the *squamous suture* with the temporal bone and part of the sphenoid, and the *coronal suture,* mentioned before, with the frontal bone.

The lower sides of the cranium and part of its floor are fashioned from two *temporal bones*. They house the middle and inner ear structures and contain the *mastoid sinuses,* notable because of the occurrence of mastoiditis, an inflammation of the mucous lining of these spaces. A description of several other temporal bone markings is included in Table 5-2, p. 96.

The *occipital bone* makes the framework of the lower, posterior part of the skull. It forms immovable joints with three other cranial bones—the parietal, temporal, and sphenoid—and a movable joint with the first cervical vertebra. Consult Table 5-2, p. 96, for a description of some of its markings.

The *sphenoid bone* resembles a bat with its wings outstretched and legs extended downward posteriorly. It constitutes the center portion of the cranial floor and forms part of the orbit's floor and sidewalls. The sphenoid bone contains fairly large mucosa-lined air-filled spaces, the *sphenoid sinuses.* Several prominent sphenoid markings are described in Table 5-2, pp. 96-97 (also see Figs. 5-8 to 5-10).

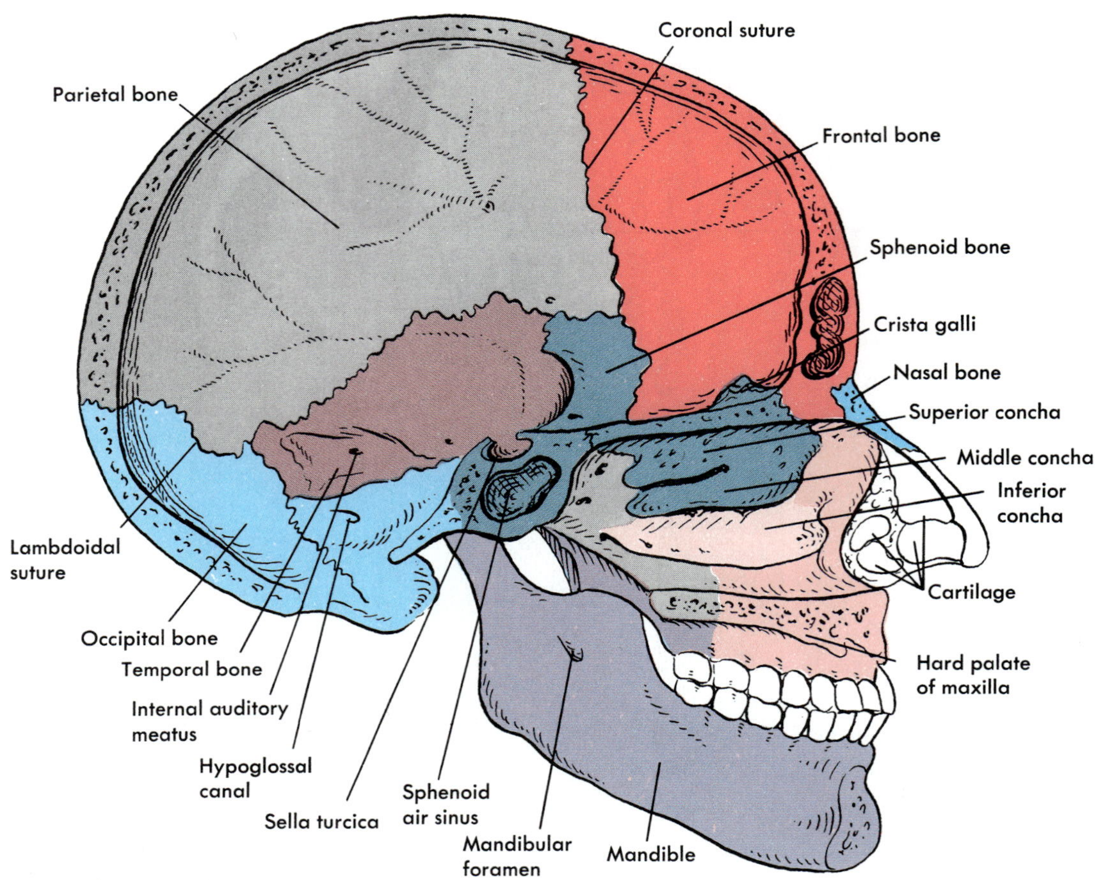

Fig. 5-10

Left half of skull viewed from within. The perpendicular plate of the ethmoid is removed to reveal the nasal concha.

The *ethmoid,* a complicated, irregular bone, lies anterior to the sphenoid but posterior to the nasal bones. It helps fashion the anterior part of the cranial floor, the medial walls of the orbits, the upper parts of the nasal septum and of the sidewalls of the nasal cavity, and the part of the nasal roof perforated by small foramina through which olfactory nerve branches reach the brain. The lateral masses of the ethmoid are honeycombed with sinus spaces. For more ethmoid markings see Table 5-2 (p. 97) (also Figs. 5-8, 5-10, and 5-14 to 5-16).

Face bones. Fourteen bones are commonly said to form the framework of the face. Actually, however, more are involved since some of the cranial bones, particularly the frontal and ethmoid, also help shape the face.

Just as the sphenoid acts as the keystone in the architecture of the cranium, so the *maxillae* serve in this capacity for the face. With the exception of the mandible, all the

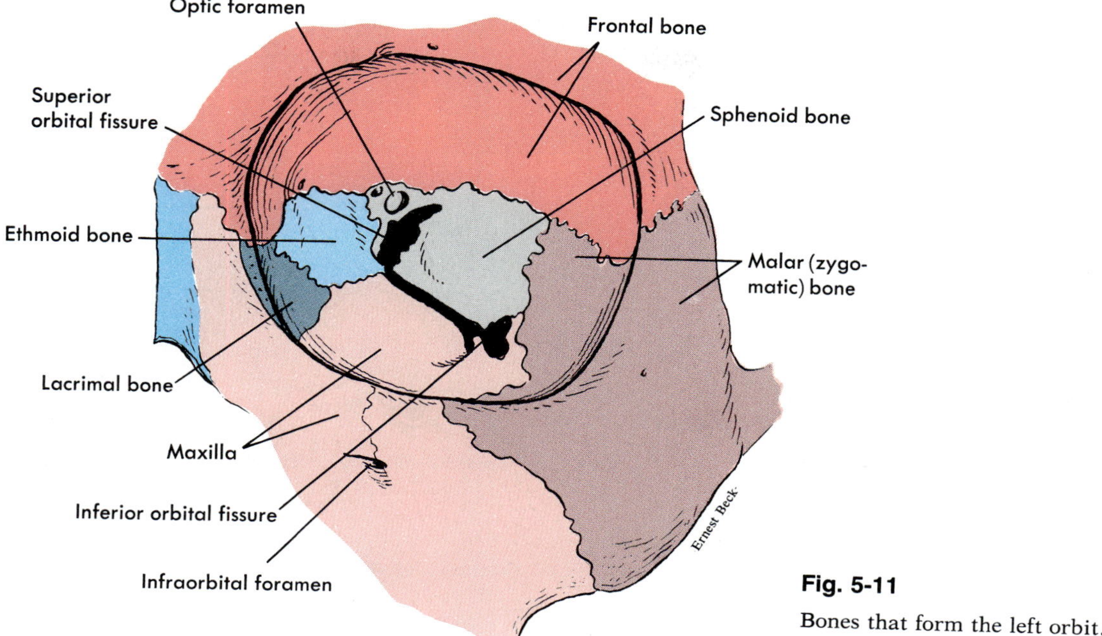

Fig. 5-11

Bones that form the left orbit.

face bones articulate with the maxillae, which also articulate with each other, in the midline. The maxillae form part of the floor of the orbits, part of the roof of the mouth, and part of the floor and sidewalls of the nose. Each maxilla contains a mucosa-lined space, the maxillary sinus or *antrum of Highmore*. This sinus is the largest of the paranasal sinuses (that is, sinuses connected by channels to the nasal cavity). For other markings of the maxillae, see Table 5-2 (p. 97).

Unlike the upper jaw, which is formed by a pair of bones, the lower jaw, because of fusion of its halves during infancy, consists of a single bone, the *mandible*. It is the largest, strongest bone of the face. It articulates with the temporal bone in the only movable joint of the skull. Its major markings are identified in Table 5-2 (p. 97).

The cheek is shaped by the underlying *zygomatic* or malar bone. This bone also forms the outer margin of the orbit and, with the zygomatic process of the temporal bone, makes the zygomatic arch. It articulates with four other face bones: the maxillae and the temporal, frontal, and sphenoid bones.

Shape is given to the nose by the two *nasal bones*, which form the upper part of the bridge of the nose, and by *cartilage*, which forms the lower part. Though small in size, the nasal bones enter into several articulations: with the perpendicular plate of the ethmoid, the cartilaginous part of the nasal septum, the frontal bone, maxillae, and with each other.

An almost paper-thin bone, shaped and sized about like a fingernail, lies just posterior and lateral to each nasal bone. It helps form the sidewall of the nasal cavity and the medial wall of the orbit. Because it contains a groove for the nasolacrimal (tear) duct, this bone is called the *lacrimal bone* (Fig. 5-11). It joins the maxilla, frontal bone, and ethmoid.

An irregular bone whose horizontal and

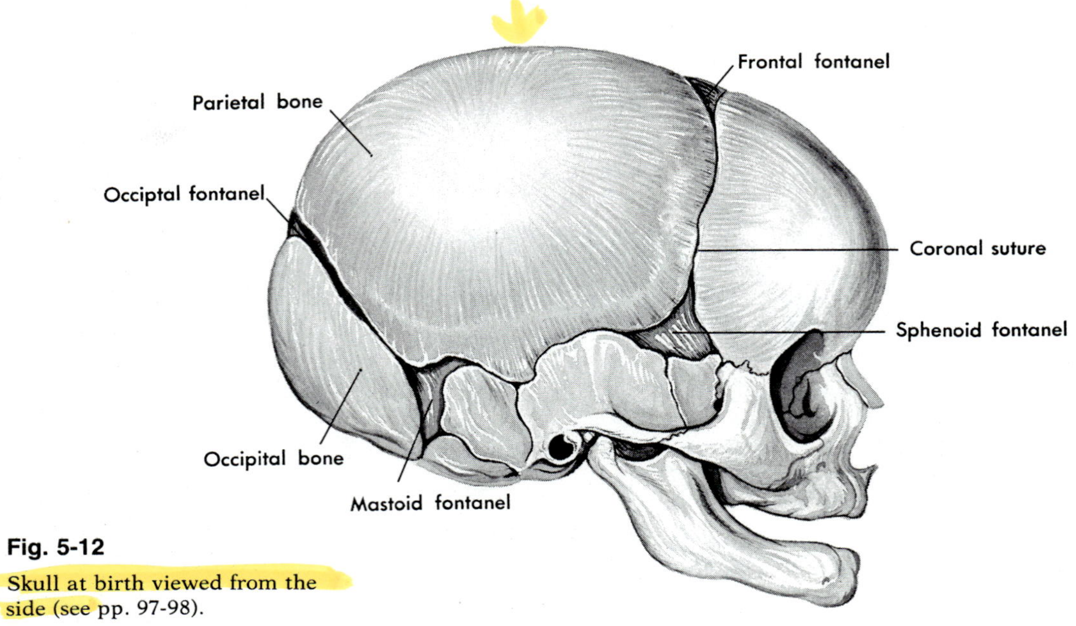

Fig. 5-12

Skull at birth viewed from the side (see pp. 97-98).

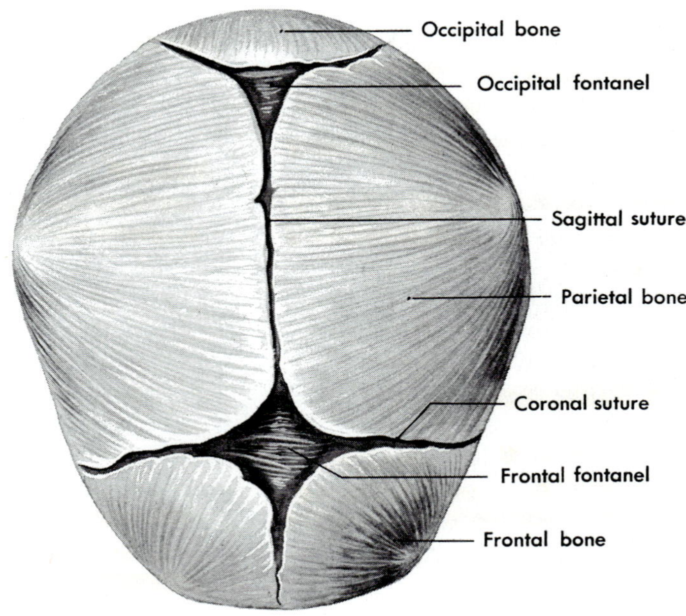

Fig. 5-13

Skull at birth viewed from above (see pp. 97-98).

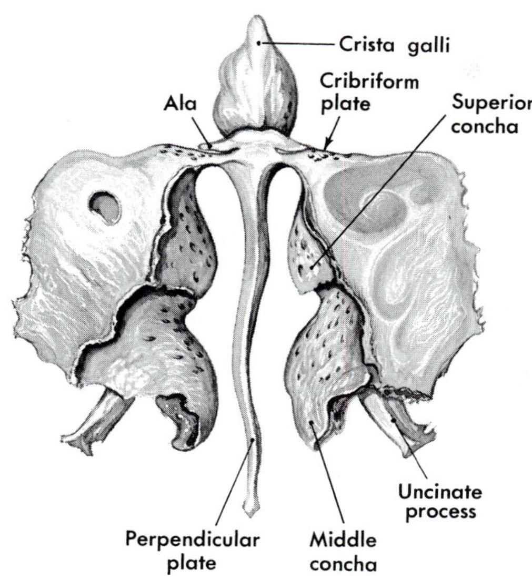

Fig. 5-14

Ethmoid bone viewed from behind.

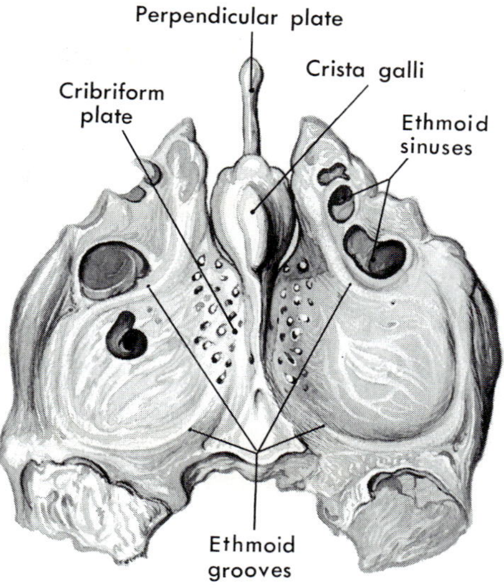

Fig. 5-15

Ethmoid bone viewed from above.

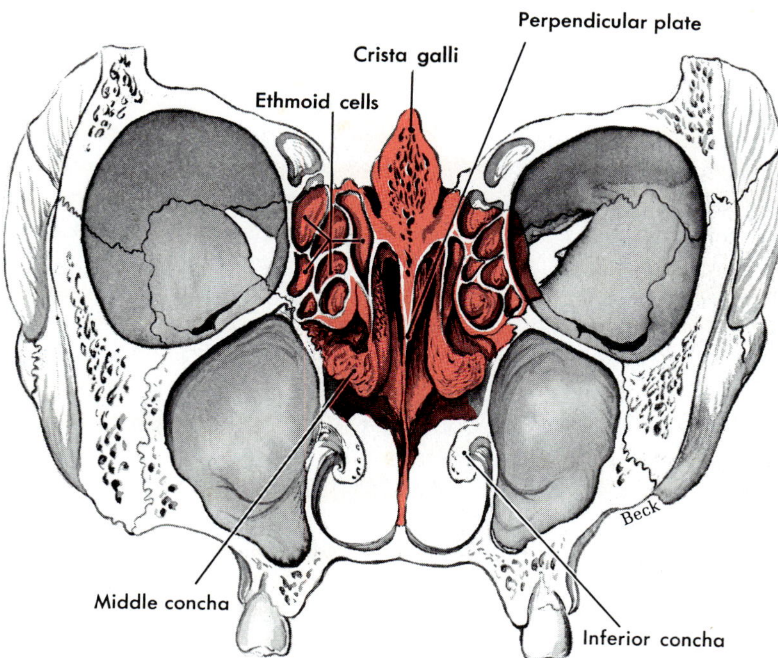

Fig. 5-16

Skull in coronal section to reveal the ethmoid bone (shown in red) as seen from in front.

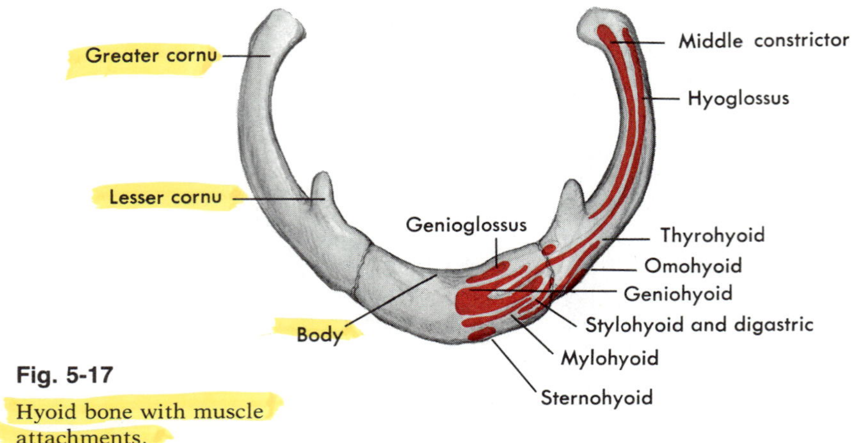

Fig. 5-17

Hyoid bone with muscle attachments.

vertical portions are joined in such a way as to make the bone roughly L shaped forms the framework for the inferior and lateral walls of the posterior part of the nasal cavity. It is known as the *palatine bone,* because it also forms the posterior part of the hard palate. In addition, it has a small upper projection that helps form the floor of the orbit. The two palatine bones are united in the midline like two L's facing each other. They articulate also with the maxillae and the sphenoid.

The *inferior nasal concha* is a scroll-like bone that forms a kind of ledge projecting into the nasal cavity from its lateral wall. In each nasal cavity there are three of these ledges, formed respectively by the superior and middle conchae (which are projections of the ethmoid) and the inferior concha (which is a separate bone). They are mucosa-covered and divide each nasal cavity into three narrow, irregular channels, the *nasal meati.* The inferior nasal conchae form immovable joints with the ethmoid, lacrimal, maxillary, and palatine bones.

Two structures that enter into the formation of the nasal septum have already been mentioned, the perpendicular plate of the ethmoid bone and the septal cartilage. One other structure, the *vomer bone,* completes the septum posteriorly. It is usually described as being shaped like a ploughshare. It forms immovable joints with four bones: the sphenoid, ethmoid, palatine, and maxillae.

Ear bones. See Table 5-1 (p. 94).

Special features. Sutures, fontanels, sinuses, orbits, nasal septum, and wormian bones are described in Table 5-2.

Hyoid bone. The hyoid bone is a single bone in the neck—a part of the axial skeleton. Its U shape may be felt just above the larynx and below the mandible where it is suspended from the styloid processes of the temporal bones. Several muscles attach to the hyoid bone. Among them are an extrinsic tongue muscle (hyoglossus) and certain muscles of the floor of the mouth (mylohyoid, geniohyoid) (Fig. 5-17). The hyoid claims the distinction of being the only bone in the body that does not articulate with any other bone.

Vertebral column. The vertebral column constitutes the longitudinal axis of the skeleton. It is a flexible rather than a rigid column because it is segmented—that is, made up of twenty-six (typical in adult) separate bones called vertebrae, so joined to each

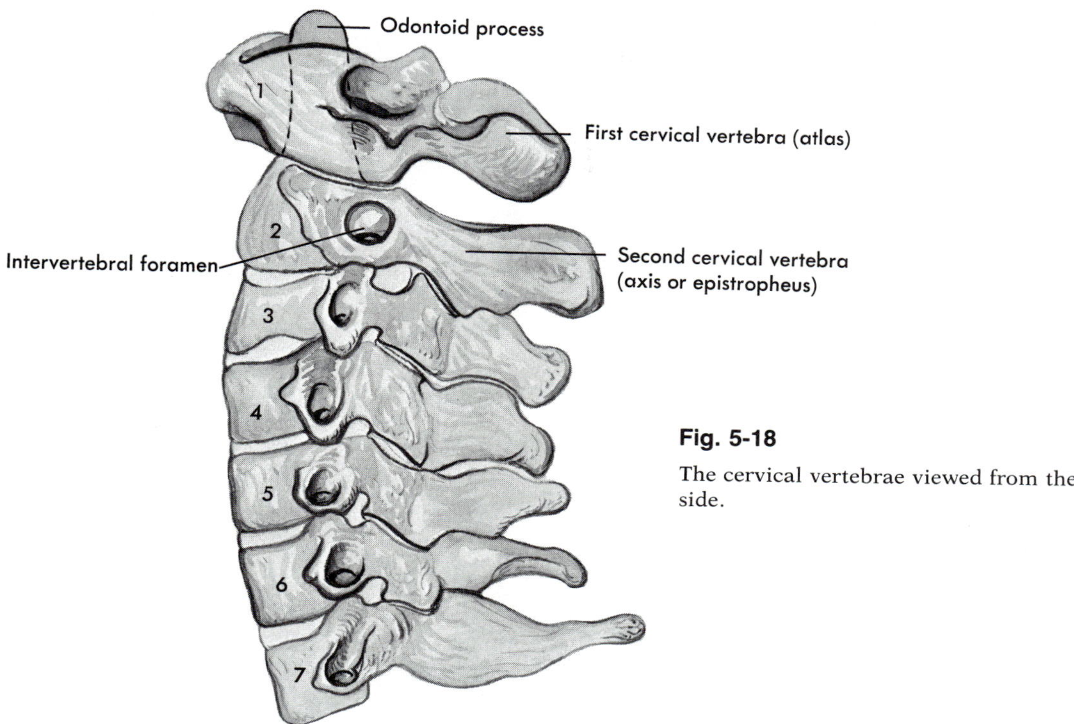

Fig. 5-18

The cervical vertebrae viewed from the side.

other as to permit forward, backward, and sideways movement of the column. The head is balanced on top of this column, the ribs and viscera are suspended in front, the lower extremities are attached below, and the spinal cord is enclosed within. It is, indeed, the "backbone" of the body.

The seven *cervical vertebrae* constitute the skeletal framework of the neck (Fig. 5-18). The next twelve vertebrae are called *thoracic vertebrae* for the obvious reason that they lie behind the thoracic cavity. The next five spinal bones, the *lumbar vertebrae,* support the small of the back. Below the lumbar vertebrae lie the *sacrum* and *coccyx*. In the adult, the sacrum is a single bone that has resulted from the fusion of five separate vertebrae, and the coccyx is a single bone that has resulted from the fusion of four or five vertebrae.

All the vertebrae resemble each other in certain features and differ in others. For example, all except the first cervical vertebra have a flat, rounded mass placed anteriorly and centrally, known as the *body,* plus a sharp or blunt *spinous process* projecting inferiorly in the posterior midline and two transverse processes projecting laterally (Figs. 5-19 to 5-23). All but the sacrum and coccyx have a central opening, the *vertebral foramen.* An upward projection (the *dens*) from the body of the second cervical vertebra furnishes an axis for rotating the head. A long, blunt spinous process that can be felt at the back of the base of the neck characterizes the seventh cervical vertebra. Each

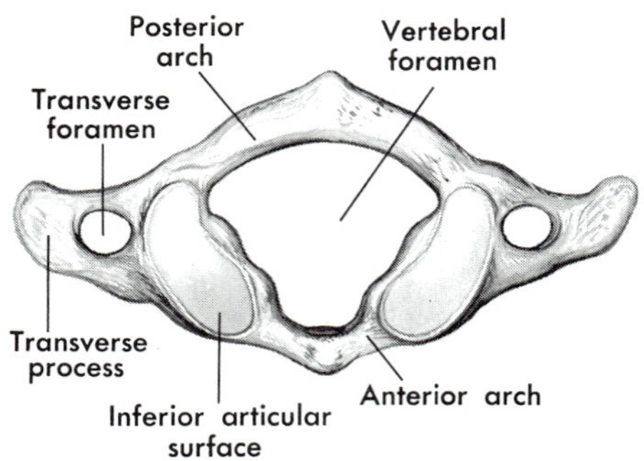

Fig. 5-19

First cervical vertebra (atlas) viewed from below.

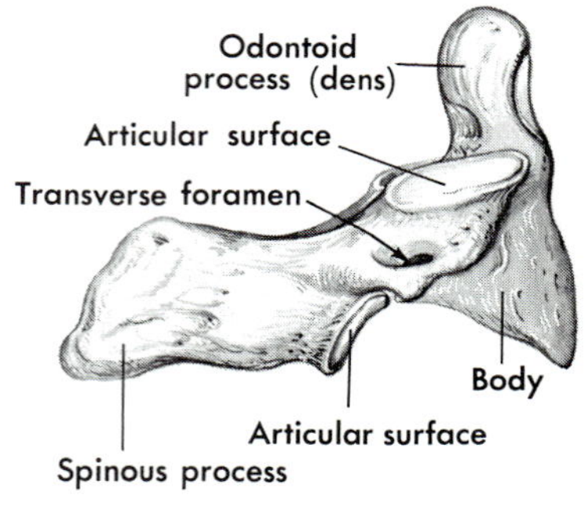

Fig. 5-20

Second cervical vertebra (axis or epistropheus) viewed from the side.

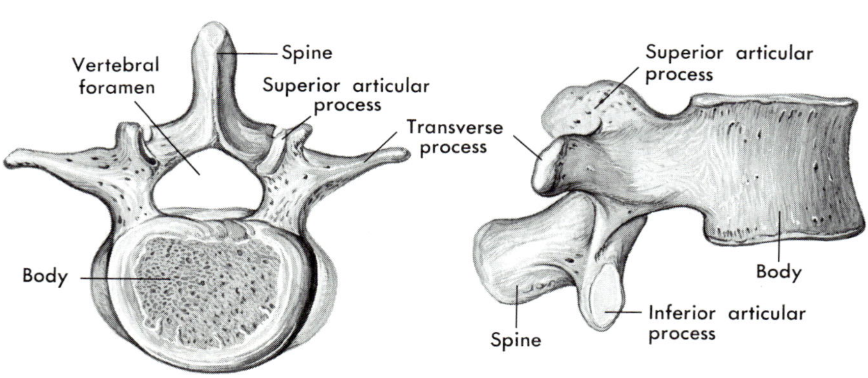

Fig. 5-21

Third lumbar vertebra viewed from above.

Fig. 5-22

Third lumbar vertebra viewed from the side.

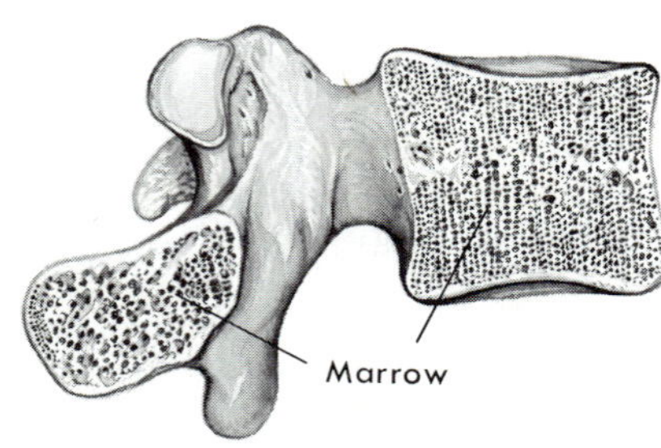

Fig. 5-23

Third lumbar vertebra sectioned.

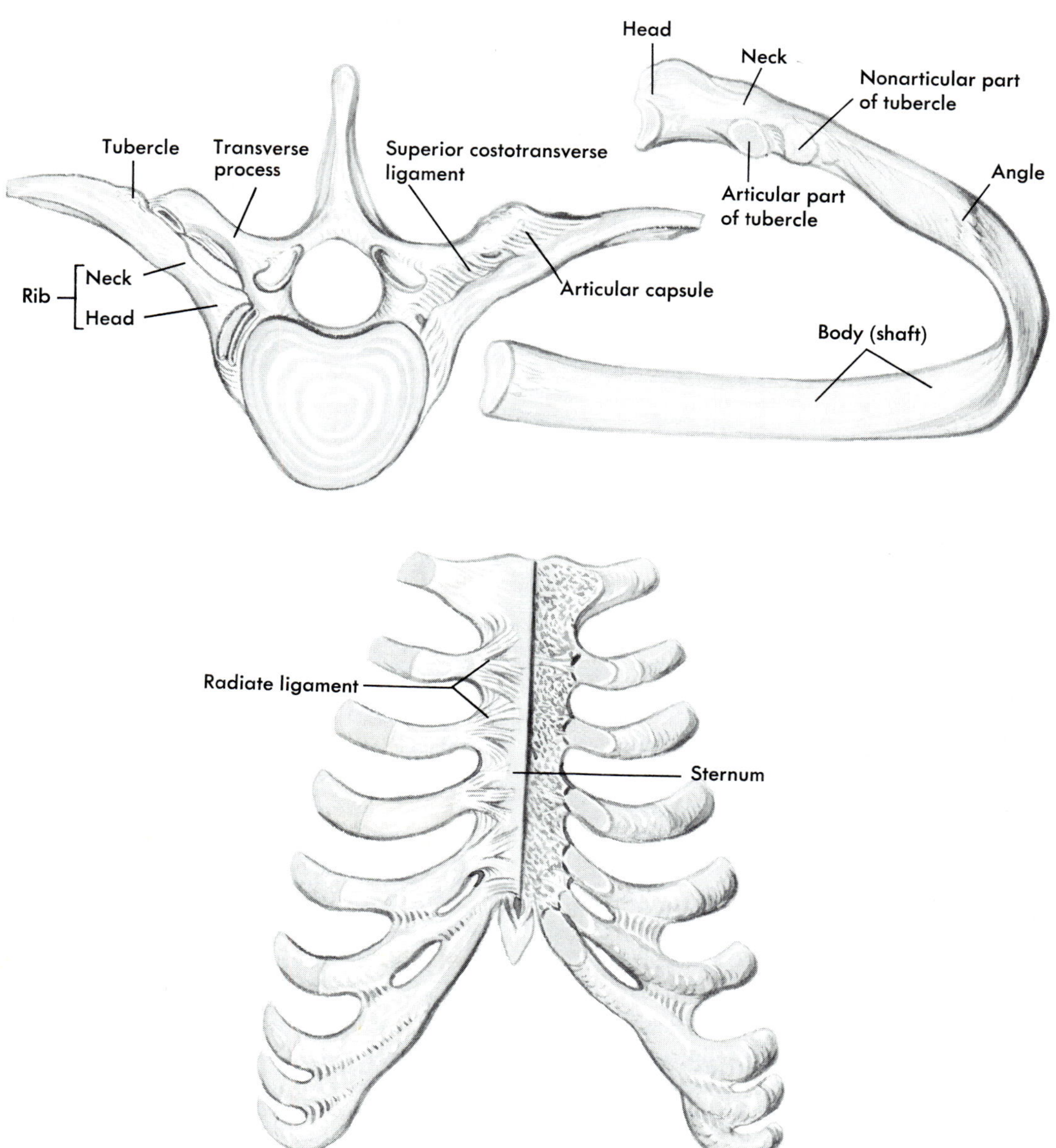

Fig. 5-24

Upper left, Articulation of ribs with a vertebra. On the left the ligaments are shown removed. *Upper right,* A central rib of the right side seen from behind. *Bottom,* The costal cartilages and their articulation with the sternum by radiate ligaments.

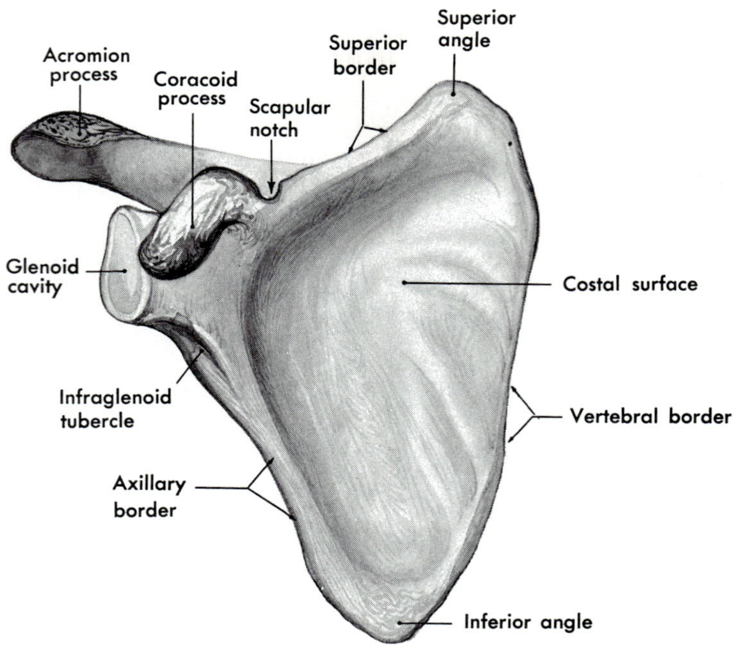

Fig. 5-25

Right scapula, anterior view.

thoracic vertebra has articular facets for the ribs. More detailed descriptions of separate vertebrae are given in Table 5-2. The vertebral column as a whole articulates with the head, ribs, and hip bones, whereas the individual vertebrae articulate with each other in joints between their bodies and between their articular processes. For a description of intervertebral joints, see Table 5-4, p. 103.

In order to increase the carrying strength of the vertebral column and to make balance possible in the upright position, the vertebral column is curved. At birth there is a continuous posterior convexity from head to coccyx. Later, as the child learns to sit and stand, secondary posterior concavities necessary for balance develop in the cervical and lumbar regions. Not uncommonly, spinal curves deviate from the normal. For example, the lumbar curve frequently shows an exaggerated concavity *(lordosis),* whereas any of the regions may have a lateral curvature *(scoliosis)*. The so-called hunchback is an exaggerated convexity in the thoracic region *(kyphosis)*.

Sternum. The *medial part* of the anterior chest wall is supported by the *sternum,* a somewhat dagger-shaped bone consisting of three parts: the upper handle part or *manubrium,* the middle blade part or *body,* and a blunt cartilaginous lower tip, the *xiphoid process.* The latter ossifies during adult life. The manubrium articulates with the clavicle and first rib, whereas the next nine ribs join the body of the sternum, either directly or indirectly, by means of the *costal cartilages.*

Ribs. *Twelve pairs of ribs,* together with the vertebral column and sternum, form the bony cage known as the *thorax.* Each rib articulates with both the body and the transverse process of its corresponding thoracic vertebra. In addition, the second through the ninth

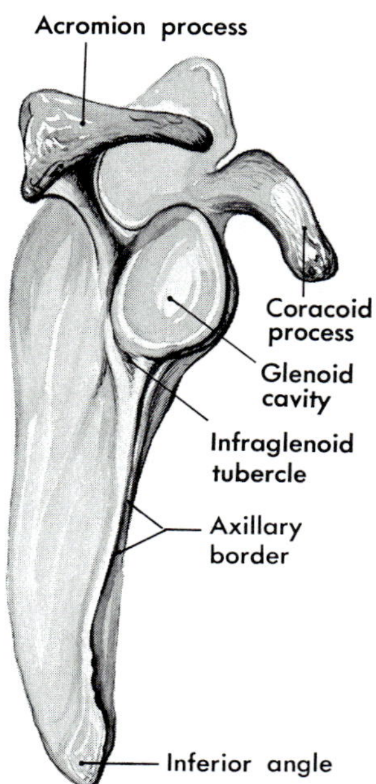

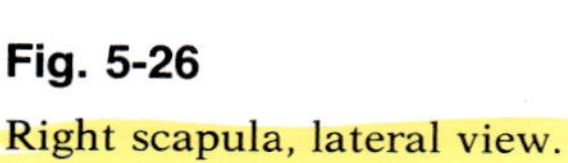
Fig. 5-26

Right scapula, lateral view.

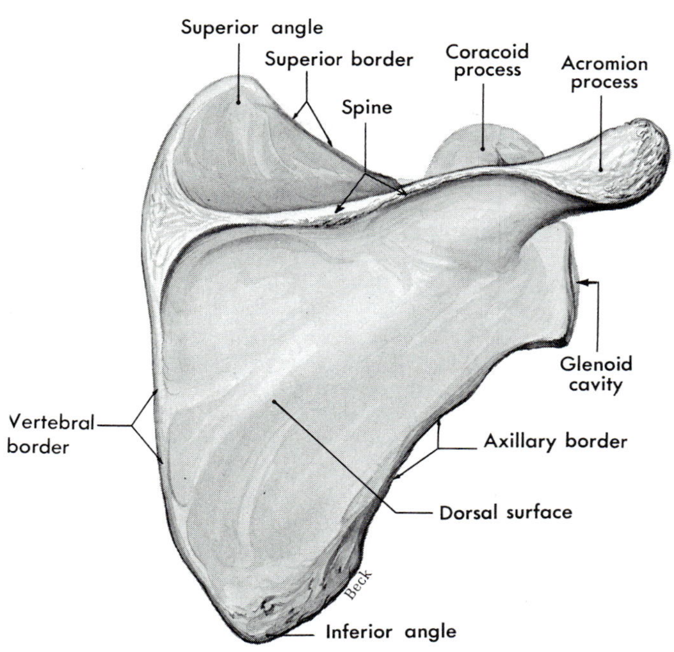

Fig. 5-27

Right scapula, posterior view.

ribs articulate with the body of the vertebra above. From its vertebral attachment each rib curves outward, then forward and downward, a mechanical fact important for breathing. Anteriorly, each of the first seven ribs joins a costal cartilage that attaches it to the sternum. Each of the costal cartilages of the next three ribs, however, joins the cartilage of the rib above to be thus indirectly attached to the sternum. Because the eleventh and twelfth ribs do not attach even indirectly to the sternum, they are designated floating ribs.

APPENDICULAR SKELETON

Upper extremity. The upper extremity consists of the bones of the shoulder girdle, upper arm, lower arm, wrist, and hand. Two bones, the *clavicle* and *scapula,* compose the *shoulder girdle.* Contrary to appearances, this girdle forms only one bony joint with the trunk: the sternoclavicular joint between the sternum and clavicle. At its outer end, the clavicle articulates with the scapula, which attaches to the ribs by muscles and tendons, not by a joint. All shoulder movements, therefore, involve the sternoclavicular joint. Various markings of the scapula are described in Table 5-2 (p. 99) (see also Figs. 5-25 to 5-27).

The *humerus* or upper arm bone, like other long bones, consists of a shaft or diaphysis and two ends or epiphyses (Figs. 5-28 and 5-29). The upper epiphysis bears several identifying structures: the head, anatomical neck, greater and lesser tubercles, intertubercular groove, and surgical neck. On the diaphysis are found the deltoid tuberosity and the radial groove. The distal epiphysis has four projections—the medial and lateral epicondyles, the capitulum, and the trochlea—and two depressions, the olecranon and coronoid fossae. For descriptions of all of these mark-

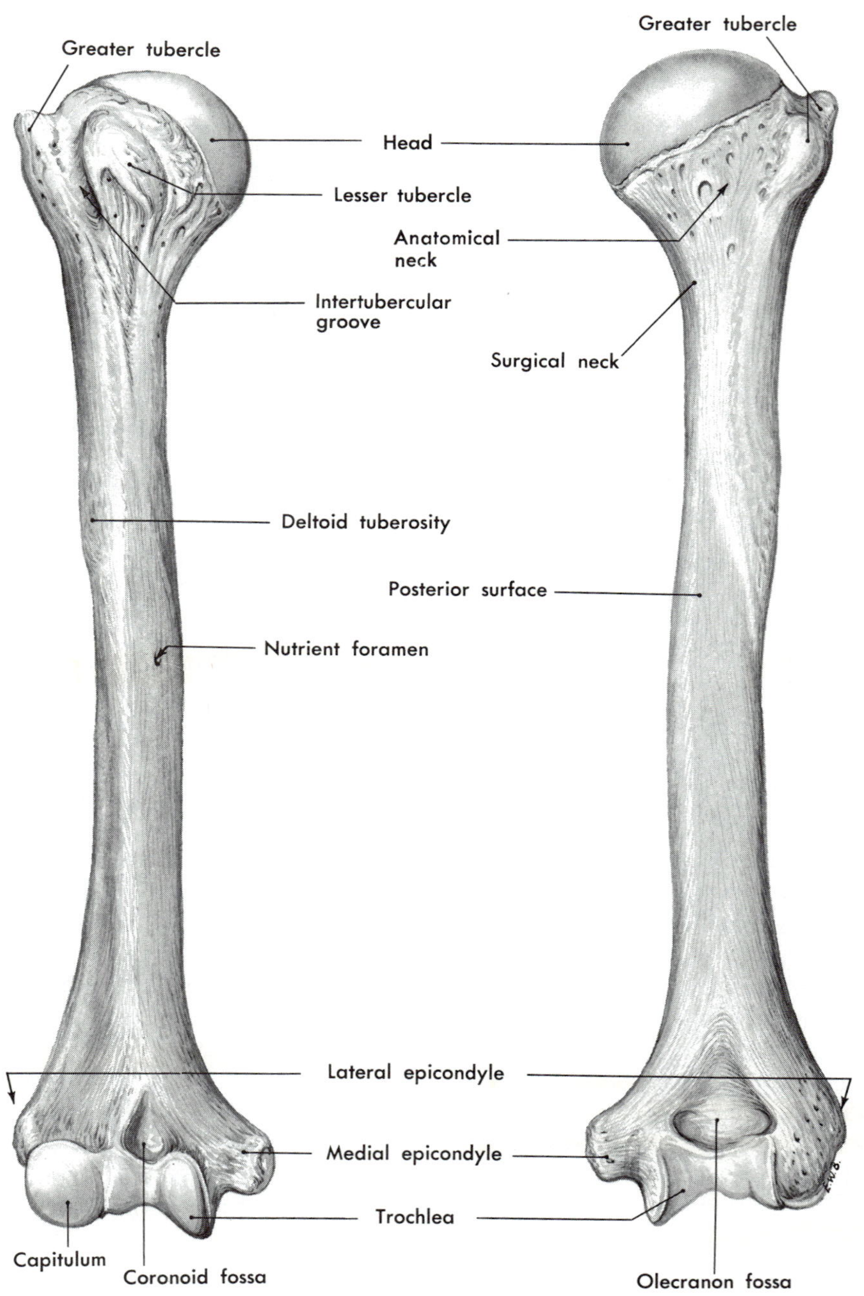

Fig. 5-28

Right humerus, anterior view.

Fig. 5-29

Right humerus, posterior view.

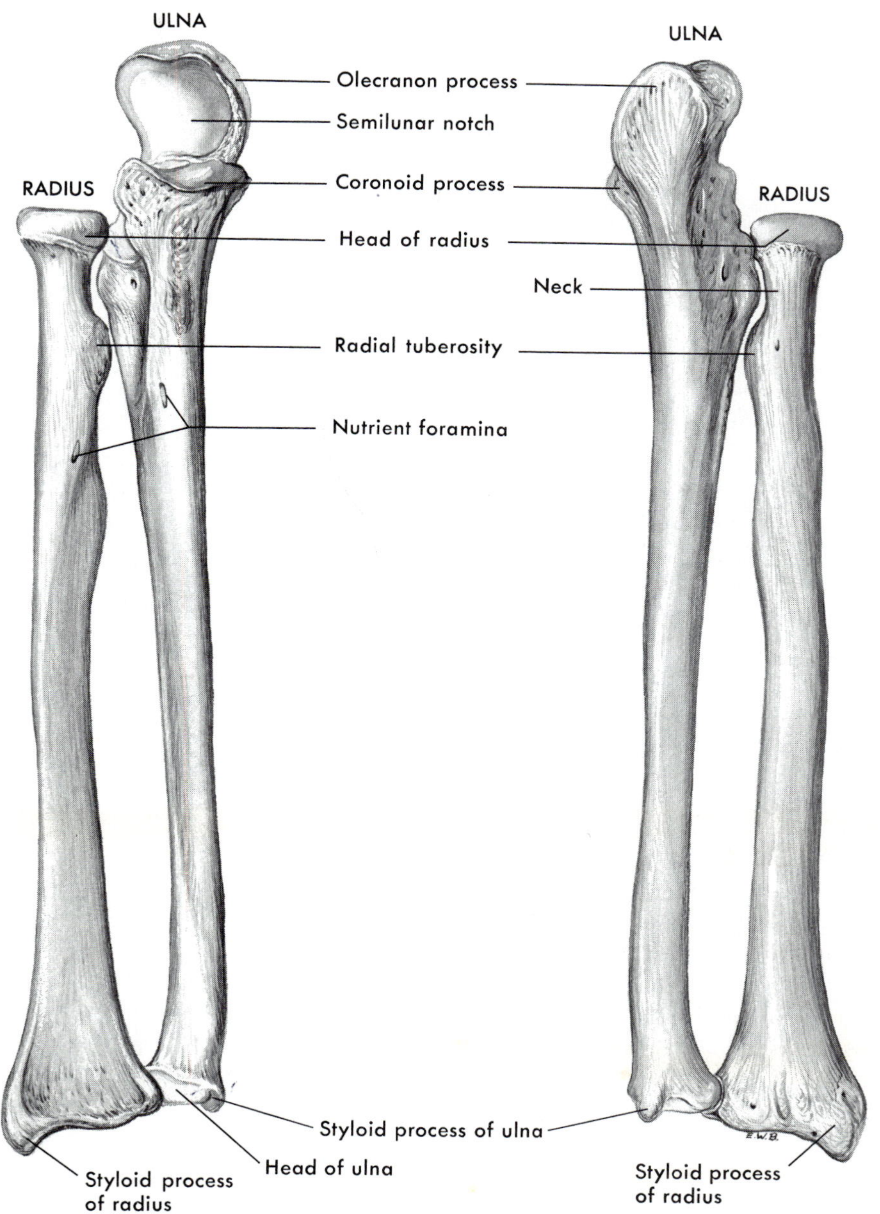

Fig. 5-30

Right radius and ulna, anterior surfaces.

Fig. 5-31

Right radius and ulna, posterior surfaces.

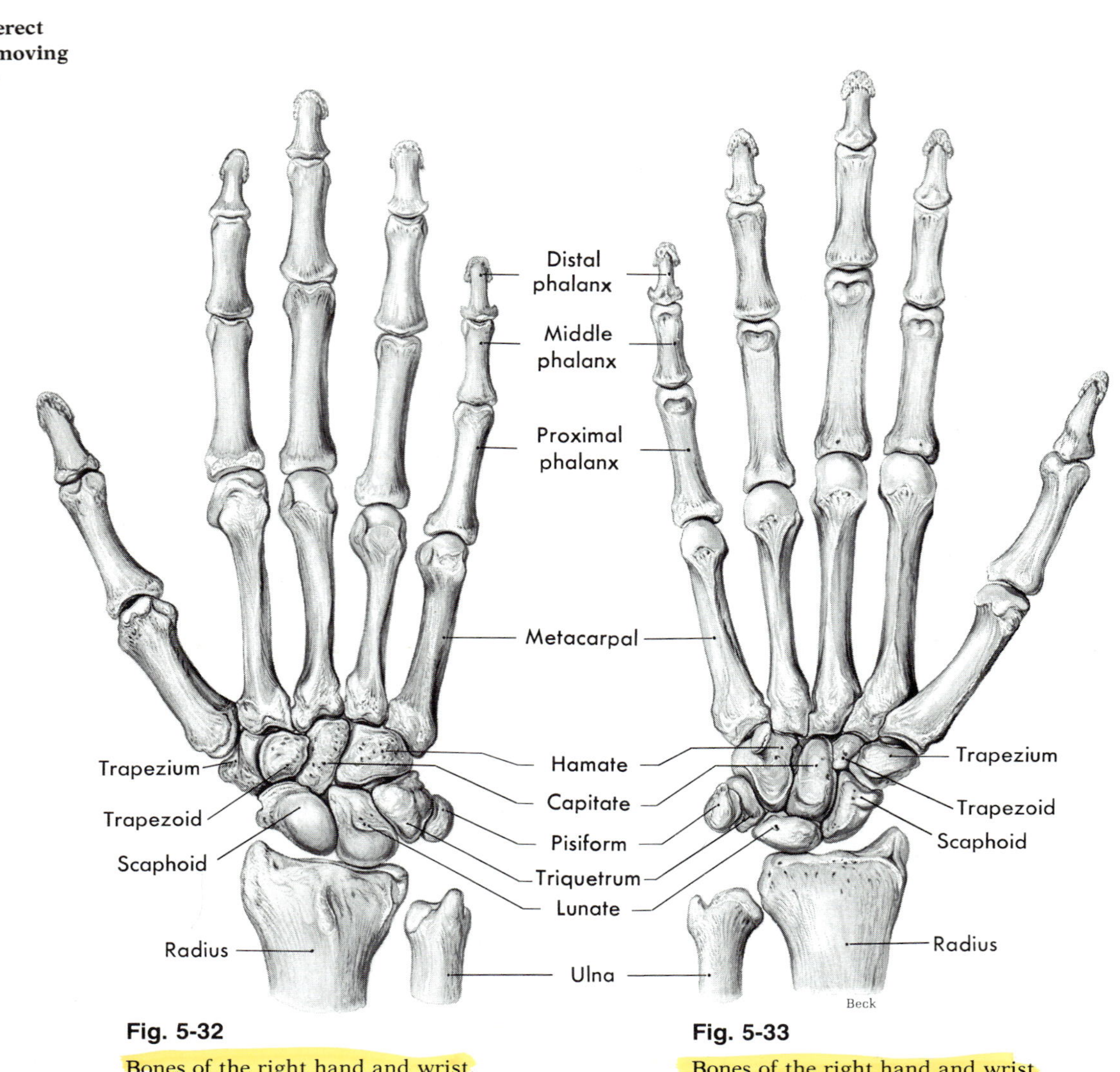

Fig. 5-32

Bones of the right hand and wrist, dorsal surface.

Fig. 5-33

Bones of the right hand and wrist, palmar surface.

ings, see Table 5-2. The humerus articulates proximally with the scapula and distally with both the radius and ulna.

Two bones form the framework for the lower arm: the *radius* on the thumb side and the *ulna* on the little finger side. At the proximal end of the ulna, the olecranon process projects posteriorly and the coronoid process anteriorly. There are also two depressions: the semilunar notch on the anterior surface and the radial notch on the lateral surface. The distal end has two projections: a rounded head and a sharper styloid process. For more detailed identification of these markings, see Table 5-2 (p. 100). The ulna articulates proximally with the humerus and radius and distally with a fibrocartilaginous disk but not with any of the carpal bones.

The radius has three projections: two at its proximal end, the head and radial tuberosity, and one at its distal end, the styloid process (Figs. 5-30 and 5-31). There are two proximal articulations: one with the capitulum of the humerus and the other with the radial notch of the ulna. The three distal articulations are with the scaphoid and lunate carpal bones and with the head of the ulna.

The eight *carpal bones* (Figs. 5-32 and 5-33)

form what most people think of as the upper part of the hand but what, anatomically speaking, is the wrist. Only one of these bones is evident from the outside, the *pisiform bone*, which projects posteriorly on the little finger side as a small rounded elevation. Ligaments bind the carpals closely and firmly together in two rows of four each: proximal row (from little finger toward thumb)—pisiform, triquetrum, lunate, and scaphoid bones; distal row—hamate, capitate, trapezoid, and trapezium bones. The joints between the carpals and the joint between the carpals and radius permit wrist and hand movements.

Of the five *metacarpal bones* that form the framework of the hand, the thumb metacarpal forms the most freely movable joint with the carpals. This fact has great significance. Because of the wide range of movement possible between the thumb metacarpal and the trapezium, particularly the ability to oppose the thumb to the fingers, the human hand has much greater dexterity than the forepaw of any animal and has enabled man to manipulate his environment effectively. The heads of the metacarpals, prominent as the proximal knuckles of the hand, articulate with the phalanges.

Lower extremity. Bones of the hip, thigh, lower leg, ankle, and foot constitute the lower extremity. Strong ligaments bind the two hip bones (*os coxae* or *os innominatum*) to the sacrum posteriorly and to each other anteriorly to form the *pelvic girdle*, a stable, circular base that supports the trunk and attaches the lower extremities to it. In early life, each innominate bone is made up of three separate bones. Later on, they fuse into a single, massive, irregular bone that is broader than any other bone in the body. The largest and uppermost of the three bones is the *ilium;* the strongest, lowermost, the *ischium;* and the anteriormost, the *pubis*. Numerous markings are present on the three bones. These are identified in Table 5-2 (also see Fig. 5-34).

The two thigh bones or *femurs* have the distinction of being the longest and heaviest bones in the body. Several prominent markings characterize them. For example, three projections are conspicuous at each epiphysis: the head and greater and lesser trochanters proximally and the medial and lateral condyles and adductor tubercle distally (Fig. 5-38). Both condyles and the greater trochanter may be felt externally. For a description of the various femur markings, see Table 5-2.

The largest sesamoid bone in the body, and the one that is almost universally present, is the *patella* or kneecap, located in the tendon of the quadriceps femoris muscle as a protection to the underlying knee joint. Although some individuals have sesamoid bones in tendons of other muscles, lists of bone names do not include them—because they are not always present, are not found in any particular tendons, and seemingly are of no importance. When the knee joint is extended, the patellar outline may be distinguished through the skin, but as the knee flexes, it sinks into the intercondylar notch of the femur and can no longer be delineated.

The *tibia* is the larger and stronger and the more medially and superficially located of the two lower leg bones. The *fibula* is smaller and more laterally and deeply placed. At its proximal end it articulates with the lateral condyle of the tibia. The proximal end of the tibia, in turn, articulates with the femur to form the knee joint, the largest and one of the most stable joints of the body. Distally, the tibia articulates with the fibula and also with the talus. The latter fits into a boxlike socket (ankle joint) formed by the medial and lateral malleoli, projections of the tibia and fibula, respectively. For other tibial markings, see Table 5-2 (p. 101) and Fig. 5-39.

Structure of the *foot* is similar to that of the hand with certain differences that adapt it for supporting weight. One example of this is the much greater solidity and the more limited

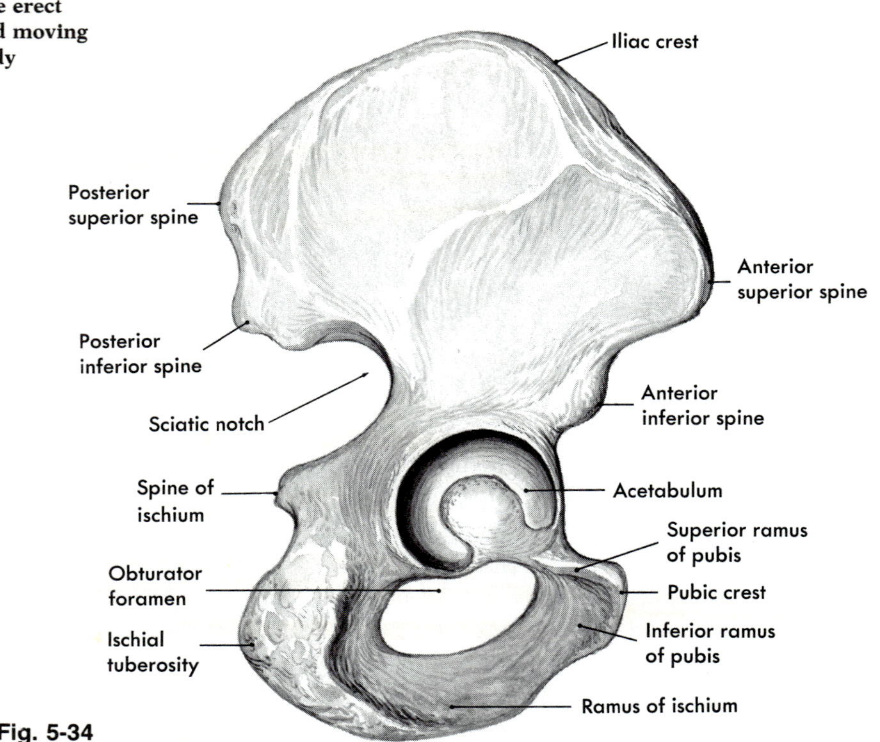

Fig. 5-34

Right hip bone disarticulated from the skeleton viewed from the side with the bone turned so as to look directly into the acetabulum.

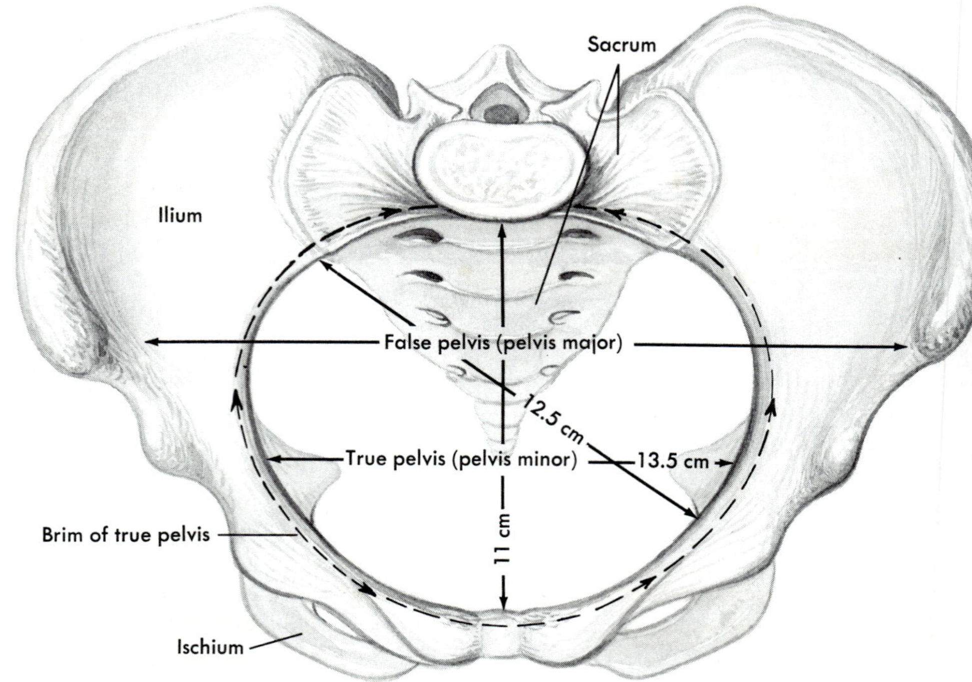

Fig. 5-35

The female pelvis viewed from above. Note the brim of the true pelvis (dotted line) that marks the boundary between the false pelvis (pelvis major) above and the true pelvis (pelvis minor) below it.

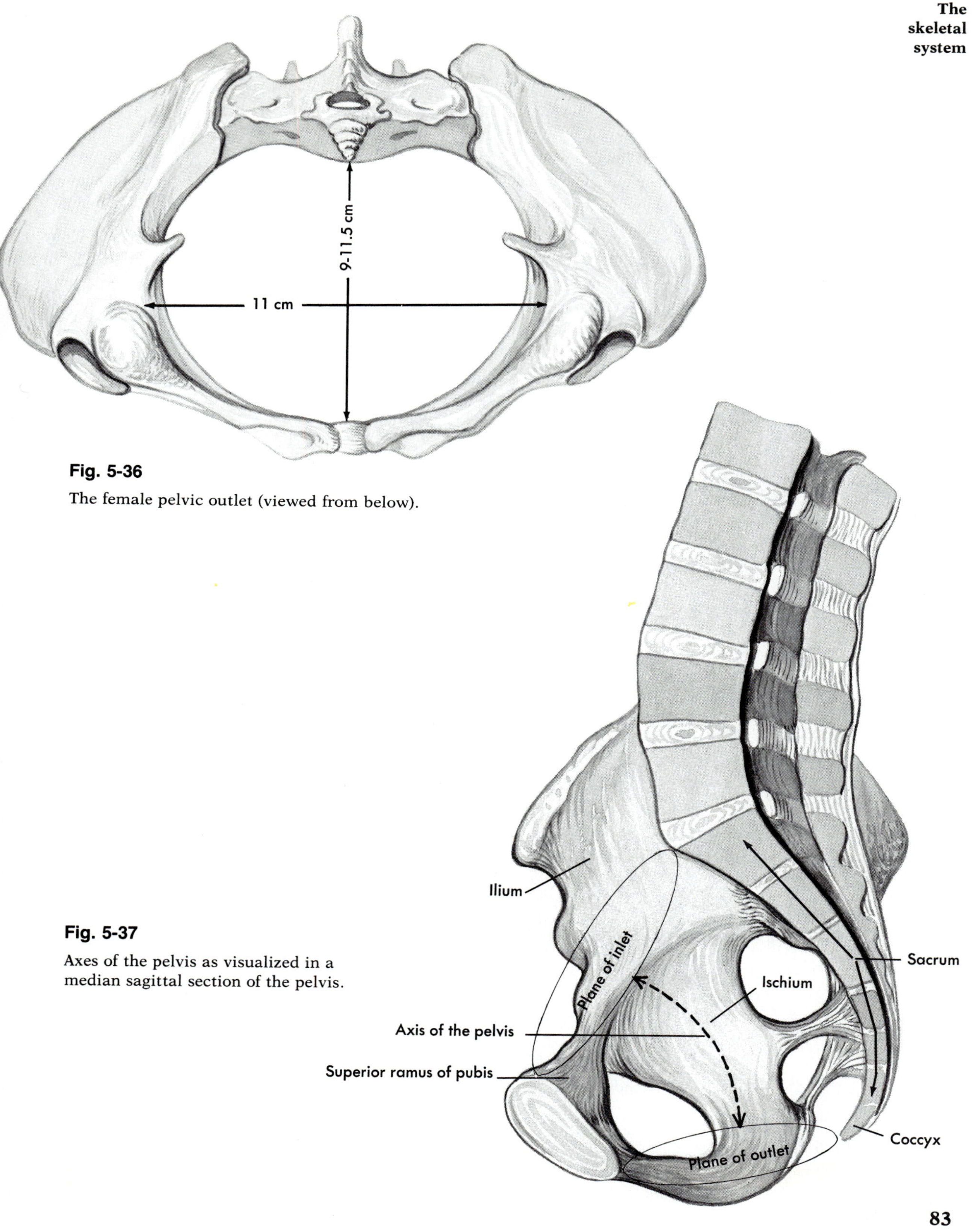

Fig. 5-36

The female pelvic outlet (viewed from below).

Fig. 5-37

Axes of the pelvis as visualized in a median sagittal section of the pelvis.

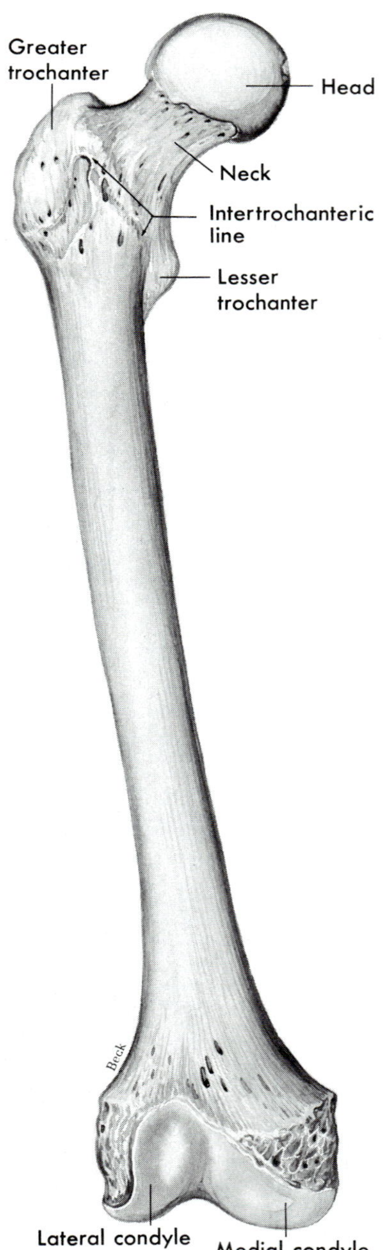

Fig. 5-38

Right femur, anterior surface.

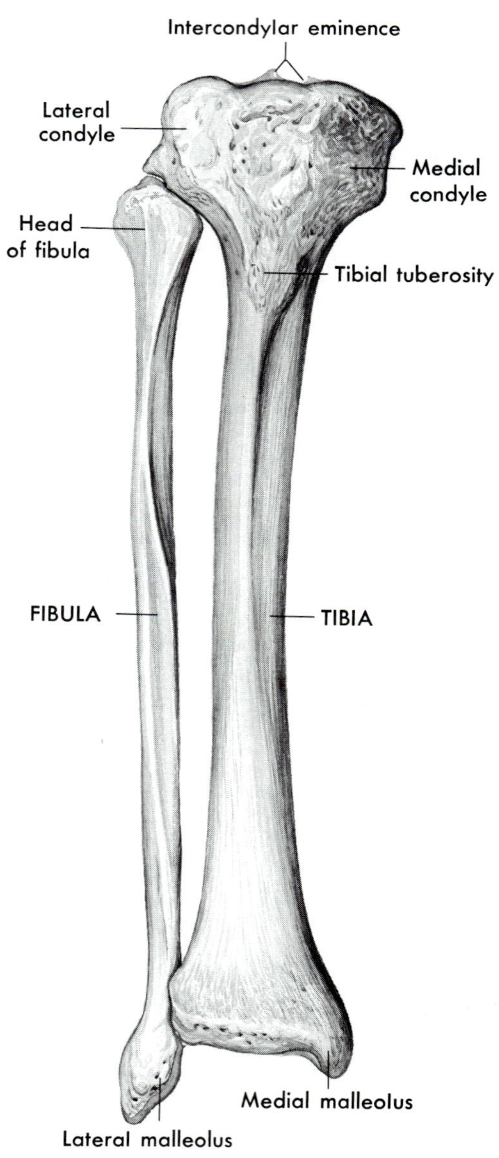

Fig. 5-39

Right tibia and fibula, anterior surface.

mobility of the great toe compared to the thumb. Then, too, the foot bones are held together in such a way as to form springy lengthwise and crosswise arches. This is architecturally sound since arches are known to furnish more supporting strength per given amount of structural material than any other type of construction. Hence, the two-way arch construction makes a highly stable base. The longitudinal arch has an inner or medial portion and an outer or lateral portion, both of which are formed by the placement of tarsals and metatarsals. Specifically, some of the tarsals (calcaneus, talus, navicular, and cuneiform) and the first three metatarsals (starting with the great toe) form the medial longitudinal arch. The calcaneus and cuboid tarsals plus the fourth and fifth metatarsals shape the lateral longitudinal arch (Figs. 5-40 to 5-42). The transverse arch results from the relative placement of the distal row of tarsals and the five metatarsals. (See Table 5-2 for specific bones of different arches.) Strong ligaments and leg muscle tendons normally hold the foot bones firmly in their arched positions, but not infrequently these weaken, causing the arches to flatten, a condition aptly called fallen arches or flatfeet. Note that the tarsals and metatarsals play the major role in the functioning of the foot as a supporting structure, with the phalanges relatively unimportant. The reverse is true for the hand. Here, manipulation is the main function rather than support. Consequently, the phalanges are all important and the carpals and metacarpals are subsidiary.

Markings

Various points on bones are labeled according to the nature of their structure. This method of identifying definite parts of different bones proves helpful when locating other structures such as muscles, blood vessels, and nerves. Table 5-2 describes many bone markings. Definitions of some of the common

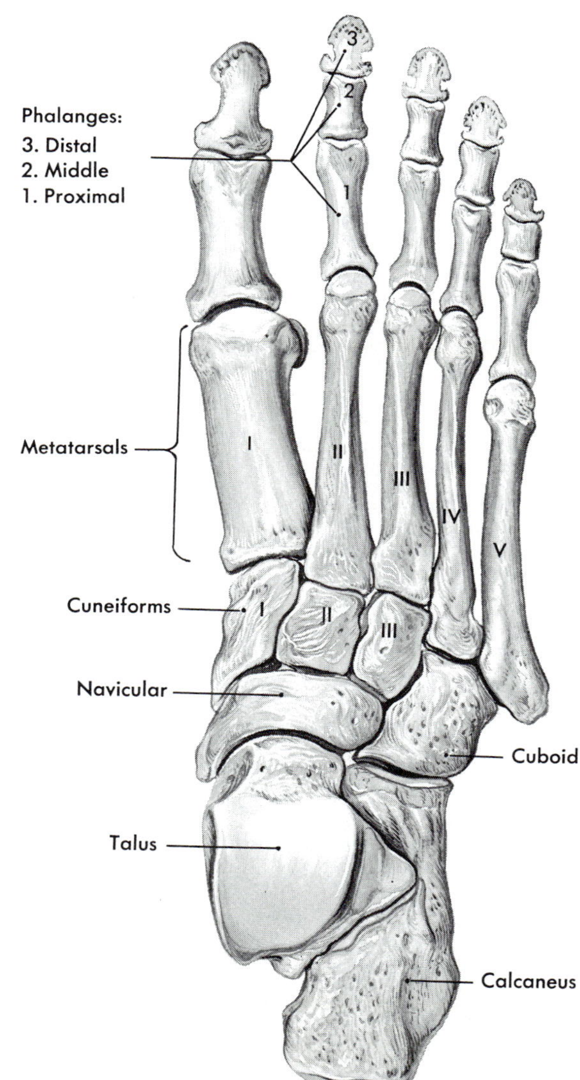

Fig. 5-40

Bones of right foot viewed from above. Tarsal bones consist of cuneiforms, navicular, talus, cuboid, and calcaneus.

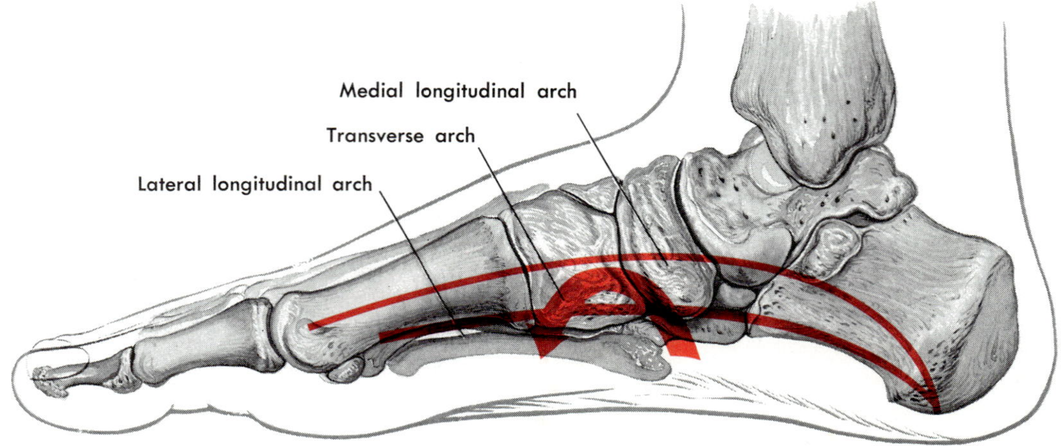

Fig. 5-41

Arches of the foot (see Table 5-2, p. 101).

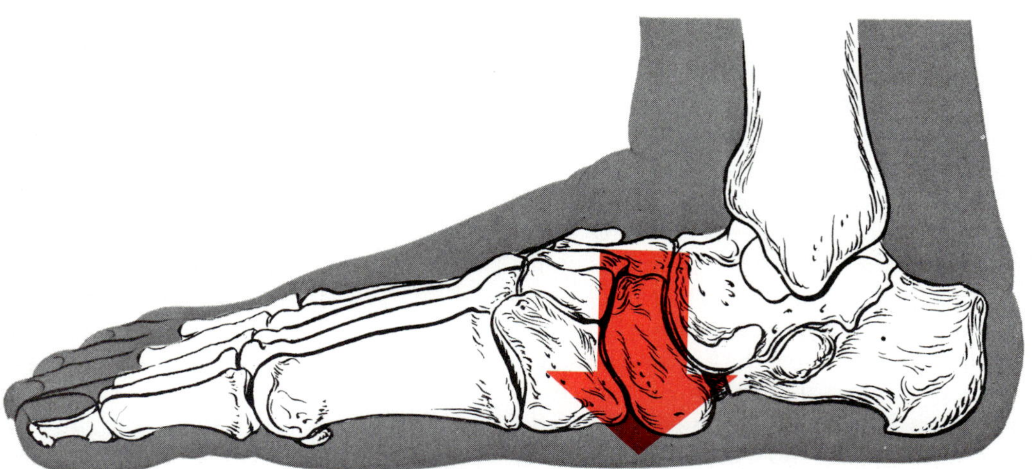

Fig. 5-42

Flatfoot results when there is a weakening of tendons and ligaments attaching to the tarsal bones. Downward pressure by the weight of the body gradually flattens out the normal arch of bones.

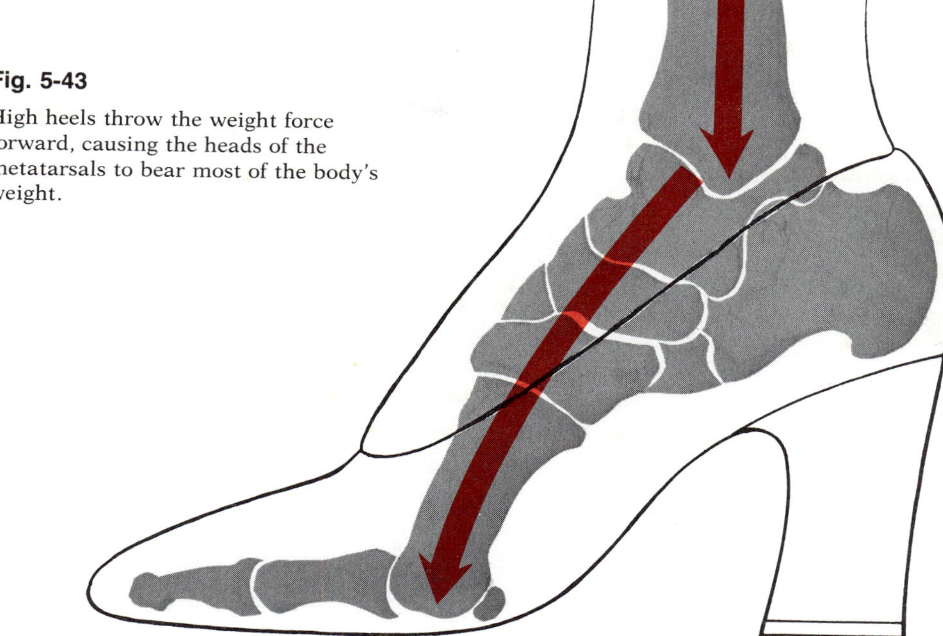

Fig. 5-43

High heels throw the weight force forward, causing the heads of the metatarsals to bear most of the body's weight.

terms applied to bone markings follow. After you have studied these, see Table 5-2, pp. 96-101, for a description of bone markings.

DEPRESSIONS AND OPENINGS

1 *Fossa*—a hollow or depression (example, mandibular fossa of the temporal bone)
2 *Sinus*—a cavity or spongelike space in a bone (example, the frontal sinus)
3 *Foramen*—a hole (example, foramen magnum of the occipital bone)
4 *Meatus*—a tube-shaped opening (example, the external auditory meatus)

PROJECTIONS OR PROCESSES

Those projections or processes that fit into joints are as follows:

1 *Condyle*—a rounded projection that enters into the formation of a joint (example, condyles of the femur)
2 *Head*—a rounded projection beyond a narrow necklike portion (example, head of the femur)

Those projections to which muscles attach include the following:

1 *Trochanter*—a very large projection (example, greater trochanter of femur)
2 *Crest*—a ridge (example, the iliac crest); a less prominent ridge is called a *line* (example, ileopectineal line)
3 *Spinous process or spine*—a sharp projection (example, anterior superior iliac spine)
4 *Tuberosity*—a large, rounded projection (example, ischial tuberosity)
5 *Tubercle*—a small, rounded projection (example, rib tubercles)

Differences between male and female skeletons

Both general and specific differences exist between male and female skeletons. The general difference is one of size and weight, the male skeleton being larger and heavier. The specific differences concern the shape of the pelvic bones and cavity. Whereas the male pelvis is deep and funnel shaped, with a narrow pubic arch (usually less than 90 degrees), the female pelvis, as Figs. 5-35 to 5-37 show, is shallow, broad, and flaring, with a wider pubic arch (usually greater than 90 degrees). The childbearing function obviously explains the necessity for these and certain other modifications of the female pelvis.

Age changes in skeleton

Changes in the skeleton as a whole from infancy to adulthood are mainly changes in the size of the bones and in the proportionate sizes between different bones. Changes in individual bones from young adulthood to old age, on the other hand, are mainly a matter of changes in the extent of calcification, in the texture, and in the contour of the margins and bone markings. Some of the major changes in the skeleton from infancy to young adulthood are as follows:

1 The head becomes proportionately smaller. Whereas the infant head is approximately one fourth the total height of the body, the adult head is only about one eighth the total height.

2 The thorax changes shape, roughly speaking, from round to elliptical.

3 The pelvis becomes relatively larger and in the female relatively wider.

4 The legs become proportionately longer and the trunk proportionately shorter.

5 The vertebral column develops two curves not present at birth—the cervical curve when the infant starts lifting up his head (at about 3 months of age) and the lumbar curve when the child begins standing (toward the end of the first year). Both of these secondary curves are concave posteriorly, whereas the primary thoracic and sacral curves are convex posteriorly.

6 The cranium shows several modifications. It grows rapidly during early childhood, enlarging its capacity from approximately 350 ml at birth to approximately 1,500 ml (about adult size) by 6 years of age. The fontanels close by about 1½ or 2 years of age, and the sutures begin to fuse in the twenties.

7 The facial bones also show several changes between infancy and adulthood. Unlike the cranial bones, their growth is slow during early childhood but rapid during the teens. Whereas the infant face compared with the entire skull bears the relationship of 1:8, the adult face bears the relationship of 1:2 to the adult skull. The sinuses are much larger in the adult. For example, at birth, only rudimentary maxillary and mastoid sinuses exist. The ethmoid and sphenoid sinuses start to appear at about 6 years of age and the frontal at about 7 years. All of the bony sinuses, but especially the frontal, grow rapidly during adolescence.

8 The epiphyses of the long bones are composed of cartilage at birth but become completely ossified (except for the thin layer of articular cartilage) by adulthood. Demonstration on x-ray films of epiphyseal cartilage between epiphyses and diaphysis indicates that skeletal growth has not ceased.

Changes in the skeleton continue to occur from adulthood to old age. Both bone margins and projections, for example, look different in old bones from those in young bones. Instead of clean-cut, distinct margins, old bones characteristically have indistinct, shaggy-appearing margins (marginal lipping and spurs)—a regrettable change because the restricted movements of old age stem partly from this piling up of bone around joint margins. Also, an increase in bone along various projections develops in old age, making ridges and processes more pronounced.

Joints between bones (articulations)

Bones are joined to one another in several ingenious ways that permit a great variety of movement. Where free movement is essential, the articulating ends of the bones are so

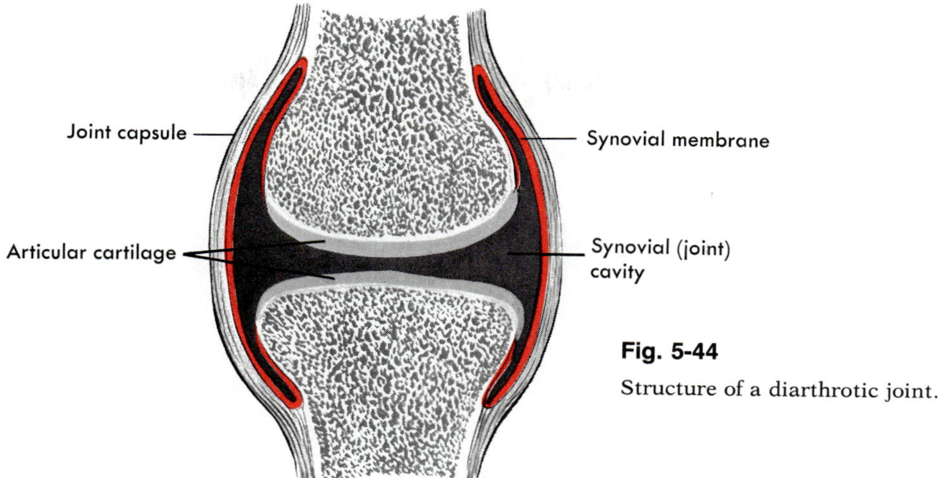

Fig. 5-44

Structure of a diarthrotic joint.

shaped and the joint so constructed as to permit and even facilitate unhampered motion. Where only slight movement is desirable, bone shape and joint structure make only slight movement possible. Where no movement between the bones is preferable, this, too, is accomplished by bone shape and joint structure. Probably most of us seldom consider how important normal joint structure and function are for the productiveness and enjoyment of life. But disease often makes this tragically clear. Joint structure often becomes so altered that crippling immobility results—sometimes only limitation of a single movement and sometimes almost complete immobilization. To inquire into joint structure and action is, therefore, an essential part of the study of anatomy and physiology. Basic information includes understanding what structural kinds of joints exist between bones and what kinds of movements each type permits.

Kinds

A confusing array of terms has grown up around the subject of joint classification. Different anatomists have used different criteria for identifying joint types and have muddled matters even more by using various names for the same kind of joint.

One of the simpler ways of classifying joints is to divide them into two main structural types: diarthroses and synarthroses.

Diarthroses are joints in which a small space, the joint cavity (Fig. 5-44), exists between the articulating surfaces of the two bones that form the joint. Because of this cavity with no tissue growing between the articulating surfaces of the bones, the surfaces are free to move against one another. Therefore, functionally, diarthrotic joints are classified as freely movable joints. By far the majority of our joints belong to this category.

Diarthroses share several characteristics besides that of having a joint cavity. A thin layer of hyaline cartilage covers the joint surfaces of the articulating bones, a sleevelike, fibrous capsule lined with smooth slippery synovial membrane encases the joint, and additional ligaments grow between the bones, lashing them firmly together. Crescent-shaped pieces of cartilage are found in some diarthrotic joints interposed between

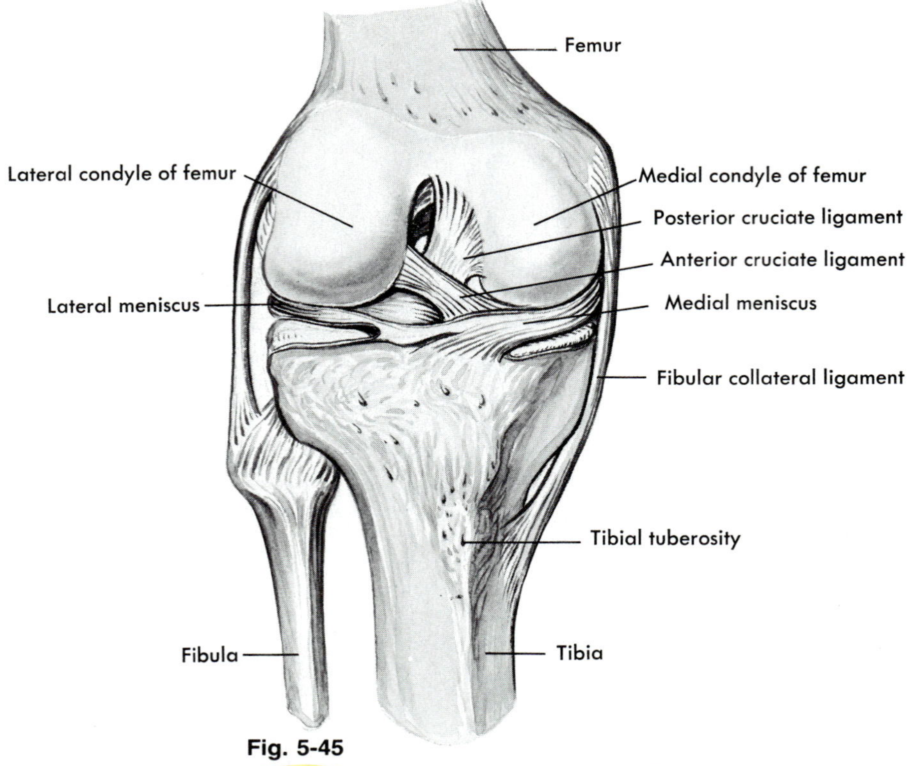

Fig. 5-45
The right knee joint viewed from the front.

the articulating ends of the two bones. Examples are the semilunar cartilages of the knee joint and the glenoid cartilages of the shoulder joint. For details about the structure of the knee joint, see Figs. 5-45 to 5-47.

Synarthroses are joints that do not have a joint cavity. Instead, tissue (fibrous cartilage, or bone) grows between the articulating surfaces of the joined bones and makes them unable to move freely against one another. Defined functionally, synarthroses are joints that do not allow free movements. In fact, they allow little or no movement.

Diarthroses and synarthroses are divided into subtypes according to such characteristics as the shape of the joint surfaces of the united bones and the type of connective tissue between them. See Fig. 5-48 and Table 5-3 (p. 102) for a summary of the main kinds of joints with examples of each. A description of individual joints is given in Table 5-4 (pp. 103-104).

Movements

Diarthrotic joints permit one or more of the following kinds of movements: (1) flexion, (2) extension, (3) abduction, (4) adduction, (5) rotation, (6) circumduction, and (7) special movements such as supination, pronation, inversion, eversion, protraction, and retraction.

Flexion. Flexion decreases the size of the angle between the anterior surfaces of articulated bones (exception: flexion of the knee and toe joints decreases the angle between the posterior surfaces of the articulated bones). Flexing movements are *bending* or

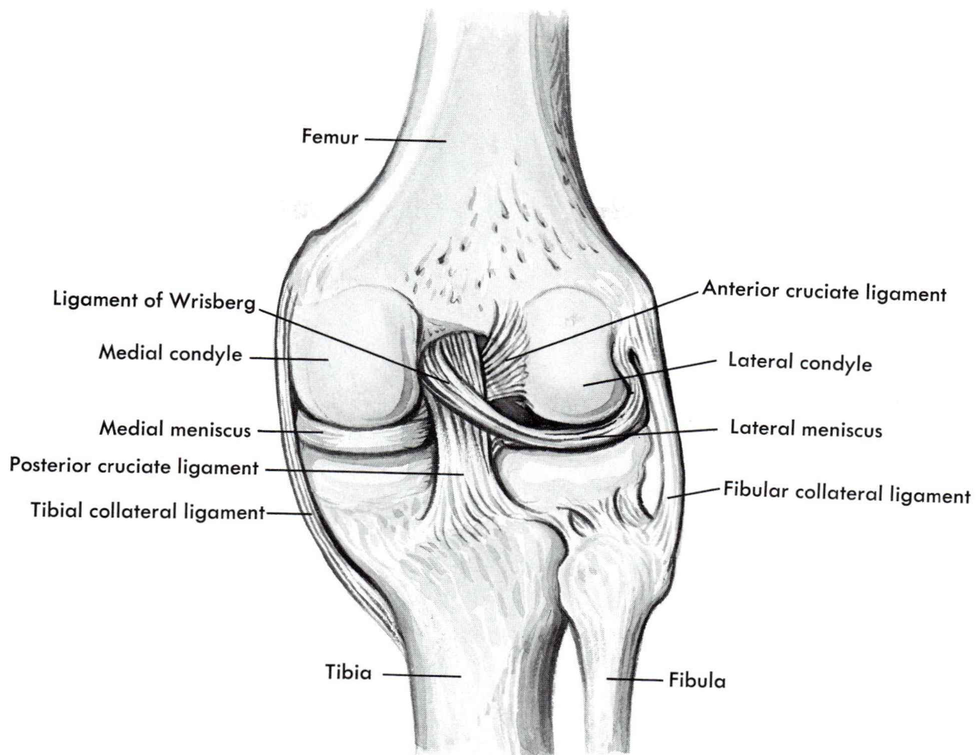

Fig. 5-46

The right knee joint viewed from behind.

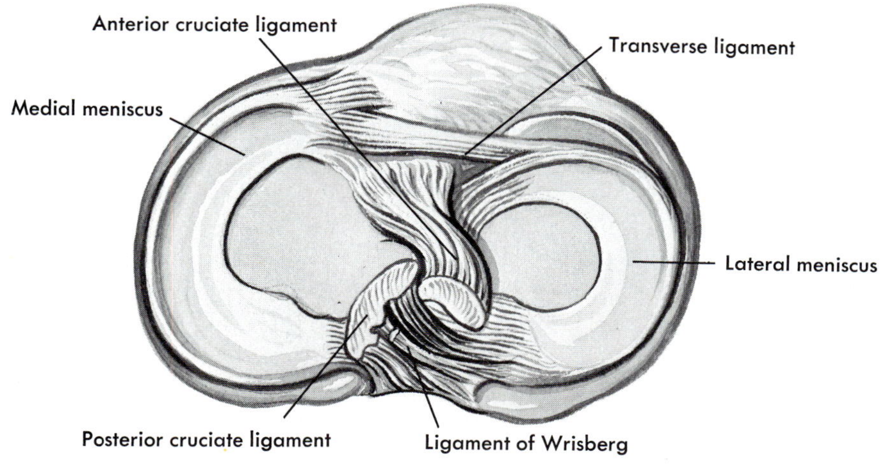

Fig. 5-47

Head of right tibia viewed from above, showing menisci and arrangement of ligaments.

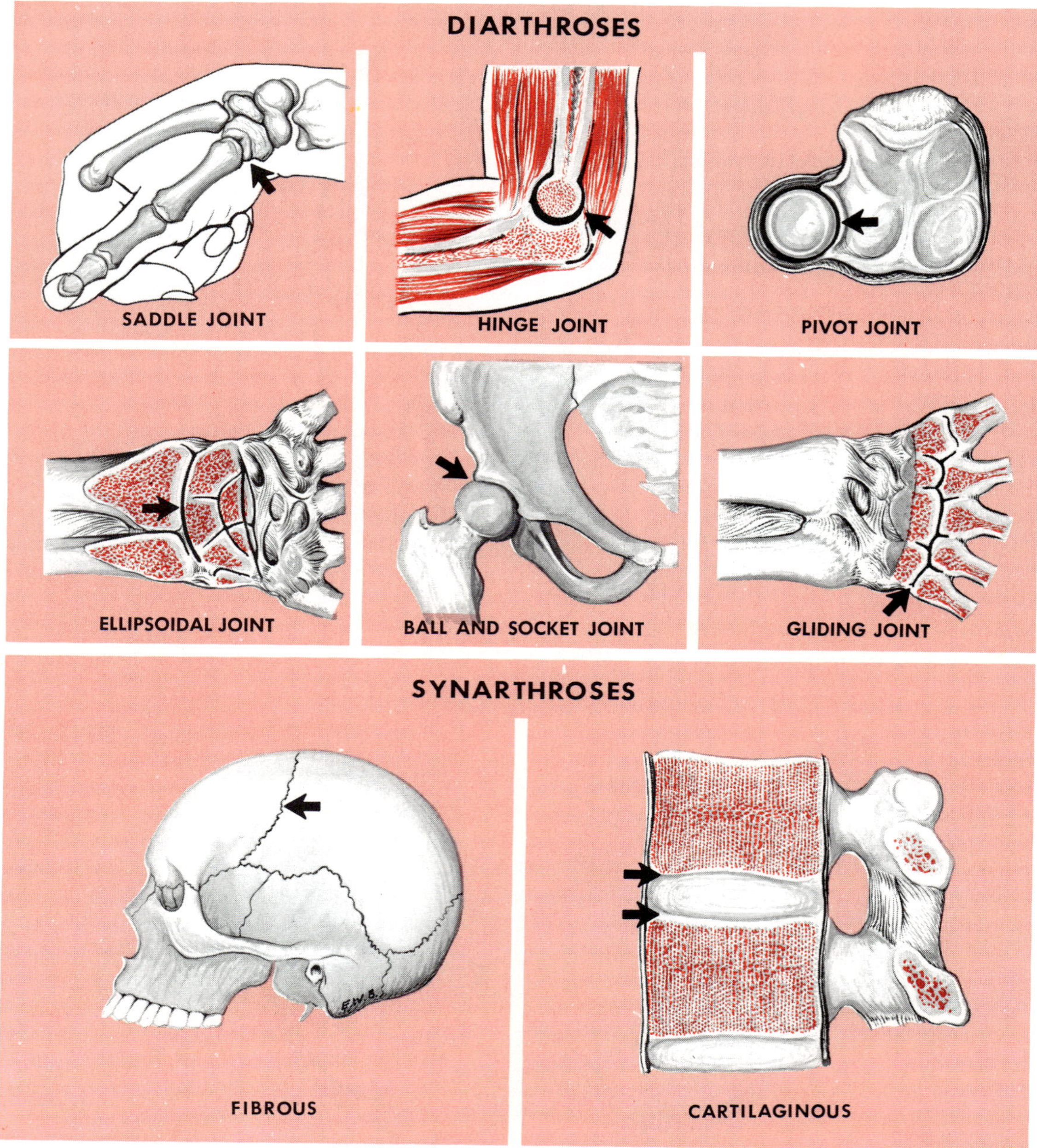

Fig. 5-48

Examples of diarthroses (movable joints with a joint cavity) and synarthroses (joints without a joint cavity). Refer to Tables 5-3 and 5-4 (pp. 102-104) for complete description and other examples.

folding movements. For example, bending the head forward is flexion of the joint between the occipital bone and the atlas, and bending the elbow is flexion of the elbow joint or of the lower arm. Flexing movements of the arms and legs may be thought of as "withdrawing" movements.

Extension. Extension is the return from flexion. Whereas bending movements are flexions, *straightening* movements are extensions. Extension restores a part to its anatomical position from the flexed position. Continuation of extension beyond the anatomical position is called *hyperextension*. Examples include flexion of the head, bending it forward as in prayer, extension of the head, returning it to the upright anatomical position from the flexed position, and hyperextension of the head, stretching it backward from the upright position. Extension of the foot at the ankle joint is commonly referred to as *plantar flexion,* while flexion of the ankle joint is called *dorsal flexion.*

Abduction. Abduction moves the bone away from the median plane of the body. An example is moving the arms straight out to the sides.

Adduction. Adduction is the opposite of abduction. It moves the part toward the median plane of the body. Examples: bringing the arms back to the sides; adduction of the fingers means moving them toward the third finger; adduction of the toes is movement toward the second toe.

Rotation. Rotation is the pivoting or moving of a bone upon its own axis somewhat as a top turns on its axis. An example is holding the head in an upright position and turning it from one side to the other.

Circumduction. Circumduction causes the bone to describe the surface of a cone as it moves. The distal end of the bone describes a circle. It combines flexion, abduction, extension, and adduction in succession. Examples are dropping the head to one shoulder, then to the chest, to the other shoulder, and backward and describing a circle with the arms outstretched.

Special movements. *Supination* is a movement of the forearm that turns the palm forward as it is in the anatomical position. *Pronation* is turning the forearm so as to bring the back of the hand forward. *Inversion* is a special movement of the ankle that turns the sole of the foot inward, while *eversion* turns it outward. *Protraction* moves a part forward, such as sticking out the jaw. *Retraction* is a reverse of protraction.

Table 5-1

Bones of skeleton

Part of body	Name of bone	Number	Description
Axial skeleton (80 bones)			Bones that form upright axis of body—skull, hyoid, vertebral column, sternum, and ribs
Skull (28 bones)			
Cranium (8 bones)			Cranium forms floor for brain to rest on and helmetlike covering over it
	Frontal	1	Forehead bone; also forms most of roof of orbits (eye sockets) and anterior part of cranial floor
	Parietal	2	Prominent, bulging bones behind frontal bone; form topsides of cranial cavity
	Temporal	2	Form lower sides of cranium and part of cranial floor; contain middle and inner ear structures
	Occipital	1	Forms posterior part of cranial floor and walls
	Sphenoid	1	Keystone of cranial floor; forms its midportion; resembles bat with wings outstretched and legs extended downward posteriorly; lies behind and slightly above nose and throat; forms part of floor and sidewalls of orbit
	Ethmoid	1	Complicated irregular bone that helps make up anterior portion of cranial floor, medial wall of orbits, upper parts of nasal septum, and sidewalls and part of nasal roof; lies anterior to sphenoid and posterior to nasal bones
Face (14 bones)	Nasal	2	Small bones forming upper part of bridge of nose
	Maxillary	2	Upper jaw bones; form part of floor of orbit, anterior part of roof of mouth, and floor of nose and part of sidewalls of nose
	Zygomatic (malar)	2	Cheekbones; form part of floor and sidewall of orbit
	Mandible	1	Lower jawbone; largest, strongest bone of face
	Lacrimal	2	Thin bones about size and shape of fingernail; posterior and lateral to nasal bones in medial wall of orbit; help form sidewall of nasal cavity; often missing in dry skull
	Palatine	2	Form posterior part of hard palate, floor, and part of sidewalls of nasal cavity and floor of orbit
	Inferior conchae (turbinates)	2	Thin scroll of bone forming kind of shell along inner surface of sidewall of nasal cavity; lies above roof of mouth
	Vomer	1	Forms lower and posterior part of nasal septum; shaped like ploughshare
Ear bones (6 bones)	Malleus (hammer) Incus (anvil) Stapes (stirrup)	2 2 2	Tiny bones referred to as auditory ossicles in middle ear cavity in temporal bones; resemble, respectively, miniature hammer, anvil, and stirrup
Hyoid bone		1	U-shaped bone in neck between mandible and upper part of larynx; claims distinction as only bone in body not forming a joint with any other bone; suspended by ligaments from styloid processes of temporal bones
Vertebral column (26 bones)			Not actually a column but a flexible segmented rod shaped like an elongated letter S; forms axis of body; head balanced above, ribs and viscera suspended in front, and lower extremities attached below; encloses spinal cord
	Cervical vertebrae	7	First or upper seven vertebrae
	Thoracic vertebrae	12	Next twelve vertebrae; twelve pairs of ribs attached to these
	Lumbar vertebrae	5	Next five vertebrae
	Sacrum	1	Five separate vertebrae until about 25 years of age; then fused to form one wedge-shaped bone
	Coccyx	1	Four or five separate vertebrae in child but fused into one in adult

Part of body	Name of bone	Number	Description
Sternum and ribs (25 bones)			Sternum, ribs, and thoracic vertebrae together form bony cage known as *thorax;* ribs attach posteriorly to vertebrae, slant downward anteriorly to attach to sternum (see description of false ribs below)
	Sternum	1	Breastbone; flat dagger-shaped bone
	True ribs	7 pairs	Upper seven pairs; fasten to sternum by costal cartilages
	False ribs	5 pairs	False ribs do not attach to sternum directly; upper three pairs of false ribs attach by means of costal cartilage of seventh ribs; last two pairs do not attach to sternum at all; therefore, called *"floating"*

Appendicular skeleton (126 bones)

Bones that are appended to axial skeleton: upper and lower extremities, including shoulder and hip girdles

Part of body	Name of bone	Number	Description
Upper extremities (including shoulder girdle) (64 bones)	Clavicle	2	Collar bones; shoulder girdle joined to axial skeleton by articulation of clavicles with sternum; scapula does not form joint with axial skeleton
	Scapula	2	Shoulder blades; scapulae and clavicles together comprise shoulder girlde
	Humerus	2	Long bone of upper arm
	Radius	2	Bone of thumb side of forearm
	Ulna	2	Bone of little finger side of forearm; longer than radius
	Carpals (scaphoid, lunate, triquetrum pisiform, trapezium, trapezoid, capitate, and hamate)	16	Arranged in two rows at proximal end of hand (Figs. 5-32 and 5-33)
	Metacarpals	10	Long bones forming framework of palm of hand
	Phalanges	28	Miniature long bones of fingers, three in each finger, two in each thumb
Lower extremities (62 bones)	Ossa coxae or innominate bones	2	Large hip bones; with sacrum and coccyx, these three bones form basinlike pelvic cavity; lower extremities attached to axial skeleton by pelvic bones
	Femur	2	Thigh bone; longest strongest bone of body
	Patella	2	Kneecap; largest sesamoid bone of body*; embedded in tendon of quadriceps femoris muscle
	Tibia	2	Shin bone
	Fibula	2	Long, slender bone of lateral side of lower leg
	Tarsals (calcaneus, talus, navicular, first, second, and third cuneiforms, cuboid)	14	Bones that form heel and proximal or posterior half of foot (Fig. 5-40)
	Metatarsals	10	Long bones of feet
	Phalanges	28	Miniature long bones of toes; two in each great toe, three in other toes
Total		206*	

*An inconstant number of small, flat, round bones known as *sesamoid bones* (because of their resemblance to sesame seeds) is found in various tendons in which considerable pressure develops. Because the number of these bones varies greatly between individuals, only two of them, the patellae, have been counted among the 206 bones of the body. Generally two of them can be found in each thumb (in flexor tendon near metacarpophalangeal and interphalangeal joints) and great toe plus several others in the upper and lower extremities. *Wormian bones,* the small islets of bone frequently found in some of the cranial sutures, have not been counted in this list of 206 bones either because of their variable occurrence.

Table 5-2

Identification of bone markings*

Marking	Description
Frontal	
Supraorbital margin	Arched ridge just below eyebrows
Frontal sinuses	Cavities inside bone just above supraorbital margin; lined with mucosa; contain air
Frontal tuberosities	Bulge above each orbit; most prominent part of forehead
Superciliary arches	Ridges caused by projection of frontal sinuses; eyebrows lie over these ridges
Supraorbital notch (sometimes foramen)	Notch or foramen in supraorbital margin slightly mesial to its midpoint; transmits supraorbital nerve and blood vessels
Glabella	Smooth area between superciliary ridges and above nose
Sphenoid	
Body	Hollow, cubelike central portion
Greater wings	Lateral projections from body; form part of outer wall of orbit
Lesser wings	Thin, triangular projections from upper part of sphenoid body; form posterior part of roof of orbit
Sella turcica (or *Turk's saddle)*	Saddle-shaped depression on upper surface of sphenoid body; contains pituitary gland
Sphenoid sinuses	Irregular air-filled mucosa-lined spaces within central part of sphenoid
Pterygoid processes	Downward projections on either side where body and greater wing unite; comparable to extended legs of bat if entire bone is likened to this animal; form part of lateral nasal wall
Optic foramen	Opening into orbit at root of lesser wing; transmits second cranial nerve
Superior orbital fissure	Slitlike opening into orbit; lateral to optic foramen; transmits third, fourth, and part of fifth cranial nerves
Foramen rotundum	Opening in greater wing that transmits maxillary division of fifth cranial nerve
Foramen ovale	Opening in greater wing that transmits mandibular division of fifth cranial nerve

*Italics suggest markings particularly useful to recognize.

Marking	Description
Temporal	
Mastoid process	Protuberance just behind ear
Mastoid air cells	Air-filled mucosa-lined spaces within mastoid process
External auditory meatus (or canal)	Opening into ear and tube extending into temporal bone
Zygomatic process	Projection that articulates with malar (or zygomatic) bone
Internal auditory meatus	Fairly large opening on posterior surface of petrous portion of bone; transmits eighth cranial nerve to inner ear and seventh cranial nerve on its way to facial structures
Squamous portion	Thin, flaring upper part of bone
Mastoid portion	Rough-surfaced lower part of bone posterior to external auditory meatus
Petrous portion	Wedge-shaped process that forms part of center section of cranial floor between sphenoid and occipital bones; name derived from Greek word for stone because of extreme hardness of this process; houses middle and inner ear structures
Mandibular fossa	Oval-shaped depression anterior to external auditory meatus; forms socket for condyle of mandible
Styloid process	Slender spike of bone extending downward and forward from undersurface of bone anterior to mastoid process; often broken off in dry skull; several neck muscles and ligaments attach to styloid process
Stylomastoid foramen	Opening between styloid and mastoid processes where facial nerve emerges from cranial cavity
Jugular fossa	Depression on undersurface of petrous portion; dilated beginning of internal jugular vein lodged here
Jugular foramen	Opening in suture between petrous portion and occipital bone; transmits lateral sinus and ninth, tenth, and eleventh cranial nerves
Carotid canal (or foramen)	Channel in petrous portion; best seen from undersurface of skull; transmits internal carotid artery

Occipital

Foramen magnum	Hole through which spinal cord enters cranial cavity
Condyles	Convex, oval processes on either side of foramen magnum; articulate with depressions on first cervical vertebra
External occipital protuberance	Prominent projection on posterior surface in midline short distance above foramen magnum; can be felt as definite bump
Superior nuchal line	Curved ridge extending laterally from external occipital protuberance
Inferior nuchal line	Less well-defined ridge paralleling superior nuchal line short distance below it
Internal occipital protuberance	Projection in midline on inner surface of bone; grooves for lateral sinuses extend laterally from this process and one for sagittal sinus extends upward from it

Ethmoid

Horizontal (cribriform) plate	Olfactory nerves pass through numerous holes in this plate
Crista galli	See Figs. 5-14 and 5-16; meninges attach to this process
Perpendicular plate	Forms upper part of nasal septum (Figs. 5-14 and 5-16)
Ethmoid sinuses	Honeycombed, mucosa-lined air spaces within lateral masses of bone (Figs. 5-15 and 5-16)
Superior and middle turbinates (conchae)	Help to form lateral walls of nose (Figs. 5-14 and 5-16)
Lateral masses	Compose sides of bone; contain many air spaces (ethmoid cells or sinuses); inner surface forms superior and middle conchae

Mandible

Body	Main part of bone; forms chin
Ramus	Process, one on either side, that projects upward from posterior part of body
Condyle (or *head*)	Part of each ramus that articulates with mandibular fossa of temporal bone
Neck	Constricted part just below condyles
Alveolar process	Teeth set into this arch
Mandibular foramen	Opening on inner surface of ramus; transmits nerves and vessels to lower teeth
Mental foramen	Opening on outer surface below space between two bicuspids; transmits terminal branches of nerves and vessels that enter bone through mandibular foramen; dentists inject anesthetics through these foramina
Coronoid process	Projection upward from anterior part of each ramus; temporal muscle inserts here
Angle	Juncture of posterior and inferior margins of ramus

Maxilla

Alveolar process	Arch containing teeth
Maxillary sinus or antrum of Highmore	Large air-filled mucosa-lined cavity within body of each maxilla; largest of sinuses
Palatine process	Horizontal inward projection from alveolar process; forms anterior and larger part of hard palate
Infraorbital foramen	Hole on external surface just below orbit; transmits vessels and nerves
Lacrimal groove	Groove on inner surface; joined by similar groove on lacrimal bone to form canal housing nasolacrimal duct

Palatine

Horizontal plate	Joined to palatine processes of maxillae to complete posterior part of hard palate

Continued.

Table 5-2

Identification of bone markings—cont'd

Marking	Description
Special features of skull	
Sutures	Immovable joints between skull bones
1 *Sagittal*	1 Line of articulation between two parietal bones
2 *Coronal*	2 Joint between parietal bones and frontal bone
3 *Lambdoidal*	3 Joint between parietal bones and occipital bone
Fontanels	"Soft spots" where ossification incomplete at birth; allow some compression of skull during birth; also important in determining position of head before delivery; six such areas located at angles of parietal bones
1 *Anterior* (or *frontal*)	1 At intersection of sagittal and coronal sutures (juncture of parietal bones and frontal bone); diamond shaped; largest of fontanels; usually closed by 1½ years of age
2 *Posterior* (or *occipital*)	2 At intersection of sagittal and lambdoidal sutures (juncture of parietal bones and occipital bone); triangular in shape; usually closed by second month
3 Anterolateral (or *sphenoid*)	3 At juncture of frontal, parietal, temporal, and sphenoid bones
4 Posterolateral (or *mastoid*)	4 At juncture of parietal, occipital, and temporal bones; usually closed by second year
Sinuses	
1 *Air* (or *bony*)	1 Spaces or cavities within bones; those that communicate with nose called *paranasal sinuses* (frontal, sphenoidal, ethmoidal, and maxillary); mastoid cells communicate with middle ear rather than nose, therefore not included among paranasal sinuses
2 *Blood*	2 Veins within cranial cavity (Figs. 13-20 and 13-21, p. 348)
Orbits formed by	
1 Frontal	1 Roof of orbit
2 Ethmoid	2 Medial wall
3 Sphenoid	3 Lateral wall
4 Lacrimal	4 Medial wall
5 Maxillary	5 Floor
6 Zygomatic	6 Lateral wall
7 Palatine	7 Floor
Nasal septum formed by	Partition in midline of nasal cavity; separates cavity into right and left halves
1 Perpendicular plate of ethmoid	1 Forms upper part of septum
2 Vomer bone	2 Forms lower, posterior part
3 Cartilage	3 Forms anterior part
Wormian bones	Small islands of bones within suture

*Italics suggest markings particularly useful to recognize.

Marking	Description
Vertebral column	
General features	Anterior part of vertebrae (except first two cervical) consists of body; posterior part of neural arch which, in turn, consists of two pedicles, two laminae, and seven processes projecting from laminae
Thoracic vertebrae	
1 *Body*	1 Main part; flat, round mass located anteriorly; supporting or weight-bearing part of vertebra
2 *Pedicles*	2 Short projections extending posteriorly from body
3 *Laminae*	3 Posterior part of vertebra to which pedicles join and from which processes project
4 *Neural arch*	4 Formed by pedicles and laminae; protects spinal cord posteriorly; together, neural arches form spinal cavity; congenital absence of one or more neural arches known as *spina bifida* (cord may protrude right through skin)
5 *Spinous process*	5 Sharp process projecting inferiorly from laminae in midline
6 *Transverse processes*	6 Right and left lateral projections from laminae
7 *Superior articulating processes*	7 Project upward from laminae
8 *Inferior articulating processes*	8 Project downward from laminae; articulate with superior articulating processes of vertebrae below
9 *Spinal foramen*	9 Hole in center of vertebra formed by union of body, pedicles, and laminae; spinal foramina, when vertebrae superimposed one upon other, form spinal cavity that houses spinal cord
Cervical vertebrae	
1 General features	1 Foramen in each transverse process for transmission of vertebral artery, vein, and plexus of nerves; short bifurcated spinous processes except on seventh vertebrae, where it is extra long and may be felt as protrusion when head bent forward; bodies of these vertebrae small, while spinal foramina large and triangular
2 *Atlas*	2 First cervical vertebra; lacks body and spinous process; superior articulating processes concave ovals that act as rockerlike cradles for condyles of occipital bone; named atlas because supports head as Atlas was thought to have supported world (Fig. 5-19)
3 *Axis* (epistropheus)	3 Second cervical vertebra, so named because atlas rotates about this bone in rotating movements of head; *dens*, or odontoid process, peglike projection upward from body of axis, forming pivot for rotation of atlas (Fig. 5-20)

Vertebral column continued

Lumbar vertebrae	Strong, massive; superior articulating processes directed inward instead of upward; inferior articulating processes, outward instead of downward; short, blunt spinous process (Figs. 5-21 to 5-23)
Sacral promontory	Protuberance from anterior, upper border of sacrum into pelvis; of obstetrical importance because its size limits anteroposterior diameter of pelvic inlet
Intervertebral foramina	Opening between vertebrae through which spinal nerves emerge
Curves	Curves have great structural importance because increase carrying strength of vertebral column, make balance possible in upright position (if column were straight, weight of viscera would pull body forward), absorb jars from walking (straight column would transmit jars straight to head), and protect column from fracture
1 *Primary*	**1** Column curves at birth from head to sacrum with convexity posteriorly; after child stands, convexity persists only in *thoracic* and *sacral* regions, which, therefore, are called primary curves
2 *Secondary*	**2** Concavities in *cervical* and *lumbar* regions; cervical concavity results from infant's attempts to hold head erect (3 to 4 months); lumbar concavity, from balancing efforts in learning to walk (10 to 18 months)
3 *Abnormal*	**3** *Kyphosis*, exaggerated convexity in thoracic region (hunchback); *lordosis*, exaggerated concavity in lumbar region, a very common condition; *scoliosis*, lateral curvature in any region

Sternum

Body	Main central part of bone
Manubrium	Flaring, upper part
Xiphoid process	Projection of cartilage at lower border of bone

Ribs

Head	Projection at posterior end of rib; articulates with corresponding thoracic vertebra and one above, except last three pairs, which join corresponding vertebra only
Neck	Constricted portion just below head
Tubercle	Small knob just below neck; articulates with transverse process of corresponding thoracic vertebra; missing in lowest three ribs
Body or shaft	Main part of rib
Costal cartilage	Cartilage at sternal end of true ribs; attaches ribs (except floating ribs) to sternum

Scapula (Figs. 5-25 to 5-27)

Borders **1** Superior **2** Vertebral **3** Axillary	 **1** Upper margin **2** Margin toward vertebral column **3** Lateral margin
Spine	Sharp ridge running diagonally across posterior surface of shoulder blade
Acromion process	Slightly flaring projection at lateral end of scapular spine; may be felt as tip of shoulder; articulates with clavicle
Coracoid process	Projection on anterior surface from upper border of bone; may be felt in groove between deltoid and pectoralis major muscles, about 1 inch below clavicle
Glenoid cavity	Arm socket

Continued.

Table 5-2

Identification of bone markings—cont'd

Humerus (Figs. 5-28 and 5-29)

Marking	Description
Head	Smooth, hemispherical enlargement at proximal end of humerus
Anatomical neck	Oblique groove just below head
Greater tubercle	Rounded projection lateral to head on anterior surface
Lesser tubercle	Prominent projection on anterior surface just below anatomical neck
Intertubercular	Deep groove between greater and lesser tubercles; long tendon of biceps muscle lodges here
Surgical neck	Region just below tubercles; so named because of its liability to fracture
Deltoid tuberosity	V-shaped, rough area about midway down shaft where deltoid muscle inserts
Radial groove	Groove running obliquely downward from deltoid tuberosity; lodges radial nerve
Epicondyles (medial and lateral)	Rough projections at both sides of distal end
Capitulum	Rounded knob below lateral epicondyle; articulates with radius; sometimes called radial head of humerus
Trochlea	Projection with deep depression through center similar to shape of pulley; articulates with ulna
Olecranon fossa	Depression on posterior surface just above trochlea; receives olecranon process of ulna when lower arm extends
Coronoid fossa	Depression on anterior surface above trochlea; receives coronoid process of ulna in flexion of lower arm

*Italics suggest markings particularly useful to recognize.

Ulna (Figs. 5-30 and 5-31)

Marking	Description
Olecranon process	Elbow
Coronoid process	Projection on anterior surface of proximal end of ulna; trochlea of humerus fits snugly between olecranon and coronoid processes
Semilunar notch	Curved notch between olecranon and coronoid, into which trochlea fits
Radial notch	Curved notch lateral and inferior to semilunar notch; head of radius fits into this concavity
Head	Rounded process at distal end; does not articulate with wrist bones but with fibrocartilaginous disk
Styloid process	Sharp protuberance at distal end; can be seen from outside on posterior surface

Radius (Figs. 5-30 and 5-31)

Marking	Description
Head	Disk-shaped process forming proximal end of radius; articulates with capitulum of humerus and with radial notch of ulna
Radial tuberosity	Roughened projection on ulnar side, short distance below head; biceps muscle inserts here
Styloid process	Protuberance at distal end on lateral surface (with forearm supinated as in anatomical position)

Os coxae (Fig. 5-34)

Marking	Description
Ilium	Upper, flaring portion
Ischium	Lower, posterior portion
Pubic bone or *pubis*	Medial, anterior section
Acetabulum	Hip socket; formed by union of ilium, ischium, and pubis
Iliac crests	Upper, curving boundary of ilium
Iliac spines	
1 *Anterior superior*	**1** Prominent projection at anterior end of iliac crest; can be felt externally as "point" of hip
2 Anterior inferior	**2** Less prominent projection short distance below anterior superior spine
3 Posterior superior	**3** At posterior end of iliac crest
4 Posterior inferior	**4** Just below posterior superior spine
Greater sciatic notch	Large notch on posterior surface of ilium just below posterior inferior spine
Gluteal lines	Three curved lines across outer surface of ilium—posterior, anterior, inferior, respectively

Os coxae continued

Iliopectineal line	Rounded ridge extending from pubic tubercle upward and backward toward sacrum
Iliac fossa	Large, smooth, concave inner surface of ilium above iliopectineal line
Ischial tuberosity	Large, rough, quadrilateral process forming inferior part of ischium; in erect sitting position body rests on these tuberosities
Ischial spine	Pointed projection just above tuberosity
Symphysis pubis	Cartilaginous, amphiarthrotic joint between pubic bones
Superior pubic ramus	Part of pubis lying between symphysis and acetabulum; forms upper part of obturator foramen
Inferior pubic ramus	Part extending down from symphysis; unites with ischium
Pubic arch	Angle formed by two inferior rami
Pubic crest	Upper margin of superior ramus
Pubic tubercle	Rounded process at end of crest
Obturator foramen	Large hole in anterior surface of os coxa; formed by pubis and ischium; largest foramen in body
Pelvic brim (or *inlet*)	Boundary of aperture leading into true pelvis; formed by pubic crests, iliopectineal lines, and sacral promontory; size and shape of this inlet has great obstetrical importance since if any of its diameters too small, infant skull cannot enter true pelvis for natural birth
True (or lesser) *pelvis*	Space below pelvic brim; true "basin" with bone and muscle walls and muscle floor; pelvic organs located in this space
False (or greater) *pelvis*	Broad, shallow space above pelvic brim, or pelvic inlet; name "false pelvis" is misleading since this space is actually part of abdominal cavity, not pelvic cavity
Pelvic outlet	Irregular circumference marking lower limits of true pelvis; bounded by tip of coccyx and two ischial tuberosities
Pelvic girdle (or bony pelvis)	Complete bony ring; composed of two hip bones (ossa coxae), sacrum, and coccyx; forms firm base by which trunk rests upon thighs and for attachment of lower extremities to axial skeleton

Femur (Fig. 5-38)

Head	Rounded, upper end of bone; fits into acetabulum
Neck	Constricted portion just below head
Greater trochanter	Protuberance located inferiorly and laterally to head
Lesser trochanter	Small protuberance located inferiorly and medially to greater trochanter
Linea aspera	Prominent ridge extending lengthwise along concave posterior surface
Gluteal tubercle	Rounded projection just below greater trochanter; rudimentary third trochanter
Supracondylar ridges	Two ridges formed by division of linea aspera at its lower end; medial supracondylar ridge extends inward to inner condyle, lateral ridge to outer condyle
Condyles	Large, rounded bulges at distal end of femur; one on medial and one on lateral surface
Adductor tubercle	Small projection just above inner condyle; marks termination of medial supracondylar ridge
Trochlea	Smooth depression between condyles on anterior surface; articulates with patella
Intercondyloid notch	Deep depression between condyles on posterior surface; cruciate ligaments that help bind femur to tibia lodge in this notch

Tibia (Fig. 5-39)

Condyles	Bulging prominences at proximal end of tibia; upper surfaces concave for articulation with femur
Intercondylar eminence	Upward projection on articular surface between condyles
Crest	Sharp ridge on anterior surface
Tibial tuberosity	Projection in midline on anterior surface
Popliteal line	Ridge that spirals downward and inward on posterior surface of upper third of tibial shaft
Medial malleolus	Rounded downward projection at distal end of tibia; forms prominence on inner surface of ankle

Continued.

Table 5-2
Identification of bone markings—cont'd

Marking	Description
Fibula (Fig. 5-39)	
Lateral malleolus	Rounded prominence at distal end of fibula; forms prominence on outer surface of ankle
Tarsals (Figs. 5-40 to 5-42)	
Calcaneus	Heel bone
Talus	Uppermost of tarsals; articulates with tibia and fibula; boxed in by medial and lateral malleoli
Longitudinal arches	Tarsals and metatarsals so arranged as to form arch from front to back of foot
1 *Inner*	1 Formed by calcaneus, navicular, cuneiforms, and three medial metatarsals
2 *Outer*	2 Formed by calcaneus, cuboid, and two lateral metatarsals
Transverse (or *metatarsal*) *arch*	Metatarsals and distal row of tarsals (cuneiforms and cuboid) so articulated as to form arch across foot; bones kept in two arched positions by means of powerful ligaments in sole of foot and by muscles and tendons

*Italics suggest markings particularly useful to recognize.

Table 5-3
Joints

Diarthroses

1 Ball and socket
Other names: spheroidal; endarthroses
Description: ball-shaped head fits into concave socket
Movement: widest range of all joints; triaxial
Examples: shoulder joint and hip joint

2 Hinge
Other name: ginglymus
Description: spool-shaped surface fits into concave surface
Movement: in one plane about single axis (uniaxial); like hinged-door movement—viz., flexion and extension
Examples: elbow, knee, ankle, and interphalangeal joints

3 Pivot
Other name: trochoid
Description: arch-shaped surface rotates about rounded or peglike pivot
Movement: rotation; uniaxial
Example: between axis and atlas; between radius and ulna

4 Ellipsoidal
Other names: condyloid, ovoid
Description: oval-shaped condyle fits into elliptical cavity
Movement: in two planes at right angles to each other—specifically, flexion, extension, abduction, and adduction; biaxial
Example: wrist joint (between radius and carpals)

5 Saddle
Description: saddle-shaped bone fits into socket that is concave-convex in opposite direction; modification of condyloid joint
Movement: same kinds of movement as condyloid joint but freer; like rider in saddle; biaxial
Example: thumb, between first metacarpal and trapezium

6 Gliding
Other name: arthrodia
Description: articulating surfaces; usually flat
Movement: gliding a nonaxial movement
Example: between carpal bones; between sacrum and ilium (sacroiliac joints)

Synarthroses

1 Cartilaginous
Other name: synchondrosis
Description: cartilage grows between two articulating surfaces (e.g., cartilage disks between bodies of vertebrae) usually reinforced by ligaments
Movement: bending and twisting or slight compression
Examples: between bodies of vertebrae; between diaphysis and epiphysis of growing bones; replaced by bone in full-grown bones

2 Fibrous
Other name: syndesmosis
Description: thin layer of fibrous tissue; continuous with periosteum; connects articulating bones
Movement: none
Example: sutures of skull; in older adults, fibrous connection replaced by bone

Table 5-4

Description of individual joints

Name	Articulating bones	Type	Movements
Atlantoepistropheal	Anterior arch of atlas rotates about dens of axis (epistropheus)	Diarthrotic (pivot type)	Pivoting or partial rotation of head
Vertebral*	Between bodies of vertebrae	Synarthrotic, cartilaginous; amphiarthrotic by other system of classifying	Slight movement between any two vertebrae but considerable motility for column as whole
	Between articular processes	Diarthrotic (gliding)	
Clavicular			
Sternoclavicular	Medial end of clavicle with manubrium of sternum; only joint between upper extremity and trunk	Diarthrotic (gliding)	Gliding; weak joint that may be injured comparatively easily
Acromioclavicular	Distal end of clavicle with acromion of scapula	Diarthrotic (gliding)	Gliding; elevation, depression, protraction, and retraction
Thoracic	Heads of ribs with bodies of vertebrae	Diarthrotic (gliding)	Gliding
	Tubercles of ribs with transverse processes of vertebrae	Diarthrotic (gliding)	Gliding
Shoulder	Head of humerus in glenoid cavity of scapula	Diarthrotic (ball and socket type)	Flexion, extension, abduction, adduction, rotation, and circumduction of upper arm; one of most freely movable of joints
Elbow	Trochlea of humerus with semilunar notch of ulna; head of radius with capitulum of humerus	Diarthrotic (hinge type)	Flexion and extension
	Head of radius in radial notch of ulna	Diarthrotic (pivot type)	Supination and pronation of lower arm and hand; rotation of lower arm on upper as in using screwdriver
Wrist	Scaphoid, lunate, and triquetral bones articulate with radius and articular disk	Diarthrotic (condyloid)	Flexion, extension, abduction, and adduction of hand
Carpal	Between various carpals	Diarthrotic (gliding)	Gliding
Hand	Proximal end of first metacarpal with trapezium	Diarthrotic (saddle)	Flexion, extension, abduction, adduction, and circumduction of thumb and opposition to fingers; motility of this joint accounts for dexterity of human hand compared with animal forepaw
	Distal end of metacarpals with proximal end of phalanges	Diarthrotic (hinge)	Flexion, extension, limited abduction, and adduction of fingers
	Between phalanges	Diarthrotic (hinge)	Flexion and extension of finger sections
Sacroiliac	Between sacrum and two ilia	Diarthrotic (gliding); joint cavity mostly obliterated after middle life	None or slight—e.g., during late months of pregnancy and during delivery

*Vertebrae not easily dislocated; held securely together by following ligaments: anterior and posterior longitudinal ligaments—between anterior and posterior surfaces of bodies of vertebrae; supraspinous ligaments (called ligamentum nuchae in cervical region)—between tops of spinous processes; interspinous ligaments—between sides of spinous processes; ligamentum flavum—between laminae.

Continued.

Table 5-4

Description of individual joints—cont'd

Name	Articulating bones	Type	Movements
Symphysis pubis	Between two pubic bones	Synarthrotic (or amphiarthrotic), cartilaginous	Slight, particularly during pregnancy and delivery
Hip	Head of femur in acetabulum of os coxa	Diarthrotic (ball and socket)	Flexion, extension, abduction, adduction, rotation, and circumduction
Knee	Between distal end of femur and proximal end of tibia; largest joint in body	Diarthrotic (hinge type)	Flexion and extension; slight rotation of tibia
Tibiofibular	Head of fibula with lateral condyle of tibia	Diarthrotic (gliding type)	Gliding
Ankle	Distal ends of tibia and fibula with talus	Diarthrotic (hinge type)	Flexion (dorsiflexion) and extension (plantar flexion)
Foot	Between tarsals	Diarthrotic (gliding)	Gliding; inversion and eversion
	Between metatarsals and phalanges	Diarthrotic (hinge type)	Flexion, extension, slight abduction, and adduction
	Between phalanges	Diarthrotic (hinge type)	Flexion and extension

Outline summary

MEANING

All bones and their joints

FUNCTIONS

1 Furnishes supporting framework
2 Affords protection
3 Movement; bones constitute levers for muscle action
4 Hemopoiesis—blood cell formation by red bone marrow (i.e., myeloid tissue)

BONE AND CARTILAGE

Microscopic structure

Bone

1 Mainly calcified matrix—cement substance impregnated with calcium salts and reinforced by collagenous fibers
2 Lamellae—concentric cylindrical layers of calcified matrix enclosing haversian canal that contains blood vessel
3 Haversian system—canal and surrounding lamellae
4 Lacunae—microscopic spaces containing osteocytes (bone cells); lie between lamellae
5 Canaliculi—microscopic canals radiating in all directions from lacunae, connecting them with haversian canals; routes by which tissue fluid reaches osteocytes
6 Compact bone (or solid bone)—no empty spaces; lamellae fit closely together
7 Cancellous bone (or spongy bone)—many spaces in matrix arranged mainly in trabeculae rather than lamellae

Cartilage

1 Similar to that of bone with following exceptions:
   a Cartilage matrix—firm gel; bone matrix—calcified cement substance
   b Cartilage matrix—no canal system, no blood vessels; bone matrix—extensive canal network

BONES

Types

1 Long (femur)
2 Short (carpals)
3 Flat (parietal)
4 Irregular (vertebrae)

Structure

1 Long bones—see Figs. 5-1 to 5-3
2 Short bones—thin layer of compact bone encasing "core" of cancellous bone
3 Flat bones—layer of cancellous bone between two plates of compact bone
4 Irregular bones—thin layer of compact bone encasing cancellous bone

Formation and growth

1 Formation
  a Skeleton preformed in hyaline cartilage and fibrous membranes; most cartilaginous or membranous structures changed into bone before birth but not complete until about 25 years of age
  b Endochondral ossification—incompletely understood process that replaces hyaline cartilage "bones" with true bones
  c Intramembranous ossification—process that replaces fibrous membrane "bones" with true bones
2 Growth
  a In length—by continual thickening of epiphyseal cartilage followed by ossification
  b In diameter—medullary cavity enlarged by osteoclasts destroying bone around it while new bone added around circumference by osteoblasts
3 Correlation with bone disease; osteoporosis—deficient synthesis of organic bone matrix by osteoblasts

Names and numbers

Total, 206 bones—Table 5-1

Axial skeleton (80 bones)

1 Skull (28 bones)
  a Cranium (8 bones)—frontal, parietal (2), temporal (2), occipital, sphenoid, and ethmoid
  b Face (14 bones)—nasal (2), maxillary (2), malar (2), mandible, lacrimal (2), palatine (2), inferior turbinates (2), and vomer
  c Ear ossicles (6 bones)—malleus (2), incus (2), and stapes (2)
2 Hyoid (1 bone)
3 Vertebral column (26 vertebrae)
  a Cervical (7)
  b Thoracic (12)
  c Lumbar (5)
  d Sacrum
  e Coccyx
4 Sternum and ribs (25 bones)—sternum, true ribs (7 pairs), false ribs (5 pairs, 2 pairs of which are floating)

Appendicular skeleton (126 bones)

1 Upper extremities (64 bones)—clavicle (2), scapula (2), humerus (2), ulna (2), radius (2), carpals (16), metacarpals (10), and phalanges (28)
2 Lower extremities (62 bones)—os coxa (2), femur (2), patella (2), tibia (2), fibula (2), tarsal (14), metatarsal (10), and phalanges (28)

Markings

Depressions and openings

1 Fossa—hollow or depression
2 Sinus—cavity of spongelike air space within bone
3 Foramen—hole
4 Meatus—tube-shaped opening

Projections or processes

1 Those that fit into joints
  a Condyle—rounded projection entering into formation of joint
  b Head—rounded projection beyond narrow neck
2 Those to which muscles attach
  a Trochanter—very large process
  b Crest—ridge
  c Spinous process or spine—sharp projection
  d Tuberosity—large, rounded projection
  e Tubercle—small rounded projection

Differences between male and female skeletons

1 Male skeleton larger and heavier
2 Male pelvis deep and funnel shaped with narrow pubic arch; female pelvis shallow, broad, and flaring with wider pubic arch and larger iliosacral notch

Age changes in skeleton

1 From infancy to young adulthood—absolute and relative sizes of bones change
2 From young adulthood to old age—texture of bones change as does contour of bone margins and markings

JOINTS BETWEEN BONES (ARTICULATIONS)

Kinds

1 Diarthroses
  a Characteristics
    1 Small space, joint cavity, present between articulating surfaces of two bones that form joint
    2 Thin layer of hyaline cartilage covers articular surfaces
    3 Fibrous, synovial-lined capsule encases joint
    4 Ligaments hold articulating bones firmly connected
    5 Free movement possible at all diarthrotic joints
  b Subtypes—see Table 5-3
2 Synarthroses
  a Characteristics
    1 No joint cavity
    2 Cartilage, fibrous tissue, or bone grows between articulating surfaces of bones
    3 Little or no movement possible at synarthrotic joints
  b Subtypes—see Table 5-3

Movements

1 Flexion—angle at joint decreases
2 Extension—angle at joint increases; return from flexion
3 Abduction—moving bone away from body's median plane
4 Adduction—moving bone back toward body's median plane
5 Rotation—pivoting bone upon its axis
6 Circumduction—describing surface of cone with moving parts
7 Special movements
  a Supination—movement of forearm that turns palm forward
  b Pronation—movement of forearm that turns back of hand forward
  c Inversion—ankle movement turning sole of foot inward
  d Eversion—ankle movement turning sole of foot outward
  e Protraction—moving part forward
  f Retraction—pulling part back; opposite of protraction

Review questions

1 What general functions does the skeletal system perform?
2 Describe the microscopic structure of bone and cartilage.
3 Describe the structure of a long bone.
4 Describe the general plan of the skeleton.
5 Name the bones of the adult skeleton.
6 Describe the structural features of diarthrotic joints that facilitate movement.
7 Give examples of several types of diarthrotic joints.
8 Explain the functions of the periosteum.
9 What joint(s) unites the shoulder girdle with the trunk? The pelvic girdle with the trunk?
10 Name the several kinds of movements possible at joints. Define each movement named.
11 Do vertebrae become dislocated easily? Substantiate your answer.
12 Name the primary and secondary curves of the spine. Describe each.
13 Name the five pairs of bony sinuses in the skull.
14 Name the bones which fuse to form the coccyx.
15 What is the true pelvis? The false pelvis? Name the boundary line between the true and false pelves.
16 Through what opening does the spinal cord enter the cranial cavity?
17 Explain the basic steps in the process of ossification according to the concept described in the text.
18 Compare osteoblasts and osteoclasts as to function.
19 What two factors presumably relate to the high incidence of osteoporosis following the menopause?
20 Define or make an identifying statement about each of the following terms: condyle, crest, diaphysis, diarthroses, endosteum, epiphysis, foramen, fossa, haversian system, kyphosis, lordosis, medullary cavity, osteoblast, osteoclast, periosteum, rotation, scoliosis, sinus, spinous process, synarthroses, trochanter, trabeculae.

6

The muscular system

Man's survival depends in large part upon his ability to adjust to the changing conditions of his environment. Movements constitute the major part of this adjustment. Whereas most of the systems of the body play some role in accomplishing movement, it is the skeletal and muscular systems acting together that actually produce movements. We have investigated the architectural plan of the skeleton and have seen how its joint structures and firm supports make movement possible. However, bones and joints cannot move themselves. They must be moved by something. Muscle tissue, because of its contractility, extensibility, and elasticity, is admirably suited to this function. Our subject then for this chapter is skeletal muscles—those muscle masses that attach to bones and move them about, the "red meat" of the body. (Cardiac muscle will be discussed in Chapter 13, and information about smooth muscle appears in several chapters.) In this chapter we shall try to discover how muscles move bones, and to do this we shall try to answer many other questions—how the structure of muscles adapts them to their function, how energy is made available for their work, how muscle activity contributes to the health and survival of the whole body—to mention only a few.

General functions

If you have any doubts about the importance of muscle function to normal life, you have only to observe a person with extensive paralysis—a victim of severe poliomyelitis, for example. Any of us possessed of normal powers of movement can little imagine life with this matchless power lost. But cardinal as it is, movement is not the only contribution muscles make to healthy survival. They also perform two other essential functions: maintenance of posture and production of a large portion of body heat.

Movement. Skeletal muscle contractions produce movements either of the body as a whole (locomotion) or of its parts.

Posture. The continued partial contraction of many skeletal muscles makes possible standing, sitting, and other maintained positions of the body.

Heat production. Muscle cells, like all cells, produce heat by the process known as catabolism (discussed in Chapter 15). But because skeletal muscle cells are both highly active and numerous, they produce a major share of total body heat. Skeletal muscle contractions, therefore, constitute one of the most important parts of the mechanism for maintaining homeostasis of temperature.

Skeletal muscle cells

Microscopic structure

Look at Fig. 6-1, *A* and *B*. As you can see there, a skeletal muscle is composed of bundles of skeletal muscle fibers, called fibers instead of cells because of their threadlike shape. Skeletal muscle fibers have many of the same structural parts as other cells. Several of them, however, bear different names in muscle fibers. For example, *sarcolemma* is the plasma membrane of a muscle fiber. *Sarcoplasm* is its cytoplasm. Muscle cells contain a network of tubules and sacs known as the *sarcoplasmic reticulum*—a structure analogous, but not identical, to the endoplasmic reticulum of other cells. Muscle fibers contain many mitochondria and, unlike other cells, they have several nuclei.

Certain structures not found in other cells are present in skeletal muscle fibers. For instance, bundles of very fine fibers—*myofibrils*—extend lengthwise of skeletal muscle fibers and almost fill their sarcoplasm. Myofibrils, in turn, are made up of still finer fibers called thick and thin filaments. The red lines in Fig. 6-1, *D*, represent thick filaments and the gray lines represent thin filaments. Find

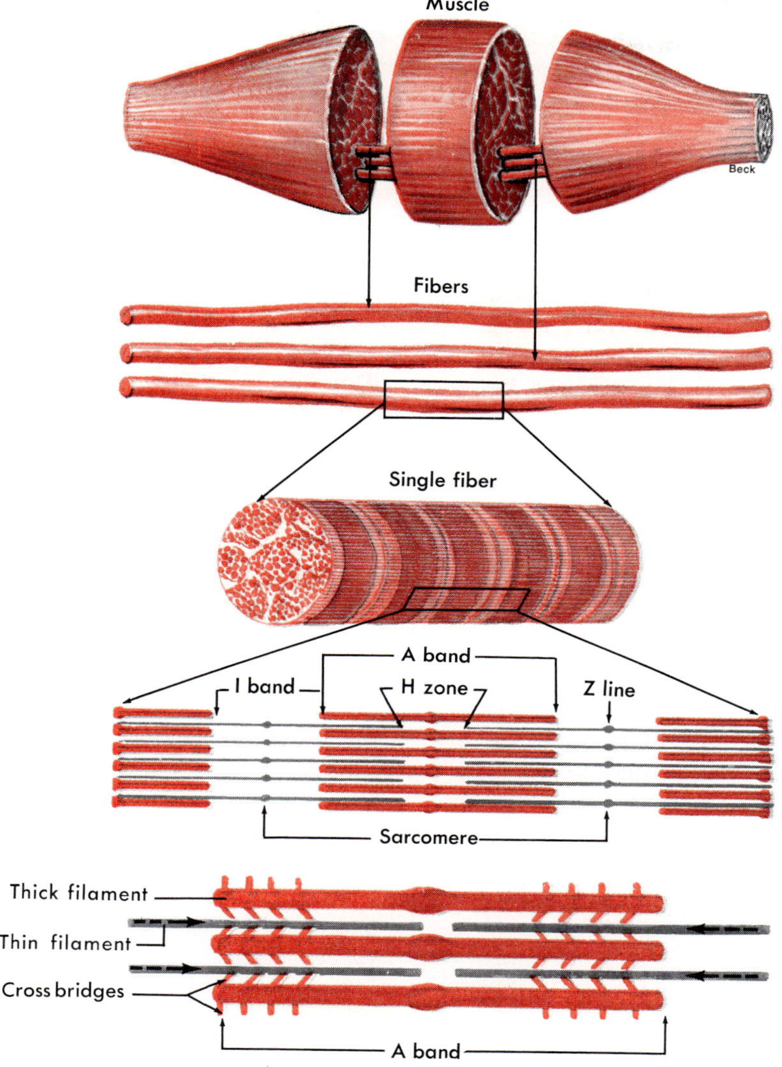

Fig. 6-1

Structure of skeletal muscle. **A,** Skeletal muscle organ, composed of bundles of muscle fibers—i.e., cells. **B,** Single fibers, enlarged. **C,** Greater magnification of single fiber showing smaller fibers—myofibrils—in its sarcoplasm. **D,** Myofibril magnified further to show thick and thin filaments composing it and producing its cross-striated appearance; dark A (anisotropic) bands alternate with light I (isotropic) bands; H zones, less dense midsections of A bands; Z lines, more dense midlines of I bands. **E,** Molecular structure of myofibril; thick filaments, myosin molecules; thin filaments, actin molecules. (Also see Fig. 6-4.)

the label sarcomere in this drawing. Note that a *sarcomere* is a segment between two successive Z lines. Each myofibril consists of a line-up of several sarcomeres, each of which functions as a contractile unit. The A bands of the sarcomeres appear as relatively wide, dark stripes (cross striae) under the microscope, and they alternate with narrower, lighter colored stripes formed by the I bands (see Fig. 6-1, *C*). Because of its cross striae, skeletal muscle is also called striated muscle.

Another structure unique to skeletal muscle cells is the *T system*. This name derives from the fact that this system consists of tubules that extend transversely into the sarcoplasm. T system tubules, as Fig. 6-2 shows, enter the sarcoplasm at the levels of the Z lines—that is, in the middle of the light I bands. Because invaginations of the sarcolemma form the T system tubules, they open to the exterior of the muscle fiber.

The sarcoplasmic reticulum is also a system of tubules in a muscle fiber. It is separate from the T system and differs from it in that the tubules of the sarcoplasmic reticulum run parallel to muscle fibers and terminate in closed sacs at the ends of each sarcomere—that is, immediately above and below each Z line. Since T tubules constitute the Z lines, the sacs of the sarcoplasmic reticulum of one sarcomere lie just above a T tubule, whereas those of the next sarcomere lie just below it. This forms a triple-layered structure (a T tubule sandwiched between sacs of the sarcoplasmic reticulum) called a *triad*.

Molecular structure

In recent years, researchers have made many discoveries about the molecular structure of muscle fibers. Here are a few of them. They have learned that four proteins—myosin, actin, tropomyosin, and troponin—are found only in muscle fibers and that these protein compounds interact to produce muscle contraction. They have established that the *thick filaments* of a myofibril consist almost entirely of myosin molecules and that there are 300 or 400 of them per filament. They know the shape of myosin molecules. Each one is a thin rod with two rounded heads at one end. They know, too, that these myosin molecules are arranged lengthwise in the thick filaments and that some of their heads point toward one end of the filament and some point toward the other end. Thus the heads of the myosin molecules jut out from the surface of both ends of the thick filaments in projections that are called *cross bridges*. A schematic diagram of these appears in Fig. 6-1, *E*.

Thin filaments consist of a complex arrangement of three kinds of protein compounds, namely, actin, tropomyosin, and troponin.* Within a myofibril, the thick and thin filaments alternate, as shown in Fig. 6-1, *D*. This arrangement is crucial for contraction. Another fact important for contraction is that the thin filaments attach to both Z lines of a sarcomere and that they extend in from the Z lines part way toward the center of the sarcomere. When the muscle fiber is relaxed, the thin filaments terminate at the outer edges of the H zones. In contrast, the thick myosin filaments do not attach to the Z lines and they extend only the length of the A bands of the sarcomeres.

Functions

Skeletal muscle fibers specialize in the function of contraction. When nerve impulses arrive at a skeletal muscle fiber, they initiate impulse conduction over its sarcolemma and inward via its T tubules. This triggers the release of calcium ions from the sacs of the sarcoplasmic reticulum into the sarcoplasm. Here, calcium ions combine with the troponin molecules in the thin filaments of the

*Murray, J. M., and Weber, A.: The cooperative action of muscle proteins, Sci. Am. **230**:59-71, Feb., 1974.

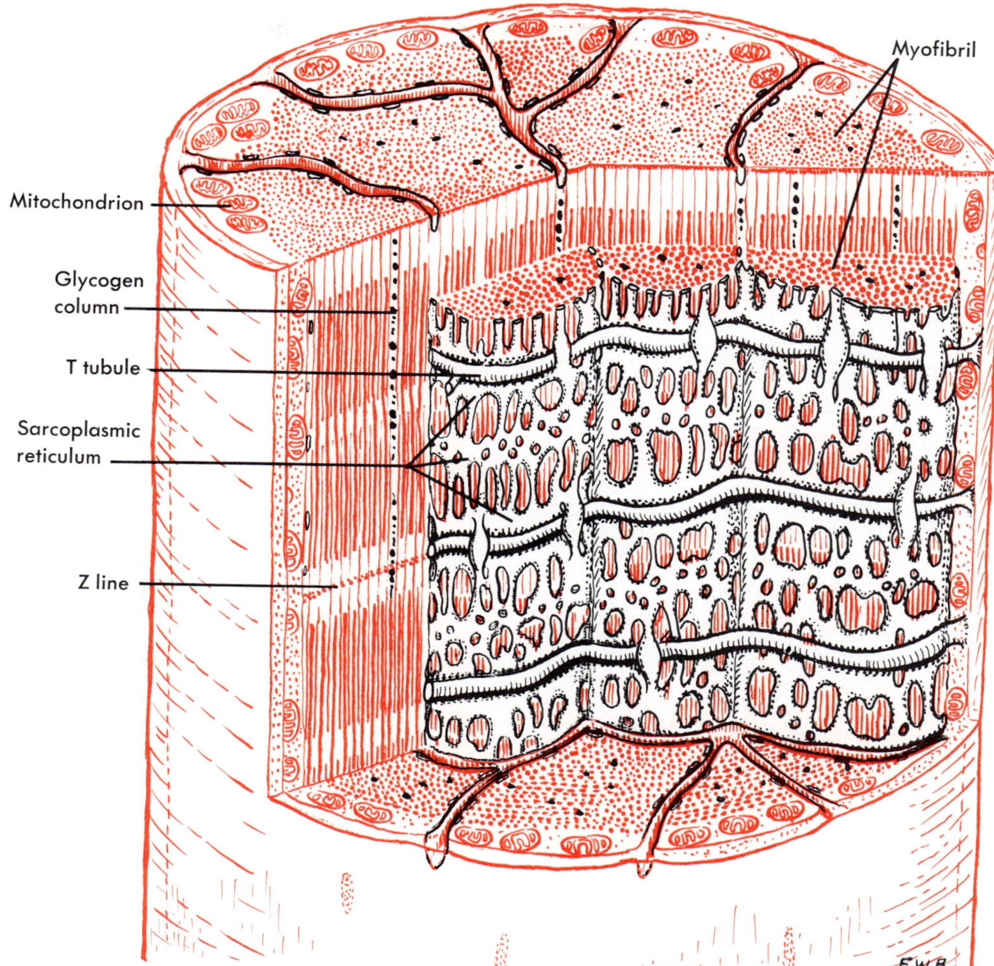

Fig. 6-2

Diagram to show sarcoplasmic reticulum and T tubules of a muscle fiber.

myofibrils. In a resting muscle fiber, troponin prevents myosin from interacting with actin. Stated differently, troponin that is not bound to calcium prevents the cross bridges of the thick filaments from attaching to the actin molecules of the thin filaments. But calcium-bound troponin permits this action. Therefore myosin interacts with actin and pulls the thin filaments toward the center of each sarcomere. This shortens the sarcomeres and thereby shortens the myofibrils and the muscle fibers that they compose. If sufficient numbers of the fibers composing a skeletal muscle organ shorten, the muscle itself shortens. In a word, it contracts.

Relaxation of a muscle fiber is now believed to be brought about by a reversal of the contraction mechanism just described. The calcium-troponin combinations separate, calcium ions reenter the sacs of the sarco-

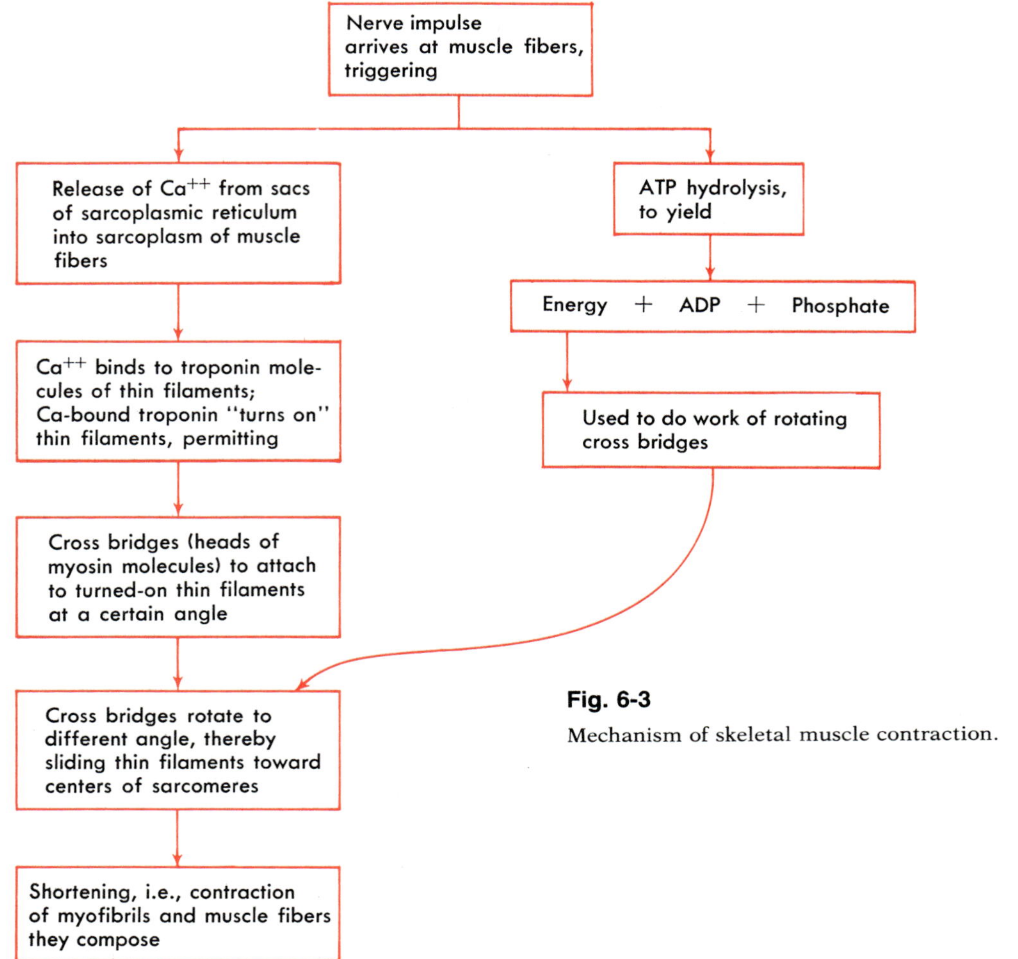

Fig. 6-3
Mechanism of skeletal muscle contraction.

plasmic reticulum, and troponin, now no longer bound to calcium, inhibits myosin-actin interaction.

Summarizing, the release of calcium ions from the sacs of the sarcoplasmic reticulum turns on muscle contraction. The withdrawal of calcium ions back into the sarcoplasmic reticulum's sacs turns off muscle contraction and turns on muscle relaxation.

Muscle cells obey the all-or-none law when they contract. This means that they either contract with all the force possible under existing conditions or they do not contract at all. However, if conditions at the time of stimulation change, then the force of the cell's contraction changes. Suppose, for example, that a particular muscle fiber receives an adequate oxygen supply at one time and an inadequate supply at another time. It will contract more forcefully with an adequate than with a deficient oxygen supply.

Energy sources for muscle contraction

The high-energy compound ATP (adenosine triphosphate) that is present in muscle fibers and all cells breaks down, releasing energy that does the work of contracting muscles. As Fig. 6-3 shows, nerve impulses arriving at a muscle fiber trigger both ATP breakdown and the release of calcium ions from the

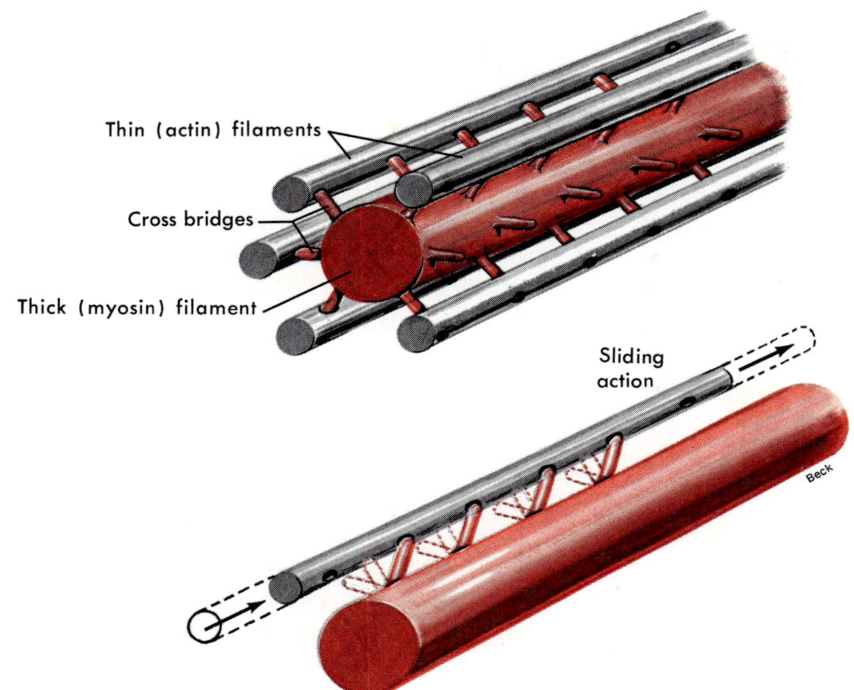

Fig. 6-4

Scheme to show how myosin interacts with actin to shorten muscle fibers. The cross bridges of a single myosin fiber are believed to point at 60-degree angles at six actin filaments. The cross bridges bond to the actin filaments and slide them toward the center of the sarcomere.

sacs of the sarcoplasmic reticulum into the sarcoplasm of the muscle fiber. The calcium ions combine with the troponin molecules of the thin filaments in myofibrils. Calcium-combined troponin acts in some way to "turn on" the thin filaments. What this means is that the actin molecules of the thin filaments become able to combine with the heads of the myosin molecules that make up the cross bridges of the thick filaments. The energy released from ATP breakdown then does the work of rotating the cross bridges to a different angle (Fig. 6-4). As the cross bridges rotate, they move the thin filaments to which they are attached in toward the centers of the

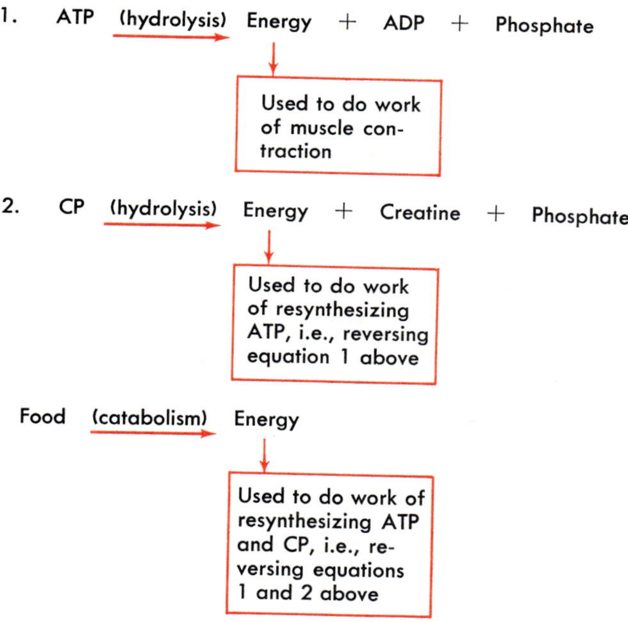

Fig. 6-5

Energy sources for muscle contraction.

sarcomeres. This necessarily shortens the sarcomeres and the myofibrils. Thus energy released from ATP breakdown does the work of muscle contraction.

Muscle fibers must continually resynthesize ATP because they can store only small amounts of it. Immediately after ATP breaks down, energy for its resynthesis is supplied by the breakdown of another high-energy compound, creatine phosphate (CP), which is also present in small amounts in muscle fibers. But ultimately, energy for both ATP and CP synthesis comes from the catabolism of foods (discussed in Chapter 15).

Skeletal muscle organs

Structure

SIZE, SHAPE, AND FIBER ARRANGEMENT

The structures called skeletal muscles are organs. They consist mainly of skeletal muscle tissue plus important connective and nervous tissue components. Skeletal muscles vary considerably in size, shape, and arrangement of fibers. They range from extremely tiny strands as, for example, the stapedius muscle of the middle ear, to large masses such as the muscles of the thigh. Some skeletal muscles are broad in shape and some narrow. Some are long and tapering and some short and blunt. Some are triangular, some quadrilateral, and some irregular. Some form flat sheets and others bulky masses.

Arrangement of fibers varies in different muscles. In some muscles, the fibers are parallel to the long axis of the muscle, in some they converge to a narrow attachment, and in some they are oblique and either pennate (like the feathers in an old-fashioned plume pen) or bipennate (double-feathered as in the rectus femoris). Fibers may even be curved, as in the sphincters of the face, for example. The direction of the fibers composing a muscle is significant because of its relationship to function. For instance, a muscle with the bipennate fiber arrangement can produce the strongest contraction.

CONNECTIVE TISSUE COMPONENTS

A fibrous connective tissue sheath *(epimysium)* envelops each muscle and extends into it as partitions between bundles of its fibers *(perimysium)* and between individual fibers *(endomysium)*. Because all three of these structures are continuous with the fibrous structures that attach muscles to bones or other structures, muscles are most firmly harnessed to the structures they pull on during contraction. The epimysium, perimysium, and endomysium of a muscle, for example, may be continuous with fibrous tissue that extends from the muscle as a *tendon*, a strong tough cord continuous at its other end with the fibrous covering of bone (periosteum). Or the fibrous wrapping of a muscle may extend as a broad, flat sheet of connective tissue *(aponeurosis)* to attach it to adjacent structures, usually the fibrous wrappings of another muscle. So tough and strong are tendons and aponeuroses that they are not often torn, even by injuries forceful enough to break bones or tear muscles. They are, however, occasionally pulled away from bones.

Tube-shaped structures of fibrous connective tissue called *tendon sheaths* enclose certain tendons, notably those of the wrist and ankle. Like the bursae, tendon sheaths have a lining of synovial membrane. Its moist smooth surface enables the tendon to move easily, almost frictionlessly, in the tendon sheath.

You may recall that a continuous sheet of loose connective tissue known as the superficial fascia lies directly under the skin. Under this lies a layer of dense fibrous connective tissue, the *deep fascia*. Plate II (in transparent Trans-Vision®) of the color insert (p. 54) shows sections of the deep fascia very clearly. Extensions of the deep fascia form the epimysium, perimysium, and endomysium of muscles and their attachments to bones and other structures and also enclose viscera, glands, blood vessels, and nerves.

NERVE SUPPLY

A nerve cell that transmits impulses to a skeletal muscle is called a *somatic motoneuron*. One such neuron plus the muscle cells in which its axon terminates constitutes a *motor unit* (Fig. 6-6). The single axon fiber of a motor unit divides, upon entering the skeletal muscle, into a variable number of branches. Those of some motor units terminate in only a few muscle fibers, whereas others terminate in numerous fibers. Consequently, impulse conduction by one motor unit may stimulate only a half dozen or so muscle fibers to contract at one time, whereas conduction by another motor unit may activate a hundred or more fibers simultaneously. This fact bears a relationship to the function of the muscle as a whole. As a general rule, the fewer the number of fibers supplied by a skeletal muscle's individual motor units, the more precise the movements that muscle can produce. For example, in certain small muscles of the hand, each motor unit includes only a few muscle fibers, and these muscles produce precise finger movements. In contrast, motor units in large abdominal muscles that do not produce precise movements are reported to include more than a hundred muscle fibers each.

The area of contact between a nerve and muscle fiber is known as the *motor end plate* or *neuromuscular junction*. When nerve impulses reach the ends of the axon fibers in a skeletal muscle, small vesicles in the axon terminals release a chemical—acetylcholine—into the neuromuscular junction. Diffusing swiftly across this microscopic trough, acetylcholine contacts the sarcolemma of the adjacent muscle fiber, stimulating the fiber to contract. In addition to the many motor nerve endings, there are also many sensory nerve endings in skeletal muscles.

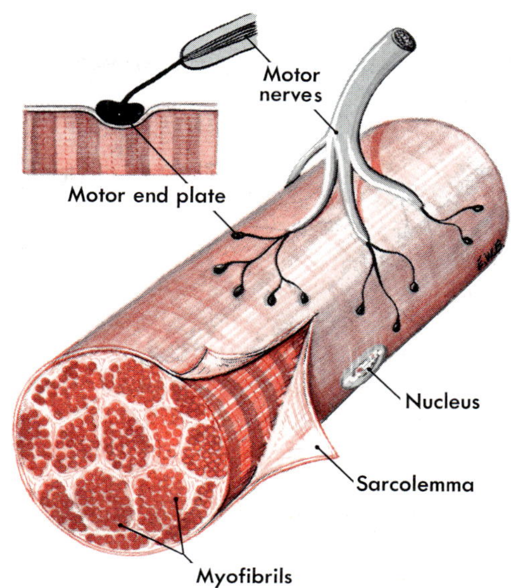

Fig. 6-6

A motor unit consists of one motoneuron and the muscle cells supplied by its axon branches. The diagram shows a motor axon ending in several unmyelinated branches. Each branch terminates in a motor end-plate embedded, as the insert shows, in a tiny trough on the surface of the muscle fiber.

AGE CHANGES

As a person grows old, his skeletal muscles undergo a process called *fibrosis*. Gradually, some of the skeletal muscle fibers degenerate, and fibrous connective tissue replaces them. With this loss of muscle fibers and increase in connective tissue comes waning muscular strength, a common finding in elderly people.* Other factors probably also contribute to decreasing muscular strength in advanced age.

Function

Several methods of study have been used to amass the present-day store of knowledge about how muscles function. They vary from the traditional and relatively simple pro-

*Shock, N. W.: The physiology of aging, Sci. Am. **206:** 100-110, Jan., 1962.

cedures, such as observing and palpating muscles in action, manipulating dissected muscles to observe movements, or deducing movements from knowledge of muscle anatomy, to the newer more complicated method of electromyography (recording action potentials from contracting muscles). As a result, there is now a somewhat overwhelming amount of knowledge about muscle functions. So perhaps we can thread our way through this maze of detail more easily if we start with general principles and then go on to the details that seem to us most useful.

BASIC PRINCIPLES

1 *Skeletal muscles contract only if stimulated*. They do not have the quality of automaticity inherent in cardiac and visceral muscle. Although nerve impulses are the natural stimuli for skeletal muscles, electrical and some other artificial stimuli can also activate them. A skeletal muscle deprived of nerve impulses by whatever cause is a functionless mass. One should, therefore, think of a skeletal muscle and its motor nerve as a physiological unit, always functioning together, either useless without the other.

2 *A skeletal muscle contraction may be any one of several types*. It may be a tonic contraction, an isotonic contraction, an isometric contraction, a twitch contraction, or a tetanic contraction. And there are also other types of contraction—called treppe, fibrillation, and convulsions.

a A *tonic contraction (tone; tonus)* is a continual, partial contraction. At any one moment, a small number of the total fibers in a muscle contract, producing a tautness of the muscle rather than a recognizable contraction and movement. Different groups of fibers scattered throughout the muscle contract in relays. Tonic contraction, or tone, is characteristic of the muscles of normal individuals when they are awake. It is particularly important for maintaining posture. A striking illustration of this fact is the following: when a person loses consciousness, his muscles lose their tone and he collapses in a heap, unable to maintain a sitting or standing posture. Muscles with less tone than normal are described as flaccid muscles and those with more than normal tone are called spastic. Impulses over stretch reflex arcs (Fig. 7-5, p. 167) maintain tone.

b An *isotonic contraction* (*iso*, same; *tonic*, pressure or tension) is a contraction in which the pressure or tension within a muscle remains the same but in which the length of the muscle changes. It shortens, producing movement.

c An *isometric contraction* is a contraction in which muscle length remains the same but in which muscle tension increases. You can observe isometric contraction by pushing your arms against a wall and feeling the tension increase in your arm muscles. Isometric contractions "tighten" a muscle but they do not produce movements nor do work. Isotonic contractions, on the other hand, both produce movements and do work.

d A *twitch contraction* is a quick, jerky contraction in response to a single stimulus. Fig. 6-7 shows a record of such a contraction. It reveals that the muscle does not shorten at the instant of stimulation, but rather a fraction of a second later, and that it reaches a peak of shortening and then gradually resumes its former length. These three phases of contraction are spoken of, respectively, as the *latent period*, the *contraction phase*, and the *relaxation phase*. The entire twitch usually lasts less than 1/10 of a second. Twitch contractions rarely occur in the body.

e A *tetanic contraction (tetanus)* is a more sustained contraction than a twitch. It is produced by a series of stimuli bombarding the muscle in rapid succession. About 30 stimuli per second, for example, evoke a tetanic contraction by a frog gastrocnemius

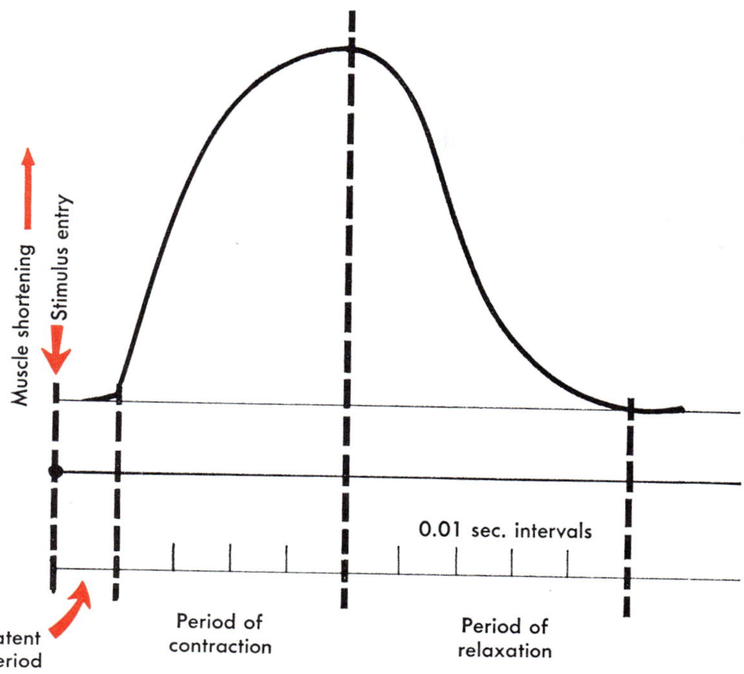

Fig. 6-7

The simple muscle twitch and its time components. (From Stacy, R. W., and Santolucito, J. A.: Modern college physiology, St. Louis, The C. V. Mosby Co.)

muscle, but the rate varies for different muscles and different conditions. Fig. 6-8 shows records of incomplete and complete tetanus. Normal movements are said to be produced by incomplete tetanic contractions.

f *Treppe (staircase phenomenon)* is a phenomenon in which increasingly stronger twitch contractions occur in response to constant strength stimuli repeated at the rate of about once or twice a second. In other words, a muscle contracts more forcefully after it has contracted a few times than when it first contracts—a principle made practical use of by athletes when they warm up but one not yet satisfactorily explained. Presumably, it relates partly to the rise in temperature of active muscles and partly to their accumulation of metabolic products. After the first few stimuli, muscle responds to a considerable number of successive stimuli with maximal contractions (Fig. 6-9). Eventually, it will respond with less and less strong contractions. The relaxation phase becomes shorter and finally disappears entirely. In other words, the muscle stays partially contracted—an abnormal state of prolonged contraction called *contracture*.

Repeated stimulation of muscle in time lessens its irritability and contractility and may result in muscle fatigue, a condition in which the muscle does not respond to the strongest stimuli. Complete muscle fatigue, however, very seldom occurs in the body but can be readily induced in an excised muscle.

g *Fibrillation* is an abnormal type of contraction in which individual fibers contract asynchronously, producing a flutter of the muscle but no effective movement. Fibrillation of the heart, for example, occurs fairly often.

h *Convulsions* are abnormal uncoordinated tetanic contractions of varying groups of muscles.

3 *Skeletal muscles contract according to the graded strength principle* (Fig. 6-10)—not according to the all-or-none principle, as do the individual muscle cells composing them. In other words, skeletal muscles contract with varying degrees of strength at differ-

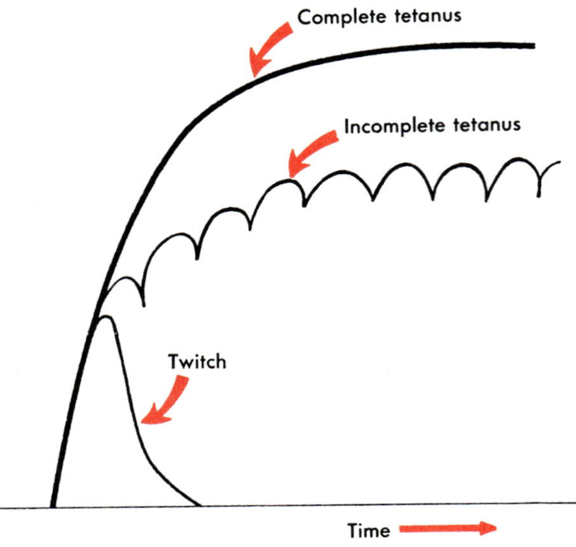

Fig. 6-8

Single twitch, incomplete tetanus, and complete tetanus of muscle. (From Stacy, R. W., and Santolucito, J. A.: Modern college physiology, St. Louis, The C. V. Mosby Co.)

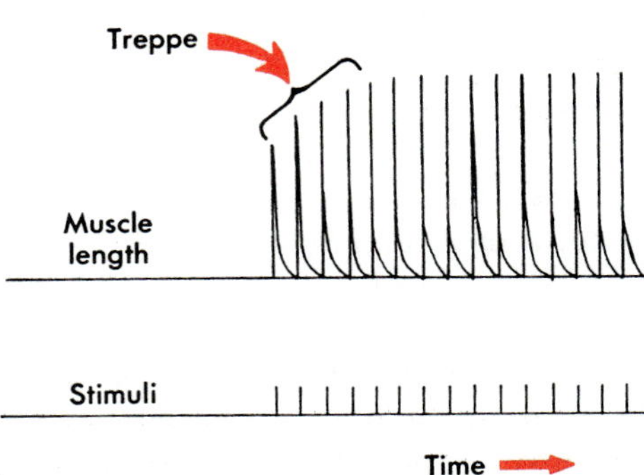

Fig. 6-9

Record of several successive muscle contractions, showing treppe occurring in the first few. (From Stacy, R. W., and Santolucito, J. A.: Modern college physiology, St. Louis, The C. V. Mosby Co.)

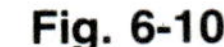

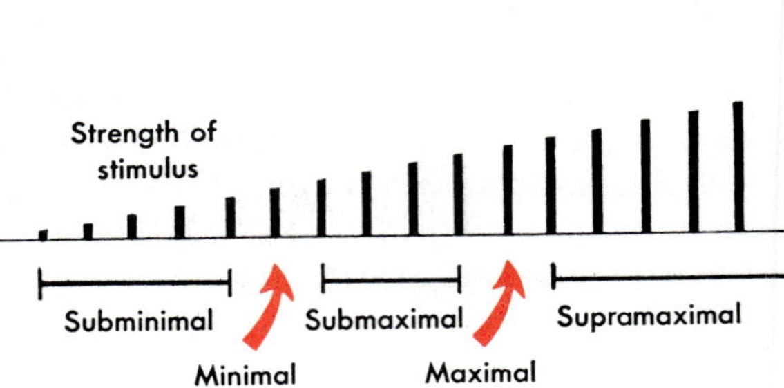

Fig. 6-10

Variation of strength of contraction of muscle with strength of stimulus. (From Stacy, R. W., and Santolucito, J. A.: Modern college physiology, St. Louis, The C. V. Mosby Co.)

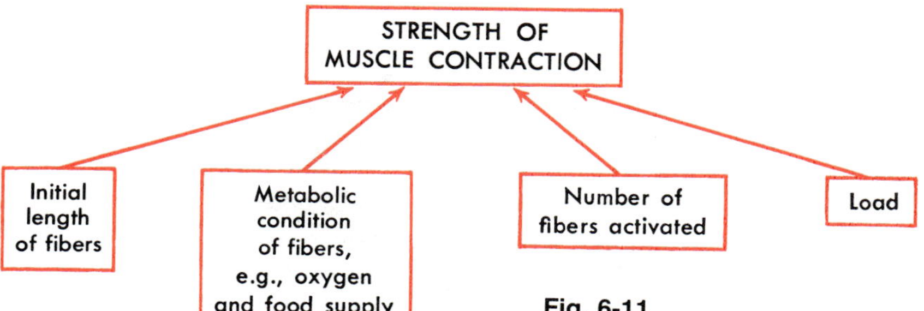

Fig. 6-11

Factors that influence the strength of muscle contraction.

ent times—a fact of practical importance. (How else, for example, could we match the force of a movement to the demands of a task?)

Several generalizations may help explain the fact of graded strength contractions. The strength of the contraction of a skeletal muscle bears a direct relationship to the initial length of its fibers, to their metabolic condition, and to the number of them contracting. If a muscle is moderately stretched at the moment when contraction begins, the force of its contraction increases. This principle, established years ago, applies experimentally to heart muscle also (Starling's law of the heart, discussed on p. 356). Outstanding among metabolic conditions that influence contraction are oxygen and food supply.

With adequate amounts of these essentials, a muscle can contract with greater force than possible with deficient amounts. The greater the number of muscle fibers contracting simultaneously, the stronger the contraction of a muscle. How large this number is depends upon how many motor units are activated, and this, in turn, depends upon the intensity and frequency of stimulation. In general, the more intense and the more frequent a stimulus, the more motor units and therefore the more fibers are activated and the stronger the contraction. Contraction strength also relates to previous contraction, the warm-up principle discussed on p. 117.

Another factor that influences the force of contraction is the size of the load imposed on the muscle. Within certain limits, the heavier the load, the stronger the contraction. Lift a pencil, for example, and then a heavy book and you can feel your arm muscles contract more strongly with the book.

The factors that influence muscle contraction are summarized in Fig. 6-11.

4 *Skeletal muscles produce movements by pulling on bones.* Most of our muscles span at least one joint and attach to both articulating bones. When they contract, therefore, their shortening puts a pull on both bones, and this pull moves one of the bones at the joint—draws it toward the other bone, much as a pull on marionette strings moves a puppet's parts. (In case you are wondering why both bones do not move since both are pulled on by the contracting muscle, the reason is that one of them is normally stabilized by contractions of other muscles or by certain features of its own that make it less mobile.)

5 *Bones serve as levers, and joints serve as fulcrums of these levers.* (By definition, a *lever* is any rigid bar free to turn about a fixed point called its *fulcrum.*) A contracting muscle applies a pulling force on a bone lever at the point of the muscle's attachment to the bone. This causes the bone (referred to as the insertion bone) to move about its joint fulcrum. We have already noted that a skeletal muscle and its motor nerve act as a functional unit. Now we can add bones and joints to this unit and can describe the physiological unit for movement as a neuromusculoskeletal unit. Disease or injury of any one of these parts of the unit—of nerve or muscle or bone or joint—can, as you might surmise, cause abnormal movements or complete loss of movement. Poliomyelitis, for example, and multiple sclerosis and hemiplegia all involve the neural part of the unit. In contrast, muscular dystrophy affects the muscular part and arthritis the skeletal part.

6 *Muscles that move a part usually do not lie over that part.* In most cases, the body of a muscle lies proximal to the part moved. Thus muscles that move the lower arm lie proximal to it—that is, in the upper arm. Applying the same principle, where would you expect muscles that move the hand to be located? Those that move the lower leg? Those that move the upper arm?

7 *Skeletal muscles almost always act in groups rather than singly.* In other words, most movements are produced by the coordinated action of several muscles. Some of the muscles in the group contract while others relax. To identify each muscle's special function in the group, the following classification is used:

a *Prime movers*—muscle or muscles whose contraction actually produces the movement

b *Antagonists*—muscles that relax while the prime mover is contracting to produce movement (exception: contraction of the antagonist at the same time as the prime mover when some part of the body needs to be held rigid, such as the knee joint when standing*)

c *Synergists*—muscles that contract at the same time as the prime mover (may help the prime mover produce its movement or may stabilize a part—hold it steady—so that the prime mover produces a more effective movement)

*Antagonistic muscles have opposite actions and opposite locations. If the flexor lies anterior to the part, the extensor will be found posterior to it. For example, the pectoralis major, the flexor of the upper arm, is located on the anterior aspect of the chest, while the latissimus dorsi, the extensor of the upper arm, is located on the posterior aspect of the chest. The antagonist of a flexor muscle is obviously an extensor muscle; that of an abductor muscle, an adductor muscle. Some frequently used antagonists are listed in Table 6-2, p. 140.

HINTS ON HOW TO DEDUCE ACTIONS

To understand muscle actions, you need first to know certain anatomical facts such as which bones muscles attach to and which joints they pull across. Then if you relate these structural facts to functional principles (for instance, those discussed in the preceding paragraphs), you may find your study of muscles more interesting and less difficult than you anticipate. Some specific suggestions for deducing muscle actions follow.

1 Start by making yourself familiar with the names, shapes, and general locations of the larger muscles, using Table 6-1, p. 139, as a guide.

2 Try to deduce which bones the two ends of a muscle attach to from your knowledge of the shape and general location of the muscle. For example, look carefully at the deltoid muscle as illustrated in Figs. 6-12 to 6-15. To what bones does it seem to attach? Check your deductions with Table 6-4, p. 142.

3 Next, make a guess as to which bone moves when the muscle shortens. (The bone moved by a muscle's contraction is its *insertion* bone; the bone that remains relatively stationary is its *origin* bone.) In many cases, you can tell by trying to move one bone and then another which one is the insertion bone. In some cases, either bone may function as the insertion. Although not all muscle attachments can be deduced as readily as those of the deltoid, they can all be learned more easily by using this deduction method than by relying on rote memory alone.

4 Deduce a muscle's actions by applying the principle that its insertion moves toward its origin. Check your conclusions with the text. Here, as in steps 2 and 3, the method of deduction is intended merely as a guide and is not adequate by itself for determining muscle actions.

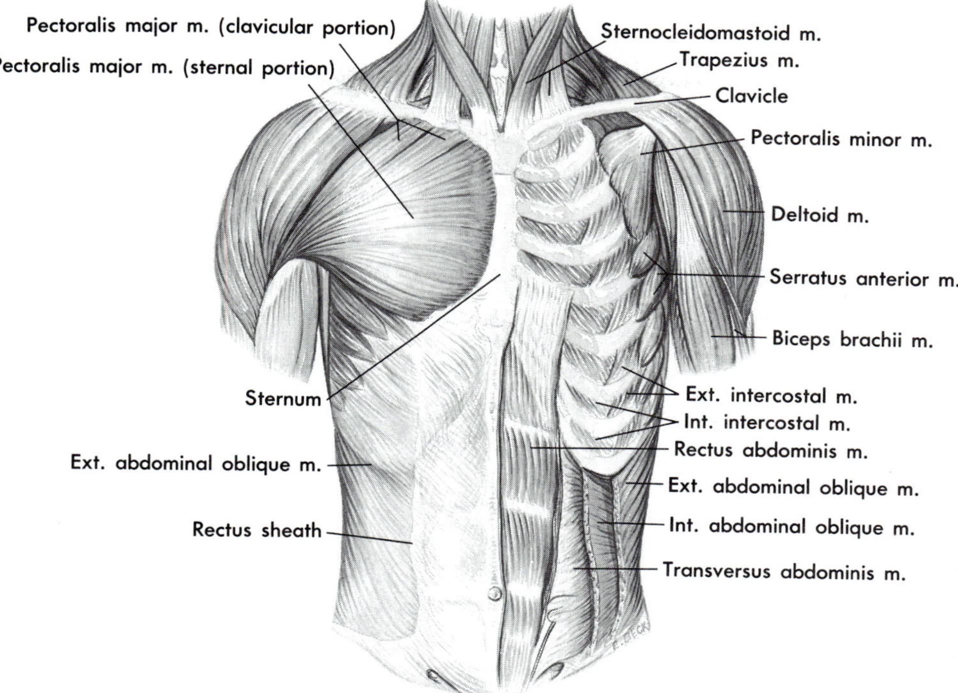

Fig. 6-12

Superficial muscles of the anterior surface of the trunk.

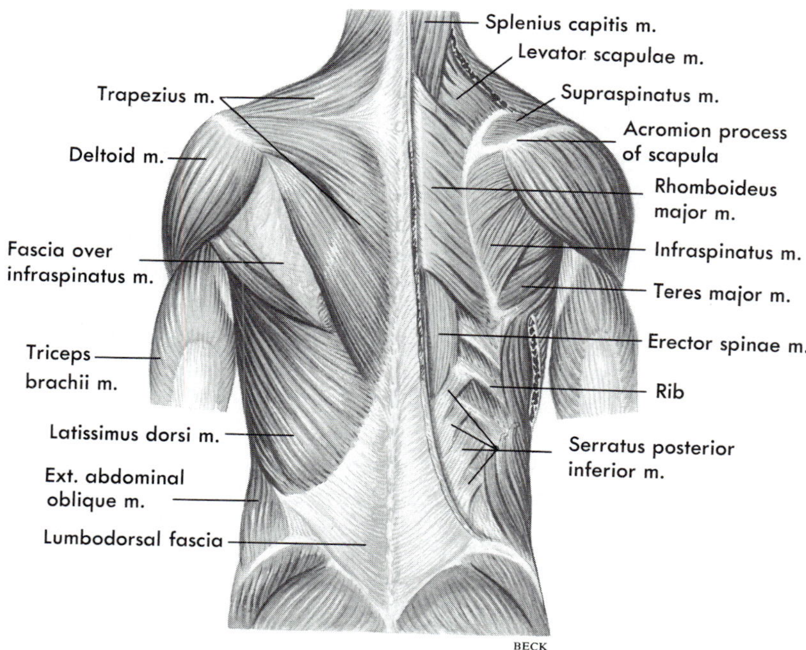

Fig. 6-13

Superficial muscles of the posterior surface of the trunk.

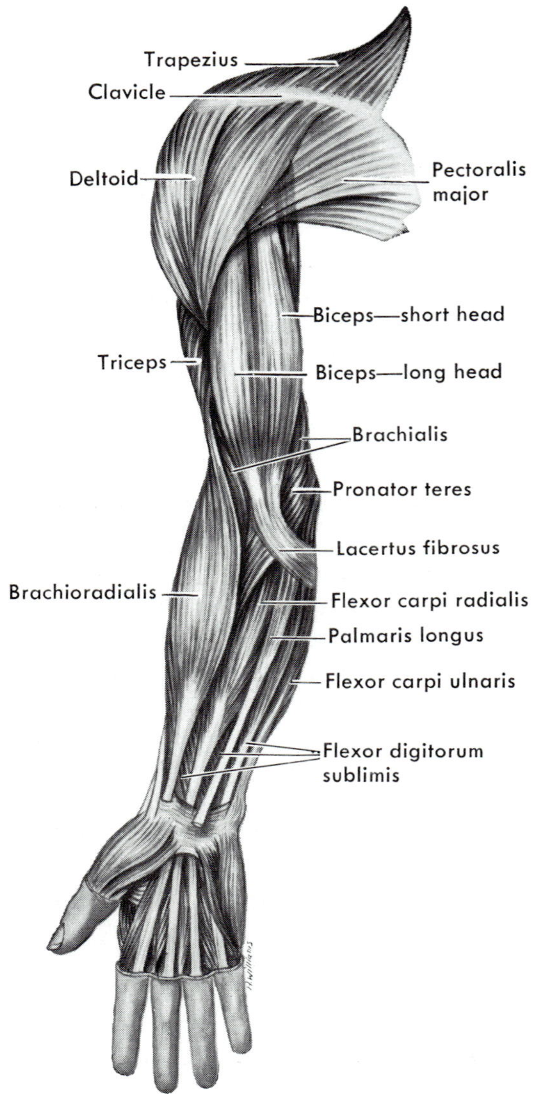

Fig. 6-14

Muscles of the flexor surface of the upper extremity.

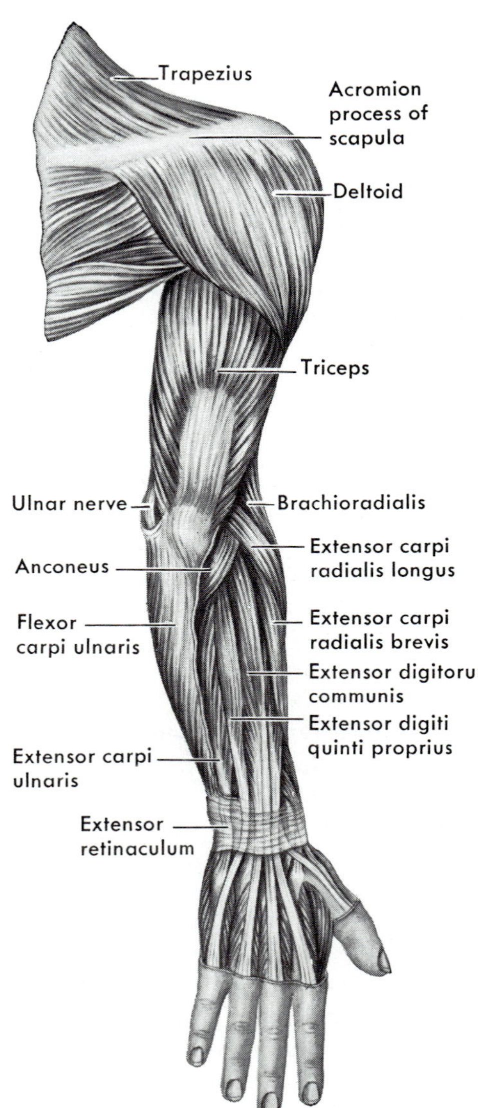

Fig. 6-15

Muscles of the extensor surface of the upper extremity.

5 To deduce which muscle produces a given action (instead of which action a given muscle produces as in step 4), start by inferring the insertion bone (bone that moves during the action). The body and origin of the muscle will lie on one or more of the bones toward which the insertion moves—often a bone or bones proximal to the insertion bone. Couple these conclusions about origin and insertion with your knowledge of muscle names and locations to deduce the muscle that produces the action.

For example, if you wish to determine the prime mover for the action of raising the upper arms straight out to the sides, you infer that the muscle inserts on the humerus since this is the bone that moves. It moves toward the shoulder—that is, the clavicle and scapula—so that probably the muscle has its origin on these bones. Because you know that the deltoid muscle fulfills these conditions, you conclude, and rightly so, that it is the muscle that raises the upper arms sidewise.

6 Do not try to learn too many details about muscle origins, insertions, and actions. Remember, it is better to start by learning a few important facts thoroughly than to half learn a mass of relatively unimportant details. Remember, too, that trying to learn too many minute facts may well result in your not retaining even the main facts.

Names

REASONS FOR NAMES

Muscle names seem more logical and therefore easier to learn when one understands the reasons for the names. Each name describes one or more of the following features about the muscle.

1 *Its action*—as flexor, extensor, adductor, etc.
2 *Direction of its fibers*—as rectus or transversus
3 *Its location*—as tibialis or femoris
4 *Number of divisions composing a muscle*—as biceps, triceps, or quadriceps
5 *Its shape*—as deltoid (triangular) or quadratus (square)
6 *Its points of attachment*—as sternocleidomastoid

A good way to start the study of a muscle is by trying to find out what its name means.

MUSCLES GROUPED ACCORDING TO LOCATION

Just as names of people are learned by associating them with physical appearance, so the names of muscles should be learned by associating them with their appearance. As you learn each muscle name, study Figs. 6-12 to 6-35 to familiarize yourself with the muscle's size, shape, and general location. To help you in this task, the names of some of the major muscles are grouped according to their location in Table 6-1, p. 139.

MUSCLES GROUPED ACCORDING TO FUNCTION

The following terms are used to designate muscles according to their main actions (see Table 6-2, p. 140, for examples).

1 *Flexors*—decrease the angle of a joint (between the anterior surfaces of the bones except in the knee and toe joints)
2 *Extensors*—return the part from flexion to normal anatomical position; increase the angle of a joint
3 *Abductors*—move the bone away from midline
4 *Adductors*—move the part toward the midline
5 *Rotators*—cause a part to pivot upon its axis
6 *Levators*—raise a part
7 *Depressors*—lower a part
8 *Sphincters*—reduce the size of an opening
9 *Tensors*—tense a part, that is, make it more rigid
10 *Supinators*—turn the hand palm upward
11 *Pronators*—turn the hand palm downward

Origins, insertions, functions, and innervations of representative muscles

Basic information about many muscles is given in Tables 6-3 to 6-15. Each table has a description of a group of muscles that move one part of the body. Muscles that, in our judgment, are the most important for beginning students of anatomy to know are set in boldface type, and the origins and insertions so judged are set in italics. Remember that the actions listed for each muscle are those for which it is a prime mover. Actually, a single muscle contracting alone rarely accomplishes a given action. Instead, muscles act in groups as prime movers, synergists, and antagonists (p. 120) to bring about movements. As you study the muscles described in Tables 6-3 to 6-15, try to follow the hints for deducing muscle actions given on p. 120.

Weak places in abdominal wall

There are several places in the abdominal wall where rupture (hernia) with protrusion of part of the intestine may occur. At these points the wall is weakened because of the presence of an interval or space in the abdominal aponeuroses. Any undue pressure on the abdominal viscera, therefore, can force a portion of the parietal peritoneum, and often a part of the intestine as well, through these nonreinforced places. The weak places are (1) the *inguinal canals*, (2) the *femoral rings*, and (3) the *umbilicus*. Hernia also occurs occasionally in the diaphragm and some other areas.

Piercing the aponeuroses of the abdominal muscles are two canals, the *inguinal canals*, one on the right and the other on the left. They lie above, but parallel to, the inguinal ligaments and are about $1\frac{1}{2}$ inches long. In the male the spermatic cords extend through the canals into the scrotum, whereas in the female the round ligaments of the uterus are in this location. The internal opening of each canal is a space in the aponeurosis of the transverse muscle known as the internal inguinal ring. The external openings or *external inguinal rings* are spaces in the aponeuroses of the external oblique muscles. They are located inferiorly and mesially to the internal rings. The fact that they are larger in the male than in the female probably explains why external inguinal hernia occurs more often in men than in women.

The *femoral rings* are openings in the groin just below the inguinal ligaments, slightly lateral to the external inguinal ring and medial to the femoral veins. They have a diameter of about $\frac{1}{2}$ inch and are usually somewhat larger in females, a fact that accounts for the greater prevalence of femoral hernia in women than in men.

Bursae

Definition

Bursae are small connective tissue sacs lined with synovial membrane and containing synovial fluid.

Locations

Bursae are located wherever pressure is exerted over moving parts—for example, between skin and bone, between tendons and bone, or between muscles, or ligaments, and bone. Some bursae that fairly frequently become inflamed (bursitis) are as follows: the subacromial bursa, between the head of the humerus and the acromion process and the deltoid muscle; the olecranon bursa, between the olecranon process and the skin; and the prepatellar bursa, between the patella and the skin. Inflammation of the prepatellar bursa is known as housemaid's knee, whereas olecranon bursitis is called student's elbow.

Function

Bursae act as cushions, relieving pressure between moving parts.

Posture

We have already discussed the major role muscles play in movement and heat production. We shall now turn our attention to a third way in which muscles serve the body as a whole—that of maintaining the posture of the body. Let us consider a few aspects of this important function.

Meaning

The term posture means simply position or alignment of body parts. "Good posture" means many things. It means body alignment that most favors function; it means position that requires the least muscular work to maintain, which puts the least strain on muscles, ligaments, and bones; it means keeping the body's center of gravity over its base. Good posture in the standing position, for example, means head and chest held high, chin, abdomen, and buttocks pulled in, knees bent slightly, and feet placed firmly on the ground about 6 inches apart.

How maintained

Since gravity pulls on the various parts of the body at all times, and since bones are too irregularly shaped to balance themselves upon each other, the only way the body can be held upright is for muscles to exert a continual pull on bones in the opposite direction from gravity. Gravity tends to pull the head and trunk forward and downward; muscles (head and trunk extensors) must therefore pull backward and upward on them. Gravity pulls the lower jaw downward; muscles must pull upward on it, etc. Muscles exert this pull against gravity by virtue of their property of tonicity. Because tonicity is absent during sleep, muscle pull does not then counteract the pull of gravity. Hence, for example, we cannot sleep standing up.

Many structures other than muscles and bones play a part in the maintenance of posture. The nervous system is responsible for the existence of muscle tone and also regulates and coordinates the amount of pull exerted by the individual muscles. The respiratory, digestive, circulatory, excretory, and endocrine systems all contribute something toward the ability of muscles to maintain posture. This is one of many examples of the important principle that all body functions are interdependent.

Importance to body as whole

The importance of posture can perhaps be best evaluated by considering some of the effects of poor posture. Poor posture throws more work on muscles to counteract the pull of gravity and therefore leads to fatigue more quickly than good posture. Poor posture puts more strain on ligaments. It puts abnormal strains on bones and may eventually produce deformities. It interferes with various functions such as respiration, heart action, and digestion. It probably is not going too far to say that it even detracts from one's feeling of self-confidence and joy. In support of this claim, consider our use of such expressions as "shoulders squared, head erect" to denote confidence and joy and "down-in-the-mouth," "long-faced," and "bowed down" to signify dejection and anxiety. The importance of posture to the body as a whole might be summed up in a single sentence: maximal health and good posture are reciprocally related—that is, each one depends upon the other.

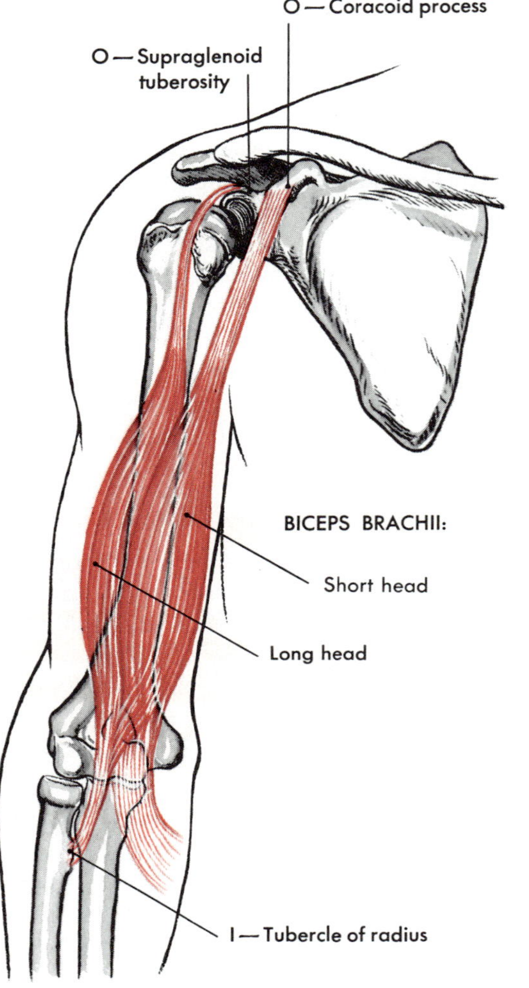

Fig. 6-16

Biceps brachii muscle. *O,* Origin. *I,* Insertion.

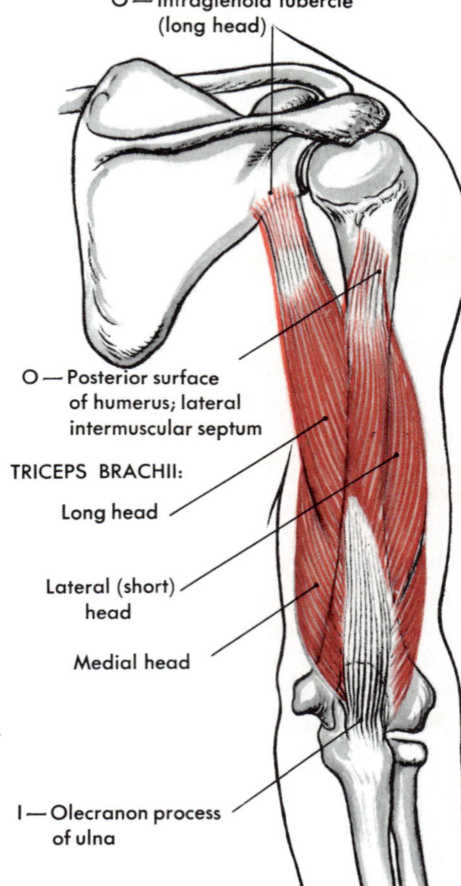

Fig. 6-17

Triceps brachii muscle. *O,* Origin. *I,* Insertion.

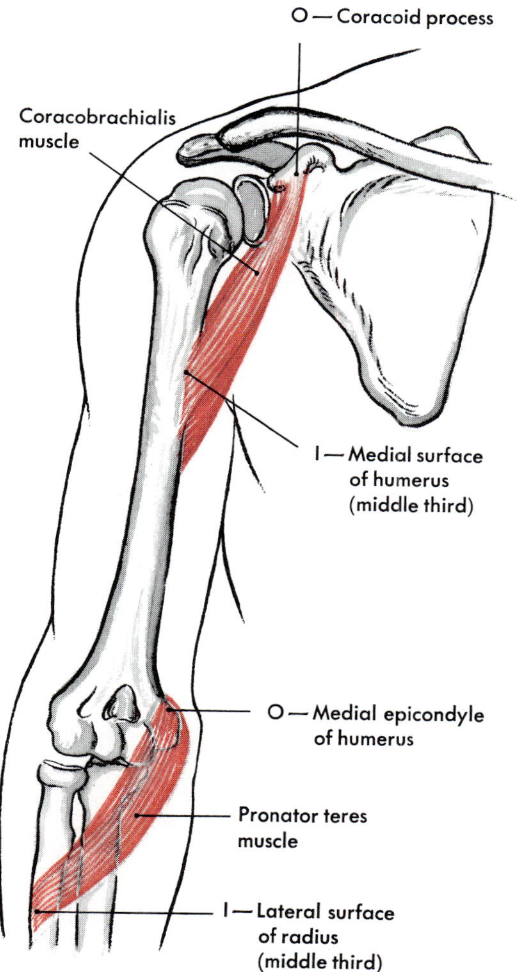

Fig. 6-18

Coracobrachialis and pronator teres muscles. *O,* Origin. *I,* Insertion.

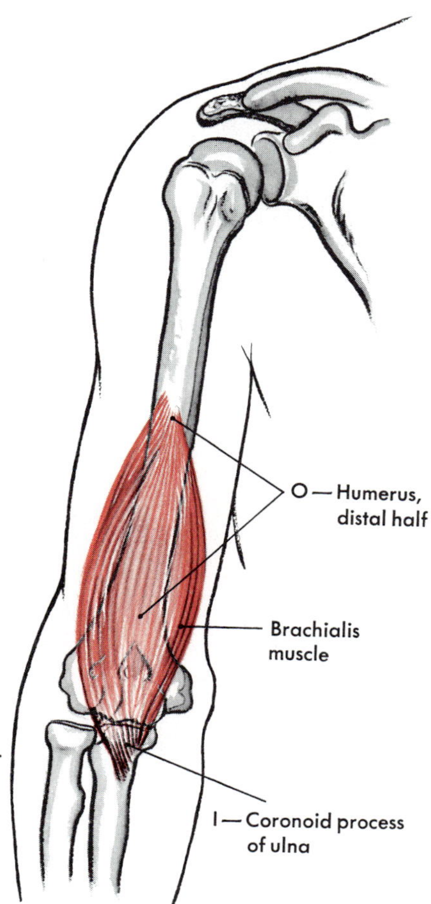

Fig. 6-19

Brachialis muscle. *O,* Origin. *I,* Insertion.

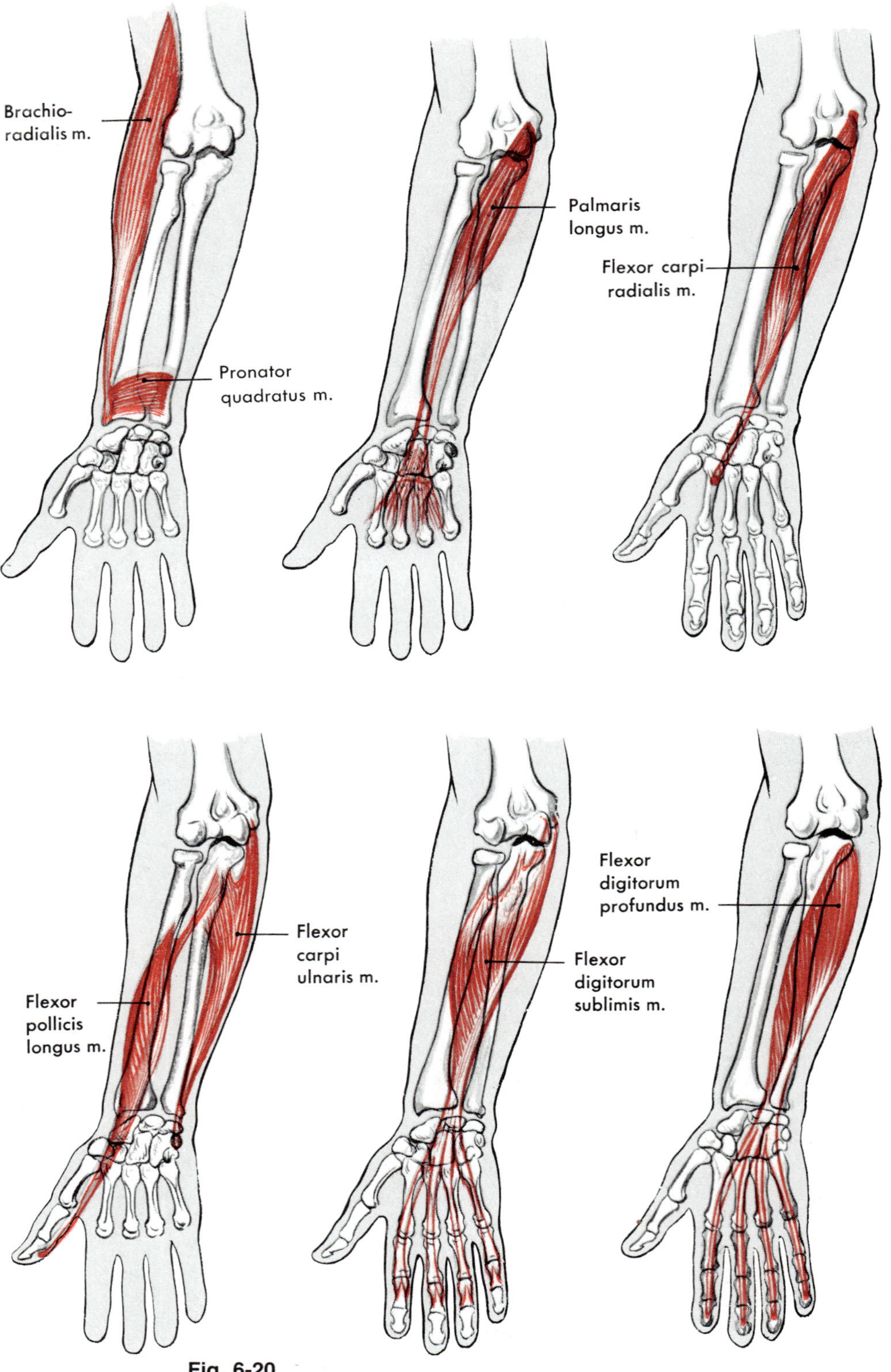

Fig. 6-20

Some muscles of the anterior (volar) aspect of the right forearm.

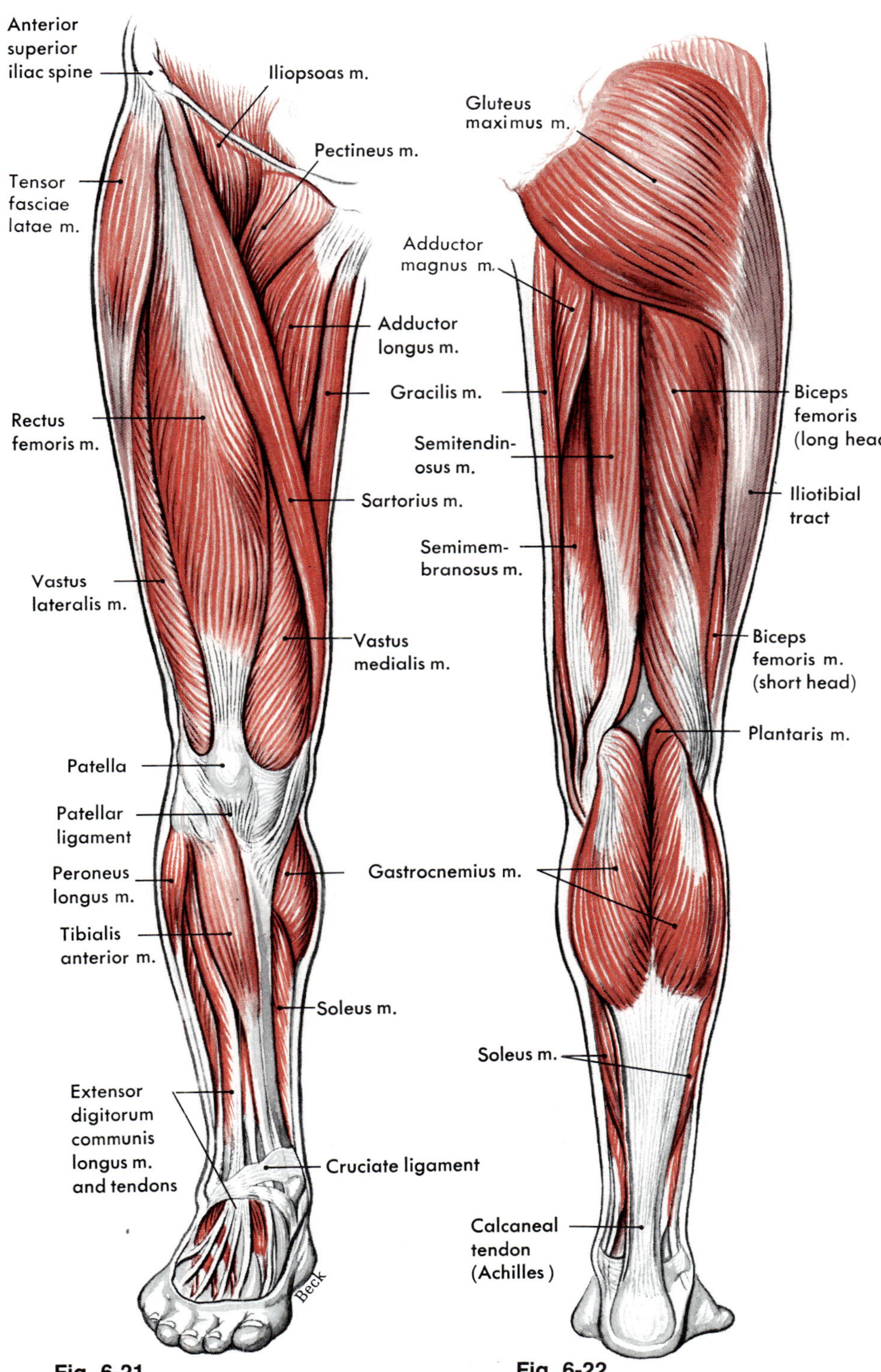

Fig. 6-21

Superficial muscles of the right thigh and leg, anterior view.

Fig. 6-22

Superficial muscles of the right thigh and leg, posterior view.

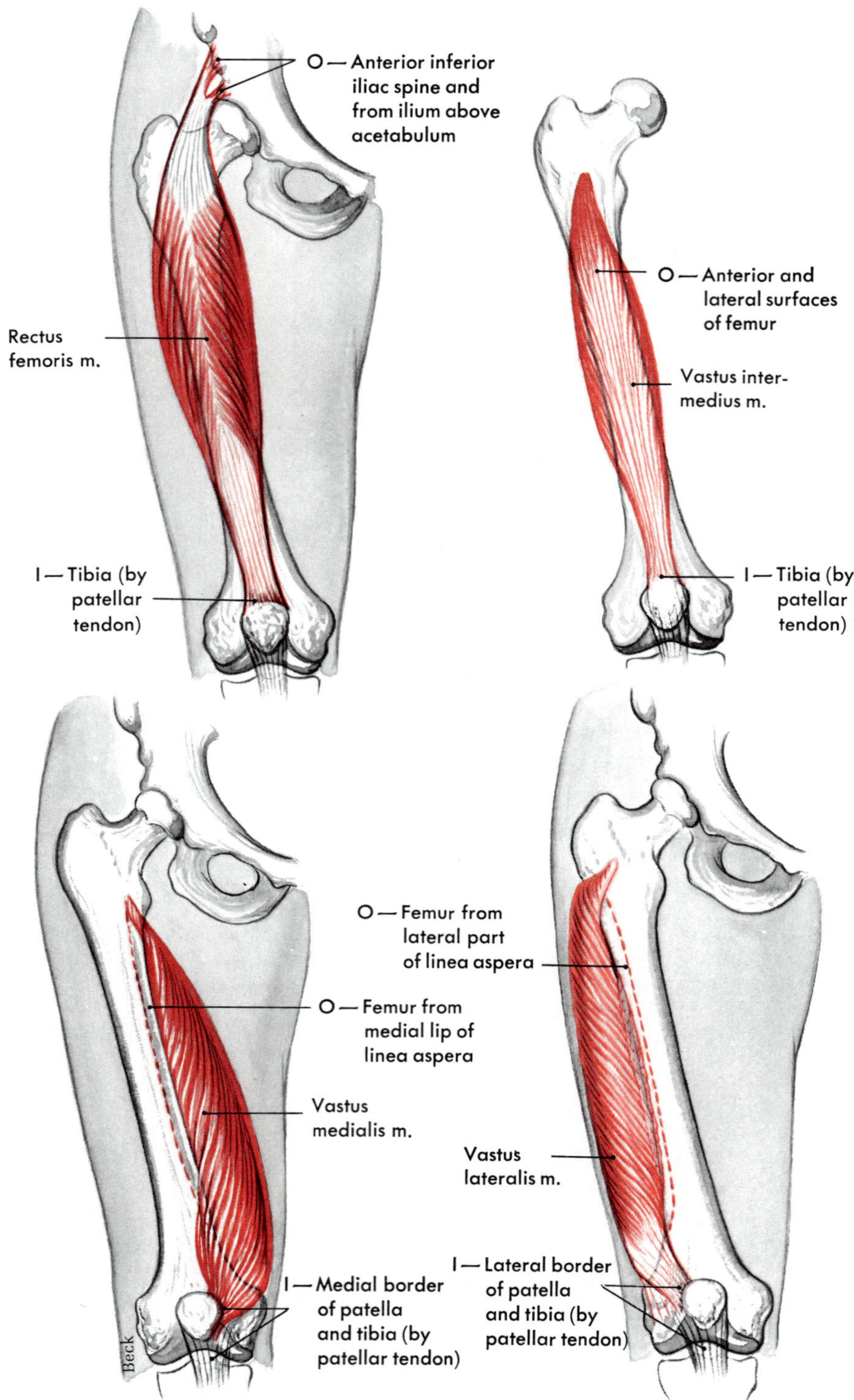

Fig. 6-23

Quadriceps femoris group of thigh muscles: rectus femoris, vastus intermedius, vastus medialis, and vastus lateralis.

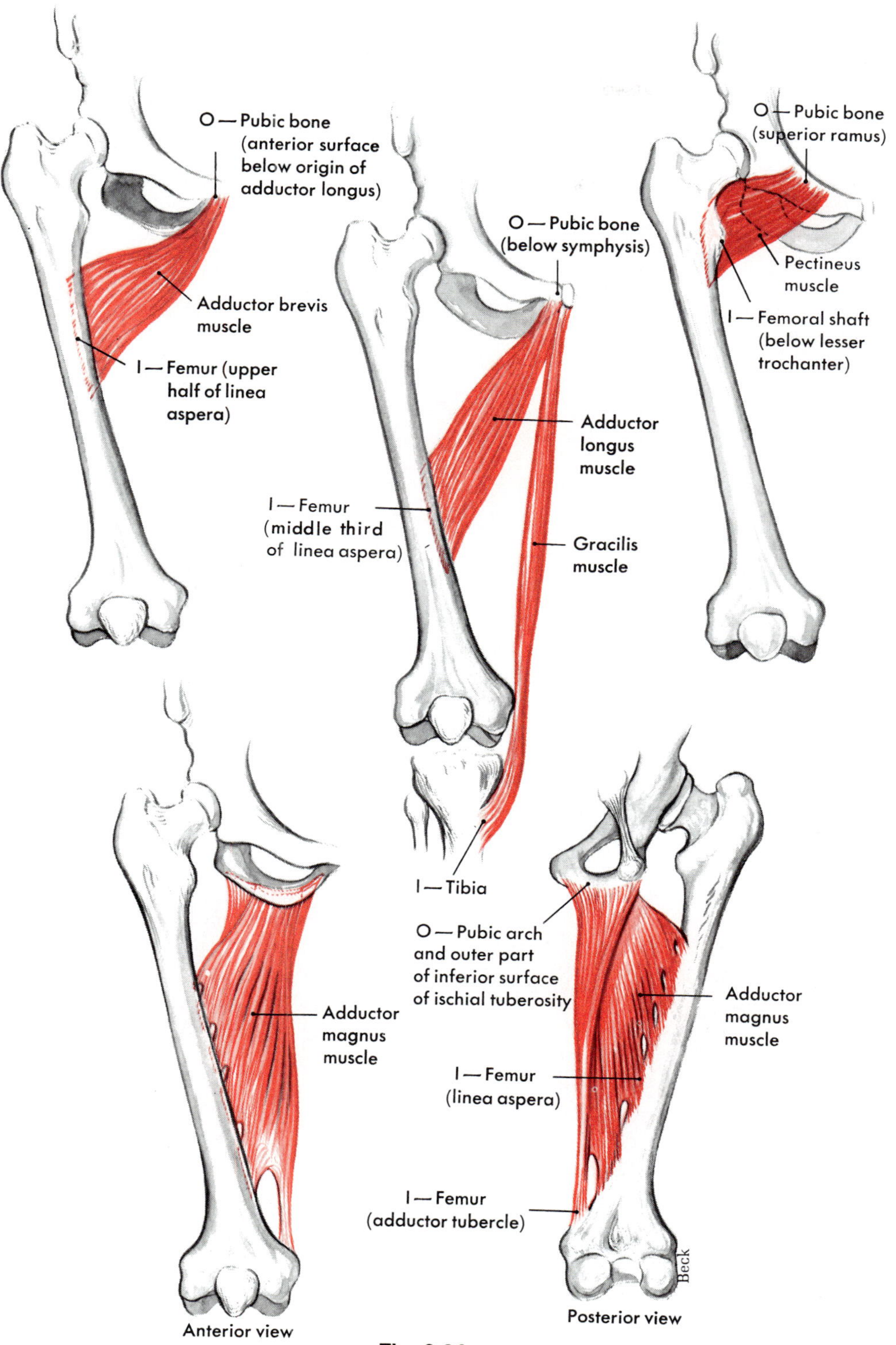

Fig. 6-24

Muscles that adduct the thigh.

O — Crest and posterior surface of ilium, sacrum, coccyx, and sacrotuberous lig.
I — Gluteal tuberosity of femur and iliotibial tract

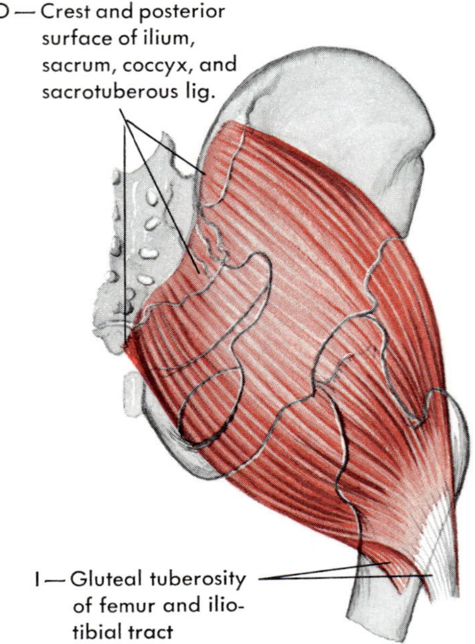

Fig. 6-25

Gluteus maximus muscle. *O*, Origin. *I*, Insertion.

O — Gluteal surface of ilium
I — Greater trochanter of femur

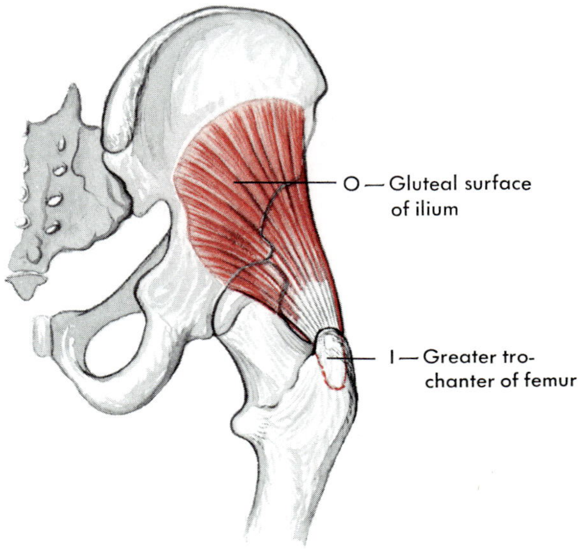

Fig. 6-26

Gluteus minimus muscle. *O*, Origin. *I*, Insertion.

O — Ilium, posterior surface of iliac crest
I — Greater trochanter

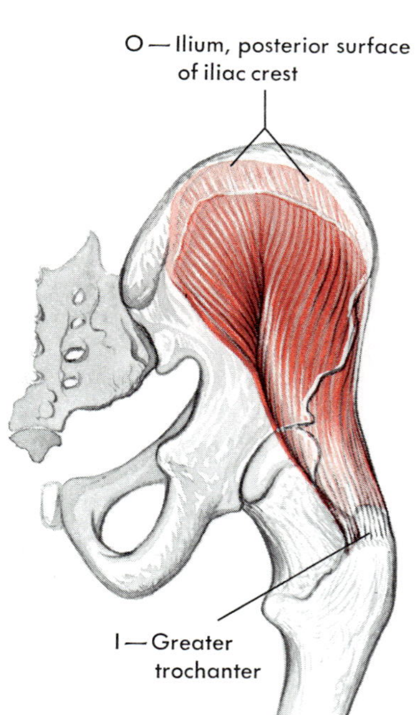

Fig. 6-27

Gluteus medius muscle. *O*, Origin. *I*, Insertion.

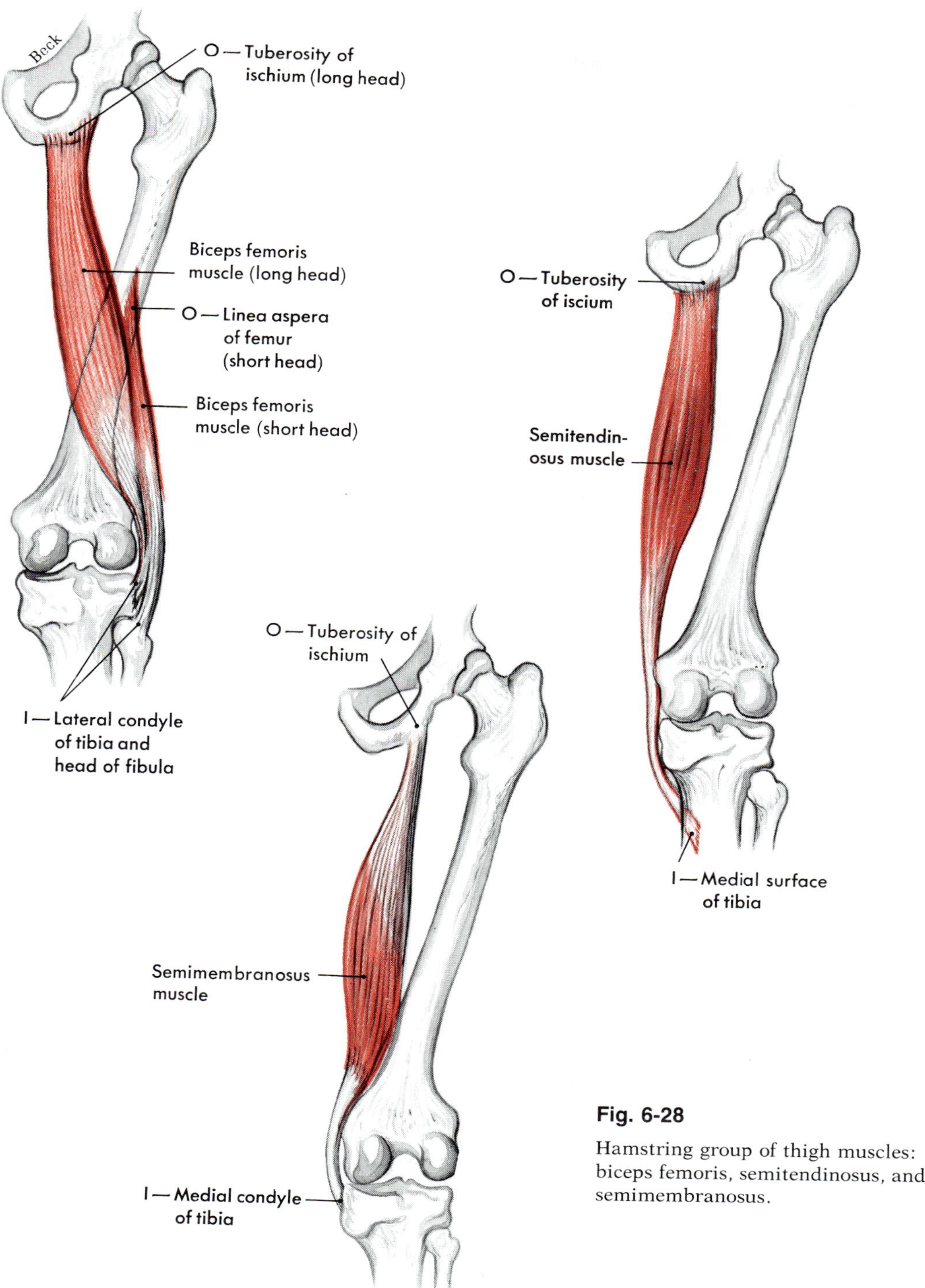

Fig. 6-28

Hamstring group of thigh muscles: biceps femoris, semitendinosus, and semimembranosus.

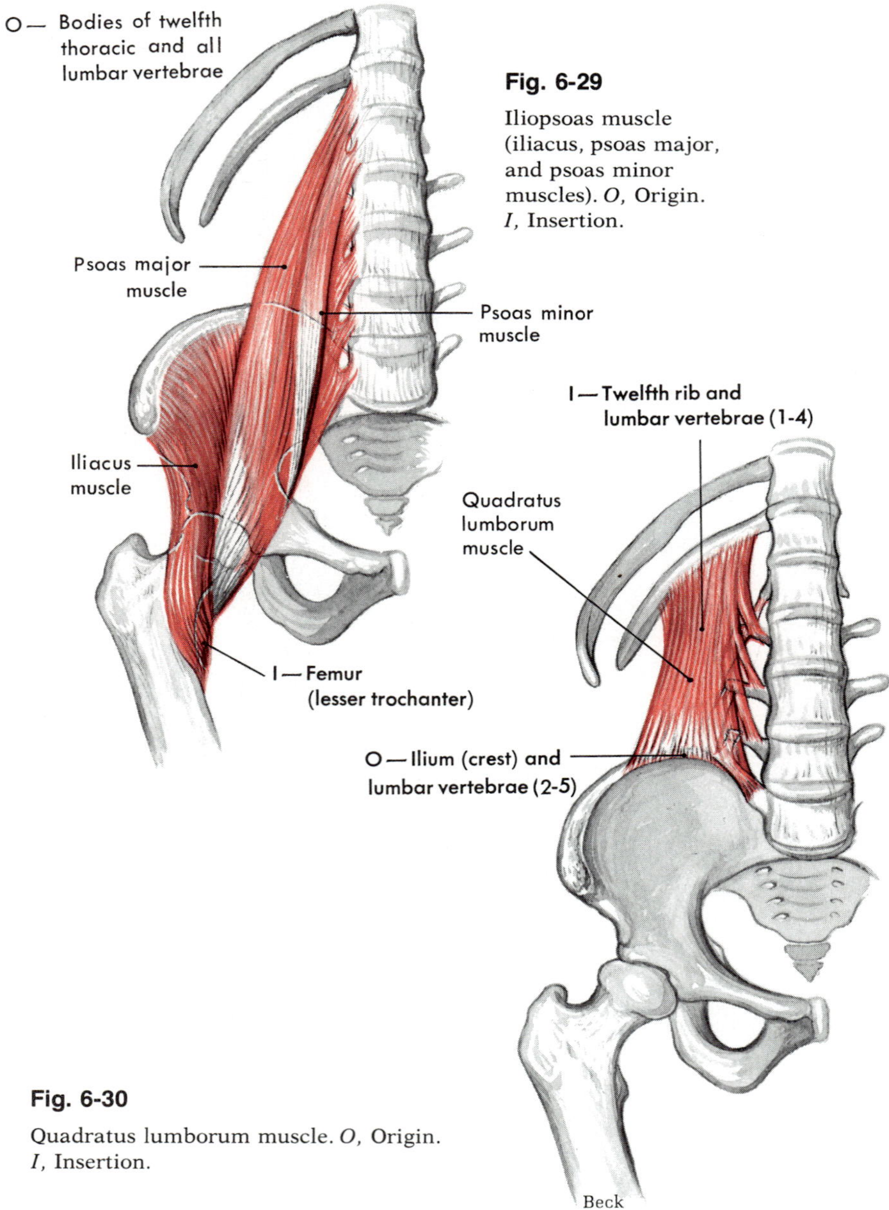

Fig. 6-29

Iliopsoas muscle (iliacus, psoas major, and psoas minor muscles). *O*, Origin. *I*, Insertion.

Fig. 6-30

Quadratus lumborum muscle. *O*, Origin. *I*, Insertion.

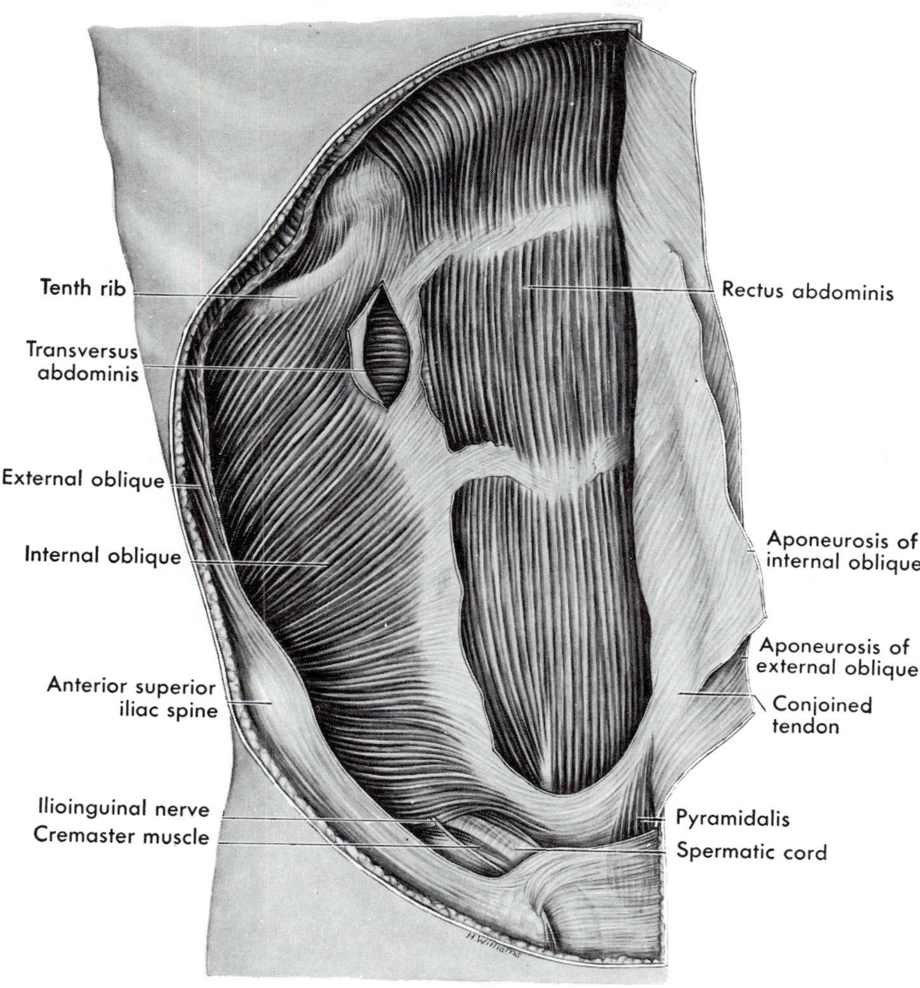

Fig. 6-31

Deep muscles of the abdominal wall. (From Francis, C. C, and Farrell, G. L.: Integrated anatomy and physiology, St. Louis, The C. V. Mosby Co.)

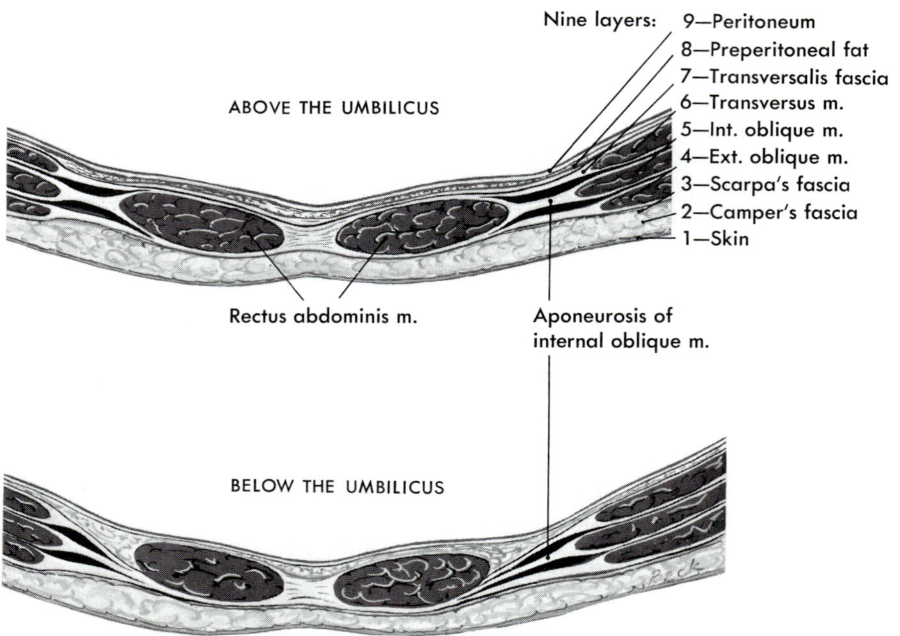

Fig. 6-32

Horizontal section of the anterolateral abdominal wall. The aponeurosis of the internal oblique muscle splits into two sections, one lying anterior and the other posterior to the rectus abdominis muscle, thereby forming an encasing sheath around this muscle.

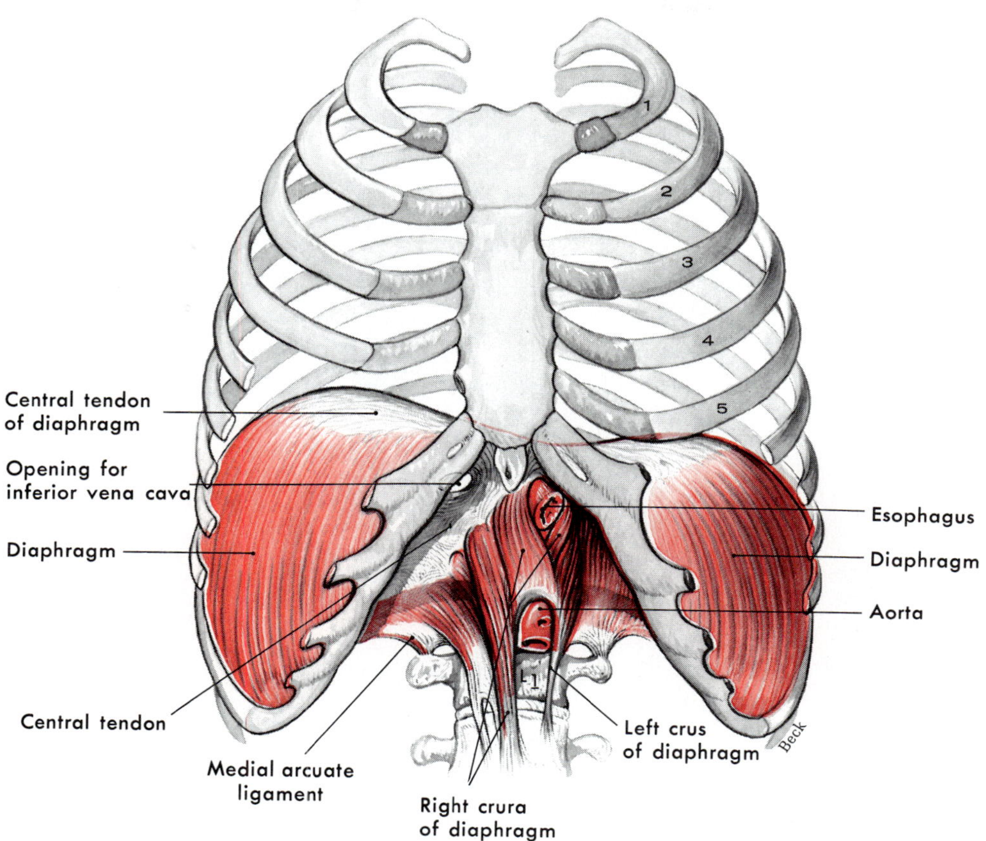

Fig. 6-33

The diaphragm as seen from the front. Note the openings in the vertebral portion for the inferior vena cava, esophagus, and aorta.

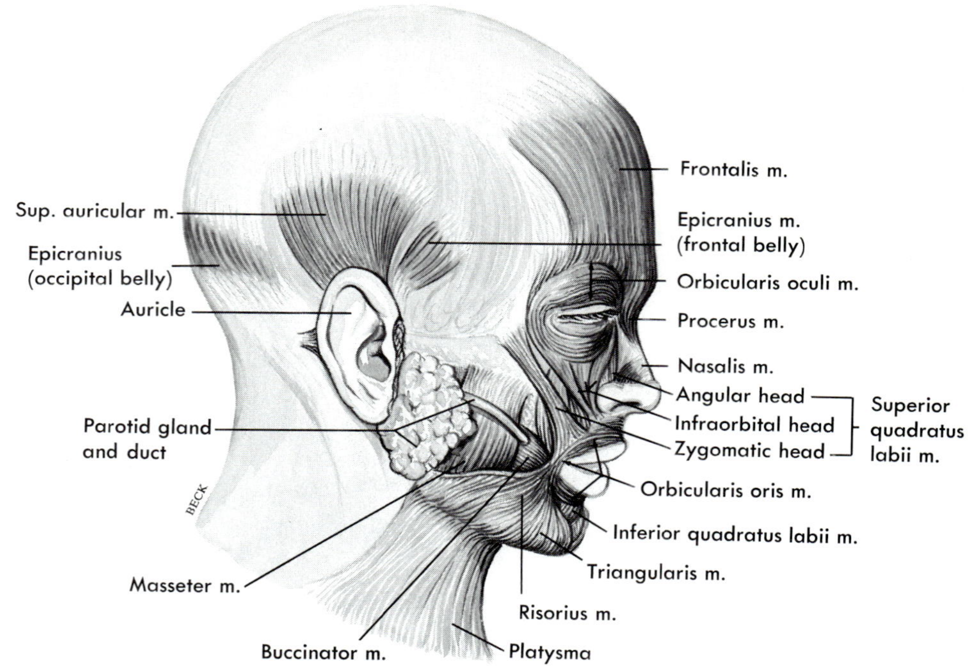

Fig. 6-34

Muscles of the head. These muscles make possible various facial expressions.

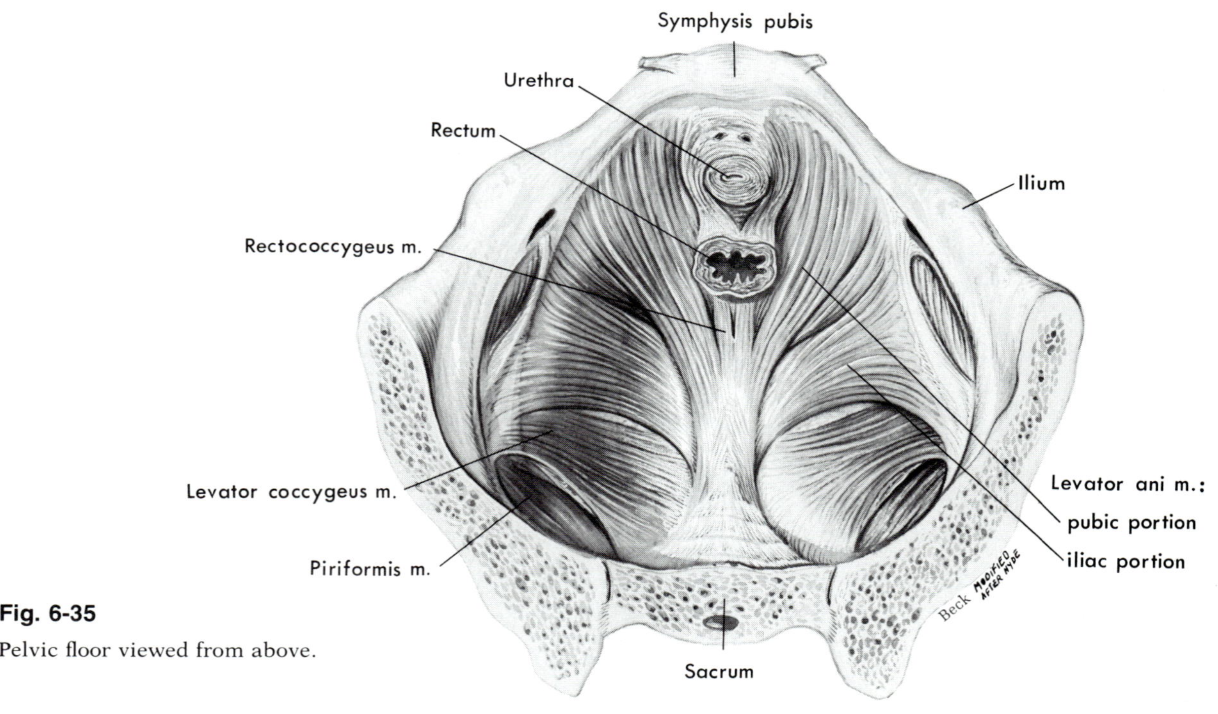

Fig. 6-35

Pelvic floor viewed from above.

Table 6-1

Muscles grouped according to location

Location	Muscles	Figures illustrating
Neck	Sternocleidomastoid	6-12
Back	Trapezius	6-12 to 6-15
	Latissimus dorsi	6-13
Chest	Pectoralis major	6-12
	Serratus anterior	6-12
Abdominal wall	External oblique	6-12, 6-13
Shoulder	Deltoid	6-12 to 6-15
Upper arm	Biceps brachii	6-14, 6-16
	Triceps brachii	6-14, 6-15, 6-17
	Brachialis	6-14, 6-19
Forearm	Brachioradialis	6-14, 6-15, 6-20
	Pronator teres	6-14, 6-18
Buttocks	Gluteus maximus	6-22, 6-25
	Gluteus minimus	6-26
	Gluteus medius	6-27
	Tensor fasciae latae	6-21
Thigh		
Anterior surface	Quadriceps femoris group	
	Rectus femoris	6-21, 6-23
	Vastus lateralis	6-21, 6-23
	Vastus medialis	6-21, 6-23
	Vastus intermedius	6-23
Medial surface	Gracilis	6-21, 6-22, 6-24
	Adductor group (brevis, longus, magnus)	6-21, 6-22, 6-24
Posterior surface	Hamstring group	
	Biceps femoris	6-22, 6-28
	Semitendinosus	6-22, 6-28
	Semimembranosus	6-22, 6-28
Leg		
Anterior surface	Tibialis anterior	6-21
Posterior surface	Gastrocnemius	6-22
	Soleus	6-22
Pelvic floor	Levator ani	6-35
	Levator coccygeus	6-35
	Rectococcygeus	6-35

Table 6-2

Muscles grouped according to function

Part moved	Example of flexor	Example of extensor	Example of abductor	Example of adductor
Head	Sternocleidomastoid	Semispinalis capitis		
Upper arm	Pectoralis major	Trapezius Latissimus dorsi	Deltoid	Pectoralis major with latissimus dorsi
Forearm	With forearm supinated: biceps brachii With forearm pronated: brachialis With semisupination or semipronation: brachioradialis	Triceps brachii		
Hand	Flexor carpi radialis and ulnaris Palmaris longus	Extensor carpi radialis, longus, and brevis Extensor carpi ulnaris	Flexor carpi radialis	Flexor carpi ulnaris
Thigh	Iliopsoas Rectus femoris (of quadriceps femoris group)	Gluteus maximus	Gluteus medius and gluteus minimus	Adductor group
Leg	Hamstrings	Quadriceps femoris group		
Foot	Tibialis anterior	Gastrocnemius Soleus	Evertors Peroneus longus Peroneus brevis	Invertor Tibialis anterior
Trunk	Iliopsoas Rectus abdominis	Sacrospinalis		

Table 6-3

Muscles that move shoulder*

Muscle	Origin	Insertion	Function	Innervation
Trapezius	*Occipital bone* (protuberance) *Vertebrae* (cervical and thoracic)	*Clavicle* *Scapula* (spine and acromion)	Raises or lowers shoulders and shrugs them Extends head when occiput acts as insertion	Spinal accessory, second, third, and fourth cervical nerves
Pectoralis minor	*Ribs* (second to fifth)	*Scapula* (coracoid)	Pulls shoulder down and forward	Medial and lateral anterior thoracic nerves
Serratus anterior	*Ribs* (upper eight or nine)	*Scapula* (anterior surface, vertebral border)	Pulls shoulder forward; abducts and rotates it upward	Long thoracic nerve

*When trying to learn the origins and insertion of the muscles listed, refer frequently to illustrations of each muscle and to the skeleton. Also, when possible, feel each muscle on your own body. Muscles judged to be most important for beginning students of anatomy to know are set in boldface type, and the origins and insertions so judged are set in italics.

Table 6-4

Muscles that move upper arm*

Muscle	Origin	Insertion	Function	Innervation
Pectoralis major	*Clavicle* (medial half) *Sternum* *Costal cartilages of true ribs*	*Humerus* (greater tubercle)	Flexes upper arm Adducts upper arm anteriorly; draws it across chest	Medial and lateral anterior thoracic nerves
Latissimus dorsi	*Vertebrae* (spines of lower thoracic, lumbar, and sacral) *Ilium* (crest) Lumbodorsal fascia†	*Humerus* (intertubercular groove)	Extends upper arm Adducts upper arm posteriorly	Thoracodorsal nerve
Deltoid	*Clavicle* *Scapula* (spine and acromion)	*Humerus* (lateral side about halfway down—deltoid tubercle)	Abducts upper arm Assists in flexion and extension of upper arm	Axillary nerve
Coracobrachialis	Scapula (coracoid process)	Humerus (middle third, medial surface)	Adduction; assists in flexion and medial rotation of arm	Musculocutaneous nerve
Supraspinatus	Scapula (supraspinous fossa)	Humerus (greater tubercle)	Assists in abducting arm	Suprascapular nerve
Teres major	Scapula (lower part, axillary border)	Humerus (upper part, anterior surface)	Assists in extension, adduction, and medial rotation of arm	Lower subscapular nerve
Teres minor	Scapula (axillary border)	Humerus (greater tubercle)	Rotates arm outward	Axillary nerve
Infraspinatus	Scapula (infraspinatus border)	Humerus (greater tubercle)	Rotates arm outward	Suprascapular nerve

*When trying to learn the origins and insertion of the muscles listed, refer frequently to illustrations of each muscle and to the skeleton. Also, when possible, feel each muscle on your own body. Muscles judged to be most important for beginning students of anatomy to know are set in boldface type, and the origins and insertions so judged are set in italics.
†Lumbodorsal fascia—extension of aponeurosis of latissimus dorsi; fills in space between last rib and iliac crest.

Table 6-5

Muscles that move lower arm*

Muscle	Origin	Insertion	Function	Innervation
Biceps brachii	*Scapula* (supraglenoid tuberosity) *Scapula* (coracoid)	*Radius* (tubercle at proximal end)	Flexes supinated forearm Supinates forearm and hand	Musculocutaneous nerve
Brachialis	*Humerus* (distal half, anterior surface)	*Ulna* (front of coronoid process)	Flexes pronated forearm	Musculocutaneous nerve
Brachioradialis	Humerus (above lateral epicondyle)	Radius (styloid process)	Flexes semipronated or semisupinated forearm; supinates forearm and hand	Radial nerve
Triceps brachii	*Scapula* (infraglenoid tuberosity) *Humerus* (posterior surface—lateral head above radial groove; medial head, below)	*Ulna* (olecranon process)	Extends lower arm	Radial nerve
Pronator teres	Humerus (medial epicondyle) Ulna (coronoid process)	Radius (middle third of lateral surface)	Pronates and flexes forearm	Median nerve
Pronator quadratus	Ulna (distal fourth, anterior surface)	Radius (distal fourth, anterior surface)	Pronates forearm	Median nerve
Supinator	Humerus (lateral epicondyle) Ulna (proximal fifth)	Radius (proximal third)	Supinates forearm	Radial nerve

*When trying to learn the origins and insertion of the muscles listed, refer frequently to illustrations of each muscle and to the skeleton. Also, when possible, feel each muscle on your own body. Muscles judged to be most important for beginning students of anatomy to know are set in boldface type, and the origins and insertions so judged are set in italics.

Table 6-6

Muscles that move hand*

Muscle	Origin	Insertion	Function	Innervation
Flexor carpi radialis	Humerus (medial epicondyle)	Second metacarpal (base of)	Flexes hand Flexes forearm	Median nerve
Palmaris longus	Humerus (medial epicondyle)	Fascia of palm	Flexes hand	Median nerve
Flexor carpi ulnaris	Humerus (medial epicondyle) Ulna (proximal two thirds)	Pisiform bone Third, fourth, and fifth metacarpals	Flexes hand Adducts hand	Ulnar nerve
Extensor carpi radialis longus	Humerus (ridge above lateral epicondyle)	Second metacarpal (base of)	Extends hand Abducts hand (moves toward thumb side when hand supinated)	Radial nerve
Extensor carpi radialis brevis	Humerus (lateral epicondyle)	Second, third metacarpals (bases of)	Extends hand	Radial nerve
Extensor carpi ulnaris	Humerus (lateral epicondyle) Ulna (proximal three fourths)	Fifth metacarpal (base of)	Extends hand Adducts hand (move toward little finger side when hand supinated)	Radial nerve

*When trying to learn the origins and insertion of the muscles listed, refer frequently to illustrations of each muscle and to the skeleton. Also, when possible, feel each muscle on your own body.

Table 6-7

Muscles that move thigh*

Muscle	Origin	Insertion	Function	Innervation
Iliopsoas (iliacus and psoas major)	*Ilium* (iliac fossa) *Vertebrae* (bodies of twelfth thoracic to fifth lumbar)	*Femur* (small trochanter)	Flexes thigh Flexes trunk (when femur acts as origin)	Femoral and second to fourth lumbar nerves
Rectus femoris	*Ilium* (anterior, inferior spine)	*Tibia* (by way of patellar tendon)	Flexes thigh Extends lower leg	Femoral nerve
Gluteal group				
Maximus	*Ilium* (crest and posterior surface) Sacrum and coccyx (posterior surface) Sacrotuberous ligament	*Femur* (gluteal tuberosity) *Iliotibial tract*†	Extends thigh—rotates outward	Inferior gluteal nerve
Medius	*Ilium* (lateral surface)	*Femur* (greater trochanter)	Abducts thigh—rotates outward; stabilizes pelvis on femur	Superior gluteal nerve
Minimus	*Ilium* (lateral surface)	*Femur* (greater trochanter)	Abducts thigh; stabilizes pelvis on femur Rotates thigh medially	Superior gluteal nerve
Tensor fasciae latae	*Ilium* (anterior part of crest)	*Tibia* (by way of *iliotibial tract*)	Abducts thigh Tightens iliotibial tract†	Superior gluteal nerve
Piriformis	Vertebrae (front of sacrum)	Femur (medial aspect of greater trochanter)	Rotates thigh outward Abducts thigh Extends thigh	First or second sacral nerves
Adductor group				
Brevis	*Pubic bone*	*Femur* (linea aspera)	Adducts thigh	Obturator nerve
Longus	*Pubic bone*	*Femur* (linea aspera)	Adducts thigh	Obturator nerve
Magnus	*Pubic bone*	*Femur* (linea aspera)	Adducts thigh	Obturator nerve
Gracilis	Pubic bone (just below symphysis)	*Tibia* (medial surface behind sartorius)	Adducts thigh and flexes and adducts leg	Obturator nerve

*When trying to learn the origins and insertion of the muscles listed, refer frequently to illustrations of each muscle and to the skeleton. Also, when possible, feel each muscle on your own body. Muscles judged to be most important for beginning students of anatomy to know are set in boldface type, and the origins and insertions so judged are set in italics.

†The iliotibial tract is part of the fascia enveloping all the thigh muscles. It consists of a wide band of dense fibrous tissue attached to the iliac crest above and the lateral condyle of the tibia below. The upper part of the tract encloses the tensor fasciae latae muscle.

Table 6-8

Muscles that move lower leg*

Muscle	Origin	Insertion	Function	Innervation
Quadriceps femoris group				
Rectus femoris	*Ilium* (anterior, inferior spine)	*Tibia* (by way of patellar tendon)	Flexes thigh Extends leg	Femoral nerve
Vastus lateralis	*Femur* (linea aspera)	*Tibia* (by way of patellar tendon)	Extends leg	Femoral nerve
Vastus medialis	*Femur*	*Tibia* (by way of patellar tendon)	Extends leg	Femoral nerve
Vastus intermedius	*Femur* (anterior surface)	*Tibia* (by way of patellar tendon)	Extends leg	Femoral nerve
Sartorius	*Os innominatum* (anterior, superior iliac spines)	*Tibia* (medial surface of upper end of shaft)	Adducts and flexes leg Permits crossing of legs tailor fashion	Femoral nerve
Hamstring group				
Biceps femoris	*Ischium* (tuberosity)	*Fibula* (head of)	Flexes leg	Hamstring nerve (branch of sciatic nerve)
	Femur (linea aspera)	*Tibia* (lateral condyle)	Extends thigh	Hamstring nerve
Semitendinosus	*Ischium* (tuberosity)	*Tibia* (proximal end, medial surface)	Extends thigh	Hamstring nerve
Semimembranosus	*Ischium* (tuberosity)	*Tibia* (medial condyle)	Extends thigh	Hamstring nerve

*When trying to learn the origins and insertion of the muscles listed, refer frequently to illustrations of each muscle and to the skeleton. Also, when possible, feel each muscle on your own body. Muscles judged to be most important for beginning students of anatomy to know are set in boldface type, and the origins and insertions so judged are set in italics.

Table 6-9

Muscles that move foot*

Muscle	Origin	Insertion	Function	Innervation
Tibialis anterior	*Tibia* (lateral condyle of upper body)	*Tarsal* (first cuneiform) Metatarsal (base of first)	Flexes foot Inverts foot	Common and deep peroneal nerves
Gastrocnemius	*Femur* (condyles)	*Tarsal* (calcaneus by way of Achilles tendon)	Extends foot Flexes lower leg	Tibial nerve (branch of sciatic nerve)
Soleus	*Tibia* (underneath gastrocnemius) *Fibula*	*Tarsal* (calcaneus by way of Achilles tendon)	Extends foot (plantar flexion)	Tibial nerve
Peroneus longus	Tibia (lateral condyle) Fibula (head and shaft)	First cuneiform Base of first metatarsal	Extends foot (plantar flexion) Everts foot	Common peroneal nerve
Peroneus brevis	Fibula (lower two thirds of lateral surface of shaft)	Fifth metatarsal (tubercle, dorsal surface)	Everts foot Flexes foot	Superficial peroneal nerve
Tibialis posterior	Tibia (posterior surface) Fibula (posterior surface)	Navicular bone Cuboid bone All three cuneiforms Second and fourth metatarsals	Extends foot (plantar flexion) Inverts foot	Tibial nerve
Peroneus tertius	Fibula (distal third)	Fourth and fifth metatarsals (bases of)	Flexes foot Everts foot	Deep peroneal nerve

*When trying to learn the origins and insertion of the muscles listed, refer frequently to illustrations of each muscle and to the skeleton. Also, when possible, feel each muscle on your own body. Muscles judged to be most important for beginning students of anatomy to know are set in boldface type, and the origins and insertions so judged are set in italics.

Table 6-10

Muscles that move head*

Muscle	Origin	Insertion	Function	Innervation
Sternocleidomastoid	*Sternum* *Clavicle*	*Temporal bone* (mastoid process)	Flexes head (prayer muscle) One muscle, alone, rotates head toward opposite side; spasm of this muscle alone or associated with trapezius called torticollis or wryneck	Accessory nerve
Semispinalis capitis	Vertebrae (transverse processes of upper six thoracic, articular processes of lower four cervical)	Occipital bone (between superior and inferior nuchal lines)	Extends head; bends it laterally	First five cervical nerves
Splenius capitis	Ligamentum nuchae Vertebrae (spinous processes of upper three or four thoracic)	Temporal bone (mastoid process) Occipital bone	Extends head Bends and rotates head toward same side as contracting muscle	Second, third, and fourth cervical nerves
Longissimus capitis	Vertebrae (transverse processes of upper six thoracic, articular processes of lower four cervical)	Temporal bone (mastoid process)	Extends head Bends and rotates head toward contracting side	

*When trying to learn the origins and insertion of the muscles listed, refer frequently to illustrations of each muscle and to the skeleton. Also, when possible, feel each muscle on your own body. Muscles judged to be most important for beginning students of anatomy to know are set in boldface type, and the origins and insertions so judged are set in italics.

Table 6-11

Muscles that move abdominal wall*

Muscle	Origin	Insertion	Function	Innervation
External oblique	*Ribs* (lower eight)	*Ossa coxae* (iliac crest and pubis by way of inguinal ligament)† *Linea alba*‡ by way of an aponeurosis§	Compresses abdomen Important postural function of all abdominal muscles is to pull front of pelvis upward, thereby flattening lumbar curve of spine; when these muscles lose their tone, common figure faults of protruding abdomen and lordosis develop	Lower seven intercostal nerves and iliohypogastric nerves
Internal oblique	*Ossa coxae* (iliac crest and inguinal ligament) *Lumbodorsal fascia*	*Ribs* (lower three) *Pubic bone* *Linea alba*	Same as external oblique	Last three intercostals; iliohypogastric and ilioinguinal nerves
Transversalis	*Ribs* (lower six) *Ossa coxae* (iliac crest, inguinal ligament) *Lumbodorsal fascia*	*Pubic bone* *Linea alba*	Same as external oblique	Last five intercostals; iliohypogastric and ilioinguinal nerves
Rectus abdominis	*Ossa coxae* (pubic bone and symphysis pubis)	*Ribs* (costal cartilage of fifth, sixth, and seventh ribs) Sternum (xiphoid process)	Same as external oblique; because abdominal muscles compress abdominal cavity, they aid in straining, defecation, forced expiration, childbirth, etc.; abdominal muscles are antagonists of diaphragm, relaxing as it contracts and vice versa Flexes trunk	Last six intercostal nerves

*When trying to learn the origins and insertion of the muscles listed, refer frequently to illustrations of each muscle and to the skeleton. Also, when possible, feel each muscle on your own body. Muscles judged to be most important for beginning students of anatomy to know are set in boldface type, and the origins and insertions so judged are set in italics.

†*Inguinal ligament* (or Poupart's)—lower edge of aponeurosis of external oblique muscle, extending between the anterior superior iliac spine and the tubercle of the pubic bone. This edge is doubled under like a hem on material. The inguinal ligament forms the upper boundary of the *femoral triangle*, a large triangular area in the thigh; its other boundaries are the adductor longus muscle mesially and the sartorius laterally.

‡*Linea alba*—literally, a white line; extends from xiphoid process to symphysis pubis; formed by fibers of aponeuroses of the right abdominal muscles interlacing with fibers of aponeuroses of the left abdominal muscles; comparable to a seam up the midline of the abdominal wall, anchoring its various layers. During pregnancy the linea alba becomes pigmented and is known as the *linea niger*.

§*Aponeurosis*—sheet of white fibrous tissue that attaches one muscle to another or attaches it to bone or other movable structures; e.g., the right external oblique muscle attaches to the left external oblique muscle by means of an aponeurosis.

Table 6-12

Muscles that move chest wall*

Muscle	Origin	Insertion	Function	Innervation
External intercostals	Rib (lower border; forward fibers)	Rib (upper border of rib below origin)	Elevate ribs	Intercostal nerves
Internal intercostals	Rib (inner surface, lower border; backward fibers)	Rib (upper border of rib below origin)	Probably depress ribs	Intercostal nerves
Diaphragm	*Lower circumference of thorax* (of rib cage)	*Central tendon of diaphragm*	Enlarges thorax, causing inspiration	Phrenic nerves

*When trying to learn the origins and insertion of the muscles listed, refer frequently to illustrations of each muscle and to the skeleton. Also, when possible, feel each muscle on your own body. Muscles judged to be most important for beginning students of anatomy to know are set in boldface type, and the origins and insertions so judged are set in italics.

Table 6-13

Muscles of pelvic floor*

Muscle	Origin	Insertion	Function	Innervation
Levator ani	Pubis—posterior surface Ischium (spine)	Coccyx	Together form floor of pelvic cavity; support pelvic organs; if these muscles are badly torn at childbirth, or become too relaxed, uterus or bladder may prolapse, i.e., drop out	Pudendal nerve
Coccygeus (posterior continuation of levator ani)	Ischium (spine)	Coccyx Sacrum	Same as levator ani	Pudendal nerve

*When trying to learn the origins and insertion of the muscles listed, refer frequently to illustrations of each muscle and to the skeleton. Also, when possible, feel each muscle on your own body.

Table 6-14

Muscles that move trunk*

Muscle	Origin	Insertion	Function	Innervation
Sacrospinalis (erector spinae)			Extend spine; maintain erect posture of trunk Acting singly, abduct and rotate trunk	Posterior rami of first cervical to fifth lumbar spinal nerves
Lateral portion:				
Iliocostalis lumborum	Iliac crest, sacrum (posterior surface), and lumbar vertebrae (spinous processes)	Ribs, lower six		
Iliocostalis dorsi	Ribs, lower six	Ribs, upper six		
Iliocostalis cervicis	Ribs, upper six	Vertebrae, fourth to sixth cervical		
Medial portion:				
Longissimus dorsi	Same as iliocostalis lumborum	Vertebrae, thoracic ribs		
Longissimus cervicis	Vertebrae, upper six thoracic	Vertebrae, second to sixth cervical		
Longissimus capitis	Vertebrae, upper six thoracic and last four cervical	Temporal bone, mastoid process		
Quadratus lumborum (forms part of posterior abdominal wall)	Ilium (posterior part of crest) Vertebrae (lower three lumbar)	Ribs (twelfth) Vertebrae (transverse processes of first four lumbar)	Both muscles together extend spine One muscle alone abducts trunk toward side of contracting muscle	First three or four lumbar nerves
Iliopsoas	See muscles that move thigh, p. 145		Flexes trunk	

*When trying to learn the origins and insertion of the muscles listed, refer frequently to illustrations of each muscle and to the skeleton. Also, when possible feel each muscle on your own body. Muscles judged to be most important for beginning students of anatomy to know are set in boldface type.

Table 6-15

Muscles of facial expression and of mastication*

Muscle	Origin	Insertion	Function	Innervation
Muscles of facial expression				
Epicranius (occipitofrontalis)	Occipital bone	Tissues of eyebrows	Raises eyebrows, wrinkling forehead, horizontally	Cranial nerve VII
Corrugator supercilii	Frontal bone (superciliary ridge)	Skin of eyebrow	Wrinkles forehead vertically	Cranial nerve VII
Orbicularis oculi	Encircles eyelid		Closes eye	Cranial nerve VII
Orbicularis oris	Encircles mouth		Draws lips together	Cranial nerve VII
Platysma	Fascia of upper part of deltoid and pectoralis major	Mandible—lower border Skin around corners of mouth	Draws corners of mouth down—pouting	Cranial nerve VII
Buccinator	Maxillae	Skin of sides of mouth	Permits smiling Blowing (e.g., as in playing a trumpet)	Cranial nerve VII
Muscles of mastication				
Masseter	Zygomatic arch	Mandible (external surface)	Closes jaw	Cranial nerve V
Temporal	Temporal bone	Mandible	Closes jaw	Cranial nerve V
Pterygoids (internal and external)	Undersurface of skull	Mandible (mesial surface)	Grate teeth	Cranial nerve V

*When trying to learn the origins and insertion of the muscles listed, refer frequently to illustrations of each muscle and to the skeleton. Also, when possible, feel each muscle on your own body.

Outline summary

GENERAL FUNCTIONS

1 Movement—sometimes locomotion, sometimes movement within given area
2 Posture
3 Heat production

SKELETAL MUSCLE CELLS
Microscopic structure

1 Muscle cells usually called muscle fibers, term descriptive of their long, narrow shape
2 Sarcolemma—cell membrane of muscle fiber
3 Sarcoplasm—cytoplasm of muscle fiber
4 Sarcoplasmic reticulum—analogous but not identical to endoplasmic reticulum of cells other than muscle fibers
5 Myofibrils—numerous fine fibers packed close together in sarcoplasm
6 Cross striae
  a Dark stripes called A bands; light H zone runs across midsection of each dark A band
  b Light stripes called I bands; dark Z line extends across center of each light I band
7 Sarcomere—section of myofibril extending from one Z line to next; each myofibril consists of several sarcomeres
8 T system—transverse tubules that extend into sarcoplasm at levels of Z lines; formed by invaginations of sarcolemma
9 Triad—triple-layered structure consisting of T tubule sandwiched between sacs of sarcoplasmic reticulum

Molecular structure

1 Proteins
  a Thick filaments—composed almost entirely of myosin molecules; heads of myosin molecules are cross bridges of thick filament
  b Thin filaments—composed of actin, tropomyosin, and troponin molecules arranged in complex fashion; thin filaments attach to Z lines, extend from them in toward center of sarcomeres; thick and thin filaments alternate in myofibrils

Functions

1 Contraction—cross bridges of thick filaments attach to thin filaments and pull them toward middle of each sarcomere (see Fig. 6-4)
2 Muscle cells obey all-or-none law when they contract—i.e., they either contract with all force possible under existing conditions or do not contract at all

Energy sources for muscle contraction

See Fig. 6-5

SKELETAL MUSCLE ORGANS
Structure

1 Size, shape, and fiber arrangement—wide variation in different muscles
2 Connective tissue components
  a Epimysium—fibrous connective tissue sheath that envelops each muscle
  b Perimysium—extensions of epimysium, partitioning each muscle into bundles of fibers
  c Endomysium—extensions of perimysium between individual muscle fibers
  d Tendon—strong, tough cord continuous at one end with fibrous wrappings (epimysium, etc.) of muscle and at other end with fibrous covering of bone (periosteum)
  e Aponeurosis—broad flat sheet of fibrous connective tissue continuous on one border with fibrous wrappings of muscle and at other border with fibrous coverings of some adjacent structure, usually another muscle
  f Tendon sheaths—tubes of fibrous connective tissue that enclose certain tendons, notably those of wrist and ankle; synovial membrane lines tendon sheaths
  g Deep fascia—layer of dense fibrous connective tissue underlying superficial fascia under skin; extensions of deep fascia form epimysium, etc. and also enclose viscera, glands, blood vessels, and nerves
3 Nerve supply
One motoneuron, together with skeletal muscle fibers it supplies, constitutes *motor unit;* number of muscle fibers per motor unit varies; in general, more precise movements produced by muscle in which motor units include fewer muscle fibers
4 Age changes
  a Fibrosis—with advancing years some skeletal muscle fibers degenerate and are replaced by fibrous connective tissue
  b Decreased muscular strength, resulting in part from fibrosis

Function

1 Basic principles
  a Skeletal muscles contract only if stimulated; natural stimulus, nerve impulses
  b Skeletal muscle contractions of several types:
    1 Tonic contraction (tone; tonus)—continual, partial contractions produced by simultaneous activation of small group of motor units, followed by relaxation of their fibers and activation of another group of motor units; all healthy muscles exhibit tone when individuals are awake
    2 Isotonic contraction—muscle shortens but its tension remains constant; isotonic contractions produce movements
    3 Isometric contraction—muscle length remains unchanged but tension within muscle increases; isometric contractions "tighten" muscles but do not produce movements
    4 Twitch contraction—quick, jerky contraction in response to single stimulus; consists of three phases—latent period, contraction phase, and relaxation phase; twitch contractions rare in normal body
    5 Tetanic contraction (tetanus)—sustained smooth contraction produced by series of stimuli bombarding muscle in rapid succession; normal movements said to be produced by incomplete tetanic contractions
    6 Treppe (staircase phenomenon)—series of increasingly stronger contractions in response to constant strength stimuli applied at rate of 1 or 2 per second; contracture—i.e., incomplete relaxation after repeated stimulation; fatigue—i.e., failure of muscle to contract in response to strongest stimuli after repeated stimulation; true muscle fatigue seldom occurs in body

7 Fibrillation—abnormal contraction in which individual muscle fibers contract asynchronously, producing no effective movement
8 Convulsions—uncoordinated tetanic contractions of varying groups of muscles

c Skeletal muscles contract according to graded-strength principle in contrast to individual muscle cells that compose them, which contract according to all-or-none law
d Skeletal muscles produce movement by pulling on insertion bones across joints
e Bones serve as levers and joints as fulcrums of these levers
f Muscles that move part usually do not lie over that part but proximal to it
g Skeletal muscles almost always act in groups rather than singly—i.e., most movements produced by coordinated action of several muscles

2 Hints of how to deduce actions
  a Deduce bones that muscle attaches to from illustrations of muscle
  b Make guess as to which bone moves (insertion)
  c Deduce movement muscle produces by applying principle that its insertion moves toward its origin

Names

1 Reasons for names—muscle names describe one or more of following features about muscle
  a Its action
  b Direction of fibers
  c Its location
  d Number of divisions composing it
  e Its shape
  f Its points of attachment
2 Muscles grouped according to location—see Table 6-1
3 Muscles grouped according to function—also see Table 6-2
  a Flexors—decrease angle of joint
  b Extensors—return part from flexion to normal anatomical position
  c Abductors—move bone away from midline of body
  d Adductors—move bone toward midline of body
  e Rotators—cause part to pivot upon its axis
  f Levators—raise part
  g Depressors—lower part
  h Sphincters—reduce size of opening
  i Tensors—tense part or make it more rigid
  j Supinators—turn hand palm upward
  k Pronators—turn hand palm downward

Origins, insertions, functions, and innervations of representative muscles

See Tables 6-3 to 6-15

WEAK PLACES IN ABDOMINAL WALL

1 Inguinal rings—right and left internal; right and left external
2 Femoral rings—right and left
3 Umbilicus

BURSAE

1 Definition—small connective tissue sacs lined with synovial membrane and containing synovial fluid
2 Locations—wherever pressure exerted over moving parts
  a Between skin and bone
  b Between tendons and bone
  c Between muscles or ligaments and bone
  d Names of bursae that frequently become inflamed (bursitis)
    1 Subacromial—between deltoid muscle and head of humerus and acromion process
    2 Olecranon—between olecranon process and skin; inflammation called student's elbow
    3 Prepatellar—between patella and skin; inflammation called housemaid's knee
3 Function—act as cushion, relieving pressure between moving parts

POSTURE

1 Meaning—position or alignment of body parts
2 How maintained—by continual pull of muscles on bones in opposite direction from pull of gravity—i.e., posture maintained by continued partial contraction of muscles, or muscle tone; therefore, indirectly dependent on many other factors—e.g., normal nervous, respiratory, and circulatory systems, health in general
3 Importance to body as whole—essential for optimal functioning of most of body—e.g., respiration, circulation, digestion, joint action, etc.; briefly, maximal health dependent upon good posture, good posture dependent upon health

Review questions

1 Differentiate between the three kinds of muscle tissue as to structure, location, and innervation.
2 Describe several physiological properties of muscle tissue.
3 What property is more highly developed in muscle than in any other tissue?
4 State a principle describing the usual relationship between a part moved and the location of muscles (insertion, body, and origin) moving the part.
5 Applying the principle stated in question 4, where would you expect muscles that move the head to be located? Name two or three muscles that fulfill these conditions.
6 Applying the principle stated in question 4, what part of the body do thigh muscles move? Name several muscles that fulfill these conditions.
7 What bone or bones serve as a lever in movements of the forearm? What structure constitutes the fulcrum for this lever?
8 Explain the meaning of the term neuromusculoskeletal unit.
9 Name the main muscles of the back, chest, abdomen, neck, shoulder, upper arm, lower arm, thigh, buttocks, leg, and pelvic floor.
10 Name the main muscles that flex, extend, abduct, and adduct the upper arm; that raise and lower the shoulder; that flex and extend the lower arm; that flex, extend, abduct, and adduct the thigh; that flex and extend the lower leg and thigh; that flex and extend the foot; that flex, extend, abduct, and adduct the head; that move the abdominal wall; that move the chest wall.
11 Discuss the chemical reactions thought to make available energy for muscle contraction.
12 What physiological reason can you give for athletes using a warming-up period before starting a game?
13 Why does an individual pant after strenuous exercise?
14 Curare preparations are often given during surgery. Would you expect this to make the patient's muscles more relaxed or more rigid? Why?
15 In general, where are bursae located? Give several specific locations.
16 Name several weak places in the abdominal wall where hernia may occur.
17 What and where are the inguinal canals? Of what clinical importance are they?
18 Good posture depends upon tonicity of the antigravity muscles, particularly of those which hold the head and trunk erect and the abdominal wall pulled in. Name several muscles that perform these functions.
19 Define the following terms: aponeurosis, bursa, contraction, contracture, elasticity, extensibility, fibrillation, insertion, motor unit, origin, oxygen debt, tetanus, tone, treppe, twitch.
20 Explain briefly the current theory about the role of calcium ions in muscle contraction and relaxation.

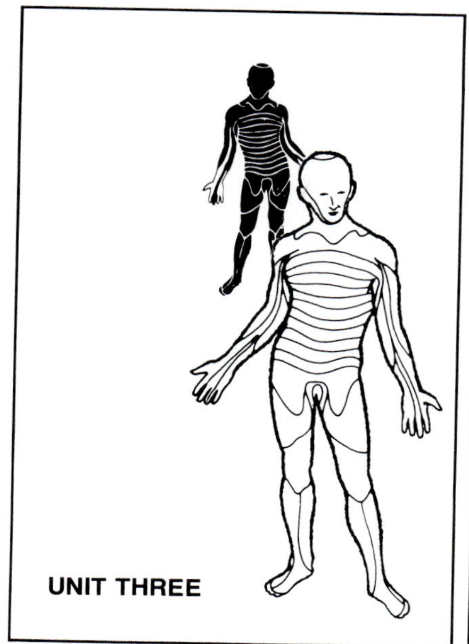

HOW THE BODY CONTROLS AND INTEGRATES ITS FUNCTIONS

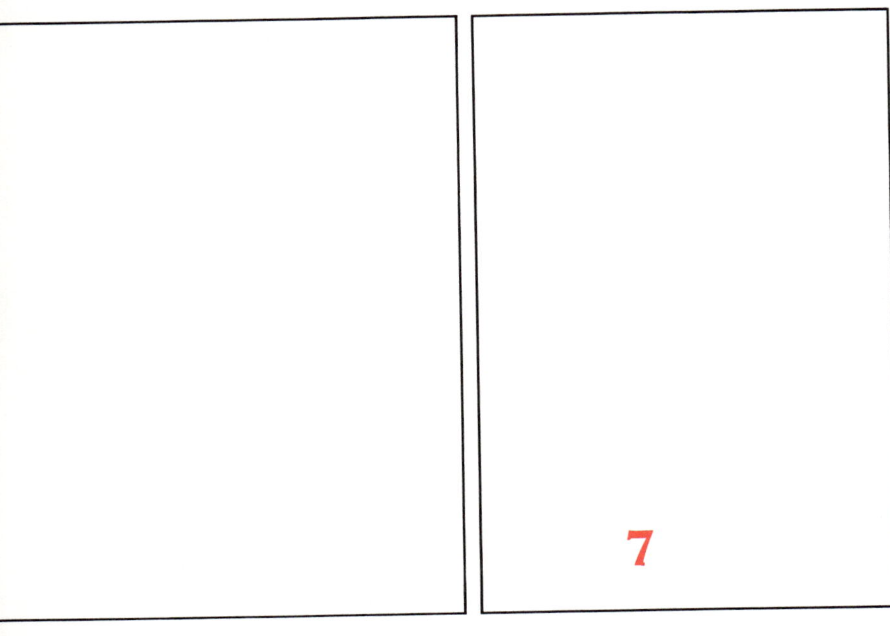

7

Nervous system cells

Cells
Neuroglia
- Types
- Structure and function

Neurons
- Types
- Structure
- Function
- Definitions

Nerve impulse (action potential)
- Definition
- Mechanism

Stimulus
- Impulse conduction routes
- Chemical conduction across synapses and neuroeffector junctions
  - Transmitters

If you want to understand the body, you need to remind yourself frequently of some principles stated in the first chapter of this book—briefly, that the body is made up of millions of smaller structures that carry on a host of different activities. But, and this is the important point, together all of these diverse activities accomplish the one big, all-encompassing function of the body—survival. Think about this for a moment. How can many parts of any kind—whether they be cells, people, or parts of a machine—be made to accomplish any one big function? How can many be made one? How can unification (integration) be achieved? Communication and control are the keys that unlock the secret of integration. Communication makes possible control, and control makes possible integration. Many familiar examples of this principle suggest themselves. A hospital, to name just one example, is a functional unit. Quite obviously, the activities of the hundreds of individuals who make up a modern hospital must be organized, coordinated, and integrated. Communication between the individual workers is what makes possible this organization, coordination, and integration. Without communication, chaos would prevail and the hospital would not survive. In no time at all it would be utterly unable to carry on its one great function of giving care to patients.

Nerve impulses and chemicals—chiefly hormones, carbon dioxide, and certain ions—constitute the two kinds of "messages" or communications sent within the body. The nervous system and the endocrine system constitute its two main communication systems. The circulatory system serves as an assistant communication system by distributing hormones and other chemical messages.

Facts, theories, and questions about the nervous system are as abundant and complex as they are fascinating. We shall approach this large body of material by considering, in this chapter, the cells of the nervous system and the mechanism of nerve impulse conduction. Chapter 8 presents information about the somatic nervous system, Chapter 9 discusses the autonomic nervous system, and Chapter 10 deals with the special senses. Discussion of the endocrine system appears in Chapter 11.

Cells

Two main kinds of cells compose nervous system structures—neurons and neuroglia. Neurons are the "specialists" of the nervous system; they specialize in impulse conduction, the function that makes possible all other nervous system functions. Neuroglia, on the other hand, perform the less specialized functions of support and protection.

Neuroglia

Types

Neuroglia are interesting cells, perhaps because their functions still remain somewhat of a mystery. Assuredly, however, some of them support neurons and anchor them to blood vessels, whereas others serve a defense or protective function. Neuroglia are important clinically because most tumors of the nervous system arise from them. Histologists identify these major types: astrocytes, oligodendroglia (or oligodendria), and microglia (Fig. 7-1).

Structure and function

Astrocytes are star-shaped cells with numerous processes. Some astrocytes lie between nerve fibers in the brain and cord, binding them to each other. Other astrocytes in these central nervous system organs lie between nerve cells and blood vessels. Little "sucker feet" attach them to adjacent blood vessels, thereby holding nerve cells close to their blood vessels.

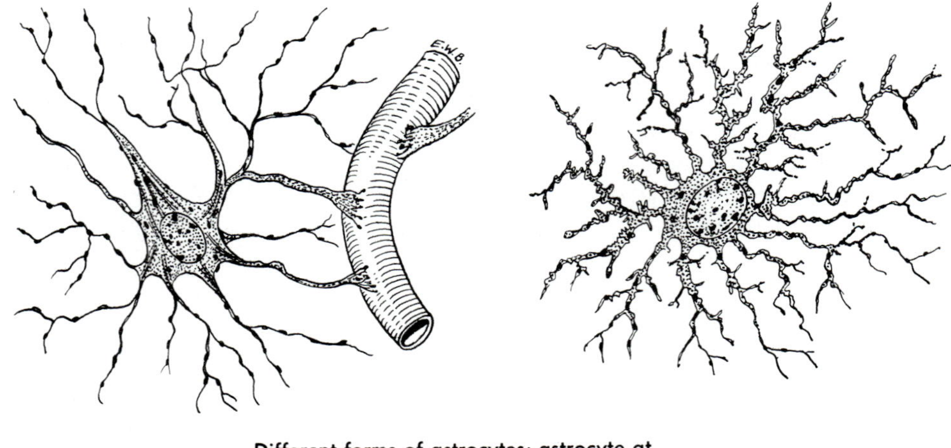

Different forms of astrocytes; astrocyte at left illustrated with footplates against blood vessel

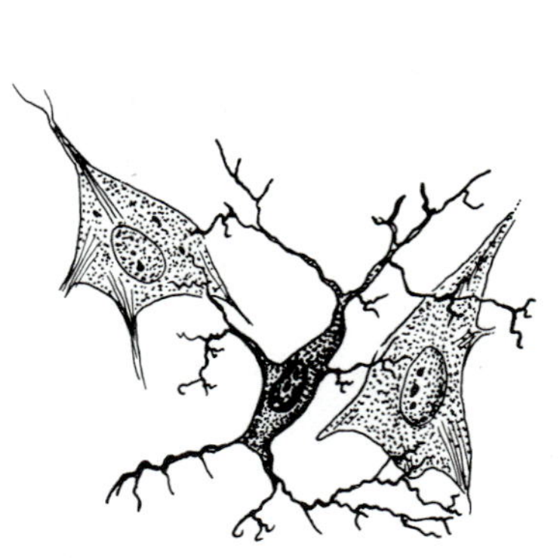

Microglia with processes extending to two nerve cell bodies

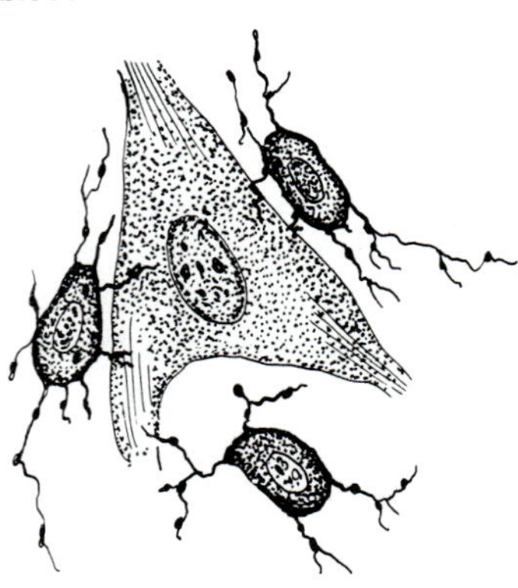

Oligodendria located near a nerve cell body

Fig. 7-1

Neuroglia, special connecting, support cells of the brain and cord.

Oligodendria are smaller cells and have fewer processes than astrocytes. Some oligodendria lie clustered around nerve cell bodies, some are arranged in rows between nerve fibers in the brain and cord. Like the astrocytes, these oligodendria help hold nerve fibers together. But oligodendria may also serve another important function. Recent evidence seems to indicate that they produce the fatty myelin sheath that envelops nerve fibers located in the brain and cord.

Microglia are small, usually stationary cells. In inflamed or degenerating brain tissue, however, microglia enlarge, move about, and carry on phagocytosis. In other words, they engulf and destroy microbes and cellular debris.

Neurons

Types

Neurons are classified according to two different criteria—the direction in which they conduct impulses and the number of processes they have.

Classified according to the direction in which they conduct impulses, there are three types of neurons: sensory, motor, and interneurons. *Sensory (afferent) neurons* transmit nerve impulses to spinal cord or brain. *Motoneurons (motor, or efferent neurons)* transmit nerve impulses away from brain or spinal cord to or toward muscle or glandular tissue. *Interneurons (internuncial or intercalated neurons)* conduct impulses from sensory to motor neurons. Interneurons lie entirely within the central nervous system (brain and spinal cord).

Classified according to the number of their processes, there are also three types of neurons: multipolar, bipolar, and unipolar. *Multipolar neurons* have only one axon but several dendrites. Most of the neurons in the brain and spinal cord are multipolar. *Bipolar neurons* have only one axon and also only one dendrite and are the least numerous kind of neuron. They are found in the retina of the eye, for example, and in the spiral ganglion of the inner ear. *Unipolar neurons* originate in the embryo as bipolar neurons. But in the course of development, their two processes become fused into one for a short distance beyond the cell body. Then they separate into clearly distinguishable axon and dendrite. Sensory neurons, like the one shown in Fig. 7-2, are usually unipolar. How would you classify the interneuron shown in this figure? Is it unipolar, bipolar, or multipolar? How would you classify the motoneuron?

Structure

All neurons consist of a cell body (also called the *soma*, or the *perikaryon*) and at least two processes: one axon and one or more dendrites. Because dendrites and axons are threadlike extensions from a neuron's soma (its cell body), they are often called *nerve fibers*.

Clusters of neuron cell bodies have a slightly gray color. *Gray matter* in the brain and spinal cord, for example, consists largely of neuron cell bodies. In many respects, the cell body, the largest part of a nerve cell, resembles other cells. It contains a nucleus, cytoplasm, and various organelles found in other cells—for example, mitochondria and a Golgi apparatus. Incidentally, Golgi first saw this apparatus in neurons. A neuron's cytoplasm extends from its cell body into its processes. A plasma membrane encloses the entire neuron.

Certain structures—dendrites, axons, neurofibrils, Nissl bodies, myelin sheath, and neurilemma—are found only in neurons. The following paragraphs describe them briefly.

Dendrites, as you can see in Fig. 7-2, branch extensively, like tiny trees. In fact, their name derives from the Greek word for tree. The distal ends of dendrites of sensory neurons are called *receptors* because they receive the stimuli that initiate conduction. Dendrites conduct impulses to the cell body of the neuron.

The *axon* of a neuron is a single process that extends out from the neuron cell body. Although a neuron has only one axon, it often has one or more side branches *(axon collaterals)*. Moreover, axons terminate in many branched filaments, and, like dendrites, they vary considerably in length. Some are as much as 3 feet or more long. Others measure only a fraction of an inch. Axons vary also in diameter—a point of interest because it relates to velocity of impulse conduction. In general, fibers with a large diameter conduct more rapidly than those with a small diameter. A neuron's axon conducts impulses away from its cell body.

Neurofibrils are very fine fibers extending through dendrites, cell bodies, and axons. The electron microscope has revealed that bundles of neurofibrils interlace to form a network in neuron cytoplasm.

Nissl bodies consist of groups of flat, mem-

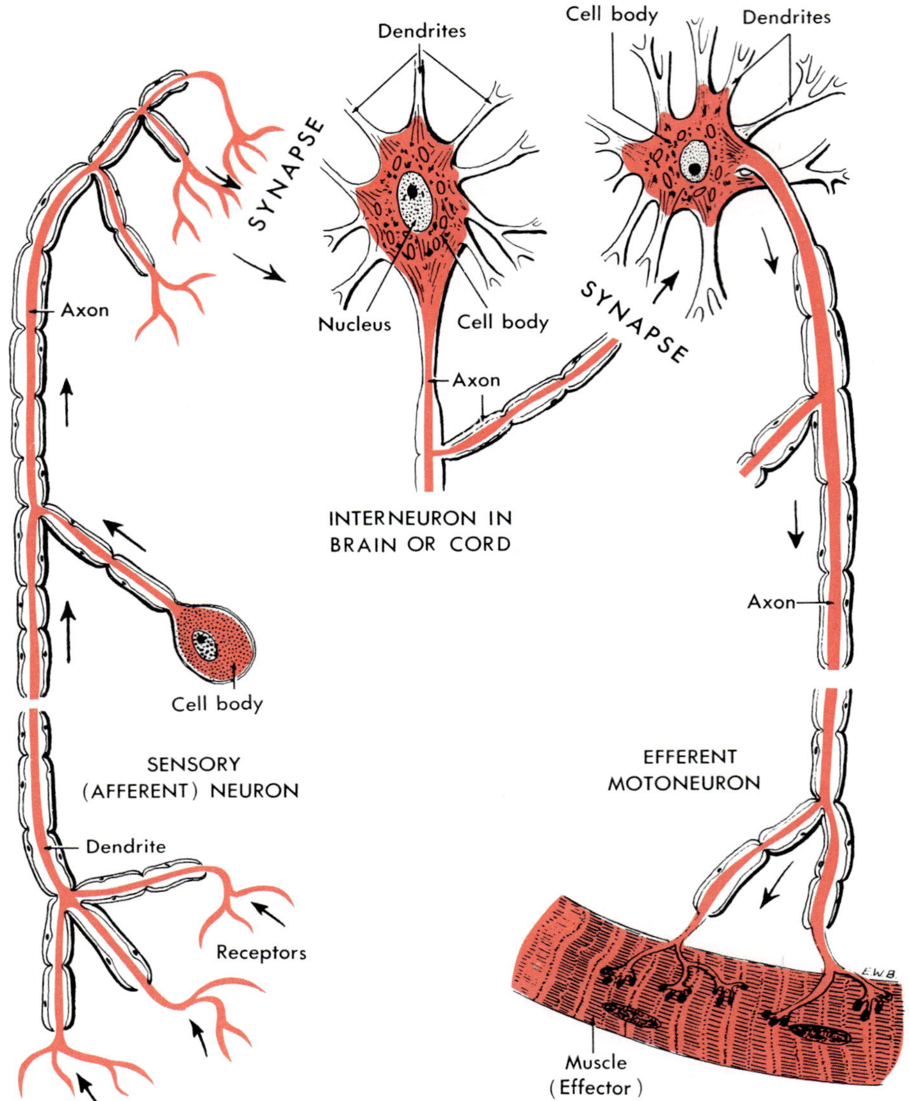

Fig. 7-2

Diagrammatic representation of structure of three kinds of neurons. Note that each neuron has three parts: a cell body and two types of extensions, dendrite(s) and an axon. Arrows indicate direction of impulse conduction. (See Fig. 7-3 for more details about coverings of processes.)

branous sacs and numerous RNA granules scattered between them. In other words, Nissl bodies constitute the rough endoplasmic reticulum of a neuron. Since this organelle specializes in protein synthesis, one would expect Nissl bodies to perform the same function. But why should neurons, whose specialty is conduction, also specialize in protein synthesis? According to one speculation, they use the protein they make for maintaining and regenerating neuron processes. Radioautographic studies have shown that proteins synthesized in neuron cell bodies quickly migrate down their axons.

The *myelin sheath* is a segmented wrapping around a nerve fiber. Note in Fig. 7-3 the small gaps called *nodes of Ranvier* between segments of the sheath. Because of the relatively high fat content in myelin, bundles of myelinated fibers appear creamy white in color. *White matter* in the brain and spinal cord consist of bundles of myelinated fibers that are called *tracts* (discussed in Chapter 8). *Nerves,* like tracts, consist of bundles of myelinated fibers and are white, but they are located outside of the brain and spinal cord, not in them as are tracts.

The *neurilemma* is a continuous sheath that encloses the segmented myelin sheath of peripheral nerve fibers. (Peripheral nerve fibers are those located outside of the central nervous system—in other words, outside of the brain and cord.) *Schwann cells* wind themselves in jelly-roll fashion around the myelin sheaths of peripheral fibers, with one Schwann cell wrapping around each segment of a myelin sheath. Thus together the several successive Schwann cells form the thin, continuous sheath called either the *sheath of Schwann* or the *neurilemma.* The neurilemma plays an essential part in peripheral nerve fiber regeneration. Brain and spinal cord fibers unfortunately do not, so far as we know, have a neurilemma, nor do they regenerate if disease or injury destroys them.* Peripheral nerve fibers, in contrast, can regenerate.

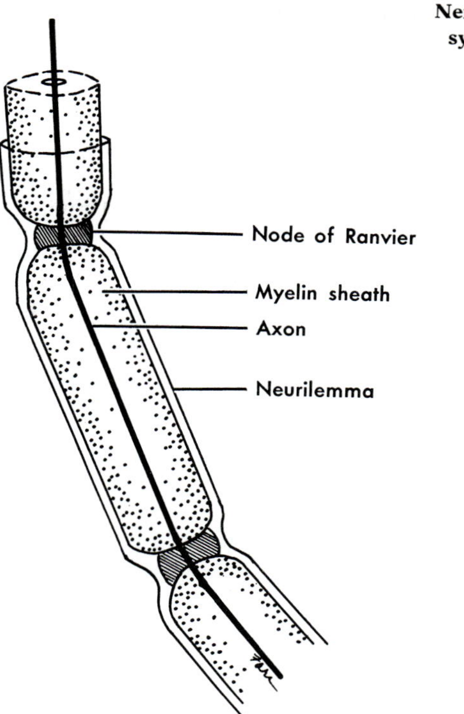

Fig. 7-3

Diagram of a nerve fiber and its coverings. Note these features of the myelin sheath. It surrounds the nerve fiber—the axon in this figure—and it is segmented. One segment is the section of myelin sheath located between two successive nodes of Ranvier. The neurilemma surrounds the myelin sheath and is continuous, not segmented.

Function

Neurons perform the specific function of conducting impulses and the general function of providing the body with its most rapid means of communication and integration. What is a nerve impulse? What produces it? These may seem like simple enough questions, but the answers to them are far from simple. Not only are they highly complex, they are also still incomplete and partly tentative. Therefore, we shall relate only some of the more widely accepted answers, but we shall try to do so as briefly, clearly, and accurately as we can. First, however, you will need to know what the following terms mean: resting state, polarized, potential difference, depolarized, and resting potential.

Definitions

When a neuron is not busy conducting impulses, it is said to be in a *resting state* and its plasma membrane is said to be polarized.

*Recently, a few scientists have presented findings that suggest that regeneration of CNS neurons may, under some circumstances, be possible. See Culliton, B. J.: Spinal cord regeneration, Sci. News **98**:337-338, Oct. 24, 1970.

A *polarized membrane* is one whose outer surface bears a different amount of electrical charge than its inner surface. The difference between two amounts of electrical charges is called a *potential difference*. Therefore another definition for polarized membrane is a membrane with a potential difference between its inner and outer surfaces. A *depolarized membrane* is the opposite, that is, one with no potential difference between its inner and outer surfaces. The potential difference that exists across the membrane of a neuron in the resting state is called, logically, the *resting potential*. The magnitude of the resting potential varies but usually ranges between 70 and 90 millivolts (mv). The inner surface of the resting neuron's membrane is 70 to 90 mv negative to its outer surface.

The complex mechanism that maintains the resting potential is not yet fully understood. A complete explanation of what is known is beyond the scope of an introductory physiology textbook. We shall, therefore, attempt only a brief explanation, knowing that it omits much of today's knowledge about this crucially important mechanism and necessarily oversimplifies it.

The direct cause of the resting potential is the difference in concentrations of positively and negatively charged ions in extracellular and intracellular fluids. As you know, the law of diffusion dictates that the net diffusion of any particular ion occurs down its own concentration gradient. Another physical law requires that ions move down an electrical gradient—that is, from a more to a less electropositive area. Net diffusion of ions across the cell membrane, therefore, should equalize the ionic concentrations of extracellular and intracellular fluids and should equalize the electrical charges on both sides of the membrane. No potential difference would then exist across it. In the normal nonconducting cell, however, these equalizations do not develop. Activity of the living cell membrane prevents them from taking place. An active process within the membrane—still not completely understood—operates continuously to pump sodium ions out of the cell. This maintains a difference between extracellular and intracellular sodium concentrations. It keeps extracellular fluid sodium concentration many times higher than that of intracellular fluid. Moreover, by maintaining an ionic concentration difference between the two fluids, the sodium pump also maintains a difference in electrical potential across the cell membrane. It keeps the outer surface of a nonconducting cell membrane positive to its inner surface. The sodium pump, in other words, polarizes cell membranes.* To deduce how important living processes are in maintaining these differences across cell membranes, you need consider only one fact—that across the membranes of dead cells, no differences in ionic concentration nor in electrical potential exist.

Nerve impulse (action potential)

Definition

According to widely accepted present-day theory, a nerve impulse is a self-propagating wave of electrical negativity that travels along the surface of a neuron's plasma membrane. One can easily demonstrate the passage of a nerve impulse by measuring the voltage on the outer surface of a neuron's membrane. During the moment the impulse passes, the voltage becomes negative. Bernstein, in 1902, postulated the reason for the change in voltage. But more than 30 years

*If you would like to delve further into the complex physiology of membrane polarization, see Mountcastle, V. B., editor: Medical physiology, ed. 13, St. Louis, 1974, The C. V. Mosby Co., pp. 3-75.

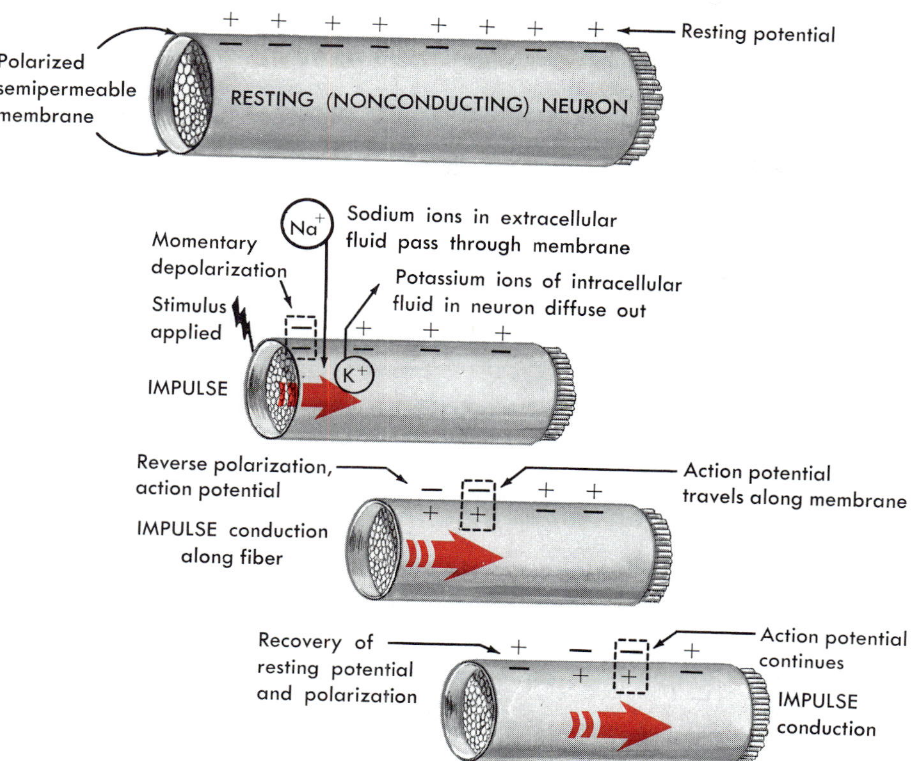

Fig. 7-4

Upper diagram represents polarized state of the membrane of a nerve fiber when it is not conducting impulses. Lower diagrams represent nerve impulse conduction—a self-propagating wave of negativity or action potential travels along membrane.

passed before the experimental methods needed to prove or disprove his theory became available and were used by Hodgkin and associates. The following paragraphs give a step-by-step description of the mechanism of the nerve impulse. As you read them, refer often to Fig. 7-4.

Mechanism

1 When an adequate stimulus is applied to a neuron, it greatly increases the membrane's permeability to sodium ions at the point of stimulation.

2 Sodium ions rush into the cell at the stimulated point. Therefore, at this point, the excess of positive ions outside rapidly dwindles to zero. Therefore, also, the membrane potential decreases to zero at this point. In other words, the stimulated point of the membrane is no longer polarized. It is *depolarized,* but only for an instant. As more sodium ions continue to stream into the cell, they almost instantaneously produce an excess of positive ions inside the cell and leave an excess of negative ions outside. In short, the influx of sodium ions reverses the resting potential, and thereby changes it into an action potential. With a typical resting potential, the inner surface of a neuron's membrane is, for example, 70 mv negative to its outer surface. In contrast, with a typical action potential, the inner surface of the neuron's membrane is about 30 mv positive to its outer surface.

3 The stimulated negatively charged point of the membrane sets up a local current with the positive point adjacent to it, and this local current acts as a stimulus. Consequently, within a fraction of a second, the adjacent point on the membrane becomes depolarized and its potential reverses from positive to negative. The action potential has thus moved from the point originally stimulated to the adjacent point on the membrane. As the cycle goes on repeating itself over and over again in rapid succession, the action potential travels point by point out the full length of the neuron. Much as a wave of water moves in from sea to shore, so an action potential moves along a neuron's membrane from its point of stimulation (on dendrites or cell body) to its axon endings. From the facts stated in this paragraph came the definition of the *action potential* (or its synonym, *nerve impulse*) as a self-propagating wave of negativity that travels along the surface of a neuron's plasma membrane.

4 By the time the action potential has moved from one point on the membrane to the next (a matter of thousandths of a second), the first point has repolarized—its resting potential has been restored. *Repolarization* results from the fact that the increased permeability to sodium induced by stimulation lasts only momentarily. It is quickly replaced by increased permeability to potassium, which therefore diffuses outward (because potassium concentration inside the cell is much greater than that outside the cell). Then the permeability of the membrane decreases and its pumping activity again becomes effective. Once again, it actively transports sodium ions out of the cell and potassium ions into it. Once again—and in much less time than it takes to tell it—more positive ions accumulate on the outer than on the inner surface of the plasma membrane. The resting potential of the cell has, in other words, been restored. The outer surface of its membrane is again electrically positive to its inner surface. (Or, stated differently, the inner surface of the neuron's membrane again becomes electrically negative to its outer surface.)

How fast does a neuron conduct impulses? It all depends—upon the diameter of its axon, the thickness of its axon's myelin sheath, and the distance between the myelin's nodes of Ranvier. The larger an axon's diameter, the faster its conduction. Fibers with a large diameter (A fibers) conduct most rapidly. Impulses travel along them at a speed of about 100 meters per second, or more than 3 miles per minute. Fibers with a small diameter (C fibers) conduct most slowly, about 0.5 meter per second, or 1 mile per hour. Fibers with a diameter of intermediate size (B fibers) conduct at speeds between those of the large, fast A fibers and the small, slow C fibers.

Heavily myelinated fibers with a greater distance between successive nodes conduct, according to another principle, many times faster than do so-called unmyelinated fibers with a lesser distance between nodes. ("Unmyelinated" fibers are not really unmyelinated. They have, we know now, a thin layer of myelin around them.)

Stimulus

A stimulus is a change in the environment. For example, common kinds of stimuli are changes in the pressure, temperature, or chemical composition of either the body's external or internal environment. When a stimulus acts on a neuron, the mechanism explained on p. 165 operates to bring about a decrease in that neuron's membrane potential. Only if its potential decreases down to a certain critical level, known as the threshold of stimulation, is impulse conduction triggered. In short, an action potential is initiated by a decrease in a neuron's membrane potential from its resting level to

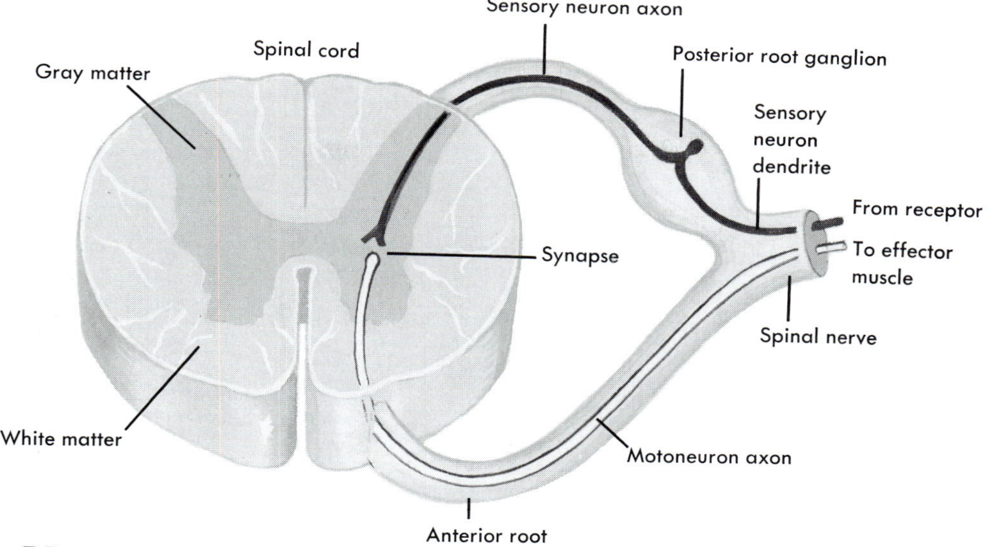

Fig. 7-5

A two-neuron ipsilateral reflex arc; also called a monosynaptic arc because impulses cross only one synapse in traversing it. In ipsilateral arcs, afferent impulses enter the cord and efferent impulses leave it on the same side. Conduction over such arcs produces stretch reflexes such as the knee jerk.

its threshold of stimulation. Once initiated, the action potential then rapidly propagates itself the full length of the neuron. Whether or not a stimulus initiates conduction depends upon the intensity or "strength" of the stimulus. A stimulus just strong enough to decrease the membrane potential to its threshold is just strong enough to initiate impulse conduction and is called a *threshold stimulus* (or liminal stimulus). Any stimulus weaker than this is a *subthreshold stimulus* (or subliminal stimulus). It decreases the membrane potential, but not down so low as the threshold level, so it does not trigger off impulse conduction. The effects of two or more subthreshold stimuli, however, can add together or "summate." By summation, they may decrease the membrane potential to its threshold level and thus trigger impulse conduction.

Impulse conduction routes

The route traveled by many nerve impulses is the one known as the *reflex arc*. Basically, the reflex arc route consists of two or more neurons, arranged in series, that conduct impulses from the periphery to the central nervous system and back out to the periphery. Periphery, as used in the preceding sentence, means any place in the body outside of the central nervous system *(spinal cord and brain)*. Impulse conduction over a reflex arc begins in receptors and ends in effectors. Look at Fig. 7-2, and you will see that receptors are the beginnings—or in technical terms, the distal ends—of a sensory neuron's dendrite. The effector shown in this figure is striated or skeletal muscle. The only other kinds of effectors are smooth (visceral) muscle and glandular tissues.

A diagram of the simplest kind of reflex arc appears in Fig. 7-5. It has two names. The first—*two-neuron ipsilateral reflex arc*—indicates that it consists of only two kinds of neurons (sensory and motor); the term ipsilateral means that both neurons terminate on the same side of the body. The second name for this kind of arc, *mono-*

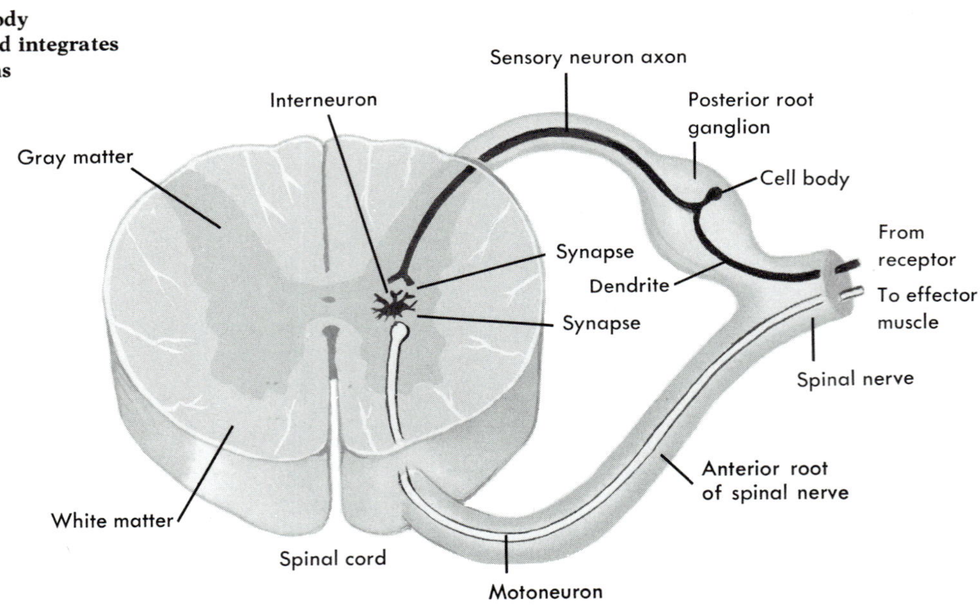

Fig. 7-6

Three-neuron ipsilateral reflex arc, consisting of a sensory neuron, an interneuron, and a motoneuron. Note the presence of two synapses in this arc: (1) between sensory neuron axon terminals and interneuron dendrites and (2) between interneuron axon terminals and motoneuron dendrites and cell bodies (located in anterior gray matter). Nerve impulses traversing such arcs produce many spinal reflexes. Example: withdrawing the hand from a hot object.

synaptic reflex arc, indicates that only one synapse occurs in the arc. A *synapse* is the place where a nerve impulse is transmitted from one neuron to another. Note the location of the synapse shown in Fig. 7-5. It lies in gray matter of the spinal cord. All the synapses in reflex arcs whose motoneurons conduct impulses to skeletal muscles lie in either spinal cord or brain gray matter.

A three-neuron reflex arc consists of three kinds of neurons—sensory neurons and motoneurons with interneurons between them. As you can see in Fig. 7-6, there are two synapses in a three-neuron arc. Now examine Fig. 7-7 to find out how a contralateral arc differs from an ipsilateral arc.

Besides simple two-neuron and three-neuron arcs, intersegmental arcs (Fig. 7-8) and even more complex multineuron, multisynaptic arcs also exist. All impulses that start in receptors, however, do not invariably travel over a complete reflex arc and terminate in effectors. Many impulses fail to be conducted across synapses. Moreover, all impulses that terminate in effectors do not invariably start in receptors. Many of them, for example, are thought to originate in the brain.

Chemical conduction across synapses and neuroeffector junctions

A description of the mechanism of impulse conduction as a self-propagating wave of electrical negativity appeared earlier in this chapter. We noted that the impulse traveled the length of a neuron's plasma membrane all the way to the branching ends of its axon. All axons terminate close by other structures. They end close to other neurons or close to effector cells (muscle or glandular cells). An

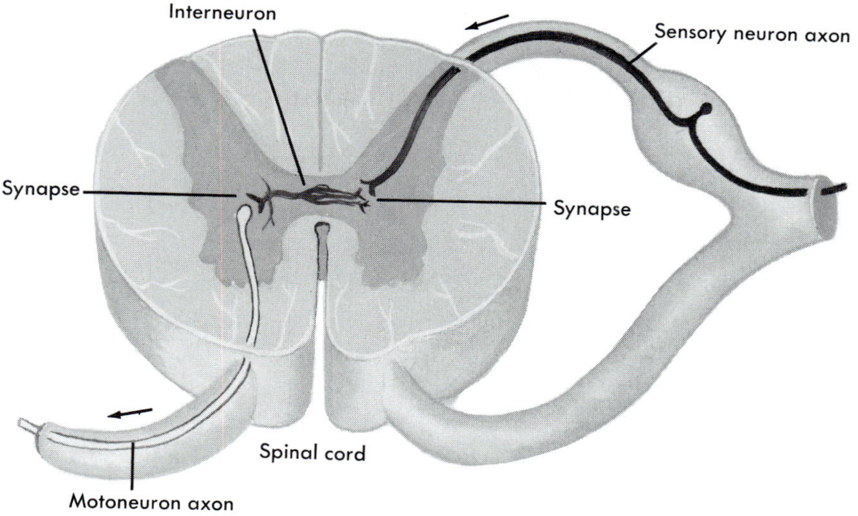

Fig. 7-7

Three-neuron contralateral reflex arc. Afferent impulses enter the cord on one side and efferent impulses leave it on the other side. Figs. 7-5 and 7-6 show ipsilateral arcs (impulses enter and leave the cord on the same side).

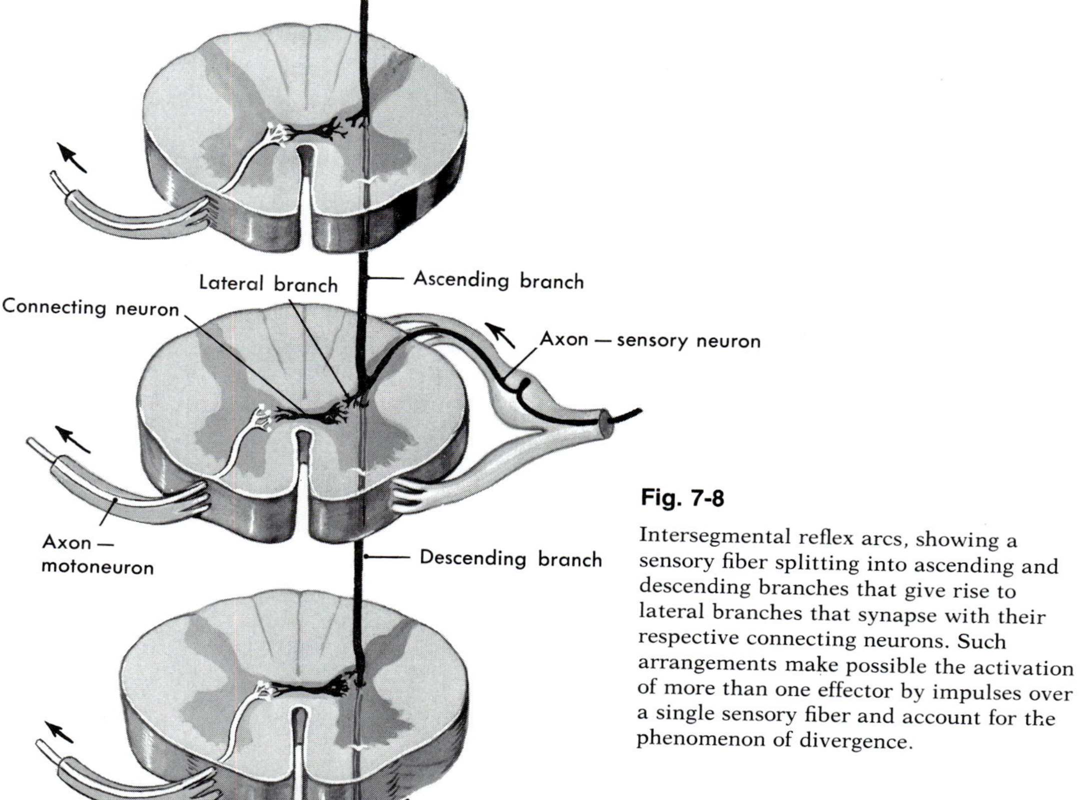

Fig. 7-8

Intersegmental reflex arcs, showing a sensory fiber splitting into ascending and descending branches that give rise to lateral branches that synapse with their respective connecting neurons. Such arrangements make possible the activation of more than one effector by impulses over a single sensory fiber and account for the phenomenon of divergence.

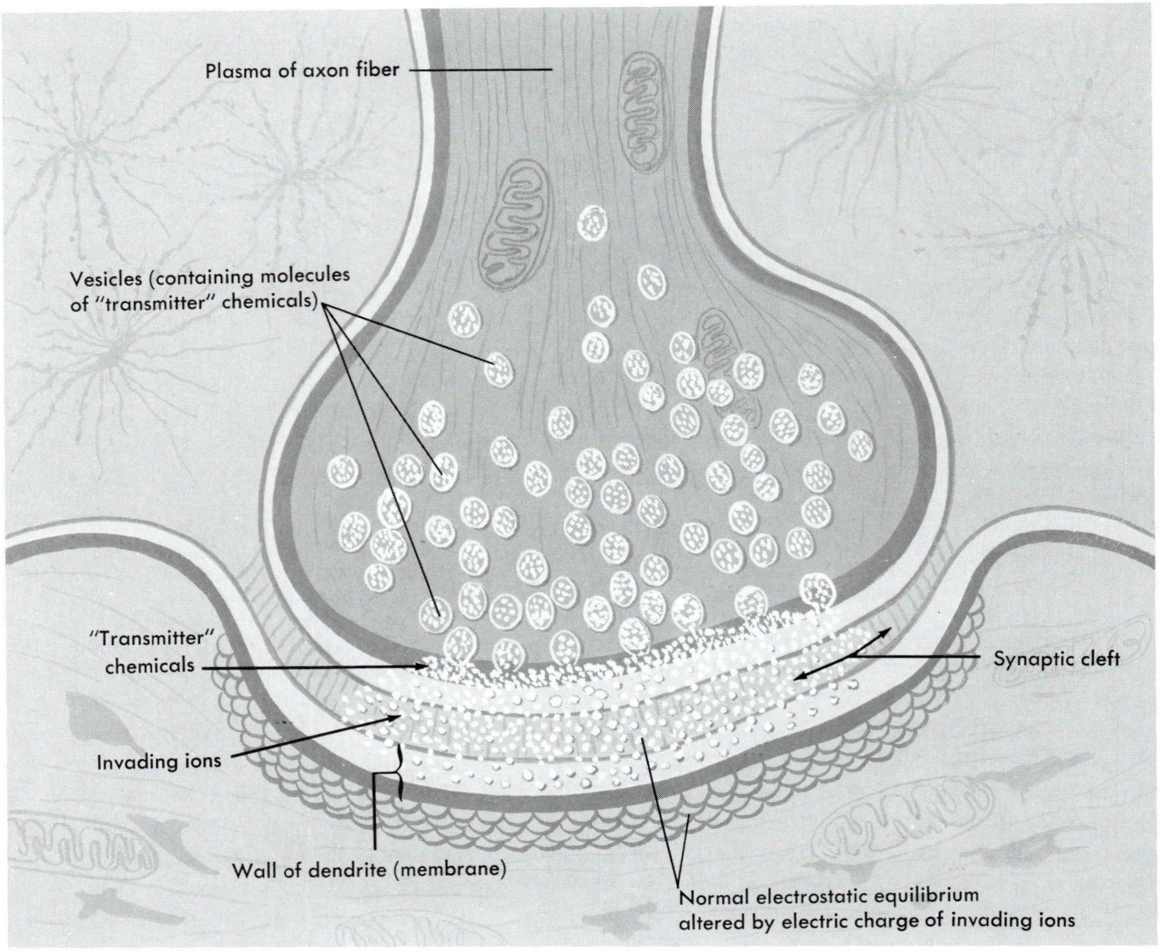

Plasma of axon fiber
Vesicles (containing molecules of "transmitter" chemicals)
"Transmitter" chemicals
Synaptic cleft
Invading ions
Wall of dendrite (membrane)
Normal electrostatic equilibrium altered by electric charge of invading ions

Fig. 7-9

Diagram showing release of transmitter molecules from vesicles in presynaptic axon terminal into synaptic cleft and moving across this microscopic space to contact the membrane of the postsynaptic dendrite.

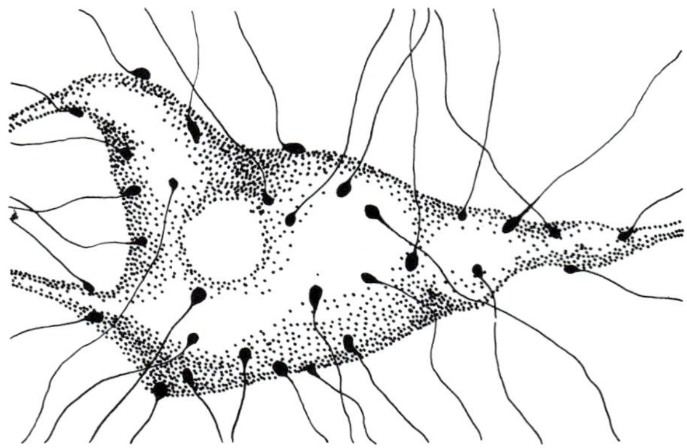

Fig. 7-10

Synaptic knobs (end feet or end buttons) on motoneuron cell body.

extremely narrow gap, the *synaptic cleft,* 200 to 300 Å or about one millionth of an inch wide, separates the branching ends of an axon from other neurons. A similar narrow gap separates the axon terminals of motoneurons from muscles or glandular cells. It is called the *neuroeffector junction.* How do impulses cross these synaptic clefts and neuroeffector junctions? Scientists have proposed various answers to this question over the years. At one time they thought that the same electrical type mechanism that conducts impulses along a single neuron also conducted them across all synapses and neuroeffector junctions. Today they know that this is not true. Conclusive evidence has shown that a chemical mechanism transmits impulses across virtually all synapses* and all neuroeffector junctions in the human body.

For our discussion of chemical conduction, let us assume that the conduction occurs from the axon of neuron A across a synapse to the dendrites or cell body of neuron B. But instead of calling them neurons A and B, we shall use their more descriptive, technical names: *presynaptic neuron* for neuron A and *postsynaptic neuron* for neuron B. The axon of the presynaptic neuron branches to form fine filaments that end in little knobs called *synaptic knobs* (or boutons). Crowded inside each knob are numerous mitochondria plus hundreds of tiny vesicles (closed sacs). A single vesicle contains a few thousand molecules of a chemical known as a neurohumor, a neurotransmitter, or, simply, a *transmitter.* When an action potential (nerve impulse) reaches the presynaptic axon terminals, some of the vesicles, as Fig. 7-9 shows, release transmitter molecules into the synaptic cleft. They diffuse almost instantaneously across its infinitesimal width and contact the postsynaptic neuron's membrane. Here, depending upon the chemical identity of the transmitter, it may produce either of two opposite effects: excitatory or inhibitory. *Excitatory transmitters* decrease the postsynaptic membrane's potential. *Inhibitory transmitters* increase its potential. In most cases, a single postsynaptic neuron receives transmitters from many presynaptic axons. Moreover, some of these axons release excitatory and some release inhibitory transmitters into the synaptic cleft. Summation of their opposite influences produces one of three changes in the postsynaptic neuron, namely, facilitation, impulse conduction, or inhibition.

Facilitation is a decrease in the postsynaptic neuron's membrane potential, changing it from the usual resting potential to what is called the *excitatory postsynaptic potential* (commonly abbreviated to EPSP). Suppose a postsynaptic neuron's resting potential is 70 mv and its threshold of stimulation is 50 mv, then 60 mv would be an example of an EPSP for that neuron. Note the relationship between the EPSP and the resting potential and threshold of stimulation. It lies somewhere between them. The EPSP is always lower than the neuron's resting potential but higher than its threshold of stimulation. And because the EPSP is higher than the threshold of stimulation, it does not trigger impulse conduction by the postsynaptic neuron. It "facilitates" it, that is, makes it ready for a weaker subsequent stimulus to initiate conduction by it. An EPSP is produced in a postsynaptic neuron when the amount of excitatory transmitter released into a synaptic cleft by presynaptic axons slightly exceeds the amount of inhibitory transmitter released by other presynaptic axons into that cleft.

Impulse conduction by a postsynaptic neuron is initiated when the amount of excitatory transmitter exceeds the amount of inhibitory transmitter released by presynaptic axons sufficiently to decrease the

*Mountcastle, V. B., editor: Medical physiology, ed. 13, St. Louis, 1974, The C. V. Mosby Co., pp. 151, 190.

postsynaptic neuron's resting potential down to its threshold of stimulation—for example, from a resting potential of 70 mv (inner surface of neuron's membrane negative to its outer surface) to a threshold of stimulation of 50 mv. Instantaneously, when the potential reaches the critical threshold level, it then drops explosively to 0 and on down until the inner surface of the membrane becomes about 30 mv positive to its outer surface. At this point, it becomes an action potential and impulse conduction by the postsynaptic neuron begins.

Inhibition of postsynaptic neurons occurs when the amount of inhibitory transmitter released by presynaptic axons exceeds the amount of excitatory transmitter released by other presynaptic axons into the same synaptic cleft. Inhibition produces an *inhibitory postsynaptic potential* (IPSP) in the postsynaptic neuron. The IPSP is higher than the usual resting potential—for example, 80 mv instead of 70 mv. When an IPSP exists in a neuron, it is said to be inhibited.

Summarizing, dozens of synaptic knobs from dozens of presynaptic axons synapse with any one postsynaptic neuron, most commonly with its dendrites or soma. Some knobs release excitatory transmitter, some release inhibitory transmitter. The algebraic sum of these antagonistic chemicals determines their effect on the postsynaptic neuron. Facilitation, impulse conduction, or inhibition of the postsynaptic neuron may result from the summation of excitatory and inhibitory transmitters in a synaptic cleft.

Whereas synapses are regions of contact between presynaptic axon terminals and postsynaptic neurons, neuroeffector junctions are regions of contact between a motoneuron's axon terminals and an effector—muscle or glandular cells. Transmitter substances released at neuroeffector junctions stimulate muscle cells to contract and glandular cells to secrete.

TRANSMITTERS

Researchers have firmly established the chemical identity of two transmitters, namely, acetylcholine (ACh) and norepinephrine (NE). Acetylcholine, for example, is the known transmitter at neuroeffector junctions with skeletal muscle cells and at some neuroeffector junctions with cardiac muscle, smooth muscle, and glandular cells. Norepinephrine is the known transmitter at some neuroeffector junctions with cardiac muscle, smooth muscle, and glandular cells. Although not yet proved, the following substances have been fairly well established as transmitters in the brains of mammals*: serotonin (5-hydroxytryptamine, 5-HT), GABA (gamma-aminobutyrate), acetylcholine, and three compounds classified as *catecholamines*, namely, norepinephrine, epinephrine, and dopamine. The next two chapters relate more information about transmitters.

*Fernstrom, J. D., and Wurtman, R. J.: Nutrition and the brain, Sci. Am. **230**:84-91, Feb., 1974.

Outline summary

CELLS

Neuroglia

1 Types—astrocytes, oligodendria, and microglia
2 Structure and function
  a Astrocytes—star shaped; numerous processes twine around neurons and attach them to blood vessels; support neurons and may carry on phagocytosis
  b Oligodendria—fewer processes than other two types of neuroglia; support neurons, connect them to blood vessels; thought to produce myelin sheath
  c Microglia—small cells but enlarge and move about in inflamed brain tissue; carry on phagocytosis

Neurons

1 Types
  a Classified according to direction of impulse conduction
    1 Sensory or afferent—conduct impulses to cord or brain
    2 Motoneurons or efferent—conduct impulses away from brain or cord to or toward muscle or glandular tissue
    3 Interneurons (internuncial or intercalated neurons) —conduct from sensory to motoneurons

b Classified according to number of processes
  1 Multipolar—one axon and several dendrites
  2 Bipolar—one axon and one dendrite
  3 Unipolar—one process comes off neuron cell body but divides almost immediately into one axon and one dendrite

2 Structure—identify parts of neurons in Figs. 7-2 and 7-3

3 Terms
  a Soma—cell body of neuron
  b Gray matter—clusters of neuron cell bodies
  c Nerve fibers—dendrites or axons
  d White matter—bundles of myelinated nerve fibers
  e Nerves—bundles of myelinated nerve fibers located peripherally, i.e., outside of brain or cord
  f Tracts—bundles of myelinated nerve fibers located in the central nervous system, i.e., in brain or cord

4 Function—respond to stimulation by conducting impulses (body's most rapid means of communication)

5 Terms
  a Resting state—neuron not conducting impulses
  b Polarized membrane—one whose outer surface bears different electrical charge than its inner surface
  c Potential difference—difference in electrical charge
  d Depolarized membrane—one with no potential difference between its outer and inner surfaces
  e Resting potential—potential difference between outer and inner surfaces of membrane of nonconducting neuron; magnitude of resting potential ranges between 70 to 90 mv with inside of membrane negative to outside

NERVE IMPULSE (ACTION POTENTIAL)

1 Definition—self-propagating wave of electrical negativity that travels along surface of neuron membrane

2 Mechanism
  a Stimulus increases permeability of neuron membrane to sodium ions
  b Rapid inward diffusion of sodium ions causes inner surface of membrane to become positive to outer surface; this reversal of resting potential marks beginning of action potential or nerve impulse

3 Rate of conduction
  a Large diameter A fibers conduct at speed of about 100 meters per second (more than 3 miles per minute)
  b Small diameter C fibers conduct at speed of about 0.5 meter per second (1 mile per hour)
  c Intermediate diameter B fibers conduct at speeds intermediate between A and C fibers
  d Heavily myelinated fibers with greater distance between successive nodes of Ranvier conduct many times faster than thinly myelinated fibers

4 Definitions
  a Resting state—state of neuron when it is not conducting impulses
  b Polarized membrane—one whose outer and inner surfaces bear unequal amounts of electrical charge
  c Potential difference—the difference between two unequal amounts of electrical charge
  d Depolarized membrane—one with no potential difference between its outer and inner surfaces
  e Resting potential—the potential that exists between the outer and inner surfaces of membrane of nonconducting neuron

STIMULUS

1 Definition—a change in either external or internal environment

2 Action—see item 2 under Nerve impulse

3 Threshold (liminal) stimulus—just strong enough to decrease potential across neuron membrane from level of resting potential (e.g., inner surface of membrane 70 mv negative to outer) down to critical level (e.g., inner surface of membrane 30 mv positive to outer) that initiates acting potential

4 Subthreshold (subliminal) stimulus—decreases membrane potential below resting level but not down to threshold level

Impulse conduction routes

1 Many, but by no means all, impulses are conducted over route known as reflex arc, i.e., two or more neurons that conduct impulses from periphery to spinal cord or brainstem and back to periphery; impulse begins in receptors and ends in effectors; receptors—distal ends of sensory neurons; effectors—muscle or glandular cells

2 Two-neuron or monosynaptic reflex arc—simplest arc possible; consists of at least one sensory neuron, one synapse, and one motoneuron (synapse is contact region between axon terminals of one neuron and dendrites or cell body of another neuron)

3 Three-neuron arc—consists of at least one sensory neuron, synapse, interneuron, synapse, and motoneuron

4 Complex, multisynaptic neural pathways also exist; many not clearly understood

Chemical conduction across synapses and neuroeffector junctions

1 Terms
  a Synaptic cleft—space about 200 Å wide; separates terminals of an axon from dendrites, soma, or axon of another neuron
  b Neuroeffector junction—space comparable to synaptic cleft but separates motoneuron axon terminals from effector cells (muscle or glandular)
  c Presynaptic neuron—conducts before synapse (region of synaptic clefts)
  d Postsynaptic neuron—conducts after synapse
  e Synaptic knobs (boutons)—small rounded ends of axon terminals; many small vesicles, each of which contains a few thousand molecules of a chemical (neurotransmitter, transmitter), are located in boutons
  f Excitatory transmitters—decrease membrane potential of postsynaptic neuron
  g Inhibitory transmitters—increase membrane potential of postsynaptic neuron

2 Release of transmitters—arrival of action potential (nerve impulse) at presynaptic axon terminals causes release of transmitter molecules into synaptic cleft

3 Summation—dozens of presynaptic axons synapse with any one postsynaptic neuron; some of these presynaptic axons release excitatory transmitter, some release inhibitory transmitter into the same synaptic cleft; adding together, or summation, of these antagonistic chemicals determines effect on postsynaptic neuron (either facilitation, inhibition, or impulse conduction)

4 Facilitation—a decrease in the postsynaptic neuron's membrane potential to a level below its resting potential

but above its threshold of stimulation; this lowered potential called the EPSP (excitatory postsynaptic potential); result of summation with slight excess of excitatory transmitter in synaptic cleft; does not initiate impulse conduction by postsynaptic neuron

5 Inhibition—an increase in the postsynaptic neuron's membrane potential above its usual resting potential; this increased potential called the IPSP (inhibitory postsynaptic potential); result of summation with excess of inhibitory transmitter in synaptic cleft

6 Impulse conduction by postsynaptic neuron initiated when summation results in an excess of excitatory transmitter sufficient to decrease postsynaptic neuron's membrane potential to its threshold of stimulation; depolarization, followed by a reverse potential of about 30 mv; when inner surface of membrane is about 30 mv positive to outer surface, it has become an action potential (nerve impulse)

Transmitters

1 At neuroeffector junctions
   - a Acetylcholine (ACh)—known transmitter at all neuroeffector junctions with skeletal muscle cells, and at some neuroeffector junctions with cardiac muscle, smooth muscle, and glandular cells
   - b Norepinephrine (NE)—known transmitter at some neuroeffector junctions with cardiac muscle, smooth muscle, and glandular cells

2 At synapses in central nervous system, following substances are fairly well established, but not yet proved to be transmitters
   - a Serotonin (5-hydroxytryptamine, 5-HT)
   - b GABA (gamma-aminobutyrate)
   - c ACh (acetylcholine)
   - d Catecholamines—norepinephrine, epinephrine, and dopamine

Review questions

1 In a word or two, what general function does the nervous system perform for the body?
2 Name another system that serves the same general function.
3 Compare neurons and neuroglia.
4 Give two names for each of the three types of neurons and make an identifying statement about each type.
5 Define briefly each of the following terms: action potential, boutons, depolarized, effector, EPSP, facilitation, gray matter, inhibition, IPSP, nerve fiber, nerve impulse, neurilemma, neuroeffector junction, neurotransmitter, polarized, potential difference, receptors, tract, synapse, white matter.
6 Name two substances that have been firmly established as transmitters at neuroeffector junctions.
7 Name five substances postulated to serve as transmitters at synapses located within the brain and cord.
8 Where in a neuron are neurotransmitters stored?
9 What structures release neurotransmitters? Dendrites? Axons?
10 Into what structures do presynaptic axons release transmitters?
11 Into what structures do the axons of motoneurons release transmitters?
12 Explain briefly what summation of impulses means and what effects it may produce.

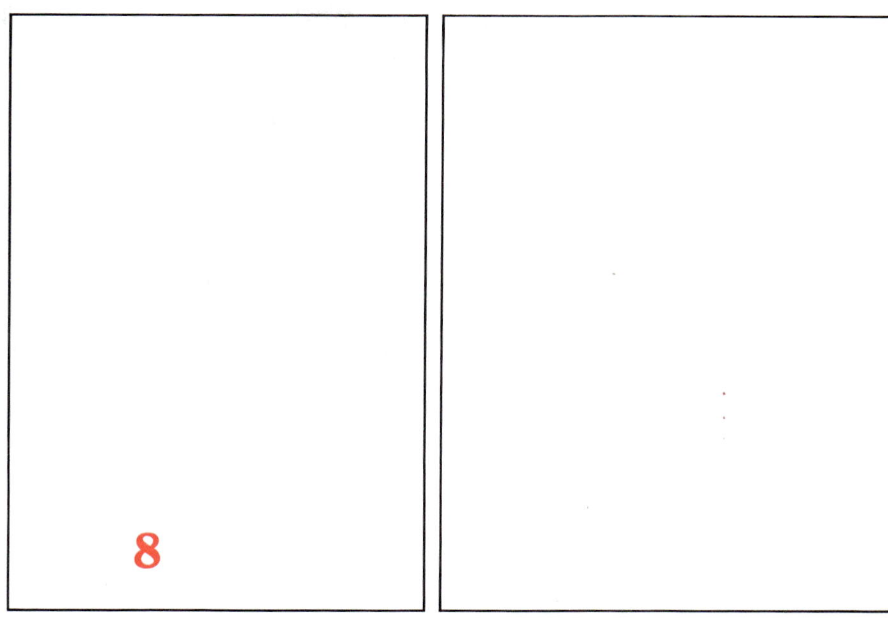

8

The somatic nervous system

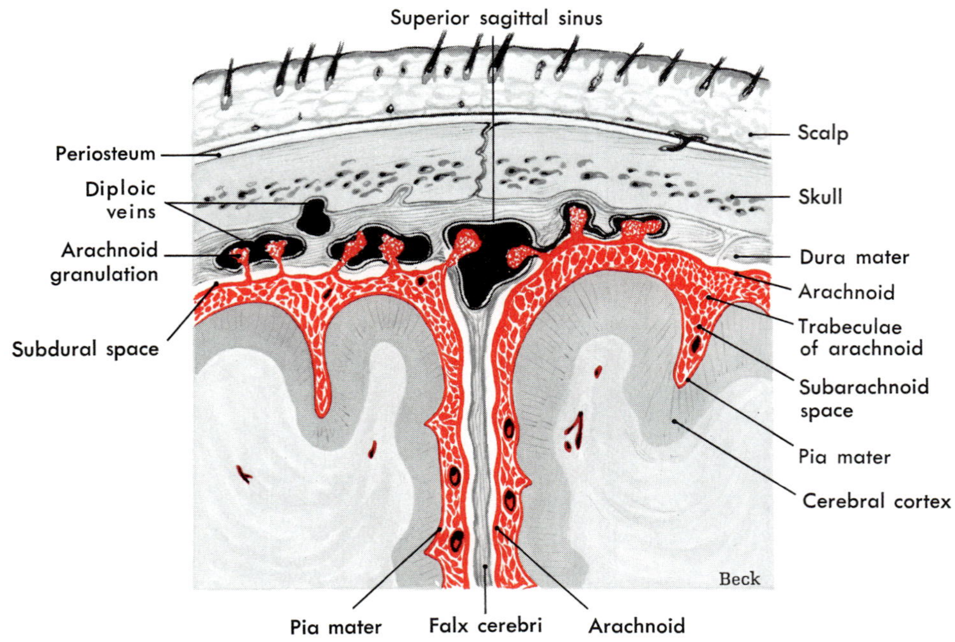

Fig. 8-1

Meninges of the brain as seen in coronal section through the skull.

From the title of this chapter you might infer that your body has more than one nervous system. It does not. It has only one nervous system, but to discuss this highly complicated set of organs, we pretty clearly must divide it up some way. One usual way is to separate it into two divisions, namely, the central nervous system, or CNS, and the peripheral nervous system, or PNS. The CNS consists of the centrally located nervous system organs, that is, the brain and spinal cord. All other nervous system organs—cranial nerves, spinal nerves, autonomic nerves, and ganglia—compose the PNS. For our discussion, we shall separate the nervous system, not into the CNS and the PNS, but into the somatic nervous system and the autonomic nervous system. The *somatic nervous system* provides communication between various parts of the body and somatic effectors (that is, skeletal muscle cells). The *autonomic nervous system,* on the other hand, conducts impulses to visceral effectors (also called autonomic effectors; specifically, smooth muscle, cardiac muscle, and glandular epithelial cells). This chapter presents basic information about the brain, spinal cord, cranial nerves, and spinal nerves, the organs that make up the somatic nervous system. But before we start our discussion of these organs, we shall describe the coverings of the brain and cord.

Brain and cord coverings

Because the brain and spinal cord are both delicate and vital, nature has provided them with two protective coverings. The outer covering consists of bone: cranial bones encase the brain and vertebrae encase the cord. The inner covering consists of membranes

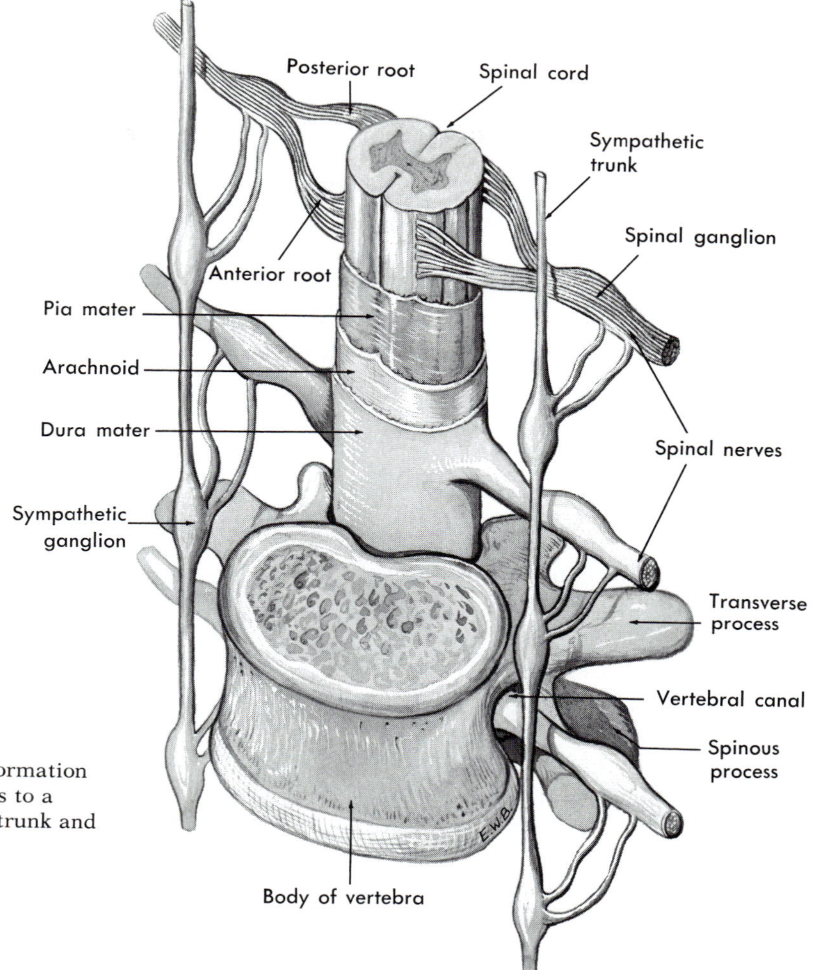

Fig. 8-2

Spinal cord showing meninges, formation of the spinal nerves, and relations to a vertebra and to the sympathetic trunk and ganglia.

known as *meninges.* Three distinct layers compose the meninges: the dura mater, the arachnoid membrane, and the pia mater. Observe their respective locations in Figs. 8-1 and 8-2. The dura mater, made of strong white fibrous tissue, serves both as the outer layer of the meninges and also as the inner periosteum of the cranial bones. The arachnoid membrane, a delicate, cobwebby layer, lies between the dura mater and the pia mater or innermost layer of the meninges. The transparent pia mater adheres to the outer surface of the brain and cord and contains blood vessels.

Three extensions of the dura mater should be mentioned: the falx cerebri, falx cerebelli, and tentorium cerebelli. The falx cerebri projects downward into the longitudinal fissure to form a kind of partition between the two cerebral hemispheres. The falx cerebelli separates the two cerebellar hemispheres. The tentorium cerebelli separates the cerebellum from the occipital lobe of the cerebrum. It takes its name from the fact that it forms a tentlike covering over the cerebellum.

Between the dura mater and the arachnoid membrane is a small space called the subdural space, and between the arachnoid and the pia mater is another space, the subarachnoid space. Inflammation of the meninges is called meningitis. It most often involves the arachnoid and pia mater or the *leptomeninges,* as they are sometimes called.

The meninges of the cord continue on down inside the spinal cavity for some distance below the end of the spinal cord. The pia mater forms a slender filament known as the filum terminale. At the level of the third segment of the sacrum, the filum terminale blends with dura mater to form a fibrous cord that disappears in the periosteum of the coccyx. This extension of the meninges beyond the cord is convenient for performing lumbar punctures. It makes it possible to insert a needle between the third and fourth or fourth and fifth lumbar vertebrae into the subarachnoid space to withdraw cerebrospinal fluid without danger of injuring the spinal cord, which ends more than an inch above that point. The fourth lumbar vertebrae can be easily located because it lies on a line with the iliac crest. Placing the patient on his side and arching his back by drawing the knees and chest together separates the vertebrae sufficiently to introduce the needle.

Brain and cord fluid spaces

In addition to the bony and membranous coverings, nature has further fortified the brain and spinal cord against injury by providing a cushion of fluid both around them and within them. The fluid is called *cerebrospinal fluid,* and the spaces containing it are as follows:

1 The subarachnoid space around the brain
2 The subarachnoid space around the cord
3 The ventricles and aqueduct inside the brain
4 The central canal inside the cord

The ventricles are cavities or spaces inside the brain. They are four in number. Two of them, the lateral (or first and second) ventricles, are located one in each cerebral hemisphere. Note in Fig. 8-3 the shape of these ventricles—roughly like the hemispheres themselves. The third ventricle is little more than a lengthwise slit in the cerebrum beneath the midportion of the corpus callosum and longitudinal fissure. The fourth ventricle is a diamond-shaped space between the cerebellum posteriorly and the medulla and pons anteriorly. Actually, it is an expansion of the central canal of the cord after the cord enters the cranial cavity and becomes enlarged to form the medulla.

Formation, circulation, and function of cerebrospinal fluid

Formation of cerebrospinal fluid occurs mainly by secretion by the choroid plexi. Choroid plexi are networks of capillaries that project from the pia mater into the lateral ventricles and into the roofs of the third and fourth ventricles. From each lateral ventricle the fluid seeps through an opening, the interventricular foramen (of Munro), into the third ventricle, then through a narrow channel, the cerebral aqueduct (or aqueduct of Sylvius), into the fourth ventricle, from which it circulates into the central canal of the cord. Openings in the roof of the fourth ventricle (the foramen of Magendie and foramina of Luschka) permit the flow of fluid into the subarachnoid space around the cord and then into the subarachnoid space around the brain. From the latter space it is gradually absorbed into the venous blood of the brain. Thus cerebrospinal fluid "circulates" from blood in the choroid plexuses, through ventricles, central canal, and subarachnoid spaces, and back into the blood.

Occasionally some condition interferes with this circuit. For example, a brain tumor may press against the cerebral aqueduct, shutting off the flow of fluid from the third to the fourth ventricle. In such an event, the fluid accumulates within the lateral and

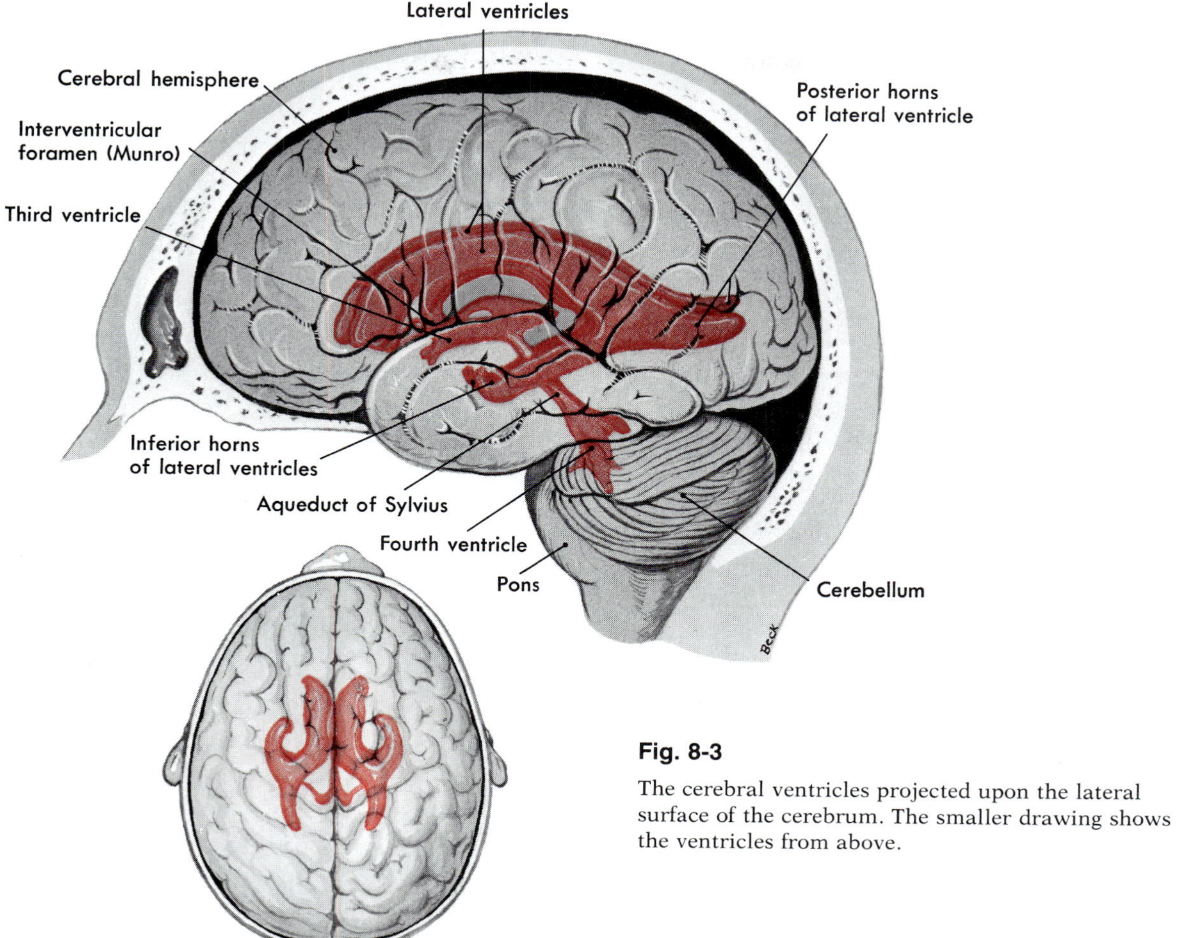

Fig. 8-3

The cerebral ventricles projected upon the lateral surface of the cerebrum. The smaller drawing shows the ventricles from above.

third ventricles because it continues to form even though its drainage is blocked. This condition is known as internal hydrocephalus. If the fluid accumulates in the subarachnoid space around the brain, external hydrocephalus results. Subarachnoid hemorrhage, for example, may lead to formation of blood clots that block drainage of the cerebrospinal fluid from the subarachnoid space. With decreased drainage, an increased amount of fluid, of course, remains in the space.

Withdrawal of some of the cerebrospinal fluid from the subarachnoid space in the lumbar region of the cord is known as a lumbar puncture.

The amount of cerebrospinal fluid in the average adult is about 140 ml (about 23 ml in the ventricles and 117 ml in the subarachnoid space of brain and cord).*

Cerebrospinal fluid serves as a protective cushion around and within the brain and cord. However, it is now known to function in other ways as well. For instance, changes in its carbon dioxide content affect neurons

*Mountcastle, V. B., editor: Medical physiology, ed. 13, St. Louis, 1974, The C. V. Mosby Co., p. 1116.

of the respiratory center in the medulla and thereby help control respirations.

Anatomy of the brain

The brain is one of the largest of adult organs. It consists of several billion neurons and presumably even more neuroglia. In most adults it weighs about 3 pounds but generally is smaller in women than in men and in older persons than in younger persons. Neurons of the brain undergo mitosis only during the prenatal period and the first few months of postnatal life. Although they grow in size after that, they do not increase in number. Malnutrition during those crucial prenatal months of neuron multiplication is reported* to hinder the process and result in fewer brain cells. The brain attains full size by about the eighteenth year but grows rapidly only during the first 9 years or so.

The brain has six major divisions: cerebrum, diencephalon, cerebellum, medulla oblongata, pons, and midbrain. The midbrain, pons, and medulla constitute the brainstem—an apt term since, viewed from the side, they look like a stem for the rest of the brain (see Fig. 8-8). Each division of the brain consists of gray matter (nuclei and centers) and white matter (tracts).

Cerebrum

The cerebrum is the largest and the most superiorly located division of the human brain.

Hemispheres, fissures, and lobes

Grooves, called fissures or sulci, dip into the surface of the cerebrum, dividing it into two hemispheres and each hemisphere into five lobes. A deep groove, the longitudinal fissure, divides the cerebrum into two halves called hemispheres. They are, however, not completely separated from each other. A structure composed of white matter and known as the corpus callosum joins them medially (Fig. 8-5). Prominent fissures, in addition to the longitudinal fissure already named, include the central fissure (of Rolando), the lateral fissure (of Sylvius), and the parieto-occipital fissure. These prominent indentations subdivide each cerebral hemisphere into four lobes, each of which bears the name of the bone lying over it: frontal lobe, parietal lobe, temporal lobe, and occipital lobe (Fig. 8-4). A fifth lobe, the insula (island of Reil), lies hidden from view in the lateral fissure. To see it, one must dissect the brain. The central fissure separates the frontal lobe from the parietal lobe. The lateral fissure separates the temporal lobe (below the fissure) from the frontal and parietal lobes above it. The parieto-occipital fissure separates the occipital lobe from the two parietal lobes.

Cerebral cortex

Each hemisphere of the cerebrum consists of external gray matter, internal white matter, and islands of internal gray matter. The cerebral cortex is the thin surface layer of the cerebrum. Gray matter only 2 to 4 mm (roughly $^{1}/_{12}$ to $^{1}/_{6}$ inch) thick composes it. But despite its thinness, the cerebral cortex consists of six layers, each one a dense network of millions of axon terminals synapsing with millions of dendrites and cell bodies of other neurons. Under the cortex lies the white matter that composes the bulk of the cerebrum's interior. Presumably, when early anatomists observed the cerebrum's outer darker layer, it reminded them of tree bark—hence their choice of the name cortex (Latin for bark) for it.

Provided one uses a little imagination, the surface of the cerebrum looks like a group of small sausages. Each "sausage" represents a convolution or gyrus. Between adjacent

*Malnutrition—effects on the brain, Sci. News **97**:70, Jan. 17, 1970.

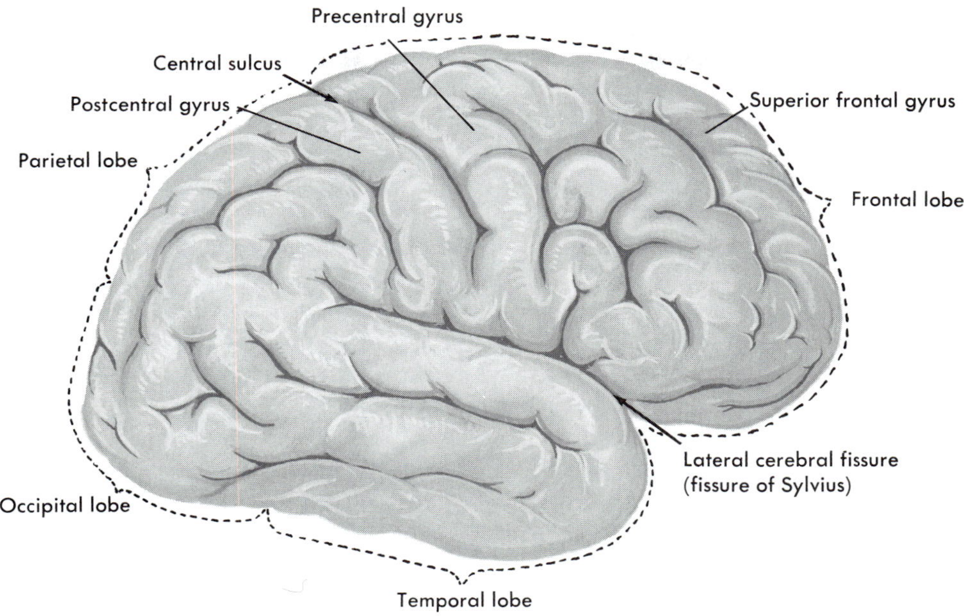

Fig. 8-4
Right hemisphere of cerebrum, lateral surface.

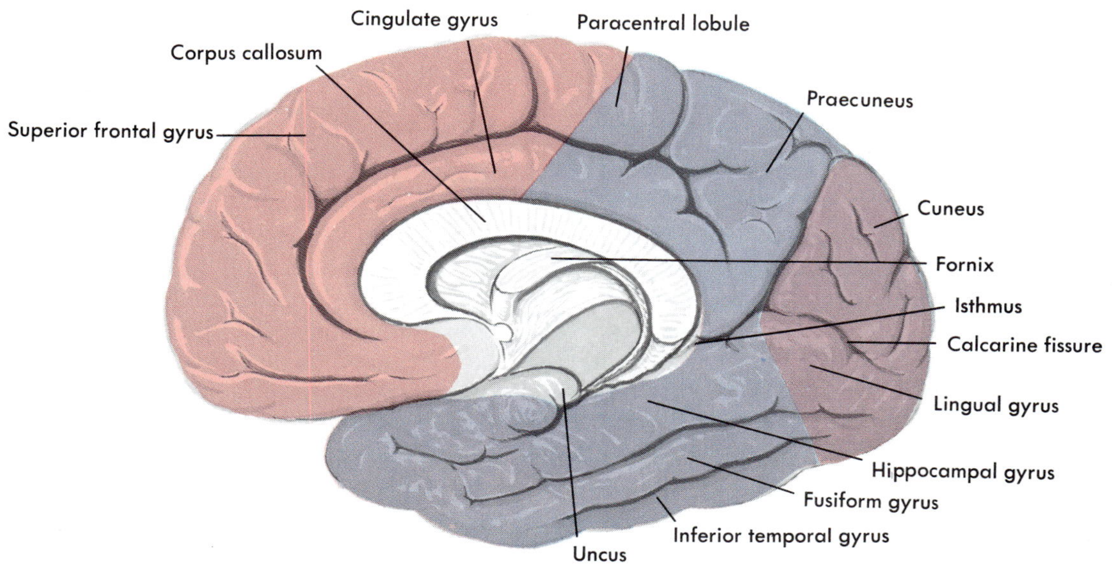

Fig. 8-5
Right hemisphere of cerebrum, medial surface.

gyri lie sulci or fissures. Identify the following structures on Figs. 8-4 and 8-5: precentral gyrus, postcentral gyrus, cuneus, cingulate gyrus, hippocampal gyrus, uncus. These are some of the structures we shall refer to later in our discussion of the physiology of the brain.

Cerebral tracts

Cerebral tracts lie interior to the cortex and are composed of great numbers of nerve fibers (axons). Tracts and nerves are comparable structures. Both consist of bundles of nerve fibers. However, whereas tracts are bundles of axons located in the brain and spinal cord, nerves are bundles of dendrites or axons, or both, located outside the brain and cord.

Tracts that conduct impulses upward are called sensory or ascending projection tracts, and those that conduct downward are referred to as motor or descending projection tracts. In one part of the interior of the cerebrum, a group of sensory and motor projection tracts forms a large irregular mass of white matter known as the internal capsule. It lies between the thalamus and the basal ganglia (Figs. 8-7 and 8-20). Some tracts are short, extending from one convolution to another in the same hemisphere. These are called association tracts.

Basal ganglia

Basal ganglia (or cerebral nuclei*) are islands of gray matter deep inside each cerebral hemisphere. Authorities agree that the most important basal ganglia are the caudate nucleus, the putamen, and the globus pallidus (pallidum). They differ somewhat, however, as to what other structures they include as basal ganglia. Look now at Fig. 8-6 to observe the size, shape, and location of the caudate nucleus and putamen. The amygdaloid body shown in this figure is also classified as one of the basal ganglia by some authorities. Next, look at Fig. 8-7 and note the location of the pallidum. The pallidum and putamen together are sometimes called the *lenticular nucleus*. The pallidum and putamen, plus two other structures, namely the caudate nucleus and internal capsule, on the other hand, are together commonly referred to as the *corpus striatum*. The term means "striped body." You can see the striped appearance of this internal region of the cerebrum in Fig. 8-7.

*When applied to the nervous system, the term nucleus means an area of gray matter in the brain or cord (composed mainly, as is all gray matter, of neuron cell bodies and dendrites). Such a cluster located outside the brain and cord is called a ganglion. So cerebral nuclei is a more accurate (but less common) name than basal ganglia.

Diencephalon

The diencephalon is the part of the brain located between the cerebrum and the mesencephalon (midbrain). Although the diencephalon consists of several structures located around the third ventricle, the main ones are the thalamus (dorsal thalamus) and the hypothalamus.

Thalamus

The right thalamus is a rounded mass of gray matter about ½ inch wide and 1½ inches long, bulging into the right lateral wall of the third ventricle. The left thalamus is a similar mass in the left lateral wall. Each thalamus consists of numerous nuclei. They are arranged in groups named the anterior, lateral, intralaminar, midline (or medial), and posterior groups of nuclei. Two important structures in the posterior group of nuclei are the medial and lateral geniculate bodies. Large numbers of axons conduct impulses into the thalamus from the cord, brainstem, cerebellum, basal ganglia, and various parts of the cerebrum. These axons terminate in thalamic nuclei where they

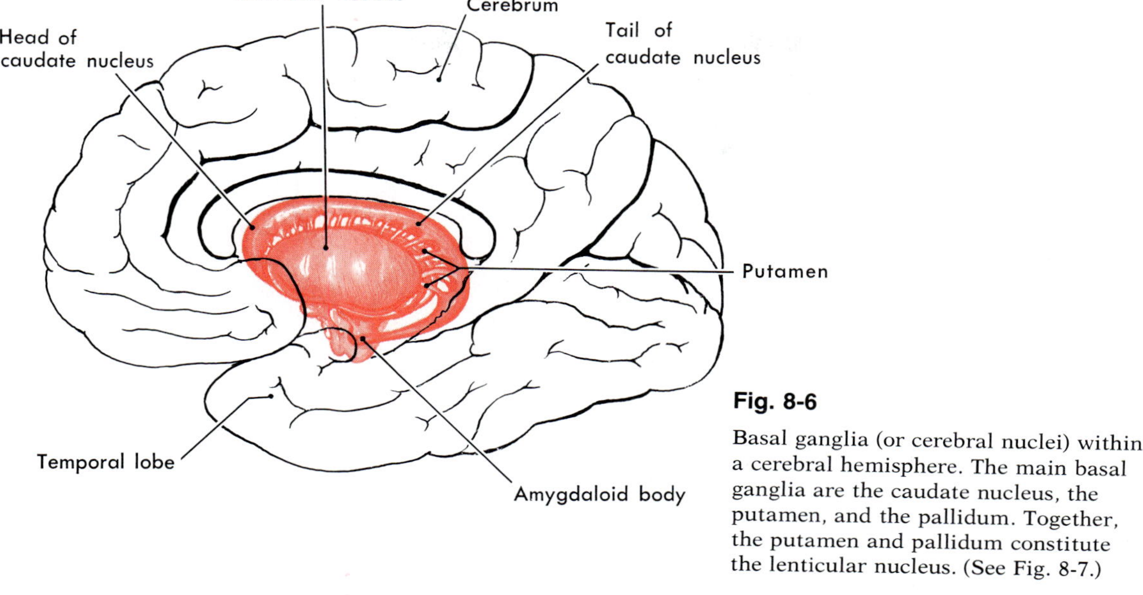

Fig. 8-6

Basal ganglia (or cerebral nuclei) within a cerebral hemisphere. The main basal ganglia are the caudate nucleus, the putamen, and the pallidum. Together, the putamen and pallidum constitute the lenticular nucleus. (See Fig. 8-7.)

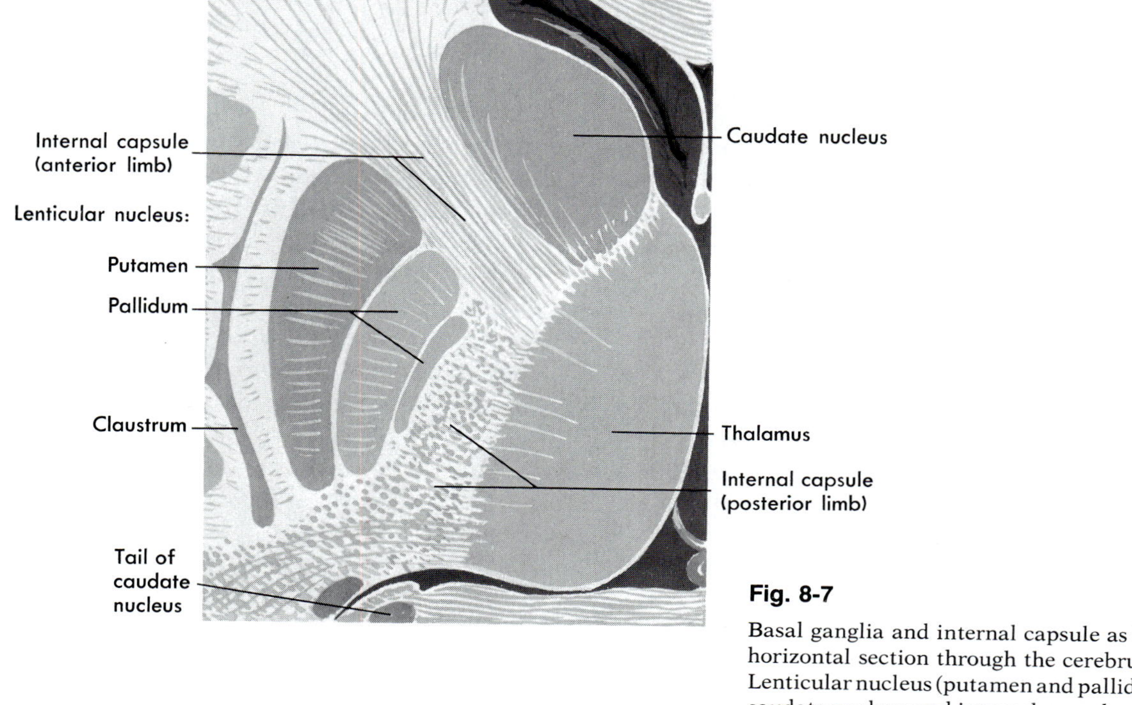

Fig. 8-7

Basal ganglia and internal capsule as seen in a horizontal section through the cerebrum. Lenticular nucleus (putamen and pallidum) plus caudate nucleus and internal capsule constitute corpus striatum.

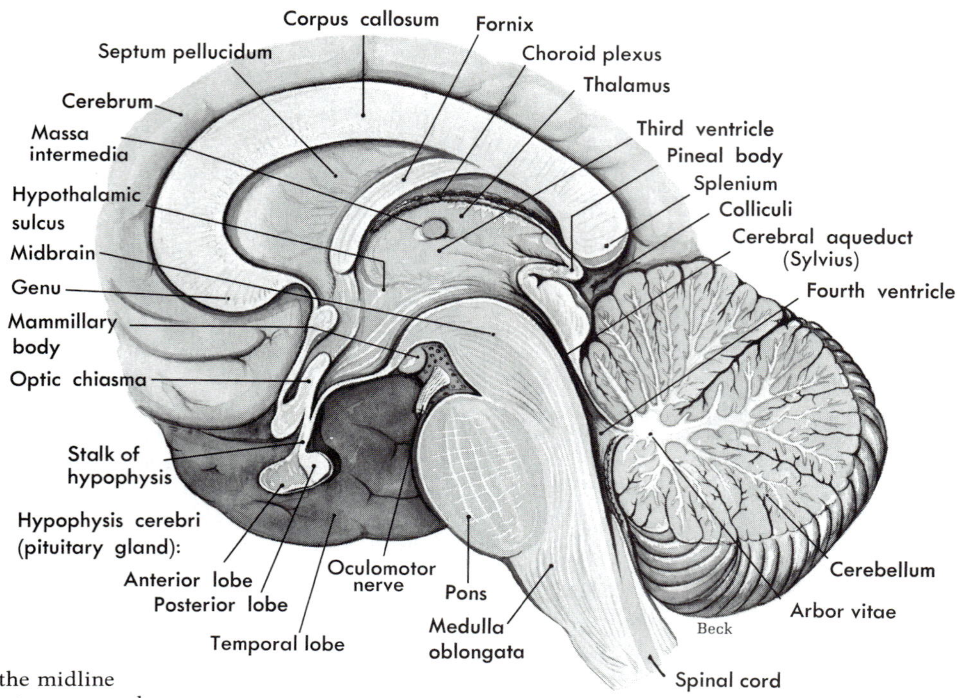

Fig. 8-8

Sagittal section through the midline of the brain showing structures around the third ventricle.

synapse with neurons whose axons conduct impulses out of the thalamus to virtually all areas of the cerebral cortex. Thus the thalamus serves as the major relay station for sensory impulses on their way to the cerebral cortex.

Hypothalamus

The hypothalamus consists of several structures that lie beneath the thalamus and form the third ventricle's floor and the lower part of its sidewall. Prominent among the structures composing the hypothalamus are the supraoptic nuclei, the paraventricular nuclei, the stalk of the hypophysis (pituitary gland), the neurohypophysis (the posterior lobe of the pituitary gland), and the mamillary bodies. Identify as many of these as you can in Figs. 8-8 and 8-9. The supraoptic nuclei consist of gray matter located just above and on either side of the optic chiasma. The paraventricular nuclei of the hypothalamus are so named because of their location close to the wall of the third ventricle. The midportion of the hypothalamus consists of the stalk of the pituitary gland and the posterior lobe of the pituitary gland (neurohypophysis). The posterior part of the hypothalamus consists mainly of the mamillary bodies, in which are located the mamillary nuclei.

Cerebellum

The cerebellum, the second largest part of the brain, is located just below the posterior portion of the cerebrum and is partially covered by it. A transverse fissure separates the cerebellum from the cerebrum. These two parts of the brain have several characteristics in common. For instance, gray matter makes up their outer portions and white

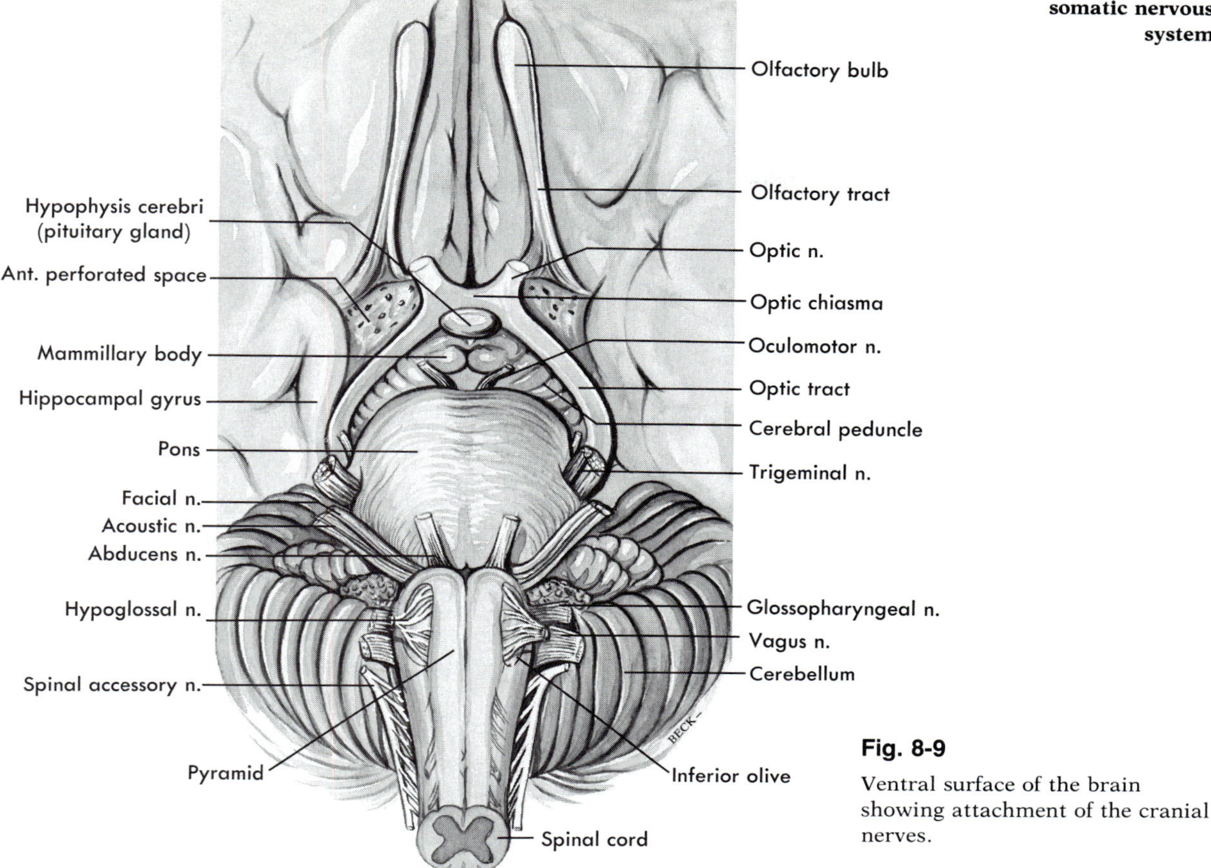

Fig. 8-9

Ventral surface of the brain showing attachment of the cranial nerves.

matter predominates in their interiors. Turn to Fig. 8-8 to observe the arbor vitae—that is, the internal white matter of cerebellum. Note its distinctive pattern, similar to the veins of a leaf. Note, too, that the surfaces of both the cerebellum and the cerebrum have numerous grooves (sulci) and convolutions (gyri). The convolutions of the cerebellum, however, are much more slender and less prominent than those of the cerebrum. The cerebellum has two large lateral masses, the cerebellar hemispheres, and a central section called the vermis because in shape it resembles a worm coiled upon itself. (For a detailed description of the several subdivisions of the cerebellum, consult a textbook on neuroanatomy.)

The internal white matter of the cerebellum is composed of some short and some long tracts. The short association tracts conduct impulses from neuron cell bodies located in the cerebellar cortex to neurons whose dendrites and cell bodies compose nuclei located in the interior of the cerebellum. The longer tracts conduct impulses from cerebellar neurons to neurons located in other parts of the brain and with the spinal cord. These enter or leave the cerebellum by way of its three pairs of peduncles as follows:

1 Inferior cerebellar peduncles (or restiform bodies)—composed chiefly of tracts into the cerebellum from the medulla and cord (notably, spinocerebellar, vestibulocerebellar, and reticulocerebellar tracts)

2 Middle cerebellar peduncles (or brachia pontis)—composed almost entirely of tracts into the cerebellum from the pons (that is, pontocerebellar tracts)

3 Superior cerebellar peduncles (or

brachia conjunctivum cerebelli)—composed principally of tracts from dentate nuclei through the red nucleus of the midbrain to the thalamus

An important pair of cerebellar nuclei are the dentate nuclei, one of which lies in each hemisphere. Tracts connect these nuclei with motor areas of the cerebral cortex (the dentatorubrothalamic tracts to the thalamus and thalamocortical tracts to the cortex). By means of these tracts, cerebellar impulses influence the motor cortex. Impulses also travel the reverse direction. Corticopontine and pontocerebellar tracts enable the motor cortex to influence the cerebellum.

Medulla oblongata

The medulla or bulb is the part of the brain that attaches to the spinal cord. It is, in fact, an enlarged extension of the cord located just above the foramen magnum. It measures only slightly more than an inch in length and is separated from the pons above by a horizontal groove. It is composed mainly of white matter (projection tracts) and reticular formation, a term that means the interlacement of gray and white matter present in the cord, brainstem, and diencephalon. Nuclei in the reticular formation of the medulla include such important centers as respiratory and vasomotor centers.

On each side of the lower posterior part of the medulla are two prominent nuclei, the nucleus gracilis and the nucleus cuneatus. Here, afferent fibers from the posterior white columns (fasciculi gracilis and cuneatus) of the cord synapse with neurons whose axons extend to the thalamus and cerebellum.

The pyramids (Fig. 8-9) are two bulges of white matter located on the anterior surface of the medulla formed by fibers of the pyramidal projection tracts.

The olive (Fig. 8-9) is an oval projection appearing one on each side of the anterior surface of the medulla. It contains the inferior olivary nucleus and two accessory olivary nuclei. Fibers from the cells of these nuclei run through the inferior cerebellar peduncles (restiform bodies) into the cerebellum. Nuclei of the ninth to the twelfth cranial nerves are also located in the medulla.

Pons

Just above the medulla lies the pons, composed like the medulla of white matter and a few nuclei. Fibers that run transversely across the pons and through the brachia pontis (middle cerebellar peduncles) into the cerebellum make up the external white matter of the pons and give it its bridgelike appearance. The reticular formation extends into the pons from the medulla. One important reticular nucleus in the pons is called the pneumotaxic center. (It functions in the control of respirations.) Nuclei of the fifth to eighth cranial nerves are located in the upper part of the pons.

Mesencephalon (midbrain)

The mesencephalon, or as it is more commonly called, the midbrain, lies below the inferior surface of the cerebrum and above the pons. It consists mainly of white matter with some internal gray matter around the cerebral aqueduct, the cavity within the midbrain. The cerebral peduncles form the ventral part of the midbrain and the corpora quadrigemina or colliculi form the dorsal part. The cerebral peduncles are two ropelike masses of white matter that extend divergently from the pons to the undersurface of the cerebral hemispheres (Fig. 8-9). In other words, the cerebral peduncles are made up of tracts that constitute the main connection between the forebrain and hindbrain. Hence, the midbrain, by function as well as by location, is well named.

The corpora quadrigemina consist of four rounded eminences, the two superior and the two inferior colliculi (Fig. 8-8), which form the dorsal part of the midbrain. Certain auditory reflex centers lie in the inferior colliculi and visual centers in the superior colliculi.

An important nucleus in the midbrain reticular formation is the red nucleus, a large gray mass ventral to the superior colliculi. Fibers from the cerebellum and from the frontal lobe of the cerebral cortex end here, whereas fibers that extend into the rubrospinal tracts of the cord have their cells of origin here. Nuclei of the third and fourth cranial nerves and the anterior part of the nucleus of the fifth cranial nerve are located deep in the midbrain. Also, as mentioned in the preceding paragraph, nuclei for certain auditory and visual reflexes lie in the colliculi of the midbrain.

Anatomy of the spinal cord

The spinal cord lies within the spinal cavity, extending from the foramen magnum to the lower border of the first lumbar vertebra (Fig. 8-10), a distance of 17 or 18 inches in the average body. The cord does not completely fill the spinal cavity—it also contains the meninges, spinal fluid, a cushion of adipose tissue, and blood vessels.

The spinal cord is an oval-shaped cylinder that tapers slightly from above downward and has two bulges, one in the cervical region and the other in the lumbar region. Two deep grooves, the *anterior median fissure* and the *posterior median sulcus,* just miss dividing the cord into separate symmetrical halves. The anterior fissure is the deeper and the wider of the two grooves—a useful fact to remember when you examine spinal cord diagrams. It enables you to tell at a glance which part of the cord is anterior and which is posterior. Gray matter composes the inner core of the cord. Although it looks like a flat letter H in cross-section views of the cord, it actually has three dimensions, since the gray matter extends the length of the cord. The limbs of the H are called either anterior, posterior, or lateral horns of gray matter or anterior, posterior, or lateral gray columns. They consist primarily of cell bodies of interneurons and motoneurons. White matter surrounding the gray matter is subdivided in each half of the cord into three *columns* (or funiculi): the anterior, posterior, and lateral white columns. These consist of large bundles of nerve fibers arranged in tracts.

Tables 8-1 and 8-2 name important tracts of each white column and give essential information about them. Also refer to Fig. 8-12.

Cranial nerves

Twelve pairs of nerves arise from the undersurface of the brain, some from each division with the exception of the cerebellum. After leaving the cranial cavity by way of small foramina in the skull, they extend to their respective destinations. Both names and numbers identify the cranial nerves. Their names suggest their distribution or function. Their numbers indicate the order in which they emerge from front to back. Some cranial nerves consist of both afferent and efferent fibers—in short, they are mixed nerves. On the other hand, some cranial nerves consist only of afferent fibers and some mainly of efferent fibers. Cell bodies of the efferent fibers lie in the various nuclei of the brainstem. Cell bodies of the afferent fibers, with few exceptions, are located in ganglia outside the brainstem—for example, the trigeminal (gasserian) ganglion of the fifth cranial nerve. The first, second, and eighth cranial nerves are purely afferent. Information about cranial nerves is summarized in Tables 8-3 and 8-5, pp. 194 and 203.

First (olfactory)

The olfactory nerves are composed of axons of neurons whose dendrites and cell bodies lie in the nasal mucosa, high up along the septum and superior conchae (turbinates).

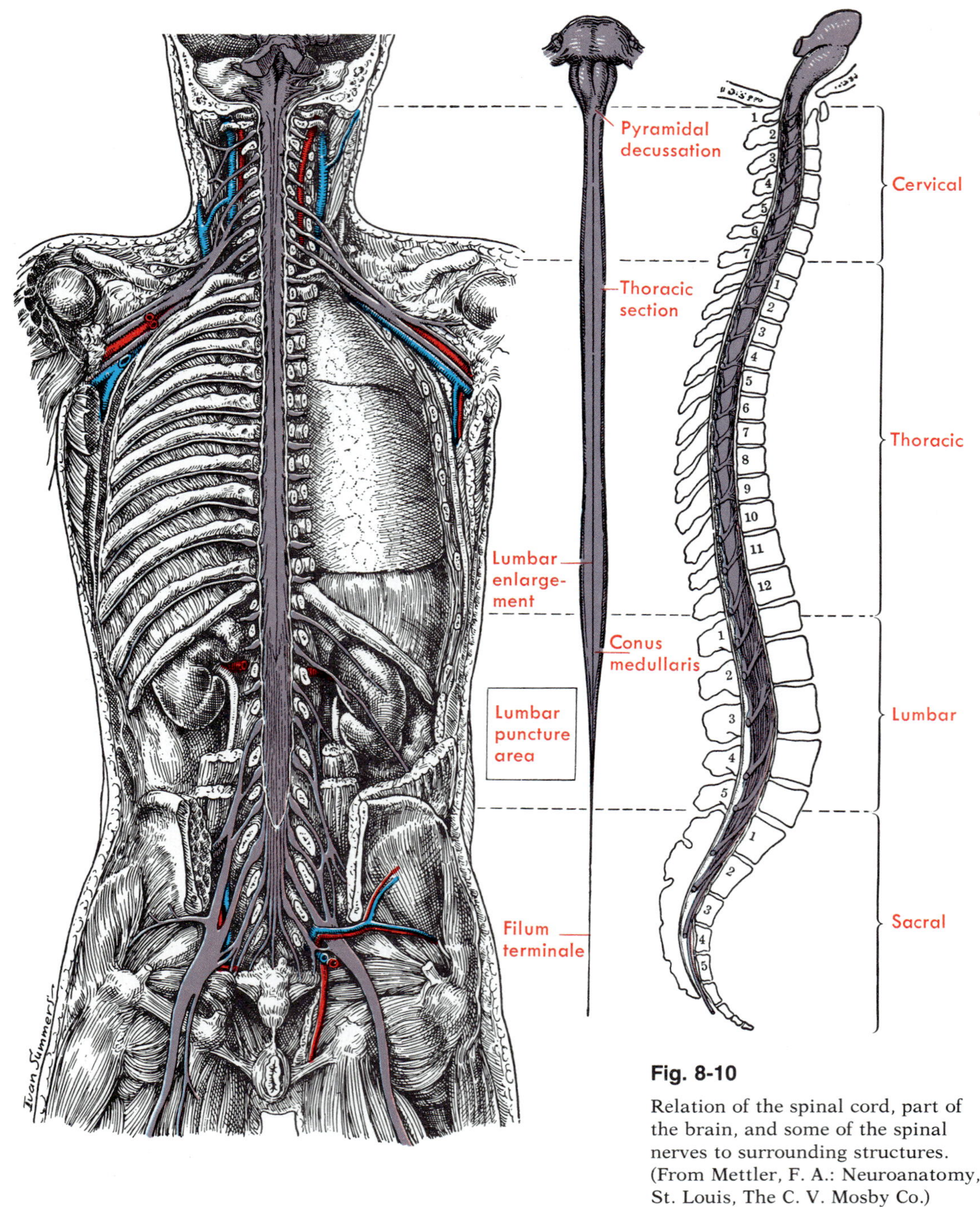

Fig. 8-10

Relation of the spinal cord, part of the brain, and some of the spinal nerves to surrounding structures. (From Mettler, F. A.: Neuroanatomy, St. Louis, The C. V. Mosby Co.)

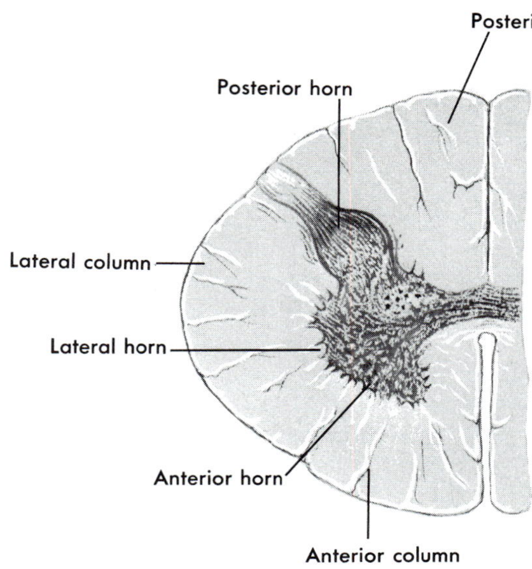

Fig. 8-11

Distribution of gray matter (horns) and white matter (columns) in a section of the spinal cord at the thoracic level.

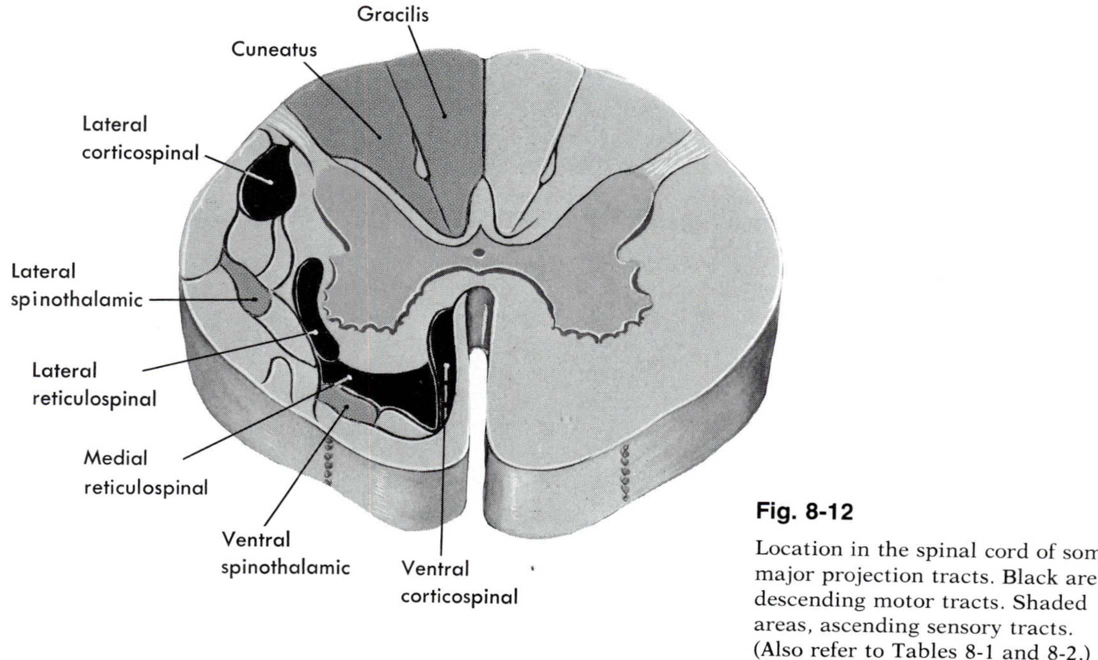

Fig. 8-12

Location in the spinal cord of some major projection tracts. Black areas, descending motor tracts. Shaded areas, ascending sensory tracts. (Also refer to Tables 8-1 and 8-2.)

Table 8-1

Major ascending tracts of spinal cord

Name	Function	Location	Origin*	Termination†
Lateral spinothalamic	Pain and temperature opposite side	Lateral white columns	Posterior gray column opposite side	Thalamus
Ventral spinothalamic	Crude touch	Anterior white columns	Posterior gray column opposite side	Thalamus
Fasciculus gracilis and cuneatus	Conscious kinesthesia, sensations of vibration, stereognosis, deep touch and pressure, two-point discrimination	Posterior white columns	Spinal ganglia same side	Medulla
Spinocerebellar	Unconscious kinesthesia	Lateral white columns	Posterior gray column	Cerebellum

*Location of cell bodies of neurons from which axons of tract arise.
†Structure in which axons of tract terminate.

Table 8-2

Major descending tracts of spinal cord

Name	Function	Location	Origin*	Termination†
Lateral corticospinal (or crossed pyramidal)	Voluntary movement, contraction of individual or small groups of muscles, particularly those moving hands, fingers, feet, and toes of opposite side	Lateral white columns	Motor areas cerebral cortex (mainly areas 4 and 6) opposite side from tract location in cord	Intermediate or anterior gray columns
Ventral corticospinal (direct pyramidal)	Same as lateral corticospinal except mainly muscles of same side	Lateral white columns	Motor cortex but on same side as tract location in cord	Intermediate or anterior gray columns
Lateral reticulospinal	Mainly facilitatory influence on motoneurons to skeletal muscles	Lateral white columns	Reticular formation midbrain, pons, and medulla	Intermediate or anterior gray columns
Medial reticulospinal	Mainly inhibitory influence on motoneurons to skeletal muscles	Anterior white columns	Reticular formation medulla mainly	Intermediate or anterior gray columns

*Location of cell bodies of neurons from which axons of tract arise.
†Structure in which axons of tract terminate.

Axons of these neurons form about twenty small fibers that pierce each cribriform plate and terminate in the olfactory bulbs, where they synapse with olfactory neurons II whose axons comprise the olfactory tracts. Summarizing:

Olfactory neurons I	
Dendrites } Cell body }	In nasal mucosa
Axons	In small fibers that extend through cribriform plate to olfactory bulb

Olfactory neurons II	
Dendrites } Cell body }	In olfactory bulb
Axons	In olfactory tracts

Second (optic)

Axons from the third and innermost layer of neurons of the retina compose the second cranial nerves. After entering the cranial cavity through the optic foramina, the two optic nerves unite to form the *optic chiasma,* in which some of the fibers of each nerve cross to the opposite side and continue in the *optic tract* of that side (Fig. 10-18). Thus each optic nerve contains fibers only from the retina of the same side, whereas each optic tract has fibers in it from both retinae, a fact of importance in interpreting certain visual disorders. Most of the optic tract fibers terminate in the thalamus (in the portion known as the lateral geniculate body). From here, a new relay of fibers runs to the visual area of the occipital lobe cortex. A few optic tract fibers terminate in the superior colliculi of the midbrain, where they synapse with motor fibers to the external eye muscles (third, fourth, and sixth cranial nerves).

Third (oculomotor)

Fibers of the third cranial nerve originate from cells in the oculomotor nucleus in the ventral part of the midbrain and extend to the various external eye muscles, with the exception of the superior oblique and the lateral rectus. Autonomic fibers whose cells lie in a nucleus of the midbrain are also contained in the oculomotor nerves. These fibers terminate in the ciliary ganglion, where they synapse with cells whose postganglionic fibers supply the intrinsic eye muscles (ciliary and iris). Still a third group of fibers are found in the third cranial nerves —namely, sensory fibers from proprioceptors in the eye muscles.

Fourth (trochlear)

Motor fibers of the fourth cranial nerve have their origin in cells in the midbrain, from which they extend to the superior oblique muscles of the eye. Afferent fibers from proprioceptors in these muscles are also contained in the trochlear nerves.

Fifth (trigeminal)

Three sensory branches (ophthalmic, maxillary, and mandibular nerves) carry afferent impulses from the skin and mucosa of the head and from the teeth to cell bodies in the semilunar or gasserian ganglion (a swelling on the nerve, lodged in the petrous portion of the temporal bone) (Fig. 8-13). Fibers extend from the ganglion to the main sensory nucleus of the fifth cranial nerve situated in the pons. A smaller motor root of the trigeminal nerve originates in the trifacial motor nucleus located in the pons just medial to the sensory nucleus. Fibers run from the motor root to the muscles of mastication by way of the mandibular nerve.

Neuralgia of the trifacial nerve, known as tic douloureux, is an extremely painful condition that can be relieved by removing the gasserian (or semilunar) ganglion, the large ganglion on the posterior root of the nerve (Fig. 8-13) containing the cell bodies of the nerve's afferent fibers. After such an operation, the patient's face, scalp, teeth, and conjunctiva on the side treated show

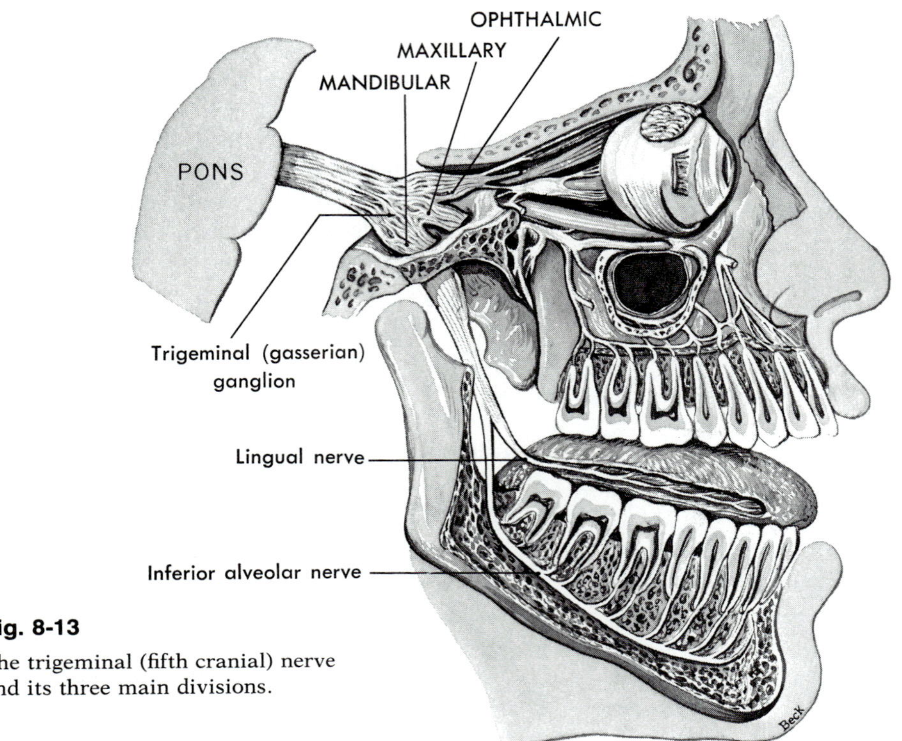

Fig. 8-13

The trigeminal (fifth cranial) nerve and its three main divisions.

anesthesia. Special care, such as wearing protective goggles and irrigating the eye frequently, is therefore prescribed. The patient is instructed also to visit his dentist regularly since he can no longer experience a toothache as a warning of diseased teeth.

Sixth (abducent)

The sixth cranial nerve is a motor nerve with fibers originating from a nucleus in the pons in the floor of the fourth ventricle and extending to the lateral rectus muscles of the eyes. It contains also some afferent fibers from proprioceptors in the lateral rectus muscles.

Seventh (facial)

The motor fibers of the seventh cranial nerve arise from a nucleus in the lower part of the pons, from which they extend by way of several branches to the superficial muscles of the face and scalp and to the submaxillary and sublingual glands. Sensory fibers from the taste buds of the anterior two thirds of the tongue run in the facial nerve to cell bodies in the geniculate ganglion, a small swelling on the facial nerve, where it passes through a canal in the temporal bone. From the ganglion, fibers extend to the nucleus solitarius in the medulla.

Eighth (vestibulocochlear)

The eighth cranial nerve has two distinct divisions: the vestibular nerve and the cochlear nerve. Both are sensory. Fibers from the semicircular canals run to the vestibular ganglion (in the internal auditory meatus), where their cell bodies are located and from which fibers extend to the vestibular nuclei in the pons and medulla. Together, these fibers constitute the *vestibular nerve.* Some of its fibers run to the cerebellum. The vestibular nerve transmits impulses that result in sensations of balance or imbalance. The *cochlear nerve* consists of fibers starting in the organ of Corti in the cochlea, have their cell bodies in the spiral ganglion in the cochlea, and terminate in the cochlear nuclei located between the medulla and pons. Conduction by the cochlear nerve results in sensations of hearing.

Ninth (glossopharyngeal)

Both sensory and motor fibers compose the ninth cranial nerve. This nerve supplies fibers not only to the tongue and pharynx, as its name implies, but also to other structures—for example, to the carotid sinus. The latter plays an important part in the control of blood pressure. Sensory fibers, with their receptors in the pharynx and posterior third of the tongue, have their cell bodies in the jugular (superior) and petrous (inferior) ganglia, located respectively in the jugular foramen and the petrous portion of the temporal bone. From these, fibers extend to the nucleus solitarius in the medulla. The motor fibers of the ninth cranial nerve originate in cells in the nucleus ambiguus in the medulla and run to muscles of the pharynx. There are also secretory fibers in this nerve, with cells of origin in the nucleus salivatorius (at the junction of the pons and medulla). These fibers run to the otic ganglion, from which postganglionic fibers extend to the parotid gland.

Tenth (vagus)

The tenth cranial nerve is widely distributed and contains both sensory and motor fibers. Its sensory fibers supply the pharynx, larynx, trachea, heart, carotid body, lungs, bronchi, esophagus, stomach, small intestine, and gallbladder. Cell bodies for these sensory dendrites lie in the jugular and nodose ganglia, located respectively in the jugular foramen and just inferior to it on the trunk of the nerve. Centrally, the sensory axons terminate in the medulla (in the nucleus solitarius) and in the pons (in the nucleus of the trigeminal nerve). Motor fibers of the vagus originate in cells in the medulla (in the dorsal motor nucleus of the vagus) and extend to various autonomic ganglia in the vagal plexus, from which postganglionic fibers run to muscles of the pharynx, larynx, and thoracic and abdominal viscera.

Eleventh (accessory)

The eleventh cranial nerve is a motor nerve. Some of its fibers originate in cells in the medulla (in the dorsal motor nucleus of the vagus and in the nucleus ambiguus) and pass by way of vagal branches to thoracic and abdominal viscera. The rest of the fibers have their cells of origin in the anterior gray column of the first five or six segments of the cervical spinal cord and extend through the spinal root of the accessory nerve to the trapezius and sternocleidomastoid muscles.

Twelfth (hypoglossal)

Motor fibers with cell bodies in the medulla (in the hypoglossal nucleus) compose the twelfth cranial nerve. They supply the muscles of the tongue. According to some anatomists, this nerve also contains sensory fibers from proprioceptors in the tongue.

■ ■ ■

The main facts about the distribution and function of each of the cranial nerve pairs are summarized in Table 8-3.

Severe head injuries often damage one or more of the cranial nerves, producing symptoms analogous to the functions of the nerve affected. For example, injury of the sixth cranial nerve causes the eye to turn in,

Table 8-3
Cranial nerves

Nerve*	Sensory fibers†			Motor fibers†		Functions‡
	Receptors	Cell bodies	Termination	Cell bodies	Termination	
I Olfactory	*Nasal mucosa*	*Nasal mucosa*	Olfactory bulbs (new relay of neurons to olfactory cortex)			*Sense of smell*
II Optic	*Retina*	*Retina*	*Nucleus in thalamus (lateral geniculate body); some fibers terminate in superior colliculus of midbrain*			*Vision*
III Oculomotor	*External eye muscles except superior oblique and lateral rectus*	?	?	**Midbrain (oculomotor nucleus and Edinger-Westphal nucleus)**	**External eye muscles except superior oblique and lateral rectus; fibers from E-W nucleus terminate in ciliary ganglion and then to ciliary and iris muscles**	**Eye movements, regulation of size of pupil, accommodation,** *proprioception (muscle sense)*
IV Trochlear	*Superior oblique*	?	?	**Midbrain**	**Superior oblique muscle of eye**	**Eye movements,** *proprioception*
V Trigeminal	*Skin and mucosa of head, teeth*	*Gasserian ganglion*	*Pons (sensory nucleus)*	**Pons (motor nucleus)**	**Muscles of mastication**	*Sensations of head and face,* **chewing movements,** *muscle sense*
VI Abducens	*Lateral rectus*			**Pons**	**Lateral rectus muscle of eye**	**Abduction of eye,** *proprioception*
VII Facial	*Taste buds of anterior two thirds of tongue*	*Geniculate ganglion*	*Medulla (nucleus solitarius)*	**Pons**	**Superficial muscles of face and scalp**	**Facial expressions, secretion of saliva,** *taste*
VIII Acoustic						
1 Vestibular branch	*Semicircular canals and vestibule (utricle and saccule)*	*Vestibular ganglion*	*Pons and medulla (vestibular nuclei)*			*Balance or equilibrium sense*
2 Cochlear or auditory branch	*Organ of Corti in cochlear duct*	*Spiral ganglion*	*Pons and medulla (cochlear nuclei)*			*Hearing*
IX Glossopharyngeal	*Pharynx; taste buds and other receptors of posterior one third of tongue*	*Jugular and petrous ganglia*	*Medulla (nucleus solitarius)*	**Medulla (nucleus ambiguus)**	**Muscles of pharynx**	*Taste and other sensations of tongue,* **swallowing movements, secretion of saliva, aid in reflex control of blood pressure and respirations**

Nerve*	Sensory fibers†			Motor fibers†		Functions‡
	Receptors	Cell bodies	Termination	Cell bodies	Termination	
IX Glossopharyngeal continued	*Carotid sinus and carotid body*	*Jugular and petrous ganglia*	*Medulla (respiratory and vasomotor centers)*	**Medulla at junction of pons (nucleus salivatorius)**	**Otic ganglion and then to parotid gland**	
X Vagus	*Pharynx, larynx, carotid body, and thoracic and abdominal viscera*	*Jugular and nodose ganglia*	*Medulla (nucleus, solitarius), pons (nucleus of fifth cranial nerve)*	**Medulla (dorsal motor nucleus)**	**Ganglia of vagal plexus and then to muscles of pharynx, larynx, and thoracic and abdominal viscera**	*Sensations* and **movements** of organs supplied; for example, **slows heart, increases peristalsis, and contracts muscles for voice production**
XI Spinal accessory	?	?	?	**Medulla (dorsal motor nucleus of vagus and nucleus ambiguus)**	**Muscles of thoracic and abdominal viscera and pharynx and larynx**	**Shoulder movements, turning movements of head, movements of viscera, voice productions,** *proprioception?*
				Anterior gray column of first five or six cervical segments of spinal cord	**Trapezius and sternocleido-mastoid muscle**	
XII Hypoglossal	?	?	?	**Medulla (hypoglossal nucleus)**	**Muscles of tongue**	**Tongue movements,** *proprioception?*

*The first letters of the words in the following sentence are the first letters of the names of the cranial nerves. Many generations of anatomy students have used this sentence as an aid to memorizing these names. It is, "On Old Olympus Tiny Tops, A Finn and German Viewed Some Hops." (There are several slightly differing versions of this mnemonic.)

†Italics indicate sensory fibers and functions. Boldface type indicates motor fibers and functions.

‡An aid for remembering the general function of each cranial nerve is the following twelve-word saying: "Some say marry money but my brothers say bad business marry money." Words beginning with S indicate sensory function. Words beginning with M indicate motor function. Words beginning with B indicate both sensory and motor functions. For example, the first, second, and eighth words in the saying start with S, which indicates that the first, second, and eighth cranial nerves perform sensory functions.

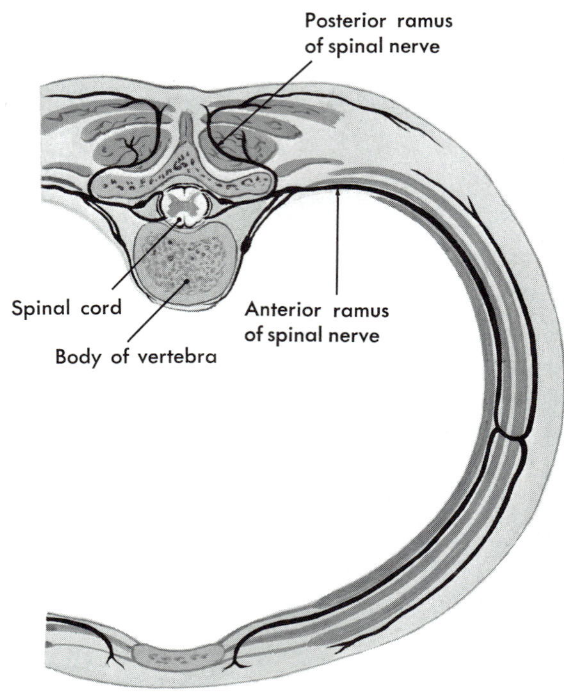

Fig. 8-14

Branchings of a spinal nerve. The anterior and posterior roots of the nerve can be seen within the vertebral foramen.

because of paralysis of the abducting muscle of the eye, whereas injury of the eighth cranial nerve produces deafness. Injury to the facial nerve results in a poker-faced expression and a drooping of the corner of the mouth from paralysis of the facial muscles.

An aneurysm of one of the middle cerebral arteries (Fig. 13-15, p. 344) with resulting pressure on the nearby oculomotor nerve (Fig. 8-9) is not uncommon. In such cases, the pupil on the same side remains dilated. In addition, the eye may turn outward and the lid may droop.

The second cranial nerve is particularly susceptible to atrophy, resulting in total blindness on the affected side.

Spinal nerves

Origin

Thirty-one pairs of nerves have their origin on the spinal cord. Unlike the cranial nerves, they have no special names but are merely numbered according to the level of the spinal column at which they emerge from the spinal cavity. Thus there are eight cervical, twelve thoracic, five lumbar, and five sacral pairs and one coccygeal pair of spinal nerves. The first cervical nerves emerge from the cord in the spaces above the first cervical vertebra (between it and the occipital bone). The rest of the cervical and all of the thoracic nerves pass out of the spinal cavity horizontally through the intervertebral foramina of their respective vertebrae. For example, the second cervical nerves emerge through the foramina above the second cervical vertebra.

Lumbar, sacral, and coccygeal nerves, on the other hand, have to descend from their point of origin at the lower end of the cord (which terminates at the level of the first lumbar vertebra) before reaching the intervertebral foramina of their respective vertebrae, through which they then emerge. This gives the lower end of the cord, with its attached spinal nerves, the appearance of a horse's tail. In fact, it bears the name *cauda equina* (Latin equivalent for horse's tail).

Each spinal nerve, instead of attaching directly to the cord, attaches indirectly by means of two short roots, anterior and posterior. The posterior roots are readily recognized by the presence on them of a swell-

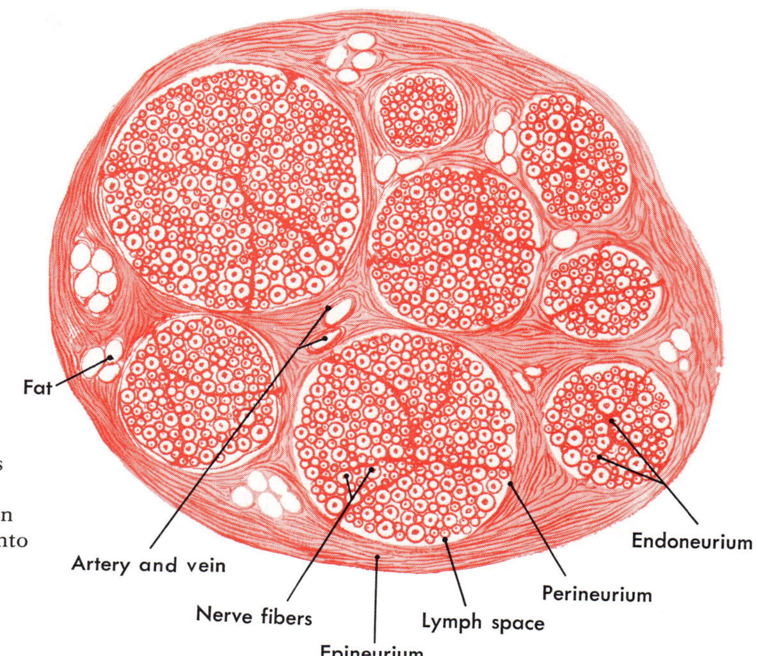

Fig. 8-15

Cross section of a nerve trunk. The fibers are medullated and are separated by a perineural sheath. The endoneurium is an extension of the perineurium that runs into the interior of the fasciculi.

ing, the posterior root ganglion or *spinal ganglion.* The roots lie within the spinal cavity with the ganglia in the intervertebral foramina (Fig. 8-14). Spinal ganglia are made up of cell bodies of sensory neurons. Fig. 8-15 shows the structure of a nerve in a cross-section view.

Distribution

After each spinal nerve emerges from the spinal cavity, it divides into anterior, posterior, and white rami. Anterior and posterior rami contain fibers belonging to the voluntary nervous system. White rami contain fibers of the autonomic nervous system. The posterior rami subdivide into lesser nerves that extend into the muscles and skin of the posterior surface of the head, neck, and trunk.

The anterior rami (except those of the thoracic nerves) subdivide and supply fibers to skeletal muscles and skin of the extremities and of anterior and lateral surfaces. Subdivisions of the anterior rami form complex networks or plexuses (Table 8-4). For example, fibers from the lower four cervical and first thoracic nerves intermix in such a way as to form a fairly definite, although apparently hopelessly confused, pattern called the *brachial plexus* (Fig. 8-16). Emerging from this plexus are smaller nerves bearing names descriptive of their locations, such as the median nerve, the musculocutaneous nerve, and the ulnar nerve. These nerves (each containing fibers from more than one spinal nerve) divide further into smaller and smaller branches, resulting ultimately in the complete innervation of the hand and most of the arm. The brachial plexus is located in the shoulder region from the neck to the axilla. It is of clinical significance since it is sometimes stretched or torn at birth, causing paralysis and numbness of the baby's arm on

Table 8-4

Spinal nerves and peripheral branches

Spinal nerves	Plexuses formed from anterior rami	Spinal nerve branches from plexuses	Parts supplied
Cervical 1, 2, 3, 4	Cervical plexus	Lesser occipital	Sensory to back of head, front of neck, and upper part of shoulder; motor to numerous neck muscles
		Great auricular	
		Cutaneous nerve of neck	
		Anterior supraclavicular	
		Middle supraclavicular	
		Posterior supraclavicular	
		Branches to numerous neck muscles	
Cervical 5, 6, 7, 8; *Thoracic (or dorsal)* 1	Brachial plexus	Suprascapular and dorsoscapular	Superficial muscles* of scapula
		Thoracic nerves, medial and lateral anterior	Pectoralis major and minor
		Long thoracic nerve	Serratus anterior
		Thoracodorsal	Latissimus dorsi
		Subscapular	Subscapular and teres major muscles
		Axillary (circumflex)	Deltoid and teres minor muscles and skin over deltoid
		Musculocutaneous	Muscles of front of arm (biceps brachii, coracobrachialis, and brachialis) and skin on outer side of forearm
		Ulnar	Flexor carpi ulnaris and part of flexor digitorum profundus; some of muscles of hand; sensory to medial side of hand, little finger, and medial half of fourth finger
		Median	Rest of muscles of front of forearm and hand; sensory to skin of palmar surface of thumb, index, and middle fingers
		Radial	Triceps muscle and muscles of back of forearm; sensory to skin of back of forearm and hand
		Medial cutaneous	Sensory to inner surface of arm and forearm
		Phrenic (branches from cervical nerves before formation of plexus; most of its fibers from fourth cervical nerve)	Diaphragm
Thoracic 2, 3, 4, 5, 6, 7, 8, 9, 10, 11, 12	No plexus formed; branches run directly to intercostal muscles and skin of thorax		

*Although nerves to muscles are considered motor, they do contain some sensory fibers that transmit proprioceptive impulses.

Spinal nerves	Plexuses formed from anterior rami	Spinal nerve branches from plexuses	Parts supplied
		Iliohypogastric } Sometimes fused	Sensory to anterior abdominal wall
		Ilioinguinal } Sometimes fused	Sensory to anterior abdominal wall and external genitalia; motor to muscles of abdominal wall
		Genitofemoral	Sensory to skin of external genitalia and inguinal region
		Lateral cutaneous of thigh	Sensory to outer side of thigh
Lumbar 1 2 3 4 5 *Sacral* 1 2 3 4 5 *Coccygeal* 1	Lumbosacral plexus	Femoral	Motor to quadriceps, sartorius, and iliacus muscles; sensory to front of thigh and to medial side of lower leg (saphenous nerve)
		Obturator	Motor to adductor muscles of thigh
		Tibial† (medial popliteal)	Motor to muscles of calf of leg; sensory to skin of calf of leg and sole of foot
		Common peroneal (lateral popliteal)	Motor to evertors and dorsiflexors of foot; sensory to lateral surface of leg and dorsal surface of foot
		Nerves to hamstring muscles	Motor to muscles of back of thigh
		Gluteal nerves, superior and inferior	Motor to buttocks muscles and tensor fasciae latae
		Posterior cutaneous nerve	Sensory to skin of buttocks, posterior surface of thigh, and leg
		Pudendal nerve	Motor to perineal muscles; sensory to skin of perineum

†Sensory fibers from the tibial and peroneal nerves unite to form the *medial cutaneous* (or sural) *nerve* that supplies the calf of the leg and the lateral surface of the foot. In the thigh, the tibial and common peroneal nerves are usually enclosed in a single sheath to form the *sciatic nerve*, the largest nerve in the body with its width of approximately ¾ of an inch. About two thirds of the way down the posterior part of the thigh, it divides into its component parts. Branches of the sciatic nerve extend into the hamstring muscles.

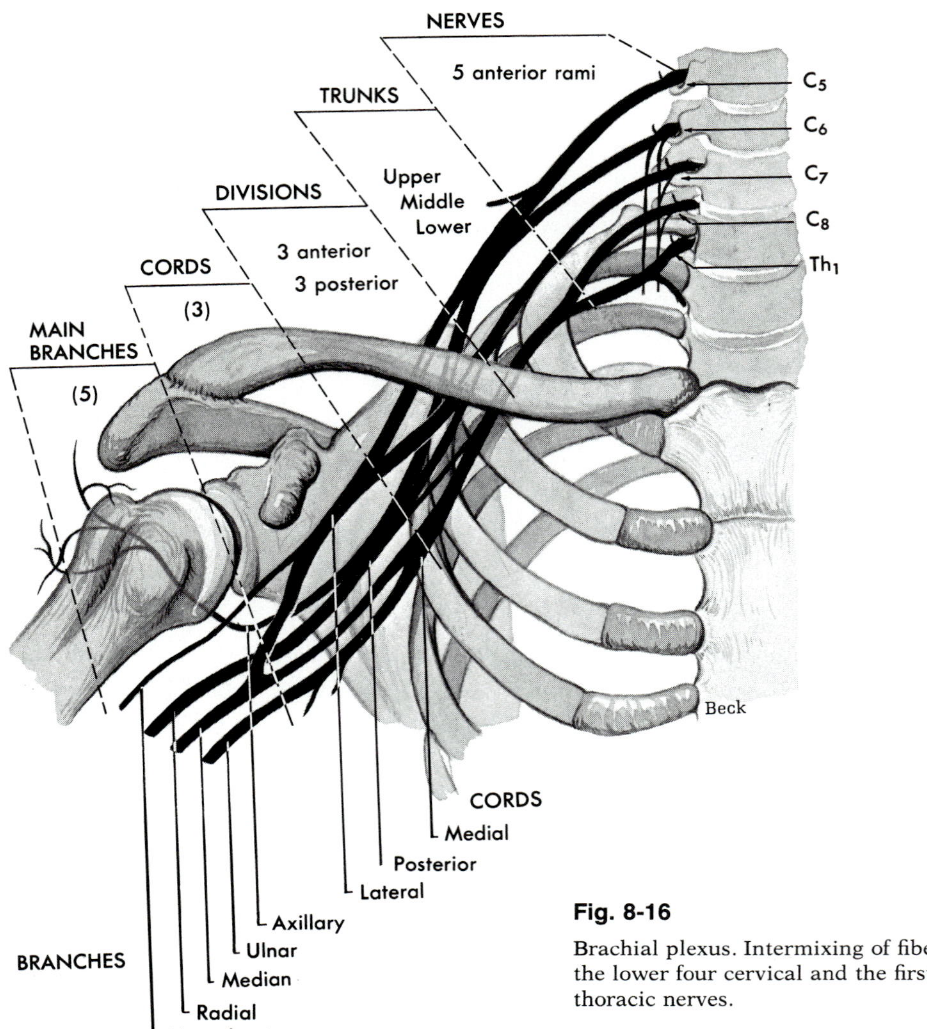

Fig. 8-16

Brachial plexus. Intermixing of fibers from the lower four cervical and the first thoracic nerves.

that side. If untreated, it results in a withered arm. Branching from this plexus are several nerves to the skin and to voluntary muscles.

The right and left phrenic nerves, whose fibers come from the third and fourth or fourth and fifth cervical spinal nerves before formation of the brachial plexus, have considerable clinical interest since they supply the diaphragm muscle. If the neck is broken in a way that severs or crushes the cord above this level, nerve impulses from the brain can, of course, no longer reach the phrenic nerves, and therefore the diaphragm stops contracting. Unless artificial respiration of some kind is provided, the patient dies of respiratory paralysis as a result of the broken neck. Poliomyelitis that attacks the cord between the third and fifth cervical segments also paralyzes the phrenic nerve and, therefore, the diaphragm.

Another spinal nerve plexus is the *lumbar plexus*, formed by the intermingling of fibers from the first four lumbar nerves. This network of nerves is located in the lumbar region of the back in the psoas muscle. The large femoral nerve is one of several nerves emerging from the lumbar plexus. It divides into many branches supplying the thigh and leg.

Fibers from the fourth and fifth lumbar nerves and the first, second, and third sacral nerves form the *sacral plexus,* located in the pelvic cavity on the anterior surface of the piriformis muscle. Among other nerves that emerge from the sacral plexus are the tibial and common peroneal nerves, which, in the thigh, form the largest nerve in the body, namely, the great sciatic nerve (Fig. 8-17). It pierces the buttocks and runs down the back of the thigh. Its many branches supply nearly all the skin of the leg, the posterior thigh muscles, and the leg and foot muscles. Sciatica or neuralgia of the sciatic nerve is a fairly common and very painful condition.

At first glance, the distribution of spinal nerves does not appear to follow an ordered arrangement. But detailed mapping of the skin surface has revealed a close relationship between the source on the cord of each spinal nerve and the vertical position of the body it innervates (Figs. 8-18 and 8-19). Knowledge of the segmental arrangement of spinal nerves has proved useful to physicians. For instance, a neurologist can identify the site of spinal cord or nerve abnormality from the area of the body insensitive to a pinprick.

Microscopic structure

All spinal nerves are *mixed nerves.* As shown in Fig. 7-5, spinal nerves are composed of both sensory dendrites and motor axons. Note also in Fig. 7-5 that the sensory dendrite shown comes from a cell body located in the ganglion on the posterior root of the spinal nerve. The motor axon shown originates from a cell body located in the anterior gray column of the cord. These neurons whose cell bodies lie in the anterior gray columns and whose axons terminate in skeletal muscle are called *somatic motoneurons.* Spinal nerves also contain axons of preganglionic autonomic motoneurons (p. 230).

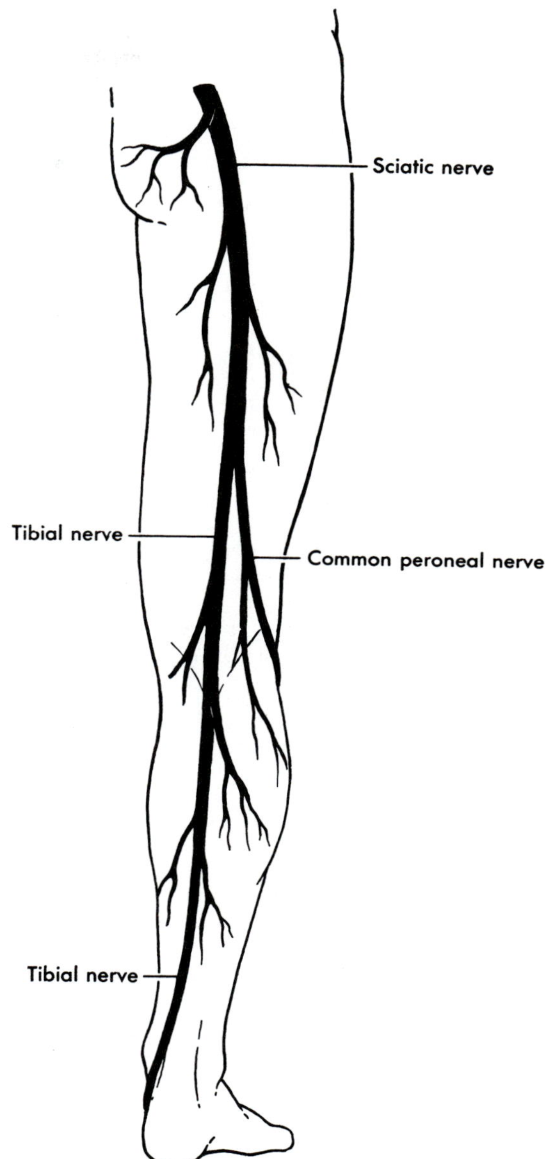

Fig. 8-17

Main nerves of the lower extremity.

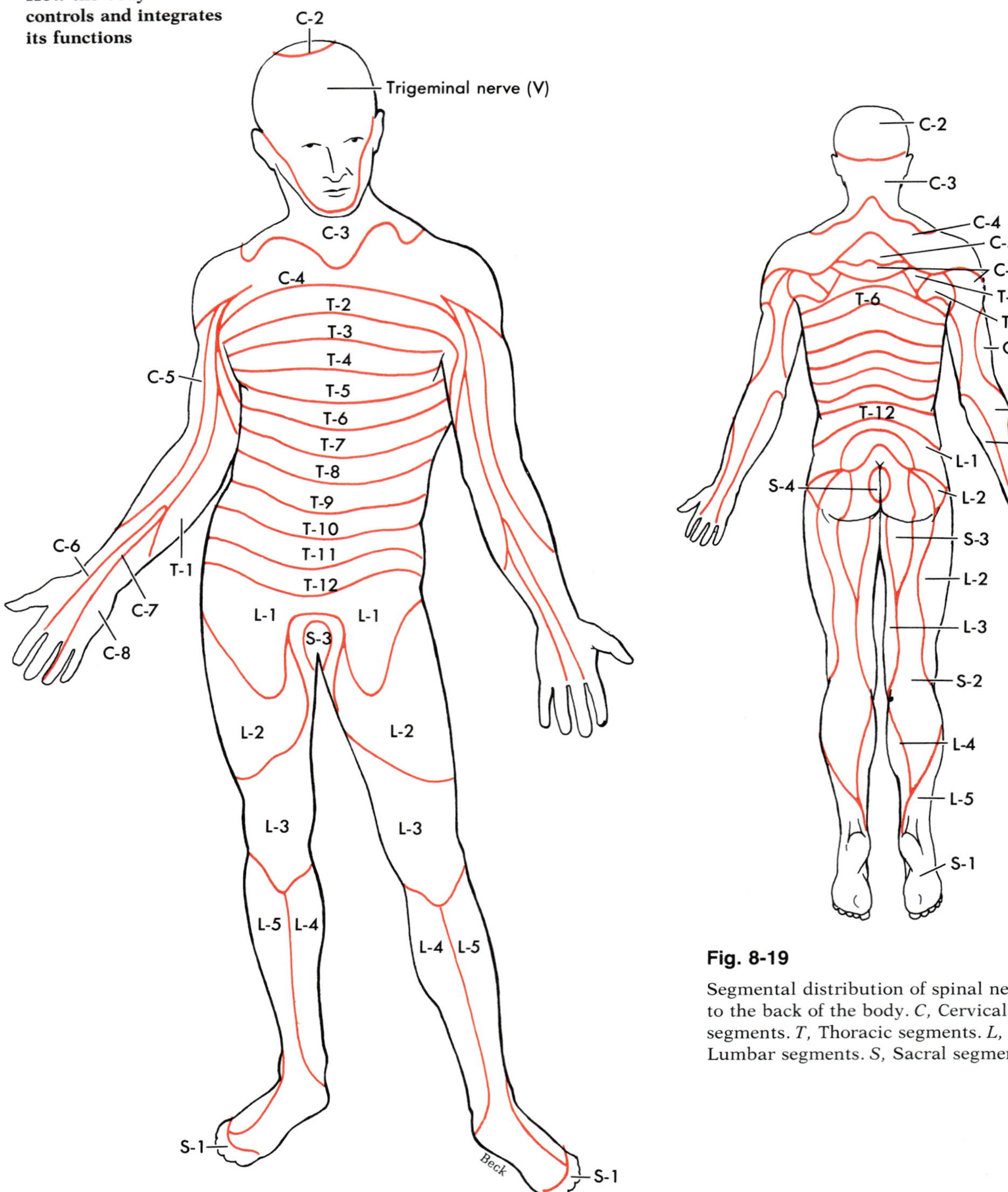

Fig. 8-18

Segmental distribution of spinal nerves to the front of the body. *C*, Cervical segments. *T*, Thoracic segments. *L*, Lumbar segments. *S*, Sacral segments.

Fig. 8-19

Segmental distribution of spinal nerves to the back of the body. *C*, Cervical segments. *T*, Thoracic segments. *L*, Lumbar segments. *S*, Sacral segments.

Table 8-5

Cranial nerves contrasted with spinal nerves

	Cranial nerves	Spinal nerves
Origin	Base of brain	Spinal cord
Distribution	Mainly to head and neck	Skin, skeletal muscles, joints, blood vessels, sweat glands, and mucosa except of head and neck
Structure	Some composed of sensory fibers only; some of both motor axons and sensory dendrites; some motor fibers belong to somatic nervous system, some to autonomic	All of them composed of both sensory dendrites and motor axons; some of latter, somatic, some autonomic
Function	Vision, hearing, sense of smell, sense of taste, eye movements, etc.	Sensations, movements, and sweat secretion

Physiology of the somatic nervous system

Because the physiology of the somatic nervous system is so vast and complex, we have organized our material about it around three topics—sensory pathways, motor pathways, and functions of the central nervous system—to correspond with the basic functions performed by the three kinds of somatic neurons.

Sensory pathways

1 Sensory neural pathways consist of relays of sensory neurons that conduct impulses from any part of the body to the spinal cord or brainstem and from these lower levels of the central nervous system up to its highest level, the cerebral cortex. To reach the cortex, impulses must be conducted over a relay of at least three sensory neurons. We shall identify them as sensory neurons I, II, and III. Some important general principles about sensory neural pathways from the periphery—any part of the body outside of the central nervous system—to the cerebral cortex follow.

a Sensory neurons I of the relay conduct from the periphery to the central nervous system. If the receptors of these neurons lie in regions supplied by spinal nerves, their dendrites lie in a spinal nerve and their axons terminate in gray matter of the cord or of the brainstem. Where are their cell bodies located? (Confirm or find your answer in Fig. 7-6.) If receptors of sensory neurons I lie in regions supplied by cranial nerves, their dendrites lie in a cranial nerve, their cell bodies lie in cranial nerve ganglia, and their axons terminate in gray matter of the brainstem. In either case, sensory neuron I axon terminals synapse with sensory neuron II dendrites or cell bodies. (See Fig. 8-20.)

b Sensory neurons II conduct from the cord or brainstem up to the thalamus. Their dendrites and cell bodies are located in cord or brainstem gray matter. Their axons ascend in ascending tracts up the cord, through the brainstem, and terminate in the thalamus. Here they synapse with sensory neuron III dendrites or cell bodies (Fig. 8-20).

c Sensory neurons III conduct from the thalamus to the postcentral gyrus of the parietal lobe. Over the years, sev-

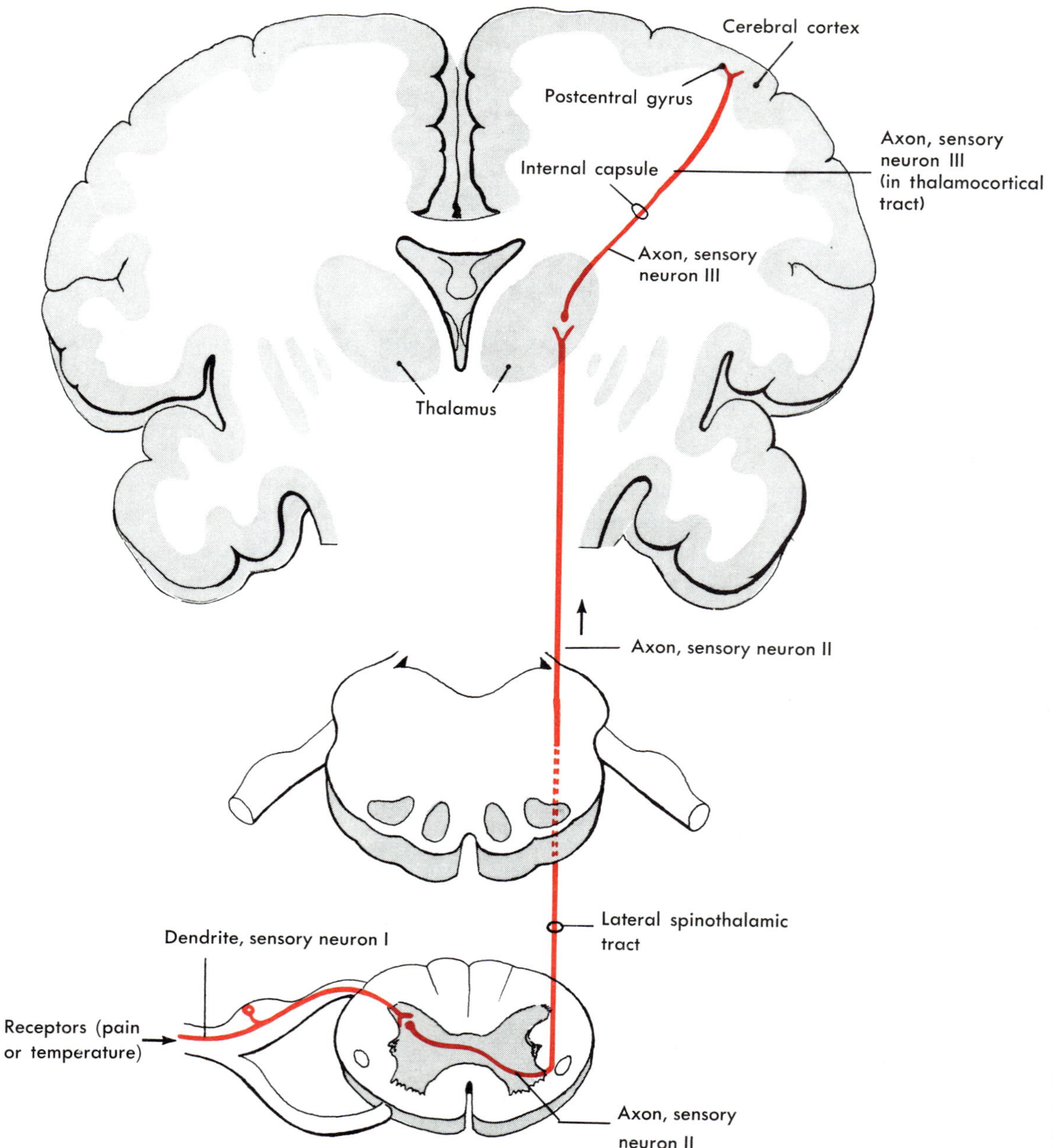

Fig. 8-20

The lateral spinothalamic tract relays sensory impulses from pain and temperature receptors up the cord to the thalamus. Thalamocortical tract fibers relay them to the somatic sensory area of the cortex (postcentral gyrus).

eral names have designated this area; *somatic sensory area* seems to be a popular name for it now. Bundles of axons of sensory neurons III form thalamocortical tracts. They extend through the portion of cerebral white matter known as the *internal capsule* to the cerebral cortex (Fig. 8-20).

2 For the most part, sensory pathways to the cerebral cortex are crossed pathways. This means that each side of the brain registers sensations from the opposite side of the body. Examine Fig. 8-20. The axon of which sensory neuron—I, II, or III—decussates in this sensory pathway? Usually it is the axon of sensory neuron II that decussates at some level in its ascent to the thalamus.

3 The neural pathway for sensations of *pain and temperature* is called the *lateral spinothalamic pathway.*

4 Two neural pathways conduct impulses that produce sensations of touch and pressure, namely, the medial lemniscal system and the ventral spinothalamic pathway. The *medial lemniscal system* consists of the tracts that make up the posterior white columns of the cord (the fasciculi cuneatus and gracilis) plus the *medial lemniscus,* a flat band of white fibers extending through the medulla, pons, and midbrain. (Derivation of the name lemniscus may interest you. It comes from the Greek word *lemniskos* meaning "woolen band." Apparently, to some early anatomist, the medial lemniscus looked like a band of woolen material running through the brainstem.)

The fibers of the medial lemniscus, like those of the spinothalamic tracts, are axons of sensory neurons II. They originate from cell bodies in the medulla, decussate, and then extend upward to terminate in the thalamus on the opposite side. The function of the medial lemniscal system is to transmit impulses that produce our more discriminating touch and pressure sensations. These include stereognosis (awareness of an object's size, shape, and texture), precise localization, two-point discrimination, weight discrimination, and sense of vibrations.

The function of the *ventral spinothalamic pathway* is to transmit impulses that result in crude touch and pressure sensations—knowing that something touches the skin, for example, but not its precise location, size, shape, or texture.

5 The neural pathway for *kinesthesia* (sense of movement and position of body parts) is also the medial lemniscal system. See Table 8-6 for a brief summary of the sensory neural pathways named in items 2, 3, and 4.

6 The first part of the cerebral cortex reached by sensory impulses conducted up spinothalamic and medial lemniscal pathways is the postcentral gyrus of the parietal lobe—areas 3, 1, and 2 in Brodman's map of the cortex (Fig. 8-21). It has several names but the most popular one now is probably *somatic sensory area.* It functions to produce our general senses—not only the common ones such as pain, heat, cold, touch, and pressure, but also those less familiar senses such as stereognosis, kinesthesia, vibratory sense, and two-point and weight discrimination. General sensations of the right side of the body are predominantly experienced by the left cerebral hemisphere's somatic sensory area. General sensations of the left side of the body are predominantly experienced by the right hemisphere's somatic sensory area.

7 Secondary somatic sensory areas have also been identified in the cerebral cortex. One such lies in the superior part of the fissure of Sylvius.

Table 8-6

Sensory neural pathways

Neurons	Gross structures in which neuron parts are located
1 Pain and temperature	
Sensory neuron I	
Receptors	In skin, mucosa, muscles, tendons, viscera
Dendrite	In spinal nerve and branch of spinal nerve
Cell body	In spinal ganglion, on posterior root of spinal nerve
Axon	In posterior root of spinal nerve; terminates in posterior gray column of cord
Sensory neuron II	
Dendrite	Posterior gray column
Cell body	Posterior gray column
Axon	Decussates and ascends in lateral spinothalamic tract (Figs. 8-12 and 8-20); terminates in thalamus
Sensory neuron III	
Dendrite	Thalamus
Cell body	Thalamus
Axon	Thalamus via thalamocortical tract in internal capsule to general sensory area of cerebral cortex, i.e., postcentral gyrus in parietal lobe
2 Crude touch stimuli	
Sensory neuron I	Same as sensory neuron I for pain and temperature stimuli
Sensory neuron II	
Dendrite	Posterior gray column; same as sensory neuron II for pain and temperature
Cell body	Posterior gray column; same as sensory neuron II for pain and temperature
Axon	In ventral spinothalamic tract (Fig. 8-12) to thalamus
Sensory neuron III	Same as sensory neuron III for pain and temperature stimuli
3 Discriminating touch (two-point discrimination, vibrations) **deep touch,** and **pressure and kinesthesia**	
Sensory neuron I	Same as sensory neuron I for pain, temperature, and crude touch stimuli, except that axon extends up cord in posterior white columns (fasciculi gracilis and cuneatus, Fig. 8-12) to nuclei gracilis or cuneatus in medulla instead of terminating in posterior gray columns of cord
Sensory neuron II	
Dendrite	In nuclei gracilis or cuneatus of medulla
Cell body	In nuclei gracilis or cuneatus of medulla
Axon	Decussates and ascends in medial lemniscus (broad band of fibers extending up through medulla and midbrain) and terminates in thalamus
Sensory neuron III	Same as sensory neuron III for pain, temperature, and crude touch stimuli

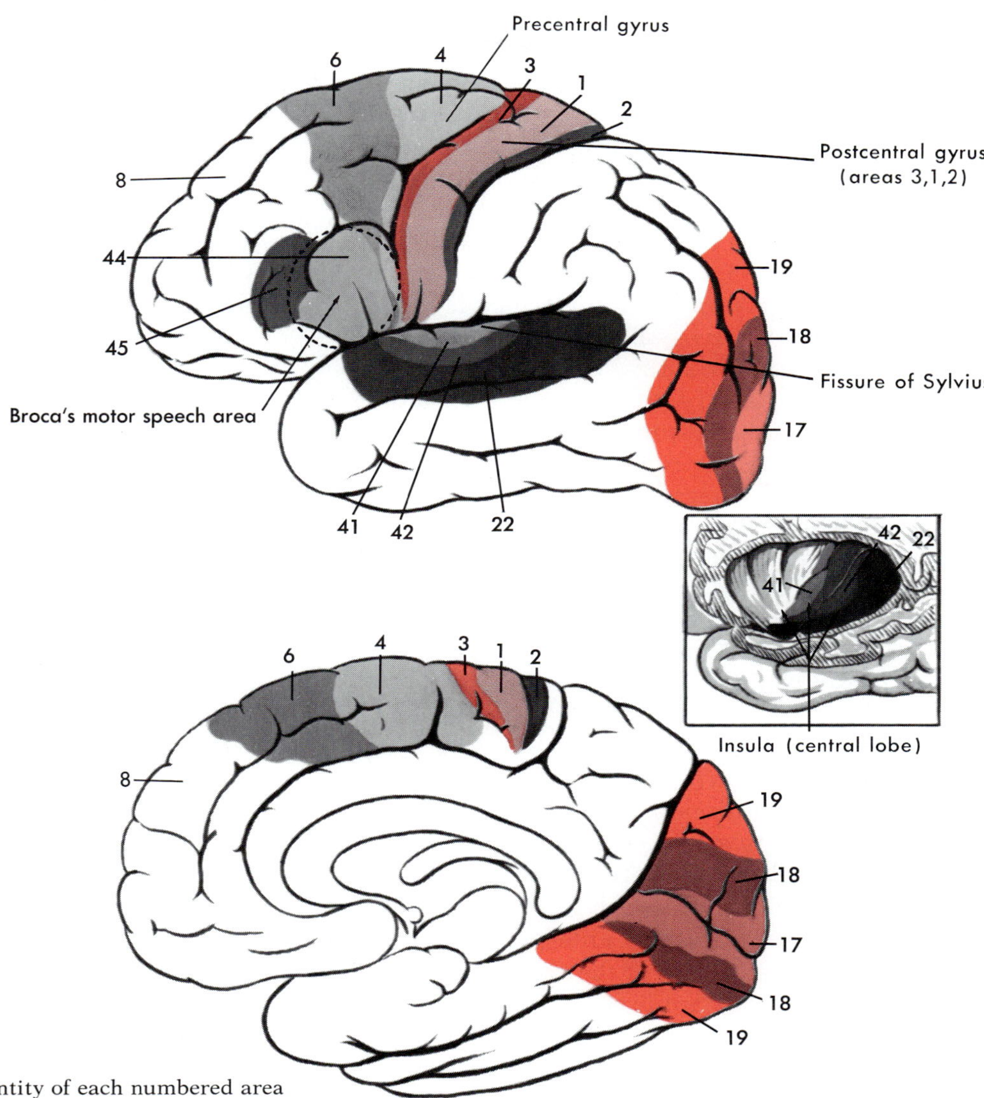

Fig. 8-21

Map of human cortex. Identity of each numbered area is determined by structural differences in neurons that compose it. Some areas whose functions are best understood are the following: areas *3*, *1*, and *2*, somatic sensory areas; area *4*, primary motor area; area *6*, secondary motor area; area *17*, primary visual area; areas *18* and *19*, secondary visual areas; areas *41* and *42*, primary auditory areas; area *22*, secondary auditory area. Area *44* and posterior part of area *45* constitute the approximate location of Broca's motor speech area. (Modified after Brodmann, K.: Feinere Anatomie des Grosshirns. In Handbuch der Neurologie, Berlin, 1910, Springer-Verlag.)

Somatic motor pathways

Somatic motor pathways consist of motoneurons that conduct impulses from the central nervous system to somatic effectors (that is, skeletal muscles). Some motor pathways are extremely complex and not at all clearly defined. Others, notably spinal cord reflex arcs, are simple and well established. You read about these in Chapter 7. Look back now at Figs. 7-5 and 7-6. From these diagrams you can derive a cardinal principle about somatic pathways—the *principle of the final common path.* It is this: only one final common path, namely, the anterior horn motoneuron, conducts impulses to skeletal muscles. Anterior horn motoneuron axons are the only ones that terminate in skeletal muscle cells. This principle of the final common path to skeletal muscles has important practical implications. For example, it means that any condition that makes anterior horn motoneurons unable to conduct impulses also makes skeletal muscle cells supplied by these neurons unable to contract. They cannot be willed to contract nor can they contract reflexly. They are, in short, paralyzed. Most famous of the diseases that produce paralysis by destroying anterior horn motoneurons is poliomyelitis. Numerous somatic motor paths conduct impulses from motor areas of the cerebrum down to anterior horn motoneurons at all levels of the cord.

Two methods are used to classify them—one based on the location of their fibers in the medulla and the other on their influence on the lower motoneurons. The first method divides them into pyramidal (Table 8-7) and extrapyramidal tracts. The second classifies them as facilitatory and inhibitory tracts.

Pyramidal tracts are those whose fibers come together in the medulla to form the pyramids, hence their name. Because axons composing the pyramidal tracts originate from neuron cell bodies located in the cerebral cortex, they also bear another name—*corticospinal tracts.* About three fourths of their fibers decussate (cross over from one side to the other) in the medulla. After decussating, they extend down the cord in the crossed corticospinal tract located on the opposite side of the cord in the lateral white column. About one fourth of the corticospinal fibers do not decussate. Instead, they extend down the same side of the cord as the cerebral area from which they came. One pair of uncrossed tracts lies in the ventral white columns of the cord, namely, the ventral corticospinal tracts. The other uncrossed corticospinal tracts form part of the lateral corticospinal tracts (Fig. 8-12). About 60% of corticospinal fibers are axons that arise from neuron cell bodies in the precentral (frontal lobe) region of the cortex.* In Fig. 8-21, these are the areas numbered 4, 6, and 8. About 40% of corticospinal fibers originate from neuron cell bodies located in postcentral areas of the cortex, areas classified as sensory; now, more accurately, they are often called sensorimotor areas.

Relatively few corticospinal tract fibers synapse directly with anterior horn motoneurons. Most of them synapse with interneurons, which in turn synapse with anterior horn motoneurons. All corticospinal fibers conduct impulses that facilitate—that is, decrease—the resting potential (p. 164) of anterior horn motoneurons. The effects of facilitatory impulses acting rapidly on any one neuron add up or summate. Each impulse, in other words, decreases the neuron's resting potential a little bit more. If sufficient numbers of impulses impinge rapidly enough on a neuron, its potential decreases to threshold level. And at that moment, the neuron starts conducting impulses. In short, it is stimulated. Stimulation of anterior horn motoneurons by corticospinal tract impulses

*Mountcastle, V. B., editor: Medical physiology, ed. 13, St. Louis, 1974, The C. V. Mosby Co., p. 748.

Table 8-7

Pyramidal path from cerebral cortex

Microscopic structures	Macroscopic structures in which neurons located
Upper motoneuron (Betz cells)	
Dendrite	Motor area of cerebral cortex
Cell body	Motor area of cerebral cortex
Axon	Motor area of cerebral cortex; descends in corticospinal (pyramidal) tract through cerebrum and brainstem; decussates in medulla and continues descent in lateral corticospinal (crossed pyramidal) tract in lateral white column; or may descend uncrossed in ventral, or direct, pyramidal tract in anterior white column and either decussate or not prior to terminating in anterior gray column
Lower motoneuron	
Dendrite	Anterior gray column
Cell body	Anterior gray column
Axon	Anterior gray column to anterior root of spinal nerve to spinal nerve and branches; terminates in somatic effector—i.e., skeletal muscle

results in stimulation of individual muscle groups (mainly of the hands and feet). Precise control of their contractions is, in short, the function of the corticospinal tracts. Without stimulation of anterior horn motoneurons by impulses over corticospinal fibers, willed movements cannot occur. This means that paralysis results whenever pyramidal corticospinal tract conduction is interrupted. For instance, the paralysis that so often follows cerebral vascular accidents ("strokes") comes from pyramidal neuron injury—sometimes of their cell bodies in the primary motor area, sometimes of their axons in the internal capsule (Fig. 8-22).

Extrapyramidal tracts are much more complex than pyramidal tracts. They consist of all motor tracts from the brain to the spinal cord anterior horn motoneurons except the corticospinal (pyramidal) tracts. Within the brain, extrapyramidal tracts consist of numerous but as yet incompletely worked-out relays of motoneurons between motor areas of the cortex, basal ganglia, thalamus, cerebellum, and brainstem. In the cord, some of the most important extrapyramidal tracts are the reticulospinal tracts.

Fibers of the *reticulospinal tracts* originate from cell bodies in the reticular formation of the brainstem and terminate in gray matter of the spinal cord, where they synapse with interneurons that synapse with lower (anterior horn) motoneurons. Some reticulospinal tracts function as facilitatory tracts, others as inhibitory tracts. Impulses over facilitatory tracts tend to decrease the lower motoneuron's resting potential. Impulses over inhibitory tracts tend to increase its potential. Summation of these opposing influences determines the lower motoneuron's response. It initiates impulse conduction only when facilitatory impulses exceed inhibitory impulses sufficiently to decrease its potential to its threshold level.

Conduction by extrapyramidal tracts plays a crucial part in producing our larger, more automatic movements because extrapyramidal impulses bring about contractions of groups of muscles in sequence or simultaneously. Such muscle action occurs, for example, in swimming and walking and, in fact, in all normal voluntary movements.

Conduction by extrapyramidal tracts plays an important part in our emotional expres-

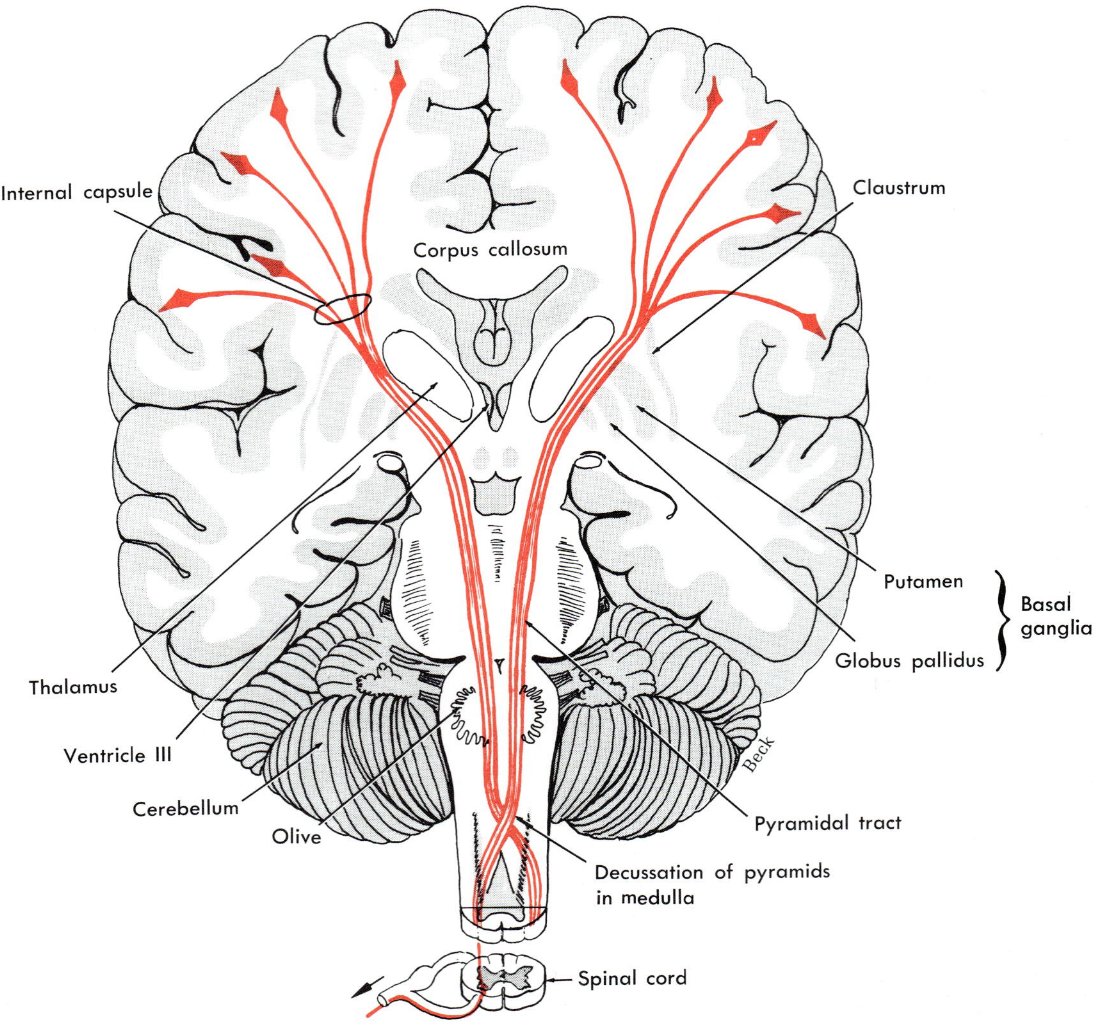

Fig. 8-22

The crossed corticospinal (pyramidal) tracts. Axons that compose pyramidal tracts come from neuron cell bodies in the cerebral cortex. After they descend through the internal capsule of the cerebrum and the white matter of the brainstem, about three fourths of the fibers decussate—cross over from one side to the other—in the medulla, as shown here. Then they continue downward in the lateral corticospinal tract (Fig. 8-12) on the opposite side of the cord. Each crossed corticospinal tract, therefore, conducts motor impulses from one side of the brain to interneurons or anterior horn motoneurons on the opposite side of the cord. Therefore, impulses from one side of the cerebrum cause movements of the opposite side of the body.

sions. For instance, most of us smile automatically at things that amuse us and frown at things that irritate us. And it is extrapyramidal impulses, not pyramidal, that produce the smiles or frowns.

Axons of many different neurons converge upon—that is, synapse—with each anterior horn motoneuron. Hence many impulses from diverse sources—some facilitatory and some inhibitory—continually bombard this final common path to skeletal muscles. Together the added or summated effect of these opposing influences determines lower motoneuron functioning. Facilitatory impulses reach these cells via sensory neurons (whose axons, you will recall, lie in the posterior roots of spinal nerves), pyramidal (corticospinal) tracts, and extrapyramidal facilitatory reticulospinal tracts. According to recent evidence, impulses over facilitatory reticulospinal fibers facilitate the lower motoneurons that supply extensor muscles. And at the same time, they reciprocally inhibit the lower motoneurons that supply flexor muscles. Hence, facilitatory reticulospinal impulses tend to increase the tone of extensor muscles and decrease the tone of flexor muscles.

Inhibitory impulses reach lower motoneurons mainly via inhibitory reticulospinal fibers that originate from cell bodies located in the *bulbar inhibitory area* in the medulla. They inhibit the lower motoneurons to extensor muscles (and reciprocally stimulate those to flexor muscles). Hence, inhibitory reticulospinal impulses tend to decrease extensor muscle tone and increase flexor muscle tone —opposite effects from facilitatory reticulospinal impulses.

Brainstem inhibitory and facilitatory areas, as indicated in Fig. 8-23, are influenced directly or indirectly by impulses from various higher motor centers. According to a recent study,* the basal ganglia and cerebellum send impulses to the thalamus, which presumably modifies this information before transmitting it to motor areas of the cortex. This concept, in other words, sees the basal ganglia and cerebellum functioning as the highest command centers for movements and relaying their orders through the cerebral motor cortex to skeletal muscles. If this is true, it implies that the function of the cerebral motor cortex may be to refine skeletal

*Evarts, E. V.: Brain mechanisms in movement, Sci. Am. **209:**96-103, July, 1973.

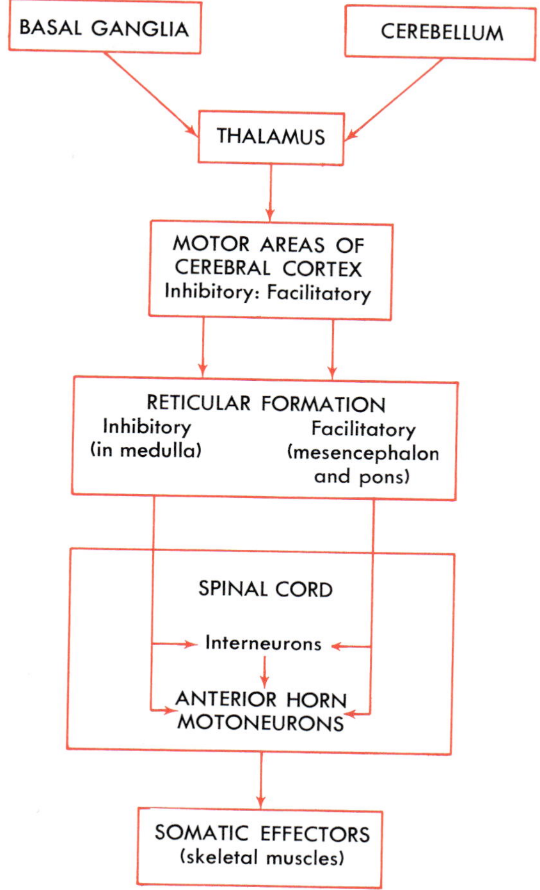

Fig. 8-23

Extrapyramidal motor paths, suggested by recent experimental studies.

muscle actions rather than to will them. The traditional view is quite different. It sees the cerebral motor cortex functioning as the highest command center for movements and relaying some of its orders through the basal ganglia and cerebellum. If this is true, it implies that the cerebral motor cortex wills skeletal muscle actions and that the basal ganglia and cerebellum refine them. The truth is that no one yet knows the truth. Physiologists still cannot tell us positively and precisely what part of the brain initiates the commands that lead to movements, nor can they describe with assurance the neural pathways through the brain from higher to lower motor centers. Whatever these pathways may be, the ratio of facilitatory and inhibitory impulses converging upon lower motoneurons (anterior horn motoneurons) is such as to maintain normal muscle tone. In other words, facilitatory impulses normally somewhat exceed inhibitory impulses. But disease sometimes alters this ratio. Parkinson's disease and "strokes," for example, may interrupt transmission by inhibitory extrapyramidal paths from basal ganglia to bulbar inhibitory centers. Facilitatory impulses then predominate, and excess muscle tone (rigidity or spasticity) develops. Injury of *upper motoneurons* (those whose axons lie in either pyramidal or extrapyramidal tracts) produces symptoms frequently referred to as "pyramidal signs," notably a spastic type of paralysis, exaggerated deep reflexes, and a positive Babinski reflex (see p. 223). Actually, pyramidal signs result from interruption of both pyramidal and extrapyramidal pathways. The paralysis stems from interruption of pyramidal tracts, whereas the spasticity (rigidity) and exaggerated reflexes come from interruption of inhibitory extrapyramidal pathways.

Injury of lower motoneurons produces different symptoms from upper motoneuron injury. Anterior horn cells or lower motoneurons, you will recall, constitute the final common path by which impulses reach skeletal muscles. This means that if they are injured, impulses can no longer reach the skeletal muscles they supply. This, in turn, means both that reflexes involving these muscles are no longer possible and that these muscles cannot be moved at the patient's will. As a result, they lose their normal tone and become soft and flabby (flaccid). In short, absence of reflexes and flaccid paralysis are the chief "lower motoneuron signs."

Functions of the cerebrum

The cerebrum above all other bodily structures deserves the title, "the most important human organ." Why? Because it can claim two distinctions not shared by any other organ. The cerebrum is the top executive of the world's most complex organization—the human body. In addition, the cerebrum provides the functions that endow us with our uniquely human qualities. During the past decade or so, research scientists in various fields—neurophysiology, neurosurgery, neuropsychiatry, and others—have added mountains of information to our knowledge about the brain. But questions come faster than answers, and clear complete understandings of the brain's mechanisms still elude us. Perhaps they forever will. Perhaps the capacity of the human brain falls short of the ability to understand its own complexity. We shall start our discussion of the cerebrum with some generalizations about its functions, and follow these with more specific information about each generalization.

1 Cerebral activity goes on as long as life itself. Only when life ceases (or moments before) does the cerebrum cease its functioning. Only then do all of its neurons stop conducting impulses. Proof of this has come from records of brain electrical potentials known as electroencephalograms, or EEG's, or "brain waves."

2 The cerebrum performs three kinds of functions: sensory functions, motor functions, and a group of activities less easily named and even less easily defined or explained. "Integrative functions" is one name for them. What most of us think of as mental activities are part, but not all, of the cerebrum's integrative functions.
3 The right and left hemispheres of the cerebrum specialize in different functions. Also, certain regions of each hemisphere play key roles in particular functions.
4 Certain chemical transmitters are unique to brain and cord synapses, that is, they are not released at neuroeffector junctions, nor at autonomic synapses.

Electroencephalograms

Electroencephalograms (EEG's) are records of the cerebrum's electrical activity. These records are usually made from a number of electrodes placed on different regions of the scalp, and they consist of waves—brain waves, as they are called. Brain waves vary as to the number of times they occur per second and as to voltage. In other words, they vary in frequency and amplitude. These variations depend on which lobe the electrodes lie over, what state of awareness the individual is in, and whether his cerebrum is functioning normally or abnormally.

Four types of brain waves are recognized: alpha, beta, delta, and theta. *Alpha waves* are moderately fast (8 to 13 per second), relatively high-voltage waves. Normally they are the dominant waves in EEG's from the parietal, occipital, and posterior parts of the temporal lobes under certain conditions—when an individual is awake, has his eyes closed, and is in a relaxed, nonattentive state, when his cerebrum is idling, so to speak. *Beta waves,* in contrast, are faster (13 to 25 per second) but lower voltage waves than alpha waves. Beta waves are prominent in EEG's made from the more anterior parts of the cerebrum, especially in those from the precentral (motor) gyrus of the frontal lobe. Beta waves normally dominate the EEG under the following conditions: when an individual is awake, has his eyes open, and is in an activated or attentive state—when his cerebrum is not idling but is busily engaged with sensory stimulation, for example, and mental activities. *Delta waves* are the slowest (0.5 to 3.5 per second) brain waves and they have a high voltage. They characterize the EEG during deep sleep. Recognizing this fact, physiologists refer to deep sleep as slow-wave sleep (or SWS). *Theta waves* are moderately slow (3 to 7 per second), low-voltage waves. They appear as drowsiness descends. Summarizing, fast waves predominate during waking states whereas slow waves predominate during drowsy or sleep states. We might coin the following nicknames for brain waves: "busy waves," "relaxed waves," "drowsy waves," and "deep sleep waves." "Busy waves" predominate when the cerebrum is busiest, or to use a more scientific term, when it is activated. Busy waves are beta waves. They are the fast, low-voltage waves that occur when we are awake, have our eyes open, and are alert and attentive. "Relaxed waves" are alpha waves. They are the moderately fast, high-voltage waves that predominate when we are relaxed and nonattentive but are awake with our eyes closed. "Drowsy waves" are theta waves—moderately slow, low-voltage waves that appear when we are drowsy, either when we are falling asleep or when we are arousing from sleep. "Deep sleep waves" are the very slow, high-voltage delta waves that identify us as being sound asleep and not easily aroused.

Scientists and laymen alike, during recent years, have become increasingly interested in brain waves. Physicians now make more frequent use of electroencephalograms as aids for diagnosing certain altered states of brain function and even for establishing death. A flat EEG indicates that the cerebrum is producing no brain waves. This is now generally considered the best criterion of death. Research physiologists also make use of electroencephalograms—for example, in their studies of the physiology of individuals who practice meditation techniques.

Sensory functions

Many of the nervous system's gross structures function to produce sensations—nerves, ganglia, cord, medulla, pons, midbrain, thalamus, and cerebral cortex. Complex discriminative sensations depend upon the cerebral cortex, especially upon its somatic sensory, visual, and auditory areas (Fig. 8-21). These regions of the cortex do more than just register separate and simple sensations. They compare them and evaluate them. They integrate them into perceptions of wholes. Suppose, for example, that someone blindfolded you and then put an ice cube in your hand. You would, of course, sense something cold touching your hand. But also, you would probably know that it was an ice cube because you would sense a total impression compounded of many sensations such as temperature, shape, size, weight, texture, and movement and position of your hand and arm.

Somatic motor functions

Mechanisms that control voluntary movements are extremely complex and are imperfectly understood. It is known, however, that for normal movements to take place, many parts of the nervous system—including certain areas of the cerebral cortex—must function. The precentral gyrus (that is, the most posterior gyrus of the frontal lobe; area 4, Fig. 8-21) constitutes the primary motor area. However, the gyrus immediately anterior to the precentral gyrus also contains motoneurons. So, too, do many other regions, including even the somatic sensory area. Neurons in the precentral gyrus are said to exert control over individual muscles, especially those that produce movements of distal joints (wrist, hand, finger, ankle, foot, and toe movements). Neurons in the gyrus just anterior to the precentral gyrus, on the other hand, are thought to activate groups of muscles simultaneously.

Integrative functions

"Integrative functions of the cerebrum" is, to say the least, a nebulous phrase, and the neural processes it designates are even more obscure. In general, they consist of all events that take place in the cerebrum between its reception of sensory impulses and its sending out of motor impulses. Integrative functions of the cerebrum include consciousness and mental activities of all kinds. Consciousness, memory, use of language, and emotions are the integrative cerebral functions that we shall discuss briefly.

Consciousness may be defined as a state of awareness of one's self and one's environment and of other beings. Very little is known about the neural mechanisms that produce consciousness. One fact known, however, is that consciousness depends upon excitation of cortical neurons by impulses conducted to them by a relay of neurons known as the reticular activating system. The *reticular activating system* consists of centers in the brainstem reticular formation that receive impulses from the cord and relay them to the thalamus and from the thalamus to all parts of the cerebral cortex. Both direct spinal reticular tracts and collateral fibers from the specialized sensory tracts (spinothalamic, lemniscal, auditory, and visual) relay impulses over the reticular activating system to the cortex. Without continual excitation of cortical neurons by reticular activating impulses, an individual is unconscious and cannot be aroused. Here then are two accepted concepts about the reticular activating system: it functions as the arousal or alerting system for the cerebral cortex, and its functioning is crucial for maintaining consciousness. Drugs known to depress the reticular activating system decrease alertness and induce sleep. Barbiturates, for example, act this way. Amphetamine, on the other hand, a drug known to have a stimulating effect on the cerebrum and to enhance alertness and produce wakefulness, probably acts by stimulating the reticular activating system.

Certain variations in the levels or states of consciousness are normal. All of us, for example, experience different levels of wakefulness. At times we are highly alert and attentive. At other times we are relaxed and nonattentive. All of us also experience different levels of sleep. Two of the best known stages are those called slow-wave sleep (SWS) and rapid eye movement (REM) sleep. Slow-wave sleep takes its name from the slow frequency, high-voltage delta waves that identify it. It is almost entirely a dreamless sleep. REM sleep, on the other hand, is associated with dreaming.

In addition to the various normal states of consciousness, altered states of consciousness (ASC) also occur under certain conditions. Anesthetic drugs produce an altered state of consciousness, namely anesthesia. Disease or injury of the brain may produce the type of ASC called coma. And lysergic acid diethylamide (LSD), a "mind-altering" drug, induces a type of ASC known in the drug culture as a "trip." Peoples of Eastern cultures have long been familiar with an altered state of consciousness called *yoga,* or *meditation.* Yoga is a waking state but differs markedly in certain respects from the usual waking state. According to an accepted definition, yoga is a "higher" or "expanded" level of consciousness. This higher consciousness is accompanied, almost paradoxically, by a high degree of both relaxation and alertness. With training in meditation techniques and practice, an individual can enter the meditative state at will and remain in it for an extended period. Undoubtedly the most widely practiced form of meditation in this country is transcendental meditation (TM) as taught by Maharishi Mahesh Yogi.* Certain physiological changes are now known to occur during meditation. We shall describe these in Chapter 9.

Memory is one of our major mental activities. The mechanisms responsible for memory, like those for consciousness, have not yet yielded their secrets. Many investigators have tried and are now trying to search out answers to explain memory. What have they learned so far? One fact they have established is that the cerebral cortex functions in memory. But they cannot yet explain how impulse conduction by cortical neurons can possibly produce memories. They know that memory does not reside in any one part of the cortex. Apparently several parts of it store memories. Dr. Wilder Penfield, a noted Canadian neurosurgeon, presented some of the earliest evidence of memory storage and recall. In the 1950's, he electrically stimulated the temporal lobes of patients undergoing brain surgery and discovered to his surprise that they responded by recalling in the most minute detail songs and events from their past. They seemed more to be reliving than remembering their experiences. Other investigators have accumulated evidence of memory storage in the occipital and parietal lobes.

More recently, researchers have reported with some certainty that the cerebrum's limbic system—the "emotional brain" as it is called—plays a key role in memory. A variety of observations support this view. To mention one, when the hippocampus (part of the limbic system) is removed, the patient

*Trotter, R. J.: Transcendental meditation, Sci. News **104**:376-378, Dec., 1973.

loses his ability to recall new information. Another idea now widely accepted is that protein synthesis constitutes a crucial part of the mechanism for long-term memory. Whether or not these proteins serve as human "memory molecules" has not yet been proved. Studies done on rats suggest that memory may also involve an increase in the number of contacts between synapsing neurons in the brain.

Speech functions consist of the use of language (speaking and writing) and the understanding of language (spoken and written). Today, these faculties are believed to depend on integrated functioning of various parts of the cerebrum. Because certain areas in the left frontal, parietal, and temporal lobes serve as focal points for this integration, they are called speech centers. Lesions in different ones of the speech centers are associated with different types of speech defects or aphasias. One authoritative source for additional information about these not uncommon language disorders is Geschwind's article, "Language and the Brain."*

Emotions—both the subjective experiencing and objective expression of them—involve functioning of the cerebrum's limbic system. The name limbic, which derives from the Latin word for border or fringe, suggests the shape of the cortical structures that make up the system. They lie on the medial surface of the cerebrum and form a curving border around the corpus callosum, the structure that connects the two cerebral hemispheres. Look now at Fig. 8-5. Here you can identify most of the structures of the limbic system. They are the cingulate gyrus, the isthmus, the hippocampal gyrus, the uncus, and the hippocampus (the extension of the hippocampal gyrus that protrudes into the floor of the inferior horn of the lateral ventricle). These limbic system structures have primary connections with various other parts of the brain, notably the septum, the amygdala (the tail of the caudate nucleus, one of the basal ganglia), and the hypothalamus. Some physiologists, therefore, include these connected structures as parts of the limbic system.

*Geschwind, N.: Language and the brain, Sci. Am. **226:** 76-83, Apr., 1972.

The limbic system (or to use its more descriptive name, the emotional brain) functions in some way to make us experience many kinds of emotions—anger, fear, sexual feelings, pleasure, and sorrow, for example. To bring about the normal expression of emotions, parts of the cerebral cortex other than the limbic system must also function. Considerable evidence exists that limbic activity without the modulating influence of other cortical areas may bring on attacks of abnormal, uncontrollable rage suffered periodically by some unfortunate individuals. Some of the evidence that suggests this has come from experiments in which the limbic cortex of monkeys was stimulated through implanted electrodes. When researchers stimulated certain limbic areas, the monkeys attacked in rage. In contrast, when they stimulated other limbic points, the animals expressed docile, affectionate behavior.*

Hemispheric specialization and functional localization within a hemisphere

The right and left hemispheres of the human cerebrum do not function identically. Each hemisphere appears to have certain functional specialties of its own. For example, the left hemisphere plays a dominant role in the production and perception of speech sounds. The left brain, so to speak, does the talking—and the understanding of spoken words.† The left hemisphere also appears to dominate the control of certain kinds of hand

*McBroom, P.: The emotional brain, Sci. News **94:**91-92, July, 1968.

†Thomsen, D. E.: Split brain and free will, Sci. News **105:**256-257, Apr. 20, 1974.

movements, notably skilled movements and gesturing movements. For instance, most people use their right hands for performing skilled movements and the left side of the cerebrum controls the muscles on the right side that execute these movements. In fact, it largely controls all muscles on the right side of the body. Also, most people gesture with their right hands when speaking; this, too, indicates left cerebral control.

Evidence that the right hemisphere of the cerebrum specializes in certain functions has rather recently been reported. As of now, it seems that one of the right hemisphere's specialties is the perception of certain kinds of auditory material. For instance, some studies have shown that the right hemisphere perceives nonspeech sounds such as melodies, coughing, crying, and laughing better than does the left hemisphere. The right hemisphere may also function better at tactual perception.*

Certain areas of the cortex in each hemisphere of the cerebrum engage predominantly in one particular function. What that function is depends upon what structures the cortical area receives impulses from or sends impulses to. For example, areas 3, 1, and 2, shown in Fig. 8-21, function mainly as general somatic sensory areas because they receive impulses from receptors stimulated by general stimuli—such as heat, cold, and touch—located in almost all parts of the body. Area 4, on the other hand, functions chiefly as the somatic motor area since it sends out impulses that eventually stimulate somatic effectors, the skeletal muscles. Other areas of the cerebral cortex that serve particular functions are the primary auditory area (areas 41 and 42 in Fig. 8-21; transverse gyrus of the temporal lobe), and the primary visual area (area 17 in the occipital lobe). It is important to remember that no part of the brain functions alone. Many structures of the central nervous system must function in order for any one part of the brain to function.

*Kimura, D.: The asymmetry of the human brain, Sci. Am. **228**:70-78, Mar., 1973; Ornstein, R. E.: The psychology of consciousness, San Francisco, 1972, W. H. Freeman and Co., chap. 2.

Chemical transmitters in brain and cord

Chemical transmitters were defined in Chapter 7 as substances released by axon terminals at synapses and neuroeffector junctions. All synapses are located in the central nervous system (spinal cord and brain) except those in the autonomic ganglia. Two substances, acetylcholine (ACh) and norepinephrine (NE), are most commonly considered to be transmitters at synapses located in the brain and cord. In addition, substantial evidence suggests that several other substances probably also serve as transmitters in the central nervous system—notably serotonin (5-hydroxytryptamine, 5-HT), dopamine (DA), and gamma-aminobutyrate (GABA). Several parts of the central nervous system have been found to contain more than one of these transmitters. Norepinephrine, serotonin, and GABA, for example, are all known to be present in the hypothalamus. Norepinephrine is especially abundant in both the hypothalamus and the cerebrum's limbic system.

Functions of the thalamus

The thalamus performs the following functions:

1 Plays two parts in the mechanism responsible for sensations:
- **a** Impulses from appropriate receptors, upon reaching the thalamus, produce conscious recognition of the cruder, less critical sensations of pain, temperature, and touch.
- **b** Neurons whose dendrites and cell bodies lie in certain nuclei of the thalamus relay all kinds of sensory impulses, except possibly olfactory, to the cerebrum.

2 Plays a part in the mechanism responsible for emotions by associating sensory impulses with feelings of pleasantness and unpleasantness.

3 Plays a part in the arousal or alerting mechanism.

4 Plays a part in mechanisms that produce complex reflex movements.

Functions of the hypothalamus

The hypothalamus is a small but functionally mighty area of the brain. It weighs little more than ¼ ounce, yet it performs many functions of the greatest importance both for survival and for the enjoyment of life. For instance, it functions as a link between the psyche (mind) and the soma (body). It also links the nervous system to the endocrine system. Certain areas of the hypothalamus function as pleasure centers or reward centers for the primary drives such as eating, drinking, and mating. The following paragraphs give a brief summary of hypothalamic functions.

1 The hypothalamus functions as a higher autonomic center or, rather, as several higher autonomic centers. By this we mean that axons of neurons whose dendrites and cell bodies lie in nuclei of the hypothalamus extend in tracts from the hypothalamus to both parasympathetic and sympathetic centers in the brainstem and cord. (Fig. 9-2 indicates these tracts in blue.) Thus impulses from the hypothalamus can simultaneously or successively stimulate or inhibit few or many lower autonomic centers. In other words, the hypothalamus serves as a regulator and coordinator of autonomic activities. It helps control and integrate the responses made by visceral effectors all over the body.

2 The hypothalamus functions as the major relay station between the cerebral cortex and lower autonomic centers. Tracts conduct impulses from various centers in the cortex to the hypothalamus (also shown in blue in Fig. 9-2). Then, via numerous synapses in the hypothalamus, these impulses are relayed to other tracts that conduct them on down to autonomic centers in the brainstem and cord and also to spinal cord somatic centers (anterior horn motoneurons). Thus the hypothalamus functions as the link between the cerebral cortex and lower centers—hence, between the psyche and the soma. It provides a crucial part of the route by which emotions can express themselves in changed bodily functions. It is the all-important relay station in the neural pathways that makes possible the mind's influence over the body—sometimes, unfortunately, even to the profound degree of producing "psychosomatic disease."

3 Neurons in the supraoptic and paraventricular nuclei of the hypothalamus synthesize the hormones secreted by the posterior pituitary gland (neurohypophysis). Because one of these hormones affects the volume of urine excreted, the hypothalamus plays an indirect but essential role in maintaining water balance (see Fig. 20-9, p. 519).

4 Some of the neurons in the hypothalamus function as endocrine glands. Their axons secrete chemicals, called releasing hormones, into blood that circulates to the anterior pituitary gland. Releasing hormones control the release of certain anterior pituitary hormones—specifically those that control hormone secretion by the sex glands, the thyroid gland, and the adrenal cortex. (Discussed in Chapter 11.) Thus, indirectly, the hypothalamus helps control the functioning of every cell in the body (see pp. 272-274).

5 The hypothalamus plays an essential role in maintaining the waking state. Presumably it functions as part of an arousal or alerting mechanism. Clinical evidence of this is that somnolence characterizes some hypothalamic disorders.

6 The hypothalamus functions as a crucial

part of the mechanism for regulating appetite and therefore the amount of food intake. Experimental and clinical findings seem to indicate the presence of a "feeding or appetite center" in the lateral part of the hypothalamus and a "satiety center" located medially. For example, an animal with an experimental lesion in the ventromedial nucleus of the hypothalamus will consume tremendous amounts of food. Similarly, a human being with a tumor in this region of the hypothalamus may eat insatiably and gain an enormous amount of weight.

7 The hypothalamus functions as a crucial part of the mechanism for maintaining normal body temperature. Hypothalamic neurons whose fibers connect with autonomic centers for vasoconstriction and dilation and sweating and with somatic centers for shivering constitute heat-regulating centers. Marked elevation of body temperature frequently characterizes injuries or other abnormalities of the hypothalamus.

Functions of the brainstem

Nuclei in the medulla contain a number of reflex centers. Some of these perform functions so necessary for survival that they are called the vital centers. They are the cardiac, vasomotor, and respiratory centers. As their names suggest, they serve as the centers for various reflexes controlling heart action, blood vessel diameter, and respirations. Because the medulla contains these centers, it is the most vital part of the entire brain—so vital, in fact, that injury or disease of the medulla often proves fatal. Blows at the base of the skull and bulbar poliomyelitis, for example, cause death if they interrupt impulse conduction by the vital respiratory centers.

The medulla contains centers for many nonvital reflexes such as vomiting, coughing, sneezing, hiccoughing, and swallowing.

All projection tracts between the cord and brain necessarily pass through the medulla. Hence it functions in a great many sensory and motor mechanisms. Fibers of the crossed corticospinal tracts decussate—that is, cross from one side to the other in the pyramids of the medulla, an anatomical fact that explains why one side of the brain is said to control the other side of the body.

The upper part of the brainstem serves as conduction pathways (projection tracts) between the cord and other parts of the brain. In addition, the pons contains the centers for reflexes mediated by the fifth, sixth, seventh, and eighth cranial nerves (see Table 8-3 for functions). And because the pons contains the pneumotaxic centers, it helps regulate respirations (discussed on p. 314).

Like the pons, the midbrain also functions as a reflex center for certain cranial nerve reflexes—for example, pupillary reflexes and eye movements, mediated by the third and fourth cranial nerves, respectively.

Functions of the cerebellum

The cerebellum performs three general functions, all of which have to do with the control of skeletal muscles. It acts with the cerebral cortex to produce skilled movements by coordinating the activities of groups of muscles. It controls skeletal muscles so as to maintain equilibrium. It helps control posture. It functions below the level of consciousness to make movements smooth instead of jerky, steady instead of trembling, and efficient and coordinated instead of ineffective, awkward, and uncoordinated (asynergic).

There have been many theories about cerebellar functions. One theory, based on comparative anatomy studies and substantiated by experimental methods, regards the cerebellum as three organs, each with a somewhat different function: synergic control of muscle action, excitation and inhibition of postural reflexes, and maintenance of equilibrium.

Synergic control of muscle action, which is ascribed to the neocerebellum (superior vermis and hemispheres), is closely associated with cerebral motor activity. Normal muscle action, you will recall, involves groups of muscles, the various members of which function together as a unit. In any given action, for example, the prime mover contracts and the antagonist relaxes but then contracts weakly at the proper moment to act as a brake, checking the action of the prime mover. Also, the synergists contract to assist the prime mover, and the fixation muscles of the neighboring joint contract. Through such harmonious coordinated group action, normal movements are smooth, steady, and precise as to force, rate, and extent. Achievement of such movements results from cerebellar activity added to cerebral activity. Impulses from the cerebrum may start the action, but those from the cerebellum synergize or coordinate the contractions and relaxations of the various muscles once they have begun. Some physiologists consider this the main, if not the sole, function of the cerebellum.

One part of the cerebellum is thought to be concerned with both exciting and inhibiting postural reflexes.

Part of the cerebellum presumably discharges impulses important to the maintenance of equilibrium. Afferent impulses from the labyrinth of the ear reach the cerebellum. Here, connections are made with the proper efferent fibers for contraction of the necessary muscles for equilibrium.

Cerebellar disease (abscess, hemorrhage, tumors, trauma, etc.) produces certain characteristic symptoms, among which ataxia (muscle incoordination), hypotonia, tremors, and disturbances of gait and equilibrium predominate. One example of ataxia is overshooting a mark or stopping before reaching it when trying to touch a given point on the body (finger-to-nose test). Drawling, scanning, or singsong speech are also examples of ataxia. Tremors are particularly pronounced toward the end of movements and with the exertion of effort. Disturbances of gait and equilibrium vary, depending upon the muscle groups involved, but the walk is often characterized by staggering or lurching and by a clumsy manner of raising the foot too high and bringing it down with a clap. Paralysis does not result from loss of cerebellar function.

Functions of the spinal cord

The spinal cord performs sensory, motor, and reflex functions. The following paragraphs describe these functions briefly.

Sensory and motor functions. Spinal cord tracts serve as two-way conduction paths between peripheral nerves and the brain. Ascending tracts conduct impulses up the cord to the brain. Descending tracts conduct impulses down the cord from the brain. Bundles of axons compose all tracts. Tracts are both structural and functional organizations of these nerve fibers. They are structural organizations in that all axons of any one tract originate from neuron cell bodies located in the same structure and all of the axons terminate in the same structure. For example, all fibers of the spinothalamic tract are axons originating from neuron cell bodies located in the spinal cord and terminating in the thalamus. Tracts are functional organizations in that all the axons that compose one tract serve one general function. For instance, fibers of the lateral spinothalamic tract serve a sensory function. They transmit the impulses that produce our sensations of pain and of temperature.

Because so many different tracts make up the white columns of the cord, we shall mention only a few that seem most important in man. Locate each tract in Fig. 8-11. Consult Tables 8-1 and 8-2 for a brief summary of information about tracts. Four important ascending or sensory tracts and their functions, stated very briefly, are as follows:

1 Lateral spinothalamic tracts—pain and temperature
2 Ventral spinothalamic tracts—crude touch
3 Fasciculi gracilis and cuneatus—discriminating touch and conscious kinesthesia
4 Spinocerebellar tracts—unconscious kinesthesia

Further discussion of the sensory neural pathways may be found on pp. 203-206.

Four important descending or motor tracts and their functions in brief are as follows:

1 Lateral corticospinal tracts—voluntary movement; contraction of individual or small groups of muscles, particularly those moving hands, fingers, feet, and toes on opposite side of body
2 Ventral corticospinal tracts—same as preceding except mainly muscles of same side of body
3 Lateral reticulospinal tracts—mainly facilitatory impulses to anterior horn motoneurons to skeletal muscles
4 Medial reticulospinal tracts—mainly inhibitory impulses to anterior horn motoneurons to skeletal muscles

Further discussion of the motor neural pathways may be found on pp. 208-212.

Reflex functions. The spinal cord functions in all reflexes except those mediated by cranial nerves. Gray matter of the cord contains innumerable reflex centers. To mention just one example, gray matter of the second, third, and fourth lumbar segments contain the centers for the knee jerk reflex. The term *reflex center* means literally the center of a reflex arc or the place in the arc at which incoming sensory impulses become outgoing motor impulses. Some reflex centers are merely synapses between sensory and motoneurons, whereas others are interneurons interposed between sensory and motoneurons.

Reflexes

Definition

The action that results from a nerve impulse passing over a reflex arc is called a *reflex*. In other words, a reflex is a response to a stimulus. It may or may not be conscious. Usually the term is used to mean only involuntary responses rather than those directly willed (that is, involving cerebral cortex activity).

A reflex consists either of muscle contraction or glandular secretion. *Somatic reflexes* are contractions of skeletal muscles. Impulse conduction over somatic reflex arcs—arcs whose motoneurons are somatic motoneurons (that is, anterior horn neurons or lower motoneurons)—produces somatic reflexes. *Autonomic* (or *visceral reflexes* consist either of contractions of smooth or cardiac muscle or secretion by glands; they are mediated by impulse conduction over autonomic reflex arcs, the motoneurons of which are autonomic neurons (discussed in the next chapter). The following paragraphs describe only somatic reflexes.

Some somatic reflexes of clinical importance

Clinical interest in reflexes stems from the fact that they deviate from normal in certain diseases. So the testing of reflexes is a valuable diagnostic aid. Physicians frequently test the following reflexes: knee jerk, ankle jerk, Babinski reflex, corneal reflex, and abdominal reflex.

The *knee jerk* or patellar reflex is an extension of the lower leg in response to tapping of the patellar tendon. The tap stretches both the tendon and its muscles, the quadriceps femoris, and thereby stimulates muscle spindles (receptors) in the muscle and initiates conduction over the following two-neuron reflex arc (Fig. 7-5):

1 *Sensory neurons*
   a Dendrites—in femoral and second, third, and fourth lumbar nerves

   b Cell bodies—second, third, and fourth ganglia
   c Axons—in posterior roots of second, third, and fourth lumbar nerves; terminate in these segments of the spinal cord; synapse directly with lower motoneurons
2 *Reflex center*—synapses in anterior gray column between axons of sensory neurons and dendrites and cell bodies of lower motoneurons
3 *Motoneurons*
   a Dendrites and cell bodies—in spinal cord anterior gray column
   b Axons—in anterior roots of second, third, and fourth lumbar spinal nerves and femoral nerves; terminate in quadriceps femoris muscle

The knee jerk can be classified in various ways as follows:

1 As a *spinal cord reflex*—because the center of the reflex arc (which transmits the impulses that activate the muscles producing the knee jerk) lies in the spinal cord gray matter
2 As a *segmental reflex*—because impulses that mediate it enter and leave the same segment of the cord
3 As an *ipsilateral reflex*—because the impulses that mediate it come from and go to the same side of the body
4 As a *stretch reflex,* or *myotatic reflex* (Greek, *mys,* muscle; + *tasis,* stretching) —because of the kind of stimulation used to evoke it
5 As an *extensor reflex*—because produced by extensors of lower leg (muscles located on an anterior surface of thigh, which extend the lower leg)
6 As a *tendon reflex*—because tapping of a tendon is the stimulus that elicits it
7 *Deep reflex*—because of the deep location (in tendon and muscle) of the receptors stimulated to produce this reflex (*Super-*

Table 8-8

Correlation of microscopic and macroscopic structures of a three-neuron cord reflex arc*

Microscopic structures	Macroscopic structures in which neurons located
Sensory neuron	
Receptor	In skin or mucosa
Dendrite	In spinal nerve and branches
Cell body	In spinal ganglion on posterior root of spinal nerve
Axon	Posterior root of spinal nerve; terminates in posterior gray column of cord
Interneuron	
Dendrite	Posterior gray column
Cell body	Posterior gray column
Axon	Central gray matter of cord, extending into anterior gray column
Motoneuron	
Dendrite	Anterior gray column
Cell body	Anterior gray column
Axon	Anterior gray column, extending into anterior root of spinal nerve, spinal nerve, and its branches
Effector	In skeletal muscles

*See Figs. 7-7 and 7-8; Fig. 7-6 shows the two-neuron cord arc.

ficial reflexes—those elicited by stimulation of receptors located in the skin or mucosa)

When a physician tests a patient's reflexes, he interprets the test results on the basis of what he knows about the reflex arcs that must function to produce normal reflexes.

To illustrate, suppose that a patient has been diagnosed as having poliomyelitis. In examining him, the physician finds that he cannot elicit the knee jerk when he taps the patient's patellar tendon. He knows that the poliomyelitis virus attacks anterior horn motoneurons. He also knows the information previously related about which cord segments contain the reflex centers for the knee jerk. On the basis of this knowledge, therefore, he deduces that in this patient the poliomyelitis virus has damaged the second, third, and fourth lumbar segments of the spinal cord. Do you think that this patient's leg would be paralyzed, that he would be unable to move it voluntarily? What neurons would not be able to function that must function to produce voluntary contractions?

Ankle jerk or Achilles reflex is an extension (plantar flexion) of the foot in response to tapping of the Achilles tendon. Like the knee jerk, it is a tendon reflex and a deep reflex mediated by two-neuron spinal arcs, but the centers for the ankle jerk lie in the first and second sacral segments of the cord.

The *Babinski reflex* is an extension of the great toe, with or without fanning of the other toes, in response to stimulation of the outer margin of the sole of the foot. Normal babies, up until they are about 1½ years old, show this positive Babinski reflex. By about this time, corticospinal fibers have become fully myelinated and the Babinski reflex becomes suppressed. Just why this is so is not clear. But at any rate it is, and a positive Babinski reflex after this age is abnormal. From then on, the normal response to stimulation of the outer edge of the sole is the *plantar reflex*. It consists of a curling under of all the toes (plantar flexion) plus a slight turning in and flexion of the anterior part of the foot. A positive Babinski reflex is one of the pyramidal signs (p. 212) and is interpreted to mean destruction of pyramidal tract (corticospinal) fibers.

The *corneal reflex* is winking in response to touching the cornea. It is mediated by reflex arcs with sensory fibers in the ophthalmic branch of the fifth cranial nerve, centers in the pons, and motor fibers in the seventh cranial nerve.

The *abdominal reflex* is drawing in of the abdominal wall in response to stroking the side of the abdomen. It is mediated by arcs with sensory and motor fibers in the ninth to twelfth thoracic spinal nerves and centers in these segments of the cord. It is classified as a superficial reflex. A decrease in this reflex or its absence occurs in lesions involving pyramidal tract upper motoneurons.

Outline summary

Brain and cord coverings

1 Bony—vertebrae around cord; cranial bones around brain
2 Membranous—called meninges and consist of three layers
- **a** Dura mater—white fibrous tissue outer layer
- **b** Arachnoid membrane—cobwebby middle layer
- **c** Pia mater—transparent; adherent to outer surface of cord and brain; contains blood vessels; therefore, nutritive layer

Brain and cord fluid spaces

1 Names
- **a** Subarachnoid space around cord
- **b** Subarachnoid space around brain
- **c** Central canal inside cord
- **d** Ventricles and cerebral aqueduct inside brain—four cavities within brain:
  - 1 First and second (lateral) ventricles—large cavities, one in each cerebral hemisphere
  - 2 Third ventricle—vertical slit in cerebrum beneath corpus callosum and longitudinal fissure
  - 3 Fourth ventricle—diamond-shaped space between cerebellum and medulla and pons; expansion of central canal of cord

2 Formation, circulation, and function of cerebrospinal fluid
- **a** Formed by plasma filtering from network of capillaries (choroid plexus) in each ventricle

b Circulates from lateral ventricles to third ventricle, cerebral aqueduct, fourth ventricle, central canal of cord, subarachnoid space of cord and brain; venous sinuses
c Function—protection of brain and cord

Anatomy of the brain

1 Cerebrum
  a Hemispheres, fissures, and lobes—longitudinal fissure divides cerebrum into two hemispheres, connected only by corpus callosum; each cerebral hemisphere divided by fissures into five lobes: frontal, parietal, temporal, occipital, and island of Reil (insula)
  b Cerebral cortex—outer layer of gray matter arranged in ridges called convolutions or gyri
  c Cerebral tracts—bundles of axons compose white matter in interior of cerebrum; ascending projection tracts transmit impulses toward or to brain; descending projection tracts transmit impulses down from brain to cord; commissural tracts transmit from one hemisphere to other; association tracts transmit from one convolution to another in same hemisphere
  d Basal ganglia (or cerebral nuclei)—masses of gray matter embedded deep inside white matter in interior of cerebrum; caudate, putamen, and pallidum; putamen and pallidum constitute lenticular nucleus
2 Diencephalon—thalamus
  a Structure and location—large rounded mass of gray matter in each cerebral hemisphere, lateral to third ventricle; composed of many nuclei
3 Diencephalon—hypothalamus
  a Gray matter around optic chiasma, pituitary stalk, posterior lobe of pituitary gland, mamillary bodies, and adjacent regions; made up of many nuclei, notably supraoptic, paraventricular, and mamillary
  b Afferent tracts conduct impulses to hypothalamus from cerebral cortex, thalamus, and basal ganglia
  c Efferent tracts conduct from hypothalamus to thalamus and to autonomic centers in brainstem and cord
4 Cerebellum
  a Second largest part of human brain
  b Has two hemispheres and center section, vermis
  c Surface grooved with sulci
  d Slightly raised, slender convolutions
  e Internal white matter arranged in pattern like veins of leaf
  f Three pairs of tracts in cerebellum—inferior, middle, and superior cerebellar peduncles
  g Most important cerebellar nucleus—dentate nucleus
5 Medulla oblongata
  a Part of brain formed by enlargement of cord as it enters cranial cavity
  b Mainly white matter (projection tracts); also reticular formation (interlacement of gray and white matter, containing many nuclei)
  c Various autonomic centers in reticular formation, e.g., cardiac and vasomotor; also respiratory, vomiting, coughing, hiccoughing, sneezing, and swallowing centers
6 Pons
  a Located just above medulla
  b White matter with few nuclei
7 Mesencephalon (midbrain)
  a Located above pons, below cerebrum and diencephalon
  b White matter with few nuclei
  c Cerebral aqueduct within midbrain
  d Cerebral peduncles are two tracts composing ventral part of midbrain and connecting pons to cerebrum
  e Corpora quadrigemina—four rounded eminences (two superior and two inferior colliculi) on dorsal surface of midbrain
  f Important nucleus—red nucleus

Anatomy of the spinal cord

1 Size, shape, location—about 1½ feet long, smaller around than spinal cavity in which it is located
2 Sulci
  a Two deep grooves, anterior median fissure and posterior median sulcus, incompletely divide cord into right and left symmetrical halves; anterior median fissure deeper and wider than posterior groove
3 Gray matter—shaped like three-dimensional letter H
4 White matter—present in columns, anterior, lateral, and posterior, composed of numerous projection tracts

Cranial nerves

1 Twelve pairs
  a I (olfactory)
  b II (optic)
  c III (oculomotor)
  d IV (trochlear)
  e V (trifacial or trigeminal)
  f VI (abducens)
  g VII (facial)
  h VIII (vestibulocochlear)
  i IX (glossopharyngeal)
  j X (vagus)
  k XI (accessory)
  l XII (hypoglossal)
2 See Table 8-3 for distribution and function

Spinal nerves

1 Thirty-one pairs
  a Eight cervical
  b Twelve thoracic
  c Five lumbar
  d Five sacral
  e One coccygeal
2 Origin—originate by anterior and posterior roots from cord, emerge through intervertebral foramina; spinal ganglion on each posterior root
3 Distribution—branches distributed to skin, mucosa, skeletal muscles; branches form plexuses, such as brachial plexus, from which nerves emerge to supply various parts
4 Microscopic structure—consist of sensory dendrites and motor axons—i.e., are mixed nerves; also contain autonomic postganglionic fibers; see Table 8-4 for peripheral branches

Physiology of the somatic nervous system

1 Sensory pathways from periphery to cerebral cortex consist of three-neuron relays as follows
  a Sensory neuron I conducts from periphery to cord or to brainstem
  b Sensory neuron II conducts from cord or brainstem to thalamus
  c Sensory neuron III conducts from thalamus to general sensory area of cerebral cortex (areas 3, 1, 2)

2 Most sensory neuron II axons decussate; so one side of brain registers mainly sensations for opposite side of body
3 Pain and temperature—lateral spinothalamic tracts to thalamus
4 Touch and pressure
 a Discriminating touch and pressure (stereognosis, precise localization, vibratory sense)—medial lemniscal system (fasciculi cuneatus and gracilis to medulla; medial lemniscus to thalamus)
 b Crude touch and pressure—ventral spinothalamic tracts
5 Conscious proprioception or kinesthesia—medial lemniscal system to thalamus (see also Table 8-6)
6 Impulses conducted up spinothalamic and medial lemniscal pathways reach somatic sensory area of cerebral cortex (postcentral gyrus or areas 3, 1, 2 on Brodman's map of cortex), the primary area for general sensations
7 Several secondary somatic sensory areas exist in cerebral cortex
8 Somatic motor pathways
 a Principle of final common path—motoneurons in anterior gray horns of cord constitute final common path for impulses to skeletal muscles; are only neurons transmitting impulses into skeletal muscles
 b Motor pathways from cerebral cortex to anterior horn cells classified according to way fibers enter cord:
  1 Pyramidal (or corticospinal) tracts—dendrites and cells in cortex; axons descend through internal capsule, pyramids of medulla, and spinal cord and synapse directly with anterior horn cells or indirectly via internuncial neurons; impulses via pyramidal tracts essential for voluntary contractions of individual muscles to produce small discrete movements; also help maintain muscle tone
  2 Extrapyramidal tracts—all pathways between motor cortex and anterior horn cells, except pyramidal tracts; upper extrapyramidal tracts relay impulses between cortex, basal ganglia, thalamus, and brainstem; reticulospinal tracts main lower extrapyramidal tracts; impulses via extrapyramidal tracts essential for large, automatic movements; also for facial expressions and movements accompanying many emotions
 c Motor pathways from cerebral cortex to anterior horn cells also classified according to influence on anterior horn cells
  1 Facilitatory tracts—have facilitating or stimulating effect on anterior horn cells; all pyramidal tracts and some extrapyramidal tracts facilitatory, notably facilitatory reticulospinal fibers
  2 Inhibitory tracts—have inhibiting effect on anterior horn cells; inhibitory reticulospinal fibers main ones
 d Principle of convergence—axons of many neurons synapse with each anterior horn motoneuron
 e Ratio of facilitatory and inhibitory impulses impinging on anterior horn cells controls activity; normally, slight predominance of facilitatory impulses maintains muscle tone
9 Functions of the cerebrum
 a Generalizations
  1 Cerebral activity goes on as long as life itself; electroencephalograms are records of cerebrum's electrical activity
  2 Cerebrum performs sensory, motor, and integrative functions
  3 Right and left cerebral hemispheres specialize in different functions; also certain regions of each hemisphere play key roles in particular functions
  4 Certain chemical transmitters are released only at synapses in brain and cord
 b Electroencephalograms (EEG's)
  1 Definition—EEG's are records of cerebrum's electrical activity
  2 Brain waves—alpha waves are moderately fast, moderately high-voltage waves; dominant in relaxed, nonattentive state; beta waves are fast, low-voltage waves, dominant in attentive waking state; delta waves are very slow, high-voltage waves, characterize deep sleep; theta waves are moderately slow, low-voltage waves, appear as drowsiness descends; absence of brain waves ("flat EEG") generally considered best criterion of death; brain wave patterns vary in altered states of consciousness
 c Sensory functions of cerebrum: comparison, evaluation, and integration of sensations to form total perceptions
 d Somatic motor functions of cerebrum: control voluntary (skeletal muscle) movements
 e Integrative functions of the cerebrum
  1 Consciousness—state of awareness of one's self, one's environment, and other beings; depends upon excitation ("arousal" or "alerting") of cortical neurons by impulses conducted to them via neurons of reticular activating system; normal variations in degree or level of consciousness—waking, dream or REM (rapid eye movement) sleep, and deep, dreamless sleep; also several kinds of altered states of consciousness (ASC), e.g., anesthesia, coma, yoga or meditative state
  2 Memory—a function of cerebral cortex neurons but mechanisms not definitely known; evidence that several parts of cortex (e.g., temporal, parietal, occipital lobes) store memories and that limbic system ("emotional brain") and protein synthesis plays key roles in memory; also, protein synthesis appears to be crucial for long-term memory
  3 Speech functions—the use of language (speaking and writing) and the understanding of language (spoken and written); depend upon widespread integrated cortical processes involving many parts of cortex but certain regions in frontal, parietal, and temporal lobes called speech centers serve as focal points for integration of speech processes; speech defects (aphasias) of different kinds result from lesions of different speech centers
  4 Emotions—both subjective experience and objective expression of emotions involve functioning of cerebrum's limbic system, i.e., part of the cortex located on medial surface of cerebrum that forms a border around the corpus callosum; limbic system made up of cingulate gyrus, isthmus, hippocampal gyrus, hippocampus, and uncus and tracts connecting these structures to several parts of brain, notably the septum, amygdala, and hypothalamus; for normal expression of emotions, other parts of the cortex must modulate limbic system activity
 f Hemispheric specialization and functional localization within a hemisphere
  1 Left hemisphere specializes in (that is, dominates) functions of producing and understanding speech sounds and also dominates control of skilled movements and gesturing movements of right hand

2 Right cerebral hemisphere specializes in perception of nonspeech sounds (e.g., melodies, coughing, crying, laughing) and in locating objects in space; right hemisphere may also specialize in tactual perception
3 Certain areas of cortex in each hemisphere play dominant role in a particular function, e.g., primary somatic sensory area (postcentral gyrus of parietal lobe) crucial for experiencing general sensations (heat, cold, touch); primary somatic motor area (precentral gyrus of frontal lobe) dominates control of somatic effectors (skeletal muscles); primary auditory area (transverse gyrus of temporal lobe) crucial for auditory sensations; primary visual area (area 17 of occipital lobe) crucial for vision; important to remember that any one function requires functioning of many parts of the nervous system

g Chemical transmitters in brain and cord synapses—acetylcholine (ACh), norepinephrine (NE), and probably also serotonin (5-hydroxytryptamine or 5-HT), dopamine (DA), and gamma-aminobutyrate (GABA), and perhaps other substances too; NE especially abundant in limbic system and hypothalamus; also present in hypothalamus in addition to NE are serotonin and GABA

10 Functions of the thalamus
a Conscious recognition of crude sensations of pain, temperature, and touch
b Relays afferent impulses to cerebral cortex
c Involved in emotional component of sensations; feelings of pleasantness or unpleasantness
d Involved in alerting or arousal mechanism
e Involved in production of complex reflex movements

11 Functions of the hypothalamus
a Higher autonomic centers help control and integrate autonomic functions
b Relay station between cerebral cortex and lower autonomic centers; crucial part of neural paths by which emotions influence bodily functions
c Synthesizes hormones secreted by posterior pituitary gland; one of these is crucial for maintaining water balance
d Hypothalamus synthesizes and secretes releasing hormones that control release of certain hormones by anterior pituitary gland
e Essential part of arousal or alerting mechanism; essential for maintaining waking state
f Crucial part of mechanism for regulating appetite and food intake
g Crucial part of mechanism for maintaining normal body temperature

12 Functions of the brainstem
a Functions of medulla
1 Helps control heartbeat, blood pressure, and respirations
2 Mediates reflexes of vomiting, coughing, hiccoughing, etc.
3 Conducts impulses between cord and brain
b Functions of pons
1 Contains projection tracts between cord and various parts of brain
2 Centers for fifth to eighth cranial nerves
c Functions of midbrain
1 Projection tracts function in sensations and movements
2 Centers for third and fourth cranial nerves, hence for pupillary reflexes and eye movements

13 Functions of the cerebellum
a Synergic control of muscle action
b Postural reflexes
c Equilibrium

14 Functions of spinal cord
a Sensory and motor conduction pathways between peripheral nerves and brain; for names and functions of tracts, see Tables 8-1 and 8-2
b Contains reflex centers for all spinal cord reflexes

15 Reflexes
a Definition—action resulting from conduction over reflex arc; reflex is response (either muscle contraction or glandular secretion) to stimulus; term reflex usually used to mean only involuntary responses
b Some reflexes of clinical importance (see also p. 221)
1 Knee jerk
2 Ankle jerk
3 Babinski reflex
4 Corneal reflex
5 Abdominal reflex

Review questions

1 What term means the membranous coverings of the brain and cord? What three layers compose this covering?
2 Where does a physician do a lumbar puncture? Why?
3 What are the cavities inside the brain called? How many are there? What do they contain?
4 Describe the circulation of cerebrospinal fluid.
5 What function does the cerebrospinal fluid serve?
6 What term means the outer layer of the cerebrum? Describe its appearance. What structure connects the two cerebral hemispheres?
7 Compare nerves and tracts.
8 Explain the system for naming individual tracts.
9 Based on the explanation you gave in question 8, what is a spinothalamic tract?
10 What are cerebral nuclei?
11 Compare the gross structures called nucleus and ganglion.
12 What is the diencephalon? Name its two main parts.
13 Name several ascending or sensory tracts in the spinal cord.
14 Name two descending or motor tracts in the spinal cord.
15 Which cranial nerves transmit impulses that result in vision? In eye movements? Hearing? Taste sensations? Slowing of the heart?
16 What is the great sensory nerve of the head?
17 Digitalis, a drug said to stimulate the vagus nerve, has been administered to a patient. What effect, if any, would you expect it to have on the patient's pulse rate?
18 Locate the dendrite, cell body, and axon of sensory neurons I, II, and III.
19 Explain each of the following: lower motoneuron, upper motoneuron, and final common path.
20 Name the tracts that transmit impulses from each of the following types of receptors to the brain and state in which column of the cord each tract is located: pain and temperature receptors, crude touch receptors, proprioceptors (name two tracts for these), discriminating touch receptors (two tracts).
21 Compare pyramidal tract and extrapyramidal tract functions.
22 Classify the following structures as somatic effectors, visceral effectors, or neither: adrenal glands, biceps femoris muscle, heart, iris, and skin.
23 Define the term electroencephalogram. What abbreviation stands for this term?
24 Identify the following kinds of brain waves according to their frequency and voltage and the level of consciousness in which they predominate: alpha, beta, delta, and theta.
25 What general functions does the cerebral cortex perform?
26 Define consciousness. Name the normal states or levels of consciousness.
27 Explain briefly what is meant by the arousal or alerting mechanism.
28 Name some altered states of consciousness.
29 Compare yoga (the meditative state) with the usual waking state.
30 Describe some of the current ideas about memory.
31 What part of the brain is called the "emotional brain"?
32 What structures make up the limbic system? Where are they located? What general functions does the limbic system perform?
33 What general functions does the thalamus perform?
34 What general functions does the hypothalamus perform?
35 What general functions does the cerebellum perform?
36 What general functions does the medulla perform?
37 What general functions does the spinal cord perform?
38 What is a reflex center?

9 The autonomic nervous system

Definition
Divisions
Macroscopic structure
Microscopic structure
Some general principles
"Fight or flight" response and physiological changes in meditation
Autonomic learning and biofeedback

The body has only one nervous system, but this one system has two major subdivisions: the somatic nervous system and the autonomic nervous system. The many structures that make up these two divisions function together as a single unit. What function do they jointly achieve? Together they integrate the multitudinous activities of the body's working parts—its somatic effectors and its visceral effectors.

Definition

By definition, the autonomic nervous system consists microscopically only of motoneurons and only those motoneurons that conduct impulses from the central nervous system to visceral effectors. The term *visceral effectors* may be defined in two ways: in terms of tissues or of organs. In terms of tissues, visceral effectors consist of cardiac muscle, smooth muscle, and glandular epithelium. In terms of organs, visceral effectors consist of the heart, blood vessels, iris, ciliary muscles, hair muscles, various thoracic and abdominal organs, and the body's many glands. Note that all of these structures innervated by the autonomic nervous system are ones we think of as involuntary. They lie beyond our conscious control. They are our automatic parts. They function without our willing them to and, for the most part, without our even being conscious of them.

Even though the autonomic nervous system consists, by definition, only of motoneurons (specifically, those that conduct impulses from the brainstem or cord to visceral effectors), nevertheless sensory neurons also take part in autonomic functioning. The autonomic nervous system functions on the reflex arc principle just as the somatic nervous system does. But any sensory neuron can theoretically function in both autonomic and somatic reflex arcs. For example, stimulation of cold receptors in the skin can initiate both an autonomic reflex (vasoconstriction of skin blood vessels) and a somatic reflex (shivering). In short, any one sensory neuron can function in both somatic and autonomic arcs, whereas any one motoneuron can function in either a somatic arc or an autonomic arc but not in both.

Divisions

Two anatomically and physiologically separate divisions compose the autonomic nervous system: the sympathetic (or thoracolumbar) division and the parasympathetic (or craniosacral) division. Both divisions consist macroscopically of ganglia and fibers and microscopically, as we have mentioned, of visceral, or autonomic, motoneurons.

Macroscopic structure

Sympathetic ganglia lie lateral to the anterior surface of the spinal column. Because short fibers extend between the sympathetic ganglia, connecting them to each other, they look a little like two chains of beads (one chain on each side of the spinal column from the level of the second cervical vertebra to the coccyx) and are often referred to as the "sympathetic chain ganglia."*

Parasympathetic ganglia lie in or near visceral effectors, not near the spinal column, as do sympathetic ganglia. One example is the ciliary ganglion located in the posterior part of the orbit near the iris and ciliary muscle.

**Sympathetic ganglia*—There are three cervical, ten or eleven thoracic, four lumbar, and four sacral ganglia in each sympathetic chain. A few sympathetic ganglia, notably the celiac, superior, and inferior mesenteric ganglia, are located a short distance from the cord and therefore are called *collateral ganglia*.

Celiac ganglia (solar plexus)—two fairly large, flat ganglia located on either side of the celiac artery just below the diaphragm.

Superior mesenteric ganglion—small ganglion located near the beginning of the superior mesenteric artery.

Inferior mesenteric ganglion—small ganglion located close to the beginning of the inferior mesenteric artery.

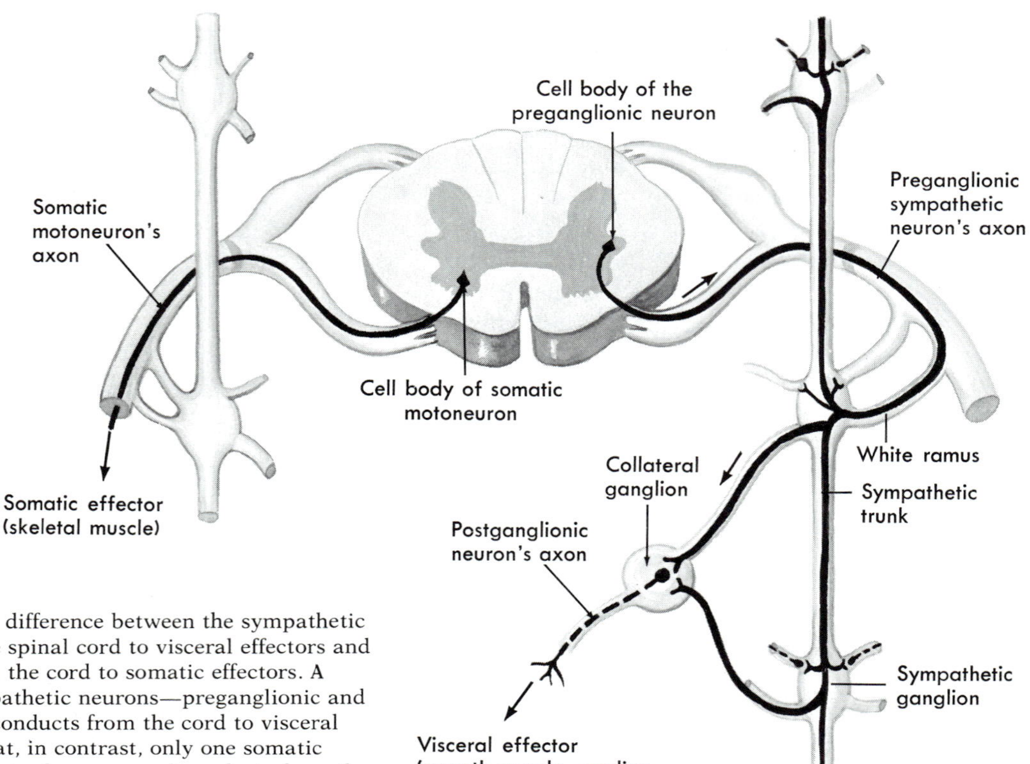

Fig. 9-1

Diagram showing difference between the sympathetic pathway from the spinal cord to visceral effectors and the pathway from the cord to somatic effectors. A relay of two sympathetic neurons—preganglionic and postganglionic—conducts from the cord to visceral effectors. Note that, in contrast, only one somatic motoneuron (anterior horn neuron) conducts from the cord to somatic effectors with no intervening synapses. Parasympathetic impulses also travel over a relay of two neurons (parasympathetic preganglionic and postganglionic) to reach visceral effectors from the central nervous system. Note the location of the sympathetic preganglionic neuron's cell body and axon. Where are the sympathetic postganglionic neuron's cell body and axon located?

Microscopic structure

Microscopically, the autonomic nervous system consists of preganglionic neurons, postganglionic neurons, and synapses between them. As their names suggest, preganglionic neurons conduct impulses before they reach a ganglion, and postganglionic neurons conduct after they reach a ganglion. For examples, look at the right side of Fig. 9-1. It shows both preganglionic and postganglionic neurons of the sympathetic system. Note the location of the preganglionic neuron's cell body in the lateral gray column of the cord. (In which column of the cord do somatic motoneuron cell bodies lie?) The axon of a preganglionic neuron terminates, as Fig. 9-1 shows, in a collateral autonomic ganglion. Here, it synapses with the dendrites or cell body of a postganglionic neuron whose axon terminates in a visceral effector. Thus it takes a relay of two neurons to conduct impulses from the central nervous system out to a visceral effector—a preganglionic neuron to conduct from either the cord or the brainstem to an autonomic ganglion and a postganglionic neuron to conduct from the ganglion to a visceral effector. How does this differ from conduction to somatic effectors from the central nervous system? Find the answer on the left side of Fig. 9-1 if you do not already know

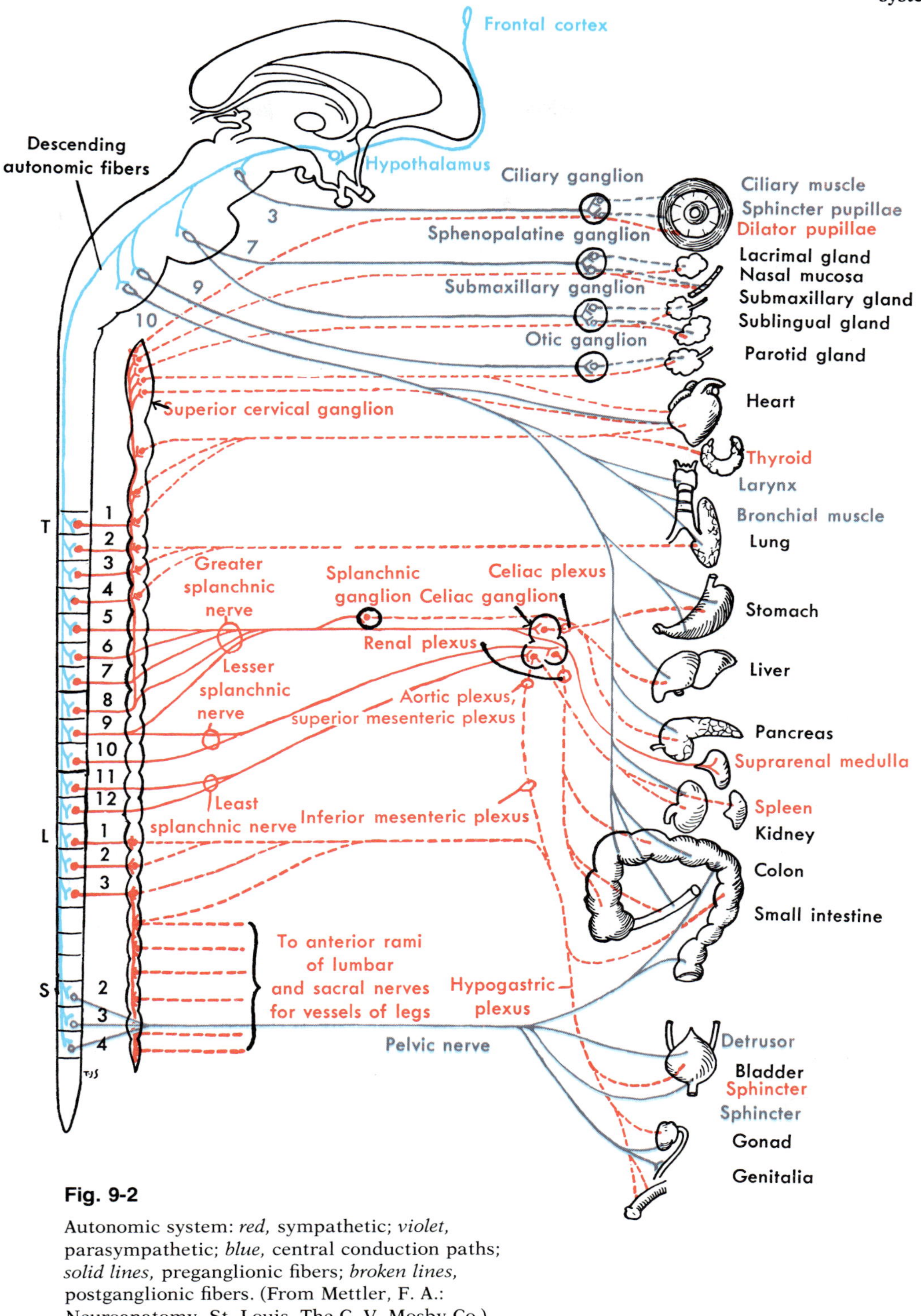

Fig. 9-2

Autonomic system: *red,* sympathetic; *violet,* parasympathetic; *blue,* central conduction paths; *solid lines,* preganglionic fibers; *broken lines,* postganglionic fibers. (From Mettler, F. A.: Neuroanatomy, St. Louis, The C. V. Mosby Co.)

it. One neuron—the anterior horn motoneuron—and not a relay of two neurons conducts from the cord to somatic effectors.

Preganglionic neurons of sympathetic system. The preganglionic neurons of the sympathetic system (Figs. 9-1 and 9-2 and Table 9-1) have their cell bodies in the lateral gray columns of the thoracic and first three or four lumbar segments of the cord. Axons from these cells extend through the anterior roots of the corresponding spinal nerves and through small side branches (the white rami) to the respective sympathetic ganglia. Here, some of them synapse with postganglionic neurons (each axon synapsing with several postganglionic neurons). Other preganglionic axons send branches up and down the sympathetic chain to terminate in ganglia above and below their point of origin. Still others extend through the sympathetic ganglia, out through the splanchnic nerves,* and terminate in the collateral ganglia.

But no matter which course a sympathetic preganglionic axon follows, it synapses with

*The three splanchnic nerves constitute the main branches from the thoracic sympathetic trunk. Since their fibers synapse in the celiac and other collateral ganglia with postganglionic neurons to abdominal structures, they form the main routes for sympathetic stimulation of abdominal viscera.

Table 9-1

Locations of autonomic neurons

Autonomic neurons (motoneurons)	Macroscopic structures in which located
Sympathetic division	
Preganglionic neurons	
Dendrites and cell bodies	In lateral gray columns (Fig. 9-1) of thoracic and first four lumbar segments of spinal cord
Axons (preganglionic fibers)	In anterior roots of spinal nerves to spinal nerves (thoracic and first four lumbar) to white rami and then in any of three following pathways: (1) through white rami to sympathetic ganglia, where they synapse with several postganglionic neurons; (2) through white rami to and through sympathetic ganglia, then up or down sympathetic trunk before synapsing in a sympathetic ganglion with several postganglionic neurons; (3) through white rami to and through sympathetic ganglia, to and through splanchnic nerves to collateral ganglia (celiac, superior, and inferior mesenteric ganglia), where they synapse with several postganglionic neurons
Postganglionic neurons	
Dendrites and cell bodies	In sympathetic ganglia or in collateral ganglia
Axons (postganglionic fibers)	In autonomic nerves that form various plexuses before supplying thoracic and abdominal viscera and blood vessels in these body cavities, or Through gray rami to spinal nerves to cutaneous blood vessels, sweat glands, and smooth muscles of hair follicles
Parasympathetic division	
Preganglionic neurons	
Dendrites and cell bodies	In midbrain, pons, or medulla (Fig. 9-2) or in lateral gray columns of sacral cord
Axons (preganglionic fibers)	From midbrain to third cranial nerve to ciliary ganglion, or From pons to seventh cranial nerve to sphenopalatine ganglion or submaxillary ganglion, or From medulla to (1) ninth cranial nerve to otic ganglion or (2) tenth and eleventh cranial nerves to cardiac and celiac ganglia
Postganglionic neurons	
Dendrites and cell bodies	In various ganglia (ciliary, sphenopalatine, submaxillary, otic, cardiac, and celiac) located in or near organs
Axons (postganglionic fibers)	In short nerve filaments to various viscera, glands, blood vessels, and intrinsic eye muscles

many postganglionic neurons, and these frequently terminate in widely separated organs. This anatomical fact explains a well-known physiological principle—sympathetic responses are usually widespread, involving many organs and not just one.

Postganglionic neurons of sympathetic system. The postganglionic neurons of the sympathetic system have their dendrites and cell bodies in the sympathetic chain ganglia or in collateral ganglia. Their axons are distributed by both spinal nerves and separate autonomic nerves. They reach the spinal nerves via small filaments (gray rami) that connect the sympathetic ganglia with the spinal nerves. They then travel in the spinal nerves to blood vessels, sweat glands, and arrector hair muscles all over the body.

The course of postganglionic axons through autonomic nerves is somewhat more complex. These nerves form complicated plexuses before fibers are finally distributed to their respective destinations. For example, postganglionic fibers from the celiac and superior mesenteric ganglia pass through the celiac plexus before reaching the abdominal viscera, those from the inferior mesenteric ganglion pass through the hypogastric plexus on their way to the lower abdominal and pelvic viscera, and those from the cervical ganglion pass through the cardiac nerves and the cardiac plexus at the base of the heart and are then distributed to the heart.

Preganglionic neurons of parasympathetic system. The preganglionic neurons of the parasympathetic system have their cell bodies in nuclei in the brainstem or in the lateral gray columns of the sacral cord. Their axons are contained in cranial nerves III, VII, IX, X, and XI and in some pelvic nerves. They extend a considerable distance before synapsing with postganglionic neurons. For example, axons arising from cell bodies in the vagus nuclei (located in the medulla) travel in the vagus nerve for a distance of a foot or more before reaching their terminal ganglia in the chest and abdomen (Fig. 9-2 and Table 9-1).

Postganglionic neurons of parasympathetic system. The postganglionic neurons of the parasympathetic system have their dendrites and cell bodies in the outlying parasympathetic ganglia and send short axons into the nearby structures. Each parasympathetic preganglionic neuron synapses, therefore, only with postganglionic neurons to a single effector. For this reason, parasympathetic stimulation frequently involves response by only one organ, as contrasted with sympathetic responses, which, as noted before, usually involve numerous organs.

Some general principles

Following are some general principles about the autonomic nervous system.

1 *Principle of autonomic functioning and homeostasis.* Both divisions of the autonomic nervous system regulate the functioning of visceral effectors, so they tend to maintain or quickly restore homeostasis. In healthy individuals, homeostasis is maintained under all but the most stressful conditions.

2 *Principle of dual autonomic innervation.* Most visceral effectors receive both sympathetic and parasympathetic fibers—notably, the following organs: eyes, heart, digestive system organs, bronchioles, and pelvic viscera (Fig. 9-2).

3 *Principle of single autonomic innervation.* Some visceral effectors are believed to receive only sympathetic fibers. Examples are the sweat glands, the smooth muscles of hairs (piloerector muscles), and most blood vessels. Only preganglionic sympathetic axons innervate the adrenal medulla. Cells of this endocrine gland are modified postganglionic sympathetic neurons.

4 *Principle of autonomic chemical transmitters.* Terminals of autonomic axons, like those of all axons, release chemicals that transmit impulses across synapses and neuroeffector junctions.* Autonomic axons fall into two classifications based on the chemical transmitters that they release. Some release acetylcholine and so are classified as cholinergic fibers. Others release norepinephrine and so are classified as adrenergic fibers.

The following are cholinergic autonomic fibers:

a All preganglionic axons

b Most, or perhaps all, parasympathetic postganglionic axons

c A few sympathetic postganglionic axons —namely, those to sweat glands, and also some of those to smooth muscle in the walls of blood vessels in skeletal muscles and in the external genitalia

The only adrenergic autonomic nerve fibers are sympathetic postganglionic axons. But not even all of these fibers are adrenergic. Sympathetic postganglionic axons to sweat glands and to blood vessels in skeletal muscles and the genitalia are cholinergic, as mentioned in **c.**

5 *Principle of tonic activity of autonomic fibers.* Many, but not all, autonomic fibers are tonically active. This means that they continually conduct impulses. Examples of tonically active autonomic fibers are adrenergic sympathetic fibers to blood vessels and both sympathetic and parasympathetic fibers to smooth muscle of the eyes and to organs of the digestive system.

6 *Principle of autonomic antagonism and summation.* Both sympathetic and parasympathetic impulses continually play on those visceral effectors that are doubly innervated and tend to produce antagonistic effects on them. Stated differently, acetylcholine and its chief antagonist, norepinephrine, are continually released at the neuroeffector junctions of doubly innervated structures. Acetylcholine tends to make them respond in one way, while norepinephrine tends to make them respond in the opposite way. Hence, the algebraic sum of these opposing chemicals determines the actual response. Here is one example of this principle of autonomic antagonism in operation: parasympathetic fibers to the heart release acetylcholine, which tends to slow the heart rate. At the same time, sympathetic fibers to the heart release norepinephrine, and this chemical tends to accelerate the heart rate. Therefore, at any one moment, the algebraic sum of these antagonists determines the actual rate of the heartbeat. If sympathetic impulses increase, the heart rate increases. If parasympathetic impulses increase, the heart rate decreases.

7 *Principle of parasympathetic dominance of digestive tract glands and smooth muscle.* Under normal conditions, parasympathetic impulses to the digestive glands and to the smooth muscle of the digestive tract dominate over sympathetic impulses to them. As

*Much of our present knowledge about chemical transmitters stems from years of research by Sweden's Dr. Ulf S. von Euler, the United States' Dr. Julius Axelrod, and England's Sir Bernard Katz. For their work, these eminent scientists shared the 1970 Nobel Prize in medicine and physiology. Their most significant findings included the following:

1 Identification of norepinephrine (NE) as the major transmitter in some brain synapses (by Dr. von Euler)

2 Identification of the enzyme catechol-O-methyl transferase (COMT) and discovery that it inactivates norepinephrine (by Dr. Axelrod)

3 Discovery that conducting cholinergic fibers rapidly release numerous packets of acetylcholine into synapses (by Sir Bernard Katz)

The above knowledge led to other discoveries and to valuable applications. For example, researchers later learned that severe psychic depression occurs when a deficit of NE exists in certain brain synapses. This finding led to the development of antidepressant drugs. Certain ones of these inhibit COMT. Because inhibited COMT does not inactivate NE, the amount of active NE in brain synapses increases and this relieves the individual's depression. The graphic names, "psychic energizers" and "mood elevators," refer to antidepressant drugs, now valuable weapons in medicine's arsenal for combating mental disease.

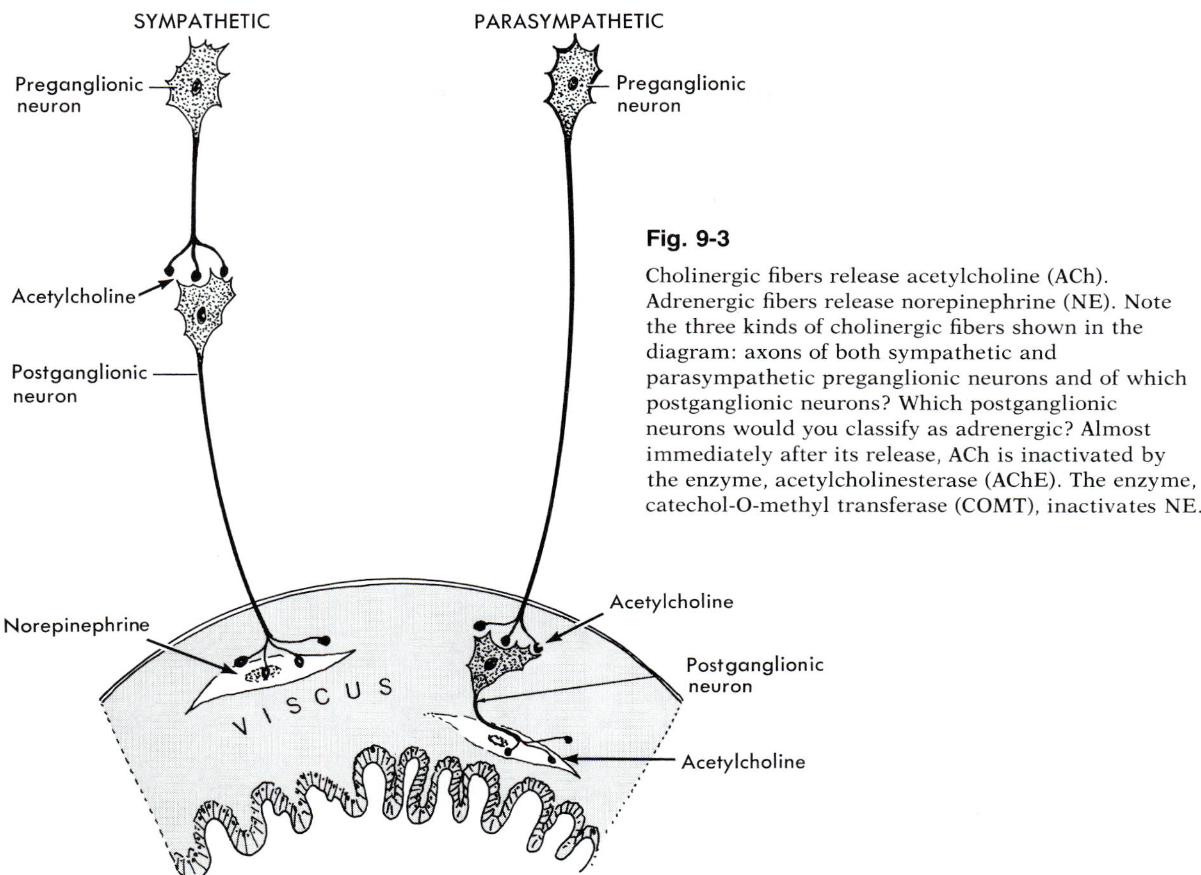

Fig. 9-3

Cholinergic fibers release acetylcholine (ACh). Adrenergic fibers release norepinephrine (NE). Note the three kinds of cholinergic fibers shown in the diagram: axons of both sympathetic and parasympathetic preganglionic neurons and of which postganglionic neurons? Which postganglionic neurons would you classify as adrenergic? Almost immediately after its release, ACh is inactivated by the enzyme, acetylcholinesterase (AChE). The enzyme, catechol-O-methyl transferase (COMT), inactivates NE.

show in Table 9-2, parasympathetic impulses tend to increase digestive gland secretions and to stimulate the smooth muscle of the digestive tract. In short, parasympathetic impulses tend to promote digestion and peristalsis. They also tend to promote elimination (defecation and also urination).

8 *Principle of sympathetic dominance under stress conditions.* Under stress conditions, from either physical or emotional causes, sympathetic impulses to most visceral effectors increase greatly. In fact, one of the very first steps in the body's complex defense mechanism against stress is a sudden and marked increase in sympathetic activity. This immediately brings about an entire set of physiological changes. All are related to making the body able to cope strenuously with the stress situation. They change the usual normal metabolic state of the body to a hypermetabolic state that enables it to put forth its greatest physical effort and expend its maximum amount of energy. Together, this group of changes, so rapidly induced by increased sympathetic activity, constitutes an "integrated response" known by such descriptive names as the sympathoadrenal response, the defense-alarm reaction, or borrowing Cannon's classic phrase, the "fight or flight" reaction. The right-hand column of Table 9-2 indicates or implies many of the physiological changes that make up the "fight

Table 9-2

Autonomic functions

Visceral effectors	Parasympathetic (cholinergic) effects	Sympathetic (adrenergic or cholinergic) effects
Cardiac muscle	Slows heart rate; decreases strength of contraction	Accelerates heart rate; increases strength of contraction
Smooth muscle of blood vessels		
Skin blood vessels	No parasympathetic fibers	Adrenergic sympathetic fibers → stimulate → constrict skin vessels
Skeletal muscle blood vessels	No parasympathetic fibers	Adrenergic sympathetic fibers → stimulate → constrict skeletal muscle vessels Cholinergic sympathetic fibers → inhibit → dilate skeletal muscle vessels
Blood vessels in cerebrum, abdominal viscera, and genitalia	Parasympathetic fibers → inhibit → dilate vessels in cerebrum, abdominal viscera, and genitalia	Adrenergic sympathetic fibers → stimulate → constrict vessels in cerebrum and abdominal viscera Cholinergic sympathetic fibers → inhibit → dilate vessels in external genitalia
Smooth muscle of hollow organs and sphincters		
Bronchi	Stimulates → bronchial constriction	Inhibits → bronchial dilation
Digestive tract	Stimulates → increased peristalsis	Inhibits → decreased peristalsis
Anal sphincter	Inhibits → opens sphincter for defecation	Stimulates → closes sphincter
Urinary bladder	Stimulates → contracts bladder	Inhibits → relaxes bladder
Urinary sphincters	Inhibits → opens sphincter for urination	Stimulates → closes sphincter
Eye		
(a) Iris	Stimulates circular fibers → constriction of pupil	Stimulates radial fibers → dilation of pupil
(b) Ciliary	Stimulates → accommodation for near vision (bulging of lens)	Inhibits → accommodation for far vision (flattening of lens)
Hairs (pilomotor muscles)	No parasympathetic fibers	Stimulates → "goose pimples" (piloerection)
Glands		
Sweat	No parasympathetic fibers	Cholinergic sympathetic fibers stimulate sweat glands
Digestive (salivary, gastric, etc.)	Stimulates secretion of saliva and gastric juice	Decreases secretion of saliva and gastric juice
Pancreas, including islets	Stimulates secretion of pancreatic juice and insulin	Decreases secretion of pancreatic juice and insulin
Liver	No parasympathetic fibers	Stimulates glycogenolysis, which tends to increase blood sugar
Adrenal medulla	No parasympathetic fibers	Stimulates epinephrine (and some norepinephrine) secretion which tends to increase blood sugar, blood pressure, and heart rate and to produce many other sympathetic effects

or flight" reaction. Particularly notable among them are a marked increase in the heart rate, blood pressure, oxygen consumption, respirations, and frequently a feeling of tenseness and being "uptight."

Sympathetic impulses usually dominate the control of most visceral effectors in times of stress, but not always. Curiously enough, parasympathetic impulses frequently become excessive to some effectors at such times. For instance, one of the first symptoms of emotional stress in many individuals is that they feel hungry and want to eat more than usual. Presumably this is partly caused by increased parasympathetic impulses to the smooth muscle of the stomach. This stimulates increased gastric contractions, which in turn may cause the feeling of hunger. The most famous disease of parasympathetic excess is peptic ulcer. In some individuals (presumably those born with certain genes), chronic stress leads to excessive parasympathetic stimulation of hydrochloric acid glands, with eventual development of peptic ulcer.

9 *Principle of nonautonomy.* The autonomic nervous system is far from being autonomous either anatomically or physiologically. Axons of many central nervous system neurons synapse with sympathetic and parasympathetic preganglionic neurons and thereby exert a major influence over their activity. Neurons located at higher and higher levels of the central nervous system function as a hierarchy in their control of autonomic activity. Lowest ranking in this hierarchy are the so-called lower autonomic centers in the spinal cord and brainstem. Here, axons from neurons located at higher levels synapse with preganglionic neurons. Highest ranking in the hierarchy of autonomic control, we now know, are certain areas of the cerebral cortex itself. Fig. 9-4 indicates the central nervous system hierarchy that regulates autonomic neuron conduction and thereby regulates the functioning of the visceral effectors they supply. Neurons in certain areas (called autonomic centers) of the cerebral cortex send impulses that either stimulate or inhibit autonomic centers in the limbic system.* Limbic system autonomic centers then relay impulses to centers in the hypothalamus, which in turn relay impulses to lower autonomic centers—to either parasympathetic centers in the brainstem or sacral segments of the spinal cord, or to sympathetic centers in the thoracolumbar segments of the cord, or to both parasympathetic and sympathetic centers. Finally, these lowest autonomic centers stimulate or inhibit conduction by parasympathetic or sympathetic preganglionic and postganglionic neurons and thereby increase or decrease visceral effector activities.

You may be wondering why the name autonomic system was ever chosen in the first place if the system is really not autonomous. Originally the term seemed appropriate. The autonomic system seemed to be self-regulating and independent of the rest of the nervous system. Common observations furnished abundant evidence of its independence from cerebral control, from direct control by the will, that is. But later, even this was found to be not entirely true. Some rare and startling exceptions were discovered. I have seen one such exception—a man who sat in a brightly lighted amphitheater in front of a class of medical students and made his pupils change from small, constricted dots (normal response to bright light) to widely dilated circles. This same man also willed gooseflesh to appear on his arms by contracting the smooth muscle of the hairs.

*See p. 216. An older name for essentially the same structures now called the limbic system was rhinencephalon (literally "nose-brain"), a name that refers to its role in olfaction.

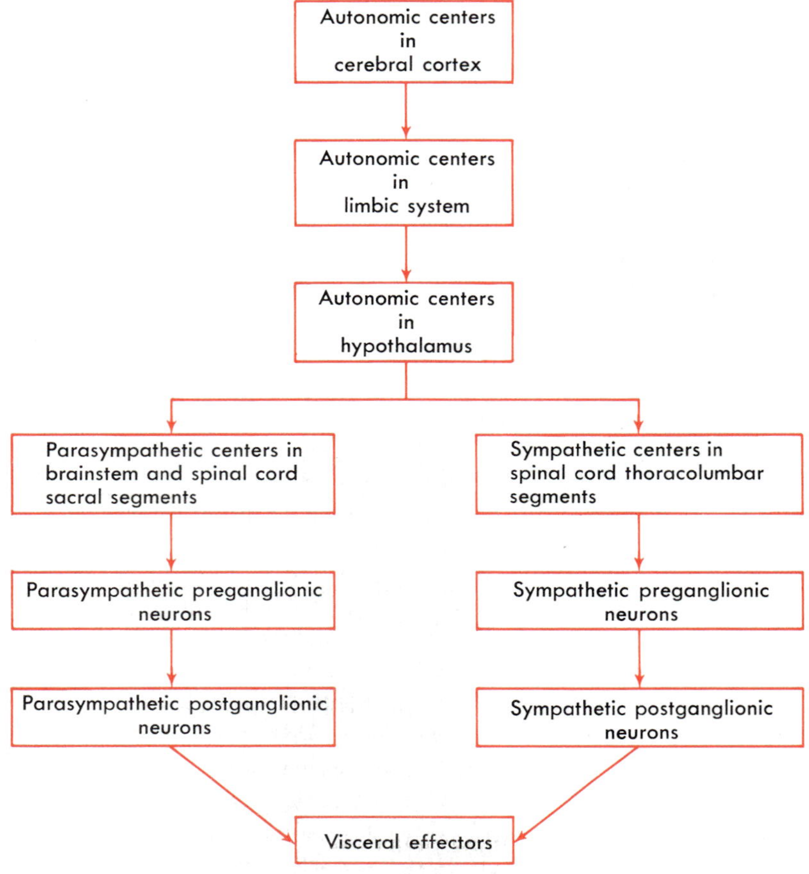

Fig. 9-4

Central nervous system hierarchy that regulates autonomic functions.

"Fight or flight" response and physiological changes in meditation

Wallace and Benson, in their major study of the physiology of meditation,* noted that a specific set of physiological changes—an "integrated response"—occurred during meditation. They observed, too, that this integrated response to meditation was a hypometabolic state and that it represented quiescence of the sympathetic nervous system. In other words, it seemed to be an integrated response that was essentially opposite to the integrated "fight or flight" response. They found that the integrated response to meditation consisted mainly of the following physiological changes: a dramatic decrease in oxygen consumption, respiratory rate and volume, a moderate slowing of the heart, and a low, almost unchanging blood pressure. The integrated "fight or flight" response, in contrast, includes opposite physiological changes: a marked increase in oxygen consumption, respiratory rate, and volume, an accelerated heart rate, and an increased blood pressure. Interestingly, the integrated

*Wallace, R. K., and Benson, H.: The physiology of meditation, Sci. Am. **226:**85-90, Feb., 1972.

response to meditation also included several physiological changes that had previously been shown to accompany a highly relaxed state of mind—intensification of slow (8 to 9 cycles per second) alpha waves in electroencephalograms of the frontal and central brain regions, a precipitous decrease in the concentration of lactate in the blood, and a rapid and marked increase in the electrical resistance of the skin. The integrated "fight or flight" response, on the other hand, is accompanied by a high degree of tenseness and often anxiety.

Autonomic learning and biofeedback

Only a few years ago, physiologists generally took for granted that learning occurred only in the somatic nervous system and that it could not take place in the autonomic nervous system. They knew, however, that conditioning could occur in the autonomic system. Pavlov had long since proved this with his dogs who salivated at the sound of a bell. But this classical or pavlovian conditioning of a visceral response was not considered true autonomic learning. True learning resulted from a process called operant conditioning, which differed somewhat from the pavlovian method. Briefly, operant conditioning consists of presenting a stimulus and a choice of several responses. When the desired response is made, it is immediately reinforced by giving a reward of some kind.

That operant condition, or true learning, can occur in the autonomic system is now no longer doubted. By the late 1960's, hard-to-refute evidence was indicating that the epithet, the "stupid autonomic nervous system," was no longer appropriate. Respected scientists in the United States and abroad were demonstrating that both animals and humans could learn to make a specific visceral response to a specific stimulus, providing they could be informed that they were making the desired visceral response and were rewarded for it. A number of so-called biofeedback instruments were designed to inform subjects when they made a desired visceral response. Various kinds of rewards were used to reenforce the response.

For example, one of the early experiments in human autonomic learning was conducted at Harvard Medical School by David Shapiro and his colleagues. In this experiment a group of young men learned to increase their blood pressure by about 4 mm Hg. Without their knowledge, their blood pressure was monitored; each time it happened to increase a light flashed (the biofeedback). After a certain number of flashes, they were rewarded with the sight of a nude pin-up picture of a pretty girl.

At the Menninger Foundation in Topeka, Kansas, patients who suffered migraine headache attacks learned to dilate blood vessels in their hands. A biofeedback instrument that detected slight temperature changes was attached to their hands and emitted a high sound each time they happened to dilate their hand blood vessels. The reward for these patients was a lessening of the migraine pain—presumably because of the shunting of blood away from the head to the hands. (Migraine headaches initially involve distention of head blood vessels.) Ninety of the first 100 subjects trained to dilate their hand blood vessels learned to control their headaches in this way. One volunteer was reported to have raised her hand temperature 10° in 2 minutes!*

Recently a good deal of interest has been focused on the spectacular feat of brain-wave training, using electroencephalograms for the biofeedback.†

*Ferguson, Marilyn: The brain revolution, New York, 1973, Taplinger Publishing Co.

†Trotter, R. J.: Listen to your head, Sci. News **100:** 314-316, Nov., 1971.

Outline summary

AUTONOMIC NERVOUS SYSTEM

1 Definition—part of nervous system that sends efferent fibers to visceral effectors; specifically, to smooth muscle, cardiac muscle, and glands
2 Divisions—sympathetic (or thoracolumbar) and parasympathetic (or craniosacral)
3 Macroscopic structure
  a Sympathetic division consists of two chains of ganglia (one on either side of backbone) and fibers that connect ganglia with each other and with thoracic and lumbar segments of cord; other fibers extend from sympathetic ganglia out to visceral effectors
  b Parasympathetic division consists of ganglia located on or near viscera with fibers between ganglia and brainstem and between ganglia and sacral region of cord; also fibers from ganglia into viscera and glands
4 Microscopic structure
  a Cell bodies of preganglionic neurons of sympathetic system in lateral gray columns of thoracic and lumbar segments of cord; lower sympathetic centers also located here
  b Cell bodies of postganglionic neurons of sympathetic system in sympathetic chain ganglia or in collateral ganglia (celiac, superior, and inferior mesenteric)
  c Cells of preganglionic neurons of parasympathetic system in various nuclei of brainstem and in gray matter of sacral segments of cord; lower parasympathetic centers also located here
  d Cells of postganglionic neurons of parasympathetic system in ganglia on or near organs innervated
5 Some general principles
  a Autonomic functioning and homeostasis—autonomic system so regulates activities of visceral effectors as to maintain or quickly restore homeostasis in healthy individuals, except under highly stressful conditions
  b Dual autonomic innervation—both sympathetic and parasympathetic fibers supply most visceral effectors
  c Single autonomic innervation—sympathetic fibers but not parasympathetic fibers to adrenal medulla, sweat glands, and probably to smooth muscles of hairs and most blood vessels
  d Autonomic chemical transmitters—autonomic axon terminals release acetylcholine at synapses and either acetylcholine or norepinephrine at neuroeffector junctions; those releasing acetylcholine called cholinergic fibers, those releasing norepinephrine called adrenergic fibers
    1 All preganglionic axons are cholinergic fibers, as are most, or perhaps all, parasympathetic postganglionic axons (to sweat glands and to smooth muscle in walls of blood vessels in skeletal muscles and external genitalia)
    2 Only adrenergic autonomic fibers are sympathetic postganglionic axons, but even few of these are cholinergic (see preceding item)
  e Principle of tonic activity of autonomic fibers—e.g., adrenergic sympathetic fibers to blood vessels, sympathetic and parasympathetic fibers to smooth muscle of eye and of digestive organs—continually conduct impulses to these visceral effectors
  f Autonomic antagonism and summation—sympathetic and parasympathetic impulses tend to produce opposite effects; algebraic sum of two opposing tendencies determines response made by doubly innervated visceral effector
  g Parasympathetic dominance of digestive tract—normally, parasympathetic impulses to glands and smooth muscle of digestive tract dominate over sympathetic impulses to them; parasympathetic impulses promote digestive gland secretion, peristalsis, and defecation
  h Sympathetic dominance in stress—under stress conditions, sympathetic impulses to most visceral effectors increase greatly, causing them to respond in ways that enable body to expend maximal amount of energy; sympathetic dominance one of first defenses against stress; parasympathetic impulses, however, may become excessive to some effectors under stress conditions
  i Nonautonomy—autonomic nervous system not autonomous with relation to rest of nervous system; partly controlled by higher autonomic centers, hierarchy of CNS structures controls autonomic nervous system—see Fig. 9-4
6 "Fight or flight" response and physiological changes in meditation—essentially opposite integrated responses

Review questions

1 Define visceral effectors. Name three kinds of visceral effectors.
2 Explain briefly the general function of the autonomic nervous system; of the sympathetic system; of the parasympathetic system.
3 Sympathetic stimulation produces massive, widespread responses, whereas reactions to parasympathetic stimulation are often highly localized. What anatomical differences between the two systems explain the physiological difference?
4 Name the transmitter released by axons of all preganglionic autonomic neurons. Where, in what structures, is it released?
5 Describe the action of the enzyme cholinesterase.
6 What transmitter substance do axons of sympathetic postganglionic neurons release at visceral neuroeffector junctions?
7 Identify COMT by name, class of compound, and action.
8 Classify the following structures as somatic effectors, visceral effectors, or neither: adrenal glands, biceps femoris muscle, heart, iris, and skin.
9 Which of the following would indicate an increase in sympathetic impulses and which might indicate an increase in parasympathetic impulses to visceral effectors—constipation, dilated pupils, dry mouth, goose pimples, "I'm always hungry and eat too much when I am upset," rapid heartbeat?
10 What is the limbic system, and what general function does it perform?
11 Compare the "fight or flight" reaction with the physiological changes of meditation.
12 Can learning in the autonomic nervous system occur? If so, explain the method for learning visceral responses.

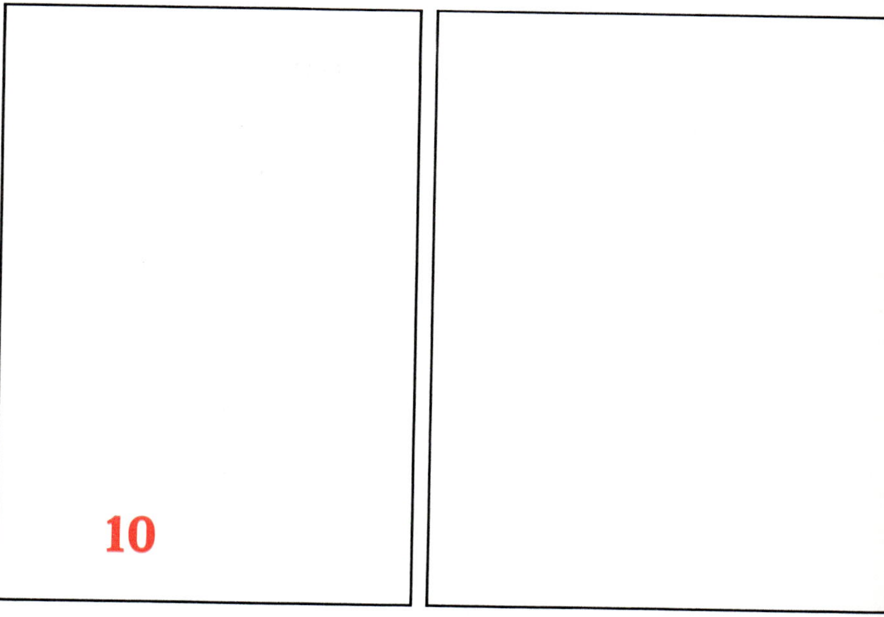

10 Sense organs

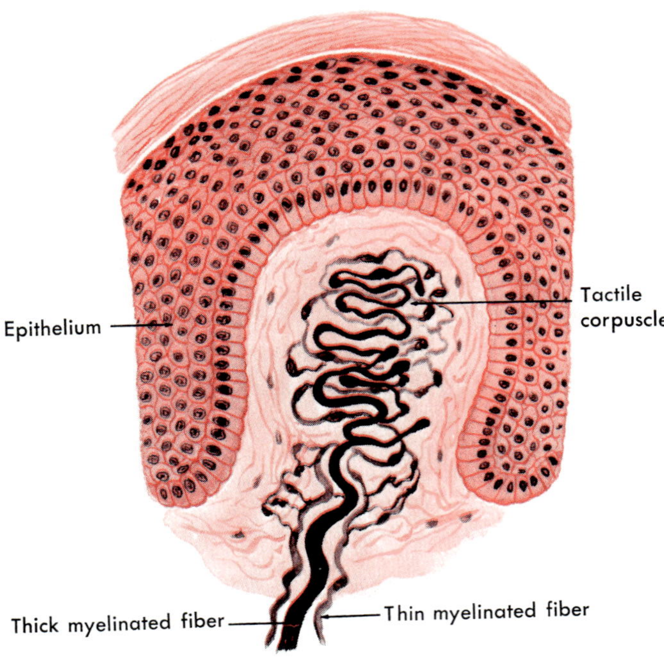

Fig. 10-1

A Meissner's corpuscle (tactile corpuscle) found in the connective tissue papillae of the skin. It is an example of an encapsulated specialized nerve ending found in the hairless portions of the skin.

Table 10-1

Receptors and sensations they mediate

Sensations	Receptors
Touch	Meissner's corpuscles (Fig. 10-1) Merkel's disks Basketlike arrangements around bases of hairs
Pressure	Vater-Pacinian corpuscles
Heat	Corpuscles of Ruffini
Cold	Krause end bulbs
Pain	Naked nerve fibers
Proprioception	Neuromuscular spindles Neurotendinous spindles Ruffini endings in joint capsules

General remarks

The body has millions of sense organs. All of its receptors—that is, the distal ends of dendrites of all its sensory neurons—are its sense organs. They serve two vital general functions—sensations and reflexes. All sensations and all reflexes result from stimulation of receptors. In short, receptors are the structures that detect changes in our external and internal environments and that initiate the responses necessary for adjusting the body to these changes so as to maintain or restore homeostasis. One more point—a matter more of interest than importance—we have more than just the "five senses," vision, hearing, taste, smell, and touch. For example, some of our other senses are warmth, cold, pain, and proprioception.

Receptors are located all over the body, inside as well as on its surfaces. Some of the main kinds of receptors are exteroceptors, visceroceptors, and proprioceptors. Exteroceptors are surface receptors of various types. Changes in the external environment stimulate exteroceptors. They are located in the skin, mucosa, eye, and ear. Visceroceptors and proprioceptors are both located internally. Visceroceptors are found, for example, in the walls of blood vessels, stomach, intestines, and various other organs. Proprioceptors are located in muscles, tendons, joints, and the internal ear.

Structurally, receptors differ considerably. Some, such as those in the eye and ear, constitute highly specialized sense organs. Others are simply naked or free nerve endings while others are encapsulated (Fig. 10-1).

According to the principle of specificity of receptors, specific kinds of receptors mediate specific sensations because they are sensitive to specific kinds of stimuli. Although generally accepted, this principle is now questioned by some investigators who think one kind of receptor may mediate more than one kind of sensation and that one kind of sensation may be mediated by more than one kind of receptor. Generally accepted ideas about which types of receptors mediate which sensations are included in Table 10-1.

Somatic, visceral, and referred pain

Because stimulation of pain receptors may give warning of potentially harmful environmental changes, pain receptors are also called *nociceptors* (L. *noceo,* to injure). Any type of stimulus, provided that it be sufficiently intense, seems to be adequate for stimulating nociceptors in the skin and mucosa. In contrast, only marked changes in pressure and certain chemicals can stimulate nociceptors located in the viscera. Sometimes this knowledge proves useful. For example, it enables a physician to cauterize the uterine cervix without giving an anesthetic but with assurance that the patient will not suffer pain from the intense heat. On the other hand, he knows that if the intestine becomes markedly distended (as it sometimes does following surgery), the patient will experience pain. So, too, will the individual whose heart becomes ischemic because of coronary occlusion. Presumably, the resulting cellular oxygen deficiency leads to the formation or accumulation of chemicals that stimulate nociceptors in the heart.

Two main types of pain are recognized: somatic and visceral.

Somatic pain may be *superficial,* as when it arises from stimulation of skin receptors, or it may be *deep,* as when it results from stimulation of receptors in the skeletal muscles, fascia, tendons, and joints.

Visceral pain results from stimulation of receptors located in the viscera. Impulses are conducted from these receptors to the cord primarily by sensory fibers in sympathetic nerves and only rarely in parasympathetic nerves.

The cerebrum does not always interpret the source of pain accurately. Sometimes it erroneously refers the pain to a surface area instead of to the region in which the stimulated receptors actually are located. This phenomenon is called *referred pain.* It occurs only as a result of stimulation of pain receptors located in deep structures (notably, viscera, joints, and skeletal muscles)—never from stimulation of skin receptors. In other words, deep somatic pain and visceral pain may be referred but not superficial somatic pain.

Pain originating in the viscera and other deep structures is generally interpreted as coming from the skin area whose sensory fibers enter the same segment of the spinal cord as the sensory fibers from the deep structure. For example, sensory fibers from the heart enter the first to fourth thoracic segments. And so do sensory fibers from the skin areas over the heart and on the inner surface of the left arm. Pain originating in the heart is referred to those skin areas, but the reason for this is not clear.

Eye

Anatomy

COATS OF EYEBALL

Approximately five sixths of the eyeball lies recessed in the orbit, protected by this bony socket. Only the small anterior surface of the eyeball is exposed. Three layers of tissues or coats compose the eyeball. From the outside in they are the sclera, the choroid, and the retina. Both the sclera and the choroid coats consist of an anterior and a posterior portion.

Tough white fibrous tissue fashions the *sclera.* Deep within the anterior part of the sclera at its junction with the cornea lies a ring-shaped venous sinus, the *canal of Schlemm* (Figs. 10-6 to 10-8). The anterior portion of the sclera is called the *cornea* and lies over the colored part of the eye (iris). The cornea is transparent, whereas the rest of the sclera is white and opaque, a fact that explains why the visible anterior surface of the sclera is usually spoken of as the "whites" of

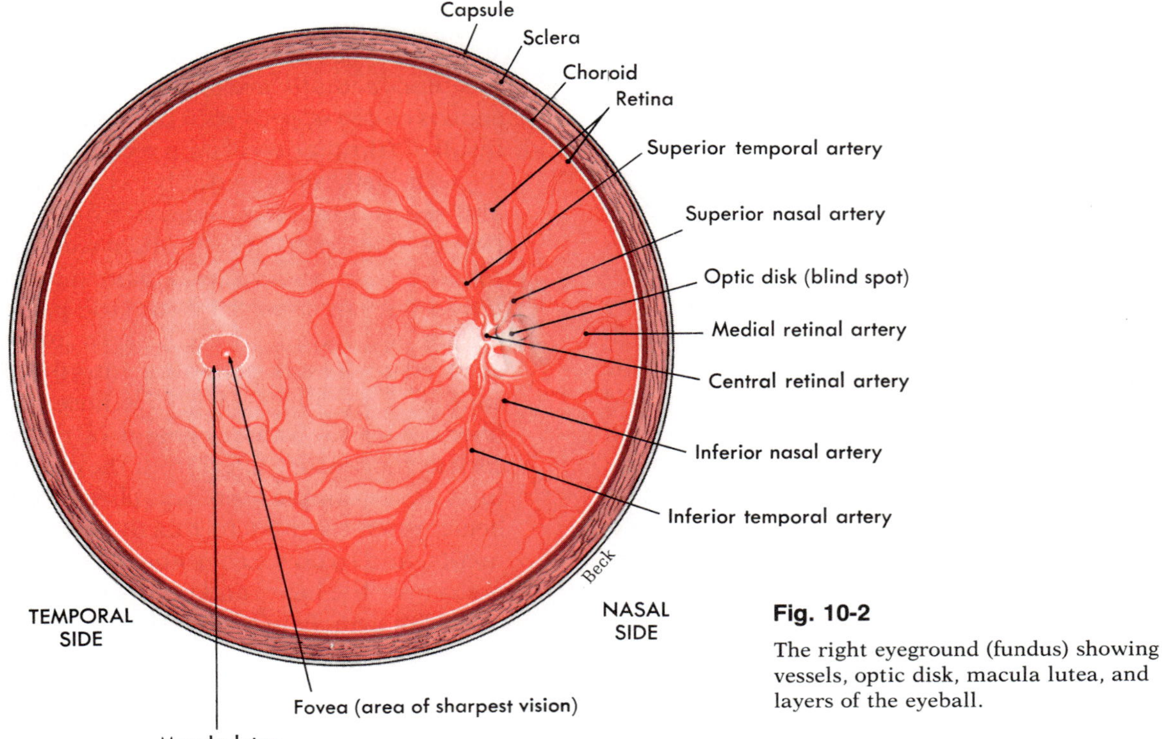

Fig. 10-2

The right eyeground (fundus) showing vessels, optic disk, macula lutea, and layers of the eyeball.

the eyes. No blood vessels are found in the cornea, in the aqueous and vitreous humors, or in the lens.

The middle or *choroid coat* of the eye contains a great many blood vessels and a large amount of pigment. Its anterior portion is modified into three separate structures: the ciliary body, the suspensory ligament, and the iris.

The *ciliary body* is formed by a thickening of the choroid and fits like a collar into the area between the anterior margin of the retina and the posterior margin of the iris. The small *ciliary muscle,* composed of both radial and circular smooth muscle fibers, lies in the anterior part of the ciliary body. Attached to the ciliary body is the *suspensory ligament,* which blends with the elastic capsule of the *lens* and holds it suspended in place.

The *iris* or colored part of the eye consists of circular and radial smooth muscle fibers arranged so as to form a doughnut-shaped structure (the hole in the middle is called the *pupil*). The iris attaches to the ciliary body.

The *retina* is the incomplete innermost coat of the eyeball—incomplete in that it has no anterior portion. It consists mainly of nervous tissue and contains three layers of neurons (Fig. 10-5). Named in the order in which they conduct impulses, they are photoreceptor neurons, bipolar neurons, and ganglion neurons. The distal ends of the dendrites of the photoreceptor neurons have been given names descriptive of their shapes. Because some look like tiny rods and others like cones, they are called *rods* and *cones,* respectively. They constitute our visual receptors, structures highly specialized for stimulation by

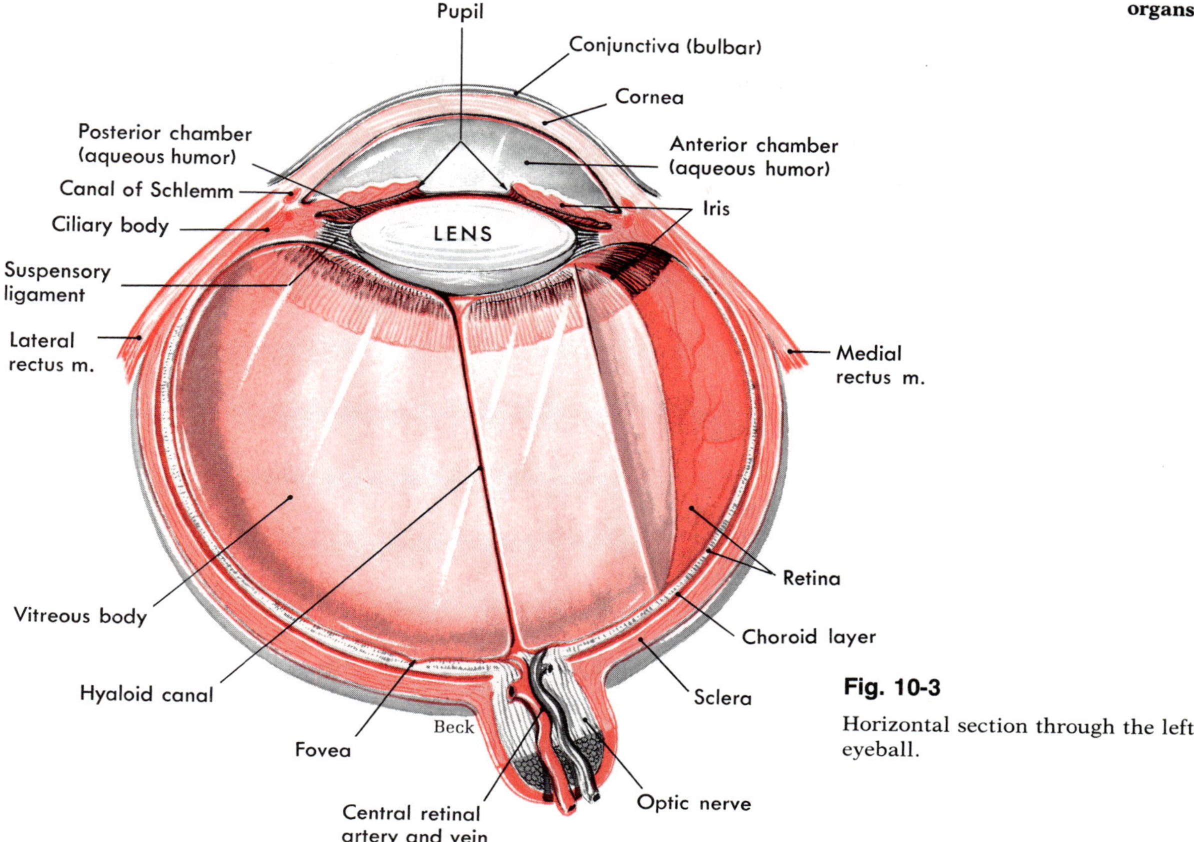

Fig. 10-3

Horizontal section through the left eyeball.

light rays (discussed on pp. 255-256). They differ as to numbers, distribution, and function. Cones are less numerous than rods and are most densely concentrated in the *fovea centralis,* a small depression in the center of a yellowish area, the *macula lutea,* found near the center of the retina. They become less and less dense from the fovea outward. Rods, on the other hand, are absent entirely from the fovea and macula and increase in density toward the periphery of the retina. How these anatomical facts relate to rod and cone functions is revealed on p. 256.

All the axons of ganglion neurons extend back to a small circular area in the posterior part of the eyeball known as the *optic disk* or papilla. This part of the sclera contains perforations through which the fibers emerge from the eyeball as the *optic nerves.* The optic disk is also called the *blind spot* because light rays striking this area cannot be seen since it contains no rods or cones, only nerve fibers.

For an outline summary of the coats of the eyeball see Table 10-2.

CAVITIES AND HUMORS

The eyeball is not a solid sphere but contains a large interior cavity that is divided into two cavities, anterior and posterior.

The *anterior cavity* has two subdivisions known as the *anterior* and *posterior chambers.* As Fig. 10-3 shows, the entire anterior cavity lies in front of the lens. The posterior chamber of the anterior cavity consists of the space directly posterior to the iris but anterior to the lens. And the anterior chamber of the anterior cavity is the small space anterior to the iris but posterior to the cornea. *Aqueous*

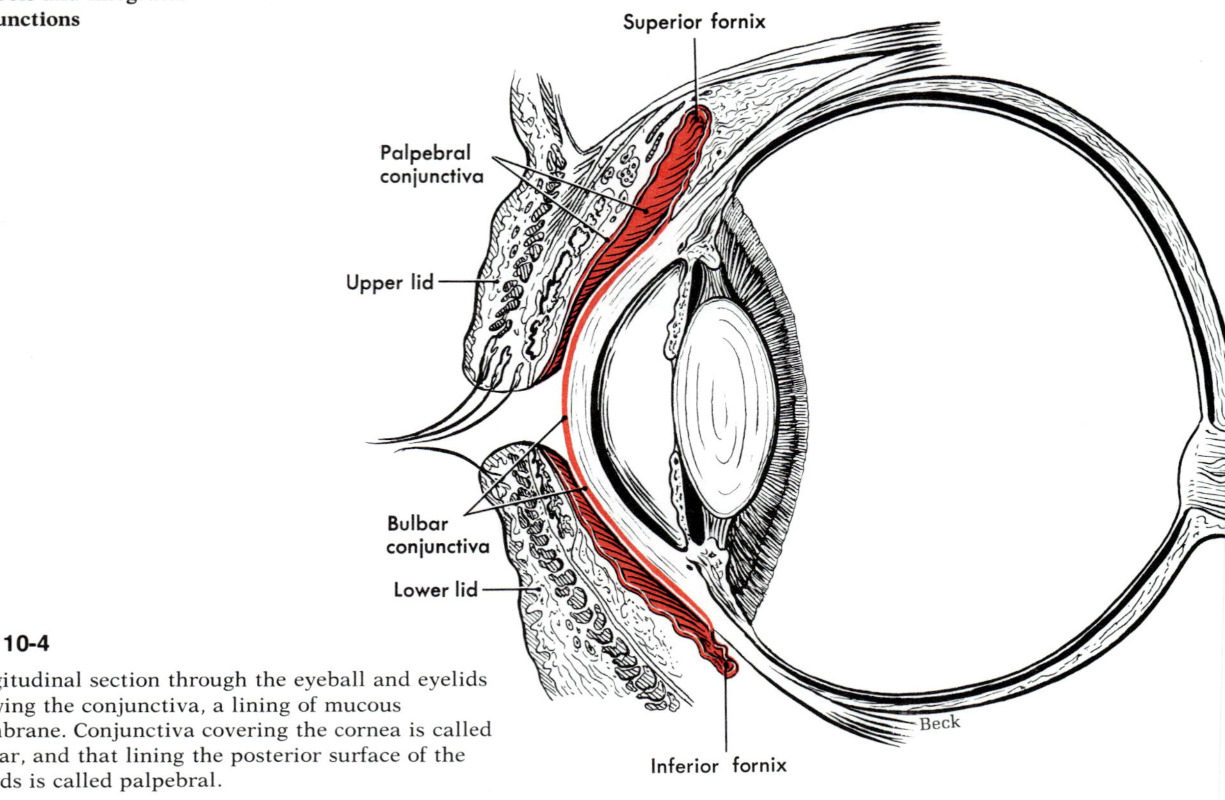

Fig. 10-4

Longitudinal section through the eyeball and eyelids showing the conjunctiva, a lining of mucous membrane. Conjunctiva covering the cornea is called bulbar, and that lining the posterior surface of the eyelids is called palpebral.

Table 10-2

Coats of the eyeball

Location	Posterior portion	Anterior portion	Characteristics
Outer coat (sclera)	Sclera proper	Cornea	Protective fibrous coat; cornea transparent; rest of coat white and opaque
Middle coat (choroid)	Choroid proper	Ciliary body; suspensory ligament; iris (pupil is hole in iris); lens suspended in suspensory ligament	Vascular, pigmented coat
Inner coat (retina)	Retina	No anterior portion	Nervous tissue; rods and cones (receptors for second cranial nerve) located in retina

Table 10-3

Cavities of the eye

Cavity	Divisions	Location	Contents
Anterior	Anterior chamber Posterior chamber	Anterior to iris and posterior to cornea Posterior to iris and anterior to lens	Aqueous humor Aqueous humor
Posterior	None	Posterior to lens	Vitreous humor

humor fills both chambers of the anterior cavity. This substance is clear and watery and often leaks out when the eye is injured.

The *posterior cavity* of the eyeball is considerably larger than the anterior since it occupies all the space posterior to the lens, suspensory ligament, and ciliary body. It contains *vitreous humor,* a substance with a consistency comparable to soft gelatin. This semisolid material helps maintain sufficient intraocular pressure to prevent the eyeball from collapsing. (An obliterated artery, the hyaloid canal, runs through the vitreous humor between the lens and optic disk.)

An outline summary of the cavities of the eye is included in Table 10-3 (also see Fig. 10-3).

Still not established is the mechanism by which aqueous humor forms. It comes from blood in capillaries (located mainly in the ciliary body). Presumably, the capillaries actively secrete aqueous humor into the posterior chamber. But also passive filtration from capillary blood may contribute to aque-

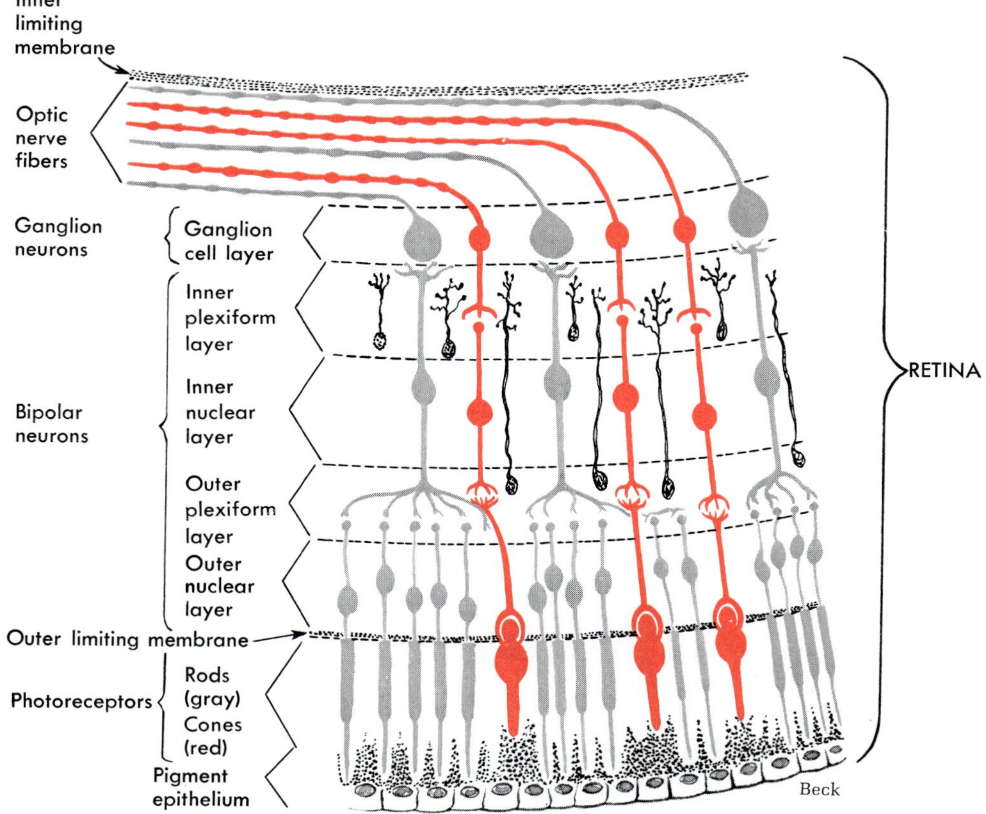

Fig. 10-5

Layers that compose the retina. The inner limiting membrane lies nearest the inside of the eyeball. It adheres to the vitreous humor. The pigment epithelium lies farthest from the inside of the eyeball. It adheres to the choroid coat. Note relay of three neurons in the retina: photoreceptor, bipolar, and ganglion neurons (named in order of impulse transmission). Light rays pass through the vitreous humor and various layers of retina to stimulate rods and cones, the receptors of the photoreceptor neurons.

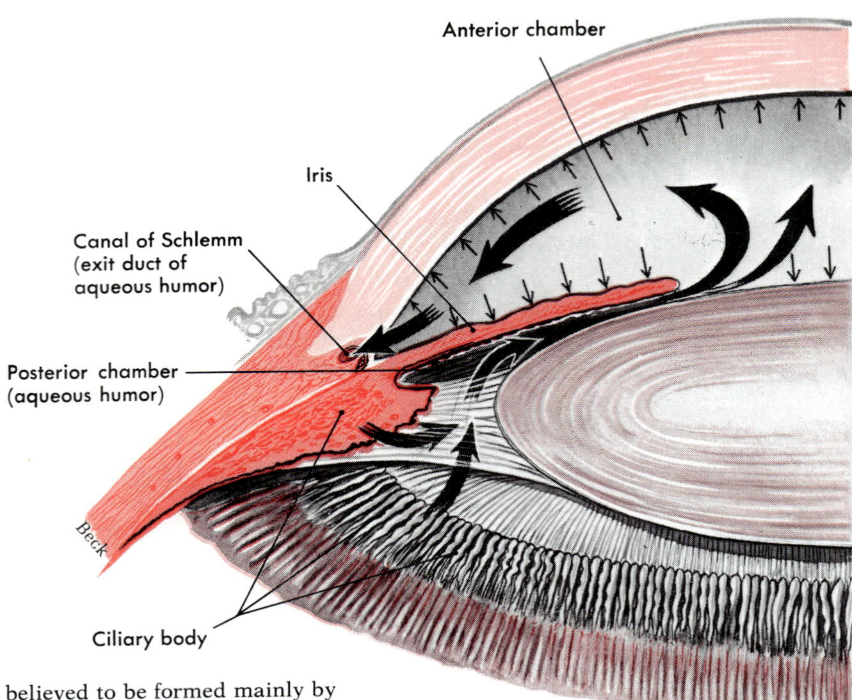

Fig. 10-6

Aqueous humor (heavy arrows) is believed to be formed mainly by secretion by the ciliary body into the posterior chamber. It passes into the anterior chamber through the pupil, from which it is drained away by the ring-shaped canal of Schlemm, and finally into the anterior ciliary veins. Small arrows indicate pressure of the aqueous humor.

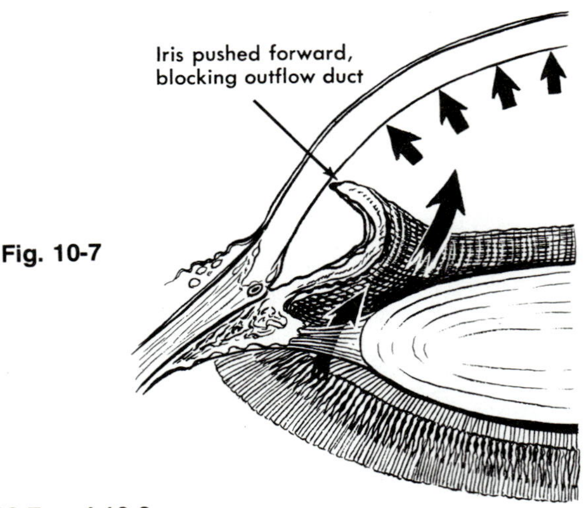

Fig. 10-7

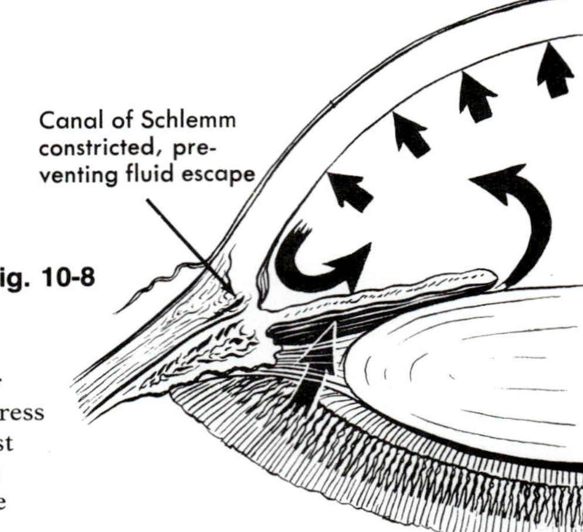

Fig. 10-8

Figs. 10-7 and 10-8

Acute glaucoma. When pressure of the aqueous humor in the anterior chamber becomes extreme, the iris is pushed forward, causing it to press upon and block the canal of Schlemm from draining the fluid. In most instances, proper eyedrop medication will dilate the duct, permitting excess fluid escape. When medication fails, a new outflow duct can be created surgically.

Table 10-4

Eye muscles

	Extrinsic muscles	Intrinsic muscles
Names	Superior rectus Inferior rectus Lateral rectus Mesial rectus Superior oblique Inferior oblique	Iris Ciliary muscle
Kind of muscle	Voluntary (striated, skeletal)	Involuntary (smooth, visceral)
Location	Attached to eyeball and bones of orbit	Modified anterior portion of choroid coat of eyeball; iris doughnut-shaped, sphincter muscle; pupil, hole in center of iris
Functions	Eye movements	Iris regulates size of pupil—therefore, amount of light entering eye; ciliary muscle controls shape of lens (accommodation)—therefore, its refractive power
Innervation	Somatic fibers of third, fourth, and sixth cranial nerves	Autonomic fibers of third and fourth cranial nerves

ous humor formation. From the posterior chamber, aqueous humor moves from the area between the iris and the lens through the pupil into the anterior chamber (Fig. 10-6). From here it drains into the canal of Schlemm and moves on into small veins. Normally, aqueous humor drains out of the anterior chamber at the same rate at which it enters the posterior chamber, so the amount of aqueous humor in the eye remains relatively constant. And so, too, does intraocular pressure. But sometimes something happens to upset this balance and intraocular pressure increases above the normal level of about 20 to 25 mm Hg pressure. The individual then has the eye disease known as glaucoma (Figs. 10-7 and 10-8). Either excess formation or, more often, decreased absorption is seen as an immediate cause of this condition, but underlying causes are unknown.

MUSCLES

Eye muscles are of two types: extrinsic and intrinsic.

Extrinsic muscles are those that attach to the outside of the eyeball and to the bones of the orbit. They move the eyeball in any desired direction and are, of course, voluntary muscles. Four of them are straight muscles and two are oblique. Their names describe their positions on the eyeball. They are the superior, inferior, mesial, and lateral rectus muscles and superior and inferior oblique muscles (Figs. 10-9 and 10-10).

Intrinsic eye muscles are those located within the eye. Their names are the iris and the ciliary muscles, and they are involuntary. Incidentally, the eye is the only organ in the body in which both voluntary and involuntary muscles are found. The iris regulates the size of the pupil. The ciliary muscle controls the shape of the lens. As the ciliary muscle contracts, it releases the suspensory ligament from the backward pull usually exerted upon it. And this allows the elastic lens, suspended in the ligament, to bulge or become more convex—a necessary accommodation for near vision (pp. 254-255). Some essential facts about eye muscles are summarized in Table 10-4.

ACCESSORY STRUCTURES

Accessory structures of the eye include the eyebrows, eyelashes, eyelids, and lacrimal apparatus.

Eyebrows and eyelashes. The eyebrows and

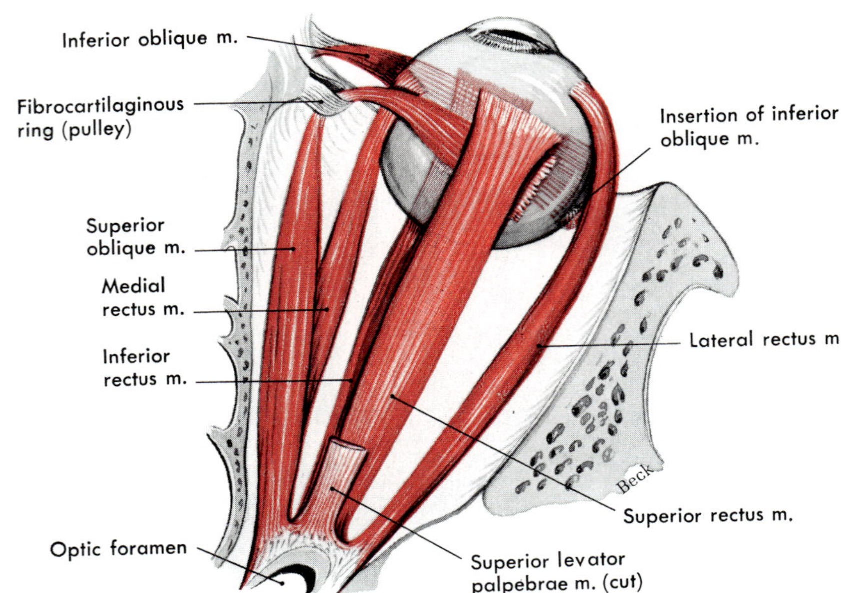

Fig. 10-9

Muscles that move the right eye as viewed from above.

eyelashes serve a cosmetic purpose and give some protection against the entrance of foreign objects into the eyes. Small glands located at the base of the lashes secrete a lubricating fluid. They frequently become infected, forming a *sty*.

Eyelids. The eyelids or palpebrae consist mainly of voluntary muscle and skin, with a border of thick connective tissue at the free edge of each lid known as the tarsal plate. One can feel the tarsal plate as a ridge when turning back the eyelid to remove a foreign object. Mucous membrane called conjunctiva lines each lid (Fig. 10-4). It continues over the surface of the eyeball, where it is modified to give transparency. Inflammation of the conjunctiva (conjunctivitis) is a fairly common infection. Because it produces a pinkish discoloration of the eye's surface, it is called pinkeye.

The opening between the eyelids bears the technical name of palpebral fissure. The height of the fissure determines the apparent size of the eyes. In other words, if the eyelids are habitually held widely opened, the eyes appear large, although actually there is very little difference in size between eyeballs of different adults. Eyes appear small if the upper eyelids droop. Plastic surgeons can correct this common aging change with an operation called blepharoplasty. The upper and lower eyelids join, forming an angle or corner known as a *canthus*, the inner canthus being the mesial corner of the eye and the outer canthus the lateral corner.

Lacrimal apparatus. The lacrimal apparatus consists of the structures that secrete tears and drain them from the surface of the eyeball. They are the lacrimal glands, lacrimal ducts, lacrimal sacs, and nasolacrimal ducts (Fig. 10-11).

The *lacrimal glands*, comparable in size and shape to a small almond, are located in a depression of the frontal bone at the upper outer margin of each orbit. Approximately a dozen small ducts lead from each gland, draining the tears onto the conjunctiva at the upper outer corner of the eye.

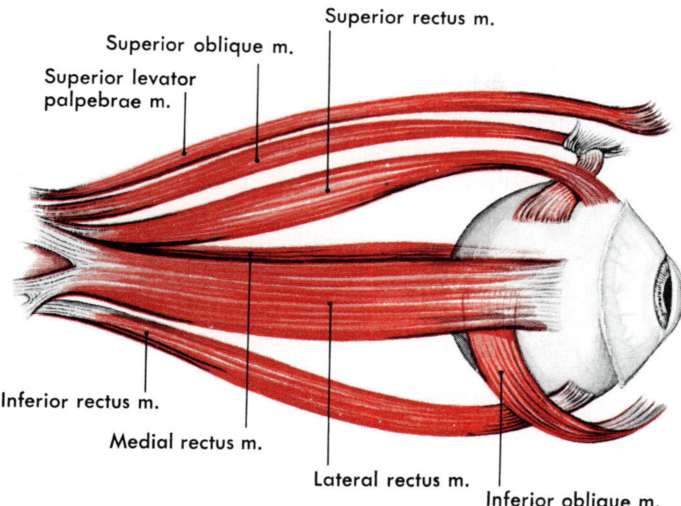

Fig. 10-10

Muscles of the right orbit as viewed from the side.

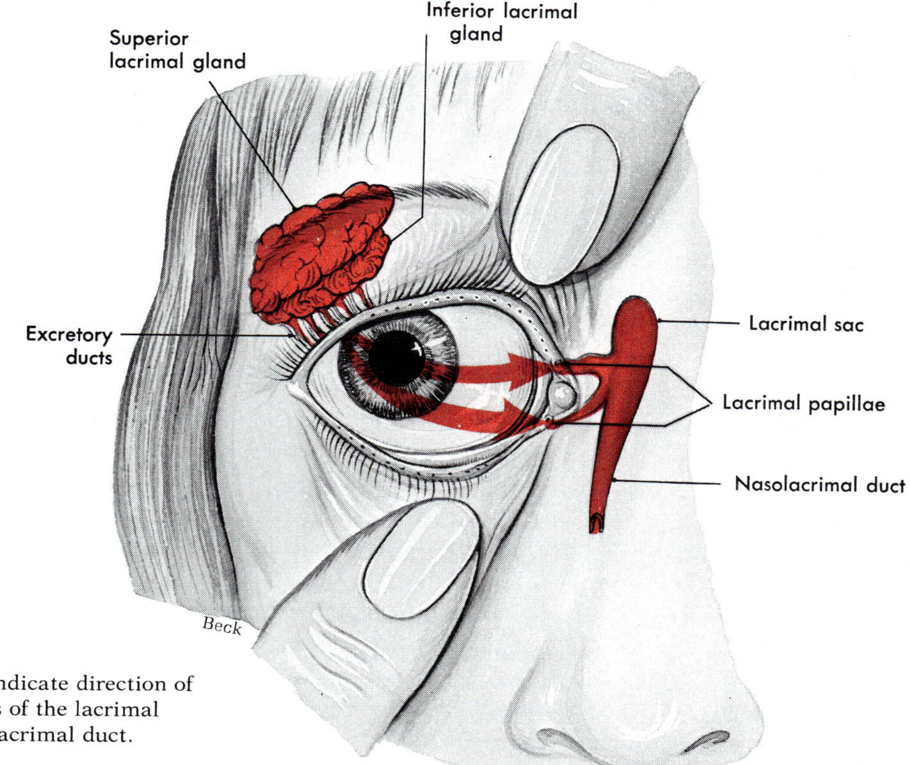

Fig. 10-11

The lacrimal apparatus. Arrows indicate direction of drainage from the excretory ducts of the lacrimal glands across the eye to the nasolacrimal duct.

The *lacrimal canals* are small channels, one above and the other below each *caruncle* (small red body at inner canthus). They empty into the lacrimal sacs. The openings into the canals are called *punctae* and can be seen as two small dots at the inner canthus of the eye. The *lacrimal sacs* are located in a groove in the lacrimal bone. The *nasolacrimal ducts* are small tubes that extend from the lacrimal sac into the inferior meatus of the nose. All the tear ducts are lined with mucous membrane, an extension of the mucosa that lines the nose. When this membrane becomes inflamed and swollen, the nasolacrimal ducts become plugged, causing the tears to overflow from the eyes instead of draining into the nose as they do normally. Hence, when we have a common cold, "watering" eyes add to our discomforts.

Physiology of vision

In order for vision to occur, the following conditions must be fulfilled: an image must be formed on the retina to stimulate its receptors (rods and cones), and the resulting nerve impulses must be conducted to the visual areas of the cerebral cortex.

FORMATION OF RETINAL IMAGE

Four processes focus light rays so that they form a clear image on the retina: *refraction* of the light rays, *accommodation* of the lens, *constriction* of the pupil, and *convergence* of the eyes.

Refraction of light rays. Refraction means the deflection or bending of light rays. It is produced by light rays passing obliquely from one transparent medium into another of different optical density, and the more convex the surface of the medium, the greater its refractive power. The refracting media of the eye are the cornea, aqueous humor, lens, and vitreous humor. Light rays are bent or refracted at the anterior surface of the cornea as they pass from the rarer air into the denser cornea, at the anterior surface of the lens as they pass from the aqueous humor into the denser lens, and at the posterior surface of the lens as they pass from the lens into the rarer vitreous humor.

When an individual goes to an ophthalmologist for an eye examination, the doctor does a "refraction." In other words, by various specially designed methods, he measures the refractory or light-bending power of that person's eyes.

In a relaxed normal (emmetropic) eye, the four refracting media together bend light rays sufficiently to bring to a focus on the retina the parallel rays reflected from an object 20 or more feet away (Fig. 10-12). Of course, a normal eye can also focus objects located much nearer than 20 feet from the eye. This is accomplished by a mechanism known as accommodation (discussed on pp. 254-255). Many eyes, however, show *errors of refraction*—that is, they are not able to focus the rays on the retina under the stated conditions. Some common errors of refraction are *nearsightedness* (myopia), *farsightedness* (hypermetropia), and *astigmatism.*

The nearsighted (myopic) eye sees distant objects as blurred images because it focuses rays from the object at a point in front of the retina (Fig. 10-13). According to one theory, this occurs because the eyeball is too long and the distance too great from the lens to the retina in the myopic eye. Concave glasses, by lessening refraction, can give clear distant vision to nearsighted individuals. Presumably, opposite conditions exist in the farsighted (hyperopic) eye (Fig. 10-14).

Astigmatism is a more complicated condition in which the curvature of the cornea or of the lens is uneven, causing horizontal and vertical rays to be focused at two different points on the retina (Figs. 10-15 and 10-16). Instead of the curvature of the cornea being a section of a sphere, it is more like that of a teaspoon with horizontal and vertical arcs un-

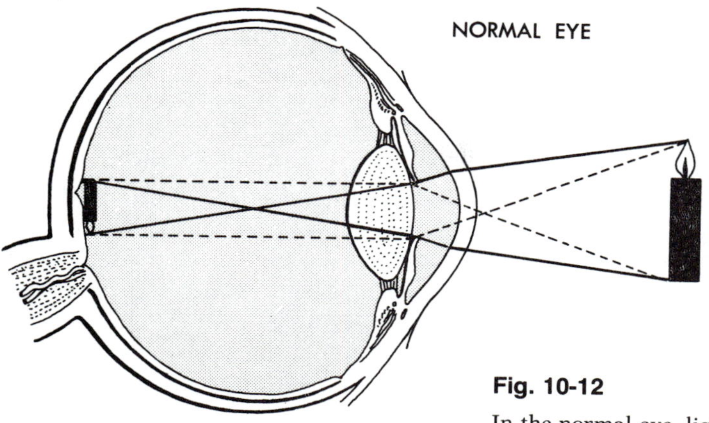

Fig. 10-12

In the normal eye, light rays from an object are refracted by the cornea, aqueous humor, lens, and vitreous humor and converge on the fovea of the retina, where an inverted image is clearly formed.

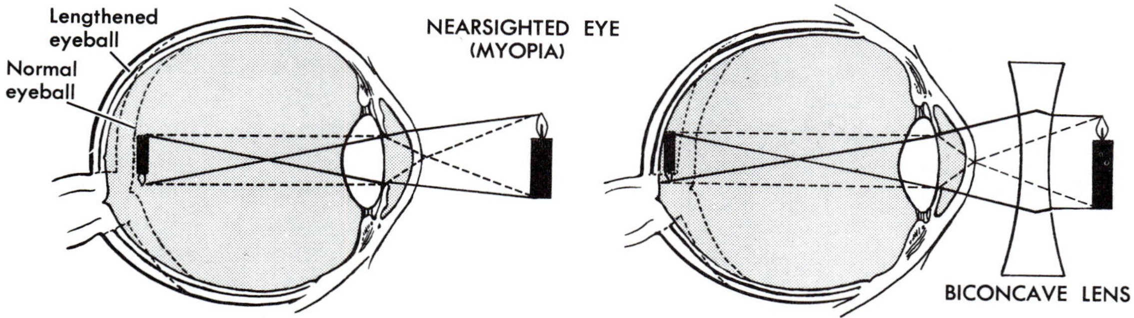

Fig. 10-13

The nearsighted or myopic eye focuses the image in front of the retina. This may occur when the eyeball is too long or the lens is too thick. Correction is by concave lens.

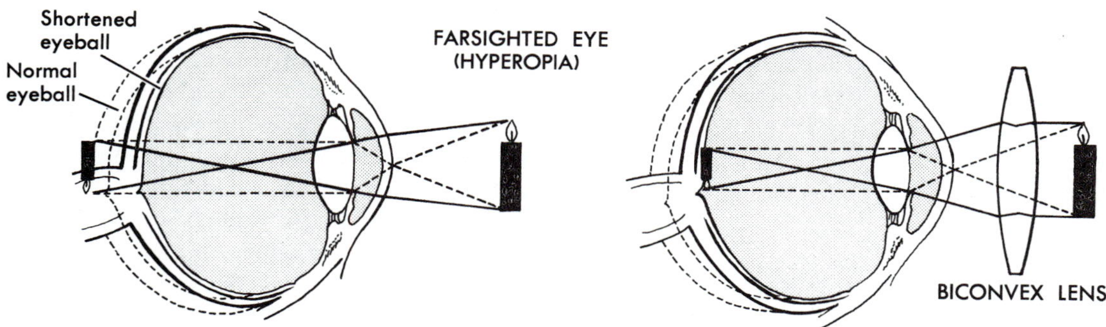

Fig. 10-14

The farsighted or hyperopic eye could only focus the image at a hypothetical distance behind the retina. This may occur when the eyeball is too short or the lens is too thin. Correction is by convex lens.

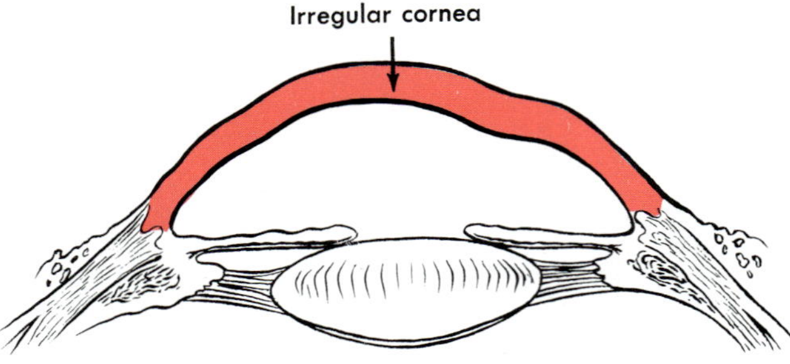

Fig. 10-15

Astigmatism results most frequently from an irregular cornea. The cornea may be only slightly flattened horizontally, vertically, or diagonally to produce distortion of vision. The compensating shape of the lens largely nullifies the irregularities of the cornea.

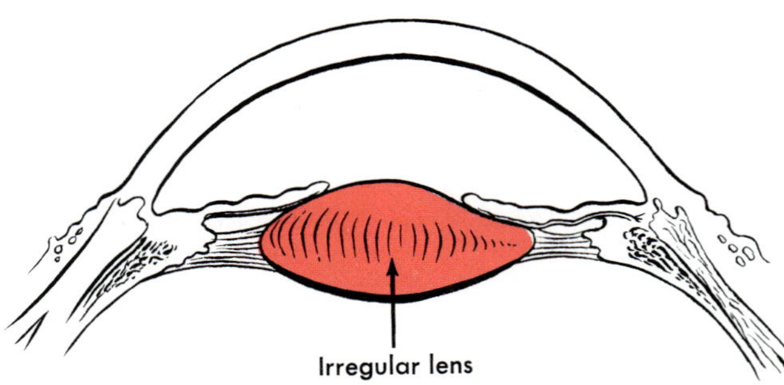

Fig. 10-16

An irregular lens causes light to be bent in such a way that it does not focus the image on the sharpest area of vision on the retina. An astigmatic eye with this defect causes distorted or blurred vision.

even. Suitable glasses correct the refraction of such an eye.

Visual acuity or the ability to distinguish form and outline clearly is indicated by a fraction that compares the distance at which an individual sees an object (usually letters of a definite size and shape) clearly with the distance at which the normal eye would see the object. Thus, if he sees clearly at 20 feet an object that the normal eye would be able to see clearly at 20 feet, his visual acuity is said to be 20/20 or normal. But if he sees an object clearly at 20 feet which the normal eye sees clearly at 30 feet, then his visual acuity is 20/30 or two thirds of normal.

As people grow older, they tend to become farsighted because lenses lose their elasticity and therefore their ability to bulge and to accommodate for near vision. This condition is called *presbyopia*.

Accommodation of lens. Accommodation for near vision necessitates three changes: increase in the curvature of the lens, constriction of the pupils, and convergence of the two eyes. Light rays from objects 20 or more feet away are practically parallel. The normal eye, as previously noted, refracts such rays sufficiently to focus them clearly on the retina. But light rays from nearer objects are divergent rather than parallel. So obviously they must be bent more acutely to bring them to a focus on the retina. Accommodation of the lens or, in other words, an increase in its curvature takes place to achieve this greater refraction. (It is a physical fact that the greater the convexity of a lens, the greater its refractive power.) Most observers accept Helmholtz' theory about the mechanism that produces accommodation of the lens. According to his theory, the ciliary muscle contracts, pulling the ciliary body and choroid forward toward the lens. This releases the tension on the suspensory ligament and therefore on the lens, which, being elastic, immediately bulges. For near vision, then, the ciliary muscle is contracted and the lens is bulging, whereas for far vision the ciliary muscle is relaxed and the lens is comparatively flat. Continual use of the eyes for near work produces eyestrain because of the prolonged contraction of the ciliary muscle. Some of the strain can be avoided by looking into the distance at intervals while doing close work.

Constriction of pupil. The muscles of the iris play an important part in the formation of clear retinal images. Part of the accommodation mechanism consists of contraction of the circular fibers of the iris, which constricts the pupil. This prevents divergent rays from the object from entering the eye through the periphery of the cornea and lens. Such peripheral rays could not be brought to a focus on the retina (because of aberration of the lens) and therefore would cause a blurred image. Constriction of the pupil for near vision is called the *near reflex* of the pupil and occurs simultaneously with accommodation of the lens in near vision. The pupil constricts also in bright light *(photopupil reflex or pupillary light reflex)* to protect the retina from too intense or too sudden stimulation.

Convergence of eyes. Single binocular vision (seeing only one object instead of two when both eyes are used) occurs when light rays from an object fall on corresponding points of the two retinas. The foveas and all points lying equidistant and in the same direction from the foveas are corresponding points. Whenever the eyeballs move in unison, either with the visual axes parallel (for far objects) or converging upon a common point (for near objects), light rays strike corresponding points of the two retinas. Convergence is the movement of the two eyeballs inward so that their visual axes come together or converge at the object viewed. The nearer the object, the greater the degree of convergence necessary to maintain single vision. A simple procedure serves to demonstrate the fact that single binocular vision results from stimulation of corresponding points of the two retinas. Gently press one eyeball out of line while viewing an object. Instead of one object, you will then see two. In order to achieve unified movement of the two eyeballs, a functional balance between the antagonistic extrinsic muscles must exist. If, for example, the right internal rectus muscle should contract more forcefully than its antagonist, the right external rectus, the right eye would be pulled in toward the nose instead of its visual axis being held parallel to that of the left eye in distant vision or converged upon the same point in near vision. Light rays from an object would then fall on noncorresponding points of the two retinas, and the object would be seen double (diplopia). Sometimes the individual can overcome the deviation of the visual axes by muscular effort (extra innervation of the weak muscle) and thereby achieve single vision, but only at the expense of muscular and nervous strain. The condition in which the imbalance of the eye muscles can be overcome by extra innervation of the weak muscle is called *heterophoria (esophoria* if the internal rectus is stronger and pulls the eye nasalward and *exophoria* if the external rectus is stronger and pulls the eye temporalward). *Strabismus* (cross-eye or squint) is an exaggerated esophoria that cannot be overcome by neuromuscular effort. An individual with strabismus usually does not have double vision, as you might expect, because he learns to suppress one of the images.

STIMULATION OF RETINA

Rods are known to contain *rhodopsin* (visual purple), a pigmented compound. As shown in Fig. 10-17, it forms by a protein opsin combining with retinal, a derivative of vitamin A. Rhodopsin is highly light-sensitive, so that when light rays strike a rod, its rhodopsin rapidly breaks down (also shown in Fig. 10-17). In some way this chemical change initi-

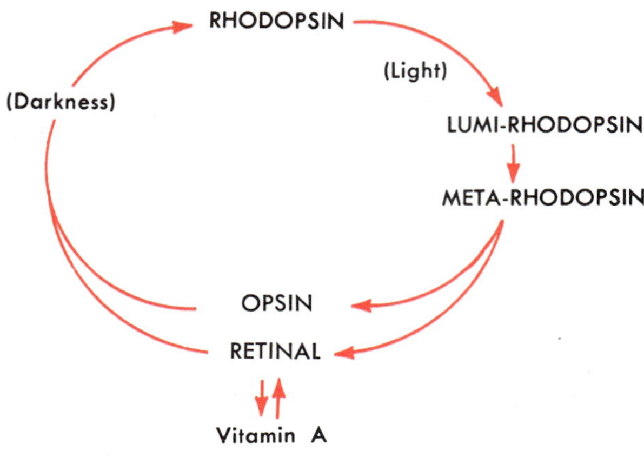

Fig. 10-17

The rhodopsin cycle.

ates impulse conduction by the rod. Then, if the rod is exposed to darkness for a short time, rhodopsin reforms from the opsin and retinal and is ready to function again. *Cones* also contain photosensitive chemicals—iodopsin and possibly others. Presumably, the cone compounds are less sensitive to light than rhodopsin. Brighter light seems necessary for their breakdown. Cones, therefore, are considered to be the receptors responsible for daylight and color visions. Rods, on the other hand, are believed to be the receptors for night vision because their rhodopsin quickly becomes almost depleted in bright light because of rapid breakdown but slow regeneration. This explains why you cannot see for a little while after you go from a bright light to darkness. But when rhodopsin has had time to reform, the rods again start functioning and dark adaptation has occurred. Or, as we say, we "can see in the dark once we get used to it." Night blindness occurs in marked vitamin A deficiency. Why? Fig. 10-17 contains a clue. The fovea contains the greatest concentration of cones and is, therefore, the point of clearest vision in good light. For this reason, when we want to see an object clearly in the daytime, we look directly at it so as to focus the image on the fovea. But in dim light or darkness we see an object better if we look slightly to the side of it, thereby focusing the image nearer the periphery of the retina where rods are more plentiful.

CONDUCTION TO VISUAL AREA

Fibers that conduct impulses from the rods and cones reach the visual cortex in the occipital lobes via the optic nerves, optic chiasma, optic tracts, and optic radiations. Look closely at Fig. 10-18. Notice that each optic nerve contains fibers from only one retina but that the optic chiasma contains fibers from the nasal portions of both retinas. Each optic tract also contains fibers from both retinas. These anatomical facts explain certain peculiar visual abnormalities that sometimes occur. Suppose a man's right optic tract were injured so that it could not conduct impulses. He would be totally blind in neither eye but partially blind in both eyes. Specifically, he would be blind in his right nasal and left temporal visual fields. Here are the reasons. The right optic tract contains fibers from the right retina's temporal area, the area that sees the right nasal visual field. And, in addition, the right optic tract contains fibers from the left retina's nasal area, the area that sees the left temporal visual field.

Auditory apparatus

Anatomy

The ears, auditory nerves, and auditory areas of the temporal lobes of the cerebrum compose the auditory apparatus. Each ear consists of three parts: external ear, middle ear, and inner ear (Fig. 10-19).

EXTERNAL EAR

The external ear has two divisions: the flap or modified trumpet on the side of the head called the *auricle* or *pinna* and the tube

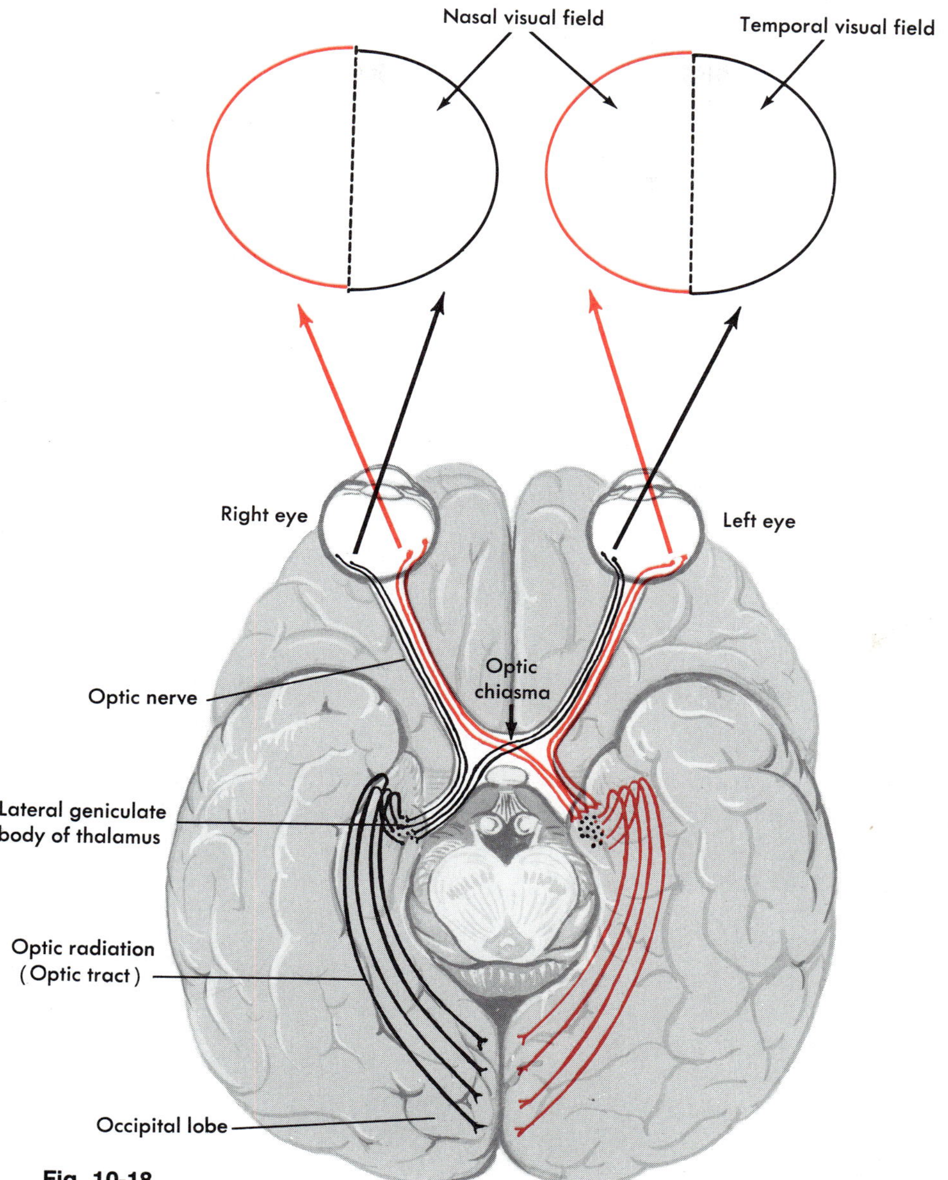

Fig. 10-18

Visual pathways. Note structures that compose each pathway: optic nerve, optic chiasma, lateral geniculate body of thalamus, optic radiations, and visual cortex of occipital lobe. Fibers from nasal portion of each retina cross over to opposite side at the optic chiasma, hence terminate in the lateral geniculate body of opposite side. The location of a lesion in the visual pathway determines the resulting visual defect. Examples: destruction of an optic nerve produces permanent blindness in the same eye. Pressure on the optic chiasma—by a pituitary tumor, for instance—produces bitemporal hemianopsia, or more simply, blindness in both temporal visual fields. Why? Because it destroys fibers from nasal sides of both retinas.

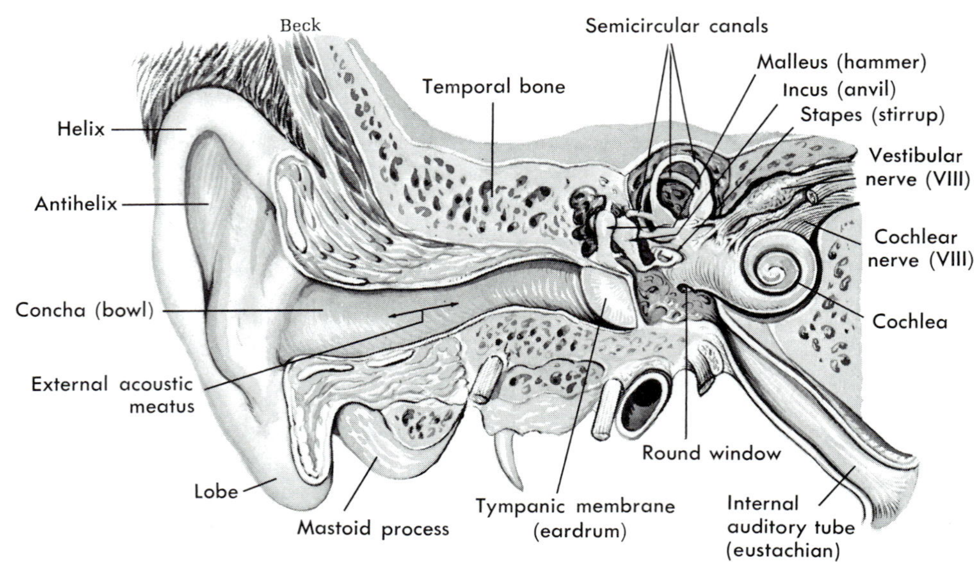

Fig. 10-19

Components of the ear. The external ear consists of the auricle (pinna), the external acoustic meatus (ear canal), and the tympanic membrane (eardrum). The middle ear includes the malleus (hammer), incus (anvil), and stapes (stirrup) and the tympanic cavity. The inner ear contains the organs of balance as well as those of hearing.

leading from the auricle into the temporal bone and named the *external acoustic meatus (ear canal)*. This canal is about 1¼ inches long and takes, in general, an inward, forward, and downward direction, although the first portion of the tube slants upward and then curves downward. Because of this curve in the auditory canal, the auricle should be pulled up and back to straighten the tube when medications are to be dropped into the ear. Modified sweat glands in the auditory canal secrete *cerumen* (waxlike substance), which occasionally becomes impacted and may cause pain and deafness. The *tympanic membrane* (eardrum) stretches across the inner end of the auditory canal, separating it from the middle ear.

MIDDLE EAR

The middle ear (tympanic cavity), a tiny epithelial-lined cavity hollowed out of the temporal bone, contains the three auditory ossicles: the malleus, incus, and stapes (Fig. 10-20). The names of these very small bones describe their shapes (hammer, anvil, and stirrup). The "handle" of the malleus is attached to the inner surface of the tympanic membrane, whereas the "head" attaches to the incus, which, in turn, attaches to the stapes. There are several openings into the middle ear cavity: one from the external auditory meatus, covered over with the tympanic membrane; two into the internal ear, the fenestra ovalis (oval window), into which the stapes fits, and the fenestra rotunda (round window), which is covered by a membrane; and one into the eustachian tube.

Posteriorly, the middle ear cavity is continuous with a number of mastoid air spaces in the temporal bone. The clinical importance of these middle ear openings is that they provide routes for infection to travel.

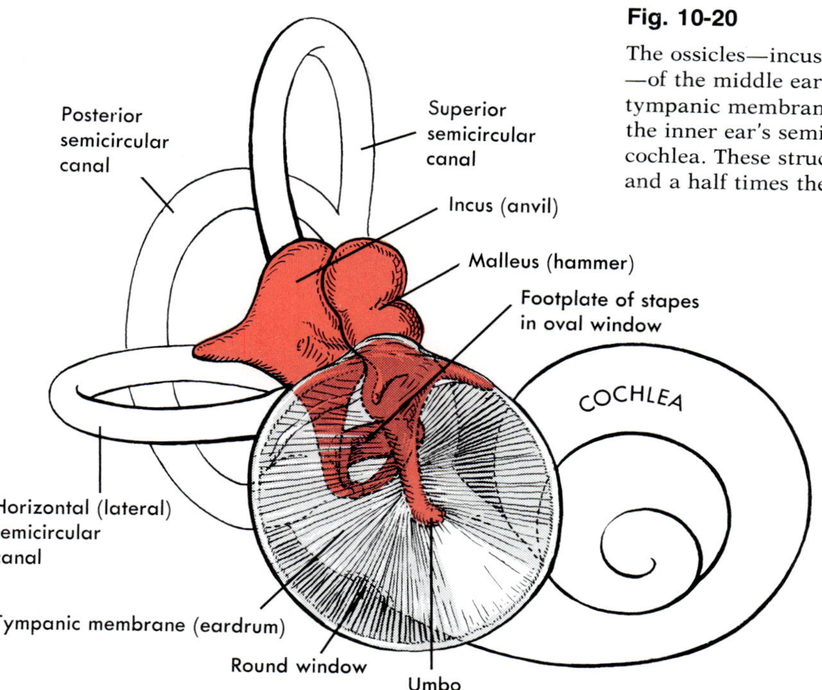

Fig. 10-20

The ossicles—incus, malleus, and stapes—of the middle ear as seen through the tympanic membrane and in relation to the inner ear's semicircular canals and cochlea. These structures are shown four and a half times their actual size.

Head colds, for example, especially in children, may lead to middle ear or mastoid infections via the nasopharynx–eustachian tube–middle ear–mastoid path.

The *eustachian* or *auditory tube* is composed partly of bone and partly of cartilage and fibrous tissue and is lined with mucosa. It extends downward, forward, and inward from the middle ear cavity to the nasopharynx (the part of the throat behind the nose).

Thus the eustachian tube provides the path by which throat infections may invade the middle ear. But the eustachian tube also serves a useful function. It makes possible equalization of pressure against inner and outer surfaces of the tympanic membrane and therefore prevents membrane rupture and the discomfort that marked pressure differences produce. The way the eustachian tube equalizes tympanic membrane pressures is this. When one swallows or yawns, air spreads rapidly through the open tube. Atmospheric pressure then presses against the inner surface of the tympanic membrane. And since atmospheric pressure is continually exerted against its outer surface, the pressures are equal. You might test this mechanism sometime when you are ascending or descending in an airplane—start chewing gum to increase your swallowing and observe whether this relieves the discomfort in your ears.

INNER EAR

The inner ear is also called the labyrinth because of its complicated shape. It consists of two main parts, a bony labyrinth and, inside this, a membranous labyrinth. The bony labyrinth consists of three parts: vestibule, cochlea, and semicircular canals. The membranous labyrinth, as Fig. 10-21 shows,

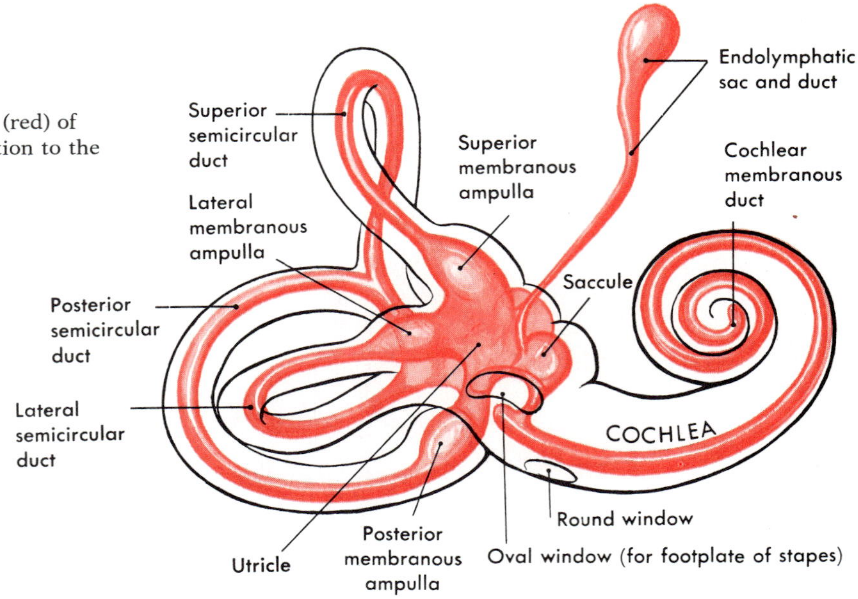

Fig. 10-21

The membranous labyrinth (red) of the inner ear shown in relation to the bony labyrinth.

consists of the utricle and saccule inside the vestibule, the cochlear duct inside the cochlea, and the membranous semicircular canals inside the bony ones.

Vestibule, utricle, and saccule. The vestibule constitutes the central section of the bony labyrinth. Into it open both the oval and round windows from the middle ear as well as the three semicircular canals of the inner ear. The utricle and saccule have membranous walls and are suspended within the vestibule. They are separated from the bony walls of the vestibule by fluid (perilymph), and both utricle and saccule contain a fluid called endolymph.

Located within the utricle (and also within the saccule) lies a small structure called the *macula*. It consists mainly of hair cells and a gelatinous membrane that contains *otoliths* (tiny ear "stones"—that is, small particles of calcium carbonate). A few delicate hairs protrude from the hair cells and are embedded in the gelatinous membrane. Receptors for the vestibular branch of the eighth cranial nerve contact the hair cells of the macula located in the utricle. Changing the position of the head causes a change in the amount of pressure on the gelatinous membrane and causes the otoliths to pull on the hair cells. This stimulates the adjacent receptors of the vestibular nerve. Its fibers conduct impulses to the brain that produce a sense of the position of the head and also a sensation of a change in the pull of gravity (for example, a sensation of acceleration). In addition, stimulation of the macula in the utricle evokes *righting reflexes,* muscular responses to restore the body and its parts to their normal position when they have been displaced. (Impulses from proprioceptors and from the eyes also activate righting reflexes. Interruption of the vestibular or visual or proprioceptive impulses that initiate these reflexes may cause disturbances of equilibrium, nausea, vomiting, and other symptoms.)

Cochlea and cochlear duct. The word cochlea, which means snail, describes the outer appearance of this part of the bony labyrinth. When sectioned, the cochlea resembles a tube wound spirally around a cone-shaped core of bone, the *modiolus*. The modiolus houses the spiral ganglion, which consists of cell bodies of the first sensory neurons in the auditory relay. Inside the cochlea lies the membranous *cochlear duct*—the only part of the inner ear concerned with hearing. This structure is shaped like a somewhat triangular tube. It forms a shelf across the inside of the bony cochlea, dividing it into upper and lower sections all along its winding course (Figs. 10-22 and 10-23). The upper section (above the cochlear duct, that is) is called the *scala vestibuli*, whereas the lower section below the cochlear duct is the *scala tympani*. The roof of the cochlear duct is known as *Reissner's membrane* or the vestibular membrane. *Basilar membrane* is the name given the floor of the cochlear duct. It is supported by bony and fibrous projections from the wall of the cochlea. Perilymph fills the scala vestibuli and scala tympani and endolymph the cochlear duct.

The hearing sense organ, the *organ of Corti*, rests on the basilar membrane throughout the whole length of the cochlear duct. The structure of the organ of Corti resembles that of the equilibrium sense organ (that is, the macula in the utricle). It consists of supporting cells plus the important *hair cells* that project into the endolymph and are topped by an adherent gelatinous membrane called the *tectorial membrane*. Dendrites of the sensory neurons whose cells lie in the spiral ganglion in the modiolus have their beginnings around the bases of the hair cells of the organ of Corti. Axons of these neurons extend in the cochlear nerve (a branch of the eighth cranial nerve) to the brain. They conduct impulses that produce the sensation of hearing.

Semicircular canals. Three semicircular canals, each in a plane approximately at right angles to the others, are found in each temporal bone. Within the bony semicircular canals and separated from them by perilymph are the membranous semicircular canals. Each contains endolymph and connects with the utricle, one of the membranous sacs inside the bony vestibule. Near its junction with the utricle the canal enlarges into an *ampulla*. Some of the receptors for the vestibular branch of the eighth cranial nerve lie in each ampulla. Like all receptors for both vestibular and auditory branches of this nerve, these receptors, too, lie in contact with hair cells in a supporting structure. Here in the ampulla, the hair cells and supporting structure together are named the *crista ampullaris*, whereas in the utricle and saccule they are called the macula and in the cochlear duct, the organ of Corti. Sitting atop the crista is a gelatinous structure called the *cupula*.

The crista with its vestibular nerve endings presumably functions as the end organ for sensations of head movements, whereas the macula in the utricle serves as the end organ for sensations of head positions. Hence, both the crista and the macula of the utricle function as end organs for the sense of equilibrium.

The function of the macula in the saccule is not known. Some evidence, however, suggests that it serves as a receptor for vibratory stimuli.

Physiology

HEARING

Hearing results from stimulation of the auditory area of the cerebral cortex (temporal lobe, Fig. 8-21, p. 207). Before reaching this area of the brain, however, sound waves must be projected through air, bone, and fluid to stimulate nerve endings and set up impulse conduction over nerve fibers.

Sound waves in the air enter the external auditory canal, probably without much aid from the pinna in collecting and reflecting them because of its smallness in man. At the inner end of the canal they strike against the tympanic membrane, setting it in vibration.

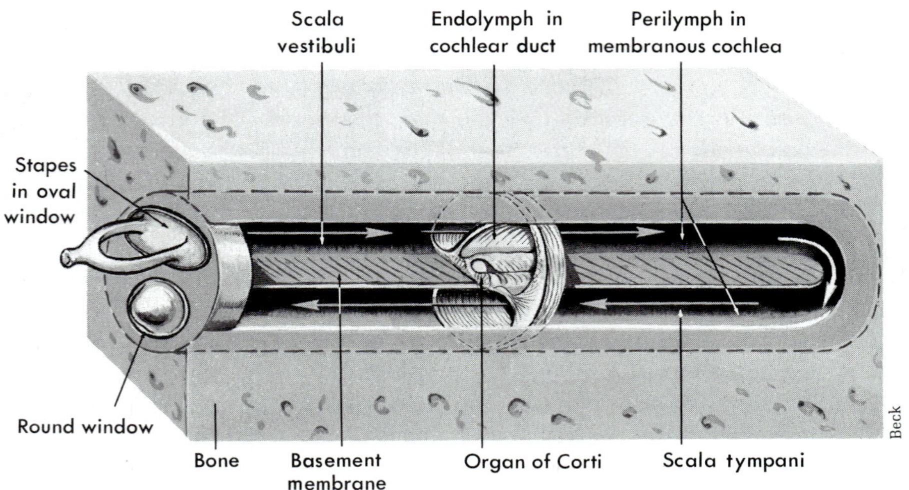

Fig. 10-22

Diagram of the bony and membranous cochlea, uncoiled. Note the end organ of Corti projecting into the endolymph contained in the cochlear duct (membranous cochlea). The perilymph indicated above the endolymph occupies the scala vestibuli. That in the lower compartment lies in the scala tympani (see also Fig. 10-23).

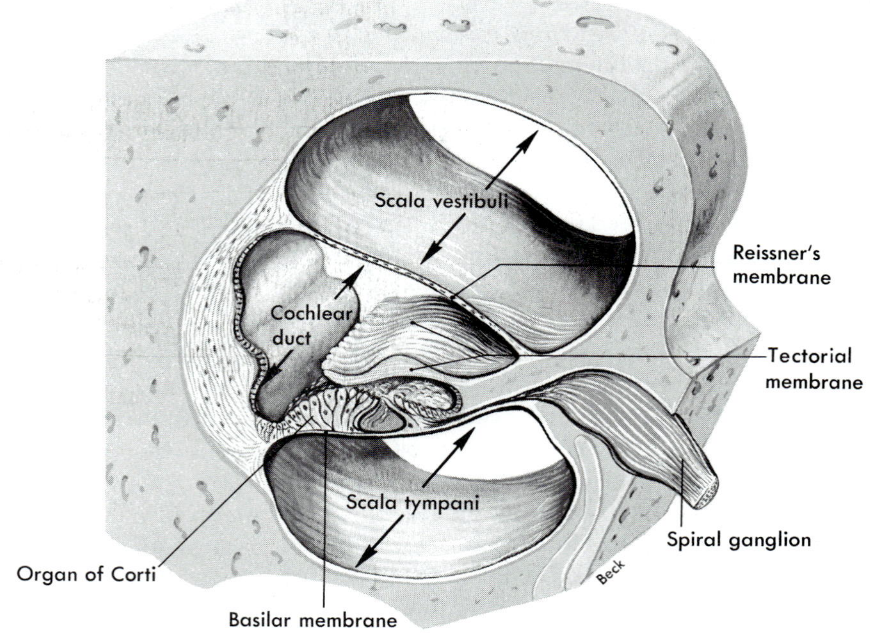

Fig. 10-23

Section through one of the coils of the cochlea. Perilymph fills the scala vestibuli and scala tympani. Endolymph fills the cochlear duct. A part of the spiral ganglion is shown (bulging stemlike structure at right of diagram).

Vibrations of the tympanic membrane move the malleus, whose handle attaches to the membrane. The head of the malleus attaches to the incus, and the incus attaches to the stapes. So when the malleus vibrates, it moves the incus, which moves the stapes against the oval window into which it fits so

Table 10-5

Divisions of internal ear (labyrinth)

Bony labyrinth (part of temporal bone)	Membranous labyrinth (inside bony labyrinth)
1. Vestibule—central section of bony labyrinth; oval and round windows are openings of middle ear into vestibule; bony semicircular canals also open into vestibule	1. Utricle—one of the two parts of the membranous labyrinth contained in bony vestibule; utricle contains fluid called endolymph and an equilibrium sense organ, the macula; macula senses both head positions and movements and acceleration and deceleration of the body; vestibular nerve (branch of eighth cranial nerve) supplies macula 2. Saccule—the other part of the membranous labyrinth within bony vestibule; contains endolymph and macula
2. Cochlea—a spiraling bony tube that resembles a snail shell in shape	3. Cochlear duct—membranous tube that forms shelf across interior of bony cochlea; contains endolymph and organ of Corti, the sense organ for hearing; cochlear nerve (branch of eighth cranial nerve) supplies organ of Corti; cochlear duct separated from bony cochlea by scala vestibuli and scala tympani, spaces that contain perilymph
3. Bony semicircular canals—three of these semicircular-shaped canals; each lies at approximately right angle to the others	4. Membranous semicircular canals—separated from bony semicircular canals by perilymph; contain endolymph and the crista, an equilibrium sense organ that senses head movements; vestibular nerve supplies crista

precisely. At this point, fluid conduction of sound waves begins. To understand this, you will probably need to refer to Figs. 10-17 to 10-23 frequently as you read the next few sentences. When the stapes moves against the oval window, pressure is exerted inward into the perilymph in the scala vestibuli of the cochlea. This starts a "ripple" in the perilymph that is transmitted through Reissner's membrane (the roof of the cochlear duct) to endolymph inside the duct and then to the organ of Corti and to the basilar membrane that supports the organ of Corti and forms the floor of the cochlear duct. From the basilar membrane the ripple is next transmitted to and through the perilymph in the scala tympani and finally expends itself against the round window—on a much reduced scale, like an ocean wave expending itself as it breaks against the shore. Dendrites (of neurons whose cell bodies lie in the spiral ganglion and whose axons make up the cochlear nerve) terminate around the bases of the hair cells of the organ of Corti, and the tectorial membrane adheres to their upper surfaces. The movement of the hair cells against the adherent tectorial membrane somehow stimulates these dendrites and initiates impulse conduction by the cochlear nerve to the brainstem. Before reaching the auditory area of the temporal lobe, impulses pass through "relay stations" in nuclei in the medulla, pons, midbrain, and thalamus.

EQUILIBRIUM

In addition to hearing, the inner ear aids in the maintenance of equilibrium. It does this by sensing head movements and positions and also by sensing acceleration or deceleration of the body (see p. 260).

Olfactory sense organs

The receptors for the fibers of the olfactory (first) cranial nerves lie in the mucosa of the upper part of the nasal cavity. Their location here explains the necessity for sniffing or drawing air forcefully up into the nose in order to smell delicate odors. The olfactory sense organ consists of hair cells and is relatively simple compared with the complex visual and auditory organs. Whereas the olfactory receptors are extremely sensitive (that is, are stimulated by even very slight odors), they are also easily fatigued—a fact that explains why odors that are at first very noticeable are not sensed at all after a short time.

Gustatory sense organs

The receptors for the taste nerve fibers (in branches of the seventh and ninth cranial nerves) are known as *taste buds*. Most of them are located on the tongue and the roof of the mouth. Taste buds are exteroceptors of the type called chemoreceptors, for the obvious reason that chemicals stimulate them. Only four kinds of taste sensations—sweet, sour, bitter, and salty—result from stimulation of taste buds. All the other flavors we sense result from a combination of taste bud and olfactory receptor stimulation. In other words, the myriads of tastes recognized are not tastes alone but tastes plus odors. For this reason, a cold that interferes with the stimulation of the olfactory receptors by odors from foods in the mouth markedly dulls one's taste sensations.

Outline summary

GENERAL REMARKS

1 Millions of receptors constitute sense organs
For receptors and sensations they mediate, see Table 10-1

SOMATIC, VISCERAL, AND REFERRED PAIN

1 Somatic—results from stimulation of pain receptors (nociceptors) in skin or in deep structures (skeletal muscles, tendons, or joints)
2 Visceral—results from stimulation of pain receptors in viscera by pressure or chemical stimuli; conducted almost exclusively by sensory fibers in sympathetic nerves
3 Referred—pain interpreted as coming from skin area when it actually originates in deep structure

EYE

Anatomy

Coats of eyeball

1 Outer coat (sclera)
2 Middle coat (choroid)
3 Inner coat (retina)
4 For outline summary, see Table 10-2

Cavities and humors

1 Anterior cavity
2 Posterior cavity
3 For outline summary, see Table 10-3

Muscles

1 Extrinsic
  a Attach to outside of eyeball and to bones of orbit
  b Voluntary muscles; move eyeball in desired directions
  c Four straight (rectus) muscles—superior, inferior, lateral, and mesial; two oblique muscles—superior and inferior
2 Intrinsic
  a Within eyeball; named iris and ciliary muscles
  b Involuntary muscles
  c Iris regulates size of pupil
  d Ciliary muscle controls shape of lens, making possible accommodation for near and far objects
3 For outline summary, see Table 10-4

Accessory structures

1 Eyebrows and eyelashes—protective and cosmetic
2 Eyelids
  a Lined with mucous membrane that continues over surface of eyeball; called conjunctiva
  b Opening between eyelids called palpebral fissure
  c Corners where upper and lower eyelids join called canthus, mesial and lateral
3 Lacrimal apparatus—lacrimal glands, lacrimal canals, lacrimal sacs, and nasolacrimal ducts

Physiology of vision

Fulfillment of following conditions results in conscious experience known as vision: formation of retinal image, stimulation of retina, and conduction to visual area

Formation of retinal image

1 Accomplished by four processes:
  a Refraction or bending of light rays as they pass through eye
  b Accommodation or bulging of lens—normally occurs if object viewed lies nearer than 20 feet from eye
  c Constriction of pupil; occurs simultaneously with accommodation for near objects and also in bright light
  d Convergence of eyes for near objects so light rays from object fall on corresponding points of two retinas; necessary for single binocular vision

Stimulation of retina

Accomplished by light rays producing photochemical change in rods and cones

Conduction to visual area

Fibers that conduct impulses from rods and cones reach visual cortex in occipital lobes via optic nerves, optic chiasma, optic tracts, and optic radiations

AUDITORY APPARATUS

Anatomy

External ear

1 Auricle or pinna
2 External acoustic meatus (ear canal)

Middle ear

1 Separated from external ear by tympanic membrane
2 Contains auditory ossicles (malleus, incus, and stapes) and openings from external acoustic meatus, internal ear, eustachian tube, and mastoid sinuses
3 Eustachian tube, collapsible tube, lined with mucosa, extending from nasopharynx to middle ear
  a Equalizes pressure on both sides of eardrum
  b Open when yawning or swallowing

Inner ear

1 Consists of bony and membranous portions, latter contained within former
2 Bony labyrinth has three divisions—vestibule, cochlea, and semicircular canals
3 Membranous cochlear duct contains receptors for cochlear branch of eighth cranial nerve (sense of hearing)
4 Utricle and membranous semicircular canals contain receptors for vestibular branch of eighth cranial nerve (sense of equilibrium)

Physiology

Hearing

Hearing results from stimulation of auditory area of temporal lobes by impulses over cochlear nerves, which are stimulated by sound waves being projected through air, bone, and fluid before reaching auditory receptors (organ of Corti in cochlear duct)

Equilibrium

Stimulation of receptors (crista in semicircular canals and utricle) leads to sense of equilibrium; also initiates righting reflexes essential for balance

OLFACTORY SENSE ORGANS

1 Receptors for olfactory (first) cranial nerve located in nasal mucosa high along septum
2 Receptors very sensitive but easily fatigued

GUSTATORY SENSE ORGANS

1 Receptors for taste nerve fibers (in branches of seventh and ninth cranial nerves) called taste buds located on tongue and roof of mouth
2 Four kinds of taste—sweet, sour, salty, and bitter
3 All other tastes result from fusion of two or more of these tastes or from olfactory stimulation

Review questions

1 What two general functions do sense organs perform?
2 Explain briefly the principle of specificity of receptors.
3 Describe briefly one theory about the mechanism of referred pain.
4 Explain briefly the mechanism for accommodation for near vision.
5 Define briefly the term *refraction*. Name the refractory media of the eye.
6 Concave glasses are prescribed for nearsighted vision. Upon what principle is this based?
7 What is the name of the receptors for vision in dim light? For bright light?
8 Distinguish between exteroceptors, proprioceptors, and visceroceptors.
9 Describe the main features of middle ear structure.
10 Name the parts of the bony and membranous labyrinths and describe the relationship of membranous labyrinth parts to those of the bony labyrinth.
11 In what ear structure(s) is the hearing sense organ located? The equilibrium sense organs?
12 What is the name of the hearing sense organ? Of the equilibrium sense organs?

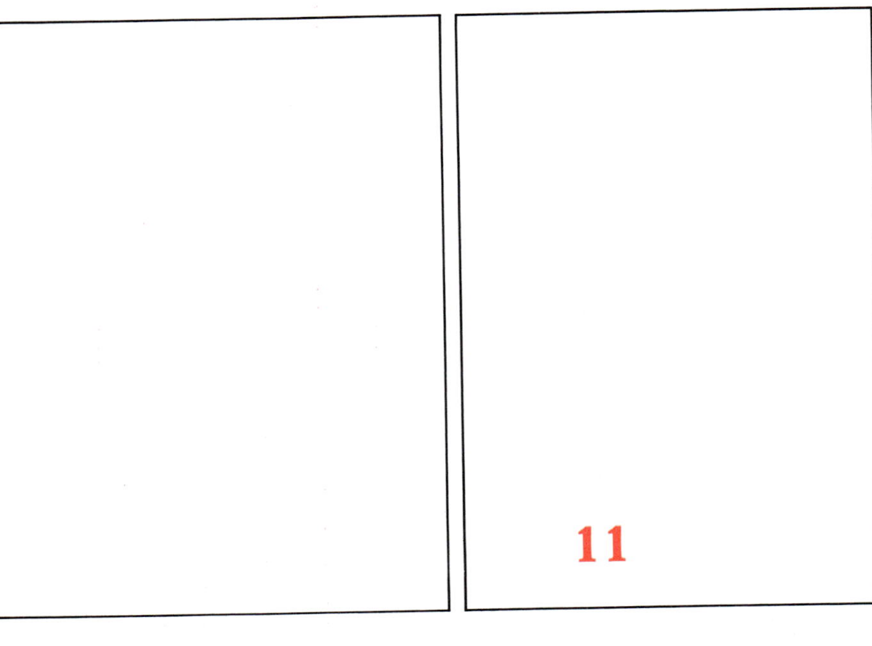

11

The endocrine system

Meaning
How hormones act

Pituitary gland (hypophysis cerebri)
Size, location, and component glands
Anterior pituitary gland (adenohypophysis)
Growth hormone
Prolactin
Tropic hormones
Melanocyte-stimulating hormone
Control of secretion
Posterior pituitary gland (neurohypophysis)
Antidiuretic hormone
Oxytocin
Control of secretion

Thyroid gland
Location and structure
Thyroid hormone and calcitonin
Effects of hypersecretion and hyposecretion

Parathyroid glands
Location and structure
Parathyroid hormone

Adrenal glands
Location and structure
Adrenal cortex
Glucocorticoids
Mineralocorticoids
Sex hormones
Control of secretion
Adrenal medulla
Epinephrine and norepinephrine
Control of secretion

Islands of Langerhans

Ovaries

Testes

Pineal gland (pineal body or epiphysis cerebri)

Placenta

Meaning

The endocrine system consists of glands that release their secretions into the blood. They have no ducts, so another name for endocrine glands is ductless glands. Exocrine or duct glands, in contrast, release their secretions into ducts. To learn the names and locations of the main endocrine glands, see Table 11-1 and Fig. 11-1.

The endocrine system and the nervous system both perform the same general functions for the body: communication, control, and integration. But they accomplish these general functions through different kinds of mechanisms and with somewhat different types of results. The mechanism of the nervous system consists of nerve impulses conducted by neurons from one specific structure to another. In contrast, the mechanism of the endocrine system consists of secretions from endocrine gland cells entering the blood and circulating to all parts of the body. Endocrine gland secretions, as you undoubtedly know, are called *hormones*. Nerve impulses and hormones do not exert exactly the same kind of control, nor do they control precisely the same structures and functions. This is fortunate. It makes for better timing and more precision of control. For instance, nerve impulses produce rapid, short-lasting responses, whereas hormones produce slower and generally longer-lasting responses. Nerve impulses control directly only two kinds of cells: muscle and gland cells. Some hormones, in contrast, exert control over all kinds of cells.

Table 11-1

Names and locations of endocrine glands

Name	Location
Pituitary gland (hypophysis cerebri) Anterior lobe (adenohypophysis) Posterior lobe (neurohypophysis)	Cranial cavity
Pineal gland (epiphysis)	Cranial cavity
Thyroid gland	Neck
Parathyroid glands	Neck
Adrenal glands Adrenal cortex Adrenal medulla	Abdominal cavity (retroperitoneal)
Islands of Langerhans	Abdominal cavity (pancreas)
Ovaries Graafian follicle Corpus luteum	Pelvic cavity
Testes (interstitial cells)	Scrotum

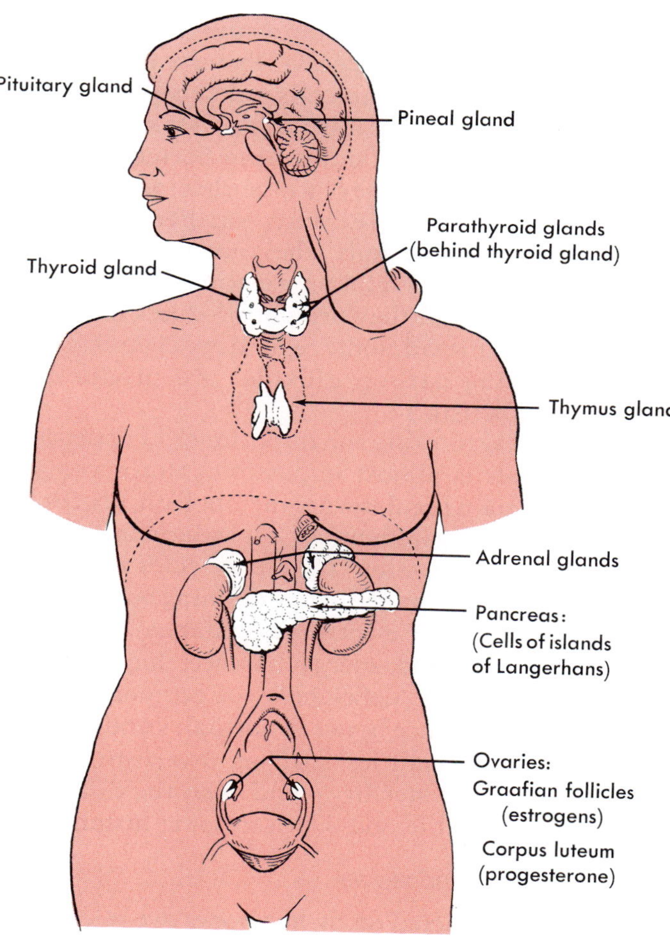

Fig. 11-1

Location of the endocrine glands in the female. Dotted line around thymus gland indicates maximum size at puberty.

Endocrine gland cells synthesize hormones by the process of anabolism. They are either protein compounds, or compounds derived from proteins, or steroid compounds. All have relatively large molecules. For instance, human growth hormone consists of a chain of 188 amino acids and has a molecular weight of 20,500.* Compared with a carbon dioxide molecule, molecular weight 44, the growth hormone molecule is indeed a giant. (Carbon dioxide is one of several so-called "regulatory chemicals." These compounds resemble hormones in two respects—they are released into the blood from cells, and they exert profound influence on various structures and functions. They differ from hormones, however, in that they are neither proteins nor steroids but are much smaller, inorganic molecules or ions, they are present in the blood in much larger amounts than are hormones, and they are not products of anabolism. Instead, some regulatory chemicals, including carbon dioxide, are products of catabolism.)

Exaggerating the importance of endocrine glands is almost impossible. Hormones are the main regulators of metabolism, of growth and development, of reproduction, and of stress responses. They play roles of the utmost importance in maintaining homeostasis—fluid and electrolyte balance, acid-base balance, and energy balance, for example. Excesses or deficiencies of hormones make the difference between normalcy and all sorts of abnormalities such as dwarfism, gigantism, and sterility—and even the difference between life and death in some instances.

How hormones act

Cells acted on by hormones are referred to descriptively as "targets." What a hormone does to its target cells to cause them to respond in particular ways has been the subject of much conjecture and research. In the late 1950's, Dr. Earl W. Sutherland isolated a chemical, cyclic AMP (abbreviation for adenosine 3′,4′-monophosphate), and suggested the part it might play in hormone action. For his brilliant work, Dr. Sutherland received the 1971 Nobel Prize for Physiology or Medicine. He conceived what is known as the "second messenger hypothesis" of hormone action. According to this concept, a hormone acts as a "first messenger," that is, it delivers its chemical message from the cells of an endocrine gland to the cells of a target organ. Upon reaching a target cell, the hormone reacts with an enzyme, adenyl cyclase, present in the cell's plasma membrane, causing it to act on ATP molecules present inside the cell. Adenyl cyclase catalyzes the conversion of ATP to cyclic AMP. Cyclic AMP serves as the "second messenger," delivering information inside the cell that causes the cell to respond by performing its specialized function. For example, cyclic AMP causes thyroid cells to respond by secreting thyroid hormones. Summarizing, hormones serve as "first messengers," providing communication between endocrine and target gland cells. Cyclic AMP then acts as the "second messenger," providing communication within a hormone's target cells.

Pituitary gland (hypophysis cerebri)

Size, location, and component glands

The pituitary gland is truly a small but mighty structure. At its largest diameter, it measures only 1.2 to 1.5 cm (less than ½ inch). By weight, it is even less impressive—only about 0.5 gm, 1/60 ounce! And yet, so crucial are the functions of the anterior lobe of this gland that it is called the "master gland."

The pituitary body has a protected location. It lies in the sella turcica and is covered over by an extension of the dura mater known as the pituitary diaphragm. The deepest part

*Mountcastle, V. B.: Medical physiology, ed. 13, St. Louis, 1974, The C. V. Mosby Co., p. 1614.

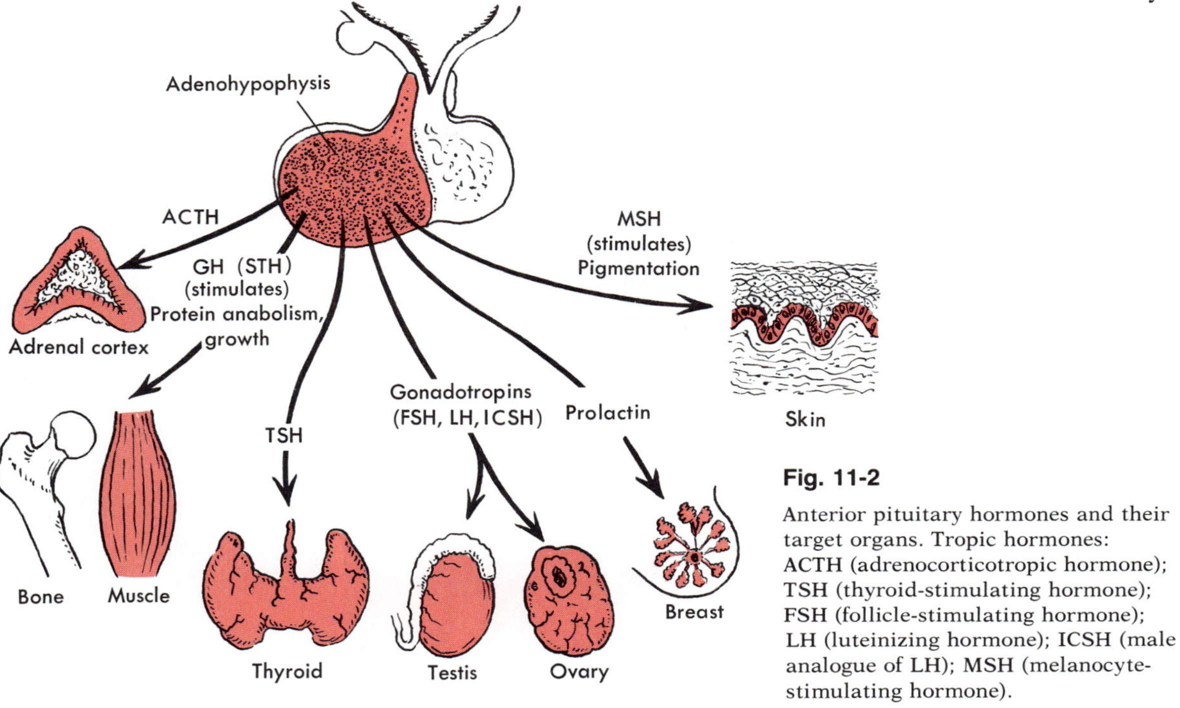

Fig. 11-2

Anterior pituitary hormones and their target organs. Tropic hormones: ACTH (adrenocorticotropic hormone); TSH (thyroid-stimulating hormone); FSH (follicle-stimulating hormone); LH (luteinizing hormone); ICSH (male analogue of LH); MSH (melanocyte-stimulating hormone).

of the sella turcica (saddle-shaped depression in the sphenoid bone) is called the pituitary fossa since the pituitary body lies in it. A stemlike portion, the pituitary stalk, juts up through the diaphragm and attaches the gland to the undersurface of the brain. More specifically, the stalk attaches the pituitary body to the hypothalamus.

Although the pituitary looks like just one gland, it actually consists of two separate glands—the adenohypophysis or anterior pituitary gland and the neurohypophysis or posterior pituitary gland. These two glands develop from different embryonic structures, have different microscopic structure, and secrete different hormones. The adenohypophysis develops as an upward projection from the embryo's pharynx, whereas the neurohypophysis develops as a downward projection from its brain. Microscopic differences between the two glands are suggested by their names—*adeno* means gland and *neuro* means nervous. The adenohypophysis has the microscopic structure of an endocrine gland, whereas the neurohypophysis has the structure of nervous tissue. Hormones secreted by the adenohypophysis serve very different functions from those secreted by the neurohypophysis.

Anterior pituitary gland (adenohypophysis)

The anterior pituitary gland consists chiefly of two main types of epithelial cells: acidophils (those that take acid stains) and basophils (those that take basic stains). Acidophils secrete growth hormone (GH; also called somatotropin or STH) and prolactin. Basophils secrete the other five hormones of the anterior lobe: thyrotropin (TH; also called thyroid-stimulating hormone or TSH), adrenocorticotropin (ACTH), two gonadotropins (follicle-stimulating hormone [FSH] and luteinizing hormone [LH]), and melanocyte-stimulating hormone (MSH).

Growth hormone

Growth hormone or somatotropin (Gr. *soma*, body; *trope*, turning) is thought to pro-

Fig. 11-3

A pituitary giant and dwarf contrasted with normal-sized men. Excessive secretion of growth hormone* by the anterior lobe of the pituitary gland during the early years of life produces giants of this type, while deficient secretion of this substance produces well-formed dwarfs. (Courtesy Dr. Edmund E. Beard, Cleveland, Ohio.)

*The structure of the human growth hormone molecule was identified by Dr. C. H. Li, Professor of Biochemistry at the University of California, in 1966. It consists of a chain of 188 amino acids with one loop of 93 subunits and another of 6 subunits. In January of 1971, Dr. Li announced that his laboratory had succeeded in synthesizing human growth hormone.

mote bodily growth indirectly by accelerating amino acid transport into cells—evidence: blood amino acid content decreases within hours after administration of growth hormone to a fasting animal. With the faster entrance of amino acid into cells, anabolism of amino acids to form tissue protein also accelerates. This in turn, tends to promote cellular growth. Growth hormone stimulates growth of both bone and soft tissues. If the anterior pituitary gland secretes an excess of growth hormone during the growth years (that is, before closure of the epiphyseal cartilages), bones grow more rapidly than normal and *gigantism* results (Fig. 11-3). Undersecretion of the growth hormone produces *dwarfism* when it occurs during the years of skeletal growth. If oversecretion of growth hormone occurs after the individual is full grown, the condition known as *acromegaly* develops. Characteristic of this disease are enlargement of the bones of the hands, feet, jaws, and cheeks and an increase, too, in their overlying soft tissues (Fig. 11-4).

In addition to its stimulating effect on protein anabolism, growth hormone also influences fat metabolism and thereby affects carbohydrate (glucose) metabolism. Briefly, growth hormone tends to accelerate both the mobilization of fats from adipose tissues and their catabolism by other tissues. In other words, it tends to cause cells to shift from glu-

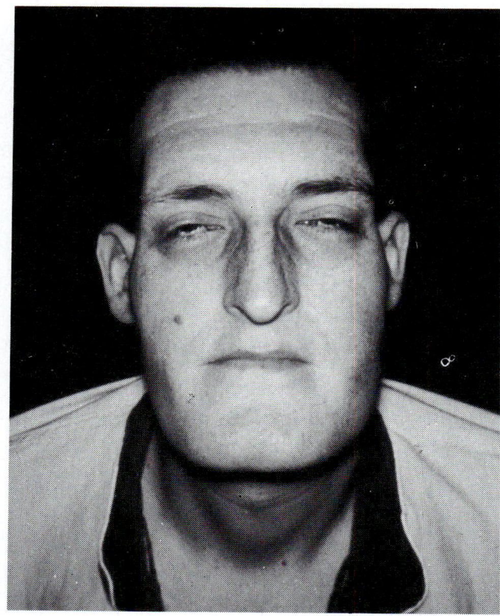

Fig. 11-4

Acromegaly. Note large head, exaggerated forward projection of the jaw, and protrusion of the frontal bone. (From Rimoin, D. L.: N. Engl. J. Med. **272:**923, 1965.)

cose catabolism to fat catabolism for their energy supply. Less glucose, therefore, leaves the blood to enter cells and more remains in the blood. Thus growth hormone tends to increase blood's concentration of glucose. In a word, it tends to have a hyperglycemic effect. Insulin produces opposite effects. It increases carbohydrate metabolism, thereby tending to decrease the blood concentration of glucose—to have a hypoglycemic effect. Whereas adequate amounts of insulin prevent diabetes mellitus, long-continued excess amounts of growth hormone produce diabetes. In short, growth hormone and insulin function as antagonists. Or, as more commonly stated, growth hormone has an anti-insulin effect, a fact of considerable clinical importance. Summarizing very briefly, growth hormone helps regulate metabolism in the following ways:

1 It promotes protein anabolism (synthesis of tissue proteins) so is essential for normal growth and for tissue repair and healing.

2 It promotes fat mobilization and catabolism and decreases glucose catabolism. By decreasing glucose catabolism, an excess of growth hormone may in time produce hyperglycemia and diabetes.

Prolactin

The anterior pituitary cells called acidophils secrete two hormones: growth hormone and prolactin. Another name for prolactin—lactogenic hormone—suggests that it "generates" (that is, initiates) milk secretion. It stimulates the mammary glands to start secreting soon after delivery of a baby. During pregnancy, it helps promote the breast development that makes possible milk secretion after pregnancy.

Tropic hormones

The anterior pituitary cells known as basophils secrete four tropic hormones—hormones that have a stimulating effect on other endocrine glands. Names of the tropic hormones are thyrotropin, adrenocorticotropin, and the two gonadotropins known as follicle-stimulating hormone and luteinizing hormone. All tropic hormones perform the same general function. Each one stimulates one other endocrine gland—stimulates it both to grow and to secrete its hormone at a faster rate. It seems to affect only this one structure as if, like a bullet, it had been aimed at a target, hence the name, "target gland."

Individual tropic hormones perform the following functions:

1 *Thyrotropin* (thyroid-stimulating hormone, TSH) promotes and maintains growth and development of its target gland, the thyroid, and stimulates it to secrete thyroid hormone.

2 *Adrenocorticotropin* (ACTH) promotes and maintains normal growth and development of the adrenal cortex and stimulates it to secrete cortisol and other glucocorticoids.

3 *Follicle-stimulating hormone* (FSH) stimulates primary graafian follicles to start growing and to continue developing to maturity, that is, to the point of ovulation. FSH also stimulates follicle cells to secrete estrogens, one type of female sex hormones. In the male, FSH stimulates development of the seminiferous tubules and maintains spermatogenesis by them.

4 *Luteinizing hormone* (LH). The name of this hormone suggests one of its functions, that of stimulating formation of the corpus luteum. Prior to this, LH acts with FSH to bring about complete maturation of the follicle; LH then produces ovulation and stimulates formation of the corpus luteum in the ruptured follicle. Finally, LH in human females stimulates the corpus luteum to secrete progesterone and estrogens. The male pituitary gland also secretes LH, but it is called *interstitial cell–stimulating hormone* (ICSH) because it stimulates interstitial cells in the testes to develop and secrete testosterone.

■ ■ ■

During childhood, the anterior pituitary gland secretes insignificant amounts of gonadotropins. Then it steps up their production gradually a few years before puberty. Just before puberty begins, presumably its secretion of these hormones takes a sudden and marked spurt. As a result, the blood concentration of gonadotropins increases markedly. This first high blood concentration of gonadotropins is the stimulus that brings on the first menstrual period and the many other changes that signal the beginning of puberty.

Melanocyte-stimulating hormone

In some species, the intermediate part of the adenohypophysis produces the melanocyte-stimulating hormone (MSH). Therefore, MSH was first named *intermedin*. However, in man, it is now thought that the anterior lobe of the pituitary gland produces most of the MSH and that both MSH and another anterior pituitary hormone (namely, ACTH) tend to produce increased pigmentation of the skin. Structurally, also, the two molecules resemble each other. Several of the same amino acids occur in the same sequence in both MSH and ACTH.

Control of secretion

Neurons in certain parts of the hypothalamus synthesize chemicals that their axons secrete into the blood in a complex of small veins known as the *pituitary portal system*. Via this portal system, the neurosecretions travel the short distance from the hypothalamus down to the anterior lobe of the pituitary gland. There they stimulate the gland to release various hormones—hence *"releasing hormones"* is the descriptive other name for neurosecretions of the hypothalamus. Here are their abbreviated names, full names, and functions:

1. GRH, or growth hormone–releasing hormone: stimulates anterior pituitary gland to release—that is, to secrete—growth hormone.
2. CRH, or corticotropin-releasing hormone: stimulates anterior pituitary secretion of ACTH.
3. TRH, or thyrotropin-releasing hormone: stimulates anterior pituitary secretion of TSH (thyroid-stimulating hormone)
4. FSH-RH, or follicle-stimulating hormone–releasing hormone: stimulates anterior pituitary secretion of FSH.
5. LH-RH, or luteinizing hormone–releasing hormone: stimulates anterior pituitary secretion of LH.

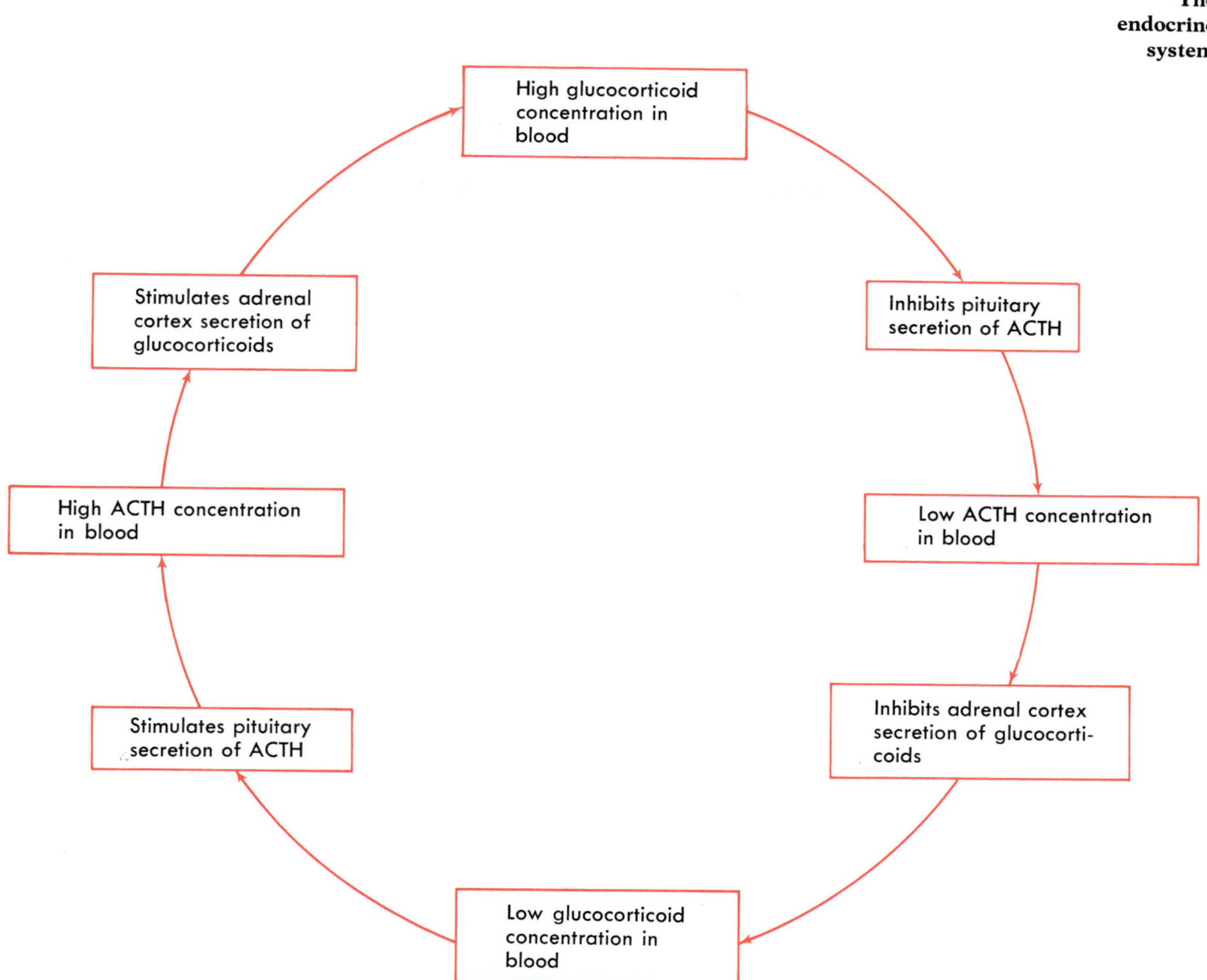

Fig. 11-5

Negative feedback control of ACTH and glucocorticoid secretion—a homeostatic mechanism that tends to keep the blood level of glucocorticoids within a narrow range.

6 P-IH, or prolactin-inhibitory hormone: inhibits anterior pituitary secretion of prolactin. Although a prolactin-releasing factor also exists, its importance in the human is not known. Prolactin secretion is normally inhibited except during pregnancy and following delivery.

Negative feedback mechanisms control the pituitary gland's secretion of tropic hormones. Such mechanisms operate on the following principles. A high blood level of a tropic hormone stimulates its target gland to increase its hormone secretion. This results in a high blood level of the target gland's hormone, which feeds back by way of the circulating blood to directly and indirectly inhibit secretion of the tropic hormone by the anterior pituitary gland. Fig. 11-5 illustrates the negative feedback mechanism that directly inhibits pituitary secretion of ACTH. Indirect inhibition of ACTH occurs as follows. A high blood glucocorticoid concentration inhibits the hypothalamus from secreting corticotropin-releasing factor. Result? A low CRF concentration in pituitary portal vein blood inhibits pituitary secretion of ACTH. Similar mechanisms regulate the anterior lobe's secretion of thyrotropin and the thyroid gland's secretion of thyroid hormone. Also, a similar mechanism controls the anterior lobe's secretion of FSH and the ovary's secretion of estrogens.

Clinical facts furnish interesting evidence about feedback control of hormone secretion by the anterior pituitary gland and its target glands. For instance, if a person has his pituitary gland removed (hypophysectomy) surgically or by radiation, he must be kept on hormone replacement therapy for the rest of his life. If not, he will develop thyroid, adrenocortical, and gonadotropic deficiencies—deficiencies, that is, of target gland hormones. Another well-known clinical fact is that estrogen deficiency develops in women between 40 and 50 years of age. By then, the ovaries seem to have tired of producing hormones and ovulating each month. Or, at any rate, they no longer respond to FSH stimulation so estrogen deficiency develops, brings about the menopause, and persists after the menopause. What, therefore, would you deduce is true of the blood concentration of FSH after the menopause? Apply the principle implied in the preceding paragraph that a low concentration of a target gland hormone stimulates tropic hormone secretion by the anterior pituitary gland.

Before leaving the subject of control of pituitary secretion, we want to call attention to another concept about the hypothalamus. It most likely functions as an important part of the body's complex machinery for coping with stress situations. For example, in severe pain or intense emotions, the cerebral cortex—especially its limbic lobe—is thought to send impulses to the hypothalamus. They stimulate it to secrete its releasing hormones into the pituitary portal veins. Circulating quickly to the anterior pituitary gland, they stimulate it to secrete more of its hormones. These, in turn, stimulate increased activity by the pituitary's target structures. In essence, what the hypothalamus does through its releasing hormones is to translate nerve impulses into hormone secretion by endocrine glands. Thus the hypothalamus links the nervous system to the endocrine system. It integrates the activities of these two great integrating systems—particularly, it seems, in times of stress. When healthy survival is threatened, the hypothalamus, via its releasing factors, can take over the command of the anterior pituitary. By so doing, it indirectly controls all of the pituitary's target glands—the thyroid, the adrenal cortex, and the gonads. Finally, by means of the hormones these glands secrete, the hypothalamus can dictate the functioning of literally every cell in our bodies.

These facts have tremendous implications. They mean that the cerebral cortex can do more than just receive impulses from all parts of the body and send out impulses to muscles and glands. They mean that the cerebral cortex—and therefore our thoughts and emotions—can, by way of the hypothalamus, influence the functioning of all our billions of cells. In short, the brain has two-way contact with every tissue of the body. Thus the state of the body can and does influence mental processes and, conversely, mental processes can and do influence the functioning of the body. In short, both somatopsychic and psychosomatic relationships exist between the body and the brain.

Posterior pituitary gland (neurohypophysis)

The posterior lobe of the pituitary gland secretes two hormones—one known as the antidiuretic hormone (ADH) and the other called oxytocin. But, strangely enough, cells of the posterior lobe do not themselves make these hormones. Neurons in the hypothalamus (supraoptic and paraventricular nuclei [Fig. 11-6]) synthesize them. From the cell bodies of these neurons the hormones pass

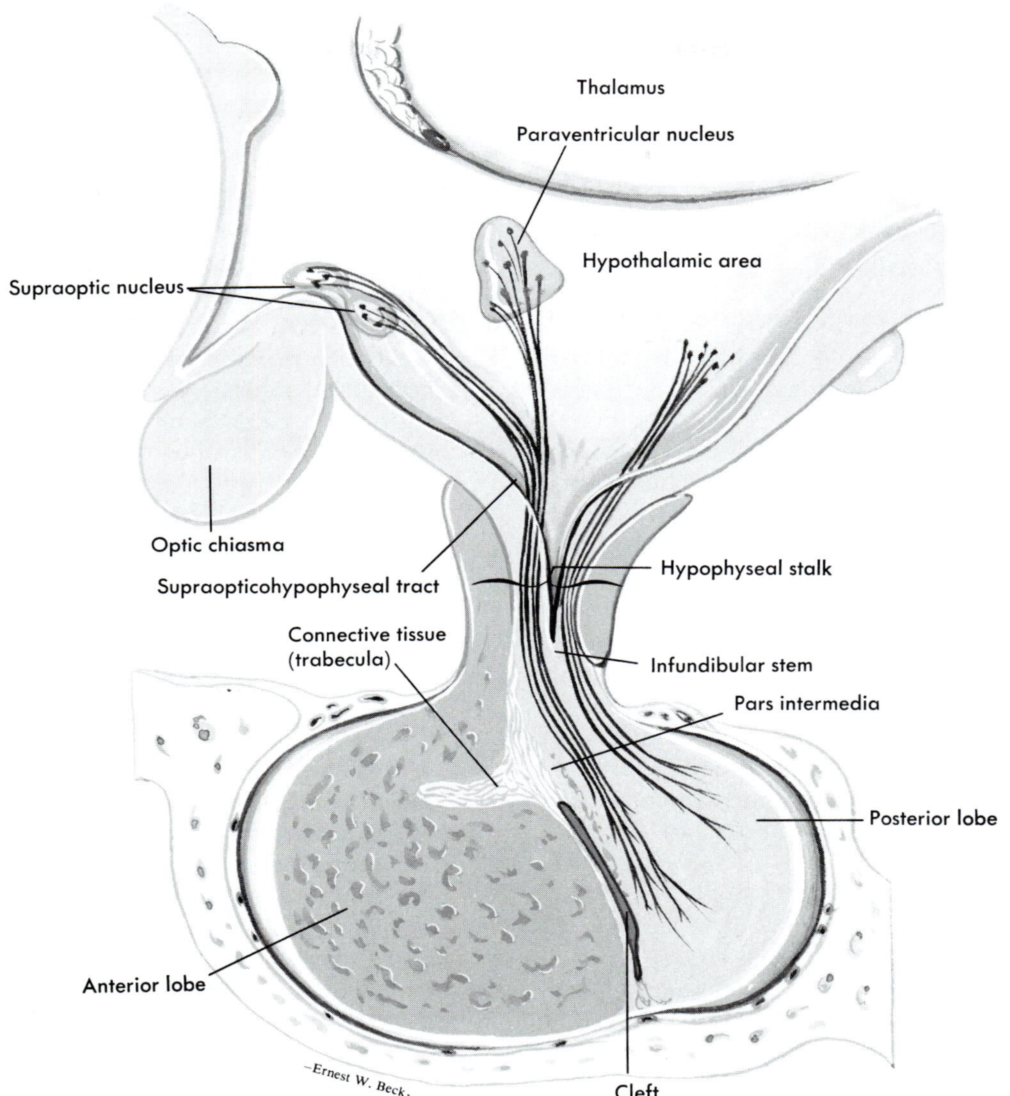

Fig. 11-6

Nerve tracts from hypothalamus to posterior lobe of pituitary gland.

down along axons (in the hypothalamohypophyseal tract) into the posterior lobe of the pituitary gland. Later, the posterior lobe secretes them into the blood.

Antidiuretic hormone

Antidiuresis means literally "against the production of a large urine volume." And this is exactly what ADH does. It prevents the formation of a large volume of urine. The one function ADH performs for the body is to decrease water loss from it. In other words, ADH is a water-retaining hormone. It performs this function by acting on cells of the distal and collecting tubules of the kidney to make them more permeable to water. This causes faster reabsorption of water from tubular urine into blood. And this, in turn, automatically produces antidiuresis (smaller urine volume). Marked diuresis—that is, abnormally large urine volume—occurs if ADH secretion is inadequate, such as occurs in the disease *diabetes insipidus.* A preparation of ADH used to treat this condition is called vasopressin or Pitressin. (The name vasopressin suggests the tendency of pharmacologic, but not physiologic, amounts of ADH to constrict blood vessels.)

Oxytocin

Oxytocin has two actions: it stimulates powerful contractions by the pregnant uterus and it causes milk ejection from the lactating breast. Under the influence of this hormone from the posterior lobe of the pituitary gland, alveoli (cells that synthesize milk) release the milk into the ducts of the breast. This is a highly important function of oxytocin because milk cannot be removed by suckling unless it has first been ejected into ducts. However, it was oxytocin's other action—its stimulating effect on contractions of the pregnant uterus—that inspired its name. The term means "swift childbirth" (Gr. *oxys,* swift; *tokos,* childbirth). Whether or not oxytocin takes part in initiating labor is still an unsettled question. Commercial preparations of oxytocin are given to stimulate uterine contractions after delivery of a baby in order to lessen the danger of hemorrhage.

Control of secretion

Details of the mechanism that controls the secretion of ADH by the posterior lobe of the pituitary gland have not been established. However, two factors play dominant roles: the osmotic pressure of the extracellular fluid and its total volume. Without going into a discussion of evidence or details, the general principles of control of ADH secretion are as follows:

1 An increase in the osmotic pressure of extracellular fluid stimulates ADH secretion. This leads to decreased urine output and tends, therefore, to increase the volume of extracellular fluid, which in turn tends to decrease its osmotic pressure back toward normal. Opposite effects result from a decrease in the osmotic pressure of extracellular fluid.

2 A decrease in the total volume of extracellular fluid acts in some way to stimulate ADH secretion and thereby to decrease urine output and increase extracellular volume back toward normal.

3 Stress, whether induced by physical or emotional factors, brings about an increased secretion of ADH. Presumably it acts in some way on the hypothalamus to stimulate ADH secretion.

■ ■ ■

About all that is known about the mechanism that controls the secretion of oxytocin by the posterior lobe of the pituitary gland is that stimulation of the nipples by the baby's nursing initiates sensory impulses that eventually reach the supraoptic and paraventricular nuclei of the hypothalamus, stimulating them to synthesize more oxytocin. The pos-

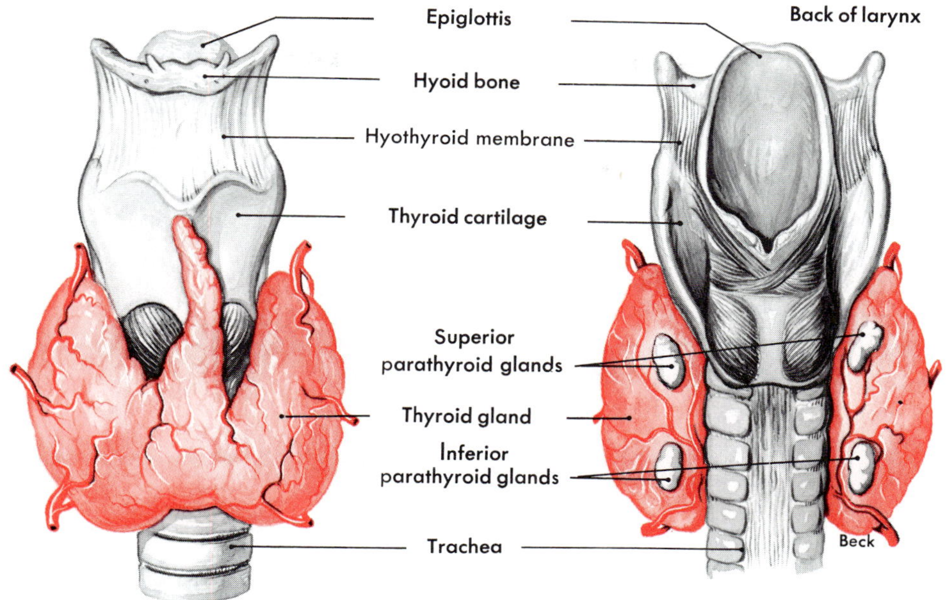

Fig. 11-7

The thyroid and parathyroid glands. Note their relations to each other and to the larynx (voice box) and trachea.

terior lobe of the pituitary gland, in turn, increases its secretion of oxytocin.

Thyroid gland

Location and structure

Two fairly large lateral lobes and a connecting portion, the isthmus, constitute the thyroid gland (Fig. 11-7). It is located in the neck just below the larynx. The isthmus lies across the anterior surface of the upper part of the trachea. Thyroid tissue contains numerous small follicles. Colloid, composed largely of an iodine-containing protein known as thyroglobulin, fills these tiny sacs.

Thyroid hormone and calcitonin

The thyroid gland secretes thyroid hormone and thyrocalcitonin. Actually, thyroid hormone consists of two hormones. Thyroxine is the name of the main one, and triiodothyronine is the name of the less abundant one. One molecule of thyroxine contains four atoms of iodine, and one molecule of triiodothyronine, as its name suggests, contains three iodine atoms. After synthesizing its hormones, the thyroid gland stores considerable amounts of them before secreting them. (Other endocrine glands do not store up their hormones.) As a preliminary to storage, thyroxine and triiodothyronine combine with a globulin in the thyroid cell to form a compound called thyroglobulin. Then thyroglobulin is stored in the colloid material in the follicles of the gland. Later, the two hormones are released from thyroglobulin and secreted into the blood as thyroxine and triiodothyronine. Almost immediately, however, they combine with a blood protein. They travel in the bloodstream in this protein-bound form. But in the tissue capillaries they are released from the protein and enter tissue cells as thyroxine and triiodothyronine.

The iodine in the protein-bound thyroxine and triiodothyronine is called protein-bound iodine (PBI). The amount of PBI can be measured by laboratory procedure and, in fact, is widely used as a test of thyroid functioning. (With a normal thyroid gland, the PBI is about 4 to 8 μg per 100 ml of plasma.)

The main physiological actions of thyroid hormone are to help regulate the metabolic rate and the processes of growth and tissue differentiation. Thyroid hormone increases the metabolic rate. How do we know this? Because oxygen consumption increases following thyroid administration. Like pituitary somatotropin, thyroid hormone stimulates growth, but unlike somatotropin, it also influences tissue differentiation and development. For example, cretins (individuals with thyroid deficiency) not only are dwarfed but also may be mentally retarded because the brain fails to develop normally. Their bones and many other tissues also show an abnormal pattern of development.

Convincing evidence indicates that the thyroid gland secretes *calcitonin.* Calcitonin acts quickly to decrease blood's calcium concentration. Presumably it produces this effect by either or both of these actions: inhibiting bone breakdown with calcium release into blood or promoting calcium deposition in bone. The function of calcitonin is to help maintain blood calcium homeostasis and prevent harmful hypercalcemia. Parathyroid hormone serves as an antagonist to calcitonin (p. 279).

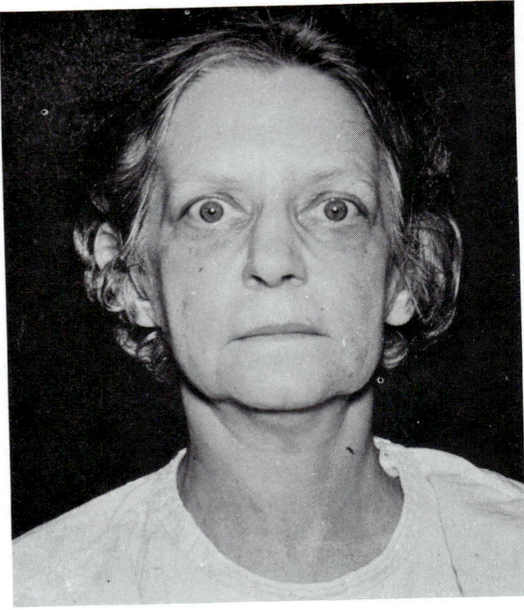

Fig. 11-8
Graves' disease caused by hypersecretion by the thyroid gland. (Courtesy Dr. William McKendree Jeffries, Case Western Reserve University School of Medicine, Cleveland, Ohio.)

Effects of hypersecretion and hyposecretion

Hypersecretion of thyroid hormone produces the disease *exophthalmic goiter* (Graves' disease, Basedow's disease, and several other names) (Fig. 11-8), characterized by an elevated PBI, an increased metabolism (+30 or more), an increased appetite, loss of weight, increased nervous irritability, and exophthalmos. Marked edema of the fatty tissue behind the eye, attributed to the high blood content of thyrotropic hormone, produces the exophthalmos.

Hyposecretion during the formative years leads to malformed dwarfism or *cretinism,* a condition characterized by a low metabolic rate, retarded growth and sexual development, and often, too, a retarded mental development. Later in life, deficient thyroid secretion produces the disease *myxedema* (Fig. 11-9). The low metabolic rate that characterizes myxedema leads to lessened mental and physical vigor, a gain in weight, loss of hair, and a thickening of the skin from an accumulation of fluid in the subcutaneous tissues. Because of a high mucoprotein content, this fluid is viscous. Therefore, it gives

Fig. 11-9

Myxedema, a condition produced by hyposecretion by the thyroid gland during the adult years. (Courtesy Dr. Edmund E. Beard, Cleveland, Ohio.)

firmness to the skin, and the skin does not pit when pressed, as it does in some other types of edema.

Parathyroid glands

Location and structure

The parathyroid glands are small round bodies attached to the posterior surfaces of the lateral lobes of the thyroid gland (Fig. 11-7). Usually there are four or five, but sometimes there are fewer and sometimes more of these glands.

Parathyroid hormone

The parathyroid glands secrete *parathyroid hormone.* Its chief function is to help maintain homeostasis of blood calcium concentration by promoting calcium absorption into the blood and thereby tending to prevent hypocalcemia. It acts as follows:

1 Parathyroid hormone acts on intestine, bones, and kidney tubules to accelerate calcium absorption from them into the blood. Hence, parathyroid hormone tends to increase the blood concentration of calcium. Its primary action on bone is to stimulate bone breakdown or resorption. This releases calcium and phosphate, which diffuse into the blood.

2 Parathyroid hormone acts on kidney tubules to accelerate their excretion of phosphates from the blood into the urine. Note, therefore, that parathyroid hormone has opposite effects on the kidney's handling of calcium and phosphate. It accelerates calcium reabsorption but phosphate excretion by the tubules. Consequently, parathyroid hormone tends to increase the blood concentration of calcium and to decrease the blood concentration of phosphate.

■ ■ ■

The maintenance of calcium homeostasis is highly important for healthy survival. Normal neuromuscular irritability, blood clotting, cell membrane permeability, and also normal functioning of certain enzymes all depend upon the blood concentration of calcium being maintained at a normal level. Neuromuscular irritability is inversely related to blood calcium concentration. In other words, neuromuscular irritability increases when the blood concentration of calcium decreases. Suppose, for example, that a parathyroid hormone deficiency develops and causes *hypocalcemia* (lower than normal blood concentration of calcium). The hypocalcemia increases neuromuscular irritability—sometimes so much that it produces muscle spasms and convulsions, a condition called tetany.

Parathyroid hormone excess produces hypercalcemia or a higher than normal blood concentration of calcium. Sometimes it causes a bone disease with the long name of osteitis fibrosa generalisata. Bone mass decreases (as a result of increased bone destruction followed by fibrous tissue replacement), decalcification occurs, and cystlike cavities appear in the bone.

Adrenal glands

Location and structure

The adrenal glands are located atop the kidneys, fitting like a cap over these organs. The outer portion of the gland is called the *cortex* and the inner substance the *medulla*. Although the adrenal cortex and adrenal medulla are structural parts of one organ, they function as two separate endocrine glands.

Adrenal cortex

Three different zones or layers of cells make up the adrenal cortex. Starting with the zone directly under the outer capsule of the gland, their names are zona glomerulosa, zona fasciculata, and zona reticularis. The outer zone of adrenal cells secretes hormones called mineralocorticoids. The middle zone secretes glucocorticoids, and the innermost zone secretes small amounts of both glucocorticoids and sex hormones. We shall now discuss briefly the functions of these three kinds of adrenal cortical hormones.

Glucocorticoids

The chief glucocorticoids secreted by the adrenal cortex are cortisol* (also called hydrocortisone) and corticosterone. Glucocor-

*Steroid compounds have the following nucleus:

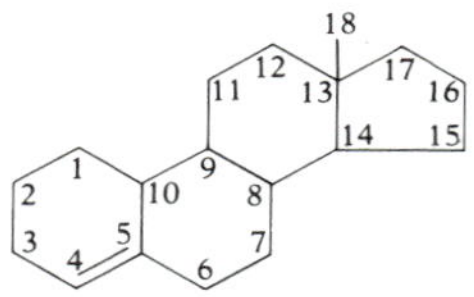

Steroid nucleus

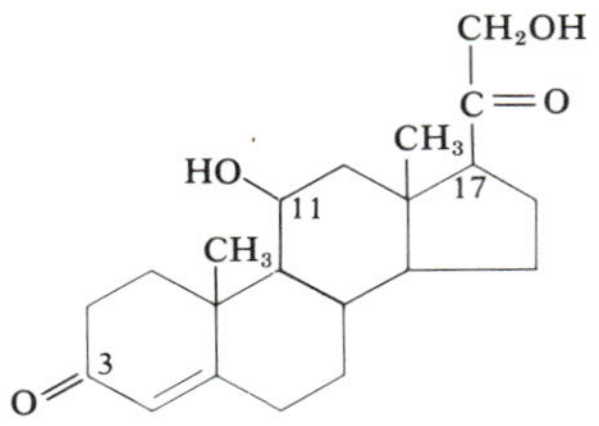

Corticosterone (compound B)

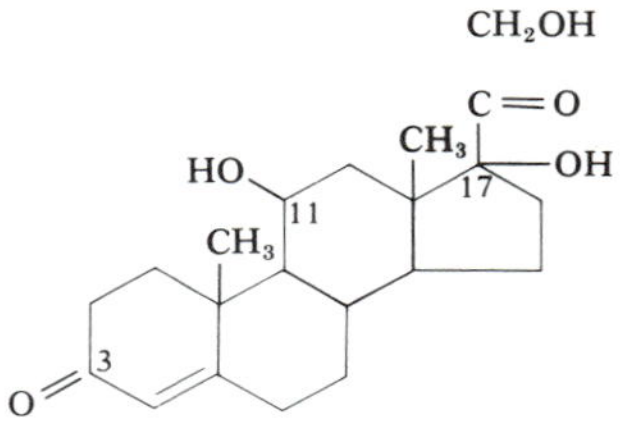

Cortisol (hydrocortisone; compound F)

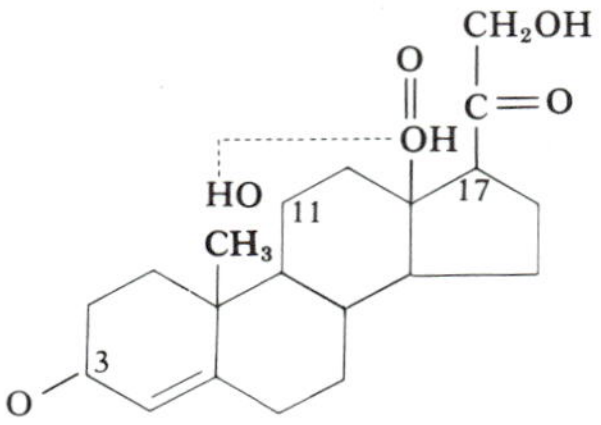

Aldosterone

Corticosterone (compound B) may be the parent substance of other corticoids.

Compound E or cortisone (chemical name, 17-hydroxy-11-dehydrocorticosterone, signifying that the molecule is the same as corticosterone with —OH instead of —H on C-17, and H on C-11).

DOC (11-desoxycorticosterone—corticosterone molecules without any oxygen at C-11).

Relation of molecular structure to function:

1 Oxygen at C-11 produces glucocorticoid effects described on pp. 281-282.
2 OH at C-17 (in addition to oxygen at C-11) enhances glucocorticoid effects.
3 Aldehyde group at C-18 (as in aldosterone) produces marked salt-retaining effect.

ticoids affect literally every cell in the body. Although the precise primary actions of glucocorticoids remain unknown, their most outstanding effects are as follows.

1 Glucocorticoids tend to accelerate the breakdown of proteins to amino acids in all cells except liver cells. These "mobilized" amino acids move out of tissue cells into blood and circulate to liver cells. Here they are changed to glucose by a process that consists of a series of chemical reactions and is called gluconeogenesis. A prolonged, high blood concentration of glucocorticoids in the blood, therefore, results in a net loss of tissue proteins (that is, a negative nitrogen balance or "tissue wasting") and a higher than normal blood glucose concentration (that is, hyperglycemia or diabetes mellitus). Summarizing, we might describe glucocorticoids as protein-mobilizing (or protein-catabolic), gluconeogenic, and hyperglycemic (or diabetogenic) hormones.

2 Glucocorticoids tend to accelerate both the mobilization of fats from adipose cells and the catabolism of fats by almost all kinds of cells. In other words, glucocorticoids tend to cause cells to "shift" for their energy supply to fat catabolism from their usual carbohydrate catabolism. But also, the fats mobilized by glucocorticoids may be used by liver cells for gluconeogenesis. Chronic excess of glucocorticoids, as in Cushing's syndrome (Fig. 11-10), results in a redistribution of body fat. It apparently accelerates fat mobilization from the arms and legs and paradoxically promotes fat deposition in the face ("moon face"), shoulders ("buffalo hump"), trunk, and abdomen.

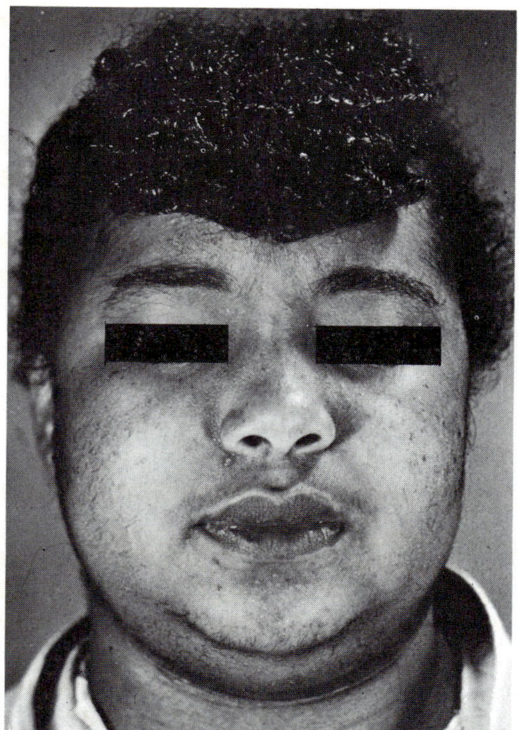

A

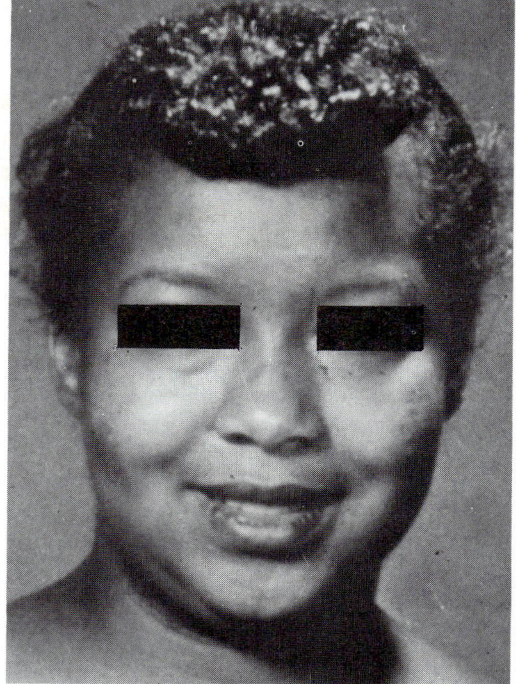

B

Fig. 11-10

Cushing's syndrome, the result of chronic excess glucocorticoids. **A,** Preoperatively. **B,** Six months postoperatively. (Courtesy Dr. William McKendree Jefferies, Case Western Reserve University School of Medicine, Cleveland, Ohio.)

3 Glucocorticoids are essential for maintaining a normal blood pressure. Without adequate amounts of glucocorticoids in the blood, the hormones norepinephrine and epinephrine cannot produce their vasoconstricting effect on blood vessels and blood pressure falls precariously.

4 Glucocorticoid secretion is known to increase during stress, particularly in stress produced by anxiety or severe injury. What is not known, after more than 30 years of study and debate, is how or even whether the increased blood glucocorticoid concentration helps the body cope successfully with factors that threaten its healthy survival.

5 A high blood concentration of glucocorticoids rather quickly causes a marked decrease in the number of eosinophils in blood (eosinopenia) and marked atrophy (decrease in size) of lymphatic tissues, particularly the thymus gland and lymph nodes. This, in turn, leads to a decrease in the number of lymphocytes and plasma cells in blood. Because of the decreased number of lymphocytes and plasma cells, antibody formation also decreases. Antibody formation is an important part of both immunity and allergy.

6 Normal physiological amounts of glucocorticoids act with epinephrine, a hormone secreted by the medullary portion of the adrenal glands, to bring about a normal recovery from injury produced by many kinds of inflammatory agents. How they accomplish this anti-inflammatory effect is still unsettled. Pharmacological amounts of glucocorticoids have been used for many years to relieve the symptoms of rheumatoid arthritis and some other inflammatory conditions.

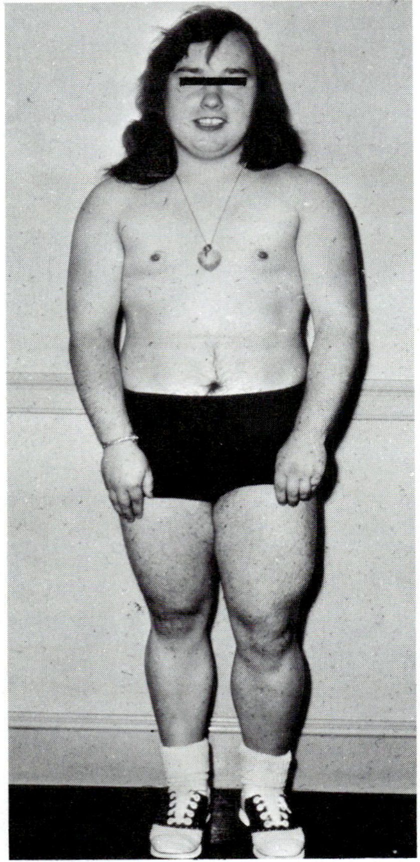

Fig. 11-11

Virilizing tumor of the adrenal cortex of a young girl. The tumor secretes excess androgens, thereby producing masculinizing adrenogenital syndrome. (Courtesy Dr. William McKendree Jefferies, Case Western Reserve University School of Medicine, Cleveland, Ohio.)

Mineralocorticoids

Mineralocorticoids, as their name suggests, play an important part in regulating mineral salt (electrolyte) metabolism. In the human, aldosterone is the only physiologically important mineralocorticoid. Its primary general function seems to be to maintain homeostasis of blood sodium concentration. It does this through its action on the distal renal tubule cells. Aldosterone stimulates them to increase their reabsorption of sodium ions from tubule urine back into blood. Because the tubule cells excrete either a potassium or a hydrogen ion in exchange for each sodium ion they reabsorb, aldosterone helps maintain a normal blood potassium concentration and a normal pH.

Moreover, as each positive sodium ion is reabsorbed, a negative ion (bicarbonate or chloride) follows along, drawn by the attraction force between ions that bear opposite electrical charges. Also, the reabsorption of electrolytes causes net diffusion of water back into blood. Briefly, then, because of its primary sodium-reabsorbing effect on kidney tubules, aldosterone tends to produce sodium and water retention but potassium and hydrogen ion loss.

Sex hormones

The normal adrenal cortex in both sexes secretes physiologically significant amounts of male hormones (androgens) and insignificant amounts of female hormones (estrogens). The androgens secreted by the cortex do not have strong masculinizing properties, except for testosterone, but the cortex secretes only trace amounts of this. New information suggests that the small amounts of androgens secreted by the female cortex probably support sexual behavior. Tumors of the adrenal cortex that secrete large amounts of androgens are known as virilizing tumors. The excessive androgens produce masculinizing effects such as those evidenced in Fig. 11-11. Extreme androgen excess in a woman may cause a beard to grow.

Control of secretion

Glucocorticoids. Mechanisms that control secretion of these hormones are discussed on pp. 273-274. hypothalamus

Mineralocorticoids (aldosterone mainly). Aldosterone secretion is largely controlled by the renin-angiotensin mechanism and by blood potassium concentration. The renin-angiotensin mechanism operates as indicated in Fig. 11-12. When blood pressure in the afferent arterioles of the kidney decreases below a certain level, it acts as a stimulus to the juxtaglomerular apparatus,* causing its cells to secrete renin into the blood and interstitial fluid. Renin is an enzyme. It catalyzes reactions that change a compound called renin substrate—a normal constituent of blood—into angiotensin I. Angiotensin I is immediately converted to angiotensin II by another enzyme normally present in blood

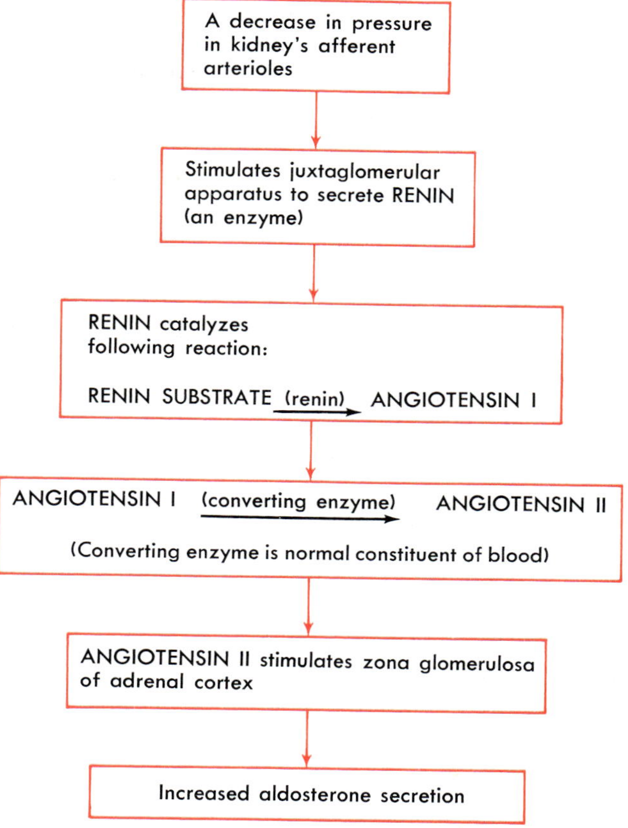

Fig. 11-12

Renin-angiotensin mechanism for regulating aldosterone secretion.

*L. *juxta*, near to. Cells located in afferent arterioles near their entry into the glomeruli constitute the juxtaglomerular apparatus.

and aptly named "converting enzyme." Angiotensin II stimulates the zona glomerulosa of the adrenal cortex to increase its secretion of aldosterone.

Blood potassium concentration also helps regulate aldosterone secretion. Specifically, a high blood potassium concentration stimulates aldosterone secretion and a low blood potassium concentration inhibits it.

Research findings of some investigators† suggest that the pineal body secretes a hormone that stimulates aldosterone secretion by the adrenal's zona glomerulosa. Therefore they named the postulated hormone adrenoglomerulotropin.

Adrenal medulla

Epinephrine and norepinephrine

The adrenal medulla secretes two catecholamines—about 80% epinephrine and the rest norepinephrine. These hormones affect smooth muscle, cardiac muscle, and glands the way sympathetic stimulation does. They serve to increase and prolong sympathetic effects.

Control of secretion

Increased epinephrine secretion by the adrenal medulla is one of the body's first responses to stress. Impulses from the hypothalamus stimulate sympathetic preganglionic neurons, which in turn stimulate the adrenal medulla to increase its output of epinephrine.

Islands of Langerhans

The beta cells of the islands of Langerhans secrete insulin, and the alpha cells secrete glucagon. Insulin tends to accelerate the movement of glucose, amino acids, and fatty acids out of blood and through the plasma membranes of cells into their cytoplasm. Hence insulin tends to lower the blood concentrations of these food compounds and to promote their metabolism (discussed in Chapter 15).

Glucagon, the hormone secreted by the alpha cells of the islands of Langerhans, tends to increase blood glucose concentration. Chapter 15 describes the mechanism by which glucagon produces this opposite effect from insulin. It also describes the mechanisms that regulate insulin and glucagon secretion.

Ovaries

The ovaries produce two kinds of steroid female hormones: estrogens (chiefly estradiol and estrone) and progesterone. The chief endocrine glands that secrete estrogens are microscopic structures located in the ovary and named ovarian or graafian follicles. Another endocrine gland in the ovary secretes mostly progesterone. Its name is the corpus luteum and it has a claim to distinction. It is a temporary structure, replaced once a month during a woman's reproductive years, except in any months when she is pregnant. In Chapter 19, we shall discuss the functions of estrogens and progesterone and the control of their secretion.

Testes

The interstitial cells of the testes secrete testosterone, a steroid hormone classed as an androgen—that is, a substance that promotes "maleness." Chapter 18 presents details about testosterone's functions and the control of its secretion.

Pineal gland (pineal body or epiphysis cerebri)

The pineal gland has long been a mystery organ. Even now its function in the human being remains a controversial matter. A small cone-shaped structure about a centimeter long, it is located in the cranial cavity behind

†Dr. Gordon L. Farrell of Case Western Reserve University, Cleveland, Ohio, and others.

the midbrain and the third ventricle (Fig. 7-8). The pineal gland may secrete only one hormone, melatonin. Recent research indicates that melatonin inhibits LH secretion.

Placenta

The placenta functions as a temporary endocrine gland. During pregnancy, it produces chorionic gonadotropins—so-called because they are secreted by cells of the chorion, the outermost fetal membrane. In addition to gonadotropins, the placenta also produces estrogens and progesterone. During pregnancy, the kidneys excrete large amounts of gonadotropins in the urine. This fact, discovered more than a half century ago by Aschheim and Zondek, led to the development of the now familiar pregnancy tests.

Outline summary

MEANING

1 Composed of glands that pour secretions into blood instead of into ducts
2 General functions and importance
  a Communication, control, integration; hormones main regulators of metabolism, reproduction, and responses to stress
  b Names and locations of endocrine glands—see Table 9-1

PITUITARY GLAND (HYPOPHYSIS CEREBRI)

Consists of two endocrine glands: anterior lobe of pituitary gland (adenohypophysis) and posterior lobe of pituitary gland (neurohypophysis)

Anterior pituitary gland (adenohypophysis)

1 Cells called acidophils secrete growth hormone (STH) and prolactin
2 Basophils secrete thyrotropin (TSH), adrenocorticotropin (ACTH), follicle-stimulating hormone (FSH), luteinizing hormone (LH), and melanocyte-stimulating hormone (MSH)
3 Growth hormone (somatotropin or somatotropic hormone)
  a Accelerates protein anabolism, so promotes growth
    1 Disorders caused by excess growth hormone: gigantism and acromegaly
    2 Disorders caused by deficient growth hormone: dwarfism and pituitary cachexia
  b Tends to accelerate fat mobilization from adipose cells and fat catabolism by other cells, thereby decreasing glucose catabolism and tending to increase blood glucose concentration (hyperglycemic effect); prolonged excess may produce diabetes
4 Prolactin (lactogenic hormone)
  a During pregnancy, promotes breast development
  b After delivery, initiates milk secretion
5 Tropic hormones
  a Thyrotropin—thyroid-stimulating hormone (TSH)
    1 Promotes growth and development of thyroid gland
    2 Stimulates thyroid gland to secrete thyroid hormone
  b Adrenocorticotropin (ACTH)
    1 Promotes growth and development of adrenal cortex
    2 Stimulates adrenal cortex to secrete cortisol and other glucocorticoids
  c Follicle-stimulating hormone (FSH)
    1 Stimulates primary graafian follicle to start growing and to develop to maturity
    2 Stimulates follicle cells to secrete estrogens
    3 In male, stimulates development of seminiferous tubules and maintains spermatogenesis by them
  d Luteinizing hormone (LH)
    1 Acts with FSH to cause complete maturation of follicle
    2 Brings about ovulation
    3 Stimulates formation of corpus luteum (luteinizing effect)
    4 Stimulates corpus luteum to secrete progesterone and estrogens
    5 In male, LH called interstitial cell–stimulating hormone (ICSH): stimulates interstitial cells in testis to develop and secrete testosterone
    6 Just before puberty, presumably sudden marked increase in secretion of gonadotropins (FSH, LH) initiates first menses
6 Melanocyte-stimulating hormone (MSH)—tends to produce increased pigmentation of skin
7 Control of secretion
  a Negative feedback mechanisms operate between target glands and anterior lobe of pituitary gland
    1 High blood concentration of target gland hormone inhibits pituitary secretion of tropic hormones either directly or indirectly (that is, via hypothalamic releasing hormones); conversely, low blood concentration of target gland hormones stimulates pituitary secretion of tropic hormones
    2 High blood concentration of tropic hormone stimulates target gland secretion of its hormone
  b Stress acts in some way to stimulate the hypothalamus to secrete releasing hormones, which in turn stimulate the anterior pituitary to increase its secretion of hormones

Posterior pituitary gland (neurohypophysis)

1 Secretes hormones (antidiuretic hormone [ADH] and oxytocin) synthesized by neurons in hypothalamus
2 ADH (vasopressin or Pitressin) stimulates water reabsorption by distal and collecting tubules and stimulates smooth muscle of blood vessels and intestine
3 Oxytocin stimulates contractions of pregnant uterus and release of milk by lactating breast
4 Control of secretion
  a ADH secretion—details of mechanism controlling secretion of this hormone not established but general principles follow
    1 Decreased ECF osmotic pressure leads to decreased ADH secretion
    2 Decreased ECF volume leads to increased ADH secretion
    3 Stress leads to increased ADH secretion

b Oxytocin—details of mechanism controlling oxytocin secretion not established but is known that stimulation of nipples by suckling leads to increased oxytocin secretion

THYROID GLAND

1 Location and structure
  a Located in neck just below larynx
  b Two lateral lobes connected by isthmus
2 Thyroid hormone—stimulates rate of oxygen consumption (metabolic rate) of all cells and thereby helps regulate physical and mental development, development of sexual maturity, and numerous other processes
3 Calcitonin—decreases blood calcium concentration, presumably by inhibiting bone breakdown with calcium release into blood—and/or increasing calcium deposition in bone
4 Effects of hypersecretion and hyposecretion
  a Hypersecretion produces exophthalmic goiter
  b Hyposecretion in early life produces malformed dwarfism or cretinism; in later life, myxedema

PARATHYROID GLANDS

1 Location and structure
  a Attached to posterior surfaces of thyroid gland
  b Small round bodies, usually four or five in number
2 Parathyroid hormone
  a Increases blood calcium by stimulating bone breakdown, releasing calcium and phosphate into blood from bone
  b Increases blood calcium by accelerating calcium absorption from intestine and kidney tubules
  c Accelerates kidney tubule excretion of phosphates from blood into urine
3 Hypersecretion causes decrease in bone mass with replacement by fibrous tissue
4 Hyposecretion produces hypocalcemia and tetany and death in few hours

ADRENAL GLANDS

1 Location and structure
  a Located atop kidneys
  b Outer portion of gland called cortex and inner portion called medulla

Adrenal cortex

1 Zones or layers (from outside in)
  a Zona glomerulosa—secretes mineralocorticoids
  b Zona fasciculata—secretes glucocorticoids
  c Zona reticularis—secretes small amounts of glucocorticoids and sex hormones
2 Glucocorticoids (mainly cortisol, smaller amounts of corticosterone)
  a Tend to accelerate tissue protein mobilization; mobilized amino acids circulate to liver cells where they are changed to glucose (process of gluconeogenesis)
  b Tend to accelerate fat mobilization and catabolism—i.e., tend to cause "shift" to fat utilization from usual carbohydrate utilization
  c Necessary for norepinephrine's vasoconstricting effect on blood vessels and therefore essential for maintaining normal blood pressure
  d Blood concentration of glucocorticoids increases during stress; value of this still controversial
  e High blood concentration of glucocorticoids causes eosinopenia and marked atrophy of thymus gland and other lymphatic tissues; this atrophy leads to a decrease in the number of lymphocytes and plasma cells and decreased antibody formation, which decreases immunity and allergic reactions
  f Glucocorticoids plus epinephrine promote normal recovery from injury by inflammatory agents (anti-inflammatory effect)
3 Mineralocorticoids (mainly aldosterone)
  a Accelerate renal tubule reabsorption of sodium ions and excretion of potassium ions (or hydrogen ions)
  b Increased renal tubule reabsorption of bicarbonate ions (or chloride ions) and water result from increased sodium reabsorption
4 Sex hormones—physiologically significant amounts of male and insignificant amounts of female hormones secreted in both sexes
5 Control of secretion
  a Glucocorticoids (see outline, p. 285)
  b Mineralocorticoids (aldosterone)
    1 Renin-angiotensin mechanism, an important regulator of aldosterone secretion; see Fig. 11-12
    2 Blood potassium concentration also helps regulate aldosterone secretion; high blood potassium concentration stimulates and low blood potassium concentration inhibits aldosterone secretion

Adrenal medulla

1 Hormones—epinephrine mainly; some norepinephrine
2 Functions of epinephrine—affects visceral effectors (smooth muscle, cardiac muscle, and glands) in same way as sympathetic stimulation of these structures; epinephrine from adrenal glands intensifies and prolongs sympathetic effects
3 Control of secretion—stress acts in some way to stimulate hypothalamus, which sends impulses to adrenal medulla via preganglionic sympathetic neurons, stimulating medulla to increase its secretion of epinephrine

ISLANDS OF LANGERHANS

1 Hormones
  a Beta cells secrete insulin
  b Alpha cells secrete glucagon
2 Functions
  a Insulin
    1 Promotes glucose transport into cells, thereby increasing glucose utilization (catabolism) and glycogenesis and decreasing blood glucose
    2 Promotes fatty acid transport into cells and fat anabolism (lipogenesis or fat deposition) in them
    3 Promotes amino acid transport into cells and protein anabolism
  b Glucagon—accelerates liver glycogenolysis so tends to increase blood glucose; in short, is insulin antagonist

OVARIES

1 Hormones
   a Ovarian (graafian) follicles secrete estrogens
   b Corpus luteum secretes progesterone
2 Functions—see Chapter 19

TESTES

1 Hormones—androgens, most important of which is testosterone
2 Functions—see Chapter 18

PINEAL GLAND (PINEAL BODY OR EPIPHYSIS CEREBRI)

Secretes melatonin, a hormone reported to inhibit LH secretion

PLACENTA

1 Temporary endocrine glands
2 Secretes estrogens, progesterone, and chorionic gonadotropin
3 Helps maintain progestational state of endometrium

Review questions

1 Name the endocrine glands and locate each one.
2 Name the hormone or hormones that help control each of the following: blood sugar level, blood calcium level, blood sodium level, blood potassium level. Explain mechanisms involved.
3 Name the hormone or hormones that help control each of the following: growth; development of secondary male characteristics and female characteristics; fluid and electrolyte balance; resistance to stress; functions of the adrenal cortex, thyroid gland, and ovaries; secretion of ACTH, TH, and FSH.
4 What hormone enhances and prolongs sympathetic effects?
5 What hormones help control protein metabolism? Explain.
6 What hormones help control fat metabolism? Explain.
7 What hormones help control carbohydrate metabolism? Explain.
8 Which gland is called the "master gland"? Why?
9 What conditions result from a deficiency of thyroid extract early in life? Later in life?
10 What disease results from too much thyroid secretion? From too much pituitary somatotropic hormone early in life? Too much later in life? Too little early in life?
11 Gigantism results from an oversupply of which endocrine secretion? Cretinism from a deficiency of which one? Acromegaly from an oversupply of which one? Myxedema from a deficiency of which one?
12 How does the hypothalamus regulate hormone secretion?
13 Describe the mechanisms that bring about increased secretion of epinephrine, ACTH, ADH, and glucocorticoids during stress.
14 Describe a mechanism by which thoughts and emotions can influence literally all body functions.
15 Make a diagram illustrating the "second messenger hypothesis" of hormone action.

Situation: A patient who has a severe form of Cushing's disease is given insulin each day.

16 Should the nurse watch this patient for signs of hypoglycemia or hyperglycemia? Explain.
17 Would this patient be likely to be "insulin-sensitive" or "insulin-resistant"? Explain.
18 Explain the meaning of the statement "hormone functions are interrelated." Give examples to support or refute this statement.
19 Compare the effects of glucocorticoids and mineralocorticoids.

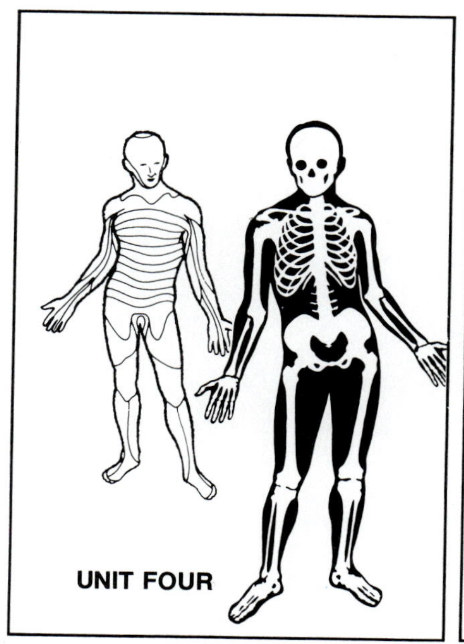

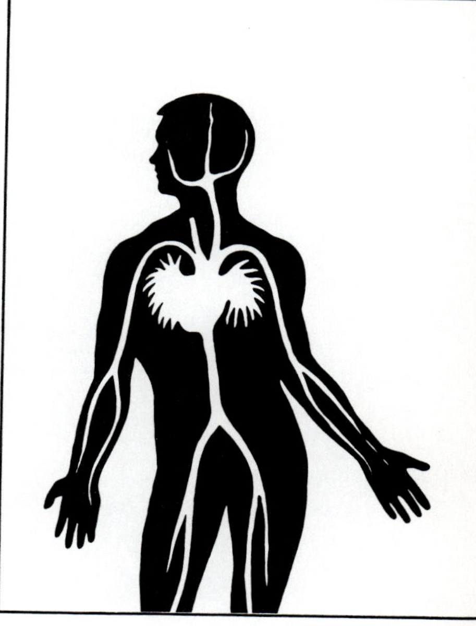

MAINTAINING THE METABOLISM OF THE BODY

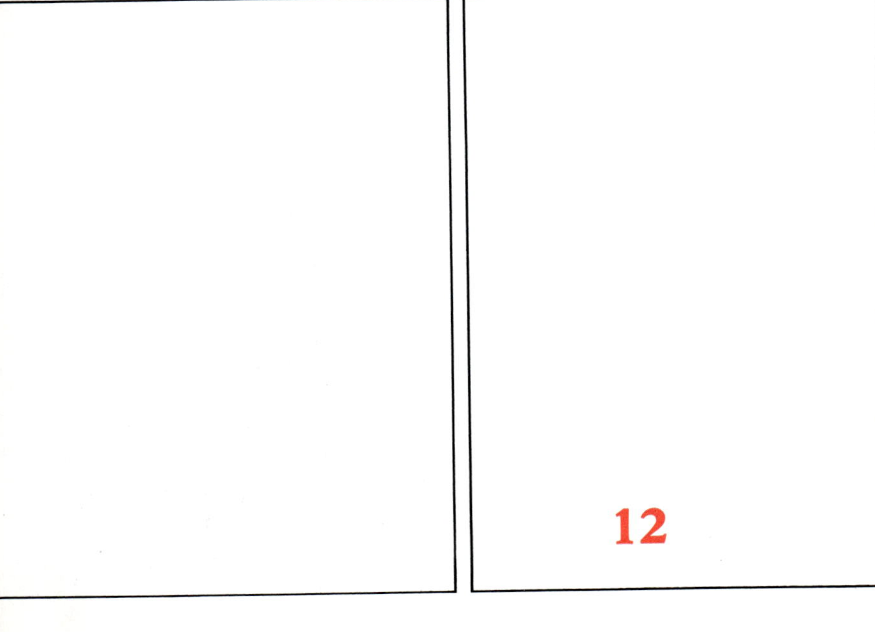

12

The respiratory system

Functions and organs

The respiratory system functions as an air distributor and gas exchanger in order that oxygen may be supplied to and carbon dioxide be removed from the body's cells. Since most of our billions of cells lie too far distant from air to exchange gases directly with it, air must first exchange gases with blood, blood must circulate, and finally blood and cells must exchange gases. These events require the functioning of two systems, namely the respiratory system and the circulatory system. All parts of the respiratory system—except its microscopic-sized sacs called alveoli—function as air distributors. Only the alveoli serve as gas exchangers.

The organs of the respiratory system are the nose, pharynx, larynx, trachea, bronchi, and lungs. Together they constitute the lifeline, the air supply line of the body. We shall first describe the structure and functions of these organs and then discuss the physiology of the respiratory system as a whole.

Organs

Nose

Structure

The nose consists of an internal and an external portion. The external portion—that is, the part that protrudes from the face—is considerably smaller than the internal portion, which lies over the roof of the mouth. The interior of the nose is hollow and is separated by a partition, the *septum* (Fig. 12-1), into a right and a left cavity. The palatine bones, which form both the floor of the nose and the roof of the mouth, separate the nasal cavities from the mouth cavity. Sometimes the palatine bones fail to unite completely, producing a condition known as *cleft palate.* When this abnormality exists, the mouth is only partially separated from the nasal cavity, and difficulties arise in swallowing.

Each nasal cavity is divided into three

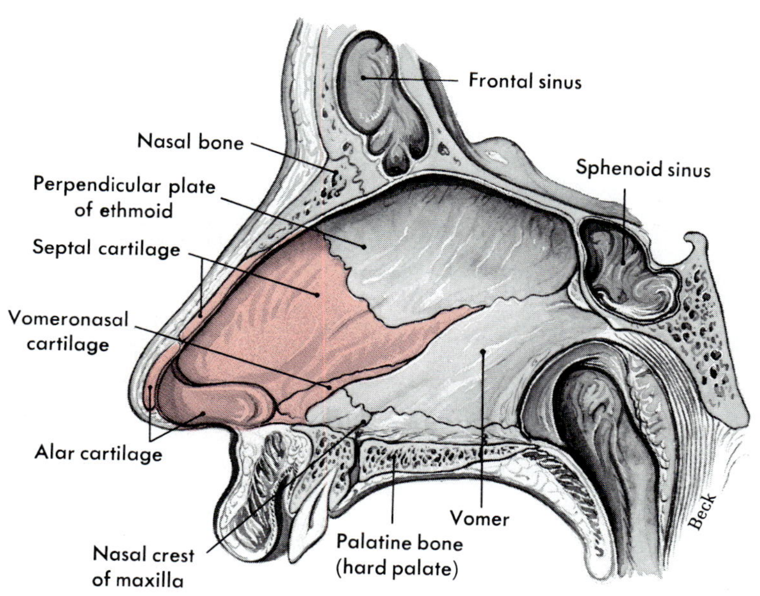

Fig. 12-1

The nasal septum consists of the perpendicular plate of the ethmoid bone, the vomer, and the septal and vomeronasal cartilages. The cartilages are shown in pink.

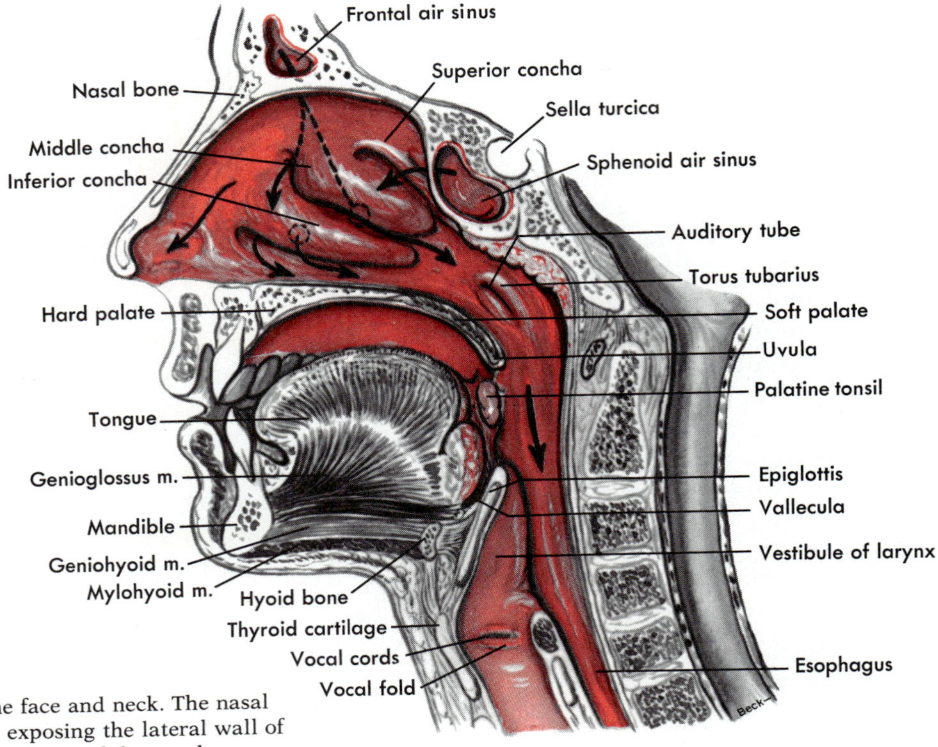

Frontal air sinus
Superior concha
Sella turcica
Nasal bone
Middle concha
Sphenoid air sinus
Inferior concha
Auditory tube
Torus tubarius
Hard palate
Soft palate
Uvula
Palatine tonsil
Tongue
Genioglossus m.
Epiglottis
Vallecula
Mandible
Vestibule of larynx
Geniohyoid m.
Mylohyoid m.
Hyoid bone
Thyroid cartilage
Vocal cords
Esophagus
Vocal fold
Beck

Fig. 12-2

Sagittal section through the face and neck. The nasal septum has been removed, exposing the lateral wall of the nasal cavity. Note the position of the conchae (turbinates).

Frontal air sinus
Ethmoid air sinus
Optic nerve
Superior concha (turbinate)
Superior meatus
Middle concha (turbinate)
Middle meatus
Maxillary sinus
Inferior concha (turbinate)
Inferior meatus
Septum

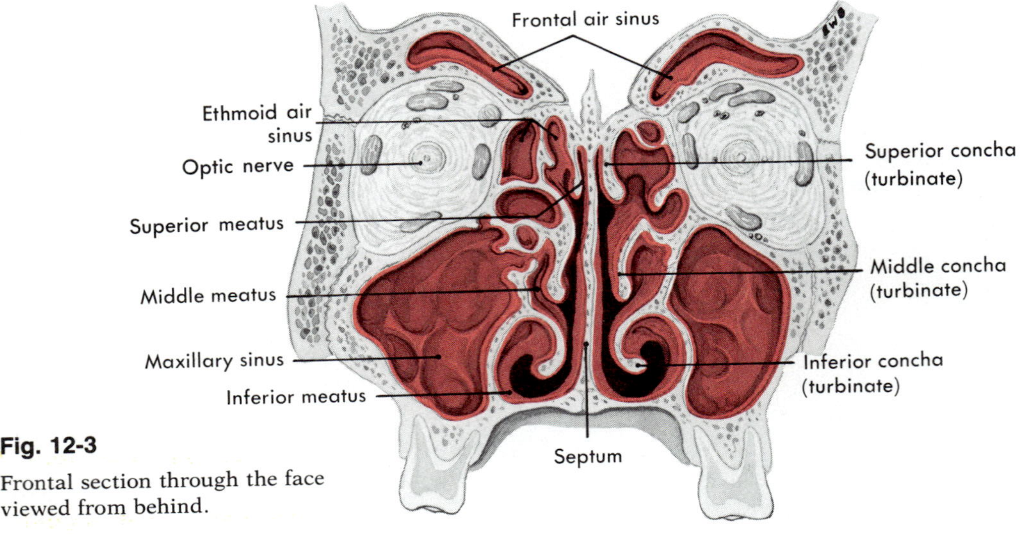

Fig. 12-3

Frontal section through the face viewed from behind.

passageways (superior, middle, and inferior meati) by the projection of the turbinates (conchae) from the lateral walls of the internal portion of the nose (Figs. 12-2 and 12-3). The superior and middle turbinates are processes of the ethmoid bone, while the inferior turbinates are separate bones.

The external openings into the nasal cavities (nostrils) have the technical name of *anterior nares*. The posterior nares (or choanae) are openings from the internal nose into the nasopharynx.

Ciliated mucous membrane lines the nose and the rest of the respiratory tract down as far as the smaller bronchioles.

Four pairs of sinuses drain into the nose. These paranasal sinuses are the frontal, maxillary, ethmoidal, and sphenoidal. They drain as follows:

1 Into the middle meatus (passageway below middle turbinate)—frontal, maxillary, and anterior ethmoidal sinuses
2 Into the superior meatus—posterior ethmoidal sinuses
3 Into the space above the superior turbinates (sphenoethmoidal recess)—sphenoid sinuses

Functions

The nose serves as a passageway for air going to and from the lungs, filtering it of impurities and warming, moistening, and chemically examining it for substances that might prove irritating to the mucous lining of the respiratory tract. It serves as the organ of smell, since olfactory receptors are located in the nasal mucosa, and it aids in phonation.

Pharynx

Structure

Another name for the pharynx is the throat. It is a tubelike structure about 5 inches long that extends from the base of the skull to the esophagus and lies just anterior to the cervical vertebrae. It is made of muscle, is lined with mucous membrane, and has three parts: one located behind the nose, the *nasopharynx;* one behind the mouth, the *oropharynx;* and another behind the larynx, the *laryngopharynx*.

Seven openings are found in the pharynx (Fig. 12-2):

1 Right and left auditory (eustachian) tubes open into the nasopharynx
2 Two posterior nares into the nasopharynx
3 The opening from the mouth, known as the *fauces*, into the oropharynx
4 The opening into the larynx from the laryngopharynx
5 The opening into the esophagus from the laryngopharynx

The *adenoids* or pharyngeal tonsils are located in the nasopharynx on its posterior wall opposite the posterior nares. If the adenoids become enlarged, they fill the space behind the posterior nares, making it difficult or impossible for air to travel from the nose into the throat. When this happens, the individual keeps his mouth open to breathe and is described as having an "adenoidy" appearance.

Two pairs of organs are found in the oropharynx: the faucial or *palatine tonsils*, located behind and below the pillars of the fauces, and the *lingual tonsils*, located at the base of the tongue. The palatine tonsils are the ones most commonly removed by a tonsillectomy. Only rarely are the lingual ones also removed.

Functions

The pharynx serves as a hallway for the respiratory and digestive tracts, since both air and food must pass through this structure before reaching the appropriate tubes. It also plays an important part in phonation. For example, only by the pharynx changing its shape can the different vowel sounds be formed.

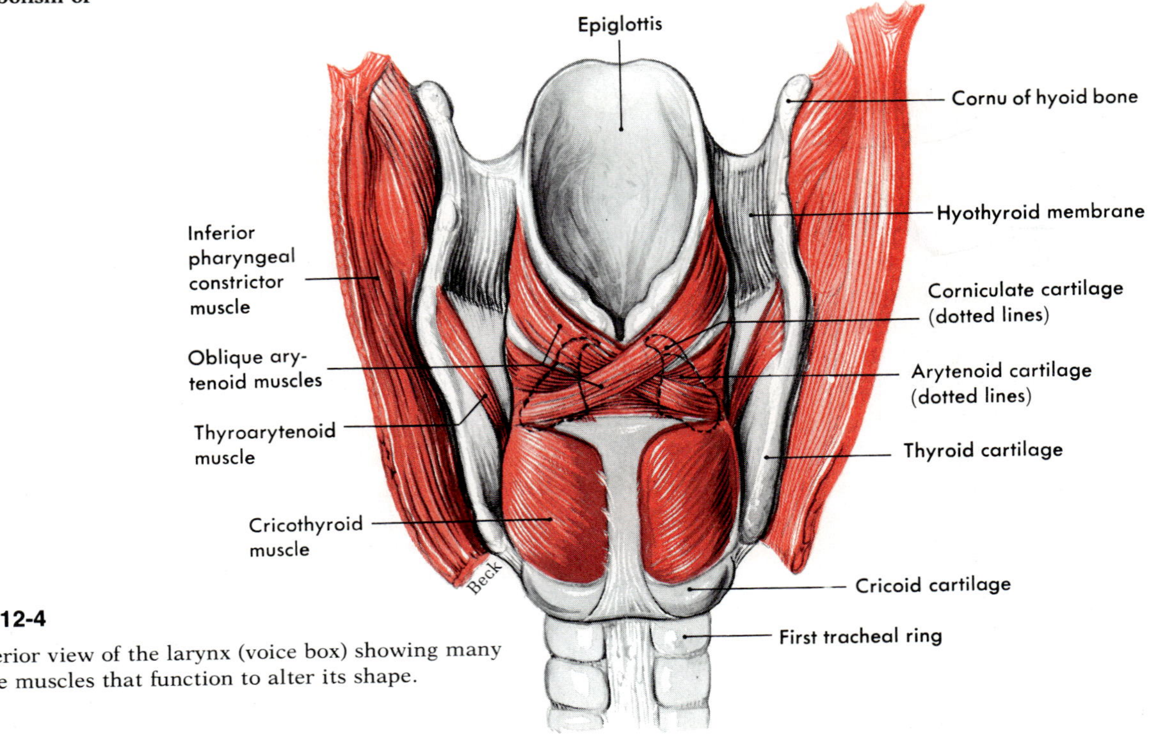

Fig. 12-4

Posterior view of the larynx (voice box) showing many of the muscles that function to alter its shape.

Larynx

Location

The larynx or voice box lies at the upper end of the trachea and just below the pharynx. It might be described as a vestibule opening into the trachea from the pharynx.

Structure

The larynx consists largely of cartilages and muscles. Nine cartilages form its framework. The three largest of these—the thyroid cartilage, the epiglottis, and the cricoid cartilage—are single structures. There are three pairs of smaller accessory cartilages, namely the arytenoid, corniculate, and cuneiform cartilages. A brief description of the three main laryngeal cartilages follows.

1 The thyroid cartilage (Adam's apple) is the largest cartilage of the larynx and is the one that gives the characteristic triangular shape to its anterior wall. It is usually larger in men than in women and has less of a fat pad lying over it—two reasons why a man's Adam's apple protrudes more than a woman's.

2 A small cartilage attached along one edge to the thyroid cartilage but free on its other borders is named the *epiglottis*.

3 The *cricoid* or signet ring cartilage, so called because its shape resembles a signet ring (turned so the signet forms part of the posterior wall of the larynx), is the most inferiorly placed of the nine cartilages.

Mucous membrane lines the larynx and forms two pairs of folds that jut inward into its cavity. The upper pair is called the false vocal folds for the rather obvious reason that they play no part in vocalization. The lower pair serves as the true vocal cords. The slitlike space between the true vocal cords—the rima glottidis or glottis—is the narrowest part of the larynx. Edema of the mucosa covering the vocal cords is a potentially lethal condition. Even a moderate amount of swelling can obstruct the glottis so that air cannot get through and asphyxiation results.

Muscles in the wall of the larynx connect its various cartilages. Their contractions play important roles in respiration, vocalization, and swallowing. During swallowing, for example, contraction of the aryepiglottic muscles (those that connect the arytenoid cartilages with the epiglottis) prevents "swallowing down the wrong throat" by squeezing shut the laryngeal inlet. Two other pairs of intrinsic laryngeal muscles function to open and close the glottis. The posterior cricoarytenoid muscles (between cricoid and arytenoid cartilages) open the glottis by abducting the true vocal cords. The lateral cricoarytenoid muscles close the glottis by adducting the cords. These events are crucial to both respiration and voice production. Certain other intrinsic muscles of the larynx function to influence the pitch of the voice by either lengthening and tensing or shortening and relaxing the vocal cords.

Functions

The larynx functions in respirations since it constitutes part of the vital airway to the lungs. It protects the airway against entrance of solids or liquids during swallowing. It also serves as the organ of voice production—hence its popular name, the voice box. Air being expired through the glottis, narrowed by partial adduction of the vocal cords, causes them to vibrate. Their vibration produces the voice. Several other structures besides the larynx contribute to the sound of the voice by acting as sounding boards or resonating chambers. Thus the size and shape of the nose, mouth, pharynx, and bony sinuses help to determine the quality of the voice.

Trachea

Structure

The trachea or windpipe is a tube about 4½ inches long that extends from the larynx in the neck to the bronchi in the thoracic cavity (Fig. 12-5). Its diameter measures about 1 inch. Smooth muscle, in which are embedded C-shaped rings of cartilage at regular intervals, fashions the walls of the trachea. The cartilaginous rings are incomplete on the posterior surface. They give firmness to the wall, tending to prevent it from collapsing and shutting off the vital airway. Often a tube is placed in the trachea (endotracheal intubation) before patients leave the operating room, especially if they have been given a muscle relaxant. The purpose of the tube is to ensure an open airway. Another procedure done frequently in today's modern hospitals is a tracheostomy—that is, the cutting of an opening into the trachea. A surgeon may do this in order that a suction device can be used to remove secretions from the bronchial tree, or he may do it so that a machine such as the intermittent positive pressure breathing (IPPB) machine can be used to improve ventilation of the lung.

Function

The trachea performs a simple but vital function—it furnishes part of the open passageway through which air can reach the lungs from the outside. Obstruction of this airway for even a few minutes causes death from asphyxiation.

Bronchi

Structure

The trachea divides at its lower end into two *primary bronchi*, of which the right bronchus is slightly larger and more vertical than the left. This anatomical fact explains why aspirated foreign objects frequently lodge in the right bronchus. In structure, the bronchi resemble the trachea. Their walls contain incomplete cartilaginous rings before the bronchi enter the lungs, but they become complete within the lungs. Ciliated mucosa lines the bronchi, as it does the trachea.

Each primary bronchus enters the lung on

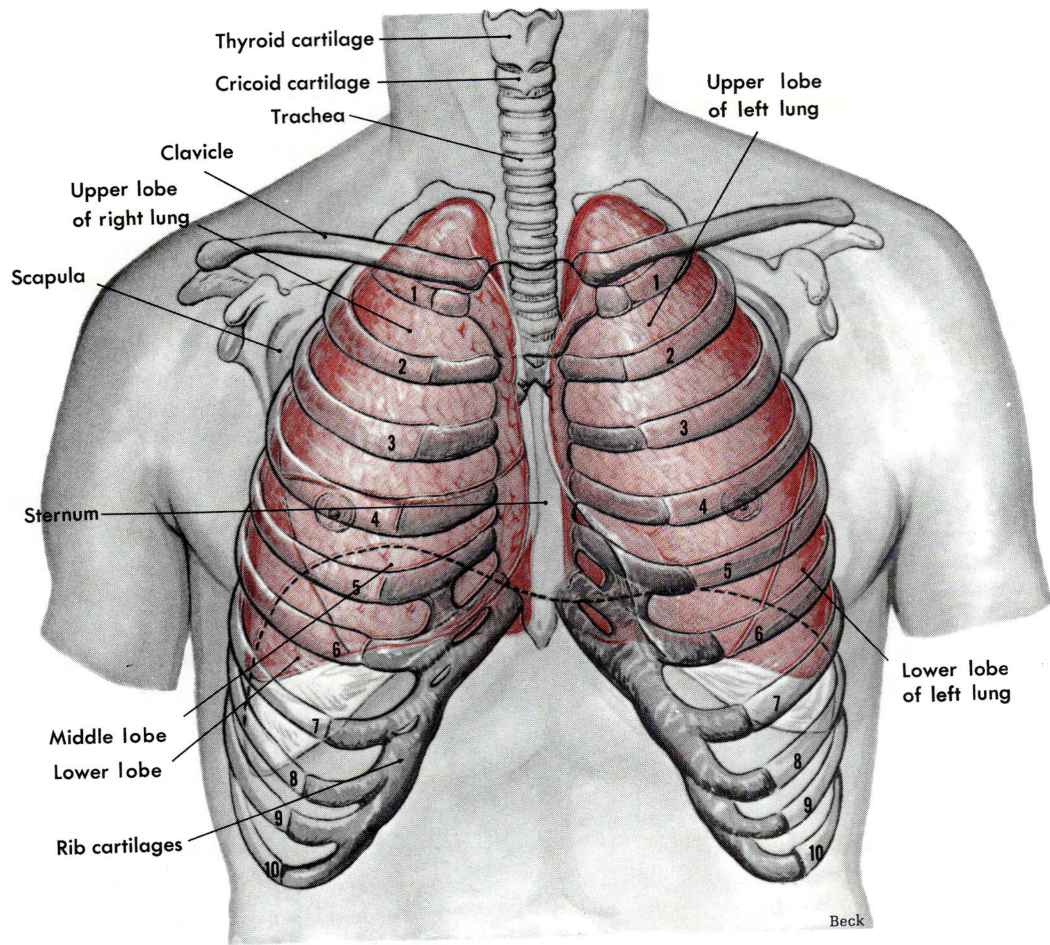

Fig. 12-5

Projection of the lungs and trachea in relation to the rib cage and clavicles. Dotted line indicates location of the dome-shaped diaphragm at the end of expiration and before inspiration. Note that apex of each lung projects above the clavicle. Ribs 11 and 12 are not visible in this view.

its respective side and immediately divides into smaller branches called *secondary bronchi.* The secondary bronchi continue to branch, forming small *bronchioles.* The trachea and the two primary bronchi and their many branches resemble an inverted tree trunk with its branches and are, therefore, spoken of as the bronchial tree. The bronchioles subdivide into smaller and smaller tubes, eventually terminating in microscopic branches that divide into *alveolar ducts,* which terminate in several alveolar sacs, the walls of which consist of numerous *alveoli* (Figs. 12-6 to 12-8). The structure of an alveolar duct with its branching alveolar sacs can be likened to a bunch of grapes—the stem represents the alveolar duct, each cluster of grapes represents an alveolar sac, and each

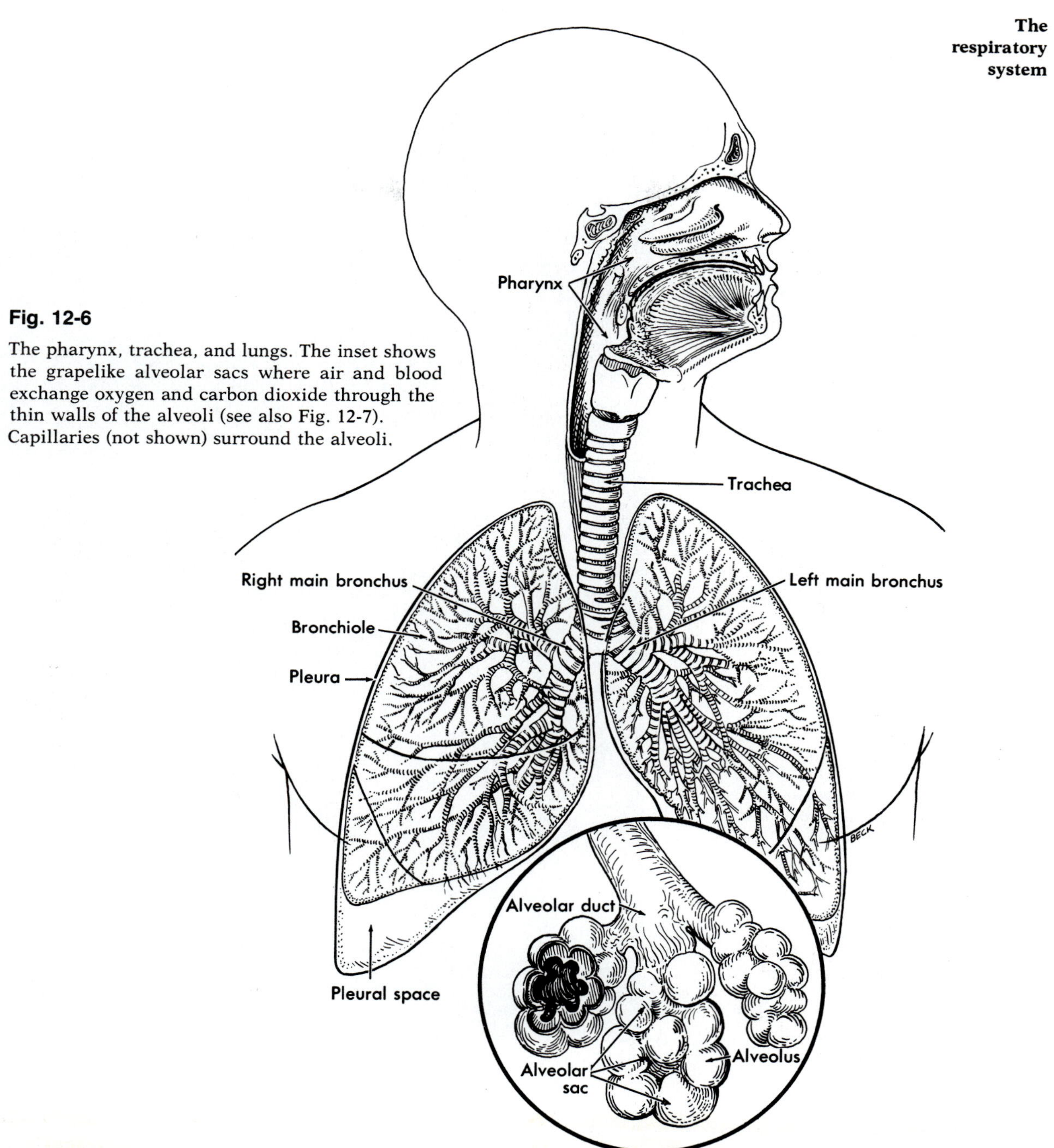

Fig. 12-6

The pharynx, trachea, and lungs. The inset shows the grapelike alveolar sacs where air and blood exchange oxygen and carbon dioxide through the thin walls of the alveoli (see also Fig. 12-7). Capillaries (not shown) surround the alveoli.

grape represents an alveolus. Some 300 million alveoli are estimated to be present in our two lungs.

The structure of the secondary bronchi and bronchioles shows some modification of the primary bronchial structure. The cartilaginous rings become irregular and disappear entirely in the smaller bronchioles. By the time the branches of the bronchial tree have dwindled sufficiently to form the alveolar ducts and sacs and the alveoli, only the internal surface layer of cells remains. In other

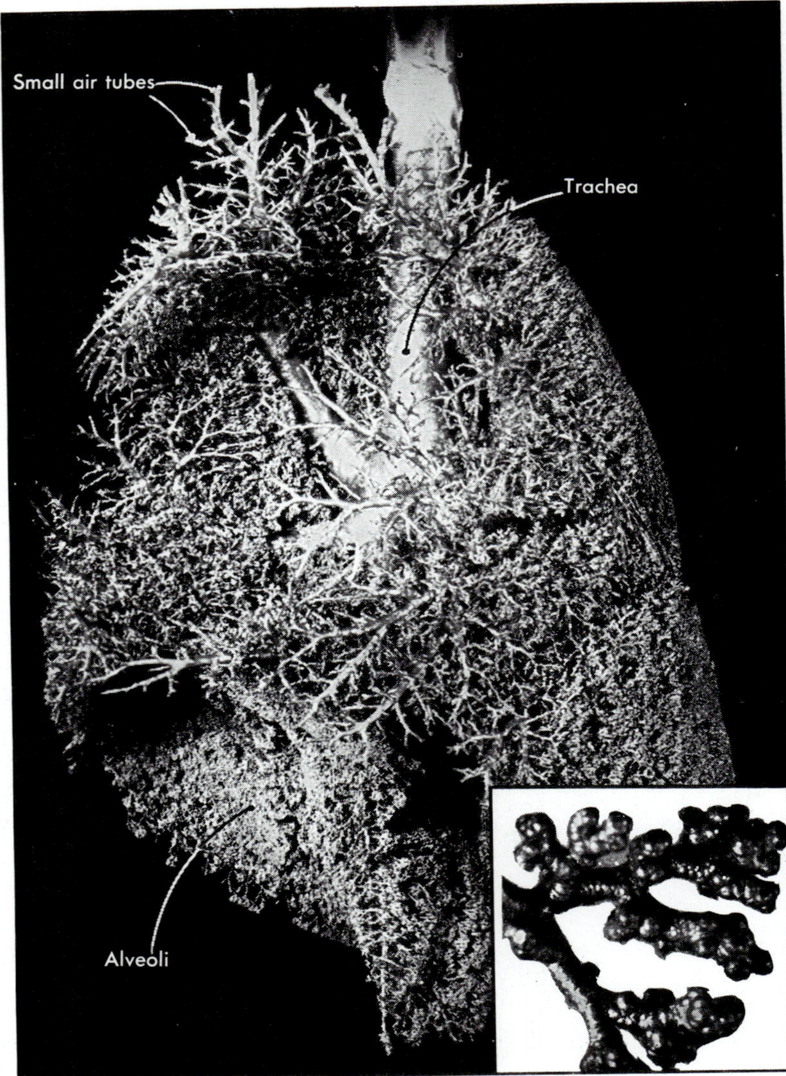

Fig. 12-7

Metal cast of air spaces of the lungs of a dog. The inset shows a cast of clusters of alveoli at the terminations of tiny air tubes. The magnification of the inset is about eleven times the actual size. (From Carlson, A. J., and Johnson, V. E.: The machinery of the body, Chicago, The University of Chicago Press.)

words, the walls of these microscopic structures consist of a single layer of simple, squamous epithelial tissue. As we shall see, this structural fact makes possible the performance of their function.

Functions

The tubes composing the bronchial tree perform the same function as the trachea—that of distributing air to the lung's interior. The alveoli, enveloped as they are by networks of capillaries, accomplish the lung's main and vital function, that of gas exchange between air and blood. Someone has observed that "the lung passages all serve the alveoli" just as "the circulatory system serves the capillaries." Certain diseases may block the passage of air through the bronchioles or

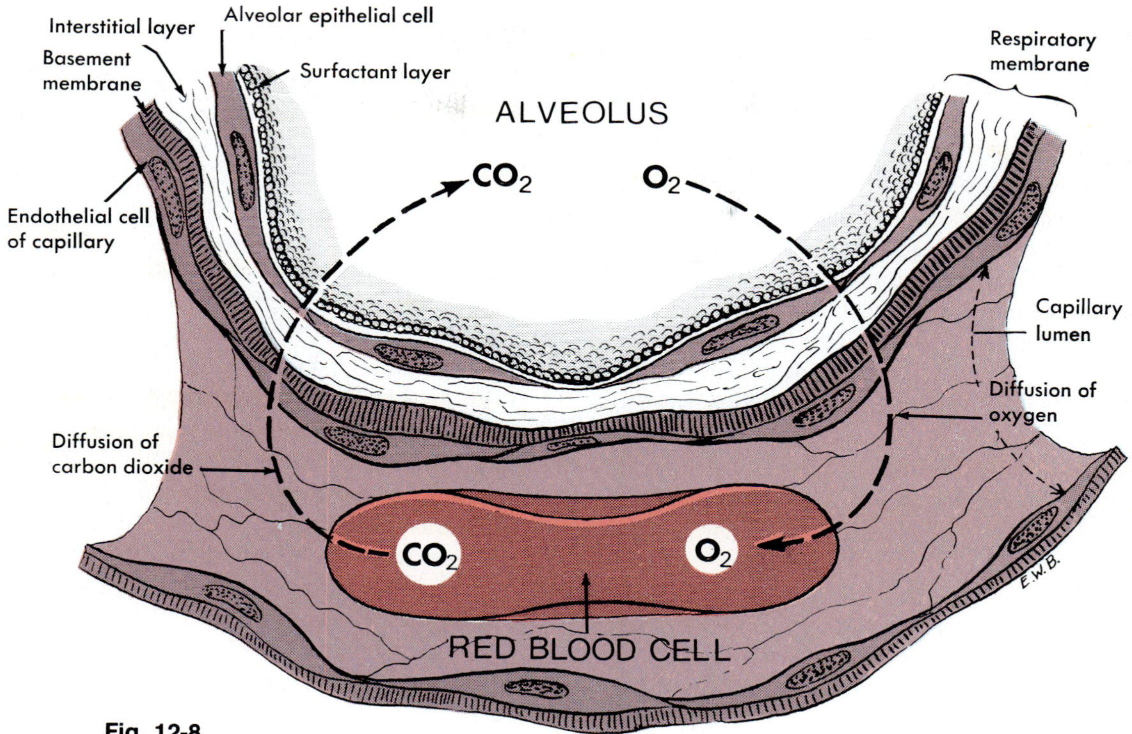

Fig. 12-8

Respiratory membrane greatly magnified to show epithelial and surfactant layers of alveolar membrane and basement membrane and endothelial layer of capillary membrane.

through the alveoli. For example, in pneumonia the alveoli become inflamed, and the accompanying wastes plug up these minute air spaces, making the affected part of the lung solid. Whether the victim survives depends largely upon the extent of the solidification.

Lungs

Structure

The lungs are cone-shaped organs, large enough to fill the pleural portion of the thoracic cavity completely. They extend from the diaphragm to a point slightly above the clavicles and lie against the ribs both anteriorly and posteriorly. The medial surface of each lung is roughly concave to allow room for the mediastinal structures and for the heart, but concavity is greater on the left than on the right because of the position of the heart. The primary bronchi and pulmonary blood vessels (bound together by connective tissue to form what is known as the *root* of the lung) enter each lung through a slit on its medial surface called the *hilum.*

The broad inferior surface of the lung,

which rests on the diaphragm, constitutes the *base*, whereas the pointed upper margin is the *apex*. Each apex projects above a clavicle.

The left lung is partially divided by fissures into two *lobes* (upper and lower) and the right lung into three lobes (superior, middle, and inferior). The interior of each lung consists of the almost innumerable tubes of dwindling diameters that make up the bronchial tree and serve as air distributors. The smallest tubes terminate in the smallest but functionally most important structures of the lung—the alveoli or "gas exchangers."

Visceral pleura covers the outer surfaces of the lungs and adheres to them much as the skin of an apple adheres to the apple.

Functions

The lungs perform two functions—air distribution and gas exchange. Air distribution to the alveoli is the function of the tubes of the bronchial tree. Gas exchange between air and blood is the joint function of the alveoli and the networks of blood capillaries that envelop them. These two structures—one part of the respiratory system and the other part of the circulatory system—together serve as highly efficient gas exchangers. Why? Because they provide an enormous surface area where the very thin-walled alveoli and equally thin-walled pulmonary capillaries come in close contact. This makes possible extremely rapid diffusion of gases between alveolar air and pulmonary capillary blood. Someone has estimated that if the lungs' 300 million or so alveoli could be opened up flat, they would form a surface about the size of a tennis court, that is, about 70 square meters,* or some 40 times the surface area of the entire body! No wonder that such large amounts of oxygen can be so quickly loaded into the blood while large amounts of carbon dioxide are rapidly being unloaded from it.

*Comroe, J. H., Jr.: The lung, Sci. Am. **214**:57-66, Feb., 1966.

Thorax (chest)

Structure

As described on p. 3, the thoracic cavity has three divisions, separated from each other by partitions of pleura. The parts of the cavity occupied by the lungs are the pleural divisions. The space between the lungs occupied mainly by the esophagus, trachea, large blood vessels, and heart is the mediastinum.

The parietal layer of the pleura lines the entire thoracic cavity. It adheres to the internal surface of the ribs and the superior surface of the diaphragm, and it partitions off the mediastinum. A separate pleural sac thus encases each lung. Since the outer surface of each lung is covered by the visceral layer of the pleura, the visceral pleura lies against the parietal pleura, separated only by a potential space (pleural space) that contains just enough pleural fluid for lubrication. Thus when the lungs inflate with air, the smooth, moist visceral pleura coheres to the smooth, moist parietal pleura. Friction is thereby avoided, and respirations are painless. In pleurisy, on the other hand, the pleura is inflamed and respirations become painful.

Functions

The thorax plays a major role in respirations. Because of the elliptical shape of the ribs and the angle of their attachment to the spine, the thorax becomes larger when the chest is raised and smaller when it is lowered. It is these changes in thorax size that bring about inspiration and expiration (discussed on p. 301). Lifting up the chest raises the ribs so that they no longer slant downward from the spine, and because of their elliptical shape, this enlarges both depth (from front to back) and width of the thorax. (If this does not sound convincing to you, examine a skeleton to see why it is so.)

Physiology

Pulmonary ventilation

Pulmonary ventilation is a technical term for what most of us call breathing. One phase of it, inspiration, moves air into the lungs and the other phase, expiration, moves air out of the lungs.

Mechanism of pulmonary ventilation

Air moves in and out of the lungs for the same basic reason that any fluid (that is, a liquid or a gas) moves from one place to another—briefly, because its pressure in one place is different from that in the other place. Or stated differently, the existence of a pressure gradient (a pressure difference) causes fluids to move. A fluid always moves down its pressure gradient. This means that a fluid moves from the area where its pressure is higher to the area where its pressure is lower. Under standard conditions, air in the atmosphere exerts a pressure of 760 mm Hg. Air in the alveoli at the end of one expiration and before the beginning of another inspiration also exerts a pressure of 760 mm Hg. This fact explains why at that moment air is neither entering nor leaving the lungs. The mechanism that produces pulmonary ventilation is one that establishes a gas pressure gradient between the atmosphere and the alveolar air.

When atmospheric pressure is greater than pressure within the lung, air flows down this gas pressure gradient. Then air moves from the atmosphere into the lungs. In other words, inspiration occurs. When pressure in the lungs becomes greater than atmospheric pressure, air again moves down a gas pressure gradient. But now, this means that it moves in the opposite direction. This time air moves out the lungs into the atmosphere. The pulmonary ventilation mechanism, therefore, must somehow establish these two gas pressure gradients—one in which intrapulmonic pressure (pressure within the lungs) is lower than atmospheric pressure to produce inspiration and one in which it is higher than atmospheric pressure to produce expiration.

These pressure gradients are established by changes in the size of the thoracic cavity, which in turn are produced by contraction and relaxation of respiratory muscles. Contraction of the diaphragm alone, or of the diaphragm and the external intercostal muscles, produces quiet inspiration. As the diaphragm contracts, it descends, and this makes the thoracic cavity longer. Contraction of the external intercostal muscles pulls the anterior end of each rib up and out. This also elevates the attached sternum and enlarges the thorax from front to back and from side to side. As the size of the thorax increases, the intrapleural (intrathoracic) pressure decreases because of operation of the familiar principle known as Boyle's law. At the end of an expiration and before the beginning of the next inspiration, intrathoracic pressure is about 3.5 cm H_2O less than atmospheric pressure (frequently written -3.5 cm H_2O). During quiet inspiration, intrathoracic pressure decreases further to -6 cm H_2O. As the thorax enlarges, it pulls the lungs along with it because of cohesion between the moist pleura covering the lungs and the moist pleura lining the thorax. Thus the lungs expand and the pressure in their tubes and alveoli necessarily decreases. Alveolar pressure decreases from an atmospheric level to a subatmospheric level—typically, to about -3.5 cm H_2O. The moment intrapulmonic pressure becomes less than atmospheric pressure, a pressure gradient exists between the atmosphere and the interior of the lungs. Air then necessarily moves into the lungs. Figs. 12-9 and 12-10 diagram the mechanisms of inspiration and expiration.

To apply some of the information just dis-

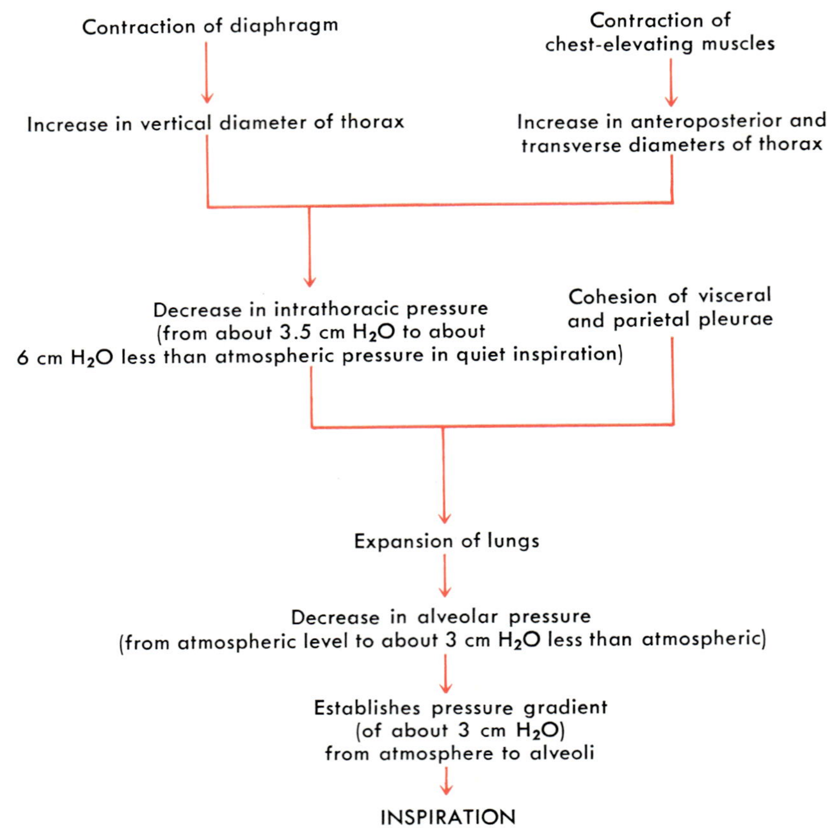

Fig. 12-9

The mechanism of normal, quiet inspiration.

cussed about the respiratory mechanism, let us suppose that a surgeon makes an incision through the chest wall into the pleural space, as he would in doing one of the dramatic, modern open-chest operations. Air would then be present in the thoracic cavity, a condition known as *pneumothorax.* What change, if any, can you deduce would take place in respirations? Compare your deductions with those in the next paragraph.

Intrathoracic pressure would, of course, immediately increase from its normal subatmospheric level to the atmospheric level. More pressure than normal would therefore be exerted upon the outer surface of the lung and would cause its collapse. It could even collapse the other lung. This is because the mediastinum is a mobile rather than a rigid partition between the two pleural sacs. This anatomical fact allows the increased pressure in the side of the chest that is open to push the heart and other mediastinal structures over toward the intact side where they exert pressure on the other lung. Pneumothorax results in many respiratory and circulatory changes. They are of great importance in determining medical and nursing care but lie beyond the scope of this book.

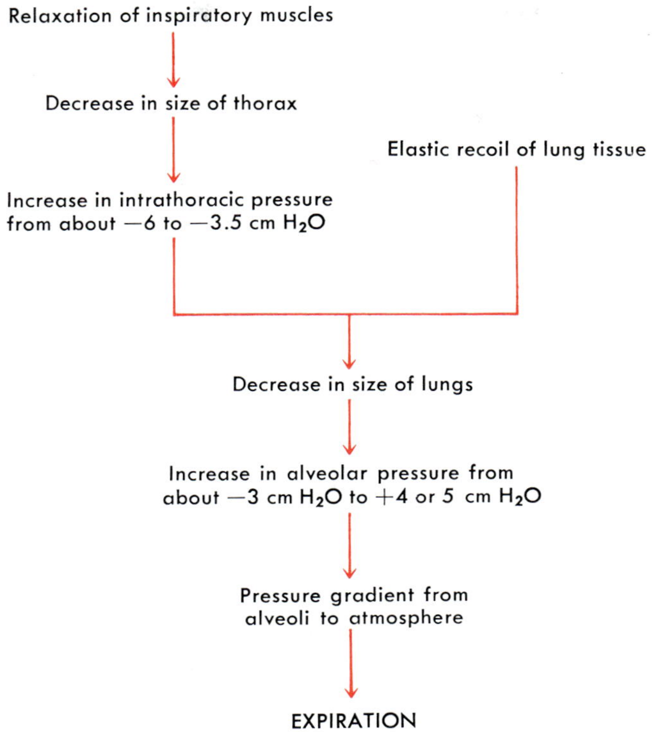

Fig. 12-10

The mechanism of normal, quiet expiration.

Volumes of air exchanged in pulmonary ventilation

The volumes of air moved in and out of the lungs and remaining in them are matters of great importance. They must be normal in order that a normal exchange of oxygen and carbon dioxide can take place between alveolar air and pulmonary capillary blood.

An apparatus called a *spirometer* is used to measure the volume of air exchanged in breathing. The volume of air exhaled normally after a normal inspiration is termed *tidal volume* (TV). As you can see in Fig. 12-11, the normal volume of tidal air for an adult is approximately 500 ml. After an individual has expired tidal air, he can force still more air out of his lungs. The largest additional volume of air that one can forcibly expire after expiring tidal air is called the *expiratory reserve volume* (ERV). An adult, as Fig. 12-11 shows, normally has an ERV of between 1,000 and 1,200 ml. *Inspiratory reserve volume* (IRV) is the amount that can be forcibly inspired over and above a normal inspiration. It is measured by having the individual exhale normally after a forced inspiration. The normal IRV is about 3.3 liters. No matter how forcefully an individual exhales, he can-

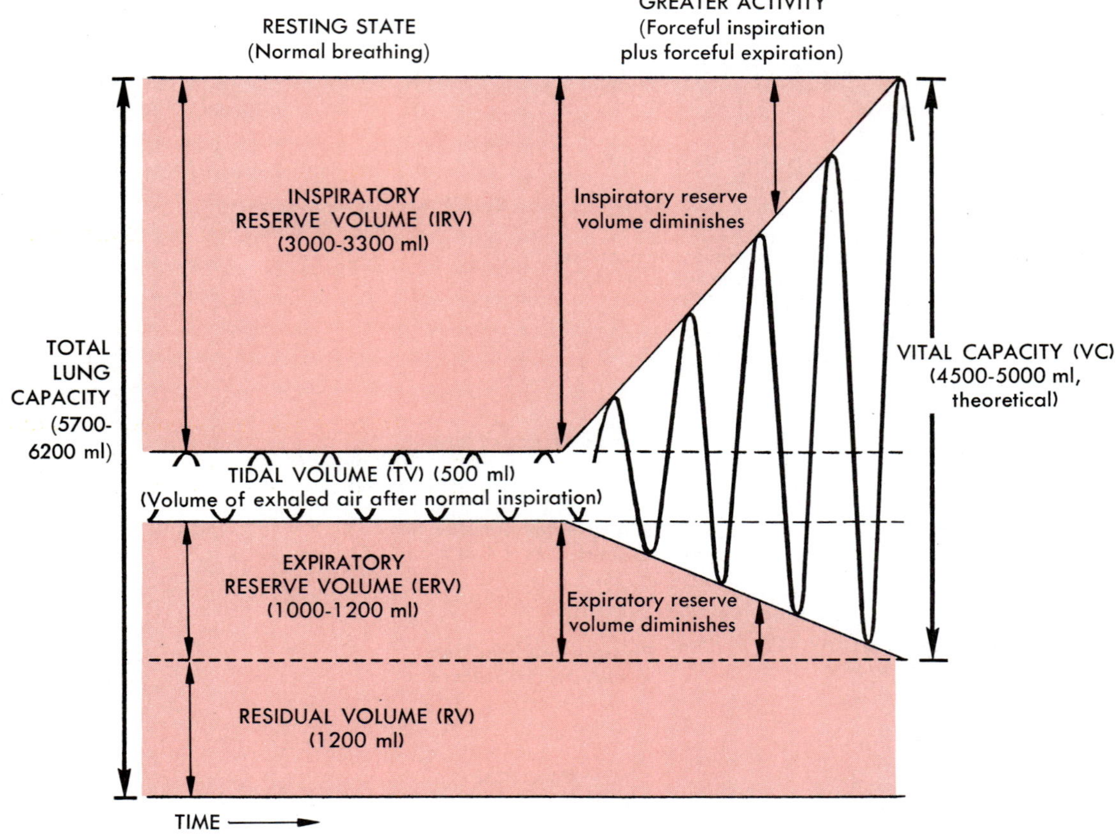

Fig. 12-11

During normal, quiet respirations, the atmosphere and lungs exchange about 500 ml of air (TV). With a forcible inspiration, about 3,300 ml more air can be inhaled (IRV). After a normal inspiration and normal expiration, approximately 1,000 ml more air can be forcibly expired (ERV). Vital capacity is the amount of air that can be forcibly expired after a maximum inspiration and indicates, therefore, the largest amount of air that can enter and leave the lungs during respiration. Residual volume is the air that remains trapped in the alveoli.

not squeeze all the air out of his lungs. Some of it remains trapped in the alveoli. This amount of air that cannot be forcibly expired is known as *residual volume* (RV) and amounts to about 1.2 liters.

Notice in Fig. 12-11 that vital capacity is the sum of IRV + TV + ERV. It represents the largest volume of air an individual can move in and out of his lungs. It is determined by measuring the largest possible expiration after the largest possible inspiration. How large a vital capacity one has depends upon many factors—the size of one's thoracic cavity, one's posture, and various other factors. In general, a larger person has a larger vital capacity than a smaller one. An individual has a larger vital capacity when he is standing erect than when he is stooped over or lying down. The volume of blood in the lungs also affects the vital capacity. If the lungs contain more blood than normal, alveolar air space is encroached upon and vital capacity accordingly decreases. This becomes a very important factor in congestive heart disease. Excess fluid in the pleural or abdominal cavities also decreases vital capacity. So too does the disease emphysema. In the latter condition, alveolar walls become stretched—that is, lose their elasticity—and are unable to recoil normally for expiration. This leads to an increased residual volume. In severe emphysema, the residual volume may increase so much that the chest occupies the inspiratory position even at rest. Excessive muscular effort is therefore necessary for inspiration, and because of the loss of elasticity of lung tissue, greater effort is required, too, for expiration.

In diagnosing lung disorders, a physician may need to know the inspiratory capacity and the functional residual capacity of the patient's lungs. *Inspiratory capacity* (IC) is the maximum amount of air an individual can inspire after a normal expiration. From Fig. 12-11, you can deduce that IC = TV + IRV. With the volumes given in the figure, how many milliliters is the inspiratory capacity? *Functional residual capacity* is the amount of air left in the lungs at the end of a normal expiration. Therefore, as Fig. 12-11 indicates, FRC = ERV + RV. With the volumes given, the functional residual capacity is 2,200 to 2,400 ml. The total volume of air a lung can hold is called the *total lung capacity.* It is, as Fig. 12-11 indicates, the sum of all four lung volumes.

The term *alveolar ventilation* means the volume of inspired air that actually reaches, "ventilates," the alveoli. Only this volume of air takes part in the exchange of gases between air and blood. (Alveolar air exchanges some of its oxygen for some of blood's carbon dioxide.) With every breath we take, part of the entering air necessarily fills our air passageways—nose, pharynx, larynx, trachea, and bronchi. This portion of air does not descend into any alveoli, so it cannot take part in gas exchange. In this sense, it is "dead air." And appropriately, the larger air passageways it occupies are said to constitute the *anatomic dead space.* One rule of thumb estimates the volume of air in the anatomic dead space as the same number of milliliters as the individual's weight in pounds. Another generalization says that the anatomic dead space approximates 30% of the tidal volume. Tidal volume minus dead space volume equals alveolar ventilation volume. Suppose you have a normal tidal volume of 500 ml and that 30% of this, or 150 ml, fills the anatomic dead space. The amount of air reaching your alveoli—your alveolar ventilation volume—is then 350 ml per breath, or 70% of your tidal volume. Emphysema and certain other abnormal conditions, in effect, increase the amount of dead space air. Consequently, alveolar ventilation decreases and this, in turn, decreases the amount of oxygen that can enter blood and the amount of carbon dioxide that can leave it. Inadequate air–blood gas exchange, therefore, is the inevitable result of inadequate alveolar ventilation. Stated differently, the alveoli must be adequately ventilated in order for an adequate gas exchange to take place in the lungs.

Types of breathing

1. *Eupnea*—normal quiet breathing
2. *Hyperpnea*—increased breathing, usually increased tidal volume with or without an increased rate of breathing
3. *Apnea*—cessation of breathing at the end of a normal expiration
4. *Apneusis*—cessation of breathing in the inspiratory position
5. *Cheyne-Stokes respirations*—gradually increasing tidal volume for several

breaths, followed by several breaths with gradually decreasing tidal volume; cycle repeats itself

6 *Biot's respirations*—repeated sequences of deep gasps and apnea

Some principles about gases

Before discussing respirations further, we need to understand the following principles.

1 *Dalton's law* (or the law of partial pressures). The term *partial pressure* means the pressure exerted by any one gas in a mixture of gases or in a liquid. The partial pressure of a gas in a mixture of gases is directly related to the concentration of that gas in the mixture and to the total pressure of the mixture. Suppose we apply this principle to compute the partial pressure of oxygen in the atmosphere. The concentration of oxygen in the atmosphere is 20.96% and the total pressure of the atmosphere is 760 mm Hg under standard conditions. Therefore:

$$\text{Atmospheric } P_{O_2} = 20.96\% \times 760 = 159.2 \text{ mm Hg}$$

The symbol used to designate partial pressure is the capital letter P preceding the chemical symbol for the gas. Examples: alveolar air P_{O_2} is about 100 mm Hg; arterial blood P_{O_2} is also about 100 mm Hg; venous blood P_{O_2} is about 37 mm Hg. The word *tension* is often used as a synonym for the term partial pressure—oxygen tension means the same thing as P_{O_2}.

2 The partial pressure of a gas in a liquid is directly determined by the amount of that gas dissolved in the liquid, which in turn is determined by the partial pressure of the gas in the environment of the liquid. Gas molecules diffuse into a liquid from its environment and dissolve in the liquid until the partial pressure of the gas in solution becomes equal to its partial pressure in the environment of the liquid. Alveolar air constitutes the environment of blood moving through pulmonary capillaries. Standing between the blood and the air are only the very thin alveolar and capillary membranes, and both of these are highly permeable to oxygen and carbon dioxide. By the time blood leaves the pulmonary capillaries as arterial blood, diffusion and approximate equilibration of oxygen and carbon dioxide across the membranes has occurred. Arterial blood P_{O_2} and P_{CO_2}, therefore, usually equal or very nearly equal alveolar P_{O_2} and P_{CO_2} (Table 12-1).

Exchange of gases in lungs

The exchange of gases in the lungs takes place between alveolar air and venous blood flowing through lung capillaries. Gases move in both directions through the alveolar-capillary membrane (Fig. 12-8). Oxygen enters blood from the alveolar air because the P_{O_2} of alveolar air is greater than the P_{O_2} of venous blood. Another way of saying this is that oxygen diffuses "down" its pressure gradient. Simultaneously, carbon dioxide molecules exit from the blood by diffusing down the carbon dioxide pressure gradient out into the alveolar air. The P_{CO_2} of venous blood is much higher than the P_{CO_2} of alveolar air. This two-way exchange of gases between alveolar air and venous blood converts venous blood to arterial blood (examine Fig. 12-12).

The amount of oxygen that diffuses into blood each minute depends upon several factors, notably upon these four:

1 The oxygen pressure gradient between alveolar air and venous blood (alveolar P_{O_2}—venous blood P_{O_2})
2 The total functional surface area of the alveolar-capillary membrane
3 The respiratory minute volume (respiratory rate per minute times volume of air inspired per respiration)
4 Alveolar ventilation (discussed on p. 305)

All four of these factors bear a direct relation to oxygen diffusion. Anything that decreases alveolar P_{O_2}, for instance, tends to decrease the alveolar-venous oxygen pressure

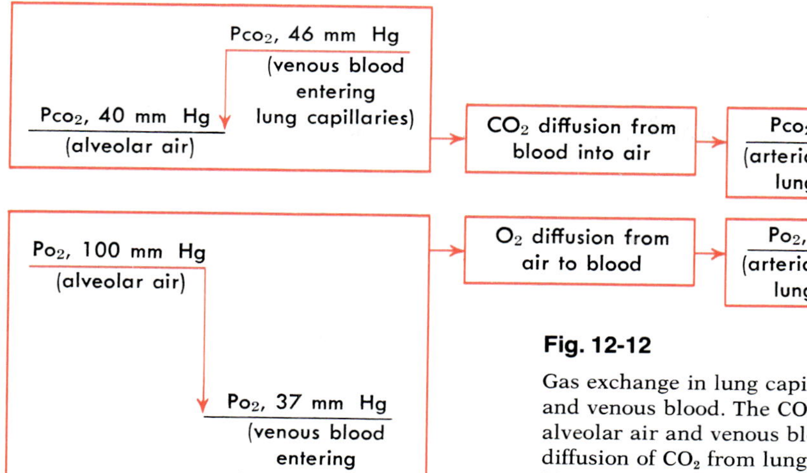

Fig. 12-12

Gas exchange in lung capillaries between alveolar air and venous blood. The CO_2 pressure gradient between alveolar air and venous blood causes outward diffusion of CO_2 from lung capillary blood; this lowers blood Pco_2 from its venous level to its arterial level. The O_2 pressure gradient between alveolar air and venous blood causes inward diffusion of O_2 into lung capillary blood; this raises blood Po_2 from its venous level up to its arterial level.

Table 12-1

Oxygen and carbon dioxide pressure gradients

	Atmosphere	Alveolar air	Arterial blood	Venous blood
Po_2	160*	100	100	37
Pco_2	0.3	40	40	46

*All figures indicate approximate mm Hg pressure under usual conditions.

gradient and therefore tends to decrease the amount of oxygen entering the blood. Application: alveolar air Po_2 decreases as altitude increases, and therefore less oxygen enters the blood at high altitudes. At a certain high altitude, alveolar air Po_2 equals venous blood Po_2. How would this affect oxygen diffusion into blood?

Anything that decreases the total functional surface area of the alveolar-capillary membrane also tends to decrease oxygen diffusion into the blood (by functional surface area is meant that which is freely permeable to oxygen). Application: in emphysema, the total functional area decreases and is one of the factors responsible for poor blood oxygenation in the condition.

Anything that decreases the respiratory minute volume also tends to decrease blood oxygenation. Application: morphine slows respirations and therefore decreases the respiratory minute volume (volume of air inspired per minute) and tends to lessen the amount of oxygen entering the blood. The main factors influencing blood oxygenation are shown in Fig. 12-13.

Several times we have stated the principle that structure determines functions. You may find it interesting to note the application of this principle to gas exchange in the lungs. Several structural facts facilitate oxygen diffusion from the alveolar air into the blood in lung capillaries:

1 The fact that the walls of the alveoli and of the capillaries together form a very thin barrier for the gases to cross (estimated at not more than 0.004 mm thick)
2 The fact that both alveolar and capillary surfaces are extremely large
3 The fact that the lung capillaries accommodate a large amount of blood at one time (about 900 ml)

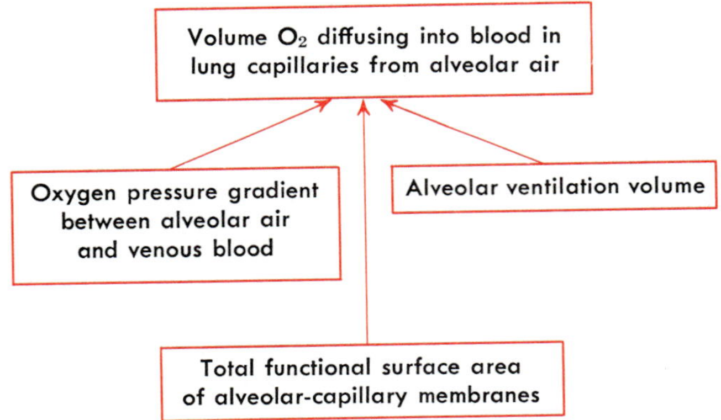

Fig. 12-13

Some major factors determining volume of oxygen entering lung capillary blood. An increase in any of the factors tends to increase oxygenation of blood. A decrease in any of them tends to decrease blood oxygenation. Alveolar ventilation volume means the volume of air that reaches the alveoli; it equals the volume of air inspired minus the volume of the anatomic dead space.

4 The fact that the blood is distributed through the capillaries in a layer so thin (equal only to the diameter of one red corpuscle) that each corpuscle comes close to alveolar air

How blood transports gases

Blood transports oxygen and carbon dioxide as solutes and as parts of molecules of certain chemical compounds. Immediately upon entering the blood, both oxygen and carbon dioxide dissolve in the plasma. But, because fluids can hold only small amounts of gas in solution, most of the oxygen and carbon dioxide rapidly form a chemical union with some other blood constituent. In this way, comparatively large volumes of the gases can be transported. With a P_{O_2} of 100 mm Hg, only 0.3 ml of oxygen dissolves in 100 ml of arterial blood. Many times that amount, however, combines with the hemoglobin in 100 ml of blood to form oxyhemoglobin. Since each gram of hemoglobin can unite with 1.34 ml of oxygen, the exact amount of oxygen in blood depends mainly upon the amount of hemoglobin present. Normally, 100 ml of blood contains about 15 gm of hemoglobin. If 100% of it combines with oxygen, then 100 ml of blood will contain 15 × 1.34, or 20.1 ml oxygen in the form of oxyhemoglobin.

Perhaps a more common way of expressing blood oxygen content is in terms of volume percent. Normal arterial blood, with P_{O_2} 100 mm Hg, contains about 20 vol.% O_2 (meaning 20 ml of oxygen in each 100 ml of blood).

Blood that contains more hemoglobin can, of course, transport more oxygen and that which contains less hemoglobin can transport less oxygen. Hence hemoglobin deficiency anemia decreases oxygen transport and may produce marked cellular hypoxia (inadequate oxygen supply).

In order to combine with hemoglobin, oxygen must, of course, diffuse from plasma into the red cells where millions of hemoglobin molecules are located. Several factors influence the rate at which hemoglobin combines with oxygen in lung capillaries. For instance, as the following equation and Fig. 12-13 show, an increasing blood P_{O_2} and a decreasing P_{CO_2} both accelerate hemoglobin association with oxygen.

$$Hb + O_2 \xrightarrow[\text{(Decreasing } P_{CO_2})]{\text{(Increasing } P_{O_2})} HbO_2$$

Decreasing P_{O_2} and increasing P_{CO_2}, on the other hand, accelerate oxygen dissociation from oxyhemoglobin—that is, the reverse of the preceding equation. Oxygen associates with hemoglobin rapidly—so rapidly, in fact, that about 97% of the

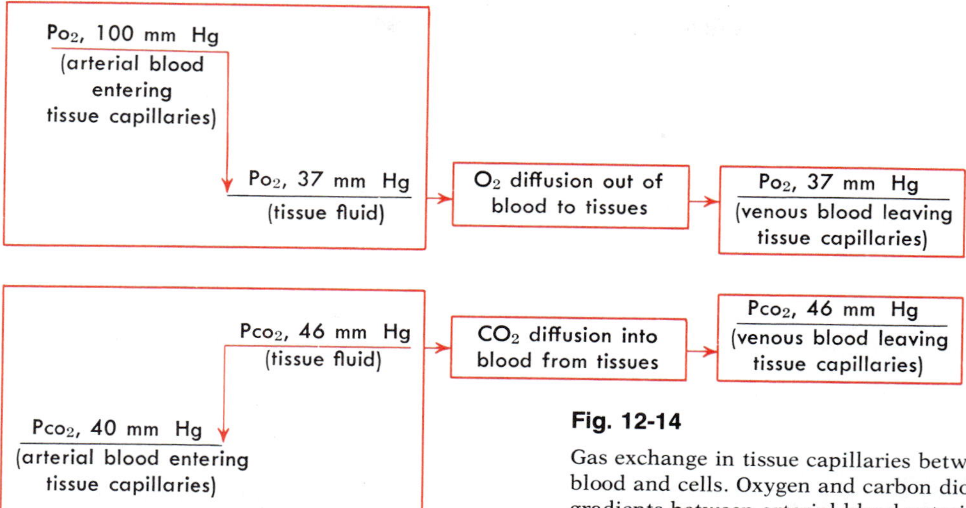

Fig. 12-14

Gas exchange in tissue capillaries between arterial blood and cells. Oxygen and carbon dioxide pressure gradients between arterial blood entering tissue capillaries and tissue fluid cause outward diffusion of oxygen from blood and inward diffusion of carbon dioxide into blood.

blood's hemoglobin has united with oxygen by the time the blood leaves the lung capillaries to return to the heart. In other words, the average *oxygen saturation* of arterial blood is about 97%.

Carbon dioxide is carried in the blood in several ways, the most important of which are described briefly as follows.

1 A small amount of carbon dioxide dissolves in plasma and is transported as a solute (dissolved carbon dioxide produces the Pco_2 of blood).

2 More than half of the carbon dioxide is carried in the plasma as bicarbonate ions (Fig. 12-17).

3 Somewhat less than one third of blood carbon dioxide unites with the NH_2 group of hemoglobin and certain other proteins to form carbamino compounds. Most of these are formed and transported in the red cells, since hemoglobin is the main protein to combine with carbon dioxide. The compound formed has a tongue-twisting name—carbaminohemoglobin. Carbon dioxide association with hemoglobin is accelerated by an increasing Pco_2 and a decreasing Po_2 and is slowed by the opposite conditions. How do the conditions that accelerate carbon dioxide association affect the rate of oxygen association with hemoglobin?

Exchange of gases in tissues

The exchange of gases in tissues takes place between arterial blood flowing through tissue capillaries and cells (Fig. 12-17). It occurs because of the principle already noted that gases move down a gas pressure gradient. More specifically, in the tissue capillaries oxygen diffuses out of arterial blood because the oxygen pressure gradient favors its outward diffusion (see Fig. 12-14). Arterial blood Po_2 is about 100 mm Hg, interstitial fluid Po_2 is considerably lower, and intracellular fluid Po_2 is still lower. Although interstitial fluid and intracellular fluid Po_2 are not definitely established, they are thought to vary considerably—perhaps from around 60 mm Hg down to about 1 mm Hg. As activity increases in any structure, its cells necessarily utilize oxygen more rapidly. This decreases intracellular and interstitial Po_2, which in turn tends to increase the oxygen pressure gradi-

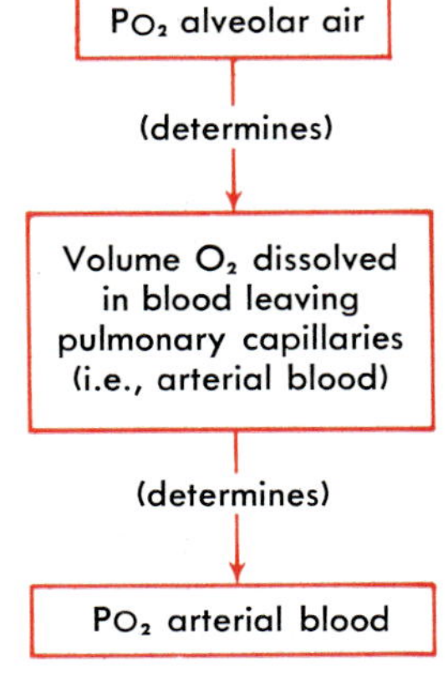

Fig. 12-15

Determinants of partial pressure of oxygen in arterial blood.

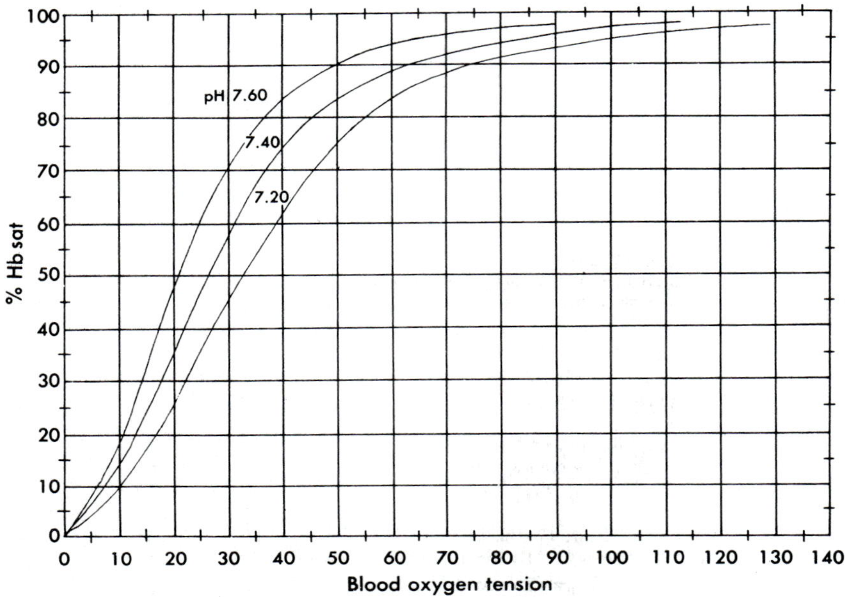

Fig. 12-16

Oxygen dissociation curve of blood at 37°C, showing variations at three pH levels. For a given oxygen tension, the higher the blood pH, the more the hemoglobin holds onto its oxygen, maintaining a higher saturation. (From Egan, D. F.: Fundamentals of respiratory therapy, ed. 2, St. Louis, 1973, The C. V. Mosby Co.)

Table 12-2

Blood oxygen

	Venous blood	Arterial blood
Po_2	37 mm Hg	100 mm Hg
Oxygen saturation	75%	97%
Oxygen content	15 ml O_2 per 100 ml blood	20 ml O_2 per 100 ml blood*

*Oxygen utilization by tissues = Difference between oxygen contents of arterial and venous blood (20-15) = 5 ml O_2 per 100 ml blood circulated per minute.

ent between blood and tissues and to accelerate oxygen diffusion out of the tissue capillaries. In this way, the rate of oxygen utilization by cells automatically tends to regulate the rate of oxygen delivery to cells. As dissolved oxygen diffuses out of arterial blood, blood Po_2 decreases, and this accelerates oxyhemoglobin dissociation to release more oxygen into the plasma for diffusion out to cells, as indicated in the following equation.

$$Hb + O_2 \xleftarrow[\text{(Increasing } Pco_2)]{\text{(Decreasing } Po_2)} HbO_2$$

Because of oxygen release to tissues from tissue capillary blood, Po_2 oxygen satura-

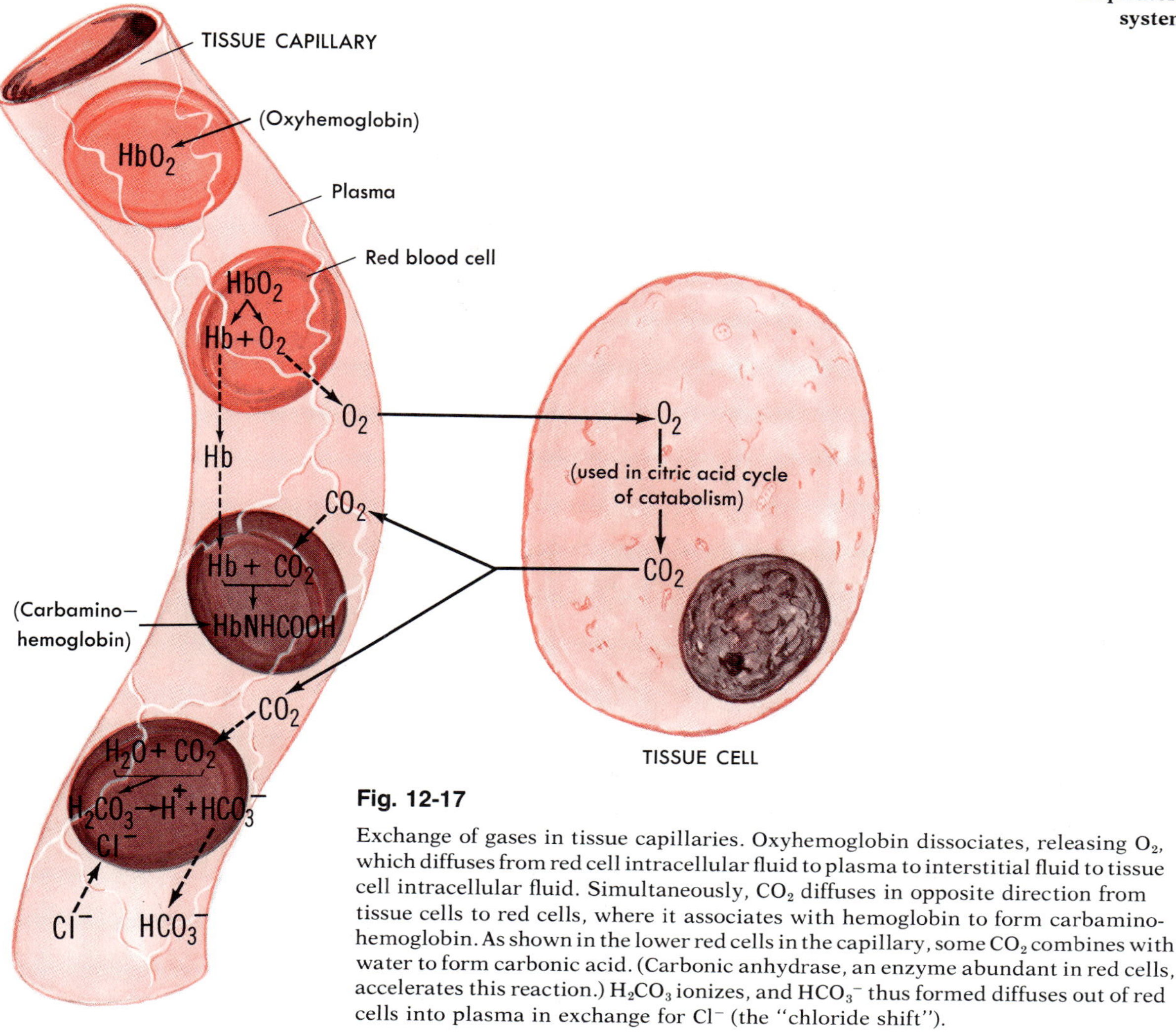

Fig. 12-17

Exchange of gases in tissue capillaries. Oxyhemoglobin dissociates, releasing O_2, which diffuses from red cell intracellular fluid to plasma to interstitial fluid to tissue cell intracellular fluid. Simultaneously, CO_2 diffuses in opposite direction from tissue cells to red cells, where it associates with hemoglobin to form carbaminohemoglobin. As shown in the lower red cells in the capillary, some CO_2 combines with water to form carbonic acid. (Carbonic anhydrase, an enzyme abundant in red cells, accelerates this reaction.) H_2CO_3 ionizes, and HCO_3^- thus formed diffuses out of red cells into plasma in exchange for Cl^- (the "chloride shift").

tion, and total oxygen content are all less in venous blood than in arterial blood, as shown in Table 12-2.

Carbon dioxide exchange between tissues and blood takes place in the opposite direction from oxygen exchange. Catabolism produces large amounts of carbon dioxide inside cells. Therefore, intracellular and interstitial P_{CO_2} are higher than arterial blood P_{CO_2}. This means that the carbon dioxide pressure gradient causes diffusion of carbon dioxide from the tissues into the blood flowing along through tissue capillaries (Fig. 12-14). Consequently, the P_{CO_2} of blood increases in tissue capillaries from its arterial level of about 40 mm Hg to its venous level of about 46 mm Hg. This increasing P_{CO_2} and decreasing P_{O_2} together produce two effects—they favor both oxygen dissociation from oxyhemoglobin and carbon dioxide association with hemoglobin to form carbaminohemoglobin (interpret the graphs in Fig. 12-16).

Regulation of respirations

The mechanism for the control of respirations has many parts (Fig. 12-18). A brief description of its main features follows.

1 The P_{CO_2}, P_{O_2}, and pH of *arterial blood* all

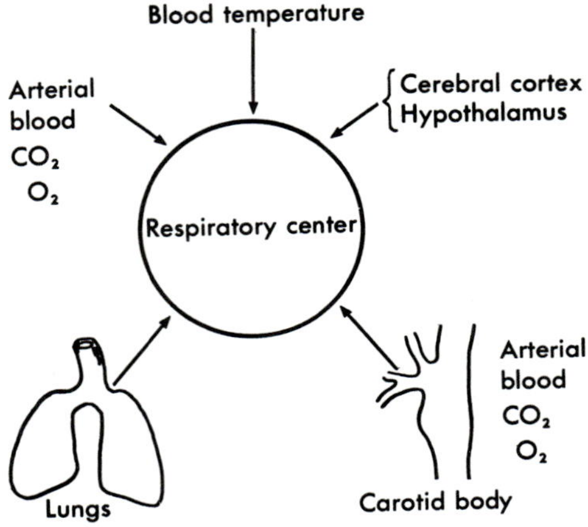

Fig. 12-18

Respiratory control mechanism. Scheme to show the main factors that influence the respiratory center and thereby control respirations. (See text for discussion.)

influence respirations. The P_{CO_2} acts on chemoreceptors located in the medulla. Chemoreceptors in this case are cells sensitive to changes in arterial blood's carbon dioxide and hydrogen ion concentrations. The normal range for arterial P_{CO_2} is about 38 to 40 mm Hg. When it increases even slightly above this, it has a stimulating effect mainly on central chemoreceptors (postulated to be present on the medulla). Large, but tolerable increases in arterial P_{CO_2} stimulate peripheral chemoreceptors present in the carotid bodies and aorta. Stimulation of chemoreceptors by increased arterial P_{CO_2} results in faster breathing, with a greater volume of air moving in and out of the lungs per minute. Decreased arterial P_{CO_2} produces opposite effects—it inhibits central and peripheral chemoreceptors, which leads to inhibition of medullary respiratory centers and slower respirations. In fact, breathing stops entirely for a few moments (apnea) when arterial P_{CO_2} drops moderately—to about 35 mm Hg, for example.

A decrease in arterial blood pH (increase in acid), within certain limits, has a stimulating effect on chemoreceptors located in the carotid and aortic bodies.

The role of *arterial blood* P_{O_2} in controlling respirations is not entirely clear. Presumably, it has little influence as long as it stays above a certain level. But neurons of the respiratory centers, like all body cells, require adequate amounts of oxygen in order to function optimally. Consequently, if they become hypoxic, they become depressed and send fewer impulses to respiratory muscles. Respirations then decrease or fail entirely. This principle has important clinical significance. For example, the respiratory centers cannot respond to stimulation by an increasing blood CO_2 if, at the same time, blood P_{O_2} falls below a critical level—a fact that may become of life and death importance during anesthesia.

However, a decrease in arterial blood P_{O_2} below 70 mm Hg but not so low as the critical level stimulates chemoreceptors in the carotid and aortic bodies and causes reflex stimulation of the inspiratory center. This constitutes an emergency respiratory control mechanism. It does not help regulate respirations under usual conditions when arterial blood P_{O_2} remains considerably higher than 70 mm Hg—the level necessary to stimulate the chemoreceptors.

2 *Arterial blood pressure* helps control respirations through the respiratory pressoreflex mechanism. A sudden rise in arterial pressure, by acting on aortic and carotid baroreceptors, results in reflex slowing of respirations. A sudden drop in arterial pressure brings about a reflex increase in rate and depth of respirations. The pressoreflex mechanism is probably not of great importance in the control of respirations. It is, however, of major importance in the control of circulation.

3 The *Hering-Breuer reflexes* help control respirations, particularly their depth and rhythmicity. They are believed to regulate the normal depth of respirations (extent of lung expansion), and therefore the volume of tidal air, in the following way. Presumably, when the tidal volume of air has been inspired, the lungs are expanded enough to stimulate baroreceptors located within them. The baroreceptors then send inhibitory impulses to the inspiratory center, relaxation of inspiratory muscles occurs, and expiration follows—the Hering-Breuer expiratory reflex. Then, when the tidal volume of air has been expired, the lungs are sufficiently deflated to inhibit the lung baroreceptors and allow inspiration to start again—the Hering-Breuer inspiratory reflex.

4 The *pneumotaxic center* in the upper part of the pons is postulated to function mainly to maintain rhythmicity of respirations. Whenever the inspiratory center is stimulated, it sends impulses to the pneumotaxic center as well as to inspiratory muscles. The pneumotaxic center, after a moment's delay, stimulates the expiratory center, which then feeds back inhibitory impulses to the inspiratory center. Inspiration, therefore, ends and expiration starts. Lung deflation soon initiates the Hering-Breuer inspiratory reflex, and inspiration starts again. In short, the pneumotaxic center and Hering-Breuer reflexes together constitute an automatic device for producing rhythmic respirations.

5 The *cerebral cortex* helps control respirations. Impulses to the respiratory center from the motor area of the cerebrum may either increase or decrease the rate and strength of respirations. In other words, an individual may voluntarily speed up or slow down his breathing rate. This voluntary control of respirations, however, has certain limitations. For example, one may will to stop breathing and do so for a few minutes. But holding the breath results in an increase in the carbon dioxide content of the blood since it is not being removed by respirations. Carbon dioxide is a powerful respiratory stimulant. So when arterial blood P_{CO_2} increases to a certain level, it stimulates the inspiratory center both directly and reflexly to send motor impulses to the respiratory muscles, and breathing is resumed even though the individual may still will contrarily. This knowledge that the carbon dioxide content of the blood is a more powerful regulator of respirations than cerebral impulses is of practical value when dealing with a child who holds his breath to force the granting of his wishes. The best treatment is to ignore such behavior, knowing that respirations will start again as soon as the amount of carbon dioxide in arterial blood increases to a certain level.

6 Miscellaneous factors also influence respirations. Among these are blood temperature and sensory impulses from skin thermal receptors and from superficial or deep pain receptors.

a *Sudden painful stimulation* produces a reflex apnea, but continued painful stimuli cause faster and deeper respirations.

b *Sudden cold stimuli* applied to the skin cause reflex apnea.

c *Afferent impulses* initiated by stretching the anal sphincter produce reflex acceleration and deepening of respirations. Use has sometimes been made of this mechanism as an emergency measure to stimulate respirations during surgery.

d *Stimulation of the pharynx or larynx* by irritating chemicals or by touch causes a temporary apnea. This is the choking reflex, a valuable protective device. It operates, for example, to prevent aspiration of food or liquids during swallowing.

Control of respirations during exercise. Respirations increase abruptly at the beginning of exercise and decrease even more markedly as it ends. This much is known. The mechanism that accomplishes this, however, is not known. It is not identical to the one that produces moderate increases in breathing. A number of studies have shown that arterial blood Pco_2, Po_2, and pH do not change enough during exercise to produce the degree of hyperpnea (faster, deeper respirations) observed. Presumably, many chemical and nervous factors and temperature changes operate as a complex, but still unknown, mechanism for regulating respirations during exercise.

Outline summary

Functions and organs

Exchange of gases between blood and air; nose, pharynx, larynx, trachea, bronchi, lungs

ORGANS

Nose

1 Structure
- **a** Portions—internal, in skull, above roof at mouth; external, protruding from face
- **b** Cavities
  - 1 Divisions—right and left
  - 2 Meati—superior, middle, and lower; named for turbinate located above each meatus
  - 3 Openings—to exterior, anterior nares; to nasopharynx, posterior nares
  - 4 Turbinates (conchae)—superior and middle, processes of ethmoid bone; inferior turbinates separate bones; divide internal nasal cavities into three passageways or meati
  - 5 Floor—formed by palatine bones that also act as roof of mouth
- **c** Lining—ciliated mucous membrane
- **d** Sinuses draining into nose (or paranasal sinuses)—frontal, maxillary (or antrum of Highmore), sphenoidal, and ethmoidal

2 Functions
- **a** Serves as passageway for incoming and outgoing air, filtering, warming, moistening, and chemically examining it
- **b** Organ of smell because olfactory receptors located in nasal mucosa
- **c** Aids in phonation

Pharnyx

1 Structure—made of muscle with mucous lining
- **a** Divisions—nasopharynx, behind nose; oropharynx, behind mouth; laryngopharynx, behind larynx
- **b** Openings—four in nasopharynx: two auditory tubes and two posterior nares; one in oropharynx: fauces from mouth; and two in laryngopharynx: into esophagus and into larynx
- **c** Organs in pharynx—adenoids or pharyngeal tonsils in nasopharynx; palatine and lingual tonsils in oropharynx

2 Functions—serves both respiratory and digestive tracts as passageway for air, food, and liquids; aids in phonation

Larynx

1 Location—at upper end of trachea, just below pharynx

2 Structure
- **a** Cartilages—nine pieces arranged in boxlike formation; thyroid largest, known as "Adam's apple"; epiglottis, "lid" cartilage; cricoid, "signet ring" cartilage
- **b** Vocal cords—false cords, folds of mucous lining; true cords, fibroelastic bands stretched across hollow interior of larynx; glottis, opening between true vocal cords
- **c** Lining—ciliated mucous membrane
- **d** Sexual differences—male larynx larger, covered with less fat, and therefore more prominent than female larynx

3 Function—expired air causes true vocal cords to vibrate, producing voice; pitch determined by length and tension of cords

Trachea

1 Structure
- **a** Walls—smooth muscle; contain C-shaped rings of cartilage at intervals, which keeps the tube open at all times; lining—ciliated mucous membrane
- **b** Extent—from larynx to bronchi; about 4½ inches long

2 Function—furnishes open passageway for air going to and from lungs

Bronchi

1 Structure—formed by division of trachea into two tubes; right bronchus slightly larger and more vertical than left; same structure as trachea; each primary bronchus branches as soon as enters lung into secondary bronchi, which branch into bronchioles, which branch into microscopic alveolar ducts, which terminate in cluster of blind sacs called alveoli; trachea and two primary bronchi and all their branches compose "bronchial tree"; alveolar walls composed of single layer of cells

2 Function—bronchi and their many branching tubes furnish passageway for air going to and from lungs; alveoli provide large, thin-walled surface area where blood and air can exchange gases

Lungs

1 Structure
- **a** Size, shape, location—large enough to fill pleural divisions of thoracic cavity; cone-shaped; extend from base, on diaphragm, to apex, located slightly above clavicle
- **b** Divisions—three lobes in right lung, two in left; root of lung consists of primary bronchus and pulmonary artery and veins, bound together by connective tissue; hilum is vertical slit on mesial surface of lung through which root structures enter lung; base is broad, inferior surface of lung; apex is pointed upper margin
- **c** Covering—visceral layer of pleura

2 Function—furnish place where large amounts of air and blood can come in close enough contact for rapid exchange of gases to occur

THORAX (CHEST)

1 Structure
  a Three divisions
    1 Right and left pleural portion—contain lungs
    2 Mediastinum—area between two lungs; contains heart, esophagus, trachea, great blood vessels, etc.
  b Lining
    1 Parietal layer of pleura lines entire chest cavity and covers superior surface of diaphragm
    2 Forms separate sac encasing each lung
    3 Separated from visceral pleura, covering lungs, only by potential space—pleural space—which contains few drops of pleural fluid
  c Shape of ribs and angle of their attachment to spine such that elevation of rib cage enlarges two dimensions of thorax, its width and depth from front
2 Function
  a Increase in size of thorax leads to inspiration
  b Decrease in size of thorax leads to expiration

PHYSIOLOGY

1 Pulmonary ventilation
  a Mechanism of pulmonary ventilation
    1 Contraction of diaphragm and chest elevating muscles enlarges thorax, thereby decreases intrathoracic pressure, which causes expansion of lungs, which decreases intrapulmonic pressure to subatmospheric level, which establishes gas pressure gradient, which causes air to move into lungs
    2 Relaxation of inspiratory muscles produces opposite effects; see Fig. 12-10
  b Volumes of air exchanged in pulmonary ventilation—directly related to gas pressure gradient between atmosphere and lung alveoli and inversely related to resistance opposing air flow
    1 Tidal volume (TV)—average amount expired after normal inspiration; approximately 500 ml or 1 pint
    2 Inspiratory reserve volume (IRV)—amount that can be forcibly inspired after normal inspiration; measured by having individual expire normally after forced inspiration
    3 Expiratory reserve volume (ERV)—additional amount of air that can be forcibly expired after a normal inspiration and expiration
    4 Residual volume (RV)—amount of air that cannot be forcibly expired from lungs
    5 Vital capacity (VC)—largest volume of air an individual can move in and out of his lungs; equals sum of inspiratory reserve volume, tidal volume, and expiratory reserve volume
    6 Anatomic dead space—volume of air that fills nose, pharynx, larynx, trachea, bronchi, and smaller tubes but does not descend into alveoli, therefore takes no part in gas exchange; typically, about 30% of tidal volume, or 150 ml, but varies under different conditions
    7 Alveolar ventilation—volume of inspired air that actually reaches alveoli; computed by subtracting dead space air volume from tidal air volume
  c Types of breathing
    1 Eupnea—normal quiet breathing
    2 Hyperpnea—abnormally increased breathing
    3 Apnea—cessation of breathing at end of normal expiration
    4 Apneusis—cessation of breathing in inspiratory position
    5 Cheyne-Stokes respirations—gradually increasing tidal volume for several breaths, followed by several breaths with gradually decreasing tidal volume; cycle repeats itself
    6 Biot's respirations—repeated sequences of deep gasps and apnea
2 Some principles about gases
  a Dalton's law—partial pressure of gas in mixture of gases directly related to concentration of that gas in mixture and to total pressure of mixture
  b Partial pressure of gas in liquid directly related to amount of gas dissolved in liquid; becomes equal to partial pressure of that gas in environment of liquid
3 Exchange of gases in lungs (between alveolar air and venous blood)
  a Where it occurs—in lung capillaries; across alveolar-capillary membrane
  b What exchange consists of—oxygen diffuses out of alveolar air into venous blood; carbon dioxide diffuses in opposite direction
  c Why it occurs—oxygen pressure gradient causes inward diffusion of oxygen; carbon dioxide pressure gradient causes outward diffusion of carbon dioxide

	Alveolar air	Venous blood
Po_2	100 mm Hg	37 mm Hg
Pco_2	40 mm Hg	46 mm Hg

  d Results of gas exchange
    1 Po_2 of blood increases to arterial blood level as blood moves through lung capillaries
    2 Pco_2 of blood decreases to arterial blood level as blood moves through lung capillaries
    3 Oxygen association with hemoglobin to form oxyhemoglobin and carbon dioxide dissociation from carbaminohemoglobin both accelerated by increasing Po_2 and decreasing Pco_2
4 How blood transports gases
  a Oxygen
    1 About 0.5 ml transported as *solute*—i.e., dissolved in 100 ml blood
    2 About 19.5 ml O_2 per 100 ml blood transported as *oxyhemoglobin* in red blood cells
    3 About 20 ml = *Total O_2 content* per 100 ml blood (100% saturation of 15 gm hemoglobin)
  b Carbon dioxide
    1 Small amount dissolves in plasma and transported as true *solute*
    2 More than half of CO_2 transported as *bicarbonate ion* in plasma
    3 Somewhat less than one third of CO_2 transported in red blood cells as *carbaminohemoglobin*

5 Exchange of gases in tissues (between arterial blood and cells)
- a Where it occurs—tissue capillaries
- b What exchange consists of—oxygen diffuses out of arterial blood into interstitial fluid and on into cells, whereas carbon dioxide diffuses in opposite direction
- c Why it occurs—O_2 pressure gradient causes outward diffusion of O_2; CO_2 pressure gradient causes inward diffusion of CO_2

	Arterial blood	Interstitial fluid
Po_2	100 mm Hg	60(?) mm Hg down to 1(?) mm Hg
Pco_2	46 mm Hg	50(?) mm Hg

- d Results of oxygen diffusion out of blood and carbon dioxide diffusion into blood
  1 Po_2 blood decreases as blood moves through tissue capillaries; arterial Po_2, 100 mm Hg, becomes venous Po_2, 40 mm Hg (figures vary)
  2 Pco_2 blood increases; arterial Pco_2, 40 mm Hg, becomes venous Pco_2, 46 mm Hg (figures vary)
  3 Oxygen dissociation from hemoglobin and carbon dioxide association with hemoglobin to form carbaminohemoglobin both accelerated by decreasing Po_2 and increasing Pco_2

6 Control of respirations (see Fig. 12-18)
- a Respiratory centers—inspiratory and expiratory centers in medulla; pneumotaxic center in pons
- b Control of respiratory centers
  1 Carbon dioxide major regulator of respirations; increased blood carbon dioxide content, up to certain level, stimulates respiration and above this level depresses respirations; decreased blood carbon dioxide decreases respirations
  2 Oxygen content of blood influences respiratory center—decreased blood O_2, down to a certain level, stimulates respirations and below this critical level depresses them; O_2 control of respirations nonoperative under usual conditions
  3 Hering-Breuer mechanism helps control rhythmicity of respirations; increased alveolar pressure inhibits inspiration and starts expiration; decreased alveolar pressure stimulates inspiration and ends expiration
  4 Pneumotaxic center acts with Hering-Breuer reflexes to produce rhythmic respirations
  5 Cerebral cortex impulses to respiratory centers provide voluntary control, within limits, of rate and depth of respirations
- c Control of respirations during exercise—not yet established

Review questions

1 What anatomical feature favors the spread of the common cold through the respiratory passages and into the middle ear and mastoid sinus?
2 How are the turbinates arranged in the nose? What are they?
3 What organs are found in the nasopharynx?
4 What tubes open into the nasopharynx?
5 Make a diagram showing the termination of a bronchiole in an alveolar duct with alveoli.
6 What kind of membrane lines the respiratory system?
7 What is the serous covering of the lungs called? Where else, besides covering the lungs, is this same membrane found?
8 The pharynx is common to what two systems?
9 Are the lungs active or passive organs during breathing? Explain.
10 What is the main inspiratory muscle?
11 How is inspiration accomplished? Expiration?
12 If an opening is made into the pleural cavity from the exterior, what happens? Why?
13 What is the pleural space? What does it contain?
14 What substance found in blood is the natural chemical stimulant for the respiratory center?
15 Normally, about what percentage of the tidal volume fills the anatomic dead space?
16 Normally, about what percentage of the tidal volume is useful air, that is, ventilates the alveoli?
17 What is the voice box? Of what is it composed? What is the Adam's apple?
18 What is the epiglottis? What is its function?
19 What are the true vocal cords? Where are they? What name is given to the opening between the cords?
20 Name the three divisions of the thorax and their contents.
21 One gram of hemoglobin combines with how many milliliters of oxygen?
22 Suppose your blood has a hemoglobin content of 15 gm per 100 ml and an oxygen saturation of 97%. How many milliliters of oxygen would 100 ml of your arterial blood contain?
23 Compare the mechanisms that accelerate respiration with those that accelerate circulation during exercise.
24 Make a generalization about the effect of a moderate increase in the amount of blood CO_2 on circulation and respiration. What advantage can you see in this effect?
25 Make a generalization about the effect of a moderate decrease in blood O_2 on circulation and respiration.
26 Define the following terms briefly: alveolus, apnea, asphyxia, complemental air, cyanosis, dyspnea, minimal air, orthopnea, Pco_2, Po_2, pleurisy, residual air, respiration, spirometer, supplemental air, thorax, tidal air, vital capacity.

13

The cardiovascular system

Functions

Blood
Blood cells
  Erythrocytes
    Appearance and size
    Structure and functions
    Formation (erythropoiesis)
    Destruction
    Erythrocyte homeostatic mechanism
  Leukocytes
    Appearance and size
    Functions
    Formation
    Destruction and life span
    Numbers
  Platelets
    Appearance and size
    Functions
    Formation and life span
Blood types (or blood groups)
Blood plasma
Blood coagulation
  Purpose
  Mechanism
  Factors that oppose clotting
  Factors that hasten clotting
  Clot dissolution
  Clinical methods of hastening clotting

Heart
Covering
  Structure
  Function
Structure
  Wall
  Cavities
  Valves and openings
  Blood supply
  Conduction system
  Nerve supply
Physiology
  Function
  Cardiac cycle

Blood vessels
  Kinds
  Structure
  Functions
Main blood vessels
  Systemic circulation
    Arteries
    Veins
  Portal circulation
  Fetal circulation

Circulation
Definitions
How to trace
Functions of control mechanisms
Principles of circulation
Local control of arterioles
Important factors influencing venous return to heart

Blood pressure
Measurement of arterial blood pressure clinically
Relation to arterial and venous bleeding

Velocity of blood

Pulse
Definition
Cause
Pulse wave
Where pulse can be felt
Venous pulse

Lymphatic system
  Definition
Lymph and interstitial fluid (tissue fluid)
  Definition
Lymphatics
  Formation and distribution
  Structure
  Function
Lymph circulation
Lymph nodes
  Structure
  Location
  Functions
Thymus

Spleen
Location
Structure
Functions

Functions

The cardiovascular or circulatory system performs a vital pickup and delivery service for the body. It is the body's only transportation system. Blood picks up foods and oxygen from the digestive and respiratory systems and delivers them to cells. From cells, it picks up wastes (produced by their metabolism of foods) and delivers them to excretory organs. Blood picks up hormones from endocrine glands and delivers them to their target cells. Directly or indirectly, the circulatory system contributes to every function of every cell and every function of the body as a whole. Our main topics in this chapter are: blood, heart, blood vessels, physiology of circulation, lymphatic system, and spleen.

Blood

Blood is much more than the simple liquid it seems to be. Not only does it consist of a fluid but also of cells—billions and billions of them. The fluid portion of blood—that is, the *plasma*—is one of the three major body fluids. (Interstitial fluid and intracellular fluid are the other two.)

How much blood does an adult body contain? We can answer that question only with some generalizations. The total blood volume varies markedly in different individuals. One of the chief variables influencing normal blood volume is the amount of body fat. Blood volume per kilogram of body weight varies inversely with the amount of excess body fat. This means that the less fat there is in your body, the more blood you have per kilogram of your body weight. According to Wennesland and associates,* healthy men average 71.4 ml of blood per kilogram of body weight. Approximately 55% of total blood volume is plasma volume, and the rest is blood cell volume. Example: A man who weighs 70 kg (154 pounds) would have an average of about 5,000 ml of blood. Of this, approximately 2,750 ml would be plasma and 2,250 ml would be blood cells. Now we shall take a rather detailed look at blood cells and then consider blood plasma.

*Wennesland, R., and others: Red cell, plasma, and blood volume in healthy men measured by radiochromium (Cr^{51}) cell tagging and hematocrit, J. Clin. Invest. **38:**1065-1077, 1959.

Blood cells

Three main kinds of blood cells are recognized: red blood cells (erythrocytes), white blood cells (leukocytes), and platelets (thrombocytes). Leukocytes are further divided as shown in the following classifications of blood cells:

1. Red blood cells or erythrocytes
2. White blood cells or leukocytes
   - **a** Granular leukocytes: basophils, neutrophils, and eosinophils
   - **b** Nongranular leukocytes: lymphocytes and monocytes
3. Platelets (thrombocytes)

Erythrocytes

Appearance and size

Facts about normal red cell size and shape hold more than academic interest. They are also clinically important. For example, an increase in red cell size characterizes one type of anemia, and a decrease in red cell size characterizes another type. Red cells are extremely small. More than 3,000 of them could be placed side by side in 1 inch since they measure only about 7 μ in diameter. A normal mature red cell has no nucleus. Just before the cell reaches maturity and enters the bloodstream from the bone marrow, the nucleus is extruded, with the result that the cell caves in on both sides. So, as you can see in Fig. 13-1, normal mature red cells are shaped like tiny biconcave disks.

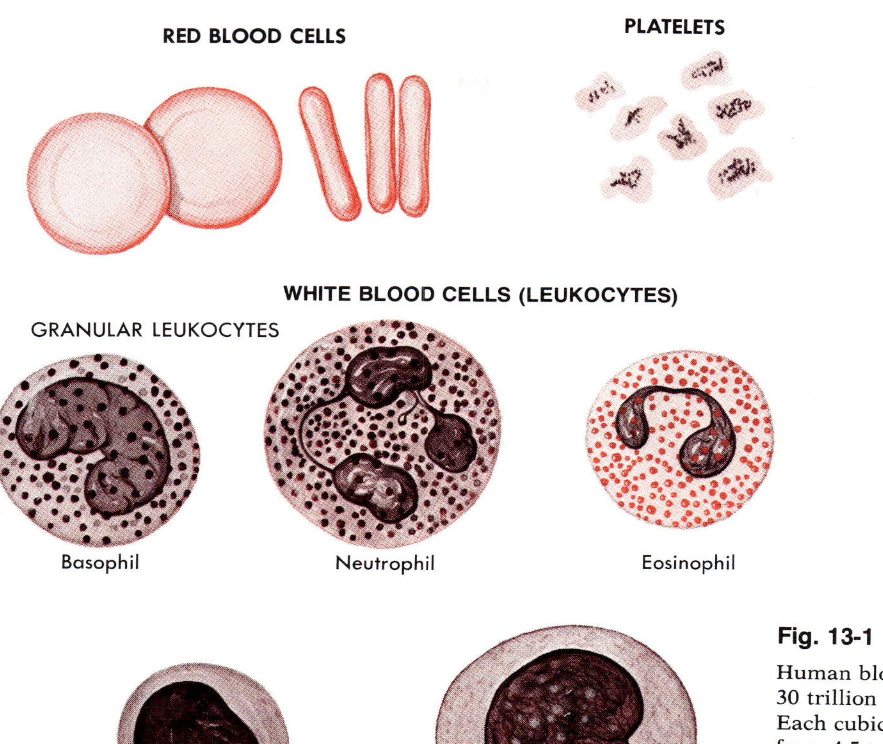

Fig. 13-1

Human blood cells. There are close to 30 trillion red blood cells in the adult. Each cubic millimeter of blood contains from 4.5 to 5.5 miilion red blood cells and an average total of 7,500 white blood cells.

Structure and functions

Red blood cell functions (specifically, the transport of oxygen and carbon dioxide) illustrate the familiar principle that structure determines function. Packed within one tiny red cell are an estimated 200 to 300 million molecules of the complex compound *hemoglobin*. One hemoglobin molecule consists of a protein molecule (globin) combined with four molecules of a pigmented compound (heme). Because each molecule of heme contains one atom of iron, one hemoglobin molecule contains four iron atoms. This structural fact enables one hemoglobin molecule to unite with four oxygen molecules to form oxyhemoglobin (a reversible reaction). Hemoglobin can also combine with carbon dioxide to form carbaminohemoglobin (also reversible). But in this reaction, the structure of the globin part of the hemoglobin molecule rather than of its heme part makes the combining possible.

A man's blood usually contains more hemoglobin than a woman's. In most normal men, 100 ml of blood contains 14 to 16 gm of hemoglobin—and the closer to 16 the more "joie de vivre," in Stewart Alsop's words.*

*Alsop, S.: Stay of execution, New York, 1973, J. B. Lippincott Co., p. 30.

The normal hemoglobin content of a woman's blood is a little less—specifically, in the range of 12 to 14 gm per 100 ml. Any adult who has a hemoglobin content of less than 12 (gm per 100 ml blood) is diagnosed as having *anemia*. An anemic person may or may not have an abnormally low red blood cell count. An adult whose blood contains more than about 17.5 gm hemoglobin per 100 ml is diagnosed as having *polycythemia*. His red blood cell count may or may not be higher than normal, although the word polycythemia means literally "many blood cells."

Formation (erythropoiesis)

Erythrocytes are formed in the red bone marrow from nucleated cells known as hemocytoblasts or stem cells. These stem cells go through several stages of development during which the nucleus becomes smaller and disappears. Newly formed red cells, at the time they leave the bone marrow and enter the blood, contain hemoglobin and a reticulum in their cytoplasm. For this reason they are called reticulocytes.

Frequently a physician needs information about the rate of erythropoiesis to help him make a diagnosis or prescribe treatment. A *reticulocyte count* gives this information. Approximately 1/2% to 1 1/2% of the red cells in normal blood are reticulocytes. A reticulocyte count of less than 1/2% of the red count usually indicates a slowdown in the process of red cell formation. Conversely, a reticulocyte count higher than 1 1/2% usually indicates an acceleration of red cell formation—as occurs, for example, following treatment of anemia.

Destruction

The life span of a red blood cell circulating in the bloodstream averages about 120 days. Reticuloendothelial cells in the lining of blood vessels, particularly in the liver, spleen, and bone marrow, phagocytose the red cells. In the process, iron is released from hemoglobin and the pigment bilirubin is formed. Both are transported to the liver, where iron is put in temporary storage and bilirubin is excreted in the bile. Eventually, the bone marrow uses most of the iron over again for new red cell synthesis, and the liver excretes the bile pigments in the bile.

Erythrocyte homeostatic mechanism

Red cells are formed and destroyed at a breathtaking rate. Normally, every hour of every day of our adult lives, over 100 million red cells are formed to replace an equal number destroyed during that brief time. Obviously, some kind of homeostatic mechanism operates to balance the number of cells formed against the number destroyed, since, in health, the number of red cells remains relatively constant at about 4.5 to 5.5 million per cubic millimeter of blood. The exact mechanism responsible for this constancy is not known. It is known, however, that the rate of red cell production soon speeds up if either the number of red cells decreases appreciably or if tissue hypoxia (oxygen deficiency) develops. Either of these conditions acts in some way to stimulate the kidneys (and perhaps some other structures) to increase their secretion of a hormone named *erythropoietin*. The resulting increased blood concentration of erythropoietin stimulates bone marrow to accelerate its production of red blood cells (Fig. 13-2). The name erythropoietin makes it easy to remember its function—erythropoietin stimulates erythropoiesis (process of red cell formation).

Note that for the red blood cell homeostatic mechanism to succeed in maintaining a normal number of red cells, the bone marrow must function adequately. To do this, the blood must supply it with adequate amounts of several substances with which to form the new red cells—vitamin B_{12}, iron, and amino acids, for example, and also copper and pos-

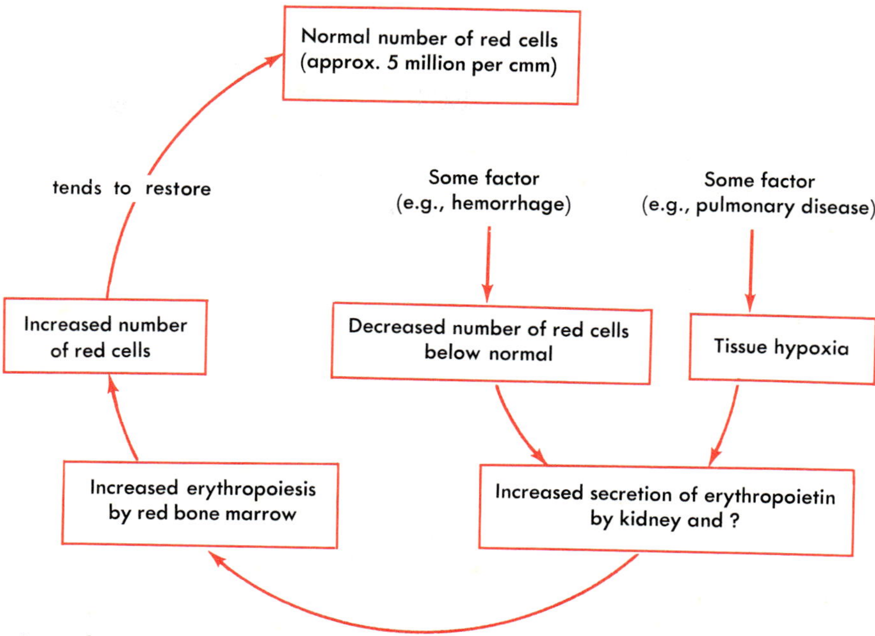

Fig. 13-2

Postulated red blood cell homeostatic mechanism—a negative feedback mechanism. A decrease in the number of circulating red blood cells "feeds back" to the red cell–forming structure (red bone marrow) to cause an increased rate of red cell formation, which in turn tends to increase the number of red cells sufficiently to restore their normal number.

sibly cobalt to serve as catalysts. In addition, the gastric mucosa must provide some unidentified intrinsic factor necessary for absorption of vitamin B_{12} (called extrinsic factor because of its external source in foods; also called antianemic principle).

Pernicious anemia develops when the gastric mucosa fails to produce the intrinsic factor needed for adequate vitamin B_{12} absorption. The marrow then produces fewer but larger red cells than normal. Many of these cells are immature with overly fragile membranes, a fact that leads to their more rapid destruction.

A clinical example of failure of red cell homeostasis seen in recent years is anemia caused by bone marrow injury by x-ray or gamma ray radiations.

Other factors may also cause marrow damage. To help diagnose this condition, a sample of marrow is removed—for example, from the sternum by means of a sternal puncture—and is studied microscopically for abnormalities. When damaged marrow can no longer keep red cell production apace with destruction, red cell homeostasis is not maintained. Instead, the red cell count falls below normal. Anemia (fewer red cells than normal) develops whenever the rates of red cell formation and destruction become unequal. Therefore, either a decrease in red cell formation (as in pernicious anemia or bone marrow injury by radiation) or an increase in red cell destruction (as in infections and malignancies) can lead to anemia.

The number of red blood cells is determined by the *"red count"* or is estimated by the hematocrit. The *hematocrit* is the volume percentage of red cells in whole blood. To be more specific, a hematocrit of 47 means that in every 100 ml of whole blood there are 47 ml of blood cells and 53 ml of fluid (plasma). Normally, the average hematocrit for a man is about 47 (±7, normal range) and for a woman about 42 (±5).

Leukocytes

Appearance and size

Consult Fig. 13-1. Note particularly the differences in color of the cytoplasmic granules and in the shapes of the nuclei of the granular leukocytes. Neutrophils take their name from the fact that they stain with neutral dyes. Because their nuclei have two to five or more lobes, neutrophils are also called *polymorphonuclear leukocytes* or, to avoid that tongue twister, simply "polys." Eosinophils stain with acid dyes. Their nuclei have two oval lobes. Basophils stain with basic dyes. Their nuclei are roughly S shaped.

Functions

White blood cells function as part of the body's defense against microorganisms. All leukocytes are motile cells. This characteristic enables them to move out of capillaries by squeezing through the intercellular spaces of the capillary wall (a process called *diapedesis*) and to migrate by ameboid movement toward microorganisms or other injurious particles that may have invaded the tissues. Neutrophils are highly motile, whereas lymphocytes, monocytes, and eosinophils are sluggishly so. Once in the tissue spaces, white cells phagocytose—ingest and digest microbes or other injurious particles. At any one time, most white blood cells are moving about in the tissues performing, or ready to perform, their function of phagocytosis. Red blood cells and platelets, in contrast, perform their functions within the blood vessels, not in the tissues.

You may recall that reticuloendothelial cells also perform the function of phagocytosis. In general, however, they do this work within more localized areas than white cells. Reticuloendothelial cells, since they do not enter the bloodstream, cannot be transported by it to any part of the body, as white cells can.

Lymphocytes play an important, though still not thoroughly understood, role in the development of immunity—the formation of antibodies. They respond to the presence of antigens (usually foreign proteins introduced into the body) in the following two ways. Some of them multiply by mitosis and undergo some changes. They differentiate, that is, become plasma cells. The plasma cells secrete antibodies into the blood that react, in the circulating blood, with the particular antigens that initiated the process that led to the formation of these particular antibodies. The reaction of an antibody with its antigen destroys that antigen or renders it harmless to the body—it "makes us immune," we say (to the particular antigens taking part in the antigen-antibody reaction).

As explained in the preceding paragraph, some lymphocytes differentiate into plasma cells that synthesize and secrete *circulating antibodies*—so-called for the obvious reason that they circulate in the bloodstream. (Another name for circulating antibodies is humoral antibodies.) Some lymphocytes, however, synthesize antibodies but do not secrete them into the blood. Instead, they remain in the cells that made them and, logically, are called *cellular antibodies.* The type of antigens that initiate cellular antibody formation are *cellular antigens*—that is, antigens present on or in cells. If, therefore, a tissue or an organ is transplanted from one person to another, its cellular antigens may initiate cellular-antibody formation by the recipient's lymphocytes. For example, lymphocytes may, and unfortunately often do, infiltrate the vicinity around a kidney transplant. There they proliferate and form cellular antibodies. These antibodies react with the kidney's cellular antigens and in the process destroy the foreign cells. In more picturesque language, "tissue-attacking lymphocytes reject" the transplanted kidney. *Tissue-attacking lymphocytes* and *graft-rejection cells* are descriptive names for the same cells—that is, for lymphocytes that synthesize cellular antibodies. Some surgeons have given antilymphocyte serum (ALS) to patients receiving an organ transplant. Rationale? To try to decrease the number of infiltrating lymphocytes and the amount of cellular antibodies formed and, thereby, prevent rejection of the transplant.

Formation

Neutrophils, eosinophils, basophils, and a few lymphocytes and monocytes originate, as do erythrocytes, in red bone marrow (myeloid tissue). Most lymphocytes and monocytes derive from hemocytoblasts in lymphatic tissue. Although many lymphocytes are found in bone marrow, presumably most were formed in lymphatic tissues and carried to the bone marrow by the bloodstream.

Myeloid tissue (bone marrow) and lymphatic tissue together constitute the hemopoietic or blood cell–forming tissues of the body. Red bone marrow is myeloid tissue that is actually producing blood cells. Its red color comes from the red cells it contains. Yellow marrow, on the other hand, is yellow because it stores considerable fat. It is not active in the business of blood cell formation so long as it remains yellow. Sometimes, however, it becomes active and red in color when an extreme and prolonged need for red cell production occurs.

Destruction and life span

The life span of white blood cells is not known. Some evidence seems to indicate that granular leukocytes may live 3 days or less, whereas other evidence suggests that they may live about 12 days. Some of them are probably destroyed by phagocytosis and some by microorganisms. Recent evidence suggests that many lymphocytes may live for years.

Numbers

A cubic millimeter of normal blood usually contains about 5,000 to 9,000 leukocytes, with different percentages of each type. Because these numbers change in certain abnormal conditions, they have clinical signif-

Table 13-1

White blood cells (leukocytes)

Class	Differential count*: Normal range (%)	Differential count*: Typical normal (%)
Those with granular cytoplasm and irregular nuclei		
Neutrophils (neutral staining)	65 to 75	65
Eosinophils (acid staining)	2 to 5	3
Basophils (basic staining)	½ to 1	1
Those with nongranular cytoplasm and regular nuclei		
Lymphocytes (large and small)	20 to 25	25
Monocytes	3 to 8	6
Total		100

*In any differential count, the sum of the percentages of the different kinds of leukocytes must, of course, total 100%.

icance. In acute appendicitis, for example, the percentage of neutrophils increases and so, too, does the total white count. In fact, these characteristic changes may be the deciding points for surgery.

The procedure in which the different types of leukocytes are counted and their percentage of the total white count is computed is known as a *differential count*. In other words, a differential count is a percentage count of white cells. The different kinds of white cells and a normal differential count are listed in Table 13-1. A decrease in the number of white blood cells is *leukopenia*. An increase in the number of white cells is *leukocytosis*. (*Leukemia* is a malignant disease characterized by a marked increase in the number of white blood cells.)

Platelets

Appearance and size

To compare platelets with other blood cells as to appearance and size, see Fig. 13-1. Note that platelets look like small fragments of cells.

Functions

Platelets help set in operation the blood-clotting mechanism (p. 327).

Formation and life span

Platelets are formed in the red bone marrow presumably by fragmentation of very large cells known as megakaryocytes. Platelets have a short life span, an average of about 10 days. A summary of the basic facts about blood cells is given in Table 13-2.

Blood types (or blood groups)

The term blood type refers to the type of antigens* present on red blood cell membranes. Antigens A, B, and Rh are the most important blood antigens as far as transfusions and newborn survival are concerned. Many other antigens have also been identified, but they are less important clinically and are too complex to discuss here. Every

*Antigen—substance capable of stimulating formation of antibodies that can react with the antigen; for example, to agglutinate or clump it.

Table 13-2

Blood cells

Cells	Number	Function	Formation (hemopoiesis)	Destruction
Red blood cells (erythrocytes)	4.5 to 5.5 million/mm³ (total of approximately 30 trillion in adult body)	Transport oxygen and carbon dioxide	Red marrow of bones (myeloid tissue)	Reticuloendothelial cells in lining of blood vessels in liver, spleen, and bone marrow phagocytose old red cells; live about 120 days in bloodstream
White blood cells (leukocytes)	Usually about 5,000 to 9,000/mm³	Play important part in producing immunity—e.g., phagocytosis by neutrophils; lymphocytes form cellular antibodies; some lymphocytes become plasma cells, cells that form circulating antibodies	Granular and nongranular leukocytes in red marrow; lymphocytes formed in thymus gland of fetus; postnatally, most lymphocytes and monocytes formed in lymph nodes and other lymphatic tissues	Not known definitely; probably some destroyed by phagocytosis
Platelets (thrombocytes)	150,000 to 300,000/mm³	Initiate blood clotting	Red marrow	Unknown

person's blood belongs to one of the four AB blood groups and, in addition, is either Rh positive or Rh negative. Blood types are named according to the antigens present on red cell membranes. Here, then, are the four AB blood types:

1 *Type A*—antigen A on red cells
2 *Type B*—antigen B on red cells
3 *Type AB*—both antigen A and antigen B on red cells
4 *Type O*—neither antigen A nor antigen B on red cells

The term *Rh-positive blood* means that Rh antigen is present on its red cells. *Rh-negative blood,* on the other hand, is blood whose red cells have no Rh antigen present on them.

Blood plasma may or may not contain antibodies that can react with red cell antigens A, B, and Rh. An important principle about this is that plasma never contains antibodies against the antigens present on its own red blood cells—for obvious reasons. If it did, the antibody would react with the antigen and thereby destroy the red cells. But (and this is an equally important principle) plasma does contain antibodies against antigen A or antigen B if they are *not* present on its red cells. Applying these two principles: In type A blood, antigen A is present on its red cells; therefore, its plasma contains no anti-A antibodies but does contain anti-B antibodies. In type B blood, antigen B is present on its red cells; therefore, its plasma contains no anti-B antibodies but does contain anti-A antibodies. What antigens do you deduce are present on the red cells of type AB blood? What antibodies, if any, does its plasma contain?*

No blood normally contains anti-Rh antibodies. However, anti-Rh antibodies can appear in the blood of an Rh-negative person provided Rh-positive red cells have at some time entered his bloodstream. One way this can happen is by giving an Rh-negative person a transfusion of Rh-positive blood. In a short time, his body makes anti-Rh antibodies, and these remain in his blood. There is one other way in which Rh-positive red cells can enter the bloodstream of an Rh-negative individual—but this can happen only to a woman. If she becomes pregnant, and if her mate is Rh positive, and if the fetus, too, is Rh positive, some of the red cells of the fetus may find their way into her blood from the fetal blood capillaries (via the placenta). These Rh-positive red cells then stimulate her body to form anti-Rh antibodies. Briefly, the only people who can ever have anti-Rh antibodies in their plasma are Rh-negative men or women who have been transfused with Rh-positive blood or Rh-negative women who have carried an Rh-positive fetus.

Anti-A, anti-B, and anti-Rh antibodies are agglutinins. *Agglutinins* are chemicals that agglutinate cells—that is, make them stick together in clumps. The danger in giving a blood transfusion is that antibodies present in the plasma of the person receiving the transfusion (the recipient's plasma) may agglutinate the donor's red blood cells. If, for example, type B blood were used for transfusing a person who had type A blood, the anti-B antibodies in the recipient's blood would agglutinate the donor's type B red cells. These clumped cells are potentially lethal. They can plug vital small vessels and cause the recipient's death.

Type O blood is referred to as *universal donor* blood, a term that implies that it can safely be given to any recipient. This, however, is not true because the recipient's plasma may contain agglutinins other than anti-A and anti-B antibodies. For this reason, the recipient's and the donor's blood—even if it is type O—should be cross-matched, that is, mixed and observed for agglutination of the donor's red cells.

*Type AB blood—antigens A and B present on red cells; no anti-A nor anti-B antibodies in plasma.

Universal recipient (type AB) blood contains neither anti-A nor anti-B antibodies, so it cannot agglutinate type A nor type B donor's red cells. This does not mean, however, that any type of donor blood may be safely given to an individual who has type AB blood without first cross-matching. Other agglutinins may be present in the so-called universal recipient blood and clump unidentified antigens in the donor's blood.

Blood plasma

Plasma is the liquid part of blood—whole blood minus its cells, in other words. In the laboratory, whole blood is centrifuged to form plasma. This consists of a rapid whirling process that hurls the blood cells to the bottom of the centrifuge tube. A clear, straw-colored fluid—blood plasma—lies above the cells. Plasma consists of 90% water and 10% solutes.* By far the largest quantity of these solutes are proteins; normally, they constitute about 6% to 8% of the plasma. Other solutes present in much smaller amounts in plasma are food substances (principally glucose, amino acids, and lipids), compounds formed by metabolism (for example, urea, uric acid, creatinine, and lactic acid), respiratory gases (oxygen and carbon dioxide), and regulatory substances (hormones, enzymes, and certain other substances).

The proteins in blood plasma consist of three main kinds of compounds: albumins, globulins, and fibrinogen. Electrophoretic methods of measuring the amounts of these compounds indicate that 100 ml of plasma contain a total of approximately 6 to 8 gm of protein. Albumins constitute about 55% of this total, globulins about 38%, and fibrinogen about 7%.

Plasma proteins are crucially important substances. Fibrinogen, for instance, and an albumin named prothrombin play key roles in the blood-clotting mechanism. Globulins function as essential components of the immunity mechanism—circulating antibodies (immune bodies) are modified gamma globulins. All plasma proteins contribute to the maintenance of normal blood viscosity, blood osmotic pressure, and blood volume. Therefore, plasma proteins play an essential part in maintaining normal circulation. Synthesis of plasma proteins takes place in liver cells. They form all kinds of plasma proteins except some of the gamma globulins—specifically, the circulating antibodies synthesized by plasma cells.

*Some of the solutes present in blood plasma are true solutes or crystalloids. Others are colloids. *Crystalloids* are solute particles less than 1 mμ in diameter (ions, glucose, and other small molecules). *Colloids* are solute particles from 1 to about 100 mμ in diameter (for example, proteins of all types). Blood solutes also may be classified as *electrolytes* (molecules ionize in solution) or *nonelectrolytes*—examples: inorganic salts and proteins are electrolytes; glucose and lipids are nonelectrolytes.

Blood coagulation

Purpose

The purpose of blood coagulation is obvious—to plug up ruptured vessels so as to stop bleeding and prevent loss of a vital body fluid.

Mechanism

Because of the function of coagulation, the mechanism for producing it must be swift and sure when needed, such as when a vessel is cut or ruptured. Equally important, however, coagulation needs to be prevented from happening when it is not needed because clots can plug up vessels that must stay open if cells are to receive blood's life-sustaining cargo of oxygen.

What makes blood coagulate? Over a period of many years, a host of investigators have searched for the answer to this question. They have tried to find out what events compose the coagulation mechanism and what sets it in operation. They have succeeded in gathering an abundance of relevant information. But still the questions about this complicated and important process outnumber the answers. We shall discuss some of the answers and not delve too far into the questions.

The blood-coagulation mechanism presumably consists of a series of chemical reactions that take place in a definite and rapid sequence. The trigger that starts the first of these changes, that initiates blood clotting, is the appearance of a "rough" spot in the lining of a blood vessel—examples: a cut edge of a vessel, or most common of all, patchlike deposits of a cholesterol-lipid substance. Within a matter of 1 or 2 seconds, clumps of platelets adhere to any portion of a blood vessel that loses its normal, perfectly smooth quality and release a variety of substances. Some accelerate coagulation* and others, notably serotonin (5-HT or 5-hydroxytryptamine) constrict blood vessels. After the platelets release these substances, a series of chemical reactions take place in rapid-fire succession. Although knowledge about these reactions is still tentative, this much is known—they cannot take place in the absence of calcium ions and certain phospholipids, and they result in the conversion of prothrombin to thrombin. Prothrombin and fibrinogen are the proteins involved in clotting that are made in the liver and are normally present in blood plasma.† Prothrombin is converted to thrombin, and thrombin then converts fibrinogen to fibrin. We can summarize these changes by the following simple equations:

*VII—serum prothrombin conversion accelerator, or proconvertin; VIII—antihemophilic factor (AHF); IX—plasma thromboplastin component (PTC or Christmas factor); X—Stuart factor; XI—plasma thromboplastin antecedent (PTA); XII—Hageman factor.

†Normal prothrombin content of plasma = 10 to 15 mg per 100 ml. Normal fibrinogen content = 350 mg per 100 ml of plasma.

$$\text{PROTHROMBIN} \xrightarrow{\text{(platelet factors, Ca}^{++}\text{ tissue factors)}} \text{THROMBIN}$$

$$\text{FIBRINOGEN} \xrightarrow{\text{(thrombin)}} \text{FIBRIN}$$

Prothrombin is normally present in plasma but is an inactive substance. Thrombin, on the other hand, is an active protein (an enzyme) and is not one of the proteins usually present in plasma. It must be formed from the normal plasma protein prothrombin. As thrombin forms, it catalyzes reactions that involve another plasma protein. Thrombin accelerates the conversion of the soluble plasma protein fibrinogen to insoluble fibrin. Fibrin appears in blood as fine threads all tangled together. Blood cells catch in the entanglement, and because most of the cells are red cells, clotted blood has a red color. The pale yellowish liquid left after a clot forms is *blood serum.* How do you think serum differs from plasma? What is plasma? To check your answers, see Fig. 13-3.

Liver cells synthesize both prothrombin and fibrinogen, as they do almost all other plasma proteins. In order for the liver to synthesize prothrombin at a normal rate, blood must contain an adequate amount of vitamin K. Vitamin K is absorbed into the blood from the intestine. Some foods contain this vitamin, but it is also synthesized in the intestine by certain bacteria (not present for a time in newborn infants). Because vitamin K is fat soluble, its absorption requires bile. If, therefore, the bile ducts become obstructed and bile cannot enter the intestine, a vitamin K deficiency develops. The liver cannot then produce prothrombin at its normal rate, and the blood's prothrombin concentration soon falls below normal. A prothrombin deficiency gives rise to a bleeding tendency. As a preoperative safeguard, therefore, patients with obstructive jaundice are generally given some kind of vitamin K preparation.

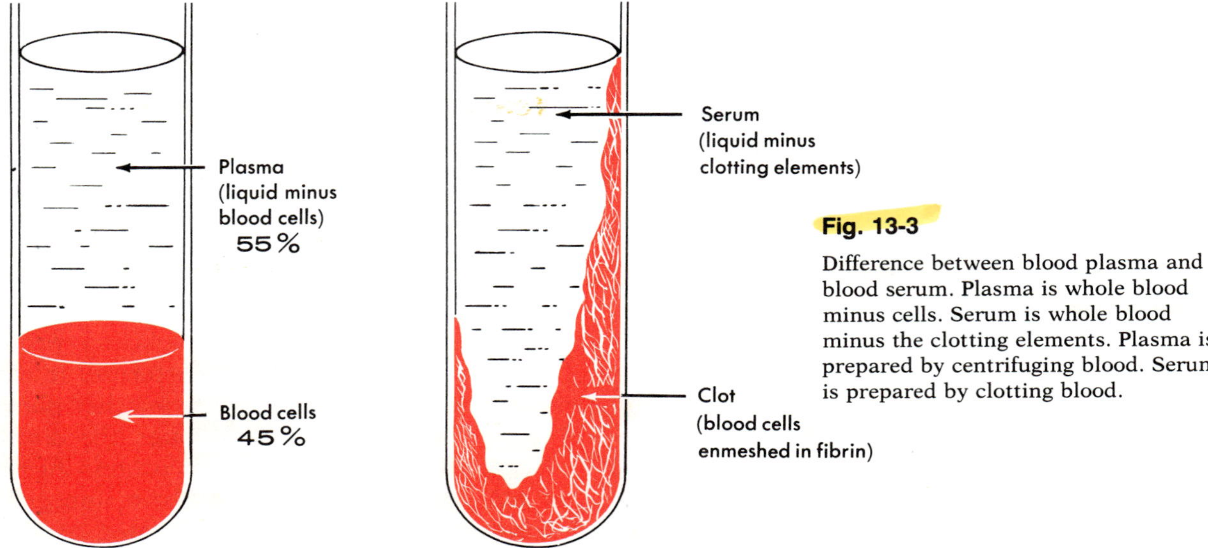

Fig. 13-3

Difference between blood plasma and blood serum. Plasma is whole blood minus cells. Serum is whole blood minus the clotting elements. Plasma is prepared by centrifuging blood. Serum is prepared by clotting blood.

Factors that oppose clotting

Although blood clotting probably goes on continuously and concurrently with clot dissolution (fibrinolysis), several factors operate to oppose clot formation in intact vessels. Most important by far is the perfectly smooth surface of the normal endothelial lining of blood vessels. Platelets do not adhere to it, consequently, they do not disintegrate and release platelet factors into the blood and, therefore, the blood-clotting mechanism does not get started in normal blood vessels. As an additional deterrent to clotting, blood contains certain substances called *antithrombins.* The name suggests their function—they oppose or inactivate thrombin. Thus antithrombins prevent thrombin from converting fibrinogen to fibrin. *Heparin,* a natural constituent of blood, acts as an antithrombin. It was first prepared from liver (hence its name), but various other organs also contain heparin. Its normal concentration in blood is too low to have much effect in keeping blood fluid. However, injections of heparin are used to prevent clots from forming in vessels. Coumarin compounds impair the liver's utilization of vitamin K and thereby slow its synthesis of prothrombin and factors VII, IX, and X. Indirectly, therefore, coumarin compounds retard coagulation.

Factors that hasten clotting

Two conditions particularly favor thrombus formation: a rough spot in the endothelium (blood vessel lining) and abnormally slow blood flow. Atherosclerosis, for example, is associated with an increased tendency toward thrombosis because of endothelial rough spots in the form of plaques of accumulated cholesterol-lipid material. Immobility, on the other hand, may lead to thrombosis because blood flow slows down as movements decrease. Incidentally, this fact is one of the major reasons why physicians insist that bed patients must either move or be moved frequently. Presumably, sluggish blood flow allows thromboplastin to accumulate sufficiently to reach a concentration adequate for clotting.

Once started, a clot tends to grow. Platelets enmeshed in the fibrin threads disintegrate, releasing more thromboplastin, which in turn causes more clotting, which enmeshes more platelets, and so on, in a vicious circle.

Clot-retarding substances, available in recent years, have proved valuable for retarding this process.

Clot dissolution

Fibrinolysis is the physiological mechanism that dissolves clots. Newer evidence indicates that the two opposing processes of clot formation and fibrinolysis go on continuously. Dr. George Fulton of Boston University has presented one bit of dramatic evidence. He took micromovies that show tiny blood vessels rupturing under apparently normal circumstances and clots forming to plug them. Blood contains an enzyme, fibrinolysin, that catalyzes the hydrolysis of fibrin, causing it to dissolve. Many other factors, however, presumably also take part in clot dissolution—for instance, substances that activate profibrinolysin (inactive form of fibrinolysin). Streptokinase, an enzyme from certain streptococci, can act this way and so can cause clot dissolution and even hemorrhage.

Clinical methods of hastening clotting

One way of treating excessive bleeding is to speed up the blood-clotting mechanism. The principle involved is apparent—to increase any of the substances essential for clotting. Application of this principle is accomplished in the following ways:

1. By applying a rough surface such as gauze, or by applying heat, or by gently squeezing the tissues around a cut vessel. Each of these procedures causes more platelets to disintegrate and release more platelet factors. This, in turn, accelerates the first of the clotting reactions.
2. By applying purified thrombin (in the form of sprays or impregnated gelatin sponges that can be left in a wound). Which stage of the clotting mechanism does this accelerate?
3. By applying fibrin foam, films, etc.

Heart

The human heart is a four-chambered muscular organ, shaped and sized roughly like a man's closed fist. It lies in the mediastinum, with approximately two thirds of its mass to the left of the midline of the body and one third to the right.

The lower border of the heart, which forms a blunt point known as the *apex*, lies on the diaphragm, pointing toward the left. To count the apical beat, one must place a stethoscope directly over the apex—that is, in the space between the fifth and sixth ribs (fifth intercostal space) on a line with the midpoint of the left clavicle.

The upper border of the heart, that is, its base, lies just below the second rib. The boundaries, which, of course, indicate its size, have considerable clinical importance since a marked increase in heart size accompanies certain types of heart disease. Therefore, when diagnosing heart disorders, the doctor charts the boundaries of the heart.

Covering

Structure

The heart has its own special covering, a loose-fitting inextensible sac called the *pericardium*. The pericardium consists of two parts: a fibrous portion and a serous portion (Fig. 13-4). The sac itself is made of tough white fibrous tissue but is lined with smooth, moist serous membrane—the parietal layer of the serous pericardium. The same kind of membrane covers the entire outer surface of the heart. This covering layer is known as the visceral layer of the serous pericardium or as the *epicardium*. The fibrous sac attaches to the large blood vessels emerging from the top of the heart but not to the heart itself. Therefore, it fits loosely around the heart with a slight space between the visceral layer ad-

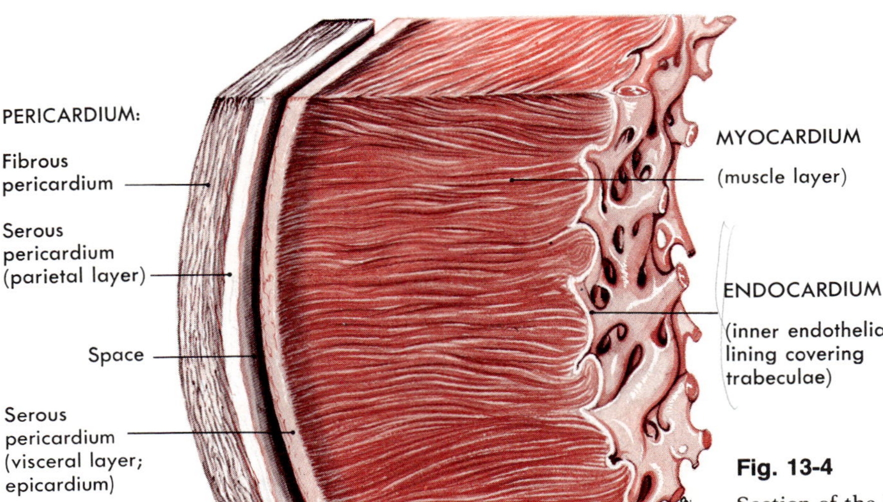

Fig. 13-4

Section of the heart wall showing the components of the outer pericardium (heart sac), muscle layer (myocardium), and inner lining (endocardium).

hering to the heart and the parietal layer adhering to the inside of the fibrous sac. This space is called the *pericardial space.* It contains a few drops of lubricating fluid secreted by the serous membrane and is called *pericardial fluid.*

The structure of the pericardium can be summarized in outline form as follows:

1 *Fibrous pericardium*—loose-fitting sac around the heart
2 *Serous pericardium*—consisting of two layers
   a *Parietal layer*—lining inside of the fibrous pericardium
   b *Visceral layer (epicardium)*—adhering to the outside of the heart; between visceral and parietal layers is a potential space, the pericardial space, which contains a few drops of pericardial fluid

Function

The fibrous pericardial sac with its smooth, well-lubricated lining provides protection against friction. The heart moves easily in this loose-fitting jacket with no danger of irritation from friction between the two surfaces so long as the serous pericardium remains normal. If, however, it becomes inflamed (pericarditis) and too much pericardial fluid or fibrin or pus develops in the pericardial space, the visceral and parietal layers may stick together. This, as you might guess, hampers heart action. In such cases, it sometimes becomes necessary to remove the fibrous pericardium with its lining of parietal serous membrane in order for the heart to continue functioning. This operation, a spectacular procedure, is called a *pericardectomy.*

Structure

Wall

Three distinct layers of tissue make up the heart wall (Fig. 13-4). The bulk of the wall consists of specially constructed muscle tissue known as cardiac muscle or the *myocardium.* Covering the myocardium on the outside and adherent to it is the visceral layer of the *serous pericardium* (or *epicardium*) already described. Lining the interior of the

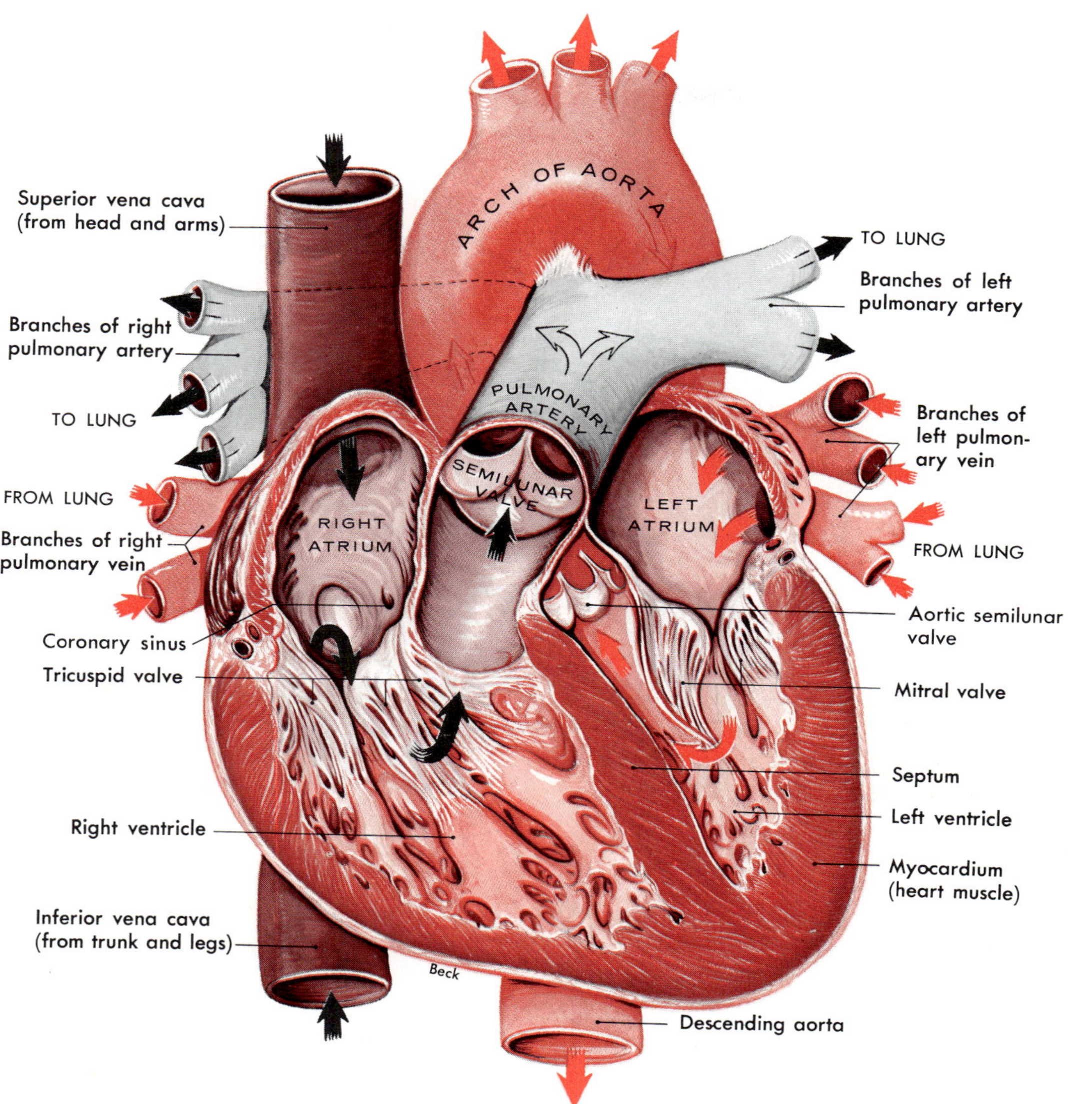

Fig. 13-5

Frontal section of the heart showing the four chambers, valves, openings, and major vessels. Arrows indicate direction of blood flow. Black arrows represent unoxygenated blood and red arrows oxygenated blood. The two branches of the right pulmonary vein extend from the right lung behind the heart to enter the left atrium.

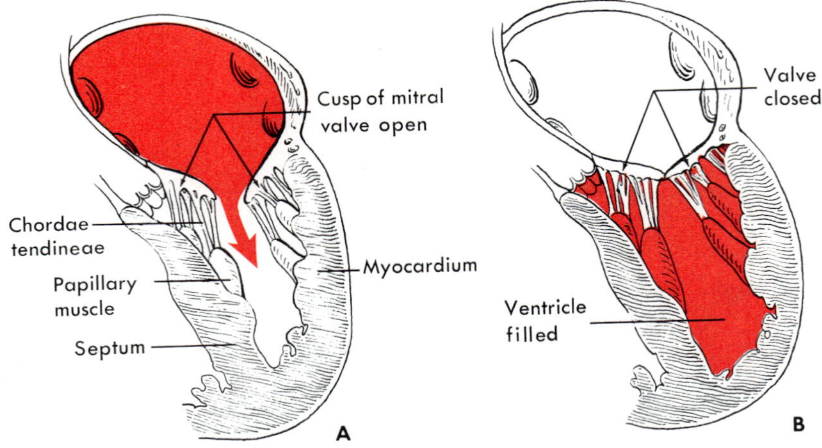

Fig. 13-6

Action of the cuspid (atrioventricular) valves. **A,** When the valves are open, blood passes freely from the atria to the ventricles. **B,** Filling of the ventricles closes the valves and prevents a backflow of blood into the atria when the ventricles contract.

myocardial wall is a delicate layer of endothelial* tissue known as the *endocardium.* On its inner surface, the myocardium is raised into ridgelike projections, the papillary muscles.

Cavities

The interior of the heart is divided into four chambers, two upper and two lower. The upper cavities are named *atria*† and the lower ones *ventricles* (Fig. 13-5). Of these, the ventricles are considerably larger and thicker walled than the atria because they carry a heavier pumping burden than the atria. Also, the left ventricle has thicker walls than the right for the same reason. It has to pump blood through all the vessels of the body, except those to and from the lungs, whereas the right ventricle sends blood only through the lungs.

*Endothelial tissue resembles simple squamous epithelial tissue in that it consists of a single layer of flat cells. It differs from epithelial tissue in that it arises from the mesoderm layer of the embryo, whereas epithelial tissue arises from the ectoderm.

†The atria are sometimes called auricles. Strictly speaking, the latter term means the earlike flaps protruding from the atria, although the two terms are often used synonymously.

Valves and openings

The heart valves are mechanical devices that permit the flow of blood in one direction only. Four sets of valves are of importance to the normal functioning of the heart (Figs. 13-5 and 13-7). Two of these, the cuspid (atrioventricular) valves, are located in the heart, guarding the openings between the atria and ventricles (atrioventricular orifices). The other two, the semilunar valves, are located inside the pulmonary artery and the great aorta just as they arise from the right and left ventricles, respectively.

The cuspid valve guarding the right atrioventricular orifice consists of three flaps of endocardium anchored to the papillary muscles of the right ventricle by several cordlike structures called *chordae tendineae.* Because

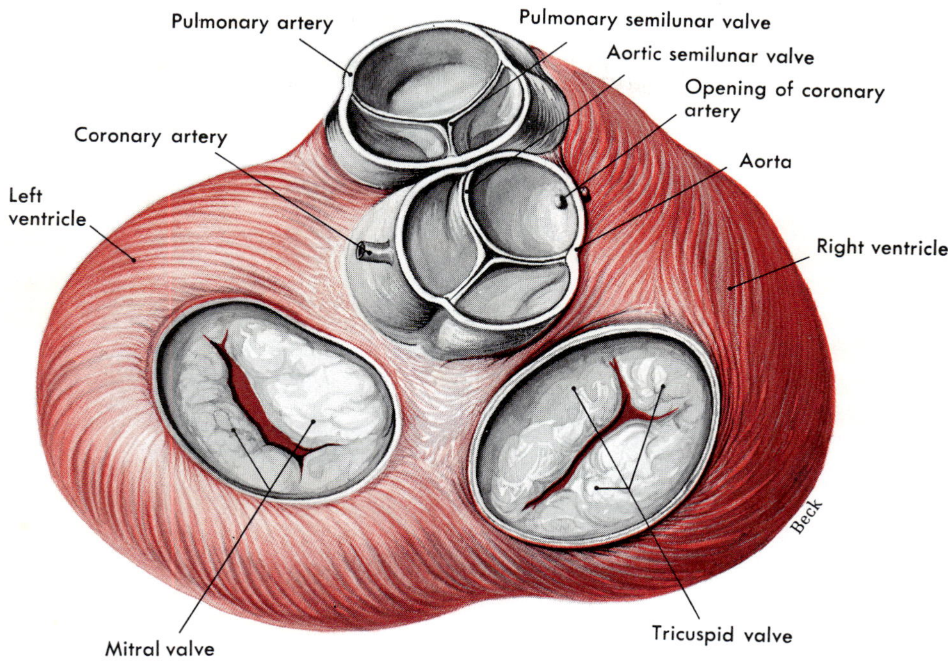

Fig. 13-7

The valves of the heart viewed from above. The atria are removed to show the mitral and tricuspid valves.

this valve has three flaps, it is appropriately named the *tricuspid valve.* The valve that guards the left atrioventricular orifice is similar in structure to the tricuspid except that it has only two flaps and is, therefore, called the *bicuspid* or, more commonly, the *mitral valve.* (An easy way to remember which valve is on the right and which is on the left is this: the names whose first letters come nearest together in the alphabet go together—thus, L and M for *l*eft side, *m*itral valve, and R and T for *r*ight side, *t*ricuspid.)

The construction of both cuspid valves allows blood to flow from the atria into the ventricles but prevents it from flowing back up into the atria from the ventricles. Ventricular contraction forces the blood in the ventricles hard against the cuspid valves, closing them and thereby ensuring the movement of the blood upward into the pulmonary artery and aorta as the ventricles contract (see Fig. 13-6).

The *semilunar valves* consist of half-moon–shaped flaps growing out from the lining of the pulmonary artery and great aorta. When these valves are closed, as in Fig. 13-7, blood fills the spaces between the flaps and the vessel wall. Each flap then looks like a tiny filled bucket. Inflowing blood smoothes the flaps against the blood vessel wall, collapsing the buckets and thereby opening the valves.

Like the cuspid valves, the semilunar valves, by preventing backflow of blood, cause it to flow forward in places where there would otherwise be considerable backflow. Whereas the cuspid valves prevent blood from flowing back up into the atria from the ventricles, the semilunar valves prevent it from flowing back down into the ventricles from the aorta and pulmonary artery.

Any one of the four valves may lose its ability to close tightly. Such a condition is known as *valvular insufficiency* or, because it permits blood to "leak" back into the part of the heart from which it came, "leakage of the heart." *Mitral stenosis* is an abnormality in which the left atrioventricular orifice becomes narrowed by scar tissue that forms as a result of disease. This hinders the passage of blood from the left atrium to the left ventricle and leads to circulatory failure. But fortunately, the marvel of open-heart surgery has made possible the correction of many valvular defects.

Blood supply

Myocardial cells receive blood by way of two small vessels, the right and left coronary arteries. Since the openings into these vitally important vessels lie behind flaps of the aortic semilunar valve, they come off of the aorta at its very beginning and are its first branches. Both right and left coronary arteries have two main branches, as shown in Table 13-3.

More than a half million Americans die every year from coronary disease and another 3.5 million or more are estimated to suffer some degree of incapacitation from this great killer.* Knowledge about the distribution of coronary artery branches, therefore, has the utmost practical importance. Here, then, are some principles about the heart's own blood supply that seem worth noting. Both ventricles receive their blood supply from branches of both the right and left coronary arteries. Each atrium, in contrast, receives blood only from a small branch of the corresponding coronary artery (Table 13-3). The most abundant blood supply of all goes to the myocardium of the left ventricle—an appropriate fact since the left ventricle does the most work and so needs the most oxygen and nutrients delivered to it. The right coronary artery is dominant in about 50% of all hearts and the left coronary artery in about 20%, and in about 30% neither right nor left coronary artery dominates.

*Effler, D. B.: Surgery for coronary disease, Sci. Am. **219**:36-43, Oct., 1968.

Table 13-3

Coronary arteries

Right coronary artery	Left coronary artery
Divides into two main branches: **1** Posterior descending artery—sends branches to both ventricles **2** Marginal artery—sends branches to right ventricle and right atrium	Divides into two main branches: **1** Anterior descending artery—sends branches to both ventricles **2** Circumflex artery—sends branches to left ventricle and left atrium

Another fact about the heart's own blood supply—and one of life-and-death importances—is the fact that only a few anastomoses exist between the larger branches of the coronary arteries. An *anastomosis* consists of one or more branches from the proximal part of an artery to a more distal part of itself or of another artery. Thus anastomoses provide detours that arterial blood can travel if the main route becomes obstructed. In short, they provide collateral circulation to a part. This explains why the scarcity of anastomoses between larger coronary arteries looms so large as a threat to life. If, for example, a blood clot plugs one of the larger coronary artery branches, as it frequently does in coronary thrombosis or embolism, too little or no blood can reach some of the heart muscle cells. They become ischemic, in other words. Deprived of oxygen, they release too little energy for their own survival. Myo-

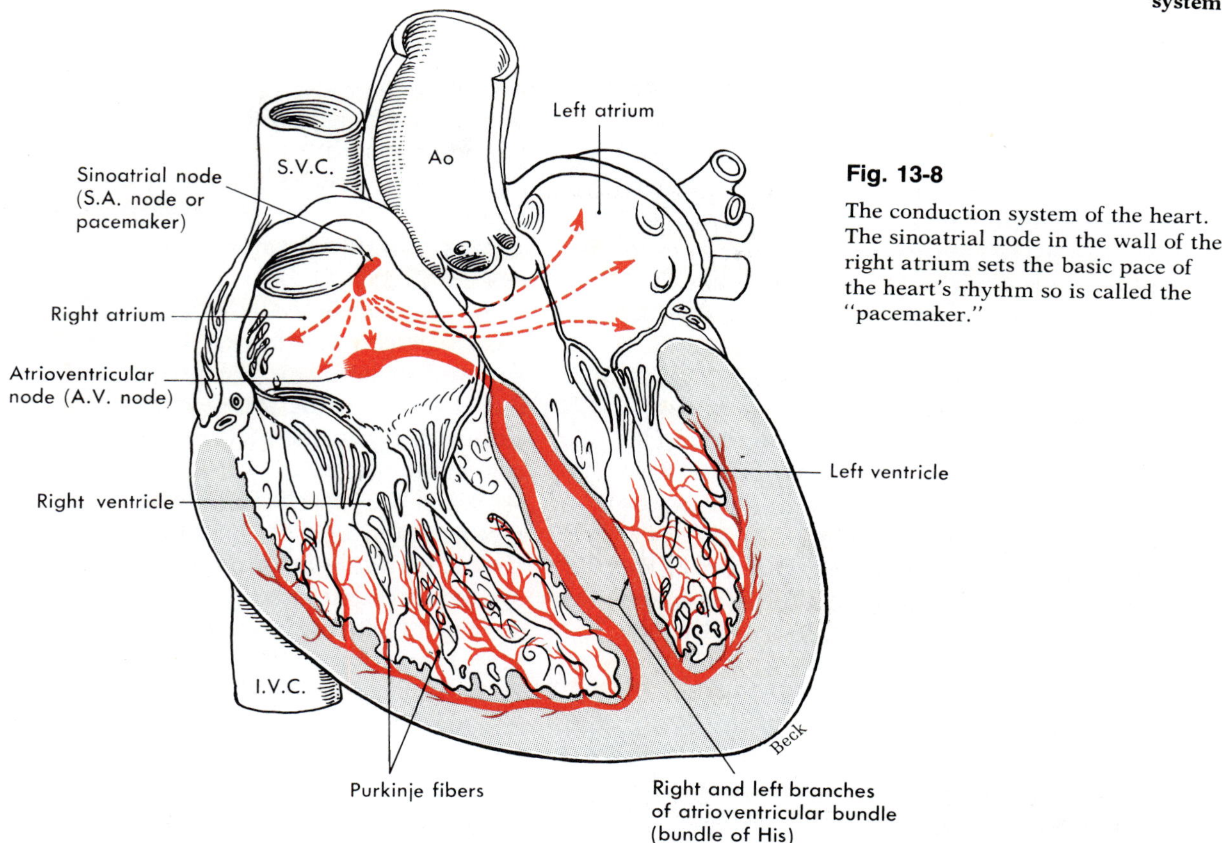

Fig. 13-8

The conduction system of the heart. The sinoatrial node in the wall of the right atrium sets the basic pace of the heart's rhythm so is called the "pacemaker."

cardial infarction (death of ischemic heart muscle cells) soon results. There is another anatomical fact, however, which brightens the picture somewhat—many anastomoses do exist between very small arterial vessels in the heart, and, given time, new ones develop and provide collateral circulation to ischemic areas. In recent years, several surgical procedures have been devised to aid this process.*

Conduction system

Four structures—the sinoatrial node, atrioventricular node, atrioventricular bundle, and Purkinje fibers—compose the conduction system of the heart (Fig. 13-8). Each of these structures consists of cardiac muscle modified enough in structure to differ in function from ordinary cardiac muscle. The main specialty of ordinary cardiac muscle is contraction. In this, it is like all muscle and, like all muscle, ordinary cardiac muscle can also conduct impulses. But conduction alone is the specialty of the modified cardiac muscle that composes the conduction system structures.

Sinoatrial node. The sinoatrial node (S.A. node, Keith-Flack node, or pacemaker) consists of hundreds of cells located in the right atrial wall near the opening of the superior vena cava (Fig. 13-8). Sinoatrial node cells possess an intrinsic rhythm. This means that without any stimulation by nerve impulses from the brain and cord, they themselves initiate impulses at regular intervals. They

*Effler, D. B.: Surgery for coronary disease, Sci. Am. **219**:36-43, Oct., 1968.

travel swiftly throughout the muscle fibers of both atria to the atrioventricular node. Thus stimulated, the atria contract as the impulses continue on their rapid way down the atrioventricular bundle to the ventricles. Here, right and left branches of the bundle fibers, and the Purkinje fibers in which they terminate, conduct the impulses throughout the muscle of both ventricles, stimulating them to contract. Thus the sinoatrial node initiates each heartbeat and sets its pace—it is the heart's own natural pacemaker. The rate it sets, however, is not an unalterable one. Various factors can and do change the rate of the heartbeat. Two major modifiers of sinoatrial node activity—and therefore of the heart rate—are the ratio of sympathetic and parasympathetic impulses conducted to the node per minute and the blood concentrations of certain hormones (notably epinephrine and thyroid hormone).

Today, everyone has heard about artificial pacemakers, devices that electrically stimulate the heart at a set rhythm. They do an excellent job of maintaining a steady heart rate and of keeping many individuals with damaged hearts alive for many years. Nevertheless, they must be judged inferior to the heart's own natural pacemaker. Why? Because they cannot speed up the heartbeat (as is necessary, for example, to make strenuous physical activity possible), nor can they slow it down again when the need has passed. The sinoatrial node, influenced as it is by autonomic impulses and hormones, can produce these changes. Artificial pacemakers can only keep the heart beating at a steady pace.

Atrioventricular node. The atrioventricular node (A.V. node or node of Tawara), a small mass of special cardiac muscle tissue, lies in the right atrium along the lower part of the interatrial septum (Fig. 13-8).

Atrioventricular bundle and Purkinje fibers. The atrioventricular bundle (bundle of His) is a bundle of special cardiac muscle fibers that originate in the atrioventricular node and that extend by two branches down the two sides of the interventricular septum. From there, they continue as the Purkinje fibers (Fig. 13-8). The latter extend out to the papillary muscles and lateral walls of the ventricles.

Impulse conduction through the heart normally starts in the sinoatrial node and spreads through atrial muscle fibers in all directions, causing atrial contraction. When impulses reach the atrioventricular node, it relays them by way of the bundle of His and Purkinje fibers to the ventricles, causing their contraction. Impulse conduction generates tiny electrical currents in the heart that spread through surrounding tissues to the surface of the body. This fact has great clinical importance. Why? Because from the skin, visible records of heart conduction can be made with the electrocardiograph or oscillograph. Skilled interpretation of these records may sometimes make the difference between life and death.

Electrocardiogram. The electrocardiogram (ECG) is a record of the heart's action currents, its conduction of impulses. It is not a record of the heart's contractions. Because electrocardiography is far too complex a science for us to attempt to explain here, * we shall merely call your attention to the appearance of a normal electrocardiogram, as shown in Fig. 13-9. Note that it is composed of a P wave, a QRS complex, and a T wave. (The letters do not stand for any words but were chosen arbitrarily.) Briefly, the P wave represents depolarization of the atria, that is, conduction through them. The QRS complex represents depolarization of the ventricles, and the T wave reflects repolarization of the ventricles.

*For more information, see Zalis, E. G., and Conover, M. H.: Understanding electrocardiography, St. Louis, 1972, The C. V. Mosby Co.

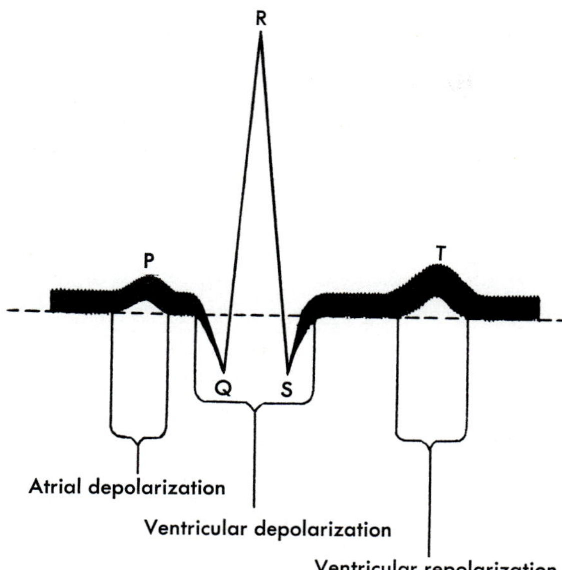

Fig. 13-9

Normal ECG deflections. Depolarization and repolarization.

Nerve supply

Both divisions of the autonomic nervous system send fibers to the heart. Sympathetic fibers (contained in the middle, superior, and inferior cardiac nerves) and parasympathetic fibers (in branches of the vagus) combine to form *cardiac plexuses* located close to the arch of the aorta. From the cardiac plexuses, fibers accompany the right and left coronary arteries to enter the heart. Here, most of the fibers terminate in the sinoatrial node, but some end in the atrioventricular node and in the atrial myocardium. Sympathetic nerves to the heart are also called accelerator nerves. Vagus fibers to the heart serve as inhibitory or depressor nerves.

Physiology

Function

The function of the heart is to pump blood in sufficient amounts to meet the varying needs of the cells of the body for the substances it transports. Mechanisms that accomplish this function of pumping different volumes of blood per minute under different conditions are discussed on pp. 354-366.

Cardiac cycle

The term *cardiac cycle* means a complete heartbeat consisting of contraction (systole) and relaxation (diastole) of both atria plus contraction and relaxation of both ventricles. The two atria contract simultaneously. Then, as they relax, the two ventricles contract and relax, instead of the entire heart contracting as a unit. This gives a kind of milking action to the movements of the heart. The atria remain relaxed during part of the ventricular relaxation and then start the cycle over again. The events occurring during the cycle are described in Table 13-4. Note the following facts:

1 The contracting force of the atria completes the emptying of blood out of the atria into the ventricles. Cuspid valves are necessarily open during this phase, the ventricles relaxed, filling with blood, and the semilunar valves closed so that blood does not flow on out into the pulmonary artery or aorta.

2 The atria relax, and blood enters them from the veins during the first part of their diastole and starts draining out into the ventricles during the latter part of it. The cuspid valves are closed during the first part of the diastole (while the ventricles are contracting, squeezing blood through the open semilunar valves into the pulmonary artery and aorta) but open as the ventricles relax, the semilunar valves close, and the ventricles start to fill with blood from the atria. About what percent of the time are the atria relaxed or resting? The ventricles? Consult Table 13-4 to find the answers.

Heart sounds during cycle. The heart makes certain typical sounds during each cycle that are described as sounding like lubb-dupp through a stethoscope. The first or

Table 13-4

The cardiac cycle (in tenths of seconds; 0.8 second for complete cycle)

Second	0.1	0.2	0.3	0.4	0.5	0.6	0.7	0.8
Atria	Contract—blood squeezed into ventricles	Relax → Blood enters from venae cavae and pulmonary veins and drains into ventricles →						
Cuspid valves	Open	Closed →		Open →				
Ventricles	Relaxed—filling with blood from atria	Contract Blood emptying into pulmonary artery and aorta →		Relax Filling with blood from atria →				
Semilunar valves	Closed	Open →		Closed →				

systolic sound is believed to be caused by the contraction of the ventricles and by vibrations of the closing cuspid valves. It is longer and lower than the second or diastolic sound, which is thought to be caused by vibrations of the closing semilunar valves.

Both of these sounds have clinical significance since they give information about the valves of the heart. Any variation from normal in the sounds indicates imperfect functioning of the valves. *Heart murmur* is one type of abnormal sound frequently heard and may signify incomplete closing of the valves (valvular insufficiency) or stenosis of them.

Blood vessels

Kinds

There are three kinds of blood vessels: arteries, veins, and capillaries. By definition an *artery* is a vessel that carries blood away from the heart. All arteries except the pulmonary artery and its branches carry oxygenated blood. Small arteries are called *arterioles*.

A *vein*, on the other hand, is a vessel that carries blood toward the heart. All of the veins except the pulmonary veins contain deoxygenated blood. Small veins are called *venules*. Both arteries and veins are macroscopic structures.

Capillaries are microscopic vessels that carry blood from small arteries to small veins—that is, from arterioles to venules. They represented the "missing link" in the proof of circulation for many years—from the time William Harvey first declared that blood circulated from the heart through arteries to veins and back to the heart until the time that microscopes made it possible to find these connecting vessels between arteries and veins. Many people rejected Harvey's theory of circulation on the basis that there was no possible way for blood to get from arteries to veins. The discovery of the capillaries formed the final proof that the blood actually does circulate from the heart into arteries to arterioles, to capillaries, to venules, to veins, and back to the heart.

Structure

Consult Figs. 13-10 and 13-11 and Table 13-5 for structure of the blood vessels.

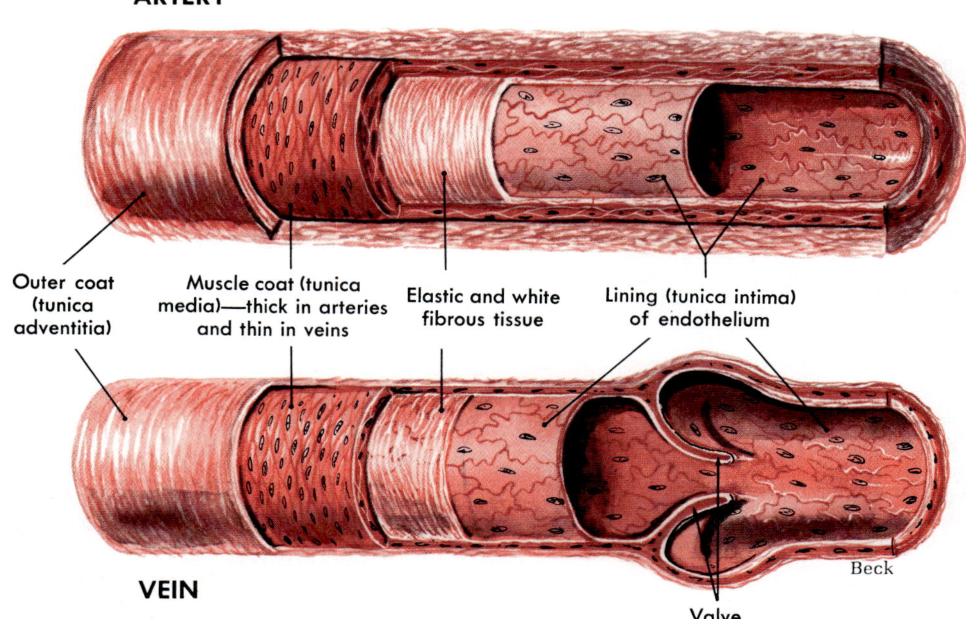

Fig. 13-10

Schematic drawings of an artery and vein showing comparative thicknesses of the three coats: outer coat (tunica adventitia), muscle coat (tunica media), and lining of endothelium (tunica intima). Note that the muscle and outer coats are much thinner in veins than in arteries and that veins have valves.

Table 13-5

Structure of blood vessels

	Arteries	Veins	Capillaries
Coats	Outer coat (tunica adventitia or externa) of white fibrous tissue; causes artery to stand open instead of collapsing when cut (see Fig. 13-10) Lining (tunica intima) of endothelium Muscle coat (tunica media) of smooth muscle, elastic, and some white fibrous tissues; this coat permits constriction and dilation	Same three coats but thinner and fewer elastic fibers; veins collapse when cut; semilunar valves present at intervals	Only lining coat present; therefore walls only one cell thick
Blood supply	Endothelial lining cells supplied by blood flowing through vessels; exchange of oxygen, etc., between cells of middle coat and blood by diffusion; outer coat supplied by tiny vessels known as *vasa vasorum* or "vessels of vessels"		
Nerve supply	Smooth muscle cells of tunica media innervated by autonomic fibers		
Abnormalities	*Atherosclerosis*—hardening of walls of arteries (arteriosclerosis) characterized by lipid deposits in tunica intima *Aneurysm*—saclike dilation of artery wall *Varicose veins*—stretching of walls, particularly around semilunar valves *Phlebitis*—inflammation of vein; "milk leg," phlebitis of femoral vein of women after childbirth		

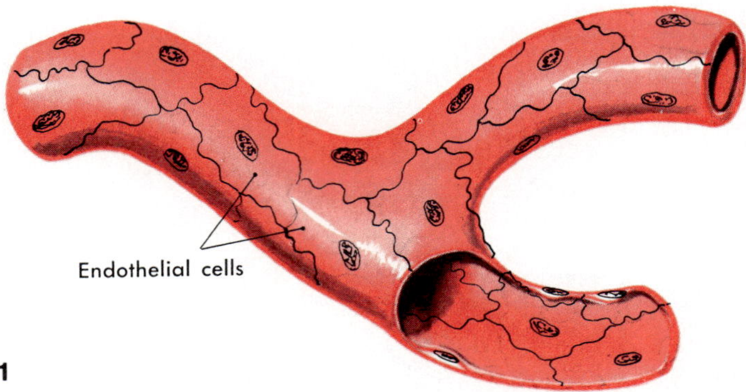

Fig. 13-11

The walls of capillaries consist of only a single layer of endothelial cells. These thin, flattened cells permit the rapid movement of substances between blood and interstitial fluid. Note that capillaries have no smooth muscle layer, elastic fibers, or surrounding adventitia.

Functions

The capillaries, though seemingly the most insignificant of the three kinds of blood vessels because of their diminutive size, nevertheless are the most important vessels functionally. Since the prime function of blood is to transport essential materials to and from the cells and since the actual delivery and collection of these substances take place in the capillaries, the capillaries must be regarded as the most important blood vessels functionally. Arteries serve merely as "distributors," carrying the blood to the arterioles. Arterioles, too, serve as distributors, carrying blood from arteries to capillaries. But in addition, arterioles perform another function, one that is of great importance for maintaining normal blood pressure and circulation. They serve as resistance vessels (discussed on p. 353). Veins function both as collectors and as reservoir vessels. Not only do they return blood from the capillaries to the heart, but they also can accommodate varying amounts of blood. This reservoir function of veins, which we will discuss later, plays an important part in maintaining normal circulation. The heart acts as a "pump," keeping the blood moving through this circuit of vessels—arteries, arterioles, capillaries, venules, and veins. In short, the entire circulatory mechanism pivots around one essential, that of keeping the capillaries supplied with an amount of blood adequate to the changing needs of the cells. All the factors governing circulation operate to this one end.

Although capillaries are very tiny (on the average, only 1 mm long or about 1/25 inch), they are so numerous as to be incomprehensible. Someone has calculated that if these microscopic tubes were joined end to end, they would extend 62,000 miles, in spite of the fact that it takes twenty-five of them to reach a single inch! According to one estimate, 1 cubic inch of muscle tissue contains over 1.5 million of these important little vessels. None of the billions of cells composing the body lies very far removed from a capillary. The reason for this lavish distribution of capillaries is, of course, apparent in view of their function of keeping the cells supplied with vital materials and rid of injurious wastes.

Table 13-6

Main arteries

Artery	Branches (only largest ones named)
Ascending aorta	Coronary arteries (two, to myocardium)
Aortic arch	Innominate artery (or brachiocephalic) Left subclavian Left common carotid
Innominate artery	Right subclavian Right common carotid
Subclavian (right and left)	Vertebral* Axillary (continuation of subclavian)
Axillary	Brachial (continuation of axillary)
Brachial	Radial Ulnar
Radial and ulnar	Palmar arches (superficial and deep arterial arches in hand formed by anastomosis of branches of radial and ulnar arteries; numerous branches to hand and fingers)
Common carotid (right and left)	Internal carotid (brain, eye, forehead, and nose)* External carotid (thyroid, tongue, tonsils, ear, etc.)
Descending thoracic aorta	Visceral branches to pericardium, bronchi, esophagus, mediastinum Parietal branches to chest muscles, mammary glands, and diaphragm
Descending abdominal aorta	Visceral branches: **1** Celiac axis (or artery), which branches into gastric, hepatic, and splenic arteries (stomach, liver, and spleen) **2** Right and left suprarenal arteries (suprarenal glands) **3** Superior mesenteric artery (small intestine) **4** Right and left renal arteries (kidneys) **5** Right and left spermatic (or ovarian) arteries (testes or ovaries) **6** Inferior mesenteric artery (large intestine) Parietal branches to lower surface of diaphragm, muscles and skin of back, spinal cord, and meninges Right and left common iliac arteries—abdominal aorta terminates in these vessels in an inverted Y formation
Right and left common iliac arteries	Internal iliac or hypogastric (pelvic wall and viscera) External iliac (to leg)
External iliac (right and left)	Femoral (continuation of external iliac after it leaves abdominal cavity)
Femoral	Popliteal (continuation of femoral)
Popliteal	Anterior tibial Posterior tibial
Anterior and posterior tibial	Plantar arch (arterial arch in sole of foot formed by anastomosis of terminal branches of anterior and posterior tibial arteries; small arteries lead from arch to toes)

*The right and left vertebral arteries extend from their origin as branches of the subclavian arteries up the neck, through foramina in the transverse processes of the cervical vertebrae, and through the foramen magnum into the cranial cavity and unite on the undersurface of the brainstem to form the *basilar artery*, which shortly branches into the right and left *posterior cerebral arteries*. The internal carotid arteries enter the cranial cavity in the midpart of the cranial floor, where they become known as the *anterior cerebral arteries*. Small vessels, the *communicating arteries*, join the anterior and posterior cerebral arteries in such a way as to form an arterial circle (the *circle of Willis*) at the base of the brain, a good example of arterial anastomosis (Fig. 13-15).

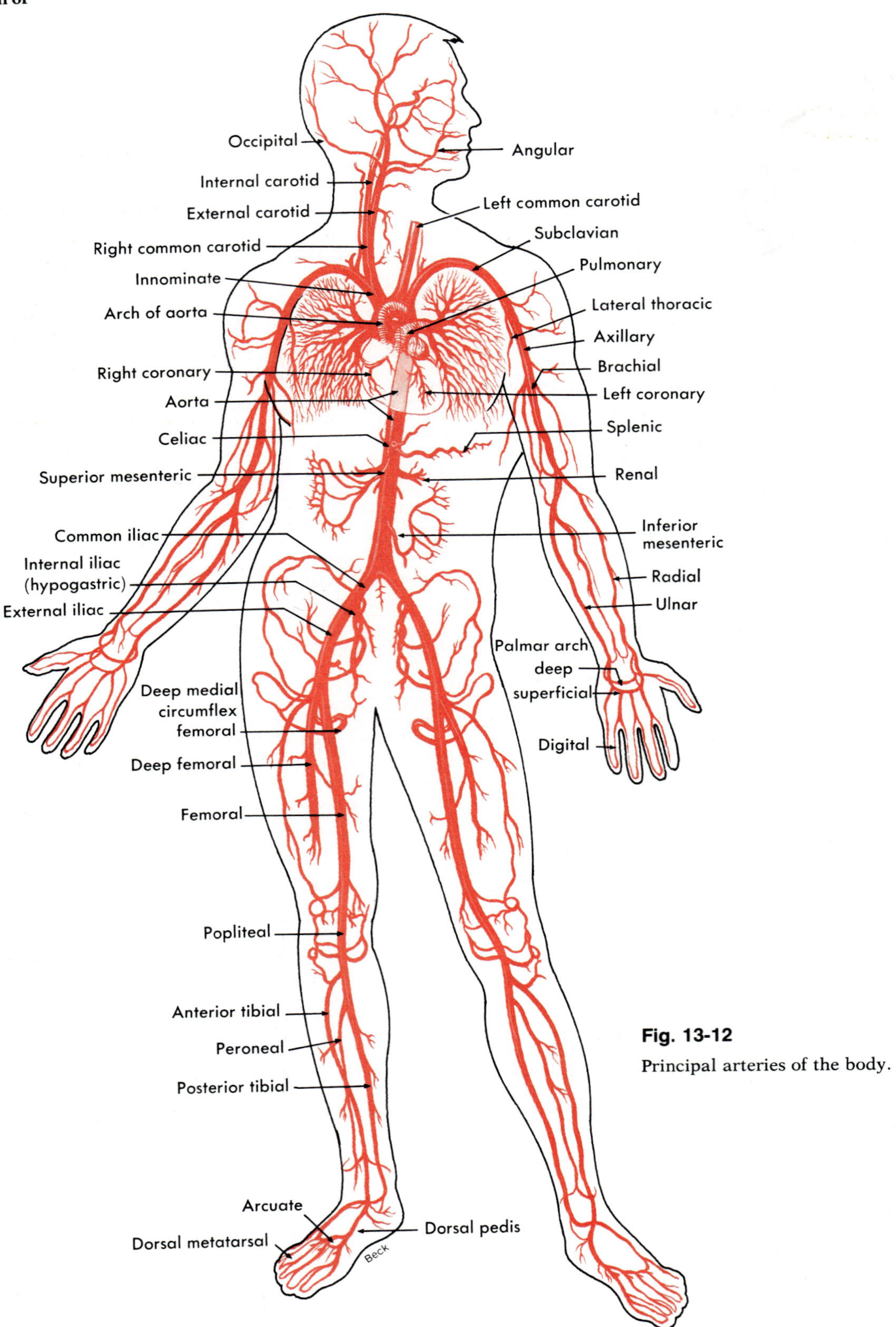

Fig. 13-12
Principal arteries of the body.

Main blood vessels

Systemic circulation

ARTERIES

Locate the arteries listed in Table 13-6 (see also Figs. 13-12 to 13-15). You may find it easier to learn the names of blood vessels and the relation of the vessels to each other from diagrams than from descriptions.

As you learn the names of the main arteries, keep in mind that these are only the major pipelines distributing blood from the heart to the various organs and that in each organ the main artery resembles a tree trunk in that it gives off numerous branches that continue to branch and rebranch, forming ever smaller vessels (arterioles), which also branch, forming microscopic vessels, the capillaries. In other words, most arteries eventually ramify into capillaries.

A few arteries open into other branches of the same or other arteries. Such a communication is termed an *arterial anastomosis.* Anastomoses, we have already noted, fulfill an important protective function in that they provide detour routes for blood to travel in the event of obstruction of a main artery. Examples of arterial anastomoses are the circle of Willis at the base of the brain and the palmar and plantar arches. Other examples are found around several joints as well as in other locations.

Fig. 13-13

Main arteries of the face and head. Superficial vessels are shown in brighter color than deep vessels. (From Francis, C. C, and Farrell, G. L.: Integrated anatomy and physiology, St. Louis, The C. V. Mosby Co.)

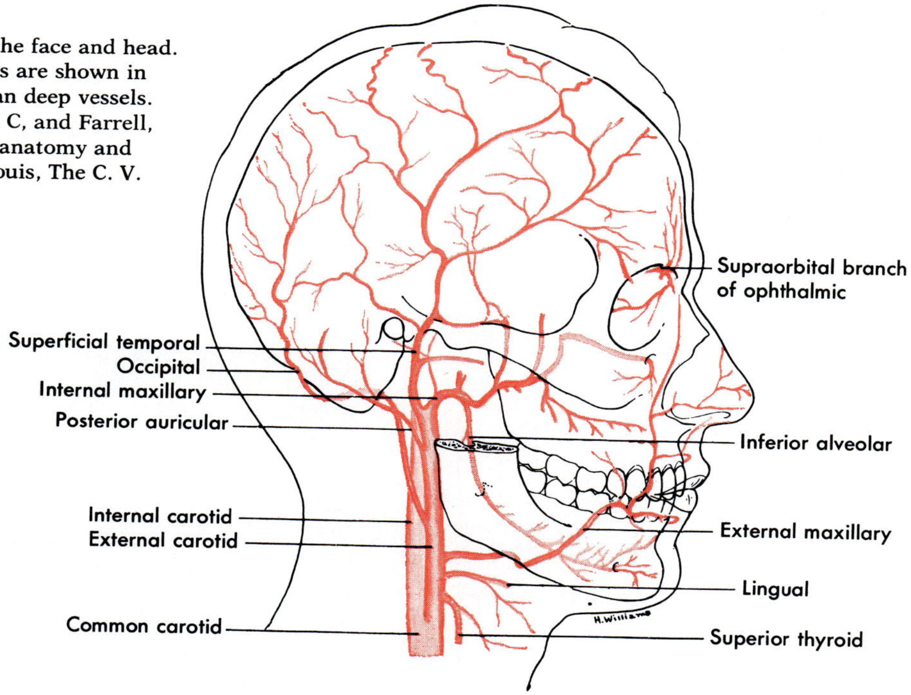

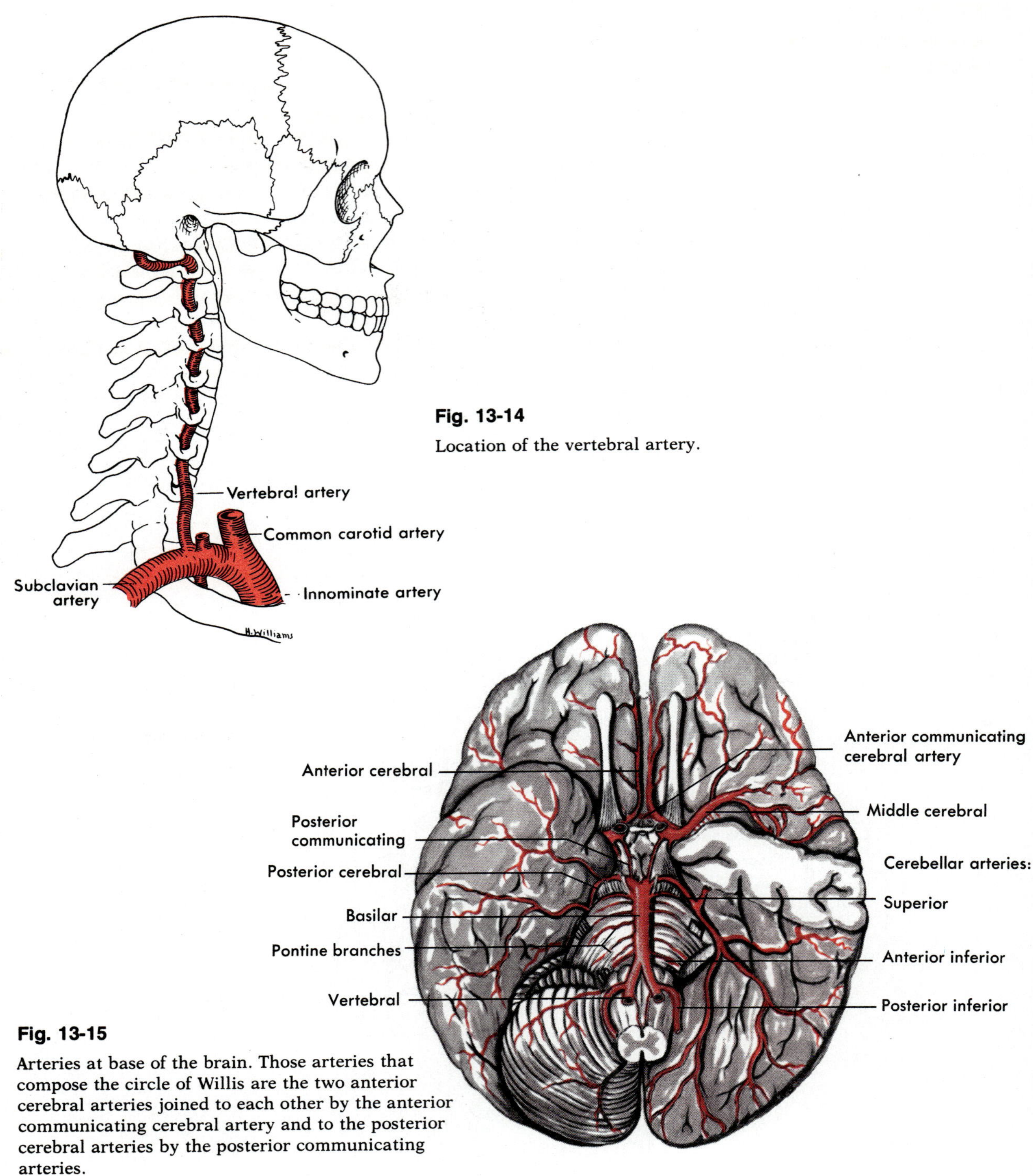

Fig. 13-14

Location of the vertebral artery.

Fig. 13-15

Arteries at base of the brain. Those arteries that compose the circle of Willis are the two anterior cerebral arteries joined to each other by the anterior communicating cerebral artery and to the posterior cerebral arteries by the posterior communicating arteries.

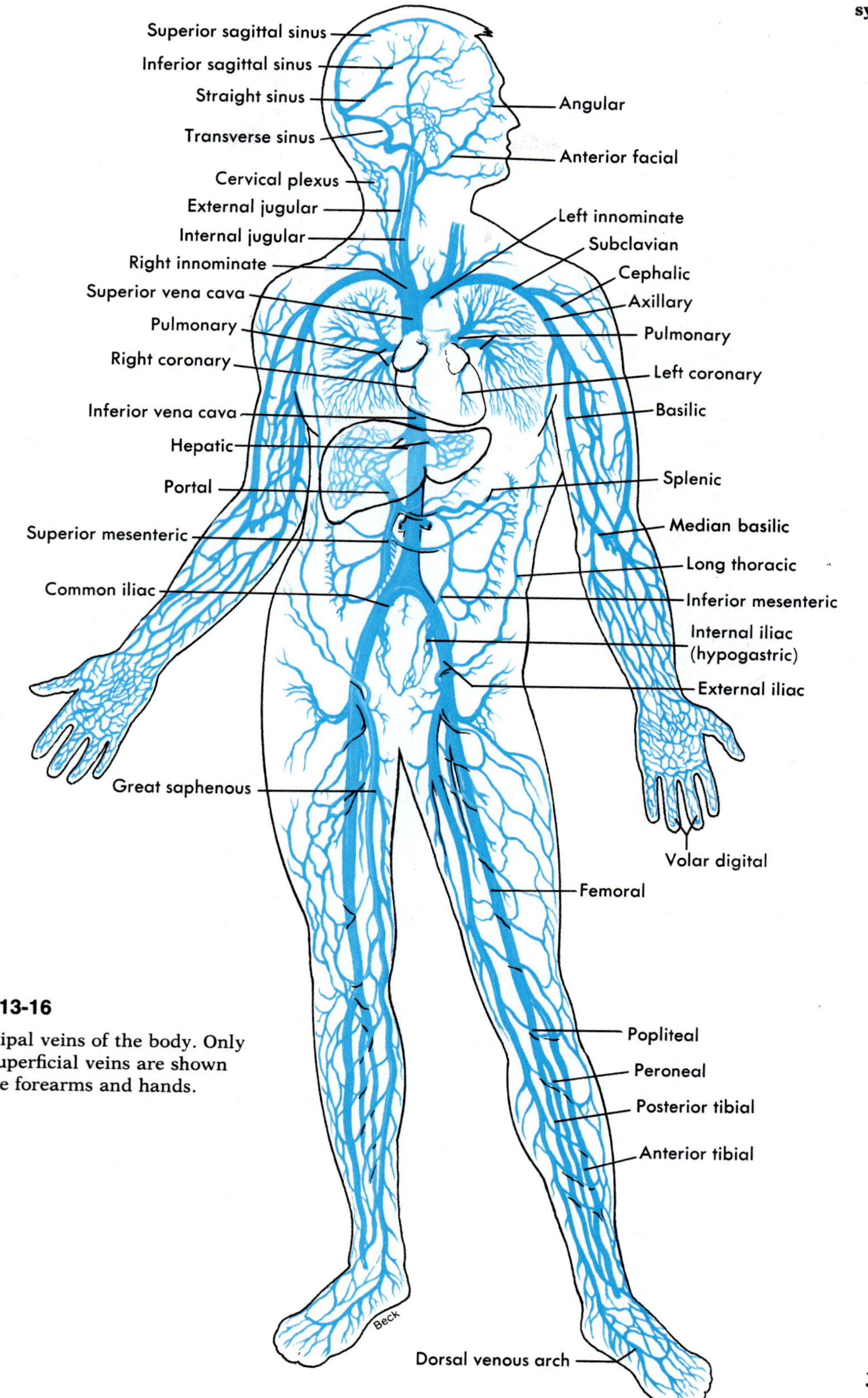

Fig. 13-16

Principal veins of the body. Only the superficial veins are shown on the forearms and hands.

VEINS

Several facts should be borne in mind while learning the names of veins.

1 Veins are the ultimate extensions of capillaries, just as capillaries are the eventual extensions of arteries. Whereas arteries branch into vessels of decreasing size to form arterioles and eventually capillaries, capillaries unite into vessels of increasing size to form venules and eventually veins.

2 Many of the main arteries have corresponding veins bearing the same name and located alongside or near the arteries. These veins, like the arteries, lie in deep, well-protected areas, for the most part close along the bones—example: femoral artery and femoral vein, both located along the femur bone.

3 Veins found in the deep parts of the body are called *deep veins* in contradistinction to *superficial veins,* which lie near the surface. The latter are the veins that can be seen through the skin.

4 The large veins of the cranial cavity, formed by the dura mater, are not called veins but *sinuses.* They should not be confused with the bony sinuses of the skull.

■ ■ ■

The following list identifies the major veins. Locate each one as named on Figs. 13-16 to 13-21.

Veins of upper extremities (Figs. 13-16 and 13-17)

Deep

Palmar (volar) arch (also superficial)
Radial (partially deep, partially superficial)
Ulnar (partially deep, partially superficial)
Brachial
Axillary (continuation of brachial)
Subclavian (continuation of axillary)

Superficial

Veins of hand from dorsal and volar venous arches, which, together with complicated network of superficial veins of lower arm, finally pour their blood into two large veins—cephalic (thumb side) and basilic (little finger side); these two veins empty into deep axillary vein

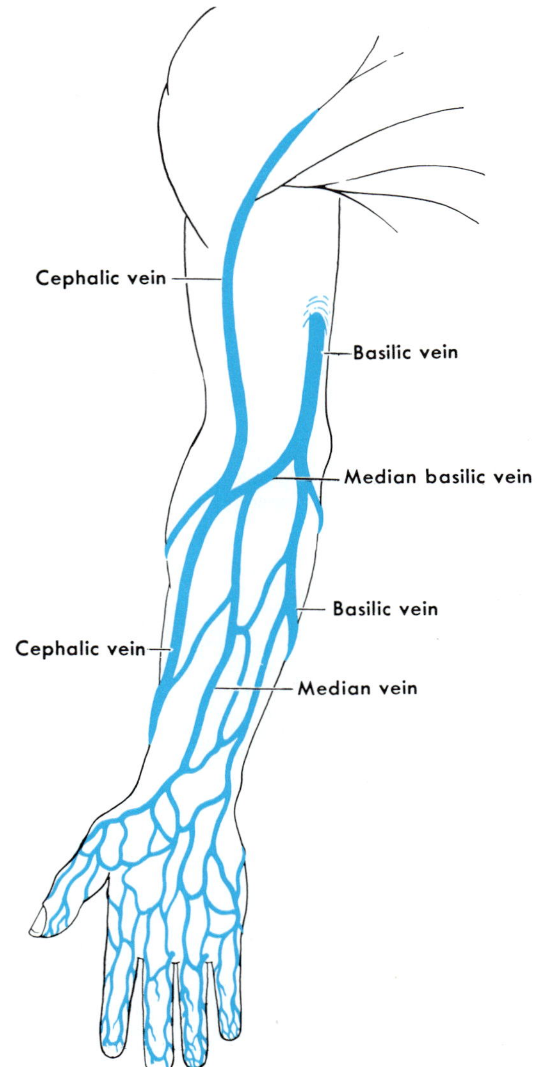

Fig. 13-17

Main superficial veins of the upper extremity, anterior view. The median basilic vein is commonly used for removing blood or giving intravenous infusions.

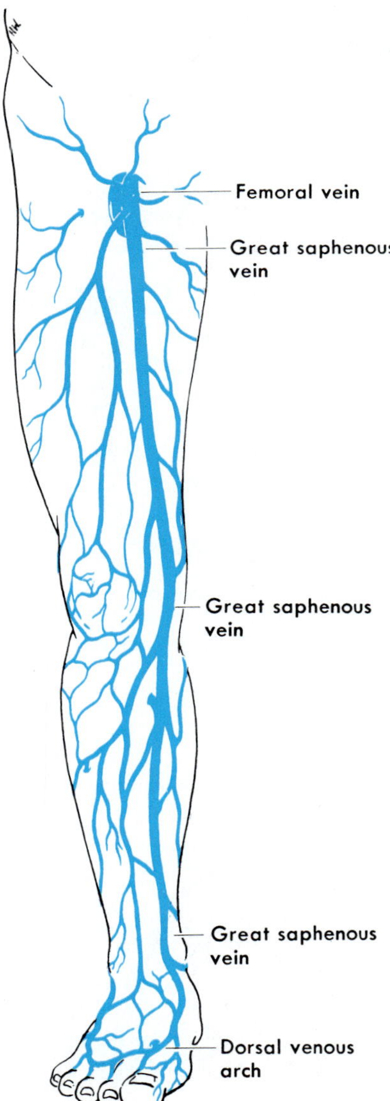

Fig. 13-18

Main superficial veins of the lower extremity, anterior view.

Fig. 13-19

Main superficial veins of the lower extremity, posterior view.

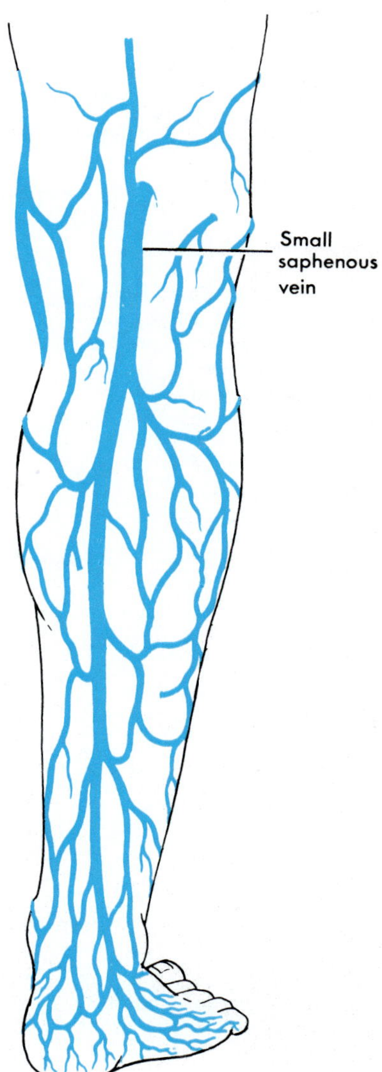

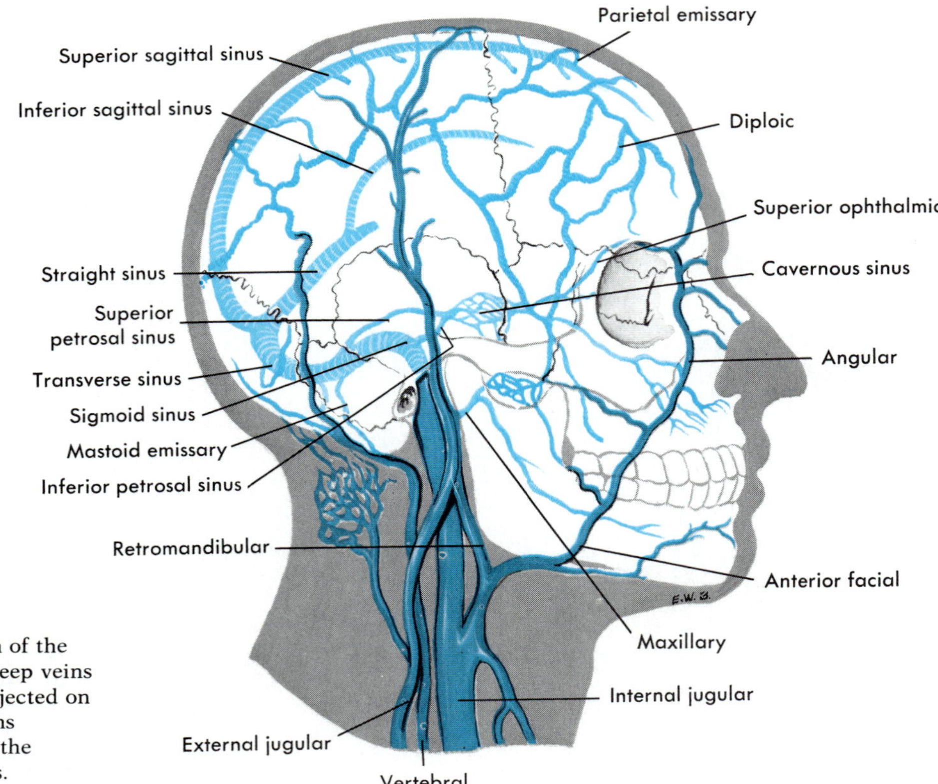

Fig. 13-20

Semischematic projection of the large veins of the head. Deep veins and dural sinuses are projected on the skull. Note connections (emissary veins) between the superficial and deep veins.

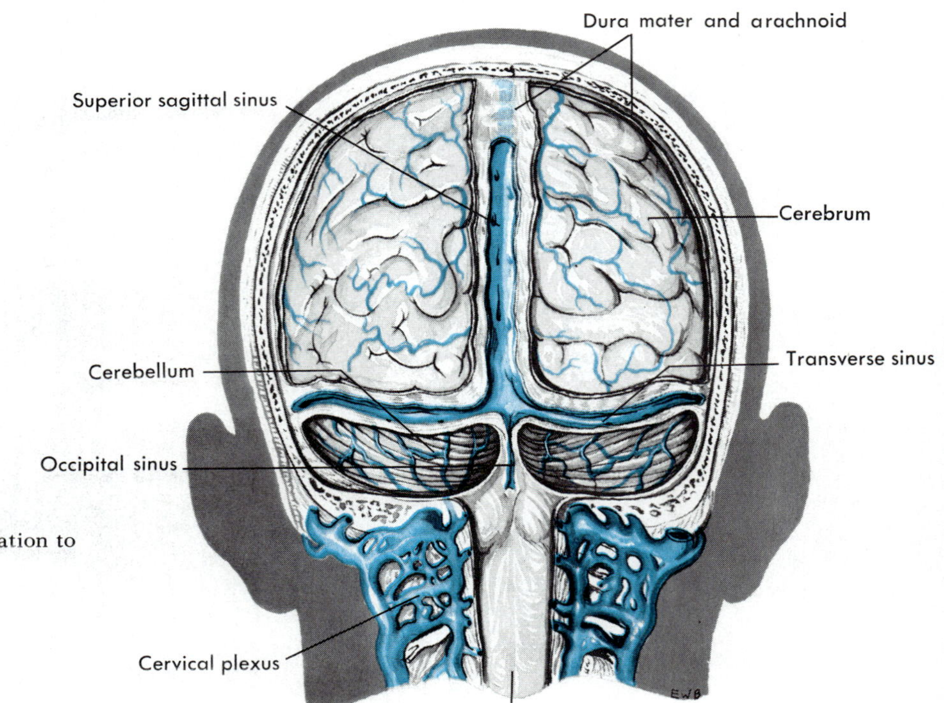

Fig. 13-21

Venous sinuses shown in relation to the brain and skull.

Veins of lower extremities (Figs. 13-16, 13-18, and 13-19)

Deep

Plantar arch
Anterior tibial
Posterior tibial
Popliteal
Femoral
External iliac

Superficial

Dorsal venous arch of foot
Great (or internal or long) saphenous
Small (or external or short) saphenous
(Great saphenous terminates in femoral vein in groin; small saphenous terminates in popliteal vein)

Veins of head and neck (Figs. 13-20 and 13-21)

Deep (in cranial cavity)

Longitudinal (or sagittal) sinus
Inferior sagittal and straight sinus
Numerous small sinuses
Right and left transverse (or lateral) sinuses
Internal jugular veins, right and left (in neck); continuations of transverse sinuses
Innominate veins, right and left; formed by union of subclavian and internal jugulars

Superficial

External jugular veins, right and left (in neck); receive blood from small superficial veins of face, scalp, and neck; terminate in subclavian veins (small emissary veins connect veins of scalp and face with blood sinuses of cranial cavity, a fact of clinical interest as a possible avenue for infections to enter cranial cavity)

Veins of abdominal organs (Figs. 13-16 and 13-22)

Spermatic (or ovarian) Renal Hepatic Suprarenal	Drain into inferior vena cava

Left spermatic and left suprarenal veins usually drain into left renal vein instead of into inferior vena cava; for return of blood from abdominal digestive organs, see discussion of portal circulation (next column); also Fig. 13-22

Veins of thoracic organs

Several small veins, such as bronchial, esophageal, pericardial, etc., return blood from chest organs (except lungs) directly into superior vena cava or into azygos vein; azygos vein lies to right of spinal column and extends from inferior vena cava (at level of first or second lumbar vertebra) through diaphragm to terminal part of superior vena cava; hemiazygos vein lies to left of spinal column, extending from lumbar level of inferior vena cava through diaphragm to terminate in azygos vein; accessory hemiazygos vein connects some of superior intercostal veins with azygos or hemiazygos vein

Correlations. Middle ear infections sometimes cause infection of the transverse sinuses with the formation of a thrombus. In such cases, the internal jugular vein may be ligated to prevent the development of a fatal cardiac or pulmonary embolism.

Intravenous injections are perhaps most often given into the median basilic vein at the bend of the elbow. Blood that is to be used for various laboratory tests is also usually removed from this vein. In an infant, however, the longitudinal sinus is more often punctured (through the anterior fontanel) because the superficial arm veins are too tiny for the insertion of a needle.

Portal circulation

Veins from the spleen, stomach, pancreas, gallbladder, and intestines do not pour their blood directly into the inferior vena cava as do the veins from other abdominal organs. Instead, they send their blood to the liver by means of the portal vein. Here, the blood mingles with the arterial blood in the capillaries and is eventually drained from the liver by the hepatic veins that join the inferior vena cava. The reason for this detouring of the blood through the liver before it returns to the heart will be discussed in the chapter on the digestive system.

Fig. 13-22 shows the plan of the portal system. In most individuals, the portal vein is formed by the union of the splenic and superior mesenteric veins, but blood from the gastric, pancreatic, and inferior mesenteric veins drains into the splenic vein before it merges with the superior mesenteric vein.

If either portal circulation or venous return from the liver is interfered with (as they often are in certain types of liver or heart disease), then venous drainage from most of the other abdominal organs is necessarily obstructed also. The accompanying increased capillary pressure accounts at least in part for the occurrence of ascites ("dropsy" of abdominal cavity) under these conditions.

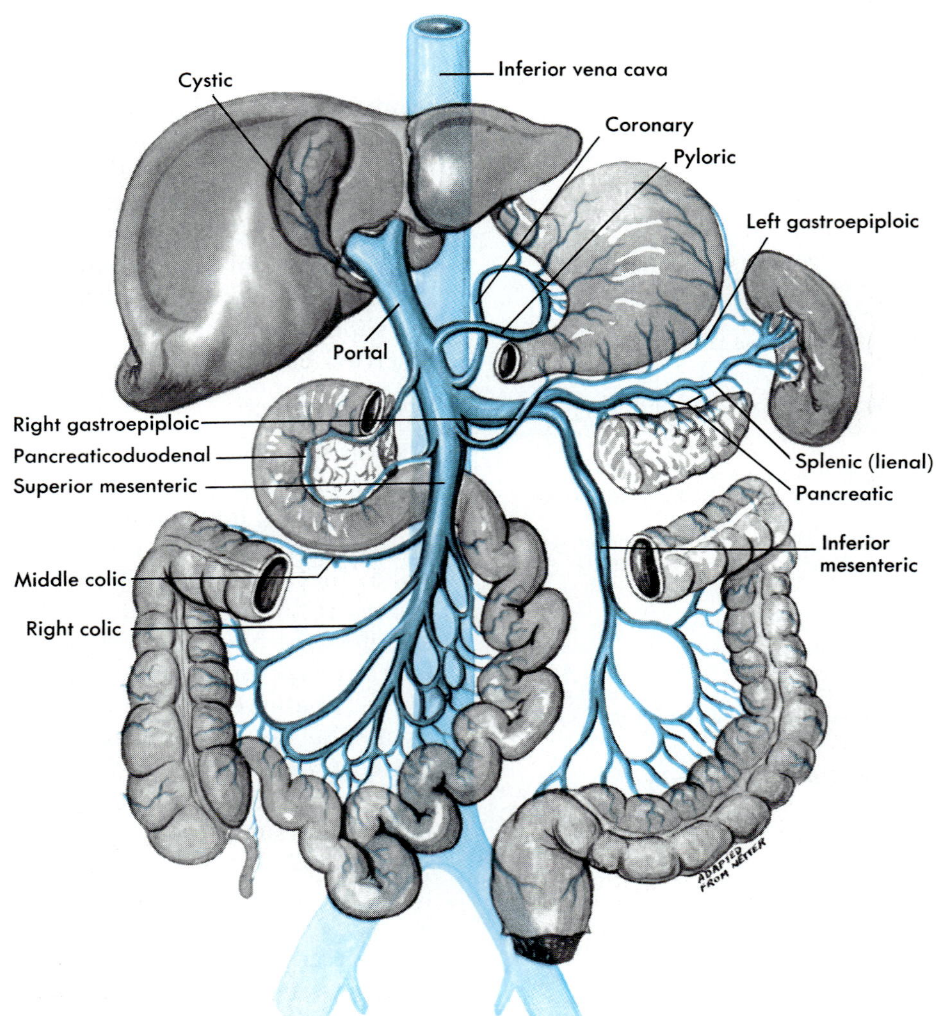

Fig. 13-22

Tributaries of the portal vein (diagrammatic).

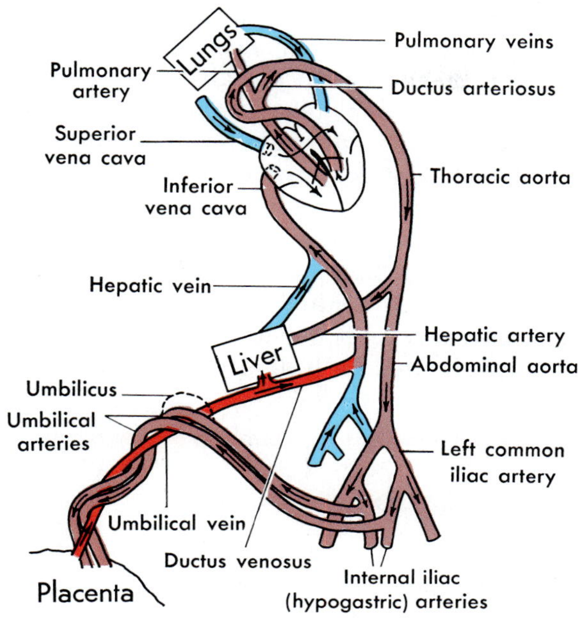

Fig. 13-23

Scheme to show the plan of fetal circulation. Note the following essential features: (1) two umbilical arteries, extensions of the internal iliac arteries, carry blood to (2) the placenta, which is attached to the uterine wall; (3) one umbilical vein returns blood, rich in oxygen and food, from the placenta; (4) the ductus venosus, a small vessel that connects the umbilical vein with the inferior vena cava; (5) the foramen ovale, an opening in the septum between the right and left atria; and (6) the ductus arteriosus, a small vessel that connects the pulmonary artery with the thoracic aorta.

Fetal circulation

Circulation in the body before birth necessarily differs from circulation after birth for one main reason—because fetal blood secures oxygen and food from maternal blood instead of from its own lungs and digestive organs, respectively. Obviously, then, there must be additional blood vessels in the fetus to carry the fetal blood into close approximation with the maternal blood and to return it to the fetal body. These structures are the two *umbilical arteries*, the *umbilical vein*, and the *ductus venosus*. Also, there must be some structure to function as the lungs and digestive organs do postnatally—that is, a place where an interchange of gases, foods, and wastes between the fetal and maternal blood can take place. This structure is the *placenta*. The exchange of substances occurs without any actual mixing of maternal and fetal bloods since each flows in its own capillaries.

In addition to the placenta and umbilical vessels, three structures located within the fetus' own body play an important part in fetal circulation. One of them (ductus venosus) serves as a detour by which most of the blood returning from the placenta bypasses the fetal liver. The other two (foramen ovale and ductus arteriosus) provide detours by which blood bypasses the lungs. A brief description of each of the six structures necessary for fetal circulation follows (also see Fig. 13-23).

1 The *two umbilical arteries* are extensions of the internal iliac (hypogastric) arteries and carry fetal blood to the placenta.

2 The *placenta* is a structure attached to the uterine wall. Exchange of oxygen and other substances between maternal and fetal blood takes place in the placenta.

3 The *umbilical vein* returns oxygenated blood from the placenta, enters the fetal body through the umbilicus, extends up to undersurface of the liver where it gives off two or three branches to the liver, and then continues on as the ductus venosus. Two umbilical arteries and the umbilical vein together constitute the *umbilical cord* and are shed at birth along with the placenta.

4 The *ductus venosus* is a continuation of the umbilical vein along the undersurface of the liver and drains into the inferior vena cava. Most of the blood returning from the placenta bypasses the liver. Only a relatively small amount enters the liver by way of the branches from the umbilical vein into the liver.

5 The *foramen ovale* is an opening in the septum between the right and left atria. A valve at the opening of the inferior vena cava into the right atrium directs most of the blood through the foramen ovale into the left atrium so that it bypasses the fetal lungs. A small percentage of the blood leaves the right atrium for the right ventricle and pulmonary artery. But even most of this does not flow on into the lungs. Still another detour, the ductus arteriosus, diverts it.

6 The *ductus arteriosus* is a small vessel connecting the pulmonary artery with the descending thoracic aorta. It therefore enables another portion of the blood to detour into the systemic circulation without going through the lungs.

■ ■ ■

Almost all fetal blood is a mixture of oxygenated and deoxygenated blood. Examine Fig. 13-23 carefully to determine why this is so. What happens to the oxygenated blood returned from the placenta via the umbilical vein? Note that it flows into the inferior vena cava.

Since the six structures that serve fetal circulation are no longer needed after birth, several changes take place. As soon as the umbilical cord is cut, the two umbilical arteries, the placenta, and the umbilical vein obviously no longer function. The placenta is shed from the mother's body as the afterbirth with part of the umbilical vessels attached. The sections of these vessels remaining in the infant's body eventually become fibrous cords which remain throughout life (the umbilical vein becomes the round ligament of the liver). The ductus venosus, no longer needed to bypass blood around the liver, eventually becomes the ligamentum venosum of the liver. The foramen ovale normally becomes functionally closed soon after a newborn baby takes his first breath and full circulation through his lungs becomes established. Complete structural closure, however, usually requires 9 months or more. Eventually, the foramen ovale becomes a mere depression (fossa ovalis) in the wall of the right atrial septum. The ductus arteriosus contracts as soon as respiration is established. Eventually, it also turns into a fibrous cord.

Circulation

Definitions

The term circulation of blood suggests its meaning—namely, blood flow through vessels arranged to form a circuit or circular pattern. Blood flow from the heart (left ventricle) through blood vessels to all parts of the body and back to the heart (to the right atrium) is spoken of as *systemic circulation.* The left ventricle pumps blood into the ascending aorta. From here it flows into arteries that carry it into the various tissues and organs of the body. Within each structure blood moves, as indicated in Fig. 13-24, from arteries to arterioles to capillaries. Here, the vital two-way exchange of substances occurs between blood and cells. Blood flows next out of each organ by way of its venules and then its veins to drain eventually into the inferior or superior vena cava. These two great veins of the body return venous blood to the heart (to the right atrium) to complete systemic circulation. But the blood has not quite come full circle back to its starting point, the left ventricle. To do this and start on its way again, it must first flow through another circuit, the *pulmonary circulation.* Observe in Fig. 13-24 that venous blood moves from the right atrium to the right ventricle to the

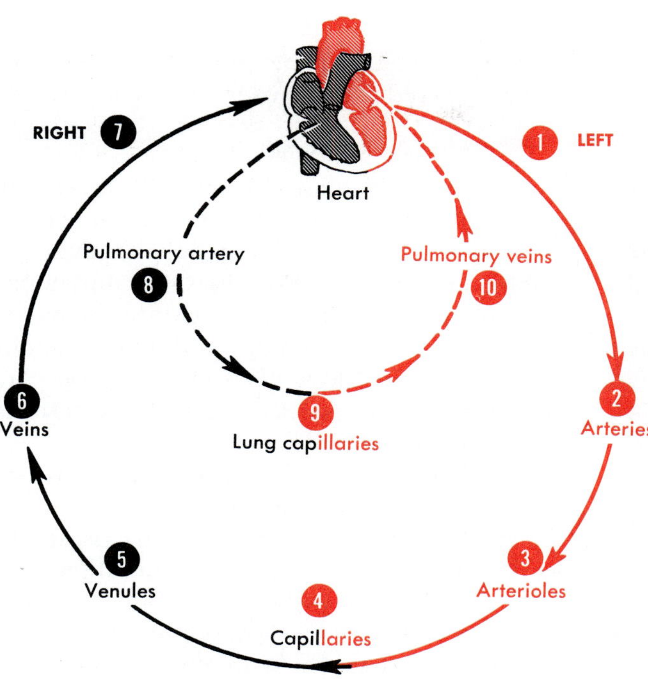

Fig. 13-24

Relationship of systemic and pulmonary circulation. As indicated by the numbers, blood circulates from the left side (ventricle) of the heart to arteries, to arterioles, to capillaries, to venules, to veins, to the right side of the heart (atrium to ventricle), to the lungs, and back to the left side of the heart, thereby completing a circuit. Refer to this diagram when tracing the circulation of blood to or from any part of the body.

pulmonary artery to lung arterioles and capillaries. Here, exchange of gases between blood and air takes place, converting venous blood to arterial blood. This oxygenated blood then flows on through lung venules into four pulmonary veins and returns to the left atrium of the heart. From the left atrium it enters the left ventricle to be pumped again through the systemic circulation.

How to trace

In order to list the vessels through which blood flows in reaching a designated part of the body or in returning to the heart from a part, one must remember the following:

1 That blood always flows in this direction —from *left ventricle* of heart to *arteries,* to *arterioles,* to *capillaries* of each body part, to *venules,* to *veins,* to *right atrium, right ventricle, pulmonary artery, lung capillaries, pulmonary veins, left atrium,* and back to left ventricle (Fig. 13-24)

2 That when blood is in capillaries of abdominal digestive organs, it must flow through portal system (Fig. 13-22) before returning to heart

3 Names of main arteries and veins of body

For example, suppose glucose were instilled into the rectum. To reach the cells of the right little finger, the vessels through which it would pass after absorption from the intestinal mucosa into capillaries would be as follows: *capillaries* into venules of large intestine into inferior mesenteric *vein,* splenic vein, portal vein, capillaries of liver, hepatic veins, inferior vena cava, *right atrium* of heart, *right ventricle, pulmonary artery, pulmonary capillaries, pulmonary veins, left atrium, left ventricle, ascending aorta,* aortic arch, innominate artery, right subclavian artery, right axillary artery, right brachial artery, right ulnar artery, arteries of palmar arch, arterioles, and *capillaries* of right little finger.

Note: The structures italicized show the direction of blood flow as described in points **1** and **2** and illustrated in Fig. 13-24. Follow this course of circulation first on Fig. 13-22 and then on Figs. 13-16 and 13-12. Try to answer question **39** of the review questions at the end of this chapter, using the plan outlined.

Functions of control mechanisms

Circulation is, of course, a vital function. It constitutes the only means by which cells can receive materials needed for their survival and can have their wastes removed. Not only is circulation necessary, but circulation of different volumes of blood per minute is also essential for healthy survival. More active cells need more blood per minute than less active cells. The reason underlying this principle is obvious. The more work cells do, the more energy they use and the more oxygen and food they need to supply this energy. Only arterial blood can deliver these energy suppliers. So the more active any part of the body is, the greater the volume of blood circulated to it per minute must be. This requires that circulation control mechanisms accomplish two functions: maintain circulation (keep blood flowing, that is) and vary the volume and distribution of the blood circulated. The greater the activity of any part of the body, the greater the volume of blood it needs circulating through it. Therefore, as any structure increases its activity, an increased volume of blood must be distributed to it—must be shifted from the less active to the more active tissues.

To achieve these two ends, a great many factors must operate together as one smooth-running though complex machine. Incidentally, this is an important physiological principle that you have no doubt observed by now—that every body function depends upon many other functions. A constellation of separate processes or mechanisms act as a single integrated mechanism. Together, they perform some one large function. For example, many mechanisms together accomplish the large function we call circulation. To try to make the complexities of circulation mechanisms a little more understandable, we shall state the major principles upon which they operate and explain each one as best we can.

Principles of circulation

1 Blood circulates for the same reason that any fluid flows—whether it be water in a river or in a garden hose or in hospital tubing or blood in vessels. A fluid flows because a pressure gradient exists between different parts of its bed. This primary fluid flow principle derives from Newton's first and second laws of motion. In essence, these laws state the following principles:

- **a** A fluid does not flow when the pressure is the same in all parts of it
- **b** A fluid flows only when its pressure is higher in one area than in another, and it flows always from its higher pressure area toward its lower pressure area

In brief, then, the primary principle about circulation is this: blood circulates from the left ventricle to the right atrium of the heart because a blood pressure gradient exists between these two structures. By blood pressure gradient, we mean the difference between the blood pressure in one structure and the blood pressure in another. For example, a typical normal blood pressure in the aorta, as the left ventricle contracts pumping blood into it, is 120 mm Hg, and as the left ventricle relaxes, it decreases to 80 mm Hg. The mean or average blood pressure, therefore, in the aorta in this instance is 100 mm Hg. Central venous pressure (pressure in the venae cavae, at their entrance into the right atrium) normally ranges between 4 and 10 cm H_2O, which is about 3 to 7 mm Hg.* The systemic blood pressure gradient equals the mean aortic pressure (at the exit of the aorta from the left ventricle) minus the central venous pressure. Using typical normal values for these pressures, the systemic blood pres-

*To convert cm H_2O to mm Hg, divide cm H_2O by 1.36. Example: 4 cm $H_2O \div 1.36 = 2.94$ mm Hg.

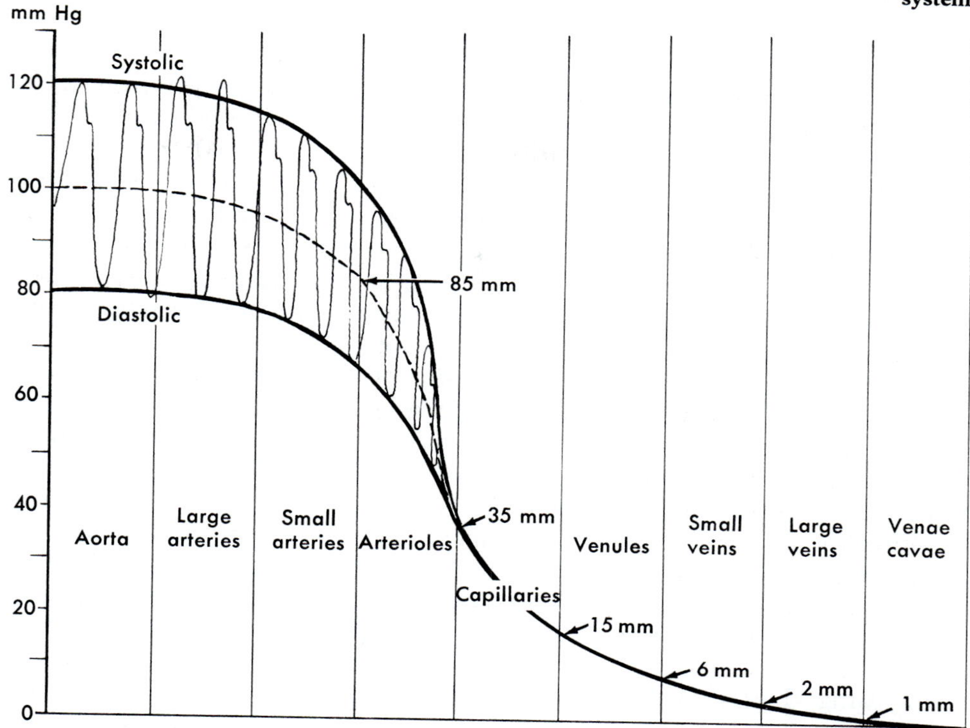

Fig. 13-25

Blood pressure gradient. Dotted line indicates the average or mean systolic pressure in arteries.

sure equals 95 mm Hg (100 mm Hg − 5 mm Hg).

The symbol ($P_1 - P_2$) is often used to stand for a pressure gradient, with P_1 the symbol for the higher pressure and P_2 the symbol for the lower pressure. Suppose, for instance, that the average capillary pressure in the capillaries of your arm muscles is 25 mm Hg and the average pressure in the arterioles from which these capillaries branch is 60 mm Hg. Which is P_1? P_2? What is this blood pressure gradient? It would cause blood to flow from the arterioles in the arm into its capillaries.

2 The primary determinant of arterial blood pressure is the volume of blood in the arteries. A direct relationship exists between arterial blood volume and arterial pressure. This means that an increase in arterial blood volume tends to increase arterial pressure and, conversely, a decrease in arterial volume tends to decrease arterial pressure.

3 Many factors together indirectly determine arterial pressure through their influence on arterial volume. Two of the most important are cardiac minute output (CMO) and peripheral resistance.

4 The cardiac minute output is determined by both the rate of the heartbeat per minute and the volume of blood pumped out of the ventricles by each beat. Contraction of the heart is called *systole*. Therefore, the volume of blood pumped by one contraction is known as *systolic discharge. Stroke volume* means the same thing, the amount of blood pumped by one stroke (contraction) of the ventricle.

Stroke volume reflects the force or strength of ventricular contraction—the stronger the contraction, the greater the stroke volume tends to be. Cardiac minute output can be computed by the following simple equation:

$$\text{Stroke volume} \times \text{Heart rate} = \text{CMO}$$

In practice, however, computing the cardiac minute output is far from simple. It requires introducing a catheter into the right side of the heart (cardiac catheterization) and working a computation known as Fick's formula.*

Since the heart's rate and stroke volume determine its output, anything that changes either the rate of the heartbeat or its stroke volume tends to change cardiac minute output, arterial blood volume, and blood pressure in the same direction. In other words, anything that makes the heart beat faster or anything that makes it beat stronger (increases its stroke volume) tends to increase cardiac minute output and, therefore, arterial blood volume and pressure. Conversely, anything that causes the heart to beat more slowly or more weakly tends to decrease cardiac minute output, arterial volume, and blood pressure. But do not overlook the word *tend* in the preceding sentences. A change in heart rate or in stroke volume does not always change the heart's output, or the amount of blood in the arteries, or the blood pressure. To see whether this is true, do the following simple arithmetic, using the simple formula for computing CMO. Assume a normal rate of 72 beats per minute and a normal stroke volume of 70 ml. Next, suppose the rate drops to 60 and the stroke volume increases to 100. Does the decrease in heart rate actually cause a decrease in cardiac minute output in this case? Clearly not—the cardiac minute output increases. Do you think it is valid, however, to say that a slower rate *tends* to decrease the heart's minute output? By itself, without any change in any other factor, would not a slowing of the heartbeat cause cardiac minute volume, arterial volume, and blood pressure to fall?

5 *Peripheral resistance* helps determine arterial blood pressure. Specifically, arterial blood pressure tends to vary directly with peripheral resistance. Peripheral resistance means the resistance to blood flow imposed by the force of friction between blood and the walls of its vessels. Friction develops partly because of a characteristic of blood—its viscosity or stickiness—and partly from the small diameter of arterioles and capillaries. Peripheral resistance helps determine arterial pressure by controlling the rate of "arteriole runoff," the amount of blood that runs out of the arteries into the arterioles. The greater the resistance, the less the arteriole runoff tends to be—and, therefore, the more blood left in the arteries and the higher the arterial pressure tends to be.

Summarized, arterial blood pressure is determined directly by arterial blood volume, which is determined by many factors but especially by the heart's output and peripheral resistance (Fig. 13-26).

6 Mechanical, neural, and chemical factors regulate the strength of the heartbeat and therefore its stroke volume. The mechanical factor that helps determine stroke volume is the length of myocardial fibers at the beginning of ventricular contraction.

Many years ago Starling described a principle later made famous as Starling's law of the heart. In this principle he stated the factor he had observed as the main regulator of heartbeat strength in experiments performed on denervated animal hearts. Starling's law of the heart, in essence, is this: within limits, the longer or more stretched the heart fibers at the beginning of contraction, the stronger is their contraction.

The factor determining how stretched the

*Mountcastle, V. B., editor: Medical physiology, ed. 13, St. Louis, 1974, The C. V. Mosby Co., pp. 906-907.

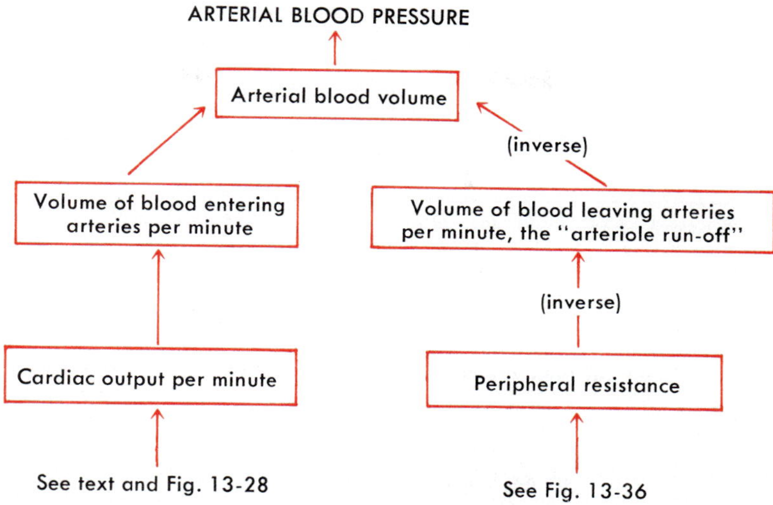

Fig. 13-26

Scheme to show how cardiac minute output and peripheral resistance affect arterial blood pressure. If cardiac minute output increases, the amount of blood entering the arteries increases and tends to increase the volume of blood in the arteries. The resulting increase in arterial volume increases arterial blood pressure. If peripheral resistance increases, it tends to increase arterial blood volume and pressure. Why? Because the increased resistance tends to decrease the amount of blood leaving the arteries, which tends to increase the amount of blood left in them. The increase in arterial volume increases arterial blood pressure.

animal hearts were at the beginning of contractions was, as you might deduce, the amount of blood in the hearts at the end of diastole. The more blood returned to the hearts per minute, the more stretched were their fibers, the stronger were their contractions, and the larger was the volume of blood they ejected with each contraction. If, however, too much blood stretched the hearts too far, beyond a certain critical point, they seemed to lose their elasticity. They then contracted less vigorously—much as a rubber band, stretched too much, rebounds with less force.

Physiologists have long accepted as fact that Starling's law of the heart operates in animals under experimental conditions. But they have questioned its validity and importance in the intact human body. Today, the prevailing opinion seems to be that Starling's law of the heart operates as a major regulator of stroke volume under ordinary conditions. Operation of Starling's law of the heart ensures that increased amounts of blood returned to the heart will be pumped out of it. It automatically adjusts cardiac output to venous return under usual conditions.

7 *Pressoreflexes* constitute the dominant heart rate control mechanism, although various other factors also influence heart rate.

Cardiac pressoreflexes. Pressure-sensitive cells called baroreceptors (or pressoreceptors) are located in the aortic arch and the carotid sinus. (The carotid sinus is a small dilation at the beginning of the internal carotid artery just above the bifurcation of the common carotid artery to form the internal and external carotid arteries.) The

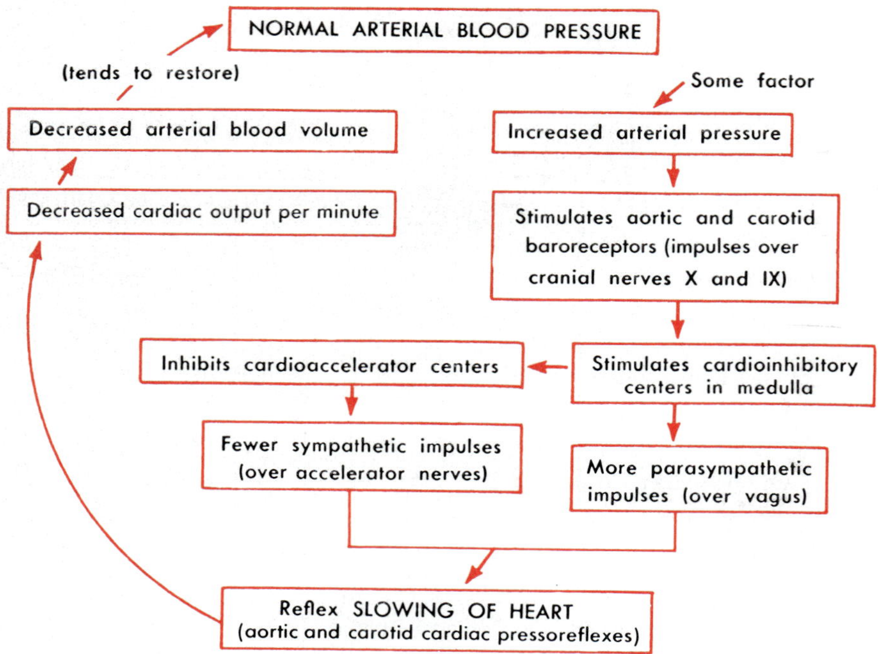

Fig. 13-27

The aortic and carotid cardiac pressoreflexes, a mechanism that tends to maintain or restore homeostasis of arterial blood pressure by regulating the rate of the heartbeat. Note that by this mechanism an increase in arterial blood pressure leads to reflex slowing of the heart and tends to lower blood pressure. The converse is also true. A decrease in blood pressure leads to reflex acceleration of the heart and tends to raise the pressure upward toward normal.

sinus lies just under the sternocleidomastoid muscle at the level of the upper margin of the thyroid cartilage. Sensory fibers extend from the aortic baroreceptors in the vagus (tenth cranial) nerve to terminate in the medulla in its cardiac and vasomotor centers. Sensory fibers from carotid sinus baroreceptors, on the other hand, run through the carotid sinus nerve (of Hering) and on through the glossopharyngeal (or ninth cranial) nerve to the cardiac and vasomotor centers.

If blood pressure within the aorta or carotid sinus increases suddenly, it stimulates the aortic or carotid baroreceptors. As shown in Fig. 13-27, this leads to stimulation of the cardioinhibitory centers and reciprocal inhibition of the accelerator centers, which in turn leads to more impulses per second over parasympathetic fibers in the vagus nerve and fewer impulses over the sympathetic fibers in the cardioaccelerator nerves to the heart. As a result, reflex slowing of the heart occurs.

On the other hand, a decrease in aortic or carotid blood pressure usually initiates reflex acceleration of the heart. The lower blood pressure decreases baroreceptors' stimulation. Hence, the cardioinhibitory center receives fewer stimulating impulses and the cardioaccelerator center fewer inhibitory

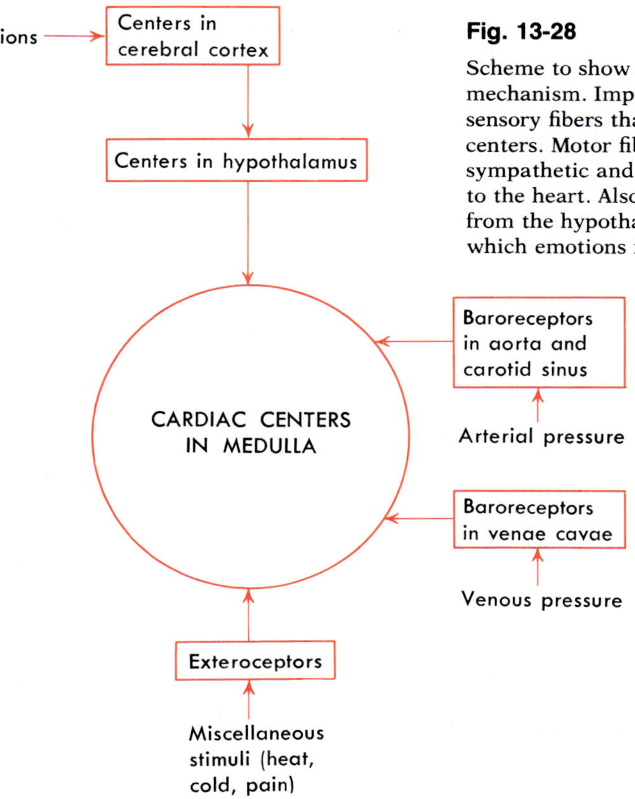

Fig. 13-28

Scheme to show some of the many parts of the heart rate control mechanism. Impulses from various receptors are conducted by sensory fibers that terminate in synapses with neurons in cardiac centers. Motor fibers from the centers relay impulses to sympathetic and parasympathetic neurons, which transmit them to the heart. Also streaming into the cardiac centers are impulses from the hypothalamus, presumably part of the pathway by which emotions influence heart rate.

impulses. Net result? The heart beats faster.

Miscellaneous factors that influence heart rate. Included in this category are such important factors as emotions, exercise, hormones, blood temperature, and stimulation of various exteroceptors. Anxiety, fear, and anger often make the heart beat faster. Grief, in contrast, tends to slow it. Presumably, emotions produce changes in the heart rate through the influence of impulses from the cerebrum via the hypothalamus, medulla, and autonomic neurons.

In exercise, the heart normally accelerates. The mechanism is not definitely known. But it is thought to include impulses from the cerebrum through the hypothalamus to cardiac centers. Epinephrine is the hormone most noted as a cardiac accelerator.

Increased blood temperature or stimulation of skin heat receptors tends to increase the heart rate, and decreased blood temperature or stimulation of skin cold receptors tends to slow it. Sudden intense stimulation of pain receptors also tends to decrease the heart rate. Fig. 13-28 shows the major factors that control the rate of the heartbeat.

8 Peripheral resistance exists mainly because of blood viscosity and the small diameter of arterioles. Viscosity is the characteristic of a fluid that results from attraction forces between its molecules or other small particles and that causes it to resist flowing. Blood viscosity stems mainly from the red cells but also partly from the protein molecules present in blood. An increase in either blood protein concentration or in the red cell count tends to increase viscosity, and a decrease in either tends to decrease it. Under normal circumstances, blood viscosity changes very little. But under certain abnormal conditions, such as marked anemia or hemorrhage, a decrease in blood viscosity may be the crucial factor lowering peripheral resistance and arterial pressure even to the point of circulatory failure.

9 Numerous factors control arteriole diameter. They might be said to constitute the vasomotor control mechanism. Like most physiological mechanisms, it consists of many parts. A change in either arterial blood pressure or in arterial blood's oxygen or carbon dioxide content sets a chemical vasomotor control mechanism in operation. A change in arterial blood pressure initiates a *vasomotor pressoreflex.*

Vasomotor pressoreflexes (Fig. 13-29). A sudden increase in arterial blood pressure stimulates aortic and carotid baroreceptors —the same ones that initiate cardiac reflexes. Not only does this lead to stimulation of cardioinhibitory centers but also to inhibition of vasoconstrictor centers. More impulses per second go out to the heart over parasympathetic vagal fibers and fewer over sympathetic fibers to blood vessels. As a result, the heartbeat slows, and arterioles and the venules of the "blood reservoirs" dilate. Since sympathetic vasoconstrictor impulses predominate at normal arterial pressures, inhibition of these is considered the major mechanism of vasodilation.

The main blood reservoirs are the venous plexuses and sinuses in the skin and abdominal organs (especially in the liver and spleen). In other words, blood reservoirs are the venous networks in most parts of the body—all but those in the skeletal muscles, heart, and brain. The term reservoir is apt, since these veins serve as storage depots for blood. It can quickly be moved out of them and "shifted" to heart and skeletal muscles when increased activity demands. The reflex vasoconstrictor mechanism that accomplishes this operates as follows.

A decrease in arterial pressure causes the aortic and carotid baroreceptors to send more impulses to the medulla's vasoconstrictor centers, thereby stimulating them. These centers then send more impulses via sympathetic fibers to the smooth muscle in the arterioles, venules, and veins of the blood reservoirs, causing their constriction. This squeezes more blood out of them, increasing the amount of venous return to the heart. Eventually, this extra blood is redistributed to more active structures such as skeletal muscles and heart because their arterioles become dilated largely from the operation of a local mechanism (p. 364). Thus the vasoconstrictor pressoreflex and the local vasodilating mechanism together serve as an important device for shifting blood from reservoirs to structures that need it more. It is an especially valuable mechanism during exercise.

Vasomotor chemoreflexes (Fig. 13-30). Chemoreceptors located in the aortic and carotid bodies are particularly sensitive to a deficiency of blood oxygen (hypoxia) and somewhat less sensitive to excess blood carbon dioxide (hypercapnia) and to decreased arterial blood pH. When one or more of these conditions stimulates the chemoreceptors, their fibers transmit more impulses to the medulla's vasoconstrictor centers, and vasoconstriction of arterioles and venous reservoirs soon follows. This stress mechanism functions as an emergency device when severe hypoxia or hypercapnia endangers survival.

The medullary ischemic reflex (Fig. 13-31). The medullary ischemic reflex mechanism is said to exert the most powerful control of all on small blood vessels. When the blood supply to the medulla becomes inadequate (ischemic), its neurons suffer from both oxygen deficiency and carbon dioxide excess. But, presumably, it is the latter, the hypercapnia, that intensely and directly stimulates the vasoconstrictor centers to bring about marked arteriole and venous constriction. If the oxygen supply to the medulla decreases below a certain level, its neurons, of course, cannot function and the medullary ischemic reflex cannot operate.

Vasomotor control by higher brain centers (Fig. 13-32). Impulses from centers in the cerebral cortex and in the hypothalamus are believed to be transmitted to the vasomotor centers in the medulla and to thereby help control vasoconstriction and dilation. One

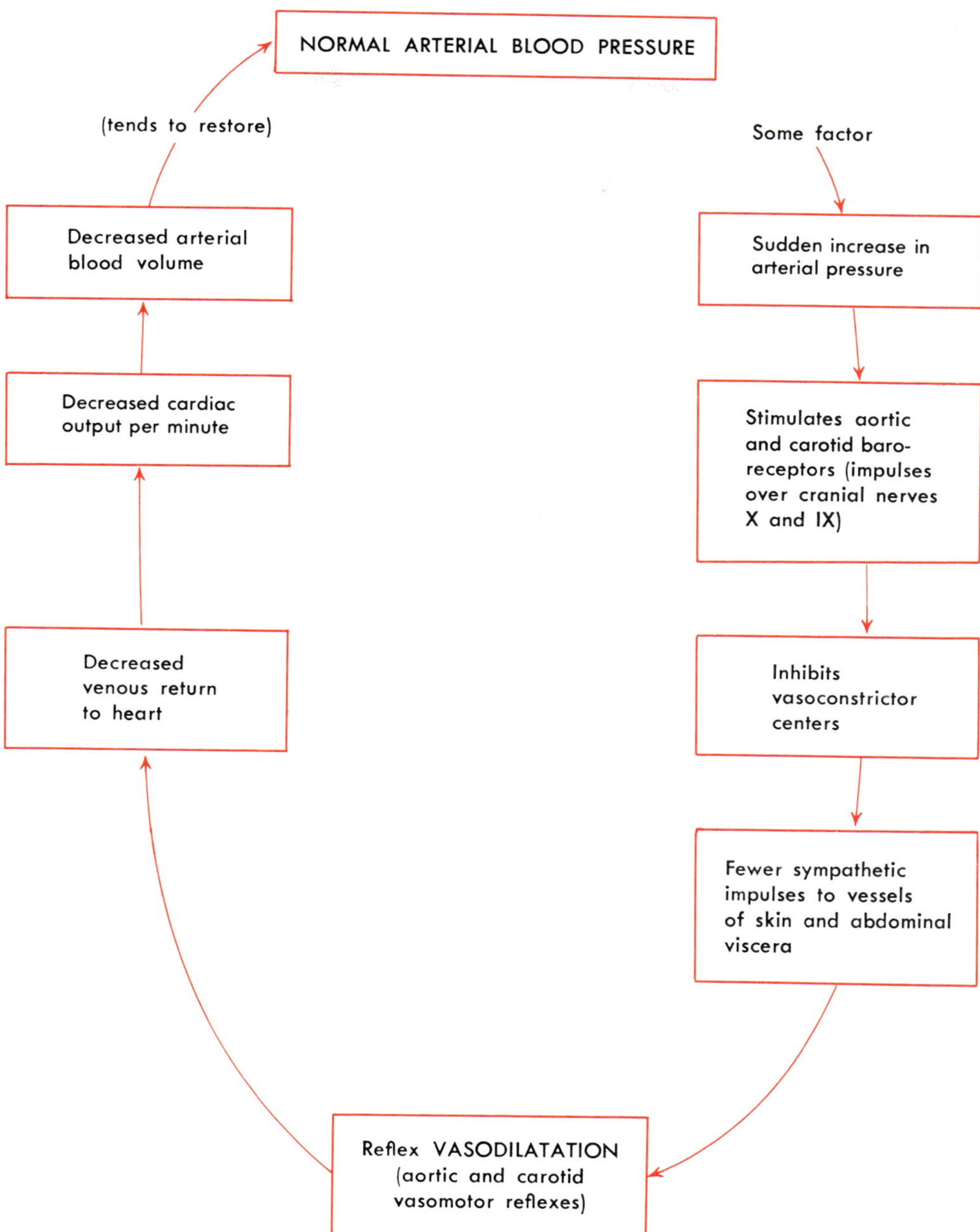

Fig. 13-29

The aortic and carotid vasomotor pressoreflex mechanism shown here is set in operation when some factor causes a sudden increase in arterial blood pressure. This mechanism and the aortic and carotid cardiac pressoreflexes (Fig. 13-27) operate simultaneously to maintain or restore homeostasis of arterial blood pressure.

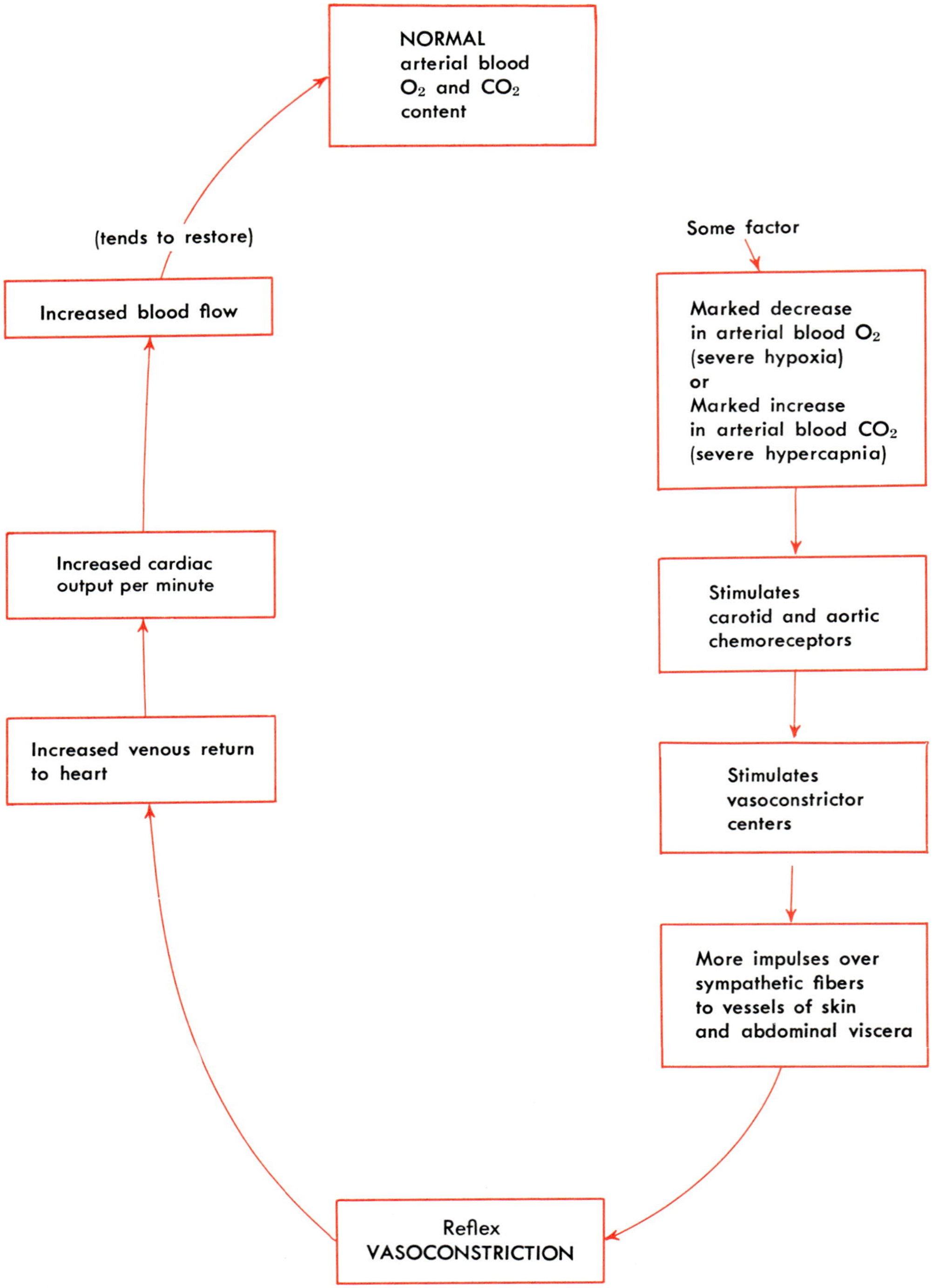

Fig. 13-30

The vasomotor chemoreflex shown here does not function under normal conditions. It operates as a response to the stress of severe hypoxia or hypercapnia and tends to restore normal blood O_2 and CO_2 levels. Because this mechanism brings about reflex vasoconstriction, it also tends to increase peripheral resistance and arterial blood pressure. (Also see Fig. 13-31.)

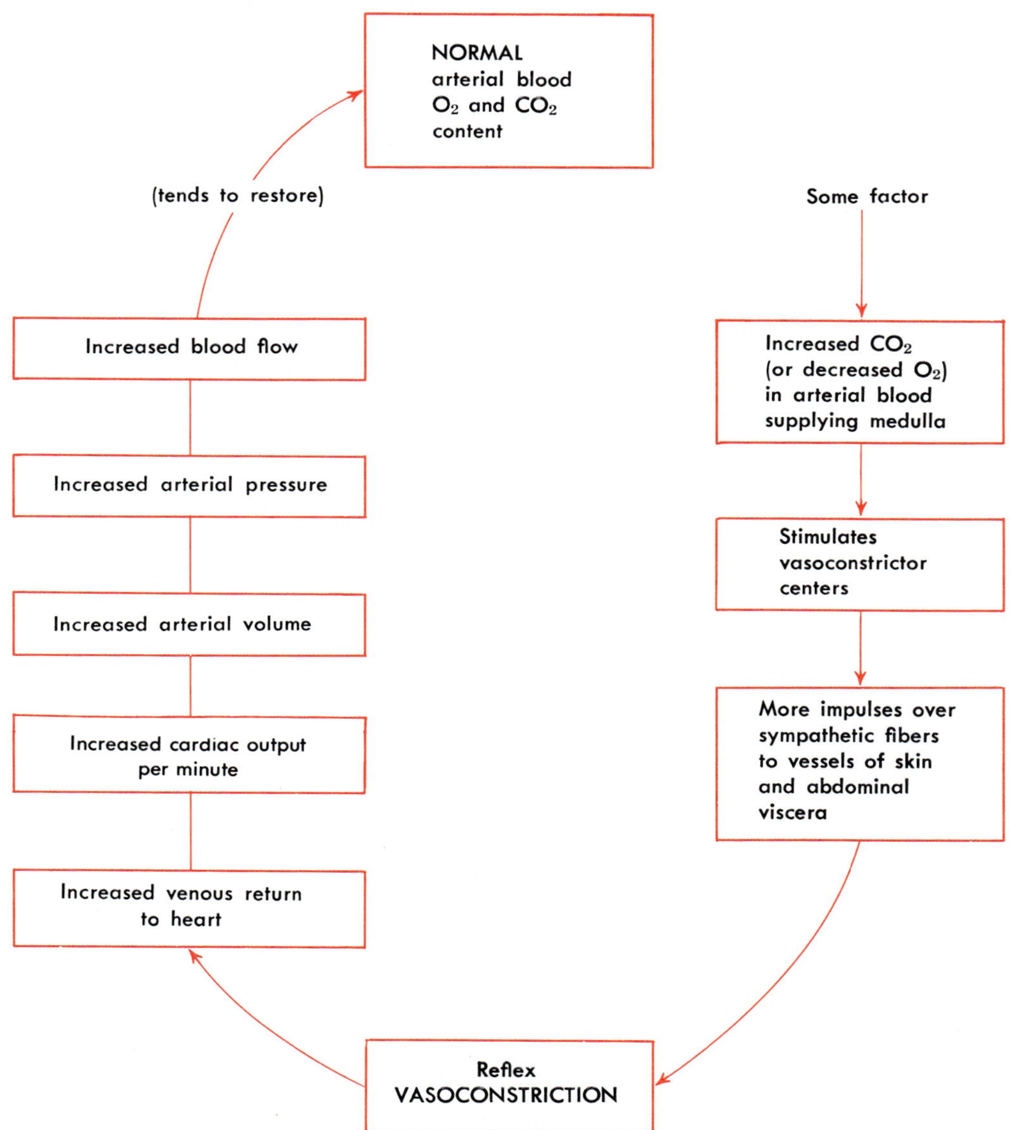

Fig. 13-31

The medullary ischemic reflex, a homeostatic mechanism that tends to restore homeostasis of blood oxygen and carbon dioxide content.

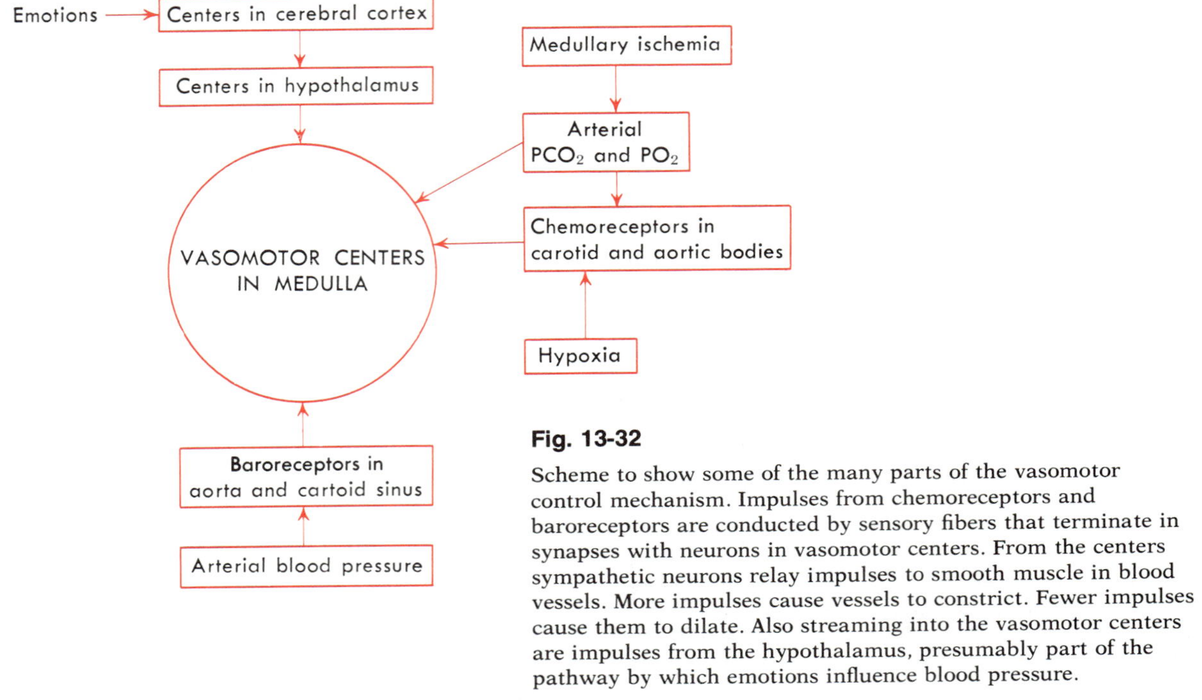

Fig. 13-32

Scheme to show some of the many parts of the vasomotor control mechanism. Impulses from chemoreceptors and baroreceptors are conducted by sensory fibers that terminate in synapses with neurons in vasomotor centers. From the centers sympathetic neurons relay impulses to smooth muscle in blood vessels. More impulses cause vessels to constrict. Fewer impulses cause them to dilate. Also streaming into the vasomotor centers are impulses from the hypothalamus, presumably part of the pathway by which emotions influence blood pressure.

evidence supporting this view, for example, is the fact that vasoconstriction and a rise in arterial blood pressure characteristically accompany emotions of intense fear or anger. Also, laboratory experiments on animals in which stimulation of the posterior or lateral parts of the hypothalamus leads to vasoconstriction support the belief that higher brain centers influence the vasomotor centers in the medulla.

Local control of arterioles

Some kind of local mechanism operates to produce vasodilation in localized areas. Although the mechanism is not clear, it is known to function in times of increased tissue activity. For example, it probably accounts for the increased blood flow into skeletal muscles during exercise. It also operates in ischemic tissues, serving as a homeostatic mechanism that tends to restore normal blood flow. Norepinephrine, histamine, lactic acid, and other locally produced substances have been suggested as the stimuli that activate the local vasodilator mechanism. Local vasodilation is also referred to as *reactive hyperemia.*

Summarizing briefly, the volume of blood circulating through the body per minute is determined by the magnitude of both the blood pressure gradient and the peripheral resistance (Figs. 13-34 to 13-36).

A nineteenth century physiologist and physicist, Poiseuille, described the relationship between these three factors—pressure gradient, resistance, and volume of fluid flow per minute—with a mathematical equation known as *Poiseuille's law.* In general, but with certain modifications, it applies to blood circulation. We can state it in a simplified form as follows: the volume of blood circulated per minute is directly related to mean arterial

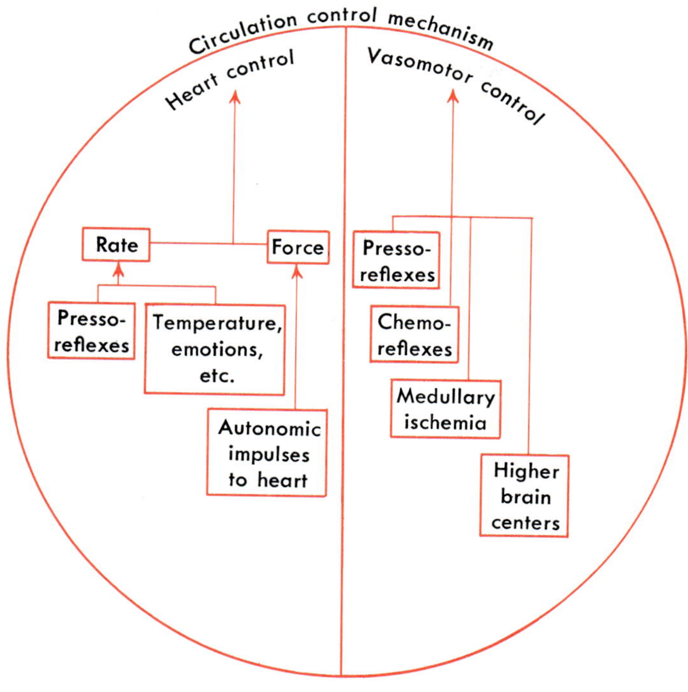

Fig. 13-33

Scheme to suggest some of the many parts of the complex circulation control mechanism. (See also Figs. 13-34 and 13-35.)

pressure minus central venous pressure and inversely related to resistance or:

$$\text{Volume of blood circulated per minute} = \frac{\text{Mean arterial pressure} - \text{Central venous pressure}}{\text{Resistance}}$$

The preceding statement and equation need qualifying with regard to the influence of peripheral resistance on circulation. For instance, according to the equation, an increase in peripheral resistance would tend to decrease blood flow. (Why? Increasing peripheral resistance increases the denominator of the fraction in the preceding equation. Increasing the denominator of any fraction necessarily does what to its value?) Increased peripheral resistance, however, has a secondary action that acts against its primary tendency to decrease blood flow. An increase in peripheral resistance hinders or decreases arteriole runoff. This, of course, tends to increase the volume of blood left in the arteries and so tends to increase arterial pressure. Note also that increasing arterial pressure tends to increase the value of the fraction in Poiseuille's equation. Therefore, it tends to increase circulation. In short, to say unequivocally what the effect of an increased peripheral resistance will be on circulation is impossible. It depends also upon arterial blood pressure—whether it increases, decreases, or stays the same when peripheral resistance increases. The clinical condition arteriosclerosis with hypertension (high blood pressure) illustrates this point. Both peripheral resistance and arterial pressure are increased in this condition. If resistance were to increase more than arterial pressure, circulation—that is, volume of blood flow per minute—would decrease. But if arterial pressure increases proportionately to resistance, circulation remains normal.

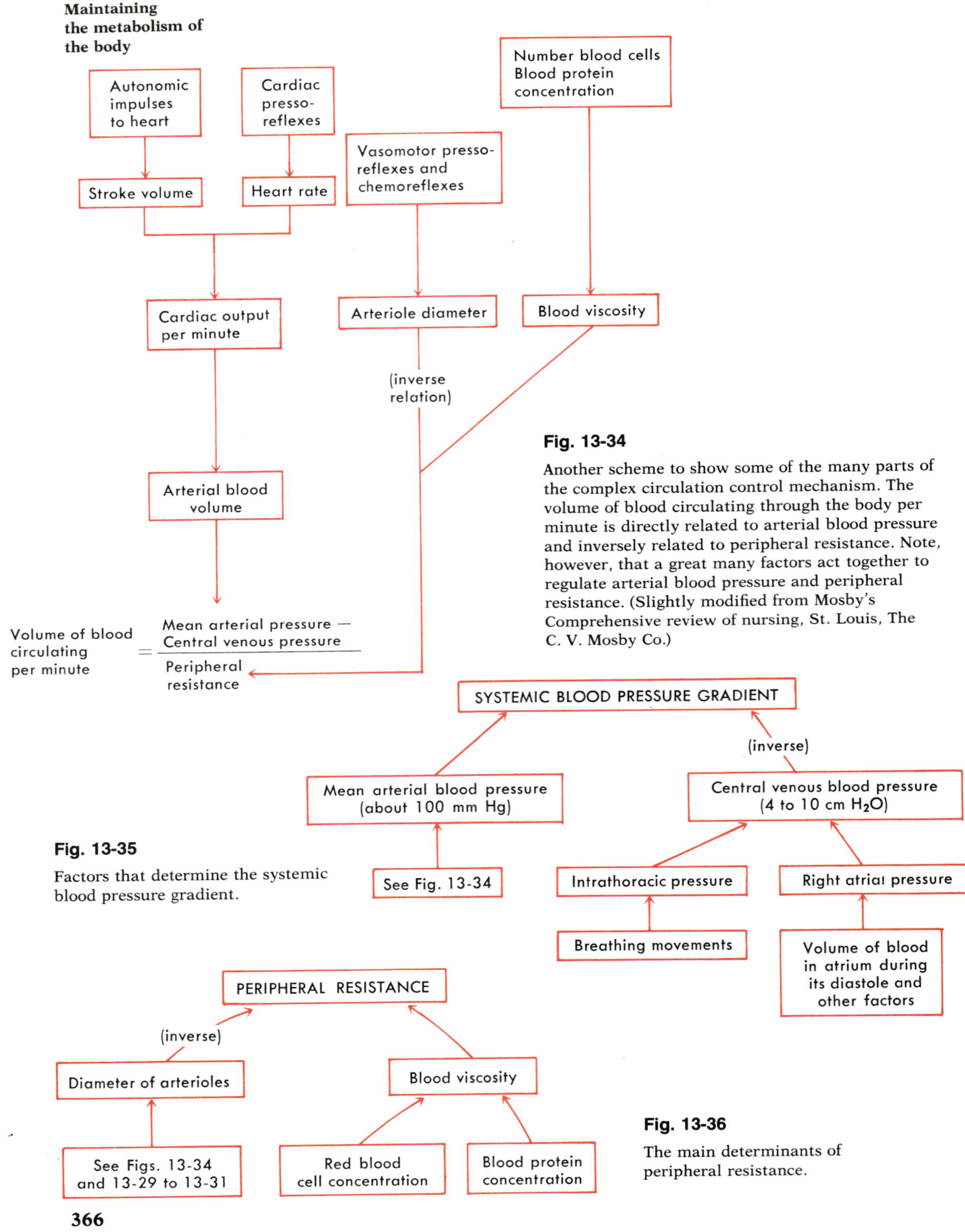

Fig. 13-34

Another scheme to show some of the many parts of the complex circulation control mechanism. The volume of blood circulating through the body per minute is directly related to arterial blood pressure and inversely related to peripheral resistance. Note, however, that a great many factors act together to regulate arterial blood pressure and peripheral resistance. (Slightly modified from Mosby's Comprehensive review of nursing, St. Louis, The C. V. Mosby Co.)

Fig. 13-35

Factors that determine the systemic blood pressure gradient.

Fig. 13-36

The main determinants of peripheral resistance.

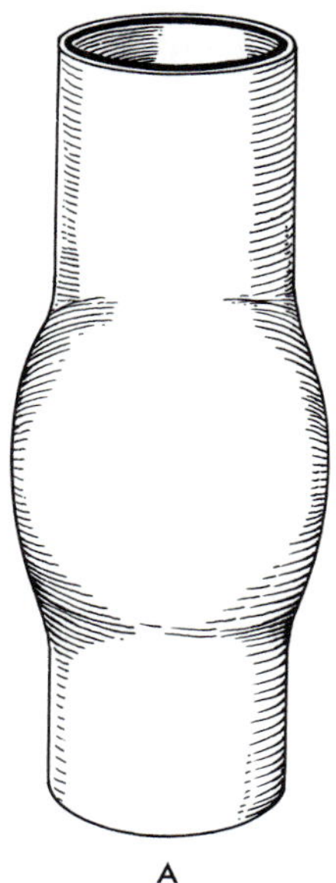

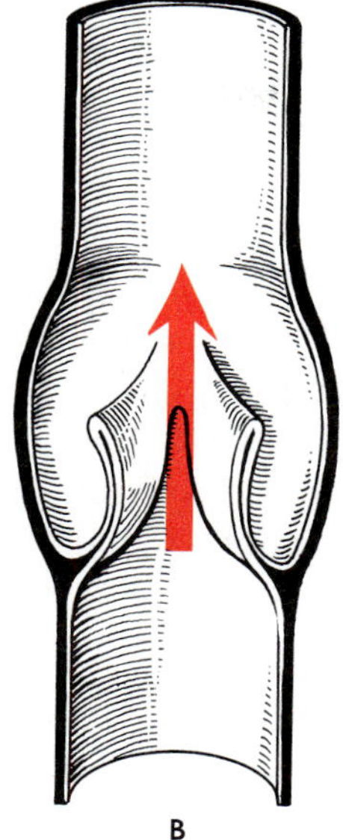

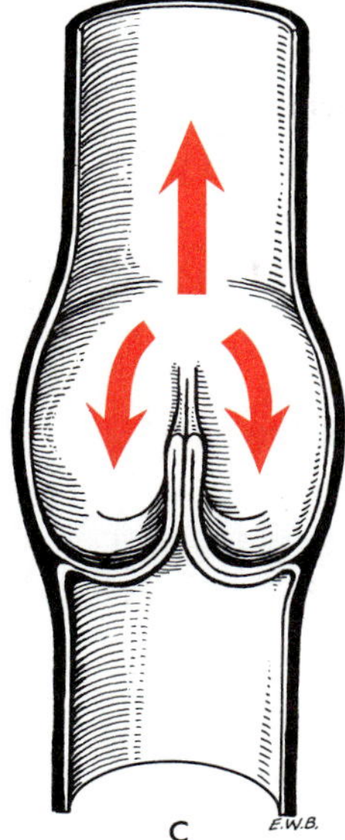

Fig. 13-37

Diagram showing the action of venous valves. **A,** External view of vein showing dilation at site of the valve. **B,** Interior of vein with the semilunar flaps in open position, permitting flow of blood through the valve. **C,** Valve flaps approximating each other, occluding the cavity and preventing the backflow of blood.

Important factors influencing venous return to heart

Two important factors that promote the return of venous blood to the heart are respirations and skeletal muscle contractions. Both produce their facilitating effect on venous return by increasing the pressure gradient between peripheral veins and venae cavae.

The process of inspiration increases the pressure gradient between peripheral and central veins by decreasing central venous pressure and also by increasing peripheral venous pressure. Each time the diaphragm contracts, the thoracic cavity necessarily becomes larger and the abdominal cavity smaller. Therefore, the pressures in the thoracic cavity, in the thoracic portion of the vena cava, and in the atria decrease, and those in the abdominal cavity and the abdominal veins increase. Deeper respirations intensify these effects and therefore tend to increase venous return to the heart more than do normal respirations. This is part of the reason why the principle is true that increased respirations and increased circulation tend to go hand in hand.

Skeletal muscle contractions serve as "booster pumps" for the heart. They promote venous return in the following way. As each skeletal muscle contracts, it squeezes the soft veins scattered through its interior, thereby milking the blood in them upward or toward the heart. The closing of the semilunar valves present in veins prevents blood from falling back as the muscle relaxes. Their flaps catch the blood as gravity pulls backward on it (Fig. 13-37). The net effect of skeletal muscle contraction plus venous valvular action, therefore, is to move venous blood toward the heart, to increase the venous return.

The value of skeletal muscle contractions in moving blood through veins is illustrated by a common experience. Who has not noticed how much more uncomfortable and tiring standing still is than walking? After several minutes of standing quietly, the feet and legs feel "full" and swollen. Blood has accumulated in the veins because the skeletal muscles are not contracting and squeezing it upward. The repeated contractions of the muscles when walking, on the other hand, keep the blood moving in the veins and prevent the discomfort of distended veins.

Blood pressure

Measurement of arterial blood pressure clinically

Blood pressure is measured with the aid of an apparatus known as a sphygmomanometer, which makes it possible to measure the amount of air pressure equal to the blood pressure in an artery. The measurement is made in terms of how many millimeters high the air pressure raises a column of mercury in a glass tube.

The sphygmomanometer consists of a rubber cuff attached by a rubber tube to a compressible bulb and by another tube to a column of mercury that is marked off in millimeters. The cuff is wrapped around the arm over the brachial artery, and air is pumped into the cuff by means of the bulb. In this way air pressure is exerted against the outside of the artery. Air is added until the air pressure exceeds the blood pressure within the artery or, in other words, until it compresses the artery. At this time no pulse can be heard through a stethoscope placed over the brachial artery at the bend of the elbow along the inner margin of the biceps muscle. By slowly releasing the air in the cuff, the air pressure is decreased until it approximately equals the blood pressure within the artery. At this point, the vessel opens slightly and a small spurt of blood comes through, producing the first sound, one with a rather sharp, taplike quality. This is followed by increasingly louder sounds that suddenly change. They become more muffled, then disappear altogether. Nurses and physicians train themselves to hear these different sounds and simultaneously to read the column of mercury since the first taplike sound represents the *systolic blood pressure*. Systolic pressure is the force with which the blood is pushing against the artery walls when the ventricles are contracting. The lowest point at which the sounds can be heard, just before they disappear, is approximately equal to the *diastolic pressure* or the force of the blood when the ventricles are relaxed. Systolic pressure gives valuable information about the force of the left ventricular contraction, and diastolic pressure gives valuable information about the resistance of the blood vessels. Clinically, diastolic pressure is considered more important than systolic pressure because it indicates the pressure or strain to which blood vessel walls are constantly subjected. It also reflects the condition of the peripheral vessels since diastolic pressure rises or falls with the peripheral resistance. If, for instance, arteries are sclerosed, peripheral resistance and diastolic pressure both increase.

Blood in the arteries of the average adult exerts a pressure equal to that required to raise a column of mercury about 120 mm (or a column of water over 5 feet) high in a glass tube during systole of the ventricles and 80 mm high during their diastole. For the sake of brevity, this is expressed as a blood pressure of 120 over 80 (120/80). The first or upper figure represents systolic pressure and the second diastolic pressure. From the figures just given, we observe that blood pressure fluctuates considerably during each heartbeat. During ventricular systole, the force is great enough to raise the mercury column 40 mm higher than during ventricular diastole. This difference between systolic and diastolic pressure is called *pulse pressure.* It characteristically increases in arteriosclerosis mainly because systolic pressure increases more than diastolic pressure does. Pulse pressure increases even more markedly in aortic valve insufficiency because of both a rise in systolic and a fall in diastolic pressure.

Relation to arterial and venous bleeding

Because blood exerts a comparatively high pressure in arteries and a very low pressure in veins, it gushes forth with considerable force from a cut artery but seeps in a slow, steady stream from a vein. As we have just seen, each ventricular contraction raises arterial blood pressure to the systolic level, and each ventricular relaxation lowers it to the diastolic level. As the ventricles contract, then, the blood spurts forth forcefully from the increased pressure in the artery, but as the ventricles relax, the flow ebbs to almost nothing because of the fall in pressure. In other words, blood escapes from an artery in spurts because of the alternate raising and lowering of arterial blood pressure but flows slowly and steadily from a vein because of the low, practically constant pressure. A uniform instead of a pulsating pressure exists in the capillaries and veins. Why? Because the arterial walls, being elastic continue to squeeze the blood forward while the ventricles are in diastole. Therefore, blood enters capillaries and veins under a steady pressure.

Velocity of blood

The speed with which blood flows (that is, distance per minute) through its vessels is governed in part by the physical principle that when a liquid flows from an area of one cross-section size to an area of larger size, its velocity slows in the area with the larger cross section. For example, a narrow river whose bed widens flows more slowly through the wide section than through the narrow. In terms of blood vascular system, the total cross-section area of all arterioles together is greater than that of the arteries. Therefore, blood flows more slowly through arterioles than through arteries. Likewise, the total cross-section area of all capillaries together is greater than that of all arterioles and, therefore, capillary flow is slower than arteriole. Venule cross-section area, on the other hand, is smaller than capillary cross-section area. Therefore, the blood velocity increases in venules and again in veins, which have a still smaller cross-section area. In short, the most rapid blood flow takes place in arteries and the slowest in capillaries. Can you think of a valuable effect stemming from the fact that blood flows most slowly through the capillaries?

Pulse

Definition

Pulse is defined as the alternate expansion and recoil of an artery.

Cause

Two factors are responsible for the existence of a pulse that can be felt:

1 Intermittent injections of blood from the heart into the aorta, which alternately increase and decrease the pressure in that vessel. If blood poured steadily out of the heart into the aorta, the pressure there would remain constant and there would be no pulse.

2 The elasticity of the arterial walls, which makes it possible for them to expand with each injection of blood and then recoil. If the vessels were fashioned from rigid material such as glass, there would still be an alternate raising and lowering of pressure within them with each systole and diastole of the ventricles, but the walls could not expand and recoil and, therefore, no pulse could be felt.

Pulse wave

Each ventricular systole starts a new pulse that proceeds as a wave of expansion throughout the arteries and is known as the pulse wave. It gradually dissipates as it travels, disappearing entirely in the capillaries. The pulse felt in the radial artery at the wrist does not coincide with the contraction of the ventricles. It follows each contraction by an appreciable interval (the length of time required for the pulse wave to travel from the aorta to the radial artery). The farther from the heart the pulse is taken, therefore, the longer that interval is.

Any nurse has only to think of the number of times she or he has counted pulses to become aware of the diagnostic importance of the pulse. It reveals important information about the cardiovascular system, about heart action, blood vessels, and circulation.

Where pulse can be felt

In general, the pulse can be felt wherever an artery lies near the surface and over a bone or other firm background. Some of the specific locations where the pulse is most easily felt are as follows:

1. *Radial artery*—at wrist
2. *Temporal artery*—in front of ear or above and to outer side of eye
3. *Common carotid artery*—along anterior edge of sternocleidomastoid muscle at level of lower margin of thyroid cartilage
4. *Facial artery*—at lower margin of lower jawbone on a line with corners of mouth and in a groove in mandible about one third of way forward from angle
5. *Brachial artery*—at bend of elbow along inner margin of biceps muscle
6. *Posterior tibial artery*—behind the medial malleolus (inner "ankle bone")
7. *Dorsalis pedis artery*—on the dorsum (upper surface) of the foot

Note: The so-called pressure points or points at which pressure may be applied to stop arterial bleeding and the points where the pulse may be felt are related. To be more specific, both are found where an artery lies near the surface and near a bone that can act as a firm background for pressure. There are six important pressure points:

1. *Temporal artery*—in front of ear
2. *Facial artery*—same place as pulse is taken
3. *Common carotid artery*—point where pulse is taken, with pressure back against spinal column
4. *Subclavian artery*—behind mesial third of clavicle, pressing against first rib
5. *Brachial artery*—few inches above elbow on inside of arm, pressing against humerus
6. *Femoral artery*—in middle of groin, where artery passes over pelvic bone; pulse can also be felt here

In trying to stop arterial bleeding by pressure, one must always remember to apply the pressure at the pressure point that lies between the bleeding part and the heart. Why? Because blood flows from the heart through the arteries to the part. Pressure between the heart and bleeding point, therefore, cuts off the source of the blood flow to that point.

Venous pulse

A pulse exists in the large veins only. It is most prominent in the veins near the heart, because of changes in venous blood pressure brought about by alternate contraction and relaxation of the atria of the heart. Venous pulse does not have as great clinical significance as arterial pulse so is less often measured.

Lymphatic system

Definition

The lymphatic system is actually part of the circulatory system since it consists of a moving fluid (lymph and tissue fluid) derived from the blood and a group of vessels (lymphatics) that return the lymph to the blood.

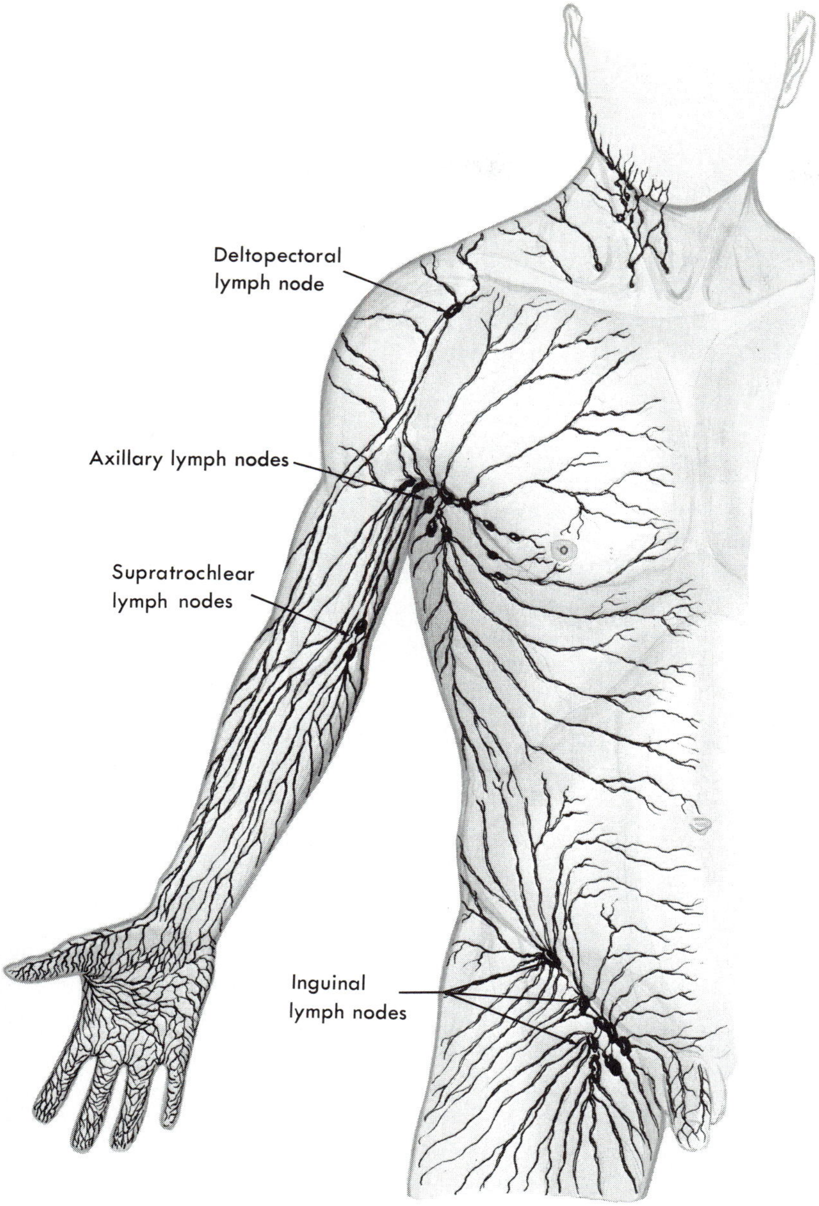

Deltopectoral lymph node

Axillary lymph nodes

Supratrochlear lymph nodes

Inguinal lymph nodes

Fig. 13-38

Superficial lymphatics of the upper extremity, anterior surface, and superficial lymphatics and lymph nodes of the front of the trunk.

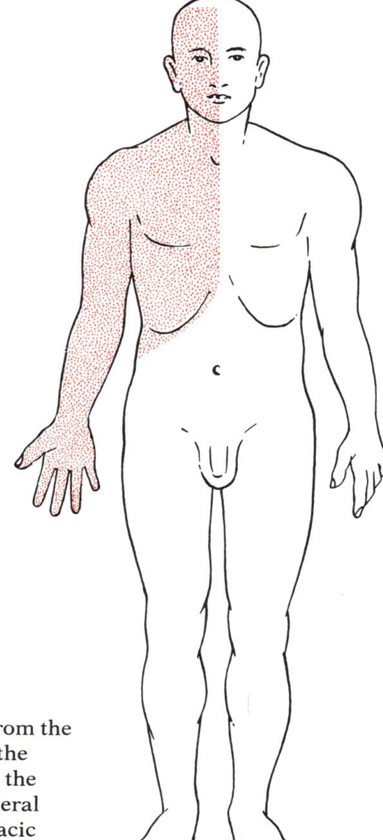

Fig. 13-39

Lymph drainage. The right lymphatic ducts drain lymph from the part of the body indicated by the stippled area. Lymph from all the rest of the body enters the general circulation by way of the thoracic duct.

Lymph and interstitial fluid (tissue fluid)

Definition

Lymph is the clear, watery-appearing fluid found in the lymphatic vessels. Interstitial fluid is also a clear, watery fluid, but it occupies the microscopic spaces between cells. In some tissues it is part of a semifluid ground substance. In others it is the bound water in a gelatinous ground substance. Interstitial fluid and blood plasma together constitute the extracellular fluid or, in the words of Claude Bernard, the "internal environment of the body"—the fluid environment of cells in contrast to the atmosphere or external environment of the body. Both lymph and interstitial fluid closely resemble blood plasma in composition. The main difference is that they contain a lower percentage of proteins than does plasma.

Lymphatics

Formation and distribution

Lymphatic vessels originate as microscopic blind-end vessels called *lymphatic capillaries* (see Plate XV of color insert). These tiny vessels are located in the intercellular spaces and are widely distributed throughout the body. As twigs of a tree join to form branches and branches join to form larger branches, and large branches join to form the tree trunk, so do lymphatic capillaries merge, forming slightly larger lymphatics that join other lymphatics to form still larger vessels, which merge to form the main lymphatic trunks: the *right lymphatic ducts* and the *thoracic duct*. Lymph from the entire body, except the upper right quadrant (Fig. 13-39), drains eventually into the thoracic duct, which drains into the left subclavian vein at the point where it joins the left internal jugular vein. Lymph from the upper right quadrant of the body empties into the right lymphatic duct (or, more commonly, into three collecting ducts) and then into the right subclavian vein. Since most of the lymph of the body returns to the bloodstream by way of the thoracic duct, this vessel is considerably larger than the other main lymph channels, the right lymphatic ducts, but is much smaller than the large veins, which it resembles in structure. It has a diameter about the size of a goose quill and a length of 15 to 18 inches. It originates as a dilated structure, the *cisterna chyli*, in the lumbar region of the abdominal cavity and ascends by a flexuous course to the root of the neck, where it joins the subclavian vein as just described (see Fig. 13-40). The presence of semilunar valves at frequent intervals along the duct gives it a somewhat varicose appearance.

Structure

Lymphatics resemble veins in structure with these exceptions:

1. Lymphatics have thinner walls
2. Lymphatics contain more valves
3. Lymphatics contain lymph nodes (or glands) located at certain intervals along their course

Lymphatics originating in the villi of the small intestine are called *lacteals*, and the milky fluid found in them after digestion is *chyle*.

Function

The function of lymphatics is to return water and proteins from the interstitial fluid to blood from which they came. Proteins can return to blood only via lymphatics. This fact has great clinical importance. For instance, if anything blocks lymphatic return, blood protein concentration and blood osmotic pressure soon fall below normal and fluid imbalance results (discussed in Chapter 20).

Lymph circulation

Water and solutes continually filter out of capillary blood into the interstitial fluid.

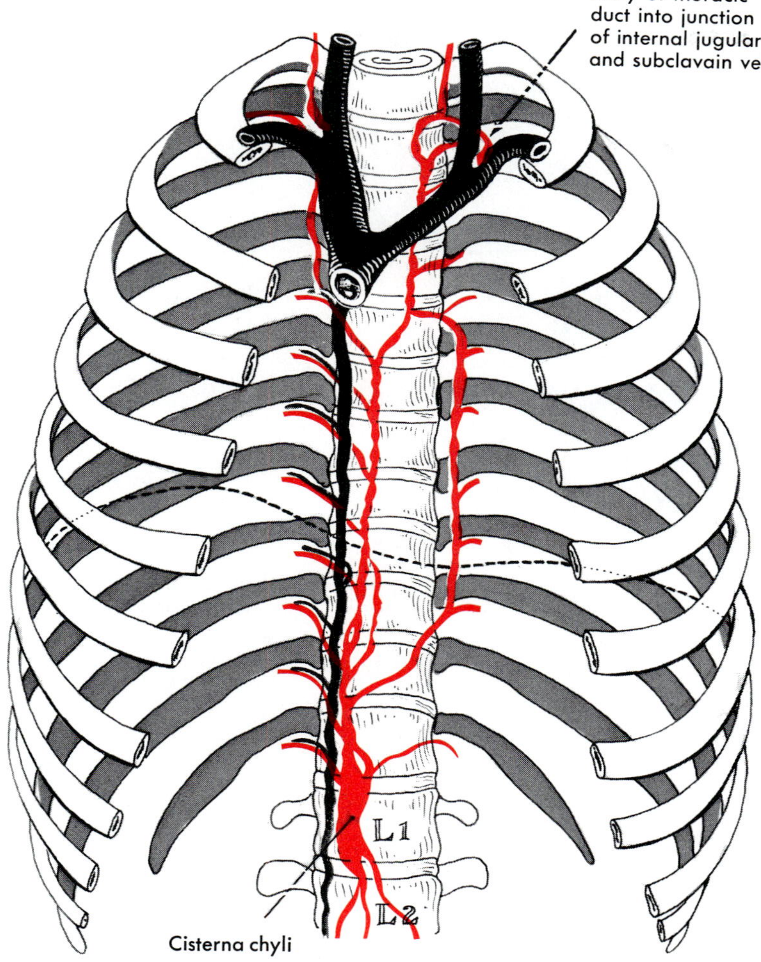

Fig. 13-40

Position of the cisterna chyli and the thoracic duct and its tributaries and the entry of the duct into the junction of the internal jugular and subclavian veins to form the innominate veins.

To balance this outflow, fluid continually reenters blood from the interstitial fluid. Experimental studies have shown that only about 40% of the fluid that filters out of blood capillaries returns to them by osmosis. The remaining 60% returns to the blood by way of the lymphatics. Also, newer evidence has disproved the old idea that healthy capillaries do not "leak" proteins. In truth, each day about 50% of the total blood proteins leak out of the capillaries into the tissue fluid and return to the blood by way of the lymphatic vessels. For more details about fluid exchange between blood and interstitial fluid, see Chapter 20. From lymphatic capillaries, lymph flows through progressively larger lymphatic vessels to eventually reenter blood at the junction of the internal jugular and subclavian veins (Fig. 13-40).

Although there is no pump connected with the lymphatic vessels to force lymph onward as the heart does blood, still lymph moves slowly and steadily along in its vessels. This occurs despite the fact that most of the flow

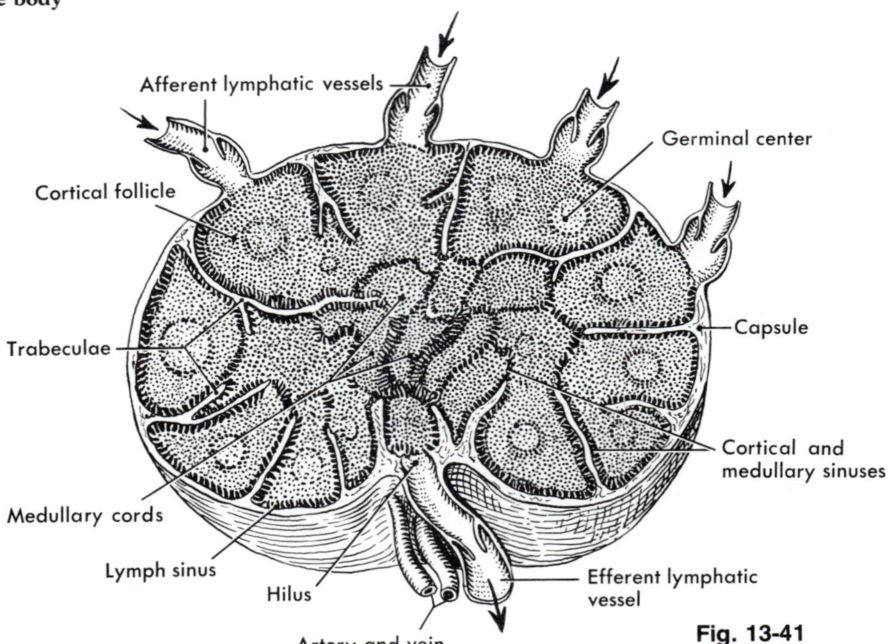

Fig. 13-41

Structure of a lymph node. Several afferent valved lymphatics bring lymph to the node. An efferent lymphatic leaves the node at the hilus. Note that the artery and vein enter and leave at the hilus.

is uphill. What mechanisms establish the pressure gradient required by the basic law of fluid flow? Two of the same mechanisms that contribute to the blood pressure gradient also establish a lymph pressure gradient. These are breathing movements and skeletal muscle contractions.

The mechanism of inspiration, resulting from the descent of the diaphragm, causes intraabdominal pressure to increase as intrathoracic pressure decreases. This simultaneously causes pressure to increase in the abdominal portion of the thoracic duct and to decrease in the thoracic portion. In other words, the process of inspiring establishes a pressure gradient in the thoracic duct that causes lymph to flow upward through it.

In addition, contracting skeletal muscles exert pressure on the lymphatics to push the lymph forward because valves within the lymphatics prevent backflow.

Lymph nodes

Structure

Lymph nodes or glands, as some people call them, are oval-shaped or bean-shaped structures. Some are as small as a pinhead and others as large as a lima bean. As shown in Fig. 13-41, lymph moves into the nodes via several afferent lymphatic vessels. Here it moves slowly through sinus channels lined with phagocytic reticuloendothelial cells and emerges usually by one efferent vessel. Lymphatic tissue, densely packed with lymphocytes, composes the substance of the node.

Location

With the exception of comparatively few single nodes, most of the lymph nodes occur in groups or clusters in certain areas. The group locations of greatest clinical importance are as follows:

1 *Submental and submaxillary groups* in the floor of the mouth—lymph from the nose, lips, and teeth drains through these nodes.

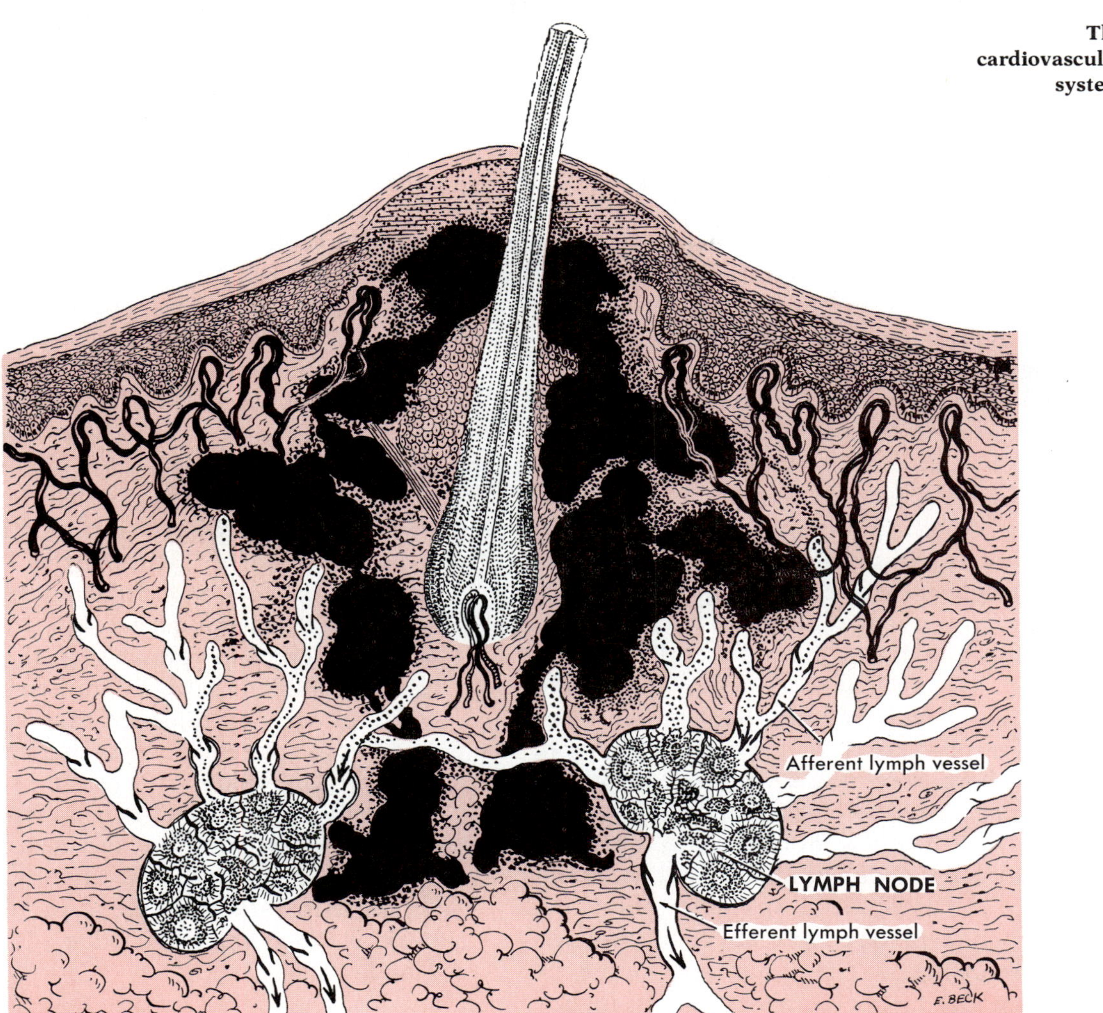

Fig. 13-42

Diagrammatic representation of a skin section in which an infection surrounds a hair follicle. The black areas represent dead and dying cells (pus). Black dots around the black areas represent bacteria. Leukocytes phagocytose many bacteria in tissue spaces. Others may enter the lymph nodes by way of afferent lymphatics. The nodes filter out those bacteria and usually destroy them all by phagocytosis.

2 *Superficial cervical glands* in the neck along the sternocleidomastoid muscle —these nodes drain lymph from the head (which has already passed through other nodes) and neck.

3 *Superficial cubital or supratrochlear nodes* located just above the bend of the elbow —lymph from the forearm passes through these nodes.

4 *Axillary nodes* (twenty to thirty large nodes clustered deep within the underarm and upper chest regions)—lymph from the arm and upper part of the thoracic wall, including the breast, drains through these nodes.

5 *Inguinal nodes* in the groin—lymph from the leg and external genitals drains through these.

Functions

Lymph nodes perform two unrelated functions, defense and hemopoiesis.

1 *Defense functions: filtration and phagocytosis.* The structure of the sinus channels within lymph nodes slows the lymph flow through them. This gives the reticuloendothelial cells that line the channels time to remove microorganisms and other injurious particles—cancer cells and soot, for example—from the lymph and phagocytose them (see Fig. 13-42). Sometimes, however, such hordes of microorganisms enter the nodes that the phagocytes cannot destroy enough of them to prevent their injuring the node. An infection of the node, adenitis, then results. Also, because cancer cells often break away from a malignant tumor and enter lymphatics, they travel to the lymph nodes where they may set up new growths. This may leave too few channels for lymph to return to the blood. For example, if tumors block axillary node channels, fluid accumulates in the interstitial spaces of the arm, causing the arm to become markedly swollen.

2 *Hemopoiesis.* The lymphatic tissue of lymph nodes forms lymphocytes and monocytes, the nongranular white blood cells, and plasma cells. Some lymphocytes, you will recall, synthesize cellular or tissue-rejecting antibodies. Some become plasma cells and produce circulating antibodies (p. 322).

Thymus

One of the body's best-kept secrets has been the function of the thymus. Before 1961, there were no significant clues as to its role. Then a young Briton, Dr. Jacques F. A. P. Miller, removed the thymus glands from newborn mice. His findings proved startling and crucial. Almost like a chain reaction, further investigations followed and led to at least a partial uncovering of the thymus' long-held secret. It now seems clear that this small structure (it weighs at most only about an ounce) plays a critical part in the body's defenses against infections—in its vital immunity mechanism.

In the mouse, and presumably in man as well, the thymus does two things. First, it serves as the source of lymphocytes before birth. (Actually, the fetal bone marrow forms lymphocytes, which then "seed" the thymus.) Many lymphocytes leave the thymus and circulate to the spleen, lymph nodes, and other lymphatic tissues. Soon after birth, the thymus is postulated to start secreting a hormone that enables lymphocytes to develop into plasma cells. Since plasma cells synthesize antibodies against foreign proteins, the thymus functions as part of the immune mechanism. It probably completes its essential work early in childhood. Its size is largest, relative to the rest of the body, when a child is about 2 years old. Its absolute size is largest at puberty. From then on, it gradually atrophies until in great old age, it may be barely recognizable. The thymus is located in the mediastinum, extending up into the neck as far as the lower edge of the thyroid gland.

Spleen

Location

The spleen is located in the left hypochondrium directly below the diaphragm, above the left kidney and descending colon, and behind the fundus of the stomach.

Structure

As Fig. 13-43 shows, the spleen is roughly ovoid in shape. Its size varies greatly in different individuals and in the same individual at different times. For example, it hypertrophies during infectious diseases and atrophies in old age. Within the spleen are numerous areas of lymphatic tissue and many venous sinuses.

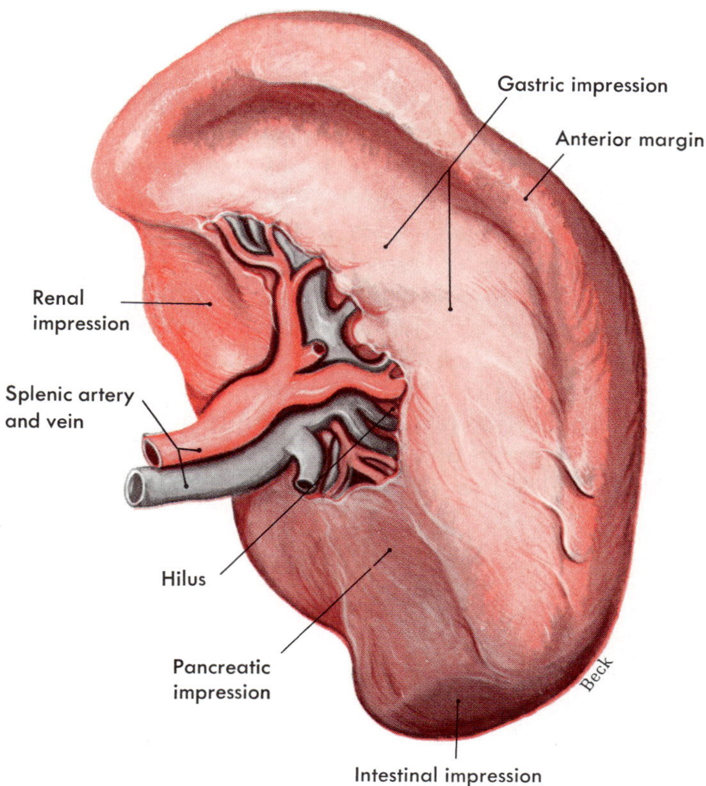

Fig. 13-43

Spleen, medial aspect. Arrangement of the vessels at the hilus is highly variable.

Functions

The spleen has long puzzled physiologists who have ascribed many and sundry functions to it. According to present-day knowledge, it performs several functions: defense, hemopoiesis, and red blood cell and platelet destruction; it also serves as a reservoir for blood.

1 *Defense.* As blood passes through the sinusoids of the spleen, reticuloendothelial cells (macrophages) lining these venous spaces remove microorganisms from the blood and destroy them by phagocytosis. Therefore, the spleen plays a part in the body's defense against microorganisms.

2 *Hemopoiesis.* Nongranular leukocytes—that is, monocytes and lymphocytes—and plasma cells are formed in the spleen. Before birth, red blood cells are also formed in the spleen, but after birth the spleen is said to form red cells only in extreme hemolytic anemia.

3 *Red blood cell and platelet destruction.* Macrophages lining the spleen's sinusoids remove worn-out red cells and imperfect platelets from the blood and destroy them by phagocytosis. They also break apart the hemoglobin molecules from the destroyed red cells and salvage their iron and globin content by returning them to the bloodstream for storage in bone marrow and liver.

4 *Blood reservoir.* The pulp of the spleen and its venous sinuses store considerable blood. Its normal volume of about 350 ml is said to decrease about 200 ml in less than a minute's time following sympathetic stimulation that produces marked constriction of its smooth capsule. This "self-transfusion" occurs, for example, as a response to the stress imposed by hemorrhage.

■ ■ ■

Although the spleen's functions make it a most useful organ, it is not a vital one. Dr. Charles Austin Doan in 1933 took the daring step of performing the first splenectomy. He removed the spleen from a 4-year-old girl who was dying of hemolytic anemia. Presumably, he justified his radical treatment on the basis of what was then merely conjecture—that is, that the spleen destroys red blood cells. The child recovered and Dr. Doan's operation proved to be a landmark. It created a great upsurge of interest in the spleen and led to many investigations of it.

Outline summary

Functions

1 Primary function—transportation of various substances to and from body cells; exchange of materials between respiratory, digestive, and excretory organs and blood and between blood and cells
2 Secondary functions—contributes to all bodily functions, for example:
   a Cellular metabolism
   b Homeostasis of fluid volume
   c Homeostasis of pH
   d Homeostasis of temperature
   e Defense against microorganisms

BLOOD

Whole blood consists of approximately 55% fluid (blood plasma) and 45% blood cells

Blood cells

Three main kinds—red blood cells (erythrocytes), white blood cells (leukocytes), platelets (thrombocytes); see Fig. 13-1

1 Erythrocytes (red blood cells)
   a Size and appearance—biconcave disks about 7μ in diameter
   b Structure and functions—millions of molecules of hemoglobin inside each red cell makes possible red cell functions of oxygen and carbon dioxide transport
   c Formation (erythropoiesis)—by myeloid tissue (red bone marrow)
   d Destruction—by fragmentation in capillaries; reticuloendothelial cells phagocytose red cell fragments and break down hemoglobin to yield iron-containing pigment and bile pigments; bone marrow reuses most of iron for new red cell synthesis; liver excretes bile pigments; life span of red cells about 120 days according to studies made with radioactive isotopes
   e Erythrocyte (red blood cell) homeostatic mechanism—stimulus thought to be tissue hypoxia; for description of mechanisms, see Fig. 13-2
   f Functions—see Table 13-2
2 Leukocytes (white blood cells)
   a Appearance and size—vary; some relatively large cells (e.g., monocytes) and some small cells (e.g., small lymphocytes); nuclei vary from spherical to S shaped to lobulated; cytoplasm of neutrophils, eosinophils, and basophils contain granules that take neutral, acid, and basic stains, respectively
   b Functions—defense or protection since they carry on phagocytosis
   c Formation—in myeloid tissue (red bone marrow) for granular leukocytes and probably monocytes; lymphocytes formed in lymphatic tissue—i.e., mainly in lymph nodes and spleen
   d Destruction and life span—some destroyed by phagocytosis; life span unknown
   e Numbers—about 5,000 to 10,000 leukocytes per cubic millimeter of blood; for differential count, see Table 13-1
   f Functions—see Table 13-2
3 Platelets
   a Appearance and size—platelets are small fragments of cells
   b Functions—blood coagulation

c Formation and life span—formed in red bone marrow by fragmentation of large cells (megakaryocytes); life span about 10 days
d Functions—see Table 13-2

Blood types (or blood groups)

1 Names—indicate type of antigen on or in red cell membranes
2 Every person's blood belongs to one of the four AB blood groups (type A, type B, type AB, or type O) and, in addition, is either Rh positive or Rh negative
3 Plasma does not normally contain antibodies against antigens present on its red cells but does normally contain antibodies against A or B antigens not present on its red cells
4 No blood normally contains anti-Rh antibodies; anti-Rh antibodies appear in blood only of an Rh-negative person and only after Rh-positive red cells have entered his bloodstream (e.g., by transfusion or by carrying an Rh-positive fetus)
5 Universal donor: type O; universal recipient: type AB but cross-matching should be done before use for safety

Blood plasma

1 Liquid part of blood or whole blood minus its cells; constitutes about 55% of total blood volume
2 Composition
  a About 90% water and 10% solutes
  b Most solutes crystalloids but some colloids
  c Most solutes electrolytes but some nonelectrolytes
  d Solutes include foods, wastes, gases, hormones, enzymes, vitamins, and antibodies and other proteins
3 Plasma proteins
  a Types—albumins, globulins, and fibrinogen
  b Total plasma proteins—6 to 8 gm per 100 ml of plasma
  c Albumins—about 55% of total plasma proteins
  d Globulins—about 38% of total plasma proteins
  e Fibrinogen—about 7% of total plasma proteins
  f Functions—contribute to blood viscosity, osmotic pressure, and volume; prothrombin (an albumin) and fibrinogen play key roles in blood clotting; modified gamma globulins are circulating antibodies, essential for immunity
  g All plasma proteins except circulating antibodies are synthesized in liver cells

Blood coagulation

1 Purpose—to plug up ruptured vessels and thus prevent fatal hemorrhage
2 Mechanism—complicated process, still incompletely understood
  a Triggered by appearance of rough spot in blood vessel lining
  b Clumps of platelets adhere to rough spot; they release "platelet factors" that initiate a series of rapidly occurring chemical reactions
  c Last two clotting reactions convert prothrombin to thrombin, an enzyme that catalyzes conversion of soluble fibrinogen to insoluble fibrin
  d Clot consists of tangled mass of fibrin threads and blood cells; blood serum is fluid left after clot forms
3 Factors that oppose clotting
  a Smoothness of endothelium that lines blood vessels prevents platelets' adherence and consequent disintegration; some platelet disintegration continuously despite this preventive
  b Blood normally contains certain anticoagulants—e.g., antithrombins, substances that inactivate thrombin so that it cannot catalyze fibrin formation
4 Factors that hasten clotting
  a Endothelial "rough" spots (e.g., lipoid plaques in atherosclerosis)
  b Sluggish blood flow
5 Pharmaceutical preparations that retard clotting—commercial heparin, bishydroxycoumarin (Dicumarol), citrates (for transfusion blood)
6 Clot dissolution—process called fibrinolysis; presumably this and opposing process (clot formation) go on all time
7 Clinical methods of hastening clotting
  a Apply rough surfaces to wound to stimulate platelets and tissues to liberate more thromboplastin
  b Apply purified thrombin
  c Apply fibrin foam, film, etc.

HEART

1 Four-chambered muscular organ
2 Lies in mediastinum with apex on diaphragm, two thirds of its bulk to left of midline of body and one third to right
3 Apical beat may be counted by placing stethoscope in fifth intercostal space on line with left midclavicular point

Covering

1 Structure
  a Loose-fitting, inextensible sac (fibrous pericardium) around heart, lined with serous pericardium (parietal layer), which also covers outer surface of heart (visceral layer or epicardium)
  b Small space between parietal and visceral layers of serous pericardium contains few drops of pericardial fluid
2 Function—protection against friction

Structure

Wall

1 Myocardium—name of muscular wall
2 Endocardium—lining
3 Pericardium—covering

Cavities

1 Upper two—atria
2 Lower two—ventricles

Valves and openings

1 Openings between atria and ventricles—atrioventricular orifices, guarded by cuspid valves, tricuspid on right and mitral or bicuspid on left; valves consist of three parts: flaps, chordae tendineae, and papillary muscle
2 Opening from right ventricle into pulmonary artery guarded by semilunar valves
3 Opening from left ventricle into great aorta guarded by semilunar valves

Blood supply

From coronary arteries; branch from ascending aorta behind semilunar valves

1 Left ventricle receives blood via both major branches of left coronary artery and from one branch of right coronary artery
2 Right ventricle receives blood via both major branches of right coronary artery and from one branch of left coronary artery

3 Each atrium receives blood only from one branch of its respective coronary artery
4 Usually only few anastomoses between larger branches of coronary arteries, so that occlusion of one of these produces areas of myocardial infarction; if not fatal, anastomoses between smaller vessels grow and provide collateral circulation

Conduction system

1 Sinoatrial node (S.A. node; pacemaker of heart)—small mass of modified cardiac muscle at junction of superior vena cava and right atrium; inherent rhythmicity of S.A. node impulses set basic rate of heartbeat; ratio of sympathetic/parasympathetic impulses to node per minute and blood concentrations of epinephrine and thyroid hormone act on node to modify its activity and alter heart rate
2 Atrioventricular node (A.V. node)—small mass of modified cardiac muscle in septum between two atria
3 Atrioventricular bundle (bundle of His)—special cardiac muscle fibers originating in A.V. node and extending down interventricular septum
4 Purkinje fibers—extension of bundle of His fibers out into walls of ventricles
5 Electrocardiogram (ECG)—a record of the heart's action currents (impulse conduction); composed of P wave, QRS complex, and T wave; interpretation complex but valuable aid for diagnosing certain heart disorders, especially disorders of heart's rhythm (e.g., fibrillation)

Nerve supply

1 Sympathetic fibers (in cardiac nerves) and parasympathetic fibers (in vagus) form cardiac plexuses
2 Fibers from plexuses terminate mainly in S.A. node
3 Sympathetic impulses tend to accelerate and strengthen heartbeat
4 Vagal impulses slow heartbeat

Physiology

1 Primary functions of circulatory system—maintain blood flow and vary rate of flow according to energy needs of cells
2 Cardiac cycle
  a Nature—consists of systole and diastole of atria and of ventricles; atria contract and as they relax, ventricles contract
  b Time required for cycle—about 4/5 second or from 70 to 80 times per minute
  c Events of cycle
    1 Atria contracted—cuspid valves open; ventricles relaxed; semilunar valves closed
    2 Atria relaxed—cuspid valves closed during first part of atrial diastole while ventricles are contracted and then open as ventricles relax; semilunar valves open during ventricular contraction
  d Heart sounds during cycle—lubb from contraction of ventricles and closure of cuspid valves; dupp from closure of semilunar valves

BLOOD VESSELS

1 Kinds
  a Arteries—vessels that carry blood away from heart; all except pulmonary artery carry oxygenated blood
  b Veins—vessels that carry blood toward heart; all except pulmonary veins carry deoxygenated blood
  c Capillaries—microscopic vessels that carry blood from small arteries (arterioles) to small veins (venules)
2 Structure—see Table 13-5
3 Functions
  a Arteries and arterioles—carry blood away from heart to capillaries
  b Capillaries—deliver materials to cells (by way of tissue fluid) and collect substances from them; vital function of entire circulatory system
  c Veins and venules—carry blood from capillaries back to heart

Main blood vessels

Systemic circulation

1 Arteries—see p. 342, Fig. 13-12, and Table 13-6
2 Veins—see pp. 345 and 348, Figs. 13-16 to 13-22

Portal circulation

See p. 348 and Fig. 13-22

Fetal circulation

See p. 350 and Fig. 13-23

CIRCULATION

1 Definitions
  a Circulation—blood flow through closed circuit of vessels
  b Systemic circulation—blood flow from left ventricle into aorta, other arteries, arterioles, capillaries, venules, and veins of all parts of body, to right atrium of heart
  c Pulmonary circulation—blood flow from right ventricle to pulmonary artery to lung arterioles, capillaries, and venules, to pulmonary veins, to left atrium
2 Functions of control mechanisms
  a Maintain circulation
  b Vary circulation; increase blood flow per minute when activity increases and decreases blood flow when activity decreases
3 Principles
  a Blood circulates because blood pressure gradient exists within its vessels; systemic blood pressure gradient (mean arterial pressure minus central venous pressure) equals about 95 mm Hg
  b Arterial blood pressure determined primarily by volume of blood in arteries; other factors remaining constant—greater arterial blood volume, greater arterial blood pressure
  c Arterial blood volume determined mainly by cardiac minute output and peripheral resistance—directly related to cardiac output and inversely related to resistance
  d Cardiac minute output determined by heart's rate of contraction and its systolic discharge and directly related to both factors
  e Arterial blood pressure tends to vary directly with peripheral resistance
  f Stroke volume under ordinary conditions determined by Starling's law of heart
  g Heart rate regulated by pressoreflexes and by many miscellaneous factors; increased arterial pressure in aorta or carotid sinus tends to produce reflex slowing of heart, whereas increased right atrial pressure tends to produce reflex cardiac acceleration
  h Peripheral resistance determined mainly by blood viscosity and by arteriole diameter; in general, less blood

viscosity, less peripheral resistance, but smaller diameter of arterioles, greater peripheral resistance

i Blood viscosity determined by concentration of blood proteins and of blood cells and directly related to both

j Arteriole diameter regulated mainly by pressoreflexes and chemoreflexes; in general, increase in arterial pressure produces reflex dilation of arterioles, whereas hypoxia and hypercapnea cause constriction of arterioles in blood reservoir organs but dilation of them in local structures, notably in skeletal muscles, heart, and brain

k Volume of blood circulating per minute determined by blood pressure gradient and peripheral resistance; according to Poiseuille's law, directly related to pressure gradient and inversely related to peripheral resistance

l Respirations and skeletal muscle contractions tend to increase venous return to heart

BLOOD PRESSURE

1 How arterial blood pressure measured clinically
  a Sphygmomanometer
  b Systolic pressure normal range—about 120 to 140 mm Hg and diastolic pressure about 80 to 90 mm Hg
2 Relation to arterial and venous bleeding
  a Arterial bleeding in spurts because of difference in amounts of systolic and diastolic pressures
  b Venous bleeding—slow and steady because of low, practically constant venous pressure

VELOCITY OF BLOOD

1 Speed with which blood flows
2 Most rapid in arteries and slowest in capillaries

PULSE

1 Definition—alternate expansion and recoil of artery
2 Cause—intermittent injections of blood from heart into aorta with each ventricular contraction; pulse can be felt because of elasticity of arterial walls
3 Pulse wave—pulse starts at beginning of aorta and proceeds as wave of expansion throughout arteries
4 Where pulse can be felt—radial, temporal, common carotid, facial, brachial, femoral, and popliteal arteries; where near surface and over firm background, such as bone; pressure points, points where bleeding can be stopped by pressure, roughly related to places where pulse can be felt
5 Venous pulse—in large veins only; caused by changes in venous pressure brought about by alternate contraction and relaxation of atria

LYMPHATIC SYSTEM

Part of circulatory system—consists of lymph, interstitial (tissue) fluid, lymphatics, and lymph nodes

Lymph and interstitial fluid (tissue fluid)

1 Definition
  a Lymph—clear, watery fluid found in lymphatic vessels
  b Interstitial fluid (tissue fluid)—clear liquid in tissue spaces

Lymphatics

1 Formation and distribution
  a Start as capillaries in tissue spaces
  b Widely distributed throughout body
  c Two or more main lymphatic ducts—thoracic duct, which drains into left subclavian vein at junction of internal jugular and subclavian, and one or more right lymphatic ducts, which drain into right subclavian vein
2 Structure
  a Similar to veins except thinner walled
  b Contain more valves and contain lymph nodes located at intervals
3 Function—return water and proteins from interstitial fluid to blood from which they came

Lymph circulation

Water and solutes from capillary blood to interstitial fluid, to lymphatics, to blood at junction of internal jugular and subclavian veins

Lymph nodes

1 Structure
  a Lymphatic tissue, separated into compartments by fibrous partitions
  b Afferent lymphatics enter each node and efferent lymphatic leaves each node
2 Location—usually in clusters (see pp. 373-374)
3 Functions
  a Defense functions—filter out injurious substances and phagocytose them
  b Hemopoiesis—formation of lymphocytes and monocytes

Thymus

1 Location—mediastinum, extends into lower neck
2 Size—relatively largest in comparison to body size at about 2 years of age; absolutely largest at puberty, after which it gradually atrophies; almost disappears by advanced old age
3 Function—forms lymphocytes before birth; postulated to secrete hormone, starting soon after birth, that permits lymphocytes to develop into plasma cells and secrete antibodies; hence thymus serves as part of body's defense against microbes and other foreign proteins

SPLEEN

1 Location—left hypochondrium
2 Structure
  a Similar to lymph nodes, ovoid in shape
  b Size varies
  c Contains numerous venous blood spaces that serve as blood reservoir
3 Functions
  a Defense—protection by phagocytosis by reticuloendothelial cells and antibody formation by some lymphocytes
  b Hemopoiesis of nongranular leukocytes (monocytes and lymphocytes) and of red cells before birth; spleen also forms plasma cells
  c Red blood cell and platelet destruction—reticuloendothelial cells phagocytose these cells
  d Blood reservoir

Review questions

1 Compare different kinds of blood cells as to appearance and size; functions; formation, destruction, and life span; number per cubic millimeter of blood.
2 "Graft-rejection cells" is a nickname for which kind of blood cells? Why?
3 What type of cells secrete "circulating" antibodies?
4 What is the function of circulating antibodies? Of cellular antibodies?
5 Since plasma cells synthesize and secrete circulating antibodies, what organelles would you expect to be prominent or abundant in plasma cells? (Review pp. 16-17 if you cannot answer this question.)
6 Suppose your doctor has told you that you have a lymphocyte count of 2,000 per cubic millimeter. Would you think this was normal or abnormal?
7 Why might a surgeon give antilymphocyte serum to a patient who had had a kidney transplant?
8 Explain what the term hematocrit means. Is it normally less than 50? More than 55?
9 What triggers blood clotting?
10 Write two equations to show the basic chemical reactions that produce a blood clot.
11 Why and how does a vitamin K deficiency affect blood clotting?
12 Explain some principles and methods by which blood clotting may be hastened.
13 Explain what these terms mean: type AB blood, Rh-negative blood.
14 Suppose you have type B blood. Which two of the following kinds of blood might be used for transfusing you: type A? type B? type AB? type O?
15 Which of the following babies would be most likely to have erythroblastosis fetalis: fourth child of Rh-positive mother and Rh-negative father? first child of Rh-negative mother and Rh-positive father? second child of Rh-negative mother and Rh-positive father? Explain your reasoning.
16 The cells of what organ synthesize most plasma proteins? What is the normal plasma protein concentration?
17 What are some of the functions served by plasma proteins?
18 Describe the pericardium, differentiating between the fibrous and serous portions.
19 Exactly where is pericardial fluid found? Explain its function.
20 Describe the heart's own blood supply. Explain why occlusion of a large coronary artery branch has serious consequences.
21 Identify, locate, and describe function of each of the following structures: S.A. node, A.V. node, bundle of His.
22 Compare arteries, veins, and capillaries as to structure and functions.
23 Differentiate between systemic, pulmonary, and portal circulation.
24 Explain the differences between fetal and postnatal circulation and the functional reasons for these differences.
25 Explain the heart control mechanism. Include control of both rate and force. Devise a diagram to indicate the different parts of the mechanism.
26 Explain the vasomotor mechanism. Devise a diagram to indicate its various parts.
27 Explain reactive hyperemia and one theory about the mechanism producing it.
28 State in your own words the basic principle of fluid flow.
29 State in your own words Poiseuille's law. Give an example of increased circulation to illustrate application of this law. Give an example of decreased circulation to illustrate application of this law.
30 What mechanisms control arterial blood pressure? Cite an example of the operation of one or more of these mechanisms to increase arterial pressure; to decrease it.
31 What mechanisms control peripheral resistance? Cite an example of the operation of one or more parts of this mechanism to increase resistance; to decrease it.
32 What two factors determine blood viscosity? What does viscosity mean? Give an example of a condition in which blood viscosity decreases. Explain its effect on circulation.
33 What effect, if any, would a respiratory stimulant drug have on circulation. Explain why it would or would not affect circulation.

34 Describe and explain the effects of exercise on circulation.
35 What and where is lymph? Describe its circulation.
36 Compare lymphatics and lymph nodes as to structure and location and function.
37 Describe the location and function of the spleen. What functions is it thought to perform?
38 If cancer cells from breast cancer were to enter the lymphatics of the breast, where do you think they might lodge and start new growths? Explain, using your knowledge of the anatomy of the lymphatic and circulatory systems.
39 Starting with the left ventricle of the heart, list the vessels through which blood would flow in reaching the small intestine; the large intestine; the liver (two ways); the spleen; the stomach; the kidneys; the suprarenal glands; the ovaries or testes; the anterior part of the base of the brain; the little finger of the right hand. List the vessels through which the blood returns from these parts to the right atrium of the heart (see Figs. 13-12, 13-16, and 13-24).
40 Name and explain the action of the heart and its valves.
41 Trace the flow of blood through the heart.
42 Give two reasons why blood is considered a protective agent against infection.
43 Describe one or more mechanisms that probably operate to return circulation to normal a short time after exercise ceases.
44 Give the general location of the following veins: longitudinal sinus, internal jugular vein, innominate vein, portal vein, great saphenous vein, inferior vena cava, basilic vein.
45 Define the following terms briefly: adenitis, anemia, aneurysm, atherosclerosis, basophils, blood pressure, diastole, differential count, embolus, endocardium, eosinophil, erythrocyte, heart block, hemophilia, leukemia, leukocyte, leukopenia, lymphocyte, monocyte, myocardium, neutrophil, pH, phagocytosis, phlebitis, plasma, peripheral resistance, pulse pressure, sphygmomanometer, thrombocyte, thrombus, systole.

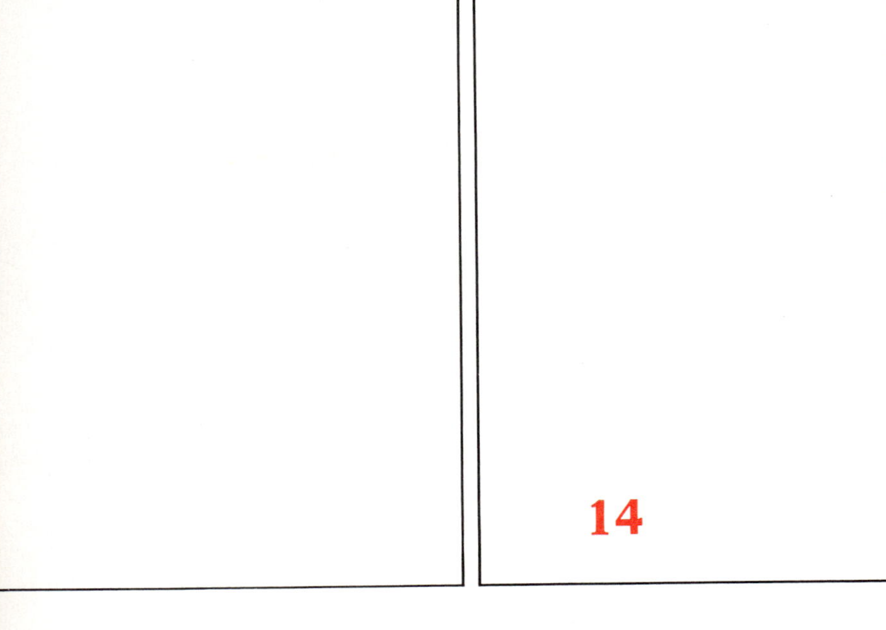

14

The digestive system

Functions and importance

Organs
Walls of organs
- Coats
- Modifications of coats

Mouth (buccal cavity)
Salivary glands
Teeth
Pharynx
Esophagus
Stomach
- Size, shape, and position
- Divisions
- Curves
- Sphincter muscles
- Coats
- Glands
- Functions

Small intestine
- Size and position
- Divisions
- Coats
- Functions

Large intestine (colon)
- Size
- Divisions
- Coats
- Functions

Liver
- Location and size
- Lobes
- Ducts
- Functions

Gallbladder
- Size, shape, and location
- Structure
- Functions
- Correlations

Pancreas
- Size, shape, and location
- Structure
- Functions

Vermiform appendix
- Size, shape, and location
- Structure

Digestion
Definition
Purpose
Kinds
Control of digestive gland secretion

Absorption
Definition
How accomplished

Functions and importance

The organs of the digestive system together perform a vital function—that of preparing food for absorption and for use by the millions of body cells. Most food when eaten is in a form that cannot reach the cells (because it cannot pass through the intestinal mucosa into the bloodstream) nor could it be used by the cells even if it could reach them. It must, therefore, be modified as to both chemical composition and physical state. This process of altering the chemical and physical composition of food so that it can be absorbed and utilized by body cells is known as digestion and is the function of the digestive system. Part of the digestive system, the large intestine, serves also as an organ of elimination, ridding the body of the wastes resulting from the digestive process.

Organs

The main organs of the digestive system (Fig. 14-1) form a tube all the way through the ventral cavities of the body. It is open at both ends. This tube is usually referred to as the *alimentary canal* (or tract) or the *gastrointestinal* or GI tract. The following organs form the gastrointestinal tract: mouth, pharynx, esophagus, stomach, and intestines. Several accessory organs are located in the main digestive organs or open into them. They are the salivary glands, teeth, liver, gallbladder, pancreas, and vermiform appendix.

Walls of organs

Coats

The alimentary canal is essentially a tube whose walls are fashioned of four layers of tissues: a mucous lining, a submucous coat of connective tissue in which are embedded the main blood vessels of the tract, a muscular coat, and a fibroserous coat.

Modifications of coats

Although the same four tissue coats form the various organs of the alimentary tract, their structure varies in different organs. Some of these modifications are listed in Table 14-1.

The parietal peritoneum, which lines the posterior wall of the abdominal cavity, projects from the lumbar region into the abdominal cavity in a double fold, shaped like a plaited fan. It is named the *mesentery* (Fig. 14-5). The loose outer edge of this great fan measures approximately 20 feet, whereas its attached posterior border has a length of from only 6 to 8 inches. Most of the small intestine is attached to its outer edge. The mesentery, then, may be defined as a fan-shaped double fold of parietal peritoneum by which the small intestine is anchored to the posterior abdominal wall. It should not be confused with the *greater omentum,** an apron-shaped double fold of peritoneum that is attached at its upper border to the first part of the duodenum, the lower edge of the stomach, and the transverse colon and hangs down loosely over the intestines. In case of a localized abdominal inflammation, such as appendicitis, the omentum envelops the inflamed area, walling it off from the rest of the abdomen. Spotty deposits of fat accumulate in the omentum, giving it a lacy appearance (Fig. 14-4).

Mouth (buccal cavity)

The following structures form the buccal cavity: the cheeks (side walls), the tongue and its muscles (floor), and the hard and soft palates (roof). Of these, only the palates and the tongue will be included in this discussion.

The *hard palate* consists of the two palatine bones and parts of the two superior maxillary

*The *lesser omentum* is a fold of peritoneum that attaches the liver to the lesser curvature of the stomach and beginning of the duodenum.

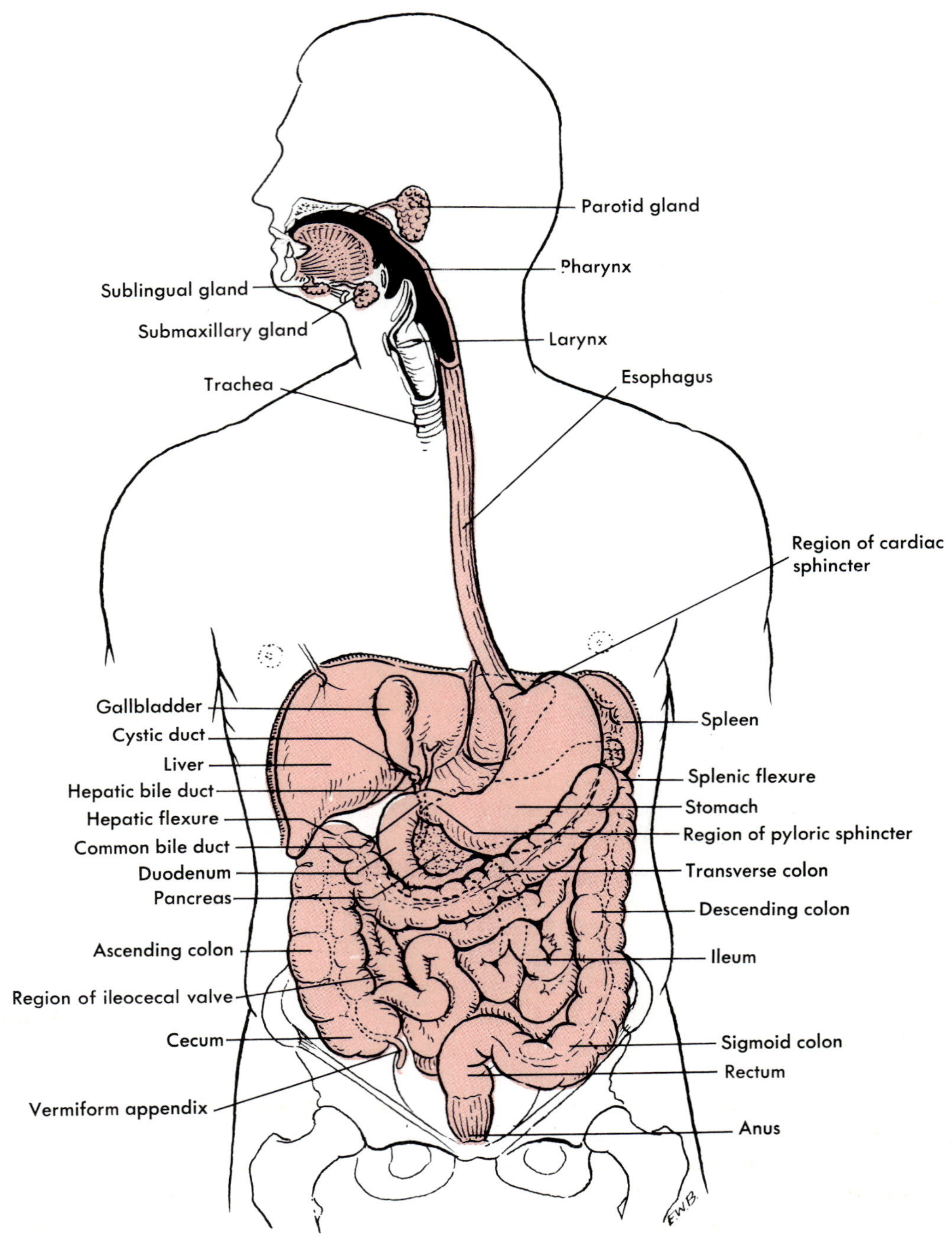

Fig. 14-1

Location of digestive system organs.

Table 14-1

Modifications of coats of digestive tract

Organ	Mucous coat	Muscle coat	Fibroserous coat
Esophagus		Two layers—inner one of circular fibers and outer one of longitudinal fibers; striated muscle in upper part and smooth in lower part of esophagus and in rest of tract	Outer coat fibrous; serous around part of esophagus in thoracic cavity
Stomach	Arranged in temporary longitudinal folds called *rugae;* allow for distention (Fig. 14-10) Contains microscopic gastric and hydrochloric acid glands	Has three layers instead of usual two—circular, longitudinal, and oblique fibers; two sphincters—cardiac at entrance of stomach and pyloric at its exit formed by circular fibers	Outer coat visceral peritoneum; hangs in double fold from lower edge of stomach over intestines, forming apronlike structure or "lace apron," *greater omentum* (Fig. 14-4)
Small intestine	Contains permanent circular folds, *valvulae conniventes* (or plica circularis) (Fig. 14-2) Microscopic fingerlike projections, *villi* (Fig. 14-3) Microscopic intestinal glands (of Lieberkühn) Microscopic duodenal (Brunner's) glands Clusters of lymph nodes, *Peyer's patches* Numerous single lymph nodes called solitary nodes	Two layers—inner one of circular fibers and outer one of longitudinal fibers	Outer coat visceral peritoneum
Large intestine	Solitary nodes Intestinal glands	Incomplete outer longitudinal coat; present only in three tapelike strips (taenia libera); small sacs (haustra) give rest of wall of large intestine puckered appearance (Fig. 14-11); internal anal sphincter formed by circular smooth fibers and external anal sphincter by striated fibers	Outer coat visceral peritoneum

bones. The *soft palate,* which forms a partition between the mouth and nasopharynx, is fashioned of muscle arranged in the shape of an arch. The opening in the arch leads from the mouth into the oropharynx and is named the *fauces,* whereas the two vertical side portions of the arch are appropriately termed the pillars of the fauces. Suspended from the midpoint of the posterior border of the arch is a small cone-shaped process, the *uvula.*

The entire buccal cavity, like the rest of the digestive tract, is lined with mucous membrane.

The salivary glands and teeth are accessory organs of the mouth.

Skeletal muscle covered with mucous membrane composes the *tongue.* Several muscles that originate on skull bones insert into the tongue. The rough elevations on the tongue's surface are called *papillae.* There are three types of them—filiform, fungiform, and circumvallate. You can readily distinguish

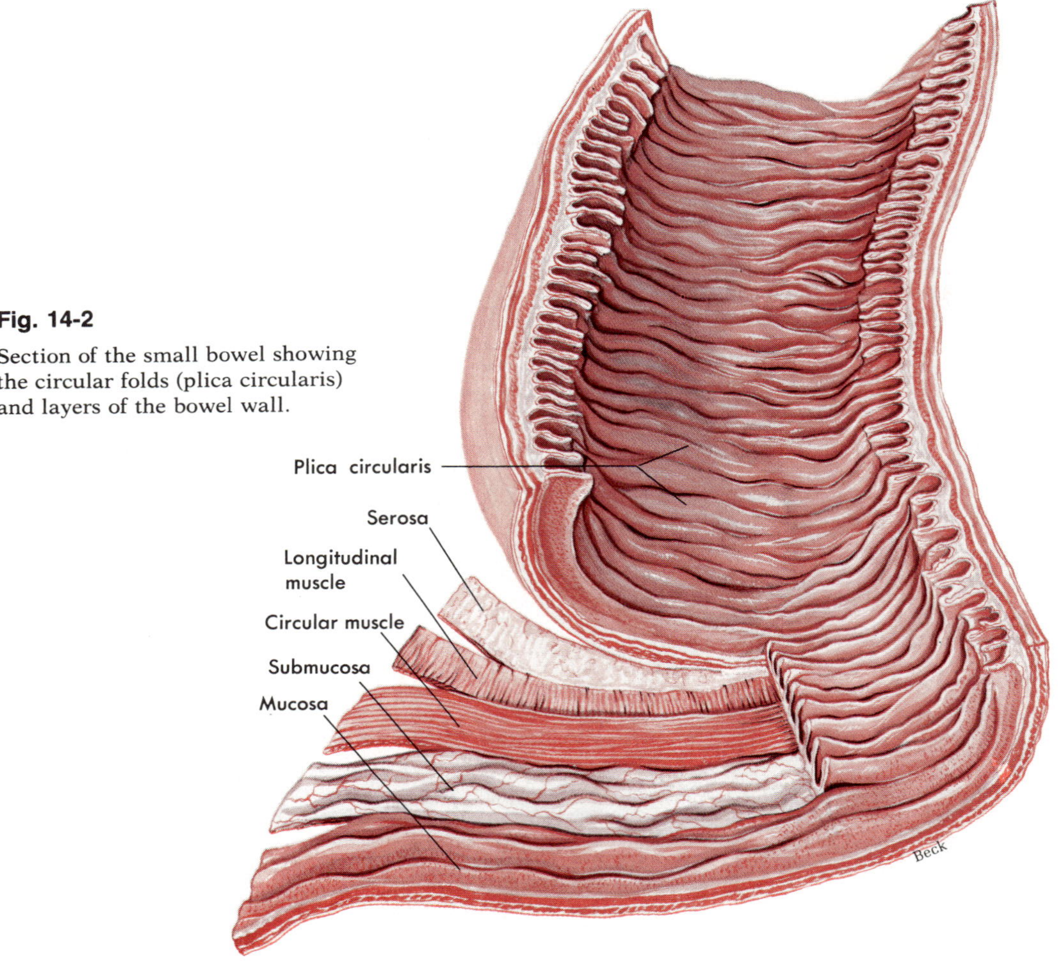

Fig. 14-2

Section of the small bowel showing the circular folds (plica circularis) and layers of the bowel wall.

them if you look at your own tongue. The filiform papillae are numerous threadlike structures distributed over the anterior two thirds of the tongue. Fungiform papillae are knoblike elevations most numerous near the edges of the tongue. Circumvallate papillae form an inverted V at the posterior part of the tongue.

The *frenum* (or frenulum) is a fold of mucous membrane in the midline of the undersurface of the tongue that helps to anchor the tongue to the floor of the mouth. If the frenum is too short for freedom of tongue movements, the individual is said to be tongue-tied, and his speech is faulty.

Salivary glands

See Table 14-2 for the names, location, and duct openings of the salivary glands.

Mumps is an acute infection of the parotid glands characterized by swelling of the glands. The act of opening the mouth causes pain because it squeezes that part of the gland that projects between the temporomandibular joint and the mastoid process.

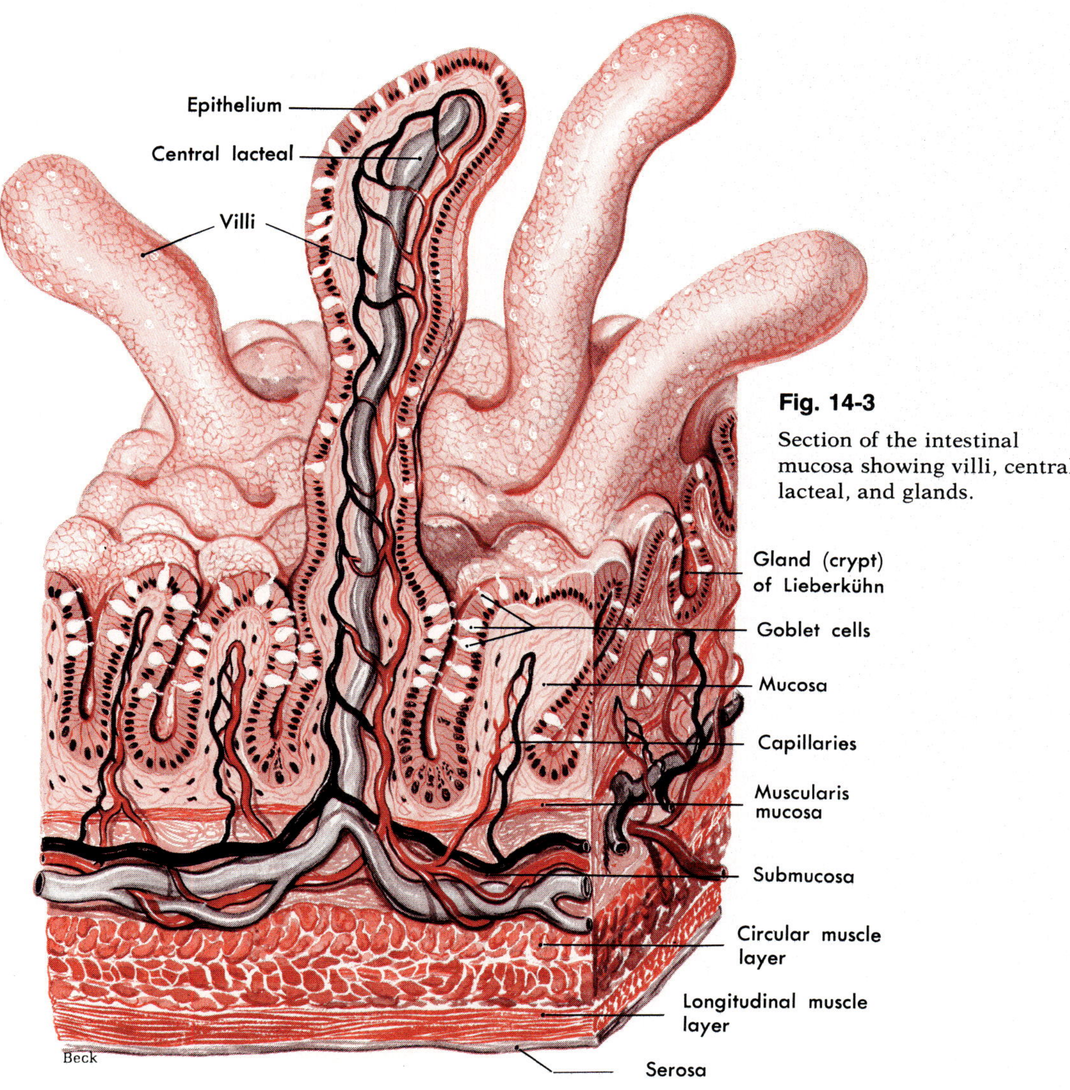

Fig. 14-3

Section of the intestinal mucosa showing villi, central lacteal, and glands.

Table 14-2

Salivary glands

Name of gland	Location	Duct openings
Parotid	Below and in front of ear	On inside of cheek, opposite upper second molar tooth; known as the parotid duct
Submandibular	Posterior part of floor of mouth	Floor of mouth, at sides of frenum; known as the submandibular duct
Sublingual	Anterior part of floor of mouth, under tongue	Several ducts open into floor of mouth

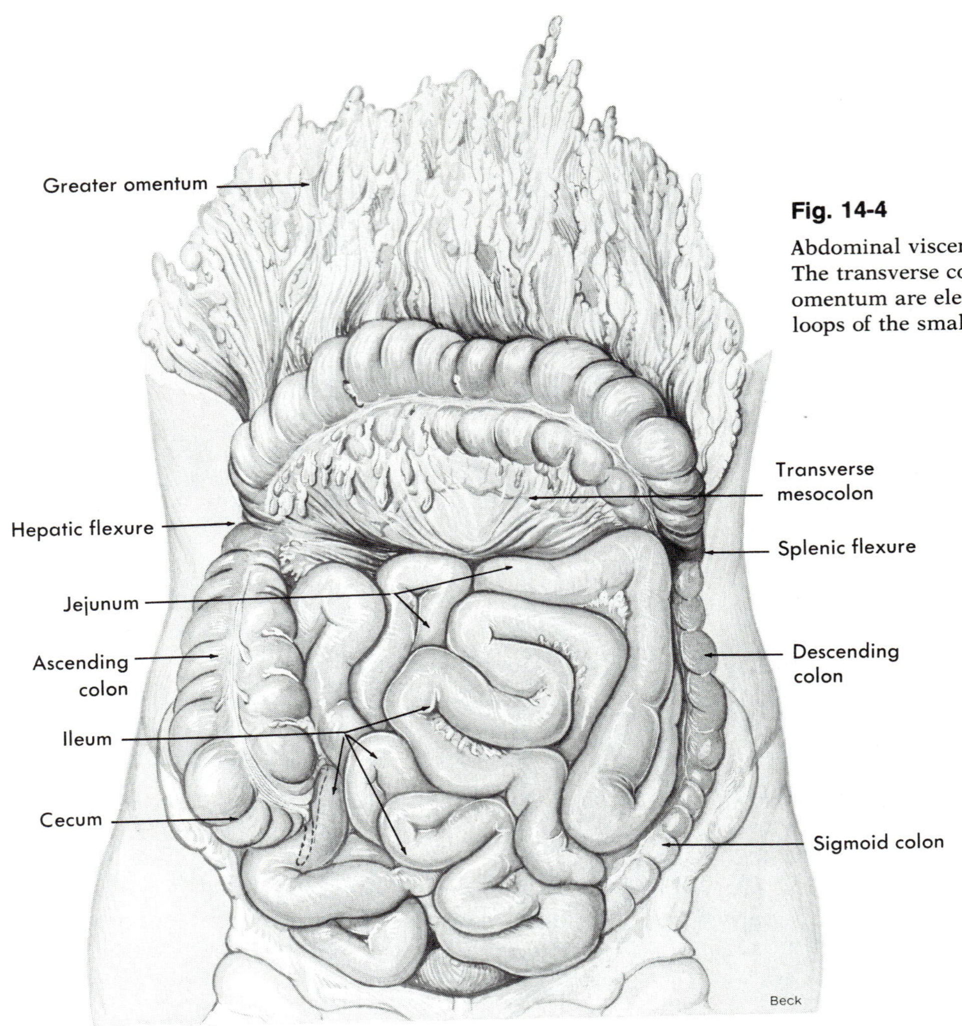

Fig. 14-4

Abdominal viscera from the front. The transverse colon and the omentum are elevated to reveal the loops of the small intestine.

Teeth

The so-called baby teeth or the set that appears first and is later shed are technically known as the *deciduous teeth* (Fig. 14-6), whereas those that replace these are the *permanent teeth* (Fig. 14-7). The names and numbers of teeth present in both sets are given in Table 14-3.

The first deciduous tooth erupts usually at the age of about 6 months. The rest follow at the rate of one or more a month until all twenty have appeared. There is, however, great individual variation in the age at which

Table 14-3

Dentition

Name of tooth	Number per jaw: Deciduous set	Number per jaw: Permanent set
Central incisors	2	2
Lateral incisors	2	2
Cuspids (canines)	2	2
Premolars (bicuspids)	0	4
First molars (tricuspids)	2	2
Second molars	2	2
Third molars	0	2
Total per jaw	10	16
Total per set	20	32

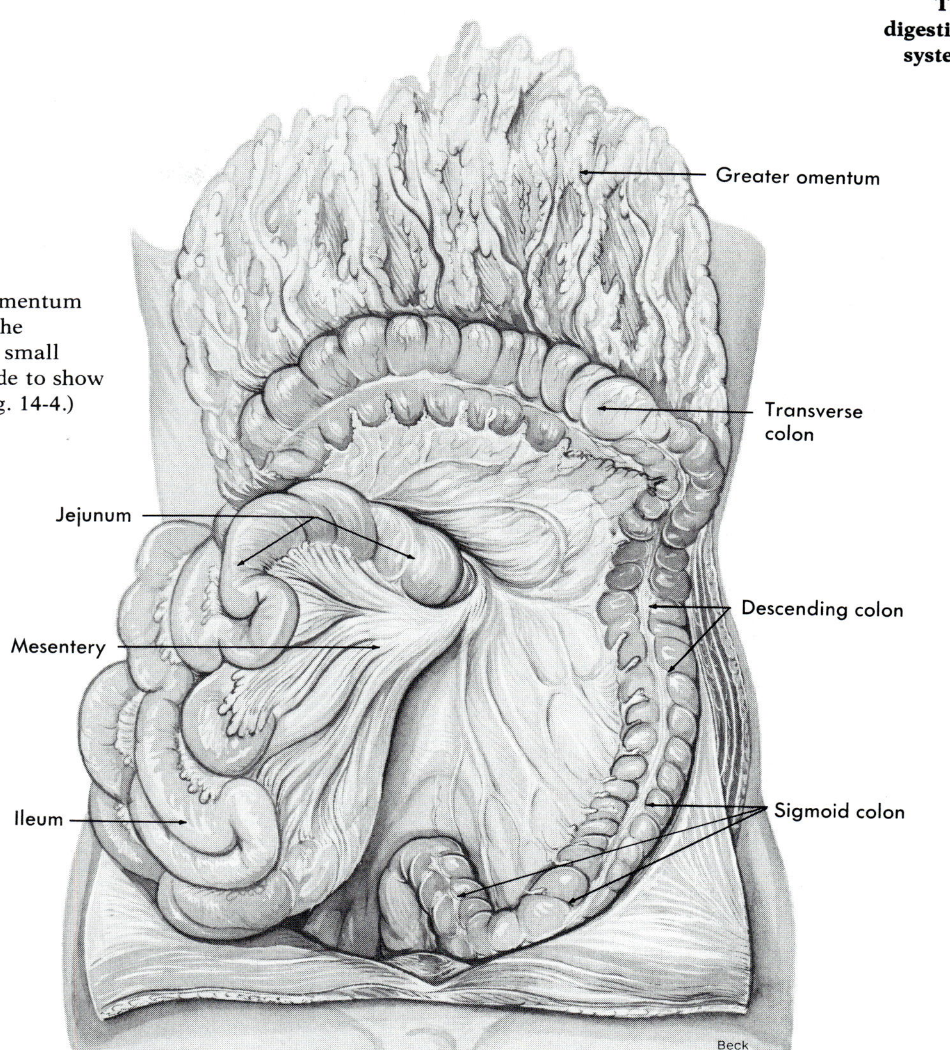

Fig. 14-5

The transverse colon and omentum are raised to demonstrate the duodenojejunal recess. The small intestine is pulled to the side to show the mesentery. (Also see Fig. 14-4.)

teeth erupt. Deciduous teeth are shed generally between the ages of 6 and 13 years. The third molars (wisdom teeth) are the last to appear, erupting usually sometime after 17 years of age.

Intact enamel resists bacterial attack, but once it is broken, the softer dentine decays. Periodontitis is an inflammation of the gums (gingivae) and periodontal membrane (Fig. 14-8).

Pharynx

For a discussion of the pharynx, see p. 293.

Esophagus

The esophagus, a collapsible tube about 10 inches long, extends from the pharynx to the stomach, piercing the diaphragm in its descent from the thoracic to the abdominal cavity. It lies posterior to the trachea and heart.

Unlike the trachea, the esophagus is a collapsible tube, its muscle walls lacking the cartilaginous rings found in the trachea.

Stomach

Size, shape, and position

Just below the diaphragm, the alimentary tube dilates into an elongated pouchlike

Fig. 14-6

The deciduous arch. Note that in the set of twenty temporary primary teeth there are no premolars (bicuspids) and there are only two pairs of molars in each jaw. Compare with permanent teeth (Fig. 14-7).

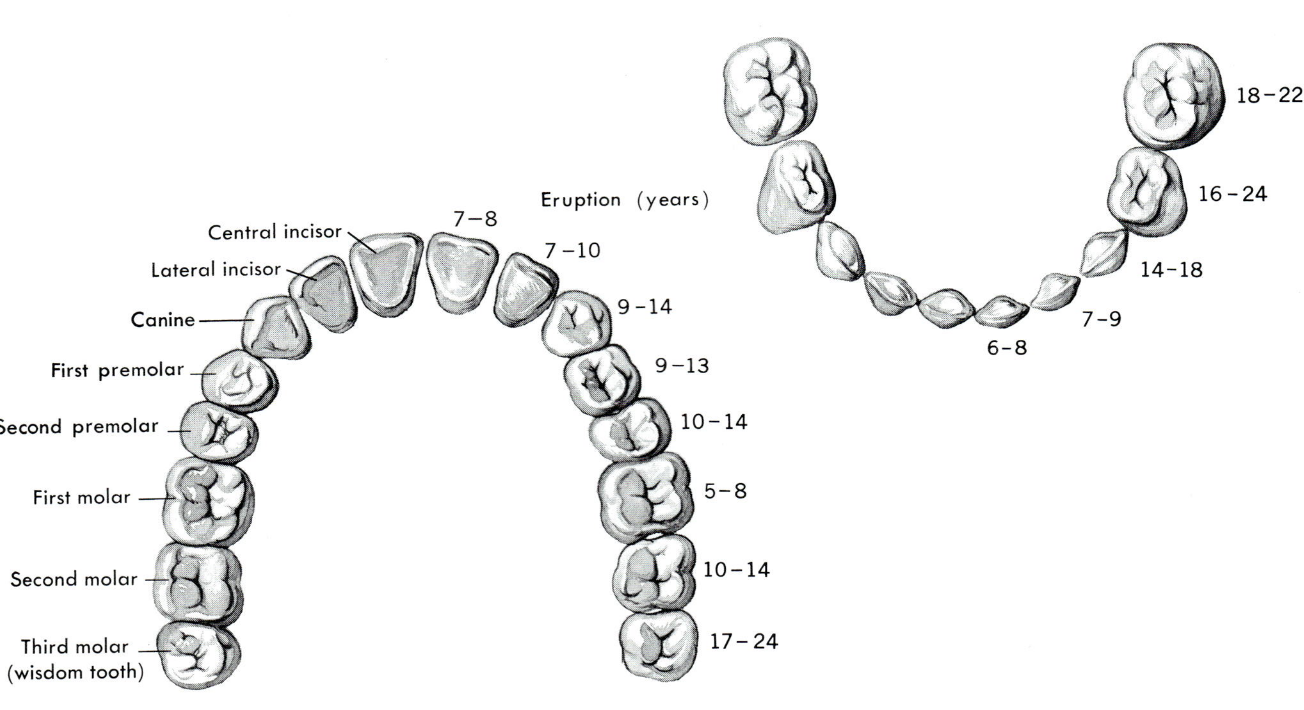

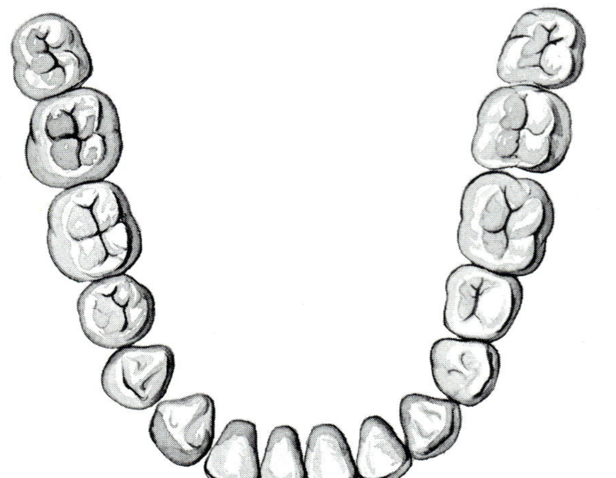

Fig. 14-7

The thirty-two permanent teeth. Generally, the lower teeth erupt before the corresponding upper teeth and all teeth usually erupt earlier in girls than in boys.

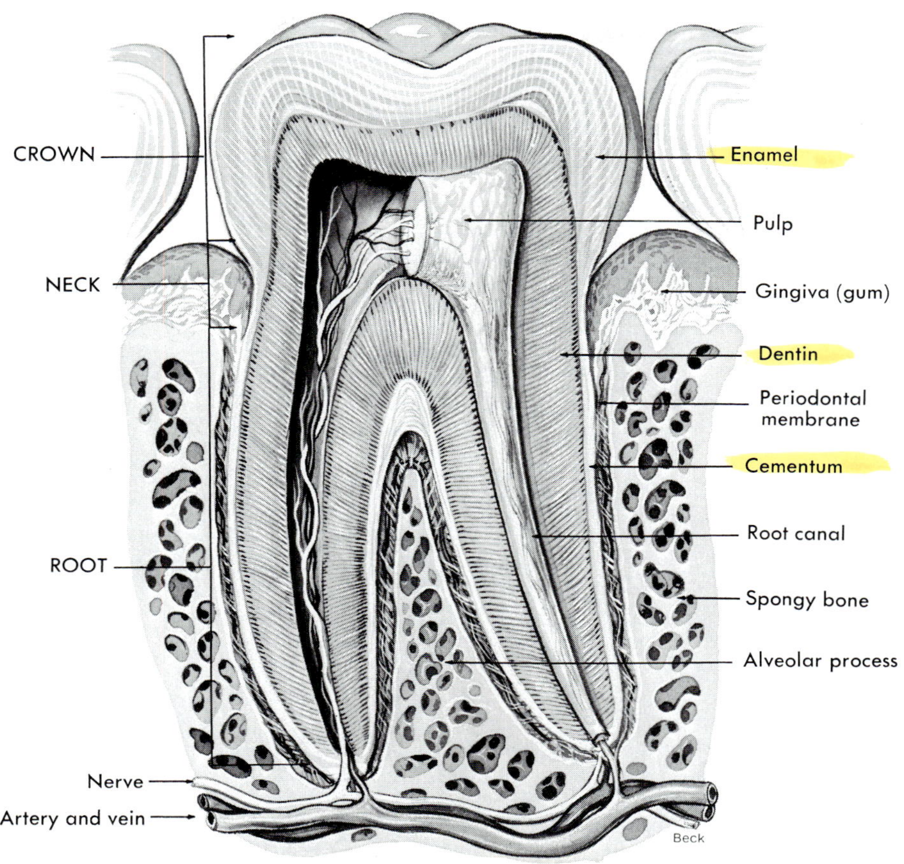

Fig. 14-8

A molar tooth sectioned to show its bony socket and details of its three main parts: crown, neck, and root. The pulp contains nerves and blood vessels.

structure, the stomach (Fig. 14-9), the size of which varies according to several factors, notably sex and the amount of distention. In general, the female stomach is usually more slender and smaller than the male stomach. For some time after a meal, the stomach is enlarged because of distention of its walls, but as food leaves, the walls partially collapse, leaving the organ about the size of a large sausage.

The stomach lies in the upper part of the abdominal cavity under the liver and diaphragm, with approximately five sixths of its mass to the left of the median line. In other words, it is described as lying in the epigastrium and left hypochondrium (Fig. 1-3, p. 6). Its position, however, alters frequently. For example, it is pushed downward with each inspiration and upward with each expiration. When it is greatly distended from

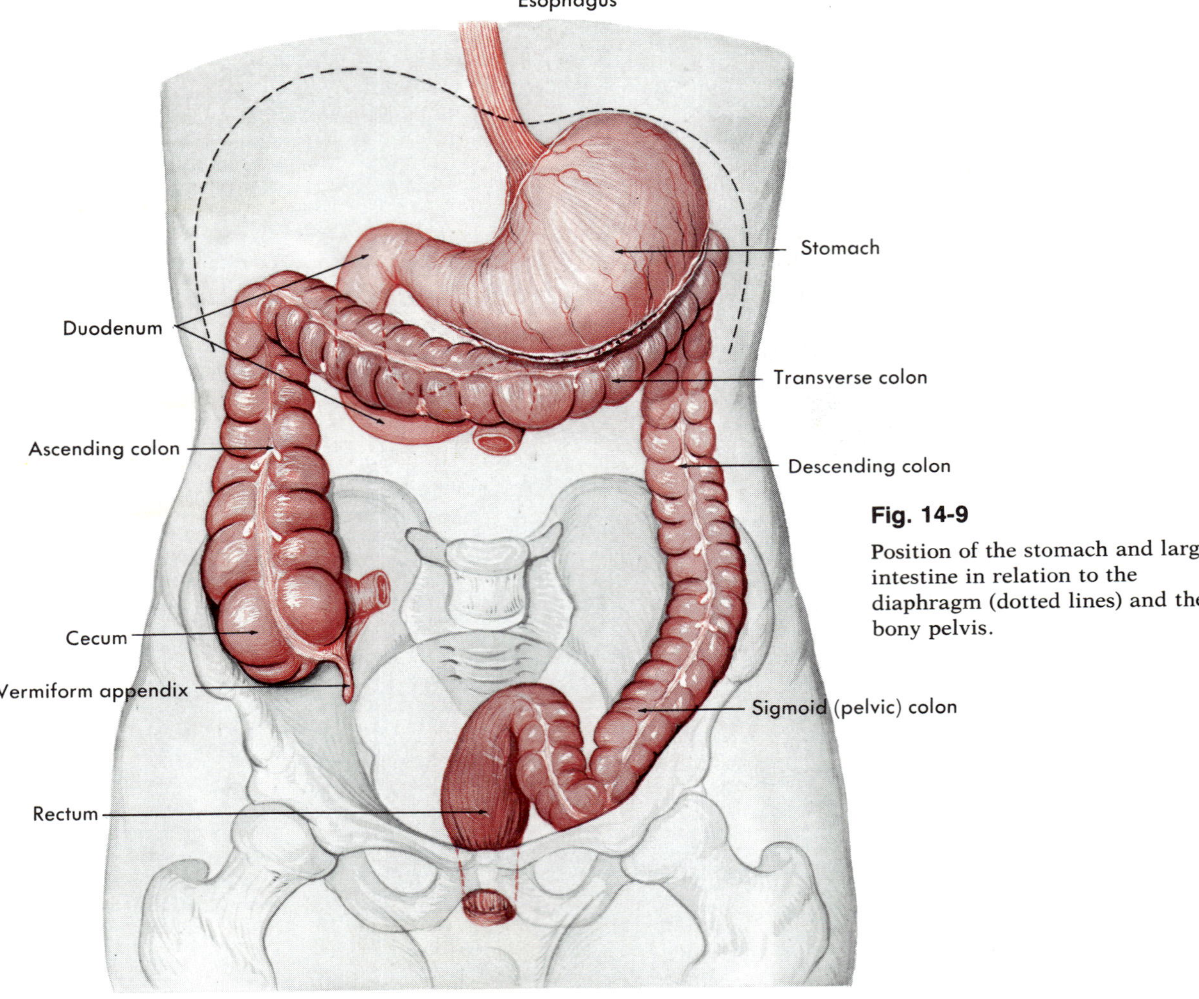

Fig. 14-9

Position of the stomach and large intestine in relation to the diaphragm (dotted lines) and the bony pelvis.

an unusually large meal, its size interferes with the descent of the diaphragm on inspiration, producing the familiar feeling of dyspnea that accompanies overeating. In this state, the stomach also pushes upward against the heart and may give rise to the sensation that the heart is being crowded.

Divisions

The *fundus,* the *body,* and the *pylorus* are the three divisions of the stomach. The fundus is the enlarged portion to the left and above the opening of the esophagus into the stomach. The body is the central part of the stomach, and the pylorus is its lower portion (Fig. 14-10).

Curves

The upper right border of the stomach presents what is known as the *lesser curvature* and the lower left border the *greater curvature.*

Sphincter muscles

Sphincter muscles guard both stomach openings. A sphincter muscle consists of circular fibers so arranged that there is an opening in the center of them (like the hole in a doughnut) when they are relaxed and no opening when they are contracted.

The *cardiac sphincter* guards the opening of the esophagus into the stomach, and the *pyloric sphincter* guards the opening from the pyloric portion of the stomach into the first part of the small intestine (duodenum). This latter muscle is of clinical importance because *pylorospasm* is a fairly common condition in babies. The pyloric fibers do not relax normally to allow food to leave the stomach, and the baby vomits his food instead of digesting and absorbing it. The condition is relieved by the administration of a drug that relaxes smooth muscle. Another abnormality of the pyloric sphincter is pyloric stenosis, an obstructive narrowing of its opening.

Coats

See Table 14-1 and Fig. 14-10 for information on coats of the stomach.

Glands

Numerous microscopic tubular glands are embedded in the gastric mucosa. Those in

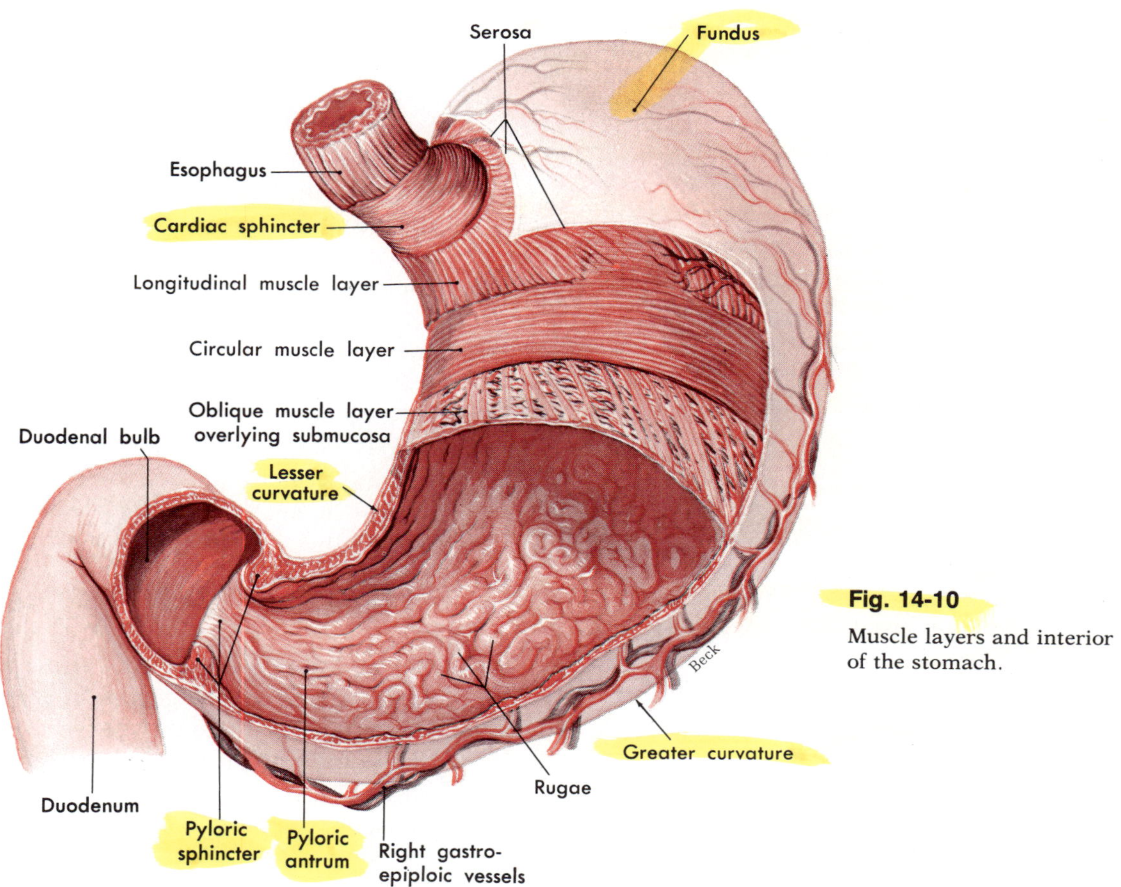

Fig. 14-10
Muscle layers and interior of the stomach.

the mucosa that lines the fundus and body of the stomach secrete most of the gastric juice, a fluid composed of mucus, enzymes, and hydrochloric acid. *Epithelial cells* that form the surface of the gastric mucosa (next to the lumen of the stomach) secrete mucus. *Parietal cells* secrete hydrochloric acid, and *chief cells* (or zymogen cells) secrete the enzymes of the gastric juice. In pernicious anemia, the gastric mucosa atrophies. As a result hydrochloric acid and a mysterious substance known as the intrinsic factor are not produced. Achlorhydria, therefore, is a characteristic finding in pernicious anemia. Absence of the intrinsic factor produces anemia. Without it, vitamin B_{12} cannot be absorbed, and without vitamin B_{12} the red bone marrow is not sufficiently stimulated to produce enough normal red blood cells.

Functions

The stomach carries on the following functions:

1 It serves as a reservoir, storing food until it can be partially digested and moved farther along the gastrointestinal tract.

2 It secretes gastric juice, one of the juices whose enzymes digest food.

3 Through contractions of its muscular coat, it churns the food, breaking it into small particles and mixing them well with the gastric juice. In due time, it moves the gastric contents on into the duodenum.

4 It secretes the intrinsic factor just mentioned.

5 It carries on a limited amount of absorption—of some water, alcohol, and certain drugs.

Small intestine

Size and position

The small intestine is a tube measuring approximately 1 inch in diameter and 20 feet in length. Its coiled loops fill most of the abdominal cavity.

Divisions

The small intestine consists of three divisions: the duodenum, the jejunum, and the ileum. The *duodenum** is the uppermost division and is the part to which the pyloric end of the stomach attaches. It is about 10 inches long and is shaped roughly like the letter C. The duodenum becomes *jejunum* at the point where the tube turns abruptly forward and downward. The jejunal portion continues for approximately the next 8 feet, where it becomes *ileum,* but without any clear line of demarcation between the two divisions. The ileum is about 12 feet long.

Coats

See Table 14-1 and Fig. 14-2 for information on the coats of the small intestine.

Functions

The small intestine carries on three main functions.

1 It completes the digestion of foods. The digestive intestinal juice contains mucus and many digestive enzymes. The glands of Lieberkühn secrete the digestive enzymes, whereas the glands of Brunner and innumerable goblet cells secrete the mucus.

2 It absorbs the end products of digestion into blood and lymph.

3 It secretes hormones—for example, some that help control the secretion of pancreatic juice, bile, and intestinal juice (pp. 409-410).

Large intestine (colon)

Size

The lower part of the alimentary canal bears the name *large intestine* because its diameter is noticeably larger than that of the small intestine. Its length, however, is

*Derivation of the word "duodenum" may interest you. It comes from words which mean 12 fingerbreadths, a distance of about 11 inches, the approximate length of the duodenum.

much less, being about 5 or 6 feet. Its average diameter is approximately $2^1/_2$ inches, but this decreases toward the lower end of the tube.

Divisions

The large intestine is divided into the cecum, colon, and rectum.

Cecum. The first 2 or 3 inches of the large intestine are named the cecum. It is located in the lower right quadrant of the abdomen (Fig. 14-9).

Colon. The colon is divided into the following portions: ascending, transverse, descending, and sigmoid (Fig. 14-1).

1 The *ascending colon* lies in the vertical position, on the right side of the abdomen, extending up to the lower border of the liver. The ileum joins the large intestine at the junction of the cecum and ascending colon, the place of attachment resembling the letter T in formation (Fig. 14-11). The ileocecal valve guards the opening of the ileum into the large intestine, permitting material to pass from the former into the latter but not in the reverse direction.

2 The *transverse colon* passes horizontally across the abdomen, below the liver and stomach and above the small intestine.

3 The *descending colon* lies in the vertical

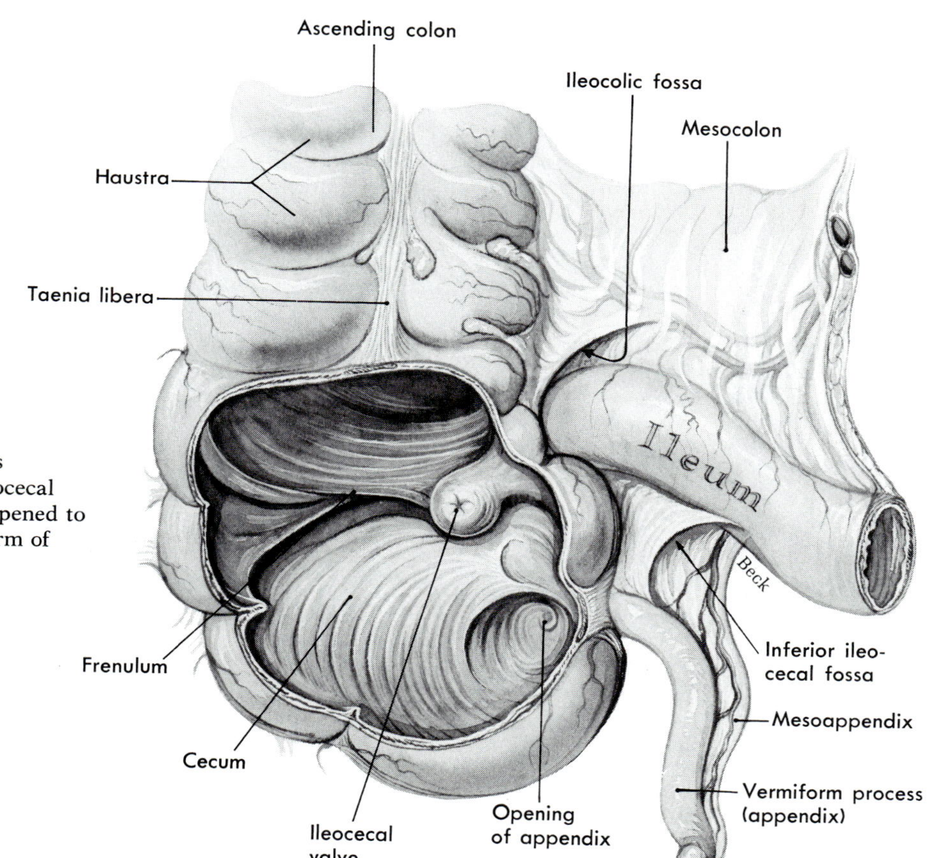

Fig. 14-11

The vermiform process (appendix) and the ileocecal region. The cecum is opened to reveal the papillary form of ileocecal sphincter.

position, on the left side of the abdomen, extending from a point below the stomach to the level of the iliac crest.

4 The *sigmoid colon* is that portion of the large intestine that courses downward below the iliac crest. It describes an S-shaped curve. The lower part of the curve, which joins the rectum, bends toward the left, the anatomical reason for placing a patient on the left side when giving an enema. In this position, gravity aids the flow of the water from the rectum into the sigmoid flexure.

Rectum. The last 7 or 8 inches of the intestinal tube is called the rectum. The terminal inch of the rectum is called the *anal canal.* Its mucous lining is arranged in numerous vertical folds known as *rectal columns,* each of which contains an artery and a vein. *Hemorrhoids* (or piles) are enlargements of the veins in the anal canal. The opening of the canal to the exterior is guarded by two sphincter muscles—an internal one of smooth muscle and an external one of striated muscle. The opening itself is called the *anus.* The general direction of the rectum is up, in, and back.

Coats

See Table 14-1 for information on the coats of the large intestine.

Functions

The main functions of the large intestine are absorption of water and elimination of the wastes of digestion.

Liver

Location and size

The liver is the largest gland in the body. It weighs between 3 and 4 pounds, lies immediately under the diaphragm, and occupies most of the right hypochondrium and part of the epigastrium (Fig. 14-12).

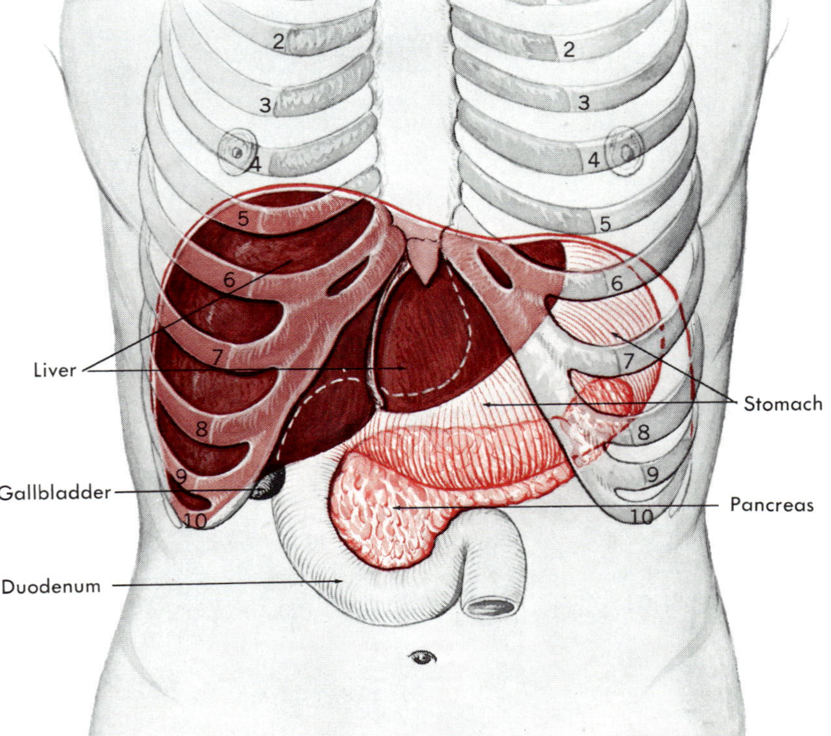

Fig. 14-12

The liver and pancreas in their normal positions relative to the rib cage, diaphragm, and stomach.

Lobes

The falciform ligament divides the liver into two main lobes, right and left, with the right lobe having three parts designated as the right lobe proper, the caudate lobe (a small four-sided area on the posterior surface), and the quadrate lobe (an approximately oblong section on the undersurface). Each lobe is divided into numerous lobules by small blood vessels and by fibrous strands that form a supporting framework (the capsule of Glisson) for them. The capsule of Glisson is an extension of the heavy connective tissue capsule that envelops the entire liver. The hepatic lobules, the anatomical units of the liver, are tiny hexagonal or pentagonal cylinders about 2 mm high and 1 mm in diameter. A small branch of the hepatic vein extends through the center of each lobule. Around this central (intralobular) vein, in columns radiating outward, are arranged the hepatic cells. Three separate sets of tiny tubes—branches of the hepatic artery, of the portal vein (interlobular veins), and of the hepatic duct (interlobular bile ducts)—are arranged around each lobule. From these, irregular branches (sinusoids) of the interlobular veins extend between the radiating columns of hepatic cells to join the central vein. Branches of the interlobular bile ducts also run between each two rows of hepatic cells.

Ducts

The small bile ducts within the liver join to form two larger ducts that emerge from the undersurface of the organ as the right and left hepatic ducts but that immediately join to form one *hepatic duct*. The hepatic duct merges with the *cystic duct* from the gallbladder, forming the *common bile duct* (Fig. 14-13), which opens into the duodenum in a small raised area, called the hepatopancreatic papilla. This papilla is located 3 to 4 inches below the pyloric opening from the stomach.

Functions

The liver is one of the most vital organs of the body. Here, in brief, are its main functions.

1. Liver cells secrete about a pint of bile a day.
2. Liver cells carry on a number of important steps in the metabolism of all three kinds of foods—proteins, fats, and carbohydrates.
3. Liver cells detoxify a variety of substances.
4. Liver cells store several substances—iron, for example, and vitamins A, B_{12}, and D.

Bile secretion by liver. Because it secretes bile into ducts, the liver qualifies as an exocrine gland. The main compounds in bile are bile salts, bilirubin, and cholesterol. Liver cells synthesize cholesterol. They secrete more than half of it into the bile. They form bile salts out of the rest of the cholesterol and secrete them into the bile. Bilirubin is formed in reticuloendothelial cells out of hemoglobin released into blood by red blood cell destruction. After circulating to the liver, bilirubin (a red pigment) is excreted into the bile by the liver cells.

Liver metabolism of proteins. Although all liver functions are important for healthy survival, some of its processes of protein metabolism are crucial for survival itself. A fairly detailed description of protein metabolism is given in Chapter 15, so now we shall merely identify the liver's protein metabolic processes. The first steps of protein catabolism take place mainly in liver cells. They consist of two sets of chemical reactions, namely, deamination and urea formation. So vital is the liver's formation of urea that if it ceases for even a few days, death ensues.

Anabolism of proteins also occurs in liver cells. They synthesize more than 95% of all the proteins present in plasma—the so-called plasma proteins. In fact, the only plasma

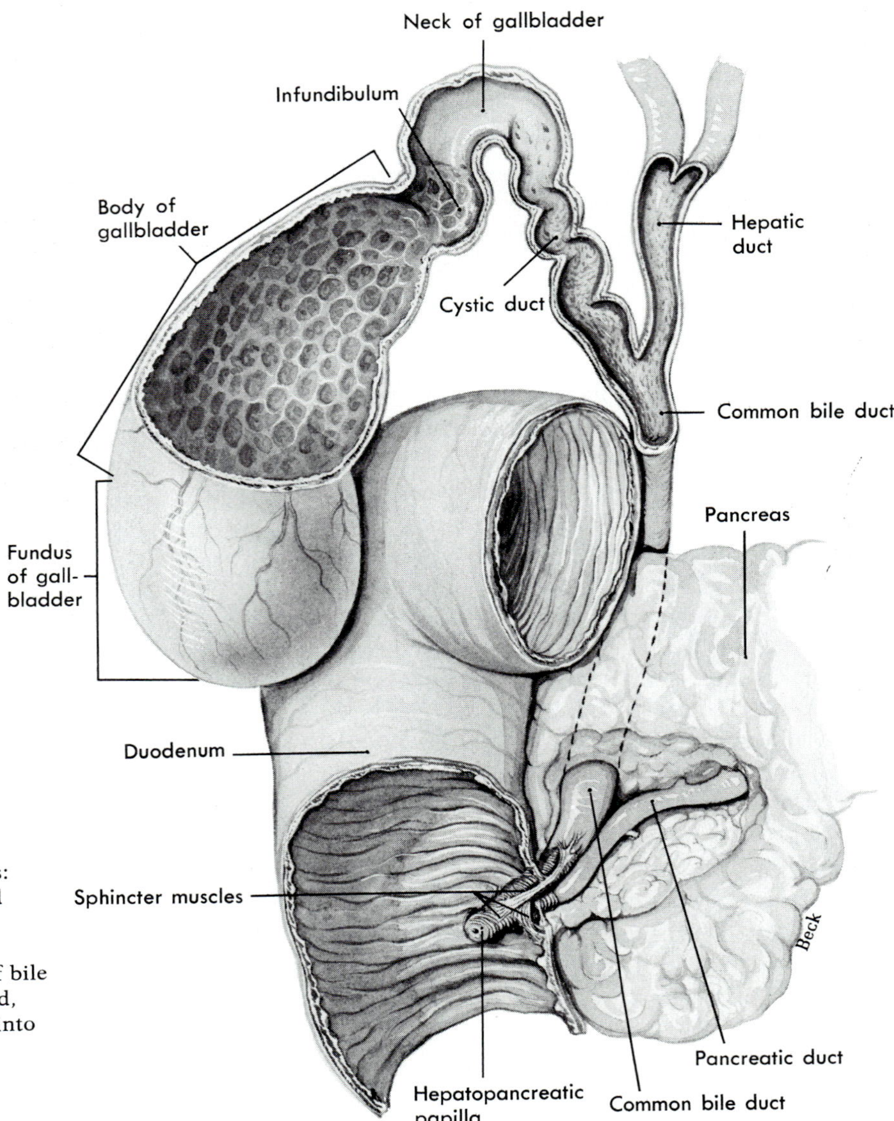

Fig. 14-13

The gallbladder and its divisions: fundus, body, infundibulum, and neck. Obstruction of either the hepatic or common bile duct by stone or spasm blocks the exit of bile from the liver, where it is formed, and prevents bile from ejecting into the duodenum.

proteins not synthesized in the liver are the gamma globulins known as antibodies. This is one of the activities that makes the liver essential for healthy survival. Consider, for instance, these vital functions of plasma proteins: prothrombin and fibrinogen play crucial parts in blood coagulation; all plasma proteins contribute to blood osmotic pressure and therefore to maintaining water balance; all of them contribute to blood viscosity and therefore are essential for maintaining normal blood pressure and circulation.

Fat metabolism in liver. Important processes of fat metabolism and fat anabolism are carried on by liver cells. For example, the first step of fat catabolism takes place mainly in liver cells. Ketogenesis is the name of this process. Liver cells also use carbohydrates and proteins to make fats, then release them into the blood to be carried to adipose cells for storage. More information about fat metabolism appears in Chapter 15.

Carbohydrate metabolism in the liver. Liver cells play an important though not vital part in carbohydrate metabolism. They carry on three processes that help maintain homeostasis of blood glucose concentration. When it rises above a certain level, the movement of glucose molecules out of blood into liver cells accelerates. Inside the liver cells, the process of glycogenesis (a series of chemical reactions) changes the glucose to glycogen. The liver cells store this glycogen in their cytoplasm until blood glucose concentration starts decreasing. Then they change it back to glucose (glycogenolysis) and release the glucose into blood. Thus liver glycogenesis and glycogenolysis function to help maintain homeostasis of blood glucose. Another liver process, namely, gluconeogenesis, can also add glucose to blood as needed. Gluconeogenesis consists of a series of chemical reactions that make "new" glucose out of proteins and fats.

Detoxication by liver cells. A number of poisonous substances enter the blood from the intestines. They circulate to the liver where, through a series of chemical reactions, they are changed to nontoxic compounds. Both ingested substances—alcohol, marijuana, and various other drugs, for example—and toxic substances formed in the intestines are detoxified in the liver.

Gallbladder

Size, shape, and location

The gallbladder is a pear-shaped sac from 3 to 4 inches long and an inch or more wide (Fig. 14-13). It lies on the undersurface of the liver and is attached to this organ by areolar tissue.

Structure

Serous, muscular, and mucous coats compose the wall of the gallbladder. The mucosal lining is arranged in rugae, similar in structure and function to those of the stomach.

Functions

The gallbladder concentrates and stores the bile that enters it by way of the hepatic and cystic ducts. Then later, when digestion is going on in the stomach and intestines, the gallbladder contracts, ejecting the concentrated bile into the duodenum.

Correlations

Inflammation of the lining of the gallbladder is called *cholecystitis. Cholecystectomy* is the surgical removal of the gallbladder. *Jaundice,* a yellow discoloration of the skin and mucosa, results whenever obstruction of the hepatic or common bile duct occurs. Bile is thereby denied its normal exit from the body in the feces. Instead, it is absorbed into the blood. As more bile pigments than normal accumulate in blood, the latter takes on a yellow hue. Without the normal amount of bile pigments, the feces become a grayish, so-called clay color.

Pancreas

Size, shape, and location

The pancreas is roughly fish shaped. It lies behind the stomach, with its head and neck in the C-shaped curve of the duodenum, its body extending horizontally across the posterior abdominal wall, and its tail touching the spleen (Fig. 14-12). According to an old anatomical witticism, the "romance of the abdomen" is the pancreas lying "in the arms of the duodenum."

This gland varies in size according to sex

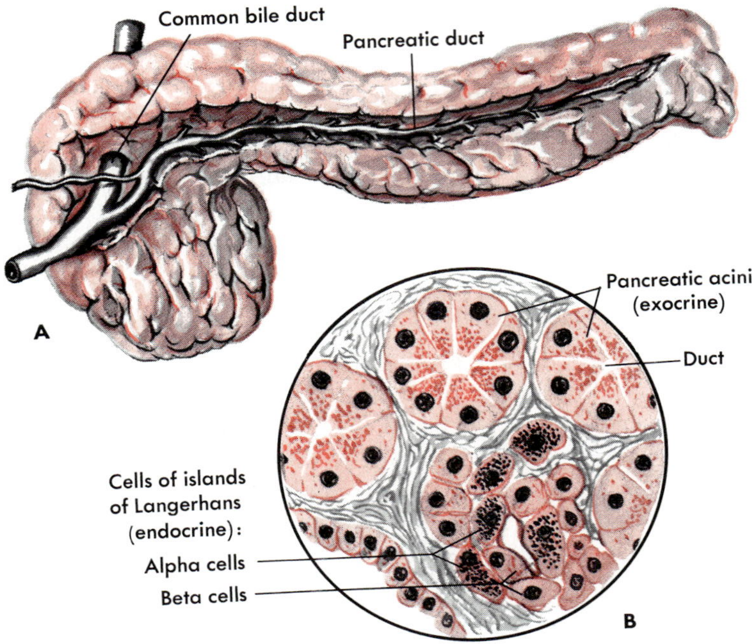

Fig. 14-14

A, Pancreas dissected to show main and accessory ducts. The main duct may join the common bile duct, as shown here, to enter the duodenum by a single opening at the hepatopancreatic papilla (see Fig. 14-13), or the two ducts may have separate openings. The accessory pancreatic duct is usually present and has a separate opening into the duodenum. **B,** Exocrine glandular cells (around small pancreatic ducts) and endocrine glandular cells of islands of Langerhans (adjacent to blood capillaries). Exocrine pancreatic cells secrete pancreatic juice, alpha endocrine cells secrete glucagon, and beta cells secrete insulin.

and individuals, being larger in men than in women. Usually its length is 6 to 9 inches, its width 1 to $1\frac{1}{2}$ inches, and its thickness $\frac{1}{2}$ to 1 inch. It weighs about 3 ounces.

Structure

The pancreas is classified as a compound tubuloacinar gland. The word compound tells us that the gland has a branching duct. The term tubuloacinar, on the other hand, tells us that some of the secreting units of the pancreas resemble tiny tubes and some tiny grapes in shape. Secreting cells constitute the walls of these tubular and acinar units. These are exocrine glands since they release their secretion into the microscopic duct within each unit (Fig. 14-14). These tiny ducts unite to form larger ducts that eventually join the main pancreatic duct which extends throughout the length of the gland from its tail to its head. It empties into the duodenum at the same point as the common bile duct—that is, at the hepatopancreatic papilla. (An accessory duct, the *duct of Santorini,* is frequently found extending from the head of the pancreas into the duodenum, about an inch above the duodenal papilla.)

In between the tubuloacinar units of the pancreas, like so many little island isolated from one another and from the ducts of the pancreas, lie clusters of cells called *islets* or *islands of Langerhans* (Fig. 14-14). Special staining techniques have revealed that two kinds of cells—alpha cells and beta cells—chiefly compose the islands of Langerhans. They are secreting cells, but their secretion passes into blood capillaries rather than into ducts. Thus the pancreas is a dual gland—an exocrine or duct gland because of the tubuloacinar units and an endocrine or ductless gland because of the islands of Langerhans.

Functions

1 The tubuloacinar units of the pancreas secrete the digestive enzymes found in pancreatic juice. Hence the pancreas plays an important part in digestion (p. 406).

2 Beta cells of the pancreas secrete *insulin,* a hormone that exerts a major control over carbohydrate metabolism (p. 422 and Fig. 15-9).

3 Alpha cells secrete *glucagon,* another hormone involved in carbohydrate metabolism (p. 422).

Vermiform appendix

Size, shape, and location

The appendix is a blind tube branching from the lower portion of the cecum (Fig. 14-11). As its name suggests, it resembles a large angle worm in size and shape, although its size varies greatly in different individuals.

Structure

The structure of the appendix is similar to that of the rest of the intestine. Its mucous lining frequently becomes inflamed, a condition well known as *appendicitis.*

Digestion

Definition

Digestion is the sum of all the changes food undergoes in the alimentary canal.

Purpose

The purpose of digestion as suggested in the dictionary definition is "the conversion of food into assimilable matter." Digestion is necessary because foods, as eaten, are too complex in physical and chemical composition to pass through the intestinal mucosa into the blood or for cells to utilize them for energy and tissue building. In other words, digestion is the necessary preliminary to both absorption and metabolism of foods.

Kinds

Since both the physical and chemical composition of ingested food makes its absorption impossible, two kinds of digestive changes are necessary, *mechanical* and *chemical.*

Mechanical digestion. Mechanical digestion consists of all those movements of the alimentary tract that bring about the following:

1. Change the physical state of ingested food from comparatively large solid pieces into minute dissolved particles, thereby facilitating chemical digestion
2. Propel the food forward along the alimentary tract, finally eliminating the digestive wastes from the body
3. Churn the intestinal contents in such a way that it becomes well mixed with the digestive juices and that all parts of it come in contact with the surface of the intestinal mucosa, thereby facilitating absorption

A list of definitions of the different processes involved in mechanical digestion, together with the organs that accomplish them, are given in Table 14-4. Swallowing, the movement of food out of the stomach, and defecation are considered in more detail below.

Swallowing consists of three steps: movement of the food through the mouth into the pharynx, through the pharynx into the esophagus, and through the esophagus into the stomach. Only the first act is voluntary—that is, can be controlled by the will. The other two actions are reflexes that occur automatically after food enters the pharynx. Stimulation of the mucosa of the back of the mouth, pharynx, or laryngeal region initiates the second step—reflex contractions of the muscular pharyngeal walls. Consequently, paralysis of the sensory nerves to the mucosa of the back of the mouth, pharynx, or laryngeal region by a drug such as procaine (Novocain) makes swallowing im-

Table 14-4

Processes in mechanics of digestion

Organ	Mechanical process	Nature of process
Mouth (teeth and tongue)	Mastication	Chewing movements—reduce size of food particles and mix them with saliva
	Deglutition	Swallowing—movement of food from mouth to stomach
Pharynx	Deglutition	
Esophagus	Deglutition	
	Peristalsis	Wormlike movements that squeeze food downward in tract; constricted ring forms first in one section, the next, etc., causing waves of contraction to spread throughout canal
Stomach	Churning	Forward and backward movement of gastric contents; peristalsis propels it forward; closed pyloric sphincter deflects it backward
	Peristalsis	Moves material through stomach and at intervals into duodenum
Small intestine	Churning (rhythmic segmentation)	Forward and backward movement within segment of intestine; purpose, to mix food and digestive juices thoroughly and to bring all digested food in contact with intestinal mucosa to facilitate absorption; purpose of peristalsis, on the other hand, to propel intestinal contents along digestive tract
	Peristalsis	
Large intestine		
Colon	Haustral churning	Churning movements within haustral sacs
	Peristalsis	
Descending colon	Mass peristalsis	Entire contents moved into sigmoid colon and rectum; usually occurs after a meal
Rectum	Defecation	Emptying of rectum, so-called bowel movement

possible. Since the pharynx serves both the digestive and respiratory systems, it has connections with the parts of the digestive tract above and below it (mouth and esophagus) and with the parts of the respiratory tract above and below (nasal cavity and larynx). When food enters the pharynx and its walls contract, theoretically its contents could be squeezed in any of these four directions—that is, back up into the nose or mouth or downward into the larynx or esophagus. That it takes only one of these routes (into the esophagus) results from the fact that the openings into the other organs become blocked off during swallowing. Elevation of the tongue (so that it presses against the roof of the mouth) bars entry into the mouth, elevation of the soft palate obstructs the passageway into the nose, and a squeezing of the larynx by contraction of certain muscles closes over the opening into this organ. An additional mechanism normally prevents our "swallowing down the wrong throat"—that is, into the larynx and lower respiratory tract. Stimulation of the mucosa of the back of the mouth, pharynx, and laryngeal region produces not only the second phase of the swallowing act but also momentary inhibition of respirations. Obviously, prevention of inspiration during swallowing greatly lessens the danger of foods entering the respiratory tract.

Stimulation of the esophageal mucosa by material entering it from the pharynx initiates the reflex of esophageal peristalsis.

Liquids and well-chewed foods are precipitated down the esophagus to the cardiac sphincter by the force of pharyngeal contraction and by gravity. They are then moved through the sphincter into the stomach by means of esophageal peristalsis. The latter is responsible for propelling large solid particles throughout the length of the esophagus.

Empyting of the stomach after a meal requires a considerable period of time (about 1 to 4 hours for the average meal). Many theories have been proposed about the mechanism regulating gastric emptying. It is now known that as small amounts of gastric contents become liquefied, they are ejected bit by bit every 20 seconds or so into the duodenum until a certain amount has accumulated there. From then on, the emptying process slows down because of operation of a mechanism known as the *enterogastric reflex.* Both nerve impulses and a hormone initiate this reflex. Fats and sugars present in the small intestine stimulate the intestinal mucosa to release a hormone (called enterogastrone) into the bloodstream. When it reaches the stomach wall via the circulation, enterogastrone has an inhibitory effect on gastric muscle, decreasing its peristalsis (the enterogastric reflex) and thereby slowing down stomach emptying. Proteins and acid also initiate the enterogastric reflex but not by means of a hormone. They initiate this reflex by stimulating vagal nerve receptors in the intestinal mucosa.

Defecation is a reflex brought about by stimulation of receptors in the rectal mucosa. Recent distention of the rectum constitutes the usual stimulus. Normally the rectum is empty until mass peristalsis moves fecal matter out of the colon into the rectum. This distends the rectum and produces the desire to defecate. Also, it stimulates colonic peristalsis and initiates reflex relaxation of the internal sphincter of the anus. Voluntary straining efforts and relaxation of the external anal sphincter may then follow as a result of the desire to defecate. Together these several responses bring about defecation. Note that this is a reflex partly under voluntary control. If one voluntarily inhibits it, the rectal receptors soon become depressed and the urge to defecate usually does not recur until about 24 hours later, when mass peristalsis again takes place. During the interim, water is absorbed from the fecal mass, producing a hardened or constipated stool.

Chemical digestion. Chemical digestion consists of all the changes in chemical composition that foods undergo in their travel through the alimentary canal. These changes result from the hydrolysis of foods. (*Hydrolysis* is a chemical process in which a compound unites with water and then splits into simpler compounds.) Numerous enzymes* present in the various digestive juices catalyze the hydrolysis of foods (Table 14-5).

*Enzymes are usually defined simply as "organic catalysts"—that is, they are organic compounds, and they accelerate chemical reactions without appearing in the final products of the reaction. Enzymes are vital substances. Without them, the chemical reactions necessary for life could not take place. So important are they that someone has even defined life as the "orderly functioning of hundreds of enzymes."

Chemical structure: Enzymes are proteins. Frequently their molecules also contain a nonprotein part called the *prosthetic group* of the enzyme molecule (if this group readily detaches from the rest of the molecule, it is spoken of as the *coenzyme*). Some prosthetic groups contain inorganic ions (Ca^{++}, Mg^{++}, Mn^{++}, etc.). Many of them contain vitamins. In fact, every vitamin of known function constitutes part of a prosthetic group of some enzyme. Nicotinic acid, thiamine, riboflavin, and other B complex vitamins, for example, function in this way.

Classification and naming: Two of the systems used for naming enzymes are as follows: suffix *-ase* is used either with the root name of the substance whose chemical reaction is catalyzed (the substrate chemical, that is) or with the word that describes the kind of chemical reaction catalyzed. Thus, according to the first method, sucrase is an enzyme that catalyzes a chemical

Footnote continued.

Table 14-5

Chemical digestion

Digestive juices and enzymes	Food enzyme digests (or hydrolyzes)	Resulting product*
Saliva		
Amylase (ptyalin)	Starch (polysaccharide or complex sugar)	Maltose (a disaccharide or double sugar)
Gastric juice		
Protease (pepsin) plus hydrochloric acid	Proteins, including casein	Proteoses and peptones (partially digested proteins)
Lipase (of little importance)	Emulsified fats (butter, cream, etc.)	*Fatty acids and glycerol*
Bile (contains no enzymes)	Large fat droplets (unemulsified fats)	Small fat droplets or emulsified fats
Pancreatic juice		
Protease (trypsin)†	Proteins (either intact or partially digested)	Proteoses, peptides, and *amino acids*
Lipase (steapsin)	Bile-emulsified fats	*Fatty acids and glycerol*
Amylase (amylopsin)	Starch	Maltose
Intestinal juice (succus entericus)		
Peptidases	Peptides	*Amino acids*
Sucrase	Sucrose (cane sugar)	*Glucose and fructose*‡ (simple sugars or monosaccharides)
Lactase	Lactose (milk sugar)	*Glucose and galactose* (simple sugars)
Maltase	Maltose (malt sugar)	*Glucose* (grape sugar)

*Substances in italics are end products of digestion or, in other words, completely digested foods ready for absorption.
†Secreted in inactive form (trypsinogen); converted to active form, trypsin, by enterokinase (now called enteropeptidase), an enzyme in the intestinal juice.
‡Glucose is also called dextrose; fructose is called levulose.

reaction in which sucrose takes part. According to the second method, sucrase might also be called hydrolase because it hydrolyzes sucrose. Enzymes investigated before these methods of nomenclature were adopted still are called by older names, such as ptyalin, pepsin, trypsin, etc.

Classified according to the kind of chemical reactions catalyzed, enzymes fall into several groups:

1 *Oxidation-reduction enzymes.* These are known as oxidases, hydrogenases, and dehydrogenases. Energy release for muscular contraction and all physiological work depends upon these enzymes.
2 *Hydrolyzing enzymes* or hydrolases. Digestive enzymes belong to this group. These are generally named after the substrate acted upon: for example, lipase, sucrase, maltase, etc.
3 *Phosphorylating enzymes.* These add or remove phosphate groups and are known as phosphorylases or phosphatases.
4 *Enzymes that add or remove carbon dioxide.* These are known as carboxylases or decarboxylases.
5 *Enzymes that rearrange atoms within a molecule.* These are known as mutases or isomerases.
6 *Hydrases.* These add water to a molecule without splitting it, as hydrolases do.

Enzymes are also classified as intracellular or extracellular, depending upon whether they act within cells or outside of them in the surrounding medium. Most enzymes act intracellularly in the body, an important exception being the digestive enzymes.

Properties: In general, the properties of enzymes are the same as those of proteins, since enzymes are pro-

Although we eat six kinds of chemical substances (carbohydrates, proteins, fats, vitamins, mineral salts, and water), only the first three named have to be chemically digested in order to be absorbed.

Different enzymes require different hydrogen ion concentrations in their environment for optimal functioning. Ptyalin, the main enzyme in saliva, functions best in the neutral to slightly acid pH characteristic of saliva. It is gradually inactivated by the marked acidity of gastric juice. In contrast, pepsin, an enzyme in gastric juice, is inactive unless sufficient hydrochloric acid is present. Therefore, in diseases characterized by gastric hypoacidity (pernicious anemia, for example), hydrochloric acid is given orally before meals.

Carbohydrate digestion (Fig. 14-15). Carbohydrates are saccharide compounds. This means that their molecules contain one or more saccharide groups ($C_6H_{10}O_5$). Polysaccharides, notably starches, contain many of these groups, disaccharides (sucrose, lactose, and maltose) contain two of them, and monosaccharides (glucose, fructose, and galactose) contain only one. Polysaccharides are hydrolyzed to disaccharides by enzymes known as *amylases* found in saliva and pancreatic juice (salivary amylase is also called ptyalin and pancreatic amylase is also known as amylopsin). Disaccharides are hydrolyzed to monosaccharides by the intestinal juice enzymes sucrase, lactase, and maltase.

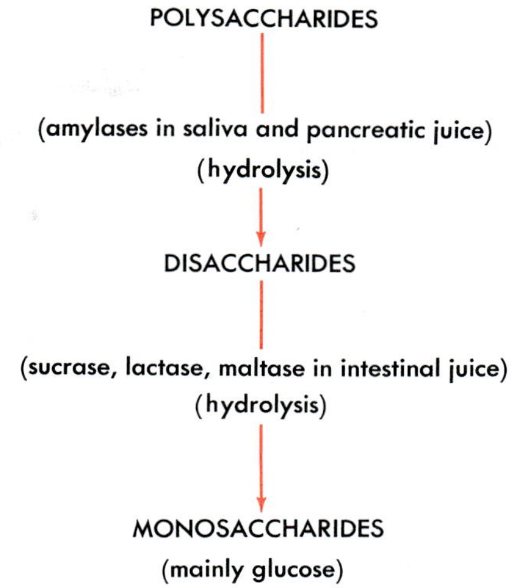

Fig. 14-15
Carbohydrate digestion.

Protein digestion (Fig. 14-16). Protein compounds have very large molecules made up of amino acids. Enzymes called *proteases*

---

teins. For example, they form colloidal solutes in water and are precipitated or coagulated by various agents, such as high temperatures and salts of heavy metals. Hence, these agents inactivate enzymes. Other important enzyme properties are as follows.

1 Most enzymes are *specific in their action*—that is, act only on a specific substrate. This is attributed to a "key-in-a-lock" kind of action, the configuration of the enzyme molecule fitting the configuration of some part of the substrate molecule.

2 Enzymes *function optimally at a specific pH* and become inactive if this deviates beyond narrow limits.

3 A *variety of physical and chemical agents inactivate or inhibit enzyme action*—for example, x-rays and radium rays (this presumably accounts for some of the ill effects of excessive radiation), certain antibiotic drugs, unfavorable pH, etc.

4 *Most enzymes catalyze a chemical reaction in both directions,* the direction and rate of the reaction being governed by the law of mass action. An accumulation of a product slows the reaction and tends to reverse it. A practical application of this fact is the slowing of digestion when absorption is interfered with and the products of digestion accumulate.

5 *Enzymes are continually being destroyed in the body* and therefore have to be continually synthesized, even though they are not used up in the reactions they catalyze.

6 *Many enzymes are synthesized* as inactive proenzymes. Substances that convert proenzymes to active enzymes are often called kinases; e.g., enterokinase (now called enteropeptidase) changes inactive trypsinogen into active trypsin, etc.

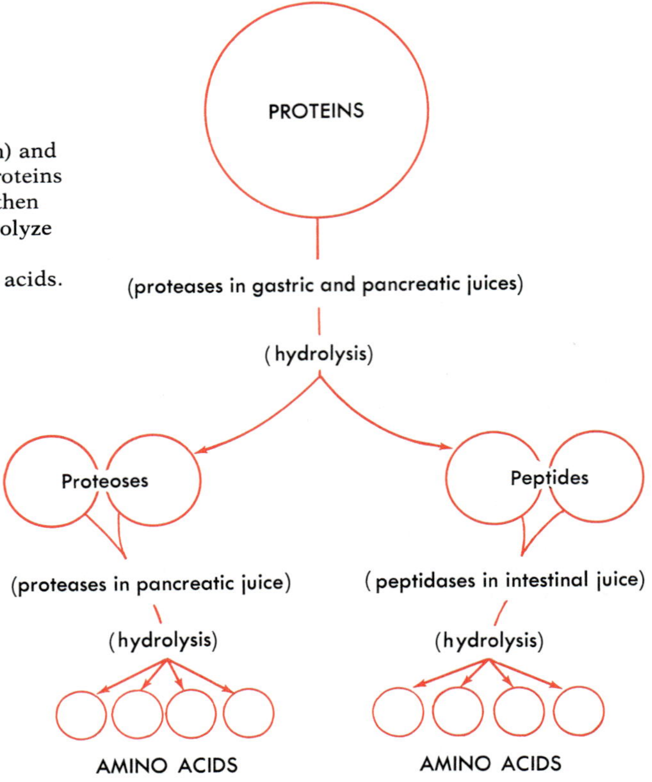

Fig. 14-16

Protein digestion. Gastric juice protease (pepsin) and pancreatic juice protease (trypsin) hydrolyze proteins to proteoses and peptides. Protein digestion is then completed by pancreatic proteases, which hydrolyze proteoses to amino acids, and intestinal juice peptidases, which hydrolyze peptides to amino acids.

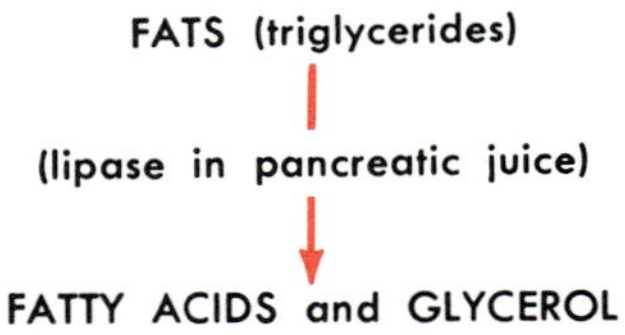

FATS (triglycerides)

(lipase in pancreatic juice)

FATTY ACIDS and GLYCEROL

Fig. 14-17

Fat digestion (hydrolysis) by lipase, facilitated first by emulsion of fats by bile.

catalyze the hydrolysis of proteins into intermediate compounds—for example, proteoses and peptides—and finally into amino acids. The main proteases are pepsin in gastric juice, trypsin in pancreatic juice, and peptidases in intestinal juice.

Fat digestion (Fig. 14-17). Because fats are insoluble in water, they must be emulsified—that is, dispersed as very small droplets—before they can be digested. Bile emulsifies fats in the small intestine. This facilitates fat digestion by providing a greater contact area between fat molecules and pancreatic lipase, the main fat-digesting enzyme (pancreatic lipase was formerly known as steapsin). For a summary of the actions of each digestive juice, see Table 14-5.

Residues of digestion. Certain components of food resist digestion and are eliminated from the intestines in the feces. These *residues of digestion* are cellulose from carbohydrates, undigested connective tissue and toxins from meat proteins, and undigested fats. In addition to these wastes, feces consists of bacteria, pigments, water, and mucus.

Control of digestive gland secretion

Digestive glands secrete when food is present in the alimentary tract or when it is seen, smelled, or imagined. Complicated reflex and chemical (hormonal) mechanisms control the flow of digestive juices in such a way that they appear in proper amounts when and for as long as needed.

Saliva. As far as is known, only reflex mechanisms control the secretion of saliva. Chemical, mechanical, olfactory, and visual stimuli initiate afferent impulses to centers in the brainstem that send out efferent impulses to salivary glands, stimulating them.

Chemical and mechanical stimuli come from the presence of food in the mouth. Olfactory and visual stimuli come, of course, from the smell and sight of food.

Gastric secretion. Stimulation of gastric juice secretion occurs in three phases controlled by reflex and chemical mechanisms. Because stimuli that activate these mechanisms arise in the head, stomach, and intestines, the three phases are known as the cephalic, gastric, and intestinal phases, respectively.

The *cephalic phase* is also spoken of as the reflex phase and as the psychic phase because a reflex mechanism controls gastric juice secretion at this time and psychic factors activate the mechanism. For example, the sight or smell or taste of food that is pleasing to an individual stimulates various head receptors and thereby initiates reflex stimulation of the gastric glands. Parasympathetic fibers in branches of the vagus nerve conduct the stimulating efferent impulses to the glands.

During the *gastric phase* of gastric juice secretion, the following chemical control mechanism dominates. Substances (such as meat extractives and products of protein digestion) in foods that have reached the pyloric portion of the stomach stimulate its mucosa to release a hormone called *gastrin* into the blood in stomach capillaries. When it circulates to the gastric glands, it greatly accelerates their secretion of gastric juice, which has a high pepsin and hydrochloric acid content (Table 14-6). Hence, this seems to be a device for ensuring that when food is in the stomach there will be enough enzymes there to digest it.

The *intestinal phase* of gastric juice secretion is less clearly understood than the other two. A chemical control mechanism, however, is believed to operate. It is known, too, that the hormone *enterogastrone* (released by intestinal mucosa when fat is in the intestine) causes a lessening of both gastric secretion and motility.

Pancreatic secretion. Chemical control of pancreatic secretion has been established. Also, reflex control is postulated. Chemical control ensures continued pancreatic secretion while food is in the duodenum. The presence in the small intestine of hydrochloric acid, protein, and fat digestion products causes the intestinal mucosa to release a hormone, *secretin,* into the blood. When

Table 14-6

Actions of some digestive hormones summarized

Hormone	Source	Action
Gastrin	Formed by gastric mucosa in presence of partially digested proteins	Stimulates secretion of gastric juice rich in pepsin and hydrochloric acid
Secretin	Formed by action of hydrochloric acid on prosecretin (chemical normally present in intestinal mucosa)	Stimulates secretion of pancreatic juice low in enzymes Stimulates secretion of bile by liver May stimulate secretion of intestinal juice
Pancreozymin	Formed by intestinal mucosa in presence of hydrochloric acid and partially digested proteins	Stimulates pancreatic cells to produce enzymes
Cholecystokinin	Formed by intestinal mucosa in presence of fats	Stimulates ejection of bile from gallbladder

secretin circulates to the pancreas, it stimulates pancreatic cells to secrete at a faster rate. But the juice they produce has a low enzyme content. Another hormone, pancreozymin, stimulates them to produce enzymes. The presence of hydrochloric acid and partially digested proteins in the intestine serves as the stimulus for pancreozymin release by the intestinal mucosa.

Secretin has an interesting claim to fame. Not only was it the first hormone to be discovered, but its discovery gave rise to the broad concept of hormonal control of body activities.

Secretion of bile. Chemical mechanisms dominate the control of bile. Secretin, the same hormone that stimulates pancreatic activity, stimulates the liver to secrete bile. The ejection of bile, however, from the gallbladder into the duodenum is largely controlled by another hormone, *cholecystokinin,* which is formed by the intestinal mucosa when fats are present in the duodenum.

Intestinal secretion. Knowledge concerning the regulation of the secretion of *intestinal juice* is still somewhat obscure. Some evidence suggests that the intestinal mucosa, stimulated by hydrochloric acid and food products, releases into the blood a hormone, *enterocrinin,* which brings about increased intestinal juice secretion. Presumably, neural mechanisms also help control the secretion of this digestive juice.

Absorption

Definition

Absorption is the passage of substances (notably digested foods, water, salts, and vitamins) through the intestinal mucosa into the blood or lymph.

How accomplished

Absorption is not entirely a passive process explainable on the basis of the physical laws of diffusion, filtration, and osmosis alone. Although these phenomena play a part in

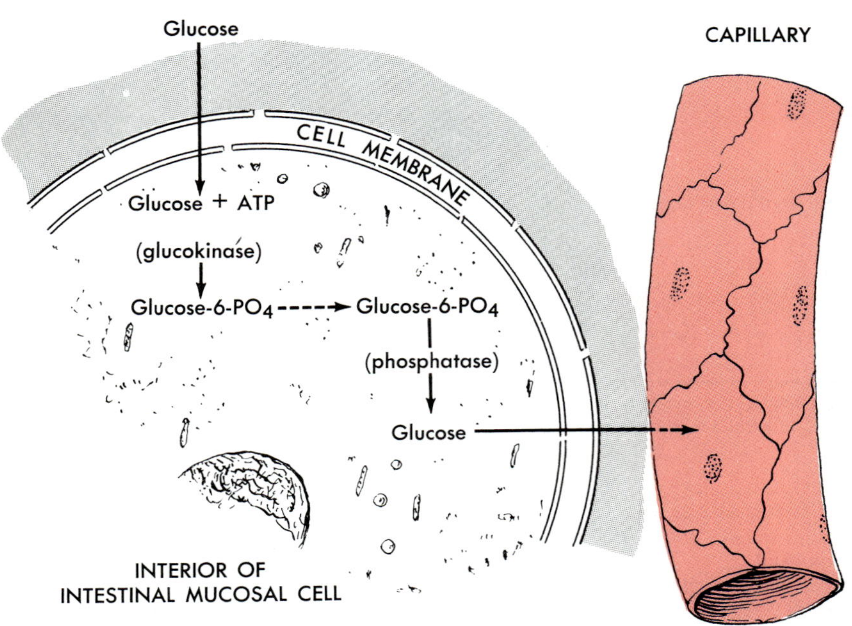

Fig. 14-18

Glucose absorption. Glucose enters mucosal cells of intestinal villi (presumably by facilitated diffusion) from the lumen of the intestine and exits from the opposite surfaces of these cells into blood capillaries.

absorption, so too do facilitated diffusion (p. 25) and active transport mechanisms. Active transport moves substances in the opposite direction from that expected according to the laws of osmosis and diffusion.

The mechanism postulated for glucose absorption (Fig. 14-18) consists of these steps:

1 Glucose moves into cells of the intestinal mucosa by the carrier-mediated process of facilitated diffusion.

2 Inside the mucosal cells, phosphorylation of the glucose molecule occurs immediately, presumably by the following reaction catalyzed by the enzyme glucokinase:

$$\text{Glucose} + \text{ATP} \xrightarrow{\text{(Glucokinase)}} \text{Glucose-6-phosphate} + \text{ADP}$$

3 Glucose-6-phosphate diffuses across the mucosal cell, and near its surface adjacent to a blood capillary, the preceding equation is reversed, but by another enzyme, phosphatase. Glucose then moves out of the cell into the blood.

Absorption of amino acids takes place by facilitated diffusion. The mechanism for fat absorption is not clear. It is thought to involve some kind of active transport. It is known, however, that some fatty acids and glycerol enter intestinal mucosal cells and there recombine to form neutral fats, which then move out of the cells into lymphatic capillaries (lacteals) in the intestinal villi. In addition, some fatty acids are absorbed as such into blood capillaries. Also, some neutral fats are absorbed without first being digested to fatty acids and glycerol.

Active transport mechanisms seem to be available for water and salt absorption, and apparently some sort of mechanism operates to prevent absorption of certain substances. Magnesium sulfate (epsom salts), for example, is not absorbed from the intestine even though its molecules are smaller than glucose molecules, which can diffuse through the intestinal mucosa. In fact, it is this nonabsorbability of magnesium sulfate that makes it an effective cathartic. (Since magnesium sulfate ions do not diffuse freely through the intestinal mucosa, do you think their presence in intestinal fluid would create an osmotic pressure gradient between the intestinal fluid and blood? If so, what effect

Table 14-7

Food absorption

Form absorbed	Structures into which absorbed	Circulation
Protein—as amino acids Perhaps minute quantities of some whole proteins absorbed—e.g., in allergic reactions	Blood in intestinal capillaries	Portal vein, liver, hepatic vein, inferior vena cava to heart, etc.
Carbohydrates—as simple sugars	Same as amino acids	Same as amino acids
Fats		
Glycerol	Lymph in intestinal lacteals	During absorption, i.e., while in epithelial cells of intestinal mucosa, glycerol and fatty acids recombine to form microscopic particles of fats (chylomicrons); lymphatics carry them by way of thoracic duct to left subclavian vein, superior vena cava, heart, etc.
Fatty acids combine with bile salts to form water-soluble substance	Lymph in intestinal lacteals	
Some finely emulsified undigested fats absorbed	Small fraction enters intestinal blood capillaries	Some fats transported by blood in form of phospholipids or cholesterol esters

would this have on water movement between these two fluids? Reread pp. 26-30 if you need help in answering these questions.)

Note that after absorption food does not pass directly into the general circulation. Instead it first travels by way of the portal system to the liver. During intestinal absorption, blood entering the liver via the portal vein contains greater concentrations of glucose, amino acids, and fats than does blood leaving the liver via the hepatic vein for the systemic circulation. Clearly, the excess of these food substances over and above the normal blood levels has remained behind in the liver.

What the liver does with them is part of the story of metabolism, our next topic for discussion.

Outline summary

Functions and importance

1 Prepare food for absorption and metabolism
2 Absorption
3 Elimination of wastes
4 Vital importance

ORGANS

1 Main organs—compose alimentary canal: mouth, pharynx, esophagus, stomach, and intestines
2 Accessory organs—salivary glands, teeth, liver, gallbladder, pancreas, and vermiform appendix

Walls of organs

1 Coats
- **a** Mucous lining
- **b** Submucous coat of connective tissue—main blood vessels here
- **c** Muscular coat
- **d** Fibroserous coat

2 Modifications of coats
- **a** Mucous lining
  - 1 Rugae and microscopic gastric and hydrochloric acid glands in stomach
  - 2 Circular folds, villi, intestinal glands, Peyer's patches, and solitary lymph nodes in small intestine
  - 3 Solitary nodes and intestinal glands in large intestine
- **b** Muscle coat
  - 1 Three layers (circular, longitudinal, oblique) in stomach instead of only two layers as in rest of tract
  - 2 Three tapelike strips make up outer, longitudinal layer and pucker large intestine into small sacs called haustra
- **c** Fibroserous coat
  - 1 Peritoneum covers stomach and intestines
  - 2 Greater omentum or lace apron—double fold of peritoneum that hangs from lower edge of stomach like an apron over intestines; should not be confused with mesentery, which also is double fold of peritoneum but fan shaped and attached at short side to posterior wall of abdominal cavity; small intestines anchored to posterior abdominal wall by means of mesentery

Mouth (buccal cavity)

1 Formed by cheeks, hard and soft palates, tongue, and muscles
2 Hard palate—formed by two palatine bones and parts of two maxillary bones
3 Soft palate—formed of muscle in shape of arch; forms partition between mouth and nasopharynx; fauces is archway or opening from mouth to oropharynx; uvula is conical-shaped process suspended from midpoint of arch
4 Tongue
- **a** Many rough elevations on tongue's surface called papillae; contain taste buds
- **b** Frenum—a fold of mucous membrane that helps anchor tongue to mouth floor

Salivary glands

1 Parotid—below and in front of ear; duct opens on inside of cheek, opposite upper second molar tooth
2 Submaxillary—posterior part of mouth floor
3 Sublingual—anterior part of mouth floor

Teeth

1 Deciduous or baby teeth—ten per jaw or twenty in set
2 Permanent—sixteen per jaw or thirty-two per set
3 Structure of typical tooth—see Fig. 14-8

Pharynx

See p. 293.

Esophagus

1 Position and extent
- **a** Posterior to trachea and heart; pierces diaphragm
- **b** Extends from pharynx to stomach, distance of approximately 10 inches

2 Structure—collapsible, muscle tube

Stomach

1 Size, shape, and position
- **a** Size varies in different individuals; also according to whether distended or not
- **b** Elongated pouch
- **c** Lies in epigastric and left hypochondriac portions of abdominal cavity

2 Divisions
- **a** Fundus—portion above esophageal opening
- **b** Body—central portion
- **c** Pylorus—constricted, lower portion

3 Curves
- **a** Lesser—upper, right border
- **b** Greater—lower, left border

4 Sphincter muscles
- **a** Cardiac—guarding opening of esophagus into stomach

b Pyloric—guarding opening of pylorus into duodenum
5 Coats—see Table 14-1
6 Glands
 a Epithelial cells of gastric mucosa secrete mucus
 b Parietal cells secrete hydrochloric acid
 c Chief cells (zymogen cells) secrete enzymes of gastric juice
7 Functions
 a Serves as food reservoir
 b Secretes gastric juice
 c Contractions break food into small particles, mix them well with gastric juice, and move contents on into duodenum
 d Secretes the antianemic intrinsic factor
 e Carries on a limited amount of absorption—some water, alcohol, and certain other drugs

Small intestine

1 Size and position
 a Approximately 1 inch in diameter and 20 feet in length
 b Its coiled loops fill most of abdominal cavity
2 Divisions
 a Duodenum
 b Jejunum
 c Ileum
3 Coats—see Table 14-1
4 Functions
 a Completes digestion of foods
 b Absorbs end products of digestion
 c Secretes hormones that help control secretion of pancreatic juice, bile, and intestinal juice

Large intestine (colon)

1 Size—approximately 2½ inches in diameter and 5 or 6 feet in length
2 Divisions
 a Cecum
 b Colon
  1 Ascending
  2 Transverse
  3 Descending
  4 Sigmoid
 c Rectum
3 Coats—see Table 14-1
4 Functions
 a Absorption of water
 b Elimination of digestive wastes

Liver

1 Location and size
 a Occupies most of right hypochondrium and part of epigastrium
 b Largest gland in body
2 Lobes
 a Right lobe, subdivided into three smaller lobes—right lobe proper, caudate, and quadrate
 b Left lobe
 c Lobes divided into lobules by blood vessels and fibrous partitions
3 Ducts
 a Hepatic duct from liver
 b Cystic duct from gallbladder
 c Common bile duct formed by union of hepatic and cystic ducts and opens into duodenum at ampulla of Vater (duodenal papilla)
4 Functions
 a Secretes bile
 b Plays essential role in metabolism of carbohydrates, proteins, and fats—e.g., carries on glycogenesis, glycogenolysis, and gluconeogenesis; also deamination and ketogenesis; synthesizes various blood proteins; detoxifies various substances; stores iron, vitamins A, B_{12}, and D

Gallbladder

1 Size, shape, and location
 a Approximately size and shape of small pear
 b Lies on undersurface of liver
2 Structure—sac of smooth muscle with mucous lining arranged in rugae
3 Functions
 a Concentrates and stores bile
 b During digestion, ejects bile into duodenum

Pancreas

1 Size, shape, and location
 a Larger in men than in women but varies in different individuals
 b Shaped something like a fish with head, body, and tail
 c Lies in C-shaped curve of duodenum
2 Structure—similar to salivary glands
 a Divided into lobes and lobules
 b Pancreatic cells pour their secretion into duct that runs length of gland and empties into duodenum at ampulla of Vater
 c Clusters of cells, not connected with any ducts, lie between pancreatic cells—called islets or islands of Langerhans—composed of alpha and beta type cells—latter thought to secrete insulin, former to secrete glucagon
3 Functions
 a Tubuloacinar units secrete pancreatic juice
 b Beta cells of islands of Langerhans secrete insulin
 c Alpha cells of islands of Langerhans secrete glucagon

Vermiform appendix

1 Size, shape, and location
 a About size and shape of large angle worm
 b Blind-end tube off cecum
2 Structure—similar to rest of intestine

DIGESTION

1 Definition—all changes food undergoes in alimentary canal
2 Purpose—conversion of foods into chemical and physical forms that can be absorbed and metabolized
3 Kinds
 a Mechanical—movements that change physical state of foods, propel them forward in alimentary tract, eliminate digestive wastes from tract and facilitate absorption (see Table 14-4 for description of processes involved in mechanical digestion)
  1 Mastication (chewing)
  2 Swallowing (deglutition)
   a Movement of food through mouth into pharynx—voluntary act

b Movement of food through pharynx into esophagus—involuntary or reflex act initiated by stimulation of mucosa of back of mouth, pharynx, or laryngeal region; paralysis of receptors here, e.g., by procaine (Novocain), makes swallowing impossible

c Movement of food through esophagus into stomach; accomplished by esophageal peristalsis—reflex initiated by stimulation of esophageal mucosa

3 Peristalsis—wormlike movements that squeeze food downward in tract; emptying of stomach or opening of pyloric sphincter regulated by enterogastric reflex; fats and sugars in intestine stimulate mucosa to release enterogastrone into blood, which in turn inhibits gastric peristalsis and slows stomach emptying; proteins and acid stimulate vagal nerve receptors in intestinal mucosa and thereby initiate reflex emptying of stomach

4 Churning

5 Mass peristalsis

6 Defecation—reflex initiated by stimulation of rectal mucosa

b Chemical—series of hydrolytic processes dependent upon specific enzymes (see Table 14-5 for description of chemical changes)

4 Control of digestive gland secretion—see Table 14-6

a Saliva—secretion is reflex initiated by stimulation of taste buds, other receptors in mouth and esophagus, olfactory receptors, and visual receptors

b Gastric secretion—controlled reflexly by same stimuli that initiate salivary secretion; also controlled chemically by hormone, gastrin, released by gastric mucosa in presence of partially digested proteins; enterogastrone (hormone just mentioned as slowing stomach emptying) also has inhibitory effect on gastric secretion

c Pancreatic secretion—controlled chemically by hormones secretin and pancreozymin formed by intestinal mucosa when hydrochloric acid enters duodenum

d Bile

1 Secretion—controlled chemically by same hormone (secretin) that regulates pancreatic secretion

2 Ejection into duodenum—controlled chemically by hormone cholecystokinin formed by intestinal mucosa when fats present in duodenum

e Intestinal secretion—control still obscure, although believed to be both reflex and chemical

ABSORPTION

1 Definition—passage of substances through intestinal mucosa into blood or lymph

2 How accomplished—mainly by facilitated diffusion and active transport

Review questions

1 Name and describe the coats that compose the walls of the esophagus, stomach, and intestines.

2 Differentiate between the peritoneum, the mesentery, and the omentum.

3 Explain what the tonsils and adenoids are and their locations.

4 Give the names and number of deciduous teeth and of permanent teeth.

5 Discuss the functions of gastric juice.

6 What juices digest proteins? Carbohydrates? Fats?

7 What functions does the liver perform? Which of these are vital functions?

8 What digestive functions does the pancreas perform?

9 What is the purpose of digestion?

10 Compare absorption of the three kinds of foods.

11 The doctor tells a patient that he has obstructive jaundice. Which duct or ducts might be obstructed to produce the symptom jaundice? Explain.

12 With obstructive jaundice, would digestion or absorption or both be affected? Explain.

13 What vitamin might the doctor prescribe for a patient with obstructive jaundice? Why?

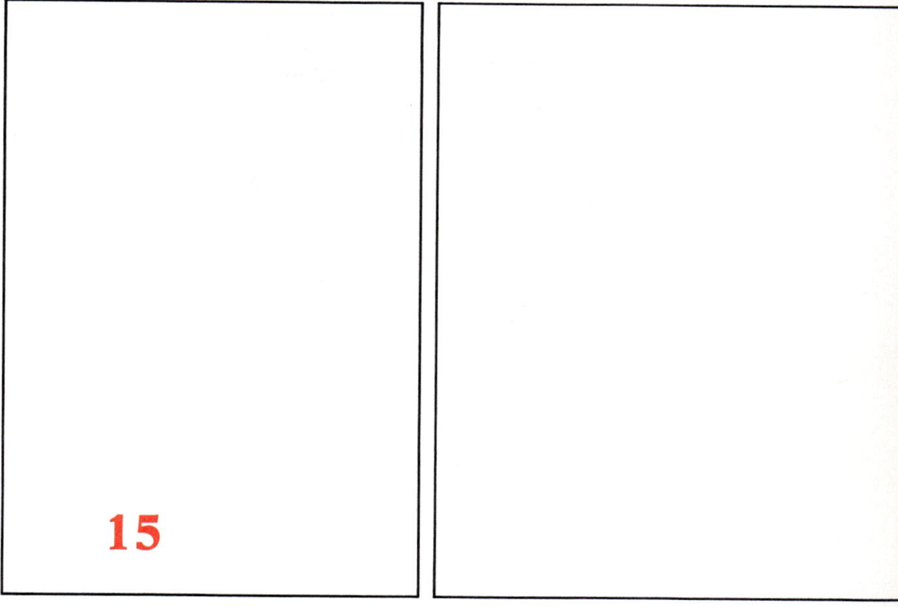

15

Metabolism

Meaning

Foods are first digested, then absorbed, and finally metabolized. Metabolism, as the word is usually defined, means the chemical changes absorbed foods undergo within the body cells. More simply, metabolism is the body's utilization of foods. Digestion and absorption are merely preliminary steps or preparations for metabolism.

Ways in which foods are metabolized

The body utilizes—that is, metabolizes—carbohydrates, proteins, and fats, to supply itself with energy and to make complex compounds that it needs. Metabolism goes on continually inside all of the body's cells. It consists of two main processes, namely catabolism and anabolism. Each of these, in turn, consists of numerous chemical reactions. *Catabolism* is the process that supplies cells with energy in a form they can use for doing their work. *Anabolism* is the process by which cells synthesize complex compounds of many different kinds—hormones, the proteins called enzymes, structural proteins, and numerous others. Anabolism is one of the several kinds of work that all cells perform. Catabolism supplies the energy that does the work of anabolism and all other kinds of cellular work. Therefore, catabolism is one of the most vital of all body processes. Without continual catabolism, cells soon cease all their activities and die.

Carbohydrate metabolism

Carbohydrate metabolism starts with the movement of glucose through cell membranes. Immediately upon reaching the interior of the cell, glucose reacts with ATP to produce glucose-6-phosphate. The reaction is referred to as glucose phosphorylation and is catalyzed by the enzyme glucokinase (or glucose hexokinase). Next, glucose-6-phosphate is either anabolized or catabolized.

Glucose catabolism

Glucose catabolism consists of two successive processes: glycolysis and the citric acid cycle (also called tricarboxylic acid cycle, Krebs' cycle, and cellular respiration).

Glycolysis is the process that changes glucose to pyruvic acid and thereby releases some of the energy stored in glucose molecules. Glycolysis takes place in the cytoplasm of cells. It consists of a series of chemical changes catalyzed by enzymes and accompanied by energy changes. Because the chemical changes of glycolysis do not use oxygen, glycolysis is referred to as the anaerobic phase of catabolism. Figs. 15-1 and 15-2 show the chemical changes of glycolysis in an abbreviated form. Stated very briefly, the chemical change produced by glycolysis is the breaking apart of 1 molecule of glucose to form 2 molecules of pyruvic acid.

The specific chemical changes, however, are not the most important facts to remember about glycolysis nor about the citric acid cycle. The most important fact is that the

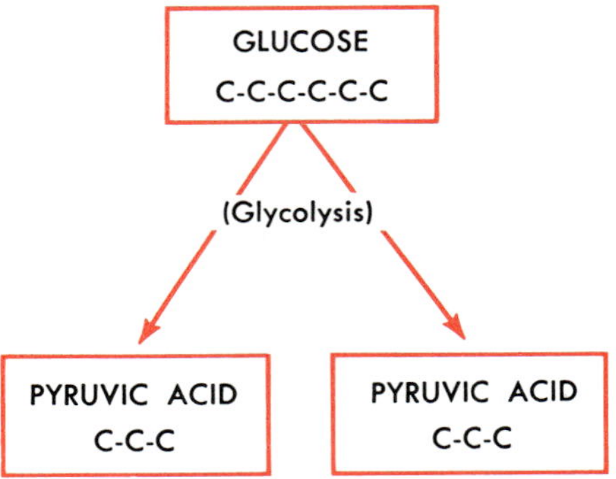

Fig. 15-1

Glycolysis. Glucose, a molecule containing 6 carbon atoms, is split by a series of chemical reactions into 2 molecules of pyruvic acid, each of which contains only 3 carbon atoms. Fig. 15-2 shows the series of intermediate reactions in brief form.

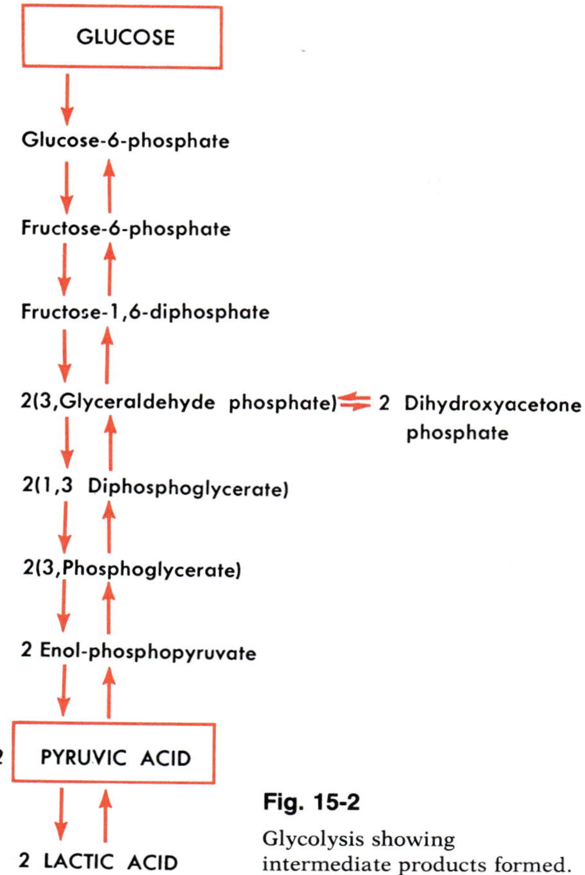

Fig. 15-2

Glycolysis showing intermediate products formed.

chemical changes of catabolism produce energy changes and that these energy changes supply all the energy that does all the work that keeps our cells and bodies alive. In essence, the energy changes produced by the chemical changes of catabolism are these (see Fig. 15-5):

1 Energy stored in the chemical bonds of food molecules is released.

2 More than half of the released energy is immediately recaptured—that is, put back into storage in chemical bonds (but in a different kind of bond in a different kind of molecule).

3 Somewhat less than half of the energy released from food molecules is transformed into heat energy. (Incidentally, this is the body's only way of producing heat—by catabolizing food molecules to transform part of their stored energy into heat energy.) In later paragraphs (pp. 418-421) more meaning will be given to the idea of energy changes during catabolism.

Citric acid cycle. The citric acid cycle (or *Krebs' cycle* as it is called in honor of the man who formulated it) is the second and final process in the catabolism of glucose. It consists of a cyclical series of chemical reactions catalyzed by enzymes and accompanied by energy changes. Because some of these chemical reactions use oxygen, the citric acid cycle is referred to as the aerobic phase of catabolism. Figs. 15-3 and 15-4 show the chemical changes of the citric acid cycle in abbreviated forms. Put in fewest words, the chemical change produced by the citric acid cycle is that pyruvic acid is broken down and oxidized to form carbon dioxide and water. Six molecules of oxygen are used up in the process. Note well this fact. Why? Because this is the only way cells utilize oxygen—to carry on the citric acid cycle—and the reason they and you must have a continuous supply of oxygen to survive.

Enzymes that catalyze the chemical reactions of the citric acid cycle are protein molecules located in the thousands of particles attached to the outer membrane of a cell's mitochondria. Investigators believe that even the placing of these enzyme molecules in the outer membrane particles is precise, that they are placed one after the other in the exact sequence of the chemical reactions they catalyze. Think of that! Even the molecules in a cell are organized so as to produce perfect timing of their functioning—a persuasive illustration of the principle: "Organization is a dominant characteristic of life." Now let us examine some of the specific chemical changes of the citric acid cycle to see how they produce energy changes.

In Fig. 15-4, chemical reaction (1) removes 1 carbon dioxide molecule and 2 hydrogen atoms from each pyruvic acid molecule, or a total of 2 carbon dioxide molecules and 4 hydrogen atoms for each glucose molecule catabolized. The 4 hydrogen atoms immediately ionize to form 4 hydrogen ions and 4 electrons.

Molecules of the coenzyme NAD* quickly accept these electrons, shuttle them across the microscopic fluid-filled space between the outer and inner membranes of the mitochondrial wall, and pass them on to enzymes presumably located in particles attached to the inner membrane. There are thousands of these tiny particles. Evidence suggests that each one consists of a series or chain of several enzymes.

Because electrons are transferred from one enzyme to another in the chain, like so many buckets of water passed along a bucket brigade, these enzymes are referred to as "electron-transfer enzymes," and the inner membrane particles that contain them are known as *"electron-transport particles."* The movement of electrons down the chain of electron-transfer enzymes releases energy. Somewhat more than half of this released energy is used to synthesize ATP (adenosine triphosphate) molecules. The rest of it is transformed to heat energy.

Finally, at the end of the electron-transfer chain, the electrons recombine with hydrogen ions to form hydrogen atoms that unite with oxygen to form water.

Briefly then, reaction (1) produces acetyl-CoA + CO_2 + H_2O + ATP + Heat. Acetyl-CoA immediately combines with oxaloacetic acid to form citric acid—reaction (2) in Fig. 15-4. Citric acid then undergoes a series of changes that eventually convert it to oxaloacetic acid, carbon dioxide, and water. Through the addition of oxygen during several reactions of the citric acid cycle, each pyruvic acid molecule entering the cycle is converted to 3 CO_2 molecules (reactions 1, 6, and 7) and 3 H_2O mole-

*Nicotinamide adenine dinucleotide phosphate.

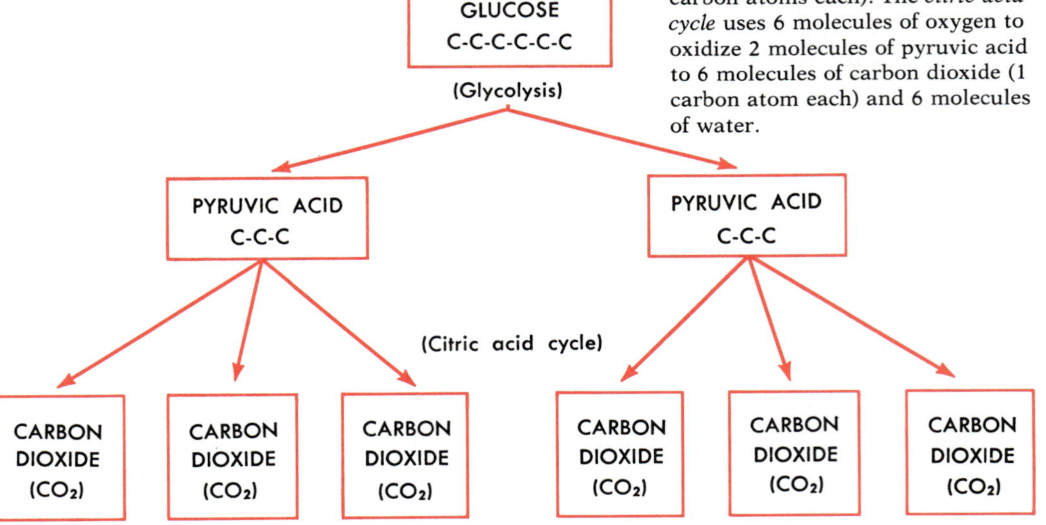

Fig. 15-3

Scheme to show that catabolism breaks larger molecules down into smaller ones. *Glycolysis* splits 1 molecule of glucose (6 carbon atoms) into 2 molecules of pyruvic acid (3 carbon atoms each). The *citric acid cycle* uses 6 molecules of oxygen to oxidize 2 molecules of pyruvic acid to 6 molecules of carbon dioxide (1 carbon atom each) and 6 molecules of water.

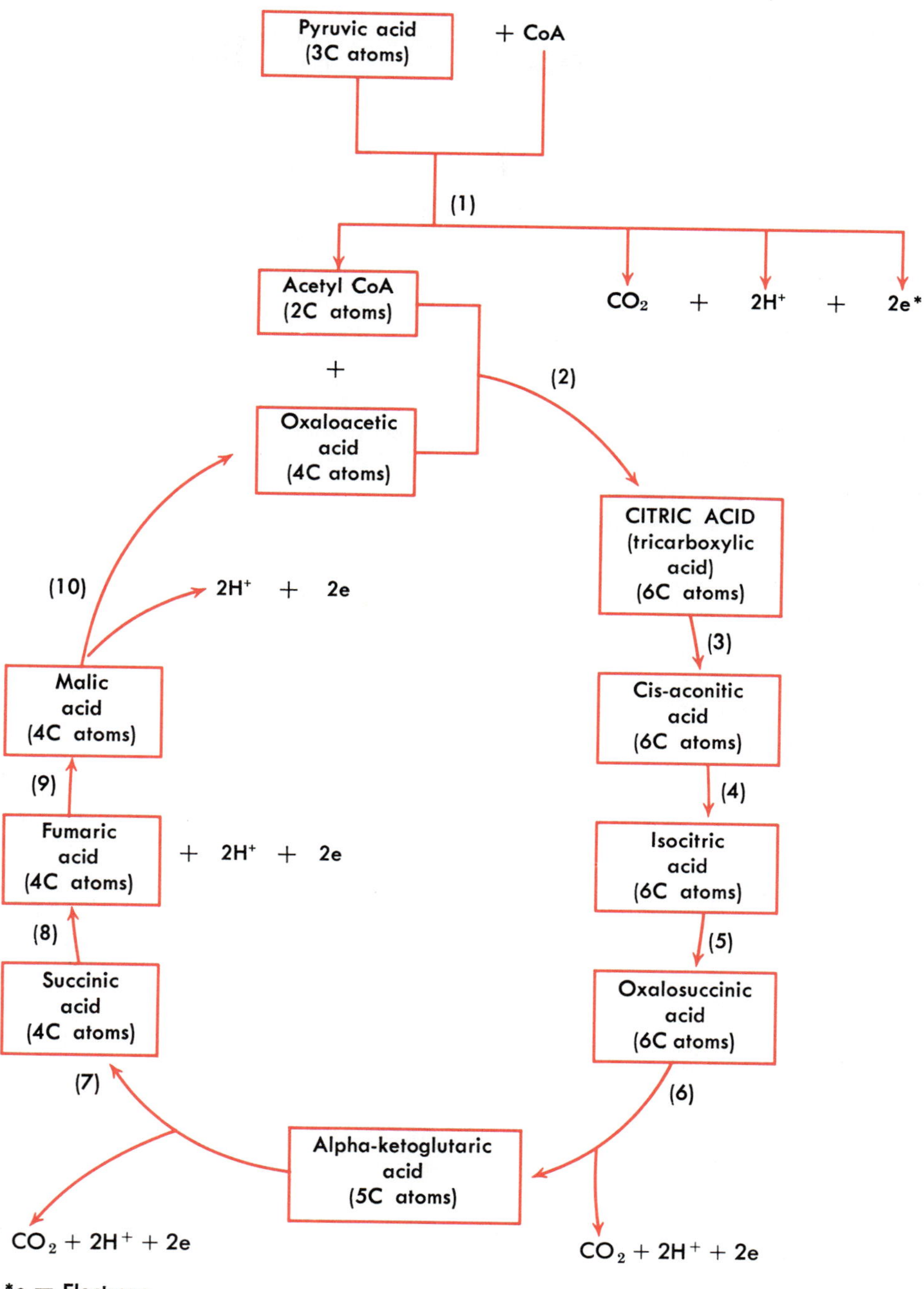

Fig. 15-4

The citric acid (Krebs') cycle. See text (pp. 418-421) for explanation.

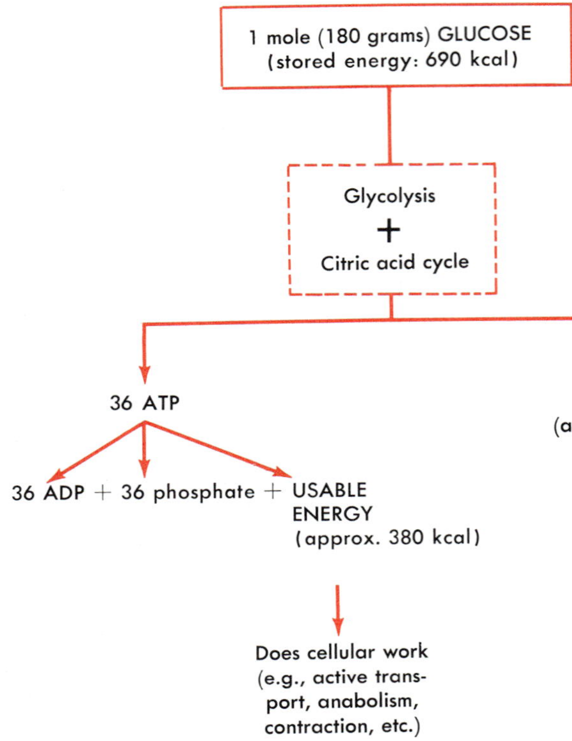

Fig. 15-5

Catabolism releases energy stored in chemical bonds of glucose molecules. 180 gm or 1 mole of glucose contains 690 kilocalories (kcal) of stored energy. Catabolism releases about 310 kcal of this energy as heat and puts about 380 kcal back in storage in high-energy bonds of 36 moles of ATP. Thus the efficiency of glucose catabolism as a mechanism for supplying cells with usable energy is about 55% (380/690).

cules (reactions 1, 8, and 10). Also, for each pyruvic acid molecule oxidized, 17 ATP molecules are synthesized, using energy released by electron transfer during reactions 1, 8, and 10. Note in Fig. 15-5 that the complete catabolism of 1 molecule of glucose yields 36* molecules of ATP—34 ATP from the citric acid cycle and a net of only 2 ATP from glycolysis.

Summarizing the changes of catabolism in equation form:

Glycolysis:

$$\text{Glucose} \rightarrow 2 \text{ pyruvic acid} + 2 \text{ ATP} + \text{Heat}$$

*Some authorities report that the complete catabolism of 1 molecule of glucose yields 38 ATP—36 ATP from the citric acid cycle and 2 ATP from glycolysis. Other authorities report fewer ATP molecules formed during the citric acid cycle.

Citric acid cycle:

$$2 \text{ pyruvic acid} + 6\ O_2 \rightarrow 6\ CO_2 + 6\ H_2O + 34 \text{ ATP} + \text{Heat}$$

We can even summarize the long series of chemical reactions that make up the process of glucose catabolism with one short equation:

$$\underset{\text{(glucose)}}{C_6H_{12}O_6} + 6\ O_2 \quad 6\ CO_2 + 6\ H_2O + 36 \text{ ATP} + \text{Heat}$$

As you read the next few paragraphs, refer often to Fig. 15-5. It summarizes the energy changes accomplished by the chemical changes of glucose metabolism.

The synthesis of ATP is a phosphorylation reaction carried on during the citric acid cycle in the thousands of tiny particles attached to the inner membrane of each mitochondrion (the same structures in which electron transfer occurs).

$$\underset{\text{(adenosine diphosphate)}}{\text{ADP}} + \text{Phosphate} + \text{ENERGY} \rightarrow \underset{\text{(adenosine triphosphat}}{\text{ATP}}$$

The energy shown in the preceding equation (in other words, the energy used to do the work of synthesizing ATP molecules) is released from the stable chemical bonds in glucose molecules by the passage of electrons down the chains of electron-transfer enzymes. Immediately the process of ATP synthesis puts some of the released energy back in storage in chemical bonds, but in unstable (labile) high-energy bonds of ATP molecules.

As its name suggests, ATP or adenosine triphosphate contains three phosphate groups. Only the last two of these are attached by labile, high-energy bonds. A high-energy bond is one that can break apart with explosive rapidity to release active energy that performs some kind of work.

$$\text{ATP} \rightarrow \text{ENERGY} + \text{ADP} + \text{Phosphate}$$

All cellular work, it is now believed, is done by energy set free from high-energy phosphate bonds, mainly those in ATP molecules. Hence, ATP is one of the most important biological compounds known. It supplies energy directly to all kinds of energy-consuming mechanisms in all kinds of living organisms from one-celled plants to billion-celled animals, including man.

Since the citric acid cycle releases most of the energy from food molecules, and since this cycle of chemical reactions takes place inside mitochondria, these tiny structures are aptly described as the "power plants" of cells.

Albert Lehninger, the biochemist who discovered that citric acid cycle and electron-transfer enzymes are located in mitochondria, writes: "If the classical engineering science of energy transformation is humbled by what is now known about the power plants of the cell, so are the newer and more glamorous branches of engineering. The technology of electronics has achieved amazing success in packaging and miniaturizing the components of a computer. But these advances still fall far short of accomplishing the unbelievable miniaturization of complex energy-transducing components that has been perfected by organic evolution in each living cell."*

*From Lehninger, A. L.: How cells transform energy, Sci. Am. **205:**62-73, Sept., 1961; copyright © 1961 by Scientific American, Inc.; all rights reserved.

Glycogenesis and glycogenolysis

Glycogenesis is a process of glucose anabolism. Simply defined, glycogenesis is the formation of glycogen from glucose. Note in Fig. 15-6 that glycogenesis consists of a series of chemical reactions, each one catalyzed by a different enzyme. Glycogen molecules do not remain permanently in the cell but from time to time are broken down or hydrolyzed back to glucose-6-phosphate or, in certain cells, all the way back to glucose. This reversal of the process of glycogenesis, this hydrolysis of glycogen, is called *glycogenolysis.* Now look at Fig. 15-8. Phosphorylase, the enzyme that catalyzes the first reaction of glycogenolysis—that is, the conversion of glycogen to glucose-1-phosphate—is presumably present in all cells. But phosphatase (glucose phosphatase), the enzyme that catalyzes the final step of glycogenolysis—the changing of glucose-6-phosphate back to glucose—is lacking in most cells. Only liver cells, intestinal mucosal cells, and kidney tubule cells contain glucose phosphatase. Therefore, the term glycogenolysis means different things in different cells. *Muscle cell glycogenolysis,* for example, means the conversion of glycogen to glucose-6-phosphate, but *liver glycogenolysis* means

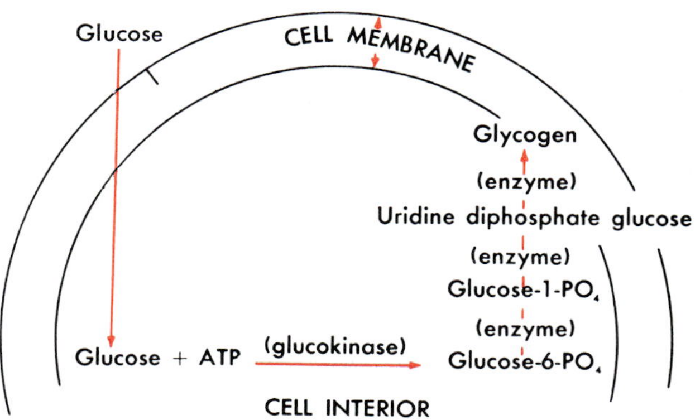

Fig. 15-6

Glycogenesis.

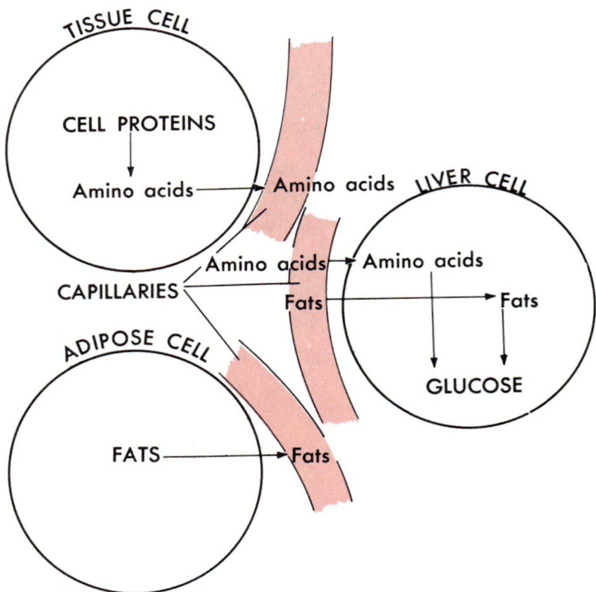

Fig. 15-7

Liver gluconeogenesis from mobilized tissue proteins and fats.

the changing of glycogen to glucose and the addition of this glucose to the bloodstream.

Gluconeogenesis

Gluconeogenesis means literally the formation of "new" glucose—"new" in the sense that it is made from proteins or fats, not from carbohydrates. The process occurs chiefly in the liver. It consists of many complex chemical reactions. The new glucose produced from fats or proteins by gluconeogenesis diffuses out of liver cells into the blood (Fig. 15-7). Gluconeogenesis, therefore, can add glucose to the blood when needed, as can the process of liver glycogenolysis. Obviously, then, the liver is a most important organ for maintaining blood glucose homeostasis.

Control of glucose metabolism

The complex mechanism that normally maintains homeostasis of blood glucose concentration consists of hormonal and neural devices. At least five endocrine glands—islands of Langerhans, anterior pituitary gland, adrenal cortex, adrenal medulla, and thyroid gland—and at least eight hormones secreted by those glands function as key parts of the glucose homeostatic mechanism.

Beta cells of the islands of Langerhans in the pancreas secrete the most famous sugar-regulating hormone of them all, *insulin.* Although no one yet knows exactly how insulin acts, several of its effects are well known. For instance, insulin is known to act in some way to accelerate glucose transport through cell membranes and to increase the activity of the enzyme glucokinase. As shown in Fig. 15-6, glucokinase catalyzes glucose phosphorylation, the reaction that must occur before either glycogenesis or glucose catabolism can take place. By applying these facts, you can deduce for yourself some of the prominent metabolic defects resulting from insulin deficiency such as occurs in diabetes mellitus. Slow glycogenesis with resulting low glycogen storage, decreased glucose catabolism, and increased blood glucose all result from the slower glucose transport into cells and the decreased glucokinase activity produced by insulin deficiency.

The islands of Langerhans secrete two sugar-regulating hormones—insulin from the beta cells and glucagon from the alpha cells. Whereas insulin tends to decrease blood glucose, glucagon tends to increase it. *Glucagon* increases the activity of the enzyme phosphorylase (Fig. 15-8), and this necessarily accelerates liver glycogenolysis and releases more of its product, glucose, into the blood.

The anterior pituitary gland (adenohypophysis) also secretes three hormones that help regulate carbohydrate metabolism. Growth hormone, adrenocorticotropic hormone (ACTH), and thyroid-stimulating hormone are their names.

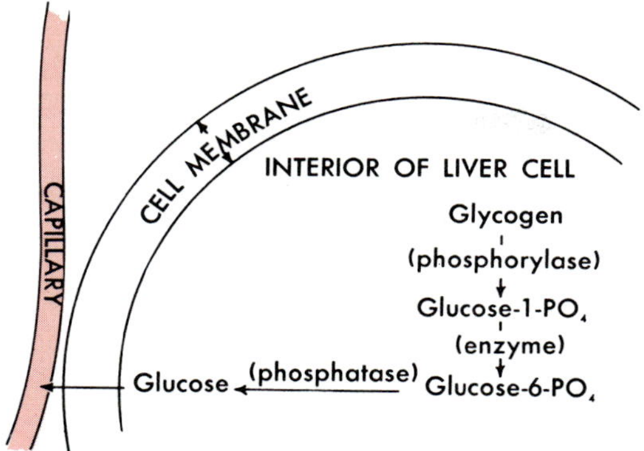

Fig. 15-8

Glycogenolysis in a liver cell.

Growth hormone tends to bring about a shift from carbohydrate catabolism to fat catabolism. Less glucose, therefore, leaves the blood to enter cells or, stated the other way around, more glucose remains in the blood. In other words, blood glucose tends to increase. Growth hormone presumably brings about this shift to fat utilization by influencing several processes. It appears to act in some way to decrease the deposition of fats in adipose tissue, to increase the mobilization of fats from adipose tissue, and to decrease carbohydrate utilization by all tissues.

ACTH also tends to increase blood glucose concentration, but it does it in a different and an even more indirect way than does growth hormone. ACTH stimulates the adrenal cortex to increase its secretion of glucocorticoids. Glucocorticoids accelerate the mobilization of proteins—that is, the breakdown or hydrolysis of tissue proteins to amino acids. More amino acids enter the circulation and are carried to the liver. Liver cells step up their production of "new" glucose from the mobilized amino acids—in technical language, gluconeogenesis accelerates. More glucose streams out of liver cells into the blood and adds to the blood glucose level. In short, growth hormone and glucocorticoids are hyperglycemic hormones. They both tend to increase blood glucose concentration, in other words. ACTH also is hyperglycemic, but only indirectly through its effects on glucocorticoid secretion.

The anterior pituitary gland's thyroid-stimulating hormone (thyrotropin) stimulates the thyroid gland to increase its secretion of thyroid hormone. Thyroid hormone accelerates catabolism—usually glucose catabolism since glucose is the body's "preferred fuel."

Epinephrine is a hormone secreted in large amounts by the adrenal medulla in times of emotional or physical stress. Like the hormone glucagon secreted by the alpha islet cells, epinephrine increases phosphorylase activity (Fig. 15-8). This makes liver glycogenolysis go on at a faster rate and causes more of its product, glucose, to enter the blood. Epinephrine accelerates both liver and muscle glycogenolysis, whereas glucagon accelerates only liver glycogenolysis.

In summary, glucose metabolism is regulated mainly by the following hormones:

1. *Insulin,* a hypoglycemic hormone—that is, it tends to decrease blood glucose
2. Glucagon, epinephrine, growth hormone, ACTH, and glucocorticoids are hyperglycemic hormones—that is, they tend to increase blood glucose
3. Thyroid-stimulating hormone and thyroid hormone—ordinarily tend to decrease blood glucose

■ ■ ■

Some important principles about normal carbohydrate metabolism may be summarized as follows.

1 *Principle of "preferred energy fuel."* All body cells catabolize glucose for their needed energy supply when the amount of glucose and the amount of insulin in the blood are adequate. Glucose is the "preferred energy fuel" of human cells—preferred to fat and

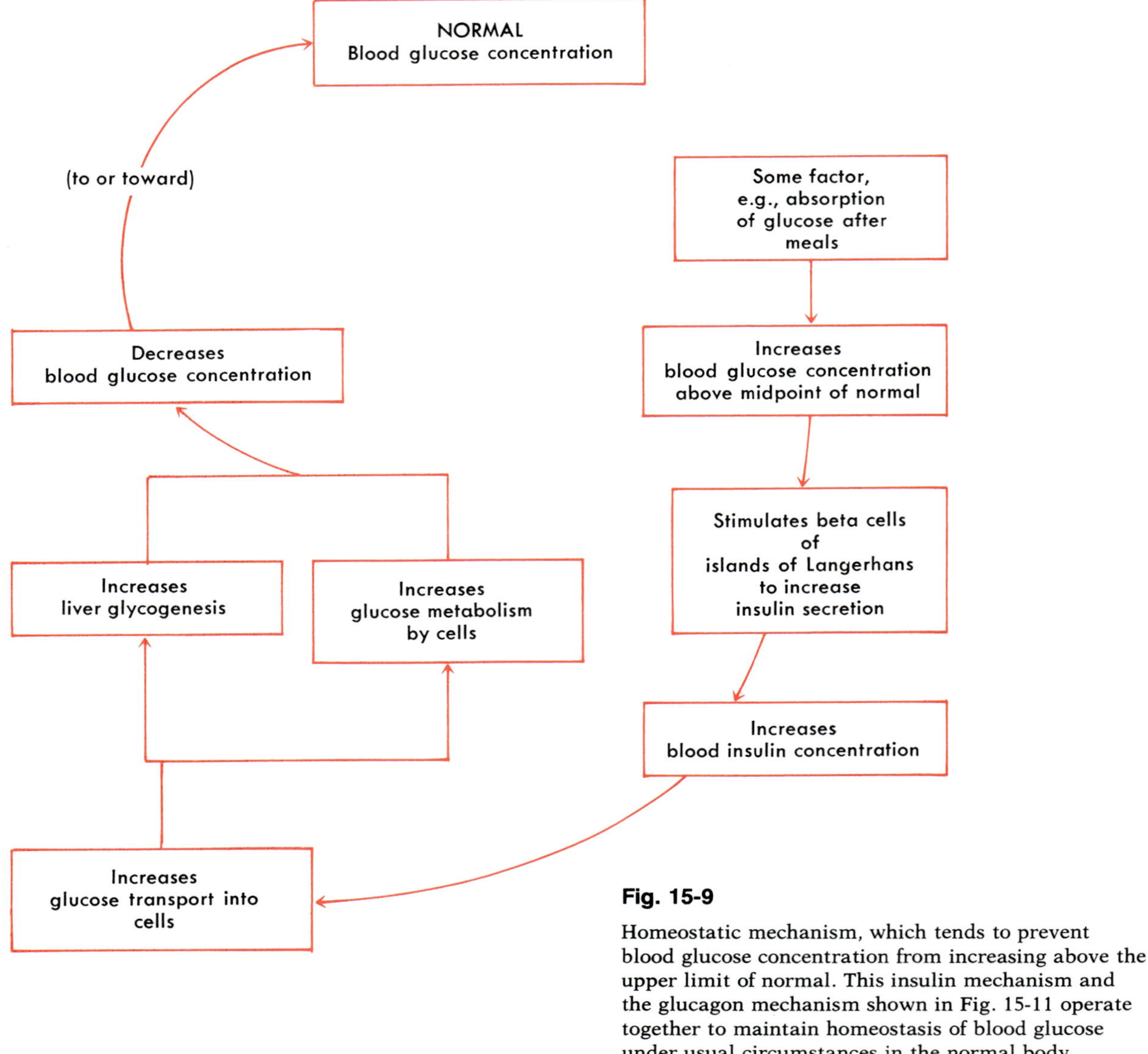

Fig. 15-9

Homeostatic mechanism, which tends to prevent blood glucose concentration from increasing above the upper limit of normal. This insulin mechanism and the glucagon mechanism shown in Fig. 15-11 operate together to maintain homeostasis of blood glucose under usual circumstances in the normal body.

protein, that is. Therefore, cells get their energy first from glucose—for as long as enough of it continues to enter them—then next from fats, and last from proteins.

2 *Principle of glycogenesis.* The process of glycogenesis is part of a homeostatic mechanism that operates when blood glucose concentration increases above the midpoint of its normal range. When blood glucose concentration increases, glycogenesis also increases. This soon decreases blood glucose, ordinarily enough to maintain its normal level. Example: soon after meals, while glucose is being absorbed rapidly, a great many glucose molecules leave the blood for storage as glycogen—mainly in liver cells but also in muscle and various other cells (Figs. 15-9 and 15-10).

3 *Principle of glycogenolysis.* The process of liver glycogenolysis is part of a homeostatic mechanism that operates when blood glucose concentration decreases below the midpoint of its normal range. When blood glucose concentration decreases, liver glycogenolysis increases (Figs. 15-10 and 15-11). Example: a few hours after meals, when blood glucose decreases below the midpoint of normal, liver glycogenolysis accelerates. This, of course, adds glucose molecules to blood and tends to raise glucose concentration back up to the midpoint of normal. However, glycogenolysis alone can probably maintain homeostasis of blood glucose concentration only a few hours since the body can store only modest amounts of glycogen.

4 *Principle of gluconeogenesis.* Liver gluconeogenesis is another part of the homeostatic mechanism that operates when blood glucose decreases below the midnormal point (Figs. 15-7 and 15-10). When this happens or when the amount of glucose entering cells is inadequate (for example, in diabetes mellitus), liver gluconeogenesis accelerates and tends to raise blood glucose concentration back to or above normal.

5 *Principle of glucose storage as fat.* If blood glucose still remains higher than normal after glycogenesis and catabolism and if the blood insulin content is adequate, the excess glucose is converted to fat (mainly by liver cells) and is stored as such in fat depots—in adipose tissue, in other words. Unfortunately from the standpoint of health and beauty, the amount of glucose the body can store as fat is virtually unlimited.

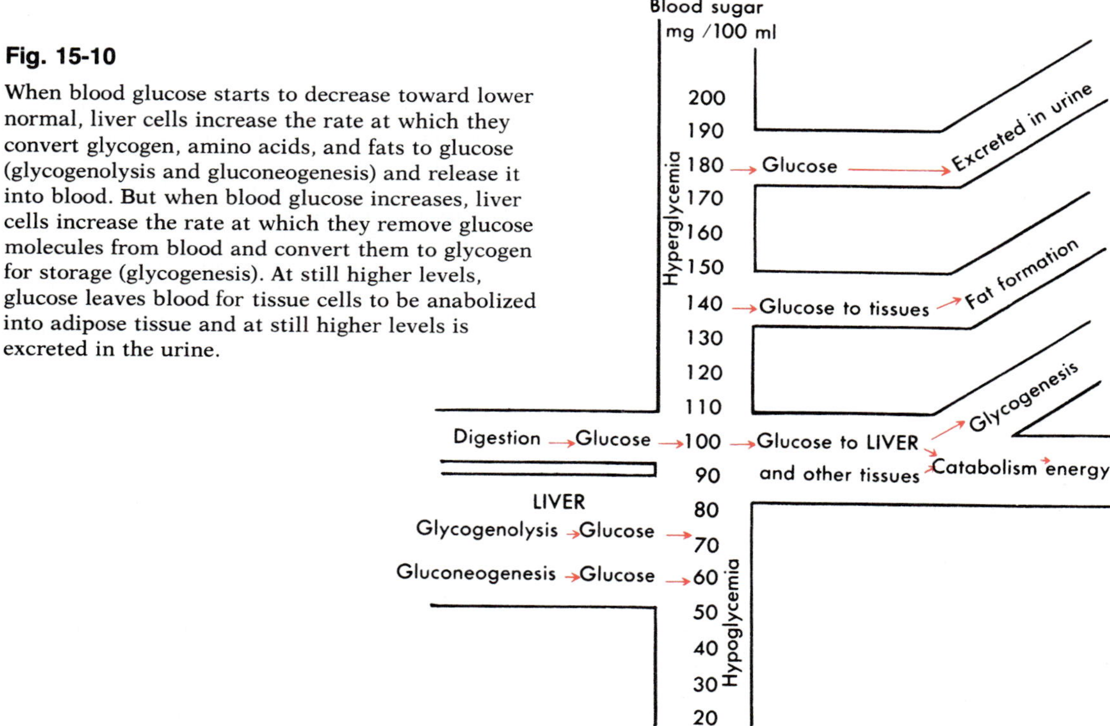

Fig. 15-10

When blood glucose starts to decrease toward lower normal, liver cells increase the rate at which they convert glycogen, amino acids, and fats to glucose (glycogenolysis and gluconeogenesis) and release it into blood. But when blood glucose increases, liver cells increase the rate at which they remove glucose molecules from blood and convert them to glycogen for storage (glycogenesis). At still higher levels, glucose leaves blood for tissue cells to be anabolized into adipose tissue and at still higher levels is excreted in the urine.

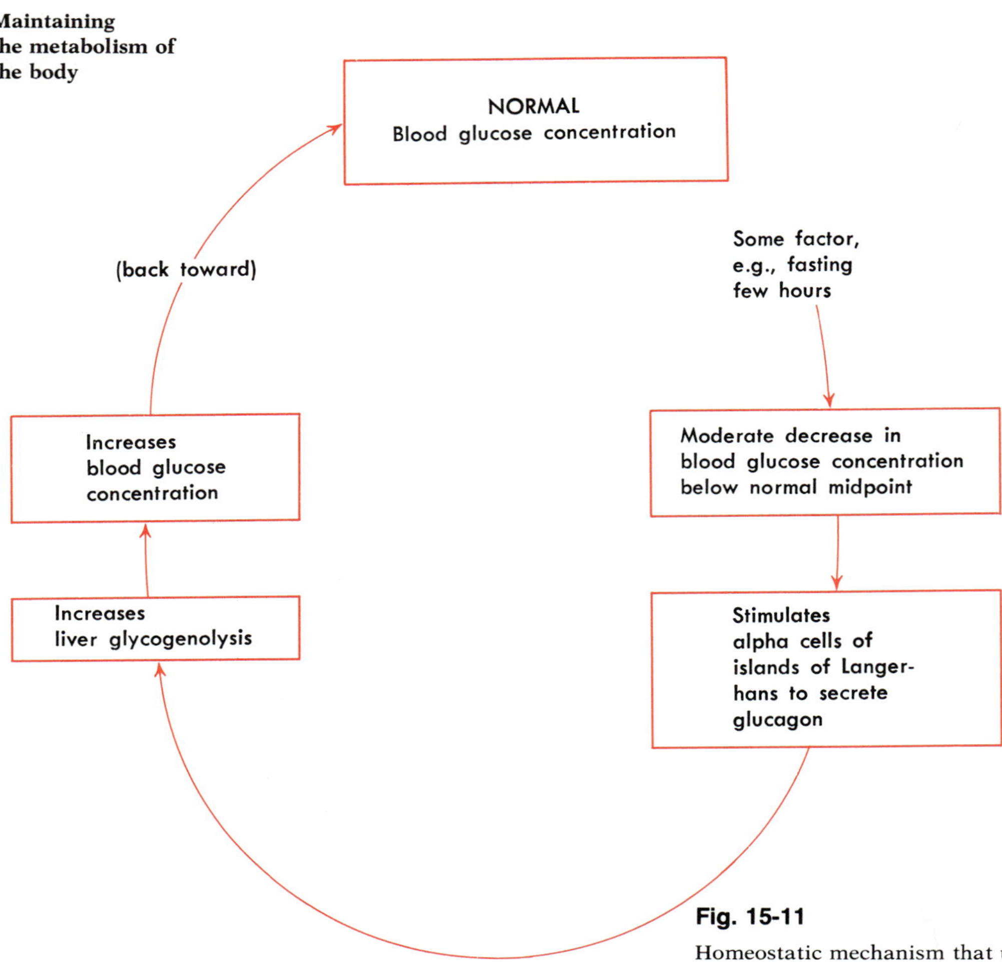

Fig. 15-11

Homeostatic mechanism that under usual conditions is chiefly responsible for preventing blood glucose from falling below the lower limit of normal. This glucagon mechanism and the insulin mechanism shown in Fig. 15-9 work together to maintain homeostasis of blood glucose in the normal body under usual circumstances.

Fat metabolism

Body cells both catabolize and anabolize fats. Fats constitute a more concentrated energy food than carbohydrates. Catabolism of 1 gm of fat yields 9 kilocalories (kcal) of heat; catabolism of 1 gm of carbohydrates yields only 4.1 kcal. *Fat catabolism,* like carbohydrate catabolism, consists of two main processes, each of which, in turn, consists of a series of chemical reactions. The first process in fat catabolism—ketogenesis—takes place chiefly in liver cells. It consists of a series of chemical reactions (known as beta oxidations) that change fatty acids to acetoacetic acid. Acetoacetic acid is classified as a ketone body and can be converted to two other ketone bodies, namely, acetone and beta-hydroxybutyric acid—hence the name ketogenesis for the first step in fat catabolism. Acetoacetic acid leaves liver cells and circulates to other tissue cells where it is changed to acetyl-CoA and oxidized via the citric acid cycle.

In other words, the final step of fat catabolism and of glucose catabolism are the same.

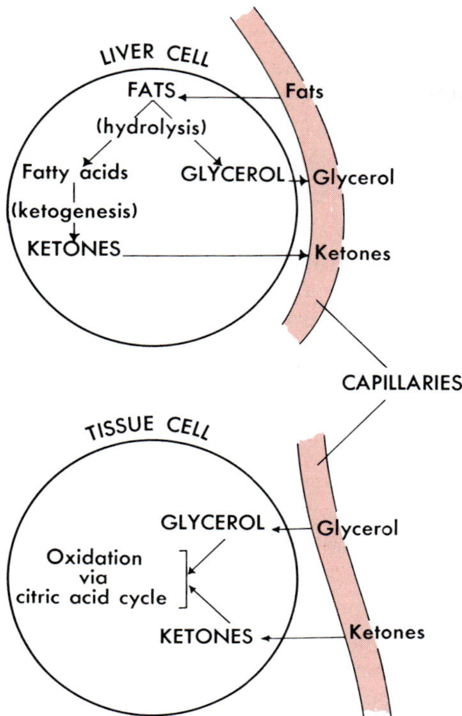

Fig. 15-12

Fat catabolism consisting, first, of ketogenesis by liver cells and, second, of oxidation by tissue and liver cells.

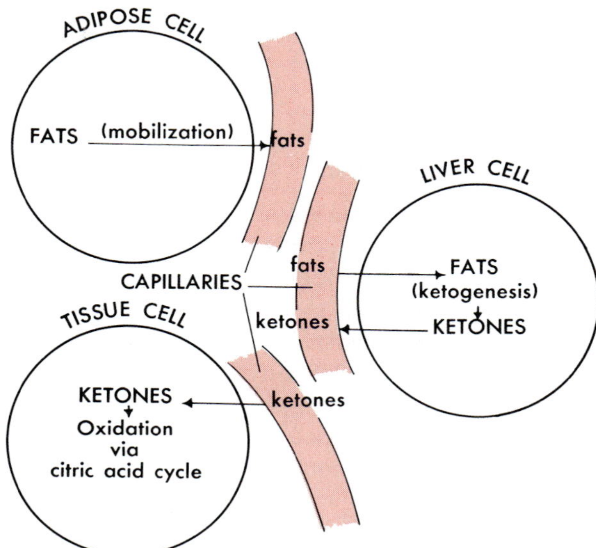

Fig. 15-13

Fat mobilization from adipose cell followed by catabolism (ketogenesis in liver cells and citric acid cycle in all cells).

Liver cells oxidize a small portion of the ketone bodies for their own energy needs, but most of them are transported by the blood to other tissue cells for the final step of catabolism. Glycerol is metabolized similarly to glucose.

Fat anabolism (fat deposition or lipogenesis) consists of the storage of fats mainly in adipose tissue. It also includes the use of fats to synthesize various complex compounds. Fats stored in the fat depots constitute the body's largest reserve energy source—too large too often, unfortunately.

The main facts about fat catabolism are summarized in Figs. 15-12 and 15-13.

Control

Fat metabolism is controlled mainly by the following hormones: insulin, growth hormone, ACTH, and glucocorticoids. You probably recall from our discussion of these hormones in connection with carbohydrate metabolism that they regulate fat metabolism in such a way that the rate of fat catabolism is inversely related to the rate of carbohydrate catabolism. If some condition such as diabetes mellitus causes carbohydrate catabolism to decrease below energy needs, increased secretion of growth hormone, ACTH, and glucocorticoids soon follows. These hormones, in turn, bring about an increase in fat catabolism. But when carbohydrate catabolism equals energy needs, fats are not mobilized out of storage and catabolized. Instead, they are spared and stored in adipose tissue. "Carbohydrates have a 'fat-sparing' effect," so says an old physiological maxim. Or, stating this truth more descriptively: "Carbohydrates have a 'fat-storing' effect."

Protein metabolism

In protein metabolism, anabolism is primary and catabolism is secondary. In carbohydrate and fat metabolism, the opposite is true—catabolism is primary and anabolism is secondary. Proteins are primarily tissue-building foods. Carbohydrates and fats are primarily energy-supplying foods.

The cellular process called protein synthesis (or protein anabolism) produces many substances—enzymes, antibodies, and some secretions, to name a few. Protein anabolism plays a major role in the growth and reproduction both of cells and of the body as a whole. Protein anabolism is also the chief process of repair. It accomplishes the healing of wounds, the formation of scar tissue, and the replacement of cells destroyed by daily wear and tear. Protein anabolism is truly "big business" in the body. Red blood cell replacement alone, for instance, runs into millions of cells per second.

A brief synopsis of the process of protein anabolism as now visualized appears on pp. 463-464.

Protein catabolism, like the catabolism of fats, consists of two processes. The first takes place mainly in liver cells and the second is the citric acid cycle and occurs in all cells. The first step in protein catabolism is known as *deamination,* a reaction in which an amino (NH_2) group is split off from an amino acid molecule to form a molecule of ammonia and one of keto acid. Most of the ammonia is converted to urea and is excreted via the urine. The keto acid may be oxidized via the citric acid cycle or may be converted to glucose (gluconeogenesis) or to fat (lipogenesis). Both protein catabolism and anabolism go on continually. Only their rates differ from time to time. With a protein-deficient diet, for example, protein catabolism exceeds protein anabolism. Various hormones, as we shall see, also influence the rates of protein catabolism and anabolism.

Usually a state of *protein balance* exists in the normal healthy adult body—that is, the rate of protein anabolism equals or balances the rate of protein catabolism. When the body is in protein balance, it is also in a state of *nitrogen balance.* For then, the amount of nitrogen taken into the body (in protein foods) equals the amount of nitrogen in protein catabolic waste products excreted in the urine, feces, and sweat. Two kinds of protein or nitrogen imbalance exist. When protein catabolism exceeds protein anabolism, the amount of nitrogen in the urine exceeds the amount of nitrogen in the protein foods ingested. The individual is then said to be in a state of *negative nitrogen balance,* or in a state of "tissue wasting"—because more of his tissue proteins are being catabolized than are being replaced by protein synthesis. Protein-poor diets, starvation, and wasting illnesses, for example, produce a negative nitrogen balance.

A *positive nitrogen balance* (nitrogen intake in foods greater than nitrogen output in urine) indicates that protein anabolism is going on at a faster rate than protein catabolism. A state of positive nitrogen balance, therefore, characterizes any condition in which large amounts of tissue are being synthesized, such as during growth, pregnancy, and convalescence from an emaciating illness.

The main facts about protein metabolism are summarized in Figs. 15-14 to 15-16. Compare them with Figs. 15-10, 15-12, and

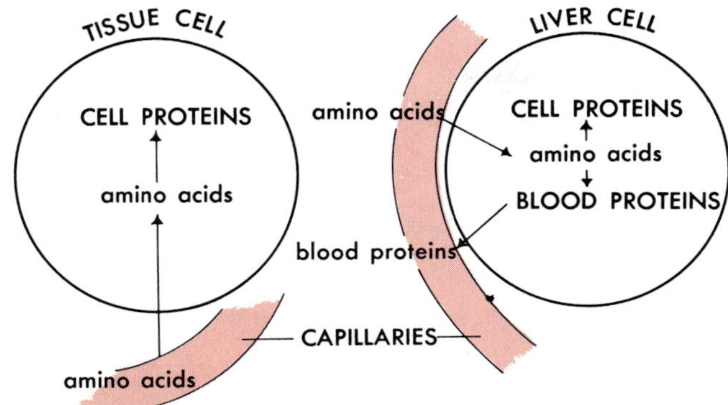

Fig. 15-14

Protein synthesis (anabolism). Growth hormone and testosterone tend to accelerate the processes shown and so are called anabolic hormones.

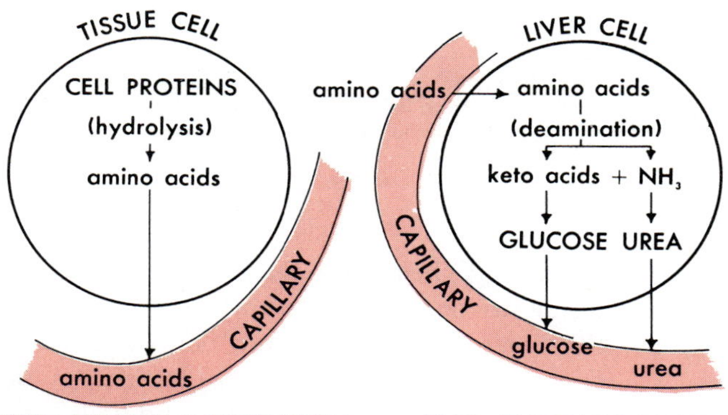

Fig. 15-15

Protein mobilization and catabolism. Glucocorticoids tend to accelerate these processes so are classed as protein catabolic hormones. (Also see Fig. 15-16.)

Fig. 15-16

Protein catabolism. First, as shown in Fig. 15-15, liver cells carry on deamination, a process that converts amino acids to keto acids and ammonia. Then keto acids may be changed to glucose by liver cells (gluconeogenesis), or liver and tissue cells may oxidize them (citric acid cycle) or convert them to fat (lipogenesis).

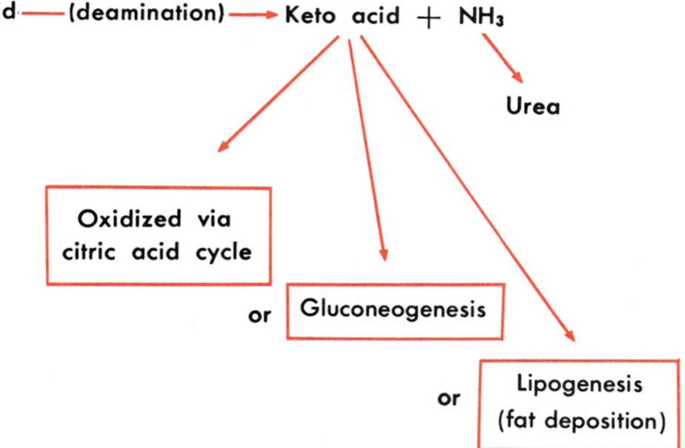

Table 15-1

Metabolism

Food	Anabolism	Catabolism
Carbohydrates	Temporary excess changed into glycogen by liver cells in presence of insulin; stored in liver and skeletal muscles until needed and then changed back to glucose (Figs. 15-7 and 15-8) True excess beyond body's energy requirements converted into adipose tissue; stored in various fat depots of body	Oxidized, in presence of insulin, to yield energy (4.1 kcal per gram) and wastes (carbon dioxide and water) $C_6H_{12}O_6 + 6\ O_2 \rightarrow$ Energy $+ 6\ CO_2 + 6\ H_2O$
Fats	Built into adipose tissue; stored in fat depots of body	Fatty acids ↓ (liver ketogenesis) Ketone bodies ↓ (tissues; citric acid cycle) Energy (9.3 kcal per gram) $+ CO_2 + H_2O$ Glycerol ↓ (liver gluconeogenesis) Glucose
Proteins	Temporary excess stored in liver and skeletal muscles Synthesized into tissue proteins, blood proteins, enzymes, hormones, etc.	Deaminated by liver forming ammonia (which is converted to urea) and keto acids (which are either oxidized or changed to glucose or fat)

15-13. Note the important part played by the liver in the metabolism of all three kinds of foods.

Control

Protein metabolism, like that of carbohydrates and fats, is controlled largely by hormones rather than by the nervous system. Growth hormone and the male hormone testosterone both have a stimulating effect on protein synthesis or anabolism. For this reason, they are referred to as anabolic hormones. Protein catabolic hormones of greatest consequence are glucocorticoids. They are thought to act in some way, still unknown, to speed up tissue protein mobilization—that is, the hydrolysis of cell proteins to amino acids, their entry into the blood, and their subsequent catabolism (Fig. 15-15). ACTH functions indirectly as a protein catabolic hormone because of its stimulating effect on glucocorticoid secretion.

Thyroid hormone is necessary for and tends to promote protein anabolism and, therefore, growth when plenty of carbohydrates and fats are available for energy production. On the other hand, under different conditions, for example, when the amount of thyroid hormone is excessive or when the energy foods are deficient, this hormone may then promote protein mobilization and catabolism.

■ ■ ■

Some of the facts about metabolism set forth in the preceding paragraphs are summarized in Table 15-1.

Metabolic rates

Meaning

The term *metabolic rate* means the amount of energy released in the body in a given time by catabolism. It represents energy expended or used for accomplishing various kinds of work. In short, metabolic rate actually means catabolic rate or rate of energy release. (It is impossible to measure the rate at which foods are anabolized but fairly easy to determine the rate at which they are catabolized—see pp. 432-434.)

Ways of expressing

Metabolic rates are expressed in either of two ways: (1) in terms of the number of kilocalories* of heat energy expended per hour or per day or (2) as normal or as a definite percentage above or below normal.

Basal metabolic rate

The basal metabolic rate (BMR) is the body's rate of energy expenditure under "basal conditions"—namely, when the individual:

1 Is awake but resting—that is, lying down and, so far as possible, not moving a muscle

2 Is in the postabsorptive state (12 to 18 hours after the last meal)

3 Is in a comfortably warm environment

*One *kilocalorie* (a so-called "large" Calorie) is the amount of heat used to raise the temperature of 1 kg (liter) of water 1° Centigrade.

Note that the basal metabolic rate is not the minimum metabolic rate. It does not indicate the smallest amount of energy that must be expended to sustain life. It does, however, indicate the smallest amount of energy expenditure that can sustain life and also maintain the waking state and a normal body temperature in a comfortably warm environment.

FACTORS INFLUENCING

The basal metabolic rate is not identical for all individuals because of the influence of various factors (Fig. 15-17), some of which are described in the following paragraphs.

Size. In computing the basal metabolic rate, size is usually considered the amount of the body's surface area. It is computed from the individual's height and weight. Per square meter of body surface, if other conditions are equal, a large individual has the same basal metabolic rate as a small one, but because a large individual has more square meters of surface area, his basal metabolism is greater than that of a small individual. For example, the BMR for a man in his twenties is about 40 kcal per square meter of body surface per hour (Table 15-2). A large man with a body surface area of

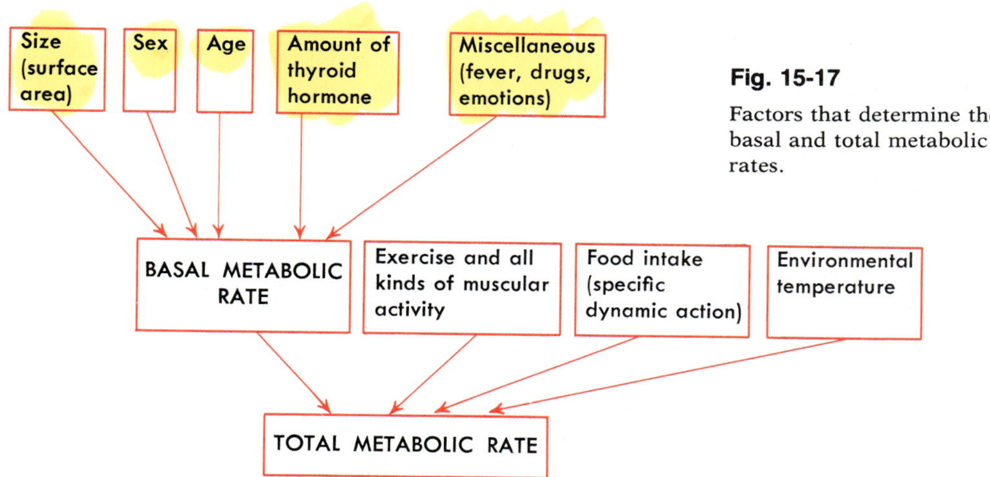

Fig. 15-17
Factors that determine the basal and total metabolic rates.

Table 15-2

Basal metabolism (Aub-DuBois)

Age (yr)	Kilocalories per hour per square meter body surface	
	Male	Female
10-12	51.5	50.0
12-14	50.0	46.5
14-16	46.0	43.0
16-18	43.0	40.0
18-20	41.0	38.0
20-30	39.5	37.0
30-40	39.5	36.5
40-50	38.5	36.0
50-60	37.5	35.0
60-70	36.5	34.0

1.9 square meters would, therefore, have a basal metabolism of 76 kcal per hour, whereas a smaller man with a surface area of perhaps 1.6 square meters would have a basal metabolism of only 64 kcal per hour. The average surface area for American adults is 1.6 square meters for women and 1.8 square meters for men.

Sex. Men oxidize their food approximately 5% to 7% faster than women. Therefore, their basal metabolic rates are about 5% to 7% higher for a given size and age. A man 5 feet, 6 inches tall, weighing 140 pounds, for example, has a 5% to 7% higher basal metabolic rate than a woman of the same height, weight, and age.

Age. That the fires of youth burn more brightly than those of age is a physiological as well as a psychological fact. In general, the younger the individual, the higher his basal metabolic rate for a given size and sex. Exception: the BMR is slightly lower at birth than a few years later. That is to say, the rate increases slightly during the first 3 to 6 years and then starts to decrease and continues to do so throughout life. The basal metabolism per hour per square meter of surface area for different age groups is given in Table 15-2.

Thyroid hormone. Thyroid hormone stimulates basal metabolism. Without a normal amount of this hormone in the blood, a normal basal metabolic rate cannot be maintained. When an excess of thyroid hormone is secreted, foods are catabolized faster, much as coal is burned faster when a furnace draft is open. Deficient thyroid secretion, on the other hand, slows the rate of metabolism.

Fever. Fever increases the basal metabolic rate. According to DuBois, metabolism increases about 13% per degree Centigrade rise in body temperature.*

Drugs. Certain drugs, such as caffeine, amphetamine, (Benzedrine), and dinitrophenol, increase the basal metabolic rate.

Other factors. Other factors, such as *emotions* and *pregnancy,* also influence basal metabolism. Both of these factors increase the basal rate.

HOW DETERMINED

Originally, basal metabolic rates were determined by a method known as direct calorimetry, a method too time consuming and costly for use on large numbers of people. Now, a rapid, inexpensive method, *indirect calorimetry,* is used in practically all hospitals, as well as in many doctors' offices and nutrition laboratories. This method of determining metabolism consists simply of an oxygen tank into which the patient breathes by means of a rubber tube leading to his mouth, his nose being clamped off. He receives oxygen from the tank and expires carbon dioxide into it, the latter being removed by soda lime contained in a tank within the oxygen tank. The amount of oxygen consumed in a given length of time is measured. Research has shown that for every liter of oxygen consumed, an average of 4.825 kcal of heat are produced under basal con-

*Mountcastle, V. B., editor: Medical physiology, ed. 13, St. Louis, 1974, The C. V. Mosby Co.

ditions. Multiplying the amount of oxygen consumed by 4.825 therefore gives the number of kilocalories produced in the given time. From that figure the number produced in 24 hours can, of course, be readily computed. This number represents the patient's basal metabolic rate per day expressed in kilocalories. Usually the basal metabolic rate is expressed as normal or as a definite percent above or below normal. The percentage is computed by comparing the actual basal metabolic rate (in kilocalories) with what is known to be an average basal metabolic rate (in kilocalories) for normal individuals of the given size, sex, and age. Statistical tables, based on research, give these normal rates for different sizes, sexes, and ages. For instance, you can compute the average BMR for a person of your own size, sex, and age in this way:

1 Start with your weight in kilograms and your height in centimeters. (Convert pounds to kilograms by dividing pounds by 2.2. Convert inches to approximate centimeters by multiplying inches by 2.5.) For example, 110 pounds = 50 kg; 5 feet, 3 inches = 158 cm.

2 Convert your weight and height to square meters, using Fig. 15-18. For example, weight 50 kg and height 158 cm = about 1.5 square meters surface area of body.

3 Find your age and sex in Table 15-2 and then multiply the number of kilocalories per square meter per hour given there by your square meters of surface area and then by 24. For example, average BMR per day for a 25-year-old female, weight 110 pounds and height 5 feet, 3 inches = 1,332 kcal (37 × 1.5 × 24).

A quick rule of thumb for estimating a young woman's BMR is to multiply her weight in pounds by twelve.

Total metabolic rate

The *total metabolic rate* is the amount of energy used or expended by the body in a

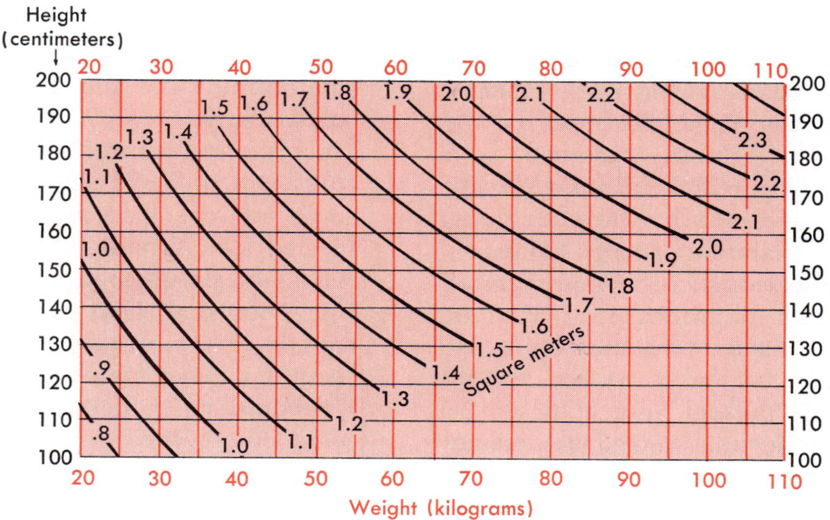

Fig. 15-18

Chart for determining surface area of man in square meters from weight in kilograms and height in centimeters according to the following formula: area (sq m) = $wt^{0.425} \times ht^{0.725} \times 7184$. (Redrawn after DuBois, D., and DuBois, E. F.: Clinical calorimetry, Arch. Intern. Med. [Chicago] **17**:863-871, 1916.)

given time. It is expressed in kilocalories per hour or per day. Most of the many factors that together determine the total metabolic rate are shown in Fig. 15-17. Of these, the main direct determinants are the following:

1 The basal metabolic rate—that is, the energy used to do the work of maintaining life under the basal conditions previously described, plus
2 The energy used to do all kinds of skeletal muscle work—from the simplest activities such as feeding oneself or sitting up in bed to the most strenuous kind of physical labor or exercise.

Energy balance and its relationship to body weight

When we say that the body maintains a state of energy balance, we mean that its energy input equals its energy output. Energy input per day equals the total calories (kilocalories) in the food ingested per day. Energy output equals the total metabolic rate expressed in kilocalories. But you may be wondering what energy intake, output, and balance have to do with body weight. "Everything" would be a fairly good one-word answer. Or, to be somewhat more explicit, the following basic principles describe the relationships between these factors:

1 Body weight remains constant (except for possible variations in water content) when the body maintains energy balance—when the total calories in the food ingested equals the total metabolic rate, that is. Example: if you have a total metabolic rate of 2,000 kcal per day and if the food you eat per day yields 2,000 kcal, your body will be maintaining energy balance and your weight will stay constant.

2 Body weight increases when energy input exceeds energy output—when the total calories of food intake per day is greater than the total calories of the metabolic rate. A small amount of the excess energy input is used to synthesize glycogen for storage in the liver and muscles. But the rest of it is used for synthesizing fat and storing it in adipose tissue. If you were to eat 3,000 kcal each day for a week and if your total metabolic rate were 2,000 kcal per day, you would gain weight. How much you would gain, you can discover by doing a little simple arithmetic:

Total energy input for week = 21,000 kcal
Total energy output for week = 14,000 kcal
Excess energy input for week = 7,000 kcal

Approximately 3,500 kcal are used to synthesize 1 pound of adipose tissue. Hence, at the end of this one week of "overeating"—of eating 7,000 kcal over and above your total metabolic rate—you would have gained about 2 pounds.

3 Body weight decreases when energy input is less than energy output—when the total number of calories in the food eaten is less than the total metabolic rate. Suppose you were to eat only 1,000 kcal a day for a week and that you have a total metabolic rate of 2,000 kcal per day. By the end of the week your body would have used a total of 14,000 kcal of energy for maintaining life and doing its many kinds of work. All 14,000 kcal of this actual energy expenditure had to come from catabolism of foods since this is the body's only source of energy. Catabolism of ingested food supplied 7,000 kcal and catabolism of stored food supplied the remaining 7,000 kcal. That week your body would not have maintained energy balance, nor would it have maintained weight balance. It would have incurred an energy deficit paid out of the energy stored in approximately 2 pounds of body fat. In short, you would have lost about 2 pounds.

▪ ▪ ▪

Anyone who wants to reduce should remember this cardinal principle: eat fewer calories than your total metabolic rate. Obey this law and you will lose weight. Ignore it and you will not lose weight. Unless caloric intake is less than caloric output (total metabolic rate), weight loss is impossible. But we ought to note one other point about this principle. There are two ways to make your caloric intake less than your total metabolic rate. The approach commonly advised to would-be reducers is simply "cut down on your calories"—a principle that proves easy to understand but difficult to apply. The other approach to weight reduction seems to be thought of or emphasized less often. It is this: increase your total metabolic rate (your caloric output, that is) and also decrease your caloric intake. Simple arithmetic shows which method produces a faster weight loss. Suppose you have a total metabolic rate of 2,000 kcal a day and that you ingest 1,500 kcal a day for a week. By the end of the week you will have taken in 3,500 kcal less than your total metabolic rate. That deficit will have been supplied by the catabolism of about 1 pound of body fat. Now suppose that the next week you still eat 1,500 kcal a day but you also exercise briskly for a half hour each day. Suppose you swim sidestroke at the rate of 1.6 miles per hour for a half hour. This will increase your total metabolic rate from 2,000 to 2,600 kcal.* Your calorie deficit will then be 2,600 minus 1,500, or 1,100 kcal a day, or 7,700 kcal for the week. Catabolism of more than 2 pounds of body fat will have supplied those 7,700 kcal. In short, you will have lost more than twice as much the second week as the first even though you ate the same amount both weeks.

Foods are stored as glycogen, fats, and tissue proteins. As you will recall, cells catabolize them preferentially in this same order: carbohydrates, fats, and proteins. If there is no food intake, almost all of the glycogen is estimated to be used up in a matter of 1 or 2 days. Then, with no more carbohydrate to act as a fat sparer, fat is catabolized. How long it takes to deplete all of this reserve food depends, of course, upon how much adipose tissue the individual has when he starts his starvation diet. Finally, with no more fat available as a protein sparer, tissue proteins are catabolized rapidly, and death soon ensues.

*See supplementary readings for this chapter, Morehouse and Miller reference.

Mechanisms for regulating food intake

Mechanisms for regulating food intake are still not clearly established. That the hypothalamus plays a part in these mechanisms, however, seems certain. A number of data seem to indicate that a cluster of neurons in the lateral hypothalamus function as an appetite center—meaning that impulses from them bring about increased food intake. Other data suggest that a group of neurons in the ventral medial nucleus of the hypothalamus functions as a satiety center—meaning that impulses from these neurons inhibit food intake. What acts directly on these centers to stimulate or depress them is still a matter of theory rather than fact. One theory (the "thermostat theory") holds that it is the temperature of the blood circulating to the hypothalamus that influences the centers. A moderate decrease in blood temperature stimulates the appetite center (and inhibits the satiety center). Result: the individual has an appetite. He wants to eat—and probably does. An increase in blood temperature produces the opposite effect, a depressed appetite (anorexia). One well-known instance of this is the loss of appetite in persons who have a fever.

Another theory (the "glucostat theory")

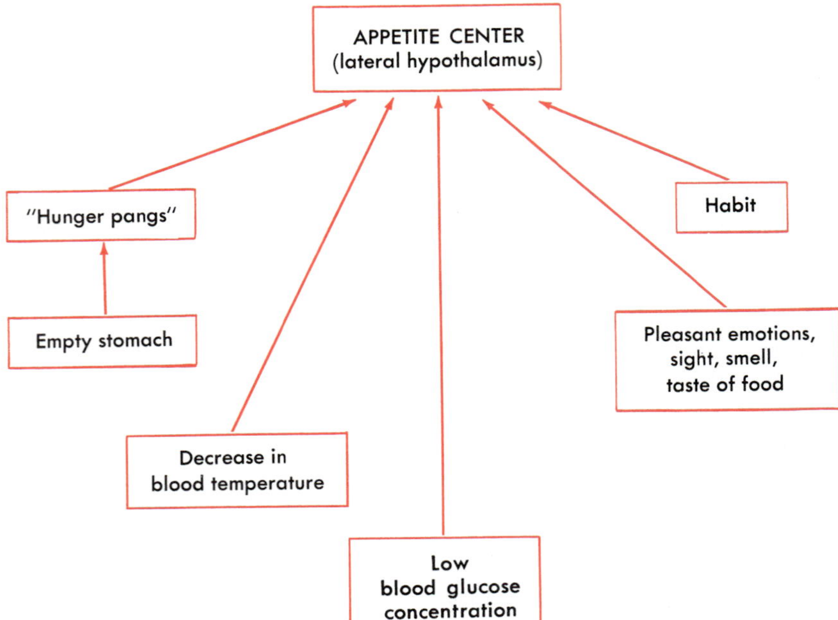

Fig. 15-19

Factors that directly or indirectly stimulate the appetite center (and inhibit the satiety center) in the hypothalamus and thereby tend to increase the amount of food eaten.

says that it is the blood glucose concentration and rate of glucose utilization that influences the hypothalamic feeding centers. A low blood glucose concentration or low glucose utilization stimulates the appetite center, whereas a high blood glucose concentration inhibits it. Unquestionably, a great many factors operate together as a complex mechanism for regulating food intake. Some of these factors are indicated in Fig. 15-19.

Homeostasis of body temperature

Warm-blood animals, such as man, maintain a remarkably constant temperature despite sizable variations in environmental temperatures.

Normally in most people, body temperature moves up and down very little in the course of a day. It hovers close to a midpoint of about 37° C., increasing perhaps to 37.6° C. by late afternoon and decreasing to around 36.2° C. by early morning. This homeostasis of body temperature is of the utmost importance. Why? Because healthy survival depends upon biochemical reactions taking place at certain rates. And these rates, in turn, depend upon normal enzyme functioning, which depends upon body temperature staying within the narrow range of normal.

In order to maintain an even temperature, the body must, of course, balance the amount of heat it produces with the amount it loses. This means that if extra heat is produced in the body, this same amount of heat must then be lost from it. Obviously if this does not occur, if increased heat loss does not follow close upon increased heat production, body temperature will climb steadily upward.

Heat production

Heat is produced by one means—catabolism of foods. Because the muscles and glands (liver, especially) are the most active tissues, they carry on more catabolism and therefore produce more heat than any of the other tissues. So the chief determinant of how much heat the body produces is the amount of muscular work it does. During exercise and shivering, for example, catabolism and heat production increase greatly. But during sleep, when very little muscular work is being done, catabolism and heat production decrease.

Heat loss

Heat is lost from the body by the physical processes of evaporation, radiation, conduction, and convection. Some 80% or more of this heat transfer occurs through the skin. The rest takes place through the mucous membranes of the respiratory, digestive, and urinary tracts.

Evaporation

Heat energy must be expended to evaporate any fluid. Evaporation of water, therefore, constitutes one method by which heat is lost from the body, especially from the skin. At moderate temperatures it accounts for about half as much heat loss as does radiation. But at high environmental temperatures, evaporation constitutes the only method by which heat can be lost from the skin. A humid atmosphere necessarily retards evaporation and therefore lessens the cooling effect derived from it—the explanation for the fact that the same degree of temperature seems hotter in humid climates than in dry ones.

Radiation

Radiation is the transfer of heat from the surface of one object to that of another without actual contact between the two. Heat radiates from the body surface to nearby objects that are cooler than the skin and radiates to the skin from those that are warmer than the skin. This is, of course, the principle of heating and cooling systems. From surfaces that have been heated to temperatures warmer than the skin, heat radiates to the skin, thereby warming it, whereas with cooled surfaces, heat radiates from the skin to them, thereby cooling the skin. The amount of heat lost by radiation from the skin is made to vary as needed by dilation of surface blood vessels when more heat needs to be lost and by vasoconstriction when heat loss needs to be decreased. In cool environmental temperatures, radiation accounts for a greater percentage of heat loss from the skin than both conduction and evaporation combined. In hot environments, on the other hand, no heat is lost by radiation but instead may even be gained by radiation from warmer surfaces to the skin.

Conduction

Conduction means the transfer of heat to any substance actually in contact with the body—to clothing or jewelry, for example, or even to cold foods or liquids ingested. This process accounts for a relatively small amount of heat loss compared to the amount lost by evaporation and radiation.

Convection

Convection is the transfer of heat away from a surface by movement of heated air or fluid particles. Usually, convection causes very little heat loss from the body's surface. But it can account for considerable heat loss —as you know from experience if you have ever stepped from your bath into even slightly moving air from an open window.

Thermostatic control of heat production and loss

The control mechanism that normally maintains homeostasis of body temperature consists of two parts:

1 A *heat-dissipating mechanism* that acts to increase heat loss when blood temperature increases above a certain point (Fig. 15-20). This mechanism, therefore, prevents body temperature from rising above normal under usual circumstances.

2 A *heat-gaining mechanism* that acts to

accelerate catabolism and thereby to increase heat production when blood temperature decreases below a certain point. Under ordinary conditions, this mechanism prevents body temperature from falling below normal.

Heat-dissipating mechanism

In the anterior part of the hypothalamus, behind the sphenoid sinuses, lies a group of cells referred to collectively as the "human thermostat." These neurons are thermal receptors—that is, they are stimulated by a very slight increase in the temperature of the blood above the point at which the human thermostat is set—normally about 37° C. In a sense, one might say that these cells of the hypothalamus take the temperature of the blood circulating to them. Whenever it increases by as little as 0.01° above 37° C* (or some other set point), these neurons send out impulses that eventually reach sweat glands and blood vessels of the skin. They stimulate the body's 2 million or more sweat glands to increase their rate of secretion, and they also cause dilation of surface blood vessels. Evaporation of the larger amount of sweat causes a greater heat loss from the skin. Also, more heat is lost by radiation from the larger quantity of blood circulating near the surface in the dilated skin vessels.

Heat-gaining mechanism

In a cold environment, the mechanism that tries to maintain homeostasis of body temperature includes two kinds of responses—those that decrease heat loss and those that increase heat production. Together, they almost always succeed in preventing a decrease in blood temperature below the lower limit of normal. Skin blood vessel constriction decreases the volume of blood circulating near the surface and so decreases heat loss by radiation. In addition, shivering and voluntary muscle contractions occur, thereby accelerating catabolism and heat production. Further details about the mechanism for preventing body temperature from falling below normal are not yet established.

*See Benzinger, T. H.: The human thermostat, Sci. Am. **204:**134-147, Jan., 1961.

Skin thermal receptors

In addition to the thermal receptors in the hypothalamus, there are many heat and cold receptors located in the skin. Impulses initiated in skin thermal receptors travel to the cerebral cortex sensory area. Here, they give rise to sensations of skin temperature and are relayed out over voluntary motor paths to produce skeletal muscle movements that affect skin temperature. For example, on a hot day you "feel hot" because of stimulation of your skin heat receptors. Often you make some sort of movements to "cool yourself off." You may start fanning yourself, or perhaps you turn on an air conditioner or go swimming. As a result, your skin temperature decreases to a more comfortable level. Thus the thermal receptors in the skin will have taken part in a conscious mechanism that helps regulate skin temperature. Convincing evidence supports the view that this is their only function and that impulses from them do not travel to the heat-regulating centers in the hypothalamus and therefore do not take part in the automatic regulation of internal body temperature.

Correlations

When a fever or higher than normal temperature exists, it is thought to result primarily from inability of the heat-dissipating mechanisms to keep pace with heat production. Some factor—perhaps chemicals from microorganisms or injured tissue cells—stimulates catabolism, thereby producing more heat in the body in a given time. Heat-dissipating mechanisms (Fig. 15-20)

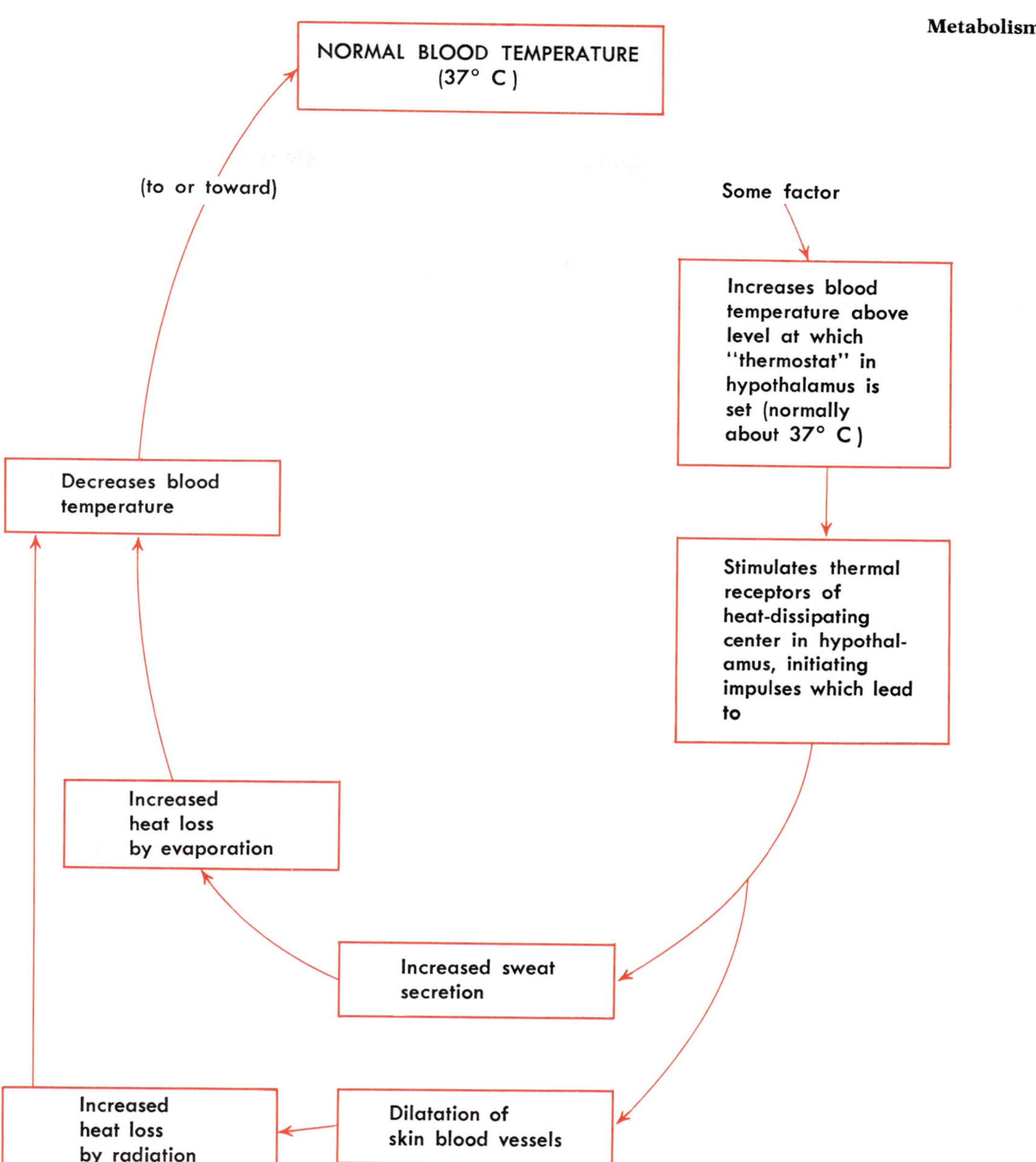

Fig. 15-20

Scheme to show how heat-dissipating mechanism operates to maintain normal body temperature. In principle, it cancels out any heat gain by bringing about an equal heat loss. Under usual circumstances, this mechanism succeeds in preventing body temperature from rising above the upper limit of normal. When it fails, fever develops. According to one theory, certain factors can increase the threshold of stimulation of hypothalamic thermal receptors—in more picturesque language, they reset the hypothalamic thermostat at a level higher than normal. Then, blood temperature rises above normal before it activates the heat-dissipating mechanism shown in this diagram.

operate in an effort to compensate for the heat gain. But presumably they cannot increase heat loss as much as heat production has increased, so body temperature necessarily increases. The patient "has a fever," in other words.

The heat-regulating centers are present at birth but do not function well for a short time after birth. Therefore, newborn babies need to be kept somewhat warmer than adults. If the baby is born prematurely, the heat-regulation centers do not function for a longer time, perhaps several weeks.

Outline summary

Meaning

Chemical changes foods undergo inside cells or utilization of foods by body cells

Ways in which foods are metabolized

1 Catabolism—breaks down food molecules to simpler compounds (carbon dioxide, water, and nitrogenous wastes), transferring some of their energy to phosphate compounds, notably ATP, and releasing some of it as heat
2 Anabolism—building up food molecules into more complex compounds—notably glycogen, enzymes and other cell proteins, hormones, etc.

Carbohydrate metabolism

1 Glucose transport through cell membranes and phosphorylation
  a Insulin promotes this transport through cell membranes
  b *Glucose phosphorylation*—conversion of glucose to glucose-6-phosphate, catalyzed by enzyme glucokinase; *insulin* increases activity of glucokinase so promotes glucose phosphorylation
2 Glucose catabolism
  a Consists of complex series of chemical reactions that take place inside cells and yield energy, carbon dioxide, and water; about half of energy released from food molecules by catabolism is put back in storage in unstable high-energy bonds of ATP molecules and the rest is transformed to heat; energy in high-energy bonds of ATP with explosive rapidity
  b Purpose to continually provide cells with utilizable energy —i.e., energy supplied instantaneously to energy-consuming mechanisms that do cellular work
  c Glycolysis—series of anaerobic (nonoxygen-utilizing) chemical reactions that convert 1 glucose molecule to 2 pyruvic acid molecules and yield slightly less than 5% of ATP produced during glucose catabolism
  d Citric acid cycle—series of aerobic chemical reactions that utilize oxygen to oxidize 2 pyruvic acid molecules to 6 carbon dioxide molecules and 6 water molecules and yield about 95% of all the ATP formed during catabolism
3 Glucose anabolism—synthesis of larger molecule compounds from glucose; an important kind of cellular work that uses some of energy made available by catabolism
4 Glycogenesis—conversion of glucose to glycogen for storage; occurs mainly in liver and muscle cells; one kind of glucose anabolism
5 Glycogenolysis
  a In muscle cells—glycogen changed back to glucose-6-phosphate, preliminary to catabolism
  b In liver cells—glycogen changed back to glucose; enzyme, glucose phosphatase, present in liver cells catalyzes final step of glycogenolysis, changing of glucose-6-phosphate to glucose; this enzyme lacking in most other cells; *glucagon* increases activity of phosphorylase so accelerates liver glycogenolysis; *epinephrine* accelerates liver and muscle glycogenolysis
6 Gluconeogenesis—sequence of chemical reactions carried on in liver cells; converts protein or fat compounds into glucose; growth hormone, ACTH, and glucocorticoids have stimulating effect on rate of gluconeogenesis
7 Control of glucose metabolism
  a By hormones secreted by islands of Langerhans in pancreas
    1 Insulin (from beta cells)—tends to accelerate glucose utilization by cells because it accelerates glucose transport through cell membranes and glucose phosphorylation; hence, insulin tends to decrease blood glucose concentration—i.e., has hypoglycemic effect
    2 Glucagon (from alpha cells)—increases activity of enzyme phosphorylase, thereby accelerating liver glycogenolysis with release of glucose into blood; hence, glucagon tends to increase blood glucose—i.e., has hyperglycemic effect
  b By hormones secreted by anterior pituitary gland, adrenal cortex, and thyroid gland
    1 Growth hormone—decreases fat deposition, increases fat mobilization and catabolism; hence, tends to bring about shift to fat utilization from "preferred" glucose utilization
    2 ACTH and glucocorticoids—ACTH stimulates adrenal cortex to increase secretion of glucocorticoids, which accelerate tissue protein mobilization and subsequent liver gluconeogenesis from mobilized proteins; therefore, ACTH and glucocorticoids tend to increase blood glucose—i.e., have hyperglycemic effect
    3 Thyrotropin—stimulates thyroid gland to increase secretion of thyroid hormone, which accelerates catabolism, usually glucose catabolism since glucose "preferred fuel"
  c By hormone secreted by adrenal medulla—epinephrine increases phosphorylase activity, so accelerates muscle and liver glycogenolysis; hence, tends to increase blood glucose—i.e., has hyperglycemic effect
8 Principles about normal carbohydrate metabolism
  a Principle of "preferred energy fuel"—cells catabolize first glucose, sparing fats and proteins; when their glucose

supply becomes inadequate, next catabolize fats, sparing proteins, and last catabolize proteins

- **b** Principle of glycogenesis—glucose in excess of about 120 to 140 mg/100 ml blood brought to liver by portal veins enters liver cells, where it undergoes glycogenesis
- **c** Principle of glycogenolysis—when blood glucose decreases below midpoint of normal, liver glycogenolysis accelerates and tends to raise blood glucose concentration back toward midpoint of normal
- **d** Principle of gluconeogenesis—when blood glucose decreases below normal or when amount of glucose entering cells inadequate, liver gluconeogenesis accelerates and tends to raise blood glucose concentration
- **e** Principle of glucose storage as fat—when blood sugar glucose higher than normal and blood insulin content adequate, excess glucose converted to fat, mainly by liver cells, and stored as such in fat depots
- **f** Principle of hyperglycemia in stress—under conditions of stress, blood glucose concentration tends to increase above normal

Fat metabolism

1 Catabolism
  - **a** Hydrolysis of fats to fatty acids and glycerol, primarily in liver cells
  - **b** Glycerol oxidized same as carbohydrates
  - **c** Fatty acids converted to ketone bodies (ketogenesis); occurs mainly in liver; largest proportion of ketones enters blood from liver cells to be transported to tissues for oxidation to carbon dioxide and water via tricarboxylic acid cycle

2 Anabolism—for tissue synthesis and for building various compounds; fats deposited in connective tissue converts it to adipose tissue

3 Fat mobilization—release of fats from adipose tissue cells, followed by their catabolism; occurs when blood contains less glucose than normal or when it contains less insulin than normal; if excessive, leads to ketosis

4 Control—by following major factors
  - **a** Rate of glucose catabolism one of main regulators of fat metabolism; in general, normal or high rates of glucose catabolism accompanied by low rates of fat mobilization and catabolism and high rates of fat deposition; converse also true
  - **b** Insulin helps control fat metabolism by its effects on glucose metabolism; in general, normal amounts of insulin and blood glucose tend to decrease fat mobilization and catabolism and to increase fat deposition; insulin deficiency increases fat mobilization and catabolism; converse also true
  - **c** Growth hormone—decreases fat deposition and increases fat mobilization and utilization—i.e., growth hormone tends to bring about shift from glucose to fat utilization
  - **d** Glucocorticoids help control fat metabolism; in general, when blood glucose lower than normal and in various stress situations, more glucocorticoids secreted and accelerate fat mobilization and gluconeogenesis from them; when blood glucose higher than normal but rate of glucose catabolism low (as in diabetes mellitus), glucocorticoids also increase fat mobilization, but it is followed by ketogenesis from them; when blood glucose higher than normal, and provided its insulin content adequate, glucocorticoids accelerate fat deposition

Protein metabolism

1 Anabolism of proteins of primary importance, their catabolism, secondary; amino acids used to synthesize all kinds of tissue (e.g., for growth and repair); also used to synthesize many other substances such as enzymes, hormones, antibodies, and blood proteins

2 Catabolism
  - **a** Deamination of amino acid molecule to form ammonia and keto acid; mainly in liver cells
  - **b** Ammonia converted to urea (mainly in liver) and excreted via urine
  - **c** Keto acids may be converted to glucose in liver, or to fat, or oxidized in liver or tissue cells via tricarboxylic acid cycle

3 Control
  - **a** Growth hormone and testosterone both have stimulating effect on protein synthesis or anabolism
  - **b** ACTH and glucocorticoids—protein catabolic hormones; accelerate tissue protein mobilization—i.e., hydrolysis of tissue proteins to amino acids and their release into blood; liver converts amino acids to glucose (gluconeogenesis) or deaminates them
  - **c** Thyroid hormone promotes protein anabolism when nutrition adequate and amount of hormone normal; therefore, adequate amounts necessary for normal growth

Metabolic rates

Meaning

Amount of heat energy expended in given time

Ways of expressing

In kilocalories or as "normal" or as a definite percentage above or below normal—e.g., +10% or −10%

Basal metabolic rate

Amount of heat produced (energy expended) in waking state when body at complete rest, 12 to 18 hours after last meal, in comfortably warm environment

1 Factors influencing
  - **a** Size—greater surface area, higher BMR (surface area computed from height and weight)
  - **b** Sex—approximately 5% higher in males
  - **c** Age—higher in youth than in aged
  - **d** Abnormal functioning of certain endocrines, particularly thyroid gland
  - **e** Fever—each Centigrade degree rise in temperature increases BMR approximately 13%
  - **f** Certain drugs (e.g., dinitrophenol) increase BMR
  - **g** Other factors (e.g., pregnancy and emotions) increase BMR

2 How determined
  - **a** Direct calorimetry—too expensive and too time consuming for wide use
  - **b** Indirect calorimetry—measures amount of oxygen inspired in given time; 4.825 kcal of heat produced for each liter of oxygen consumed

Total metabolic rate

Amount of heat produced by body in average 24 hours; equal to basal rate plus number of kilocalories produced chiefly by muscular work and adjusting to cool temperatures; expressed in kilocalories per 24 hours

Energy balance and its relationship to body weight

1 Energy balance means that energy input (total kilocalories in food ingested) equals energy output (i.e., total metabolic rate expressed in kilocalories)
2 In order for body weight to remain constant (except for variations in water content), energy balance must be maintained—total kilocalories ingested must equal total metabolic rate
3 Body weight increases when energy input exceeds energy output—when total kilocalories ingested greater than total metabolic rate
4 Body weight decreases when energy input less than energy output—when total kilocalories ingested less than total metabolic rate; no diet will reduce weight unless it contains fewer kilocalories than the total metabolic rate of the individual eating diet

Mechanisms for regulating food intake

1 Thermostat theory holds that moderate decrease in blood temperature acts as stimulant to appetite center so increases appetite; increase in blood temperature (fever) produces opposite effect
2 Glucostat theory postulates that low blood glucose concentration or low rate of glucose utilization (e.g., diabetes mellitus) acts as stimulant to appetite center so increases appetite; high blood glucose level produces opposite effect

Homeostasis of body temperature

In order to maintain homeostasis of body temperature heat production must equal heat loss

1 Heat production—by catabolism of foods in skeletal muscles and liver especially
2 Heat loss
   a By physical processes of evaporation, radiation, conduction, and convection
   b About 80% of heat loss occurs through skin; rest takes place through mucosa of respiratory, digestive, and urinary tracts
3 Thermostatic control of heat production and loss
   a Heat-dissipating mechanism—see Fig. 15-21
   b Heat-gaining mechanism
      1 Details not established but mechanism activated by decrease in blood temperature
      2 Responses—skin blood vessel constriction, shivering, and voluntary muscle contractions
4 Skin thermal receptors—stimulation of skin thermal receptors gives rise to sensations of heat or cold; also initiates voluntary movements to reduce these sensations—e.g., fanning oneself to cool off or exercising to warm up

Review questions

1 What is metabolism?
2 What two processes make up the process of metabolism?
3 Contrast the function or purpose of digestion with the function or purpose of metabolism.
4 Describe carbohydrate catabolism briefly.
5 Describe protein catabolism briefly.
6 Describe fat catabolism briefly.
7 Compare proteins, carbohydrates, and fats as to functions they serve in the body.
8 Differentiate between digestive and metabolic wastes.
9 What does the term *metabolic rate* mean?
10 Differentiate between basal and total metabolic rates.
11 Discuss possible effects of advanced liver disease on digestion, absorption, and metabolism.

Situation: Mrs. A., who weighs 160 pounds, is 5 feet, 5 inches tall, and is 50 years of age, has been given a reducing diet by her doctor. She complains to you that she "knows it won't work" because "it isn't what she eats that makes her fat." As proof, she says that both her husband and son "eat twice as much" as she and so does her younger sister.

12 How would you answer Mrs. A.?
13 Suppose that Mrs. A. has a total metabolic rate of 2,200 kcal per day and that her diet contains 1,700 kcal. How many pounds can she expect to lose per week?
14 Rapid bicycle riding for 15 minutes expends about 100 kcal. This alone, without decreasing your calorie intake at all, would cause you to lose a pound in how many days? In 50 weeks, how much could you lose this way?
15 Explain two mechanisms postulated to regulate the amount of food a person eats.
16 Explain briefly the mechanism for maintaining homeostasis of body temperature.

16 The urinary system

Organs
Kidneys
- Gross anatomy
  - Size, shape, and location
  - External structure
  - Internal structure
- Microscopic anatomy
- Physiology
  - Functions
  - How kidneys excrete urine
  - Mechanisms that control volume of urine excreted
  - Influence of kidneys on blood pressure

Ureters
- Location and structure
- Function
- Correlations

Bladder
- Structure and location
- Functions

Urethra
- Structure and location
- Functions

Urine
Physical characteristics
Chemical composition
Definitions

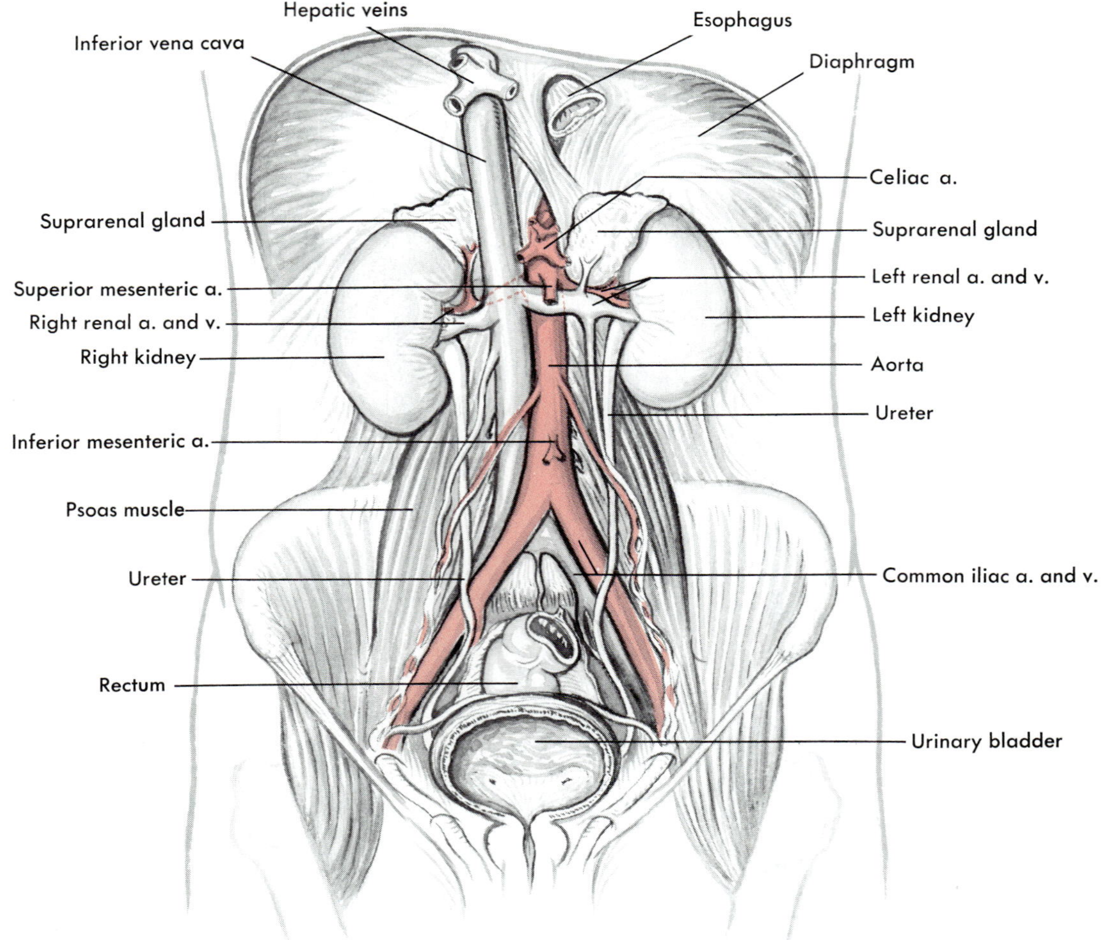

Fig. 16-1

Location of urinary system organs.

The urinary system consists of those organs that produce urine and eliminate it from the body. They are two kidneys, two ureters, one bladder, and one urethra (Fig. 16-1). The excretion of urine and its elimination from the body are vital functions since together they constitute one of the most important mechanisms for maintaining homeostasis. As one individual has phrased it: "The composition of the blood (and internal environment) is determined not by what the mouth ingests but by what the kidney keeps."*

The substances excreted from the kidneys and other excretory organs are listed in Table 16-1.

*From Smith, H. W.: Lectures on the kidney, Lawrence, Kan., 1943, University of Kansas, p. 3.

Table 16-1

Excretory organs of the body

Excretory organ	Substance excreted
Kidneys	Nitrogenous wastes (from protein catabolism)
	Toxins (e.g., from bacteria)
	Water (from ingestion and from catabolism)
	Mineral salts
Skin (sweat glands)	Water
	Mineral salts
	Small amounts of nitrogenous wastes
Lungs	Carbon dioxide (from catabolism)
	Water
Intestine	Wastes from digestion (cellulose, connective tissue, etc.)
	Some metabolic wastes (e.g., bile pigments; also salts of calcium and other heavy metals)

Organs

Kidneys

Gross anatomy

SIZE, SHAPE, AND LOCATION

The kidneys resemble lima beans in shape. An average-sized kidney measures approximately 4½ inches in length, 2 to 3 inches in width, and 1 inch in thickness. Usually the left kidney is slightly larger than the right.

The kidneys lie behind the parietal peritoneum, against the posterior abdominal wall, at the level of the last thoracic and first three lumbar vertebrae (or, as you can see in Fig. 16-1 just above the waistline). The liver pushes the right kidney down to a level somewhat lower than the left. A heavy cushion of fat normally keeps the kidneys up in position. Very thin individuals may suffer from ptosis (dropping) of one or both of these organs. Connective tissue (renal fasciae) anchors the kidneys to surrounding structures and helps to maintain their normal position.

EXTERNAL STRUCTURE

The mesial surface of each kidney presents a concave notch called the *hilum.* Structures enter the kidneys through this notch just as they enter the lung through its hilum. A tough white fibrous capsule encases each kidney.

INTERNAL STRUCTURE

When a coronal section is made through the kidney, two kinds of substances are seen composing its interior: an outer layer, the *cortex,* and an inner portion, the *medulla* (Fig. 16-2). The latter is divided into a dozen or more triangular wedges, the *renal pyramids.* The bases of the pyramids face the cortex, and their apices or *renal papillae* face the center of the kidney. The pyramids have a striated appearance, as contrasted with the smooth texture of the cortical substance. The cortex extends inward between each two pyramids, forming the *renal columns.*

Microscopic anatomy

Microscopic examination reveals the kidney to be composed of peculiarly shaped structures resembling tiny funnels with proportionately long convoluted stems. As you read about these anatomical funnels, identify each of their parts in Fig. 16-3. Note that *Bowman's capsule* consists of two layers of flat epithelial cells with a space between the layers. Observe also that Bowman's capsule has invaginated in it a cluster of capillaries designated as a *glomerulus.* A Bowman's capsule and its partially encased glomerulus is named a *renal corpuscle* (or *malpighian corpuscle*).

Blood flows into each glomerulus by way of an afferent arteriole and out of it by way of an efferent arteriole—a unique arrangement. Blood usually flows out of capillaries into what kind of vessels? Extending from each Bowman's capsule is a renal tubule. It consists of several sections. The first is known as the *proximal convoluted tubule:* proximal be-

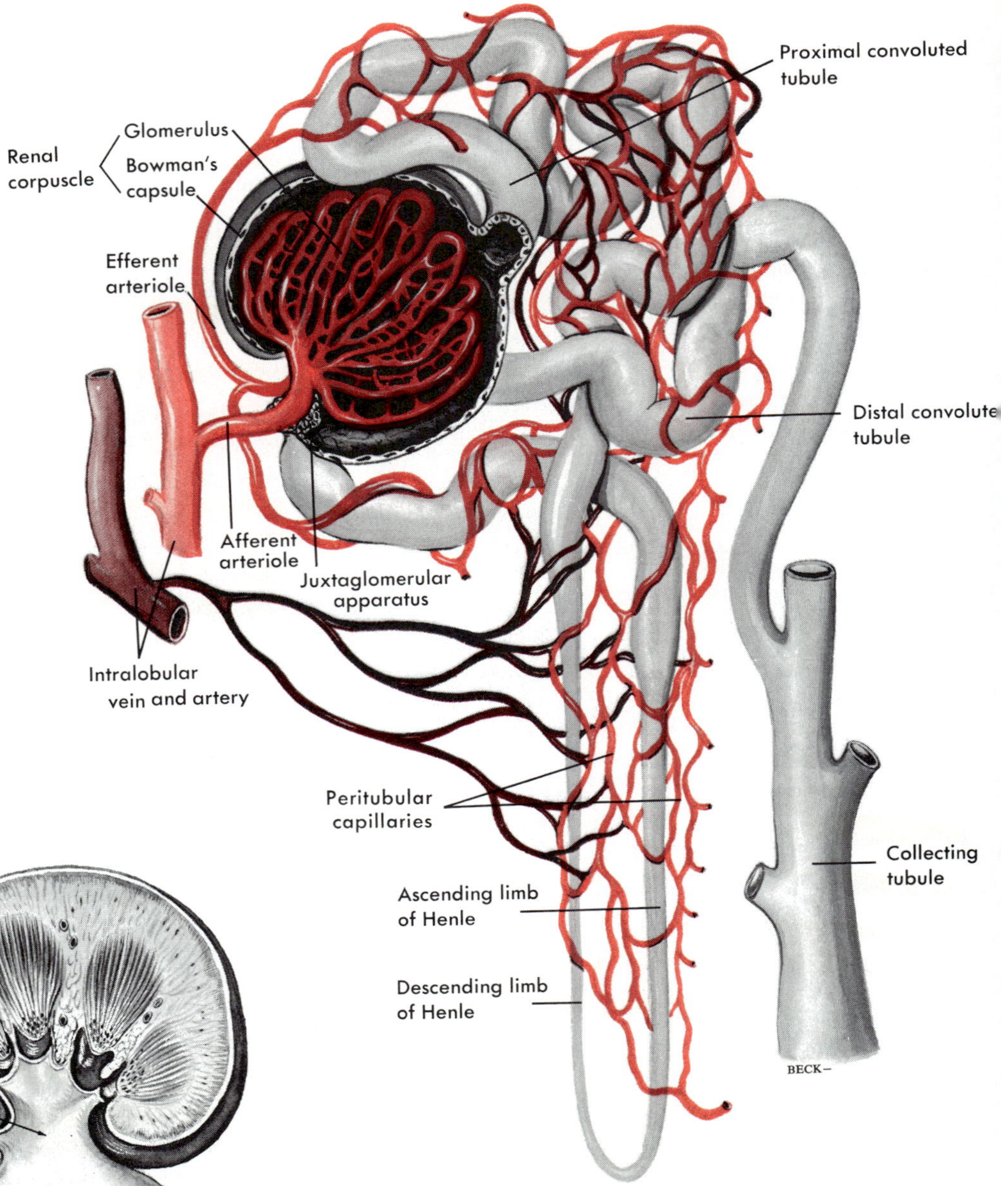

Fig. 16-3

The nephron unit with its blood vessels. Blood flows through nephron vessels as follows: intralobular artery → afferent arteriole → glomerulus → efferent arteriole → peritubular capillaries (around the tubules) → venules → intralobular vein.

Fig. 16-2

Coronal section through the right kidney.

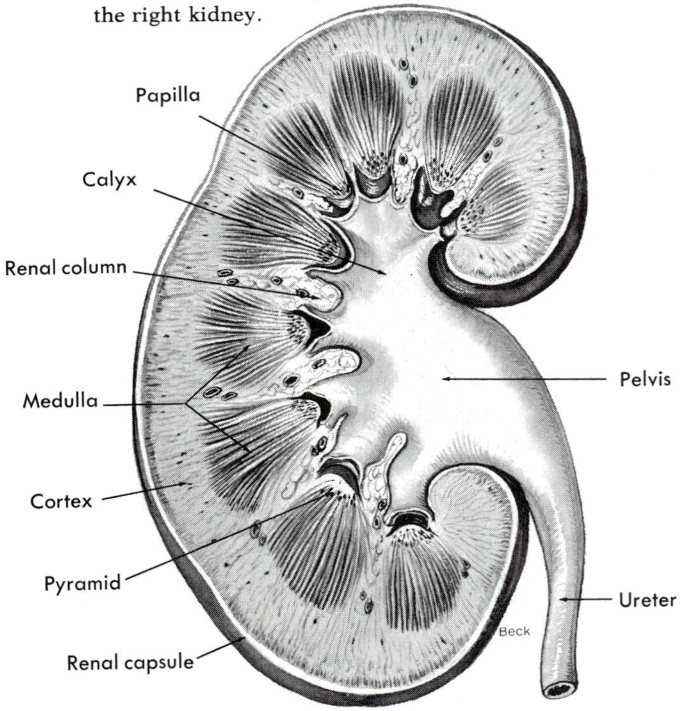

cause it is the segment nearest the tubule's origin from the Bowman's capsule and convoluted because it pursues a tortuous rather than a straight course. The proximal tubule becomes the *descending limb* of the *loop of Henle,* which becomes the *ascending limb,* which becomes the *distal convoluted tubule,* which terminates in a straight or *collecting tubule.* Now look at Fig. 16-4. The fact that the cells composing different parts of the renal tubule vary in shape suggests that they perform different functions. Whether or not they do you will discover in later paragraphs.

Collecting tubules join larger tubules, and all the larger collecting tubules of one renal pyramid converge to form one tube that opens at a renal papilla into one of the small calyces. Bowman's capsules and both convoluted tubules lie in the cortex of the kidney, whereas the loops of Henle and collecting tubules lie in its medulla.

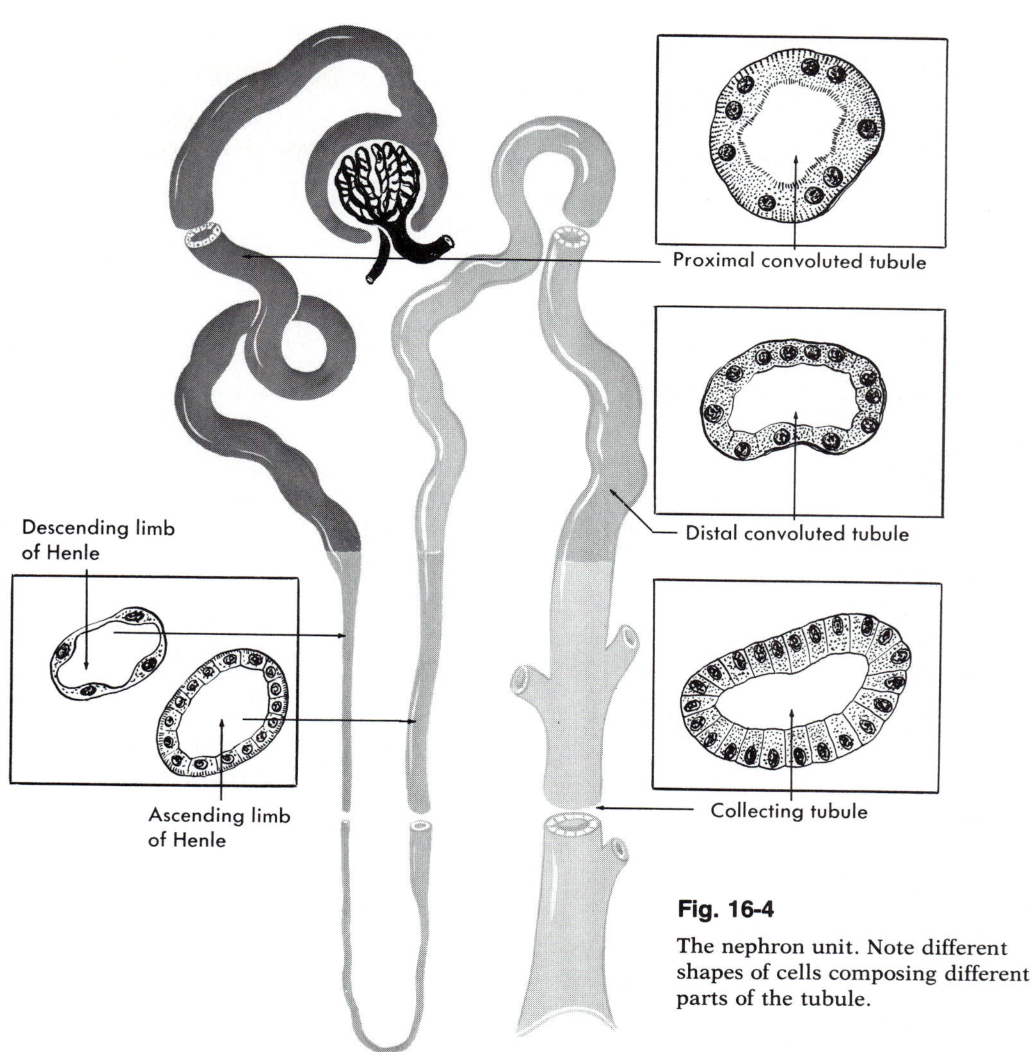

Fig. 16-4

The nephron unit. Note different shapes of cells composing different parts of the tubule.

A glomerulus, Bowman's capsule, and its tubule—proximal, convoluted, loop of Henle, and distal convoluted portions—together constitute a *nephron*, the structural and functional unit of the kidney. *Gray's Anatomy* says that there are about 1.25 million nephrons in each kidney.

Physiology

FUNCTIONS

The function of the kidneys is to excrete urine, a life-preserving function because homeostasis depends upon it. More than any other organ in the body, the kidneys can adjust the amounts of water and electrolytes leaving the body so that they equal the amounts of these substances entering the body. The vital conditions of fluid and electrolyte balance and acid-base balance, therefore, depend most of all upon adequate kidney functioning. Here are just a few of the blood constituents that cannot be held to their normal concentration ranges if the kidneys fail: sodium, potassium, chloride, and nitrogenous wastes from protein metabolism such as urea. In short, kidney failure means homeostasis failure and, if not relieved, inevitable death.

In addition to excreting urine, the kidneys are now known to influence blood pressure.

HOW KIDNEYS EXCRETE URINE

Three processes—glomerular filtration, tubular reabsorption, and tubular secretion

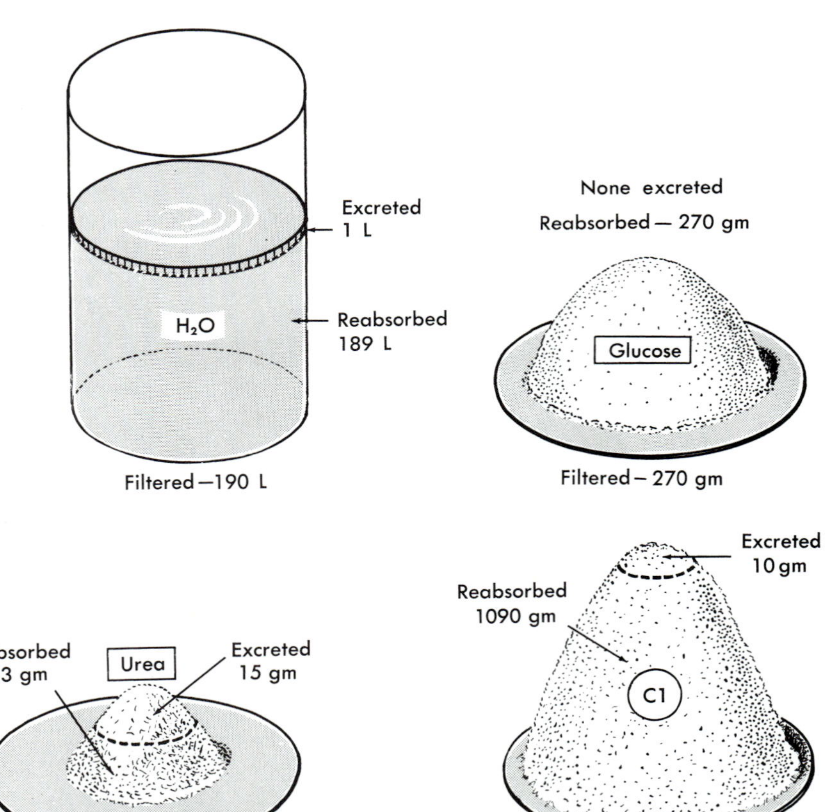

Fig. 16-5

Glomerular filtration and tubular reabsorption volumes. Note the enormous volume of water filtered out of glomerular blood per day—190 liters or many times the total volume of blood in the body. Only a very small proportion of this, however, is excreted in the urine. More than 99% of it (189 liters) is reasorbed into tubular blood. One other substance shown is also filtered and reabsorbed in large amounts. Which one? Another is normally entirely reabsorbed. Which one? Somewhat more than half of the filtered amount of which substance is reabsorbed?

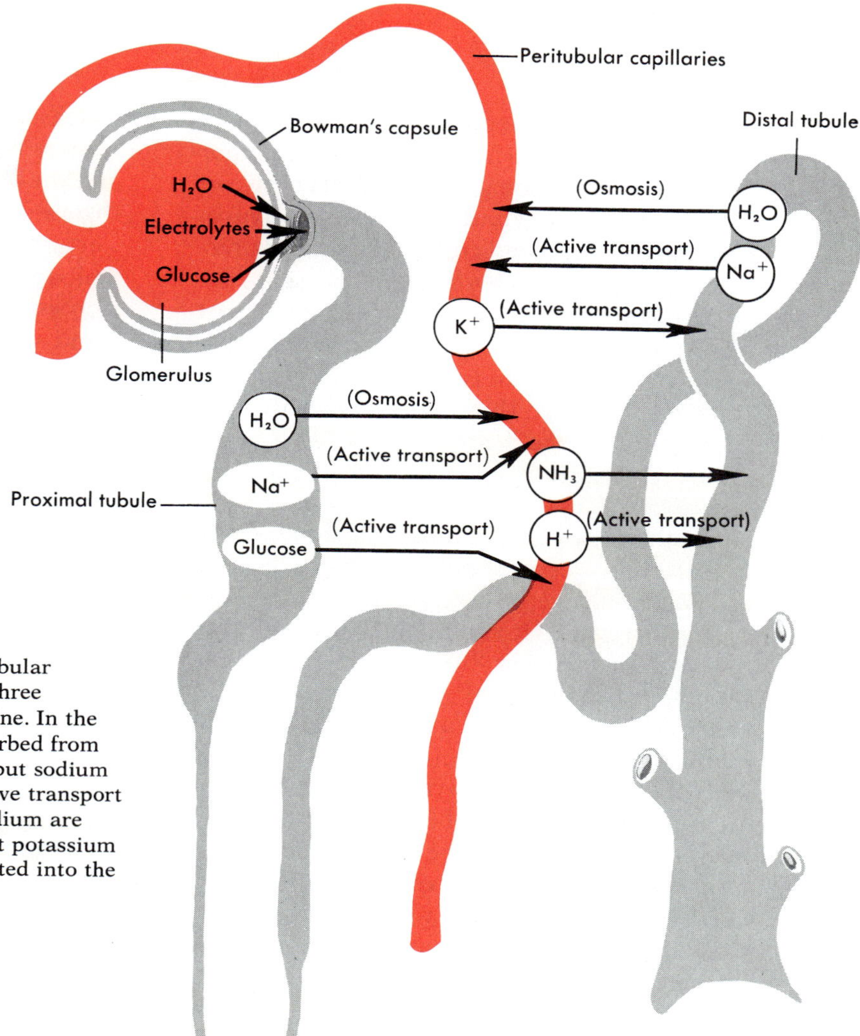

Fig. 16-6

Diagram showing glomerular filtration, tubular reabsorption, and tubular secretion—the three processes by which the kidneys excrete urine. In the proximal tubule, note that water is reabsorbed from the tubular filtrate into blood by osmosis but sodium and glucose are reabsorbed mainly by active transport mechanisms. Note, too, that water and sodium are also reabsorbed from the distal tubule. But potassium and hydrogen ions and ammonia are secreted into the distal tubule from the blood.

—together accomplish the kidneys' function of urine excretion (Figs. 16-5 and 16-6).

Glomerular filtration. Filtration, the first step in the formation of urine, takes place from the blood in the glomeruli out into the Bowman's capsules. Water and solutes filter out of the glomeruli even faster than out of ordinary capillaries. At least two structural features make renal corpuscles especially effective filtration membranes. For one thing, glomeruli have many more pores than other capillaries. The fact that the efferent arteriole is smaller in diameter than the afferent arteriole makes for a higher resistance to blood flow out of glomeruli than out of other capillaries and, therefore, for a higher blood pressure (hydrostatic pressure, that is) in glomeruli than in ordinary capillaries. Glomerular hydrostatic pressure, for instance, averages about 70 mm Hg, whereas capillary hydrostatic pressure averages only about 30 mm Hg.

Fluid moves out of the glomeruli into the Bowman's capsules for the same reason that it moves out of other capillaries into interstitial fluid or moves from any area to another —because a pressure gradient exists between the two areas. Normally, glomerular hydrostatic pressure, blood colloidal osmotic pressure, and capsular hydrostatic pressure together determine the pressure gradient (commonly spoken of as the *effective filtration pressure* or EFP) between glomeruli and capsules. Of these three pressures, the main driving force, the main determinant of the effective filtration pressure, is glomerular hydrostatic pressure. It tends to move fluid out of the glomeruli. In contrast, the *capsular* hydrostatic pressure and the blood colloidal osmotic pressure both exert force in the opposite direction. (If you do not recall why osmotic pressure is a "water-pulling" rather than a "water-pushing" force, reread p. 27.) Suppose that there is a glomerular hydrostatic pressure of 70 mm Hg and that it is opposed by a capsular hydrostatic pressure of 20 and a blood colloidal osmotic pressure of 30. There would then be a net or effective filtration pressure of 20 mm Hg (70 − 20 − 30). In other words, there would be a pressure gradient of 20 mm Hg, causing fluid to filter out of the glomeruli into the capsules.

Another factor, *capsular* colloidal osmotic pressure, operates when disease has increased glomerular permeability enough to allow blood protein molecules to diffuse out of the blood into Bowman's capsule. Under these circumstances the capsular filtrate exerts an osmotic pressure in opposition to blood osmotic pressure. Capsular osmotic pressure tends to draw water out of the blood and so constitutes another force to be added to glomerular hydrostatic pressure in determining the effective filtration pressure. By using the following formula and figures given in Table 16-2, you will be able to figure out for yourself how changes in the various pressures cause changes in the glomerular filtration rate.

$$\text{Effective filtration pressure} = \left(\text{Glomerular hydrostatic pressure} + \text{Capsular osmotic pressure}\right) - \left(\text{Glomerular osmotic pressure} + \text{Capsular hydrostatic pressure}\right)$$

For example, glomerular hydrostatic pressure may decrease sharply under stress conditions such as following severe hemorrhage. Table 16-2 gives pressures that might exist under these circumstances. Using these figures, what do you compute the effective glomerular filtration pressure to be? Do you think it is true, after doing this computation, that glomerular filtration ceases entirely when glomerular hydrostatic pressure falls below a certain critical level?

Kidney disease, as previously noted, sometimes causes a loss of blood proteins in the urine. Blood protein concentration then decreases and causes blood colloidal osmotic pressure to decrease. Suppose that blood colloidal osmotic pressure were to decrease to 20 mm Hg while the other normal figures given in Table 16-2 remained the same. Use these figures to compute the effective filtration pressure. If, based on this example, you were to formulate a principle, would you say that a decrease in blood protein concentration tends to increase or decrease the glomer-

Table 16-2

Normal and abnormal glomerulocapsular pressures

	Hydrostatic pressure	Colloidal osmotic pressure
Normal		
Glomerular blood	70 mm Hg	30 mm Hg
Capsular filtrate	20 mm Hg	—
Abnormal		
Glomerular blood	44 mm Hg	28 mm Hg
Capsular filtrate	20 mm Hg	4 mm Hg

ular filtration rate? (Check your answer in the paragraph after the next one.)

Glomerular hydrostatic pressure is regulated by mechanisms that change the size of the afferent and efferent arterioles and is also influenced by changes in systemic blood pressure. For example, sympathetic impulses cause constriction of both afferent and efferent arterioles. But with intense sympathetic stimulation, the afferent arteriole becomes much more constricted than the efferent. Consequently, glomerular hydrostatic pressure falls. Sometimes under severe stress conditions, it drops to a level too low to maintain filtration, and the kidneys "shut down" completely. In more technical language, renal suppression occurs (for example, see Table 16-2). Glomerular hydrostatic pressure and filtration are directly related to systemic blood pressure. By this, we mean that an increase in blood pressure tends to produce an increase in glomerular pressure and in the filtration rate. The converse is also true.

Glomerular filtration is inversely related to blood colloidal osmotic pressure and blood protein concentration. Specifically, a decrease in blood protein concentration causes a decrease in blood osmotic pressure, which in turn causes an increase in glomerular filtration. Normally the glomerular filtration rate averages about 125 ml per minute in men and somewhat less in women.

Tubular reabsorption (Fig. 16-6). The second step in urine formation is reabsorption of substances needed by the body—that is, of most (usually from 97% to 99%) of the water and part of the solutes from the glomerular filtrate back into the blood. Reabsorption is the function of the cells composing the walls of the convoluted tubules, loop of Henle, and collecting tubules. In the execution of this function these cells display astonishing powers of selection and discrimination. For example, they absorb most efficiently the substances that the body needs most vitally, such as Na^+, Cl^-, HCO_3^-, H_2O, and glucose.

Reabsorption is not merely a physical matter of diffusion and osmosis but consists also of somewhat obscure active transport mechanisms that require energy release by tubule cells. Because reabsorption is achieved partly by active transport mechanisms, it is one of the first powers diminished in kidney disease. Practical use is made of this fact in tests that measure the concentration of urine at different times of the day and thereby evaluate the active transport powers of the tubules.

Let us consider first the mechanism for reabsorption of solutes. Active transport mechanisms absorb glucose, amino acids, and other nutrient substances and at least some electrolytes (salts). Proximal tubule cells transport glucose and other nutrients. So these vitally important materials, after filtering out of the glomeruli, return to the blood from the first part of the renal tubule. The transport mechanisms function so effectively that normally no glucose at all is lost in the urine (Fig. 16-5). If, however, blood glucose exceeds a certain threshold amount (often around 150 mg per 100 ml of blood), the glucose transport mechanism cannot reabsorb all of it, and the excess remains in the urine. In other words, this mechanism has a maximum capacity for moving glucose molecules back into the blood. Occasionally this capacity is greatly reduced, in which case glucose appears in the urine (glycosuria or glucosuria), even though the blood sugar may be normal. This condition is known as renal diabetes or renal glycosuria.

Electrolytes are reabsorbed partly by active transport mechanisms and partly by diffusion. Sodium ions, for example, are known to be actively transported from all parts of the tubule. This has important effects on the reabsorption of certain other ions and on water reabsorption. Sodium ions are positive ions (cations). Therefore, when they diffuse from the proximal tubules into the peritubular blood, the latter momentarily becomes electropositive to the tubular filtrate. This attracts negative ions (notably chloride) and causes an equal number of them to diffuse out of the tubule. Sodium movement out of the tubule also produces a momentary disequilibrium between the tubular filtrate and peritubular blood osmotic pressures. It subtracts sodium ions from the filtrate but adds them to the blood—hence, the osmotic pressure of peritubular blood increases above the level of the filtrate osmotic pressure. Water, obeying the law of osmosis, follows rapidly along to reestablish osmotic equilibrium between the filtrate in the proximal tubules and blood. In short, the active transport of sodium out of the tubule is the main factor causing osmosis of water out of it. Therefore, the water reabsorption from the proximal tubule by this mechanism has been described as "obligatory water reabsorption"—obligatory because it is demanded by the law of osmosis. The distal and collecting tubules, in contrast, have the ability or faculty to vary the volume of water they reabsorb. Hence the term "facultative water reabsorption" indicates a function of distal and collecting tubules.

Some 20 years ago an ingenious hypothesis was first proposed to explain the mechanism by which the kidneys' tubules were able to produce either a concentrated or a dilute urine (that is, a urine that was hyperosmotic or hyposmotic to blood plasma). The hypothesis is known as the *countercurrent mechanism.* As fluid moves down the thin descending limb (Figs. 16-4 and 16-6), it becomes hyperosmotic to plasma. The countercurrent hypothesis offers this explanation: Sodium is actively transported out of the fluid in the ascending limb into interstitial fluid, but water does not follow the sodium because the ascending limb is almost impermeable to water. Hence the interstitial fluid between the ascending and descending limbs (in the kidney's medulla) becomes hyperosmotic to the fluid in both ascending and descending limbs. As a result, a net diffusion of sodium takes place from the interstitial fluid into the descending limb. Consequently the fluid in the descending limb becomes hyperosmotic while that in the ascending limb becomes hyposmotic. If antidiuretic hormone (ADH) is not present, the tubular fluid remains hyposmotic as it moves from the ascending limb and through the distal and collecting tubules, and it is excreted as hyposmotic, or in other words dilute urine. If, on the other hand, ADH is present, a concentrated urine is excreted.

This is how it happens. ADH increases the distal tubule's permeability to water. Therefore, as sodium is actively transported out of distal tubule fluid into blood, water osmoses out of the distal tubule fluid into the interstitial fluid, which is hypertonic to blood. This causes the distal tubule fluid to become isosmotic to the interstitial fluid, which is hypertonic to blood. Hence, the fluid leaving the distal tubules—urine—is also hypertonic to blood. In short, a concentrated urine is excreted.

The name antidiuretic hormone is appropriate because this hormone works against diuresis (large volume of urine). In other words, ADH decreases the amount of urine produced and, as explained in the preceding paragraphs, it also acts to concentrate it. How important the ADH-tubule mechanism is can be appreciated from the fact that water balance depends upon it and cannot be maintained if it fails.

Certain adrenal cortex hormones, classified as mineralocorticoids (M-C's), regulate the reabsorption of electrolytes—of ions, that is. Of the natural mineralocorticoids, aldosterone is the most potent. See p. 282 for a discussion of its action.

Tubular secretion. In addition to reabsorption, tubule cells also secrete certain substances. Tubular secretion (or tubular excretion, as it is also called) means the movement of substances out of the blood into the filtrate in the kidney tubules. Tubular reabsorption, you recall, means the movement of substances in the opposite direction—that is, out of the tubule filtrate into the blood. In both reabsorption and secretion, some substances are moved by active transport and some by passive mechanisms (diffusion and osmosis). For example, sodium ions and glucose molecules are reabsorbed and potassium and hydrogen ions are secreted by active transport mechanisms. But water is reabsorbed by osmosis and ammonia is secreted by diffusion, both passive processes. Tubule cells also secrete certain drugs—for example, penicillin and para-aminohippuric acid (PAH), two clinically important substances. By administering a known amount of PAH and then measuring the amount excreted in the urine, tubular secretion can be evaluated. If the tubules are impaired, they, of course, cannot move as much PAH out of the blood, and less appears in the urine within a given time.

The reabsorption and secretion mechanisms are summarized in Table 16-3. Table 16-4 and Fig. 16-6 summarize the functions of the different parts of the nephron in forming urine.

Table 16-3

Reabsorption and secretion mechanisms (for moving substances out of and into renal tubules)

Substance	Moved out of	By mechanism of	Into
Electrolytes, e.g.: Sodium ions	Blood in glomeruli	*Filtration*	Bowman's capsules
	Filtrate in tubules (all parts)	*Active transport* (reabsorption)	Blood in peritubular capillaries
Potassium ions	Blood in glomeruli	*Filtration*	Bowman's capsules
	Blood in peritubular capillaries	*Active transport* (secretion)	Filtrate in tubules
Chloride ions	Blood in glomeruli	*Filtration*	Bowman's capsules
	Filtrate in tubules	*Diffusion* (secondary to Na^+-transport)	Blood in peritubular capillaries
Nutrients, e.g.: Glucose	Blood in glomeruli	*Filtration*	Bowman's capsules
	Filtrate in proximal tubules	*Active transport* (reabsorption)	Blood in peritubular capillaries
Wastes, e.g.: Urea (most abundant solute in urine)	Blood in glomeruli	*Filtration*	Bowman's capsules
	Filtrate in tubules	*Diffusion*	Blood in peritubular capillaries
Water	Blood in glomeruli	*Filtration*	Bowman's capsules
	Filtrate in proximal tubules	*Osmosis* (obligatory water reabsorption)	Blood in peritubular capillaries
	Filtrate in distal and collecting tubules	*Osmosis* (facultative water reabsorption; ADH-controlled)	Blood in peritubular capillaries
Hydrogen ions	Blood in peritubular capillaries	*Active transport* (secretion)	Filtrate in distal tubules
Ammonia	Distal tubule cells	*Diffusion*	Filtrate in distal tubules

Table 16-4

Functions of different parts of nephron in urine formation

Part of nephron	Function	Substance moved
Glomerulus	Filtration	Water All solutes except colloids such as blood proteins
Proximal tubule and loop of Henle	Reabsorption by active transport	Na^+ and probably some other ions; nutrients—glucose and amino acids
	Reabsorption by diffusion (secondary to active transport)	Cl^-, HCO_3^-, and probably some other ions
	Obligatory water reabsorption by osmosis	Water
Distal and collecting tubules	Reabsorption by active transport	Na^+ and probably some other ions
	Facultative water reabsorption by osmosis (ADH-controlled)	Water
	Secretion by diffusion	Ammonia
	Secretion by active transport	K^+, H^+, and some drugs

MECHANISMS THAT CONTROL VOLUME OF URINE EXCRETED

The amount of urine excreted is regulated chiefly by ADH and aldosterone, hormones that influence the amount of water reabsorbed by the distal tubule cells. It is thought that only under pathological conditions does the rate of glomerular filtration change enough to alter the volume of urine produced.

The total volume of extracellular fluid is known to influence the volume of urine secreted, presumably by increasing or decreasing ADH secretion. The exact mechanism however, by which it accomplishes this is still unproved. It may operate through its effect on ECF sodium concentration. An increase in ECF volume tends to decrease blood sodium concentration, and a decrease in ECF volume tends to increase blood sodium concentration. For the effect of the latter on ADH secretion, see Fig. 20-9 (p. 519). The common experience of increased output following rapid ingestion of a large amount of fluid attests to the fact that an increased ECF volume is soon followed by increased urinary output. Evidence of the second principle is the oliguria or even anuria in dehydrated patients.

Another factor that helps regulate the amount of urine secreted is the total amount of solutes excreted through the kidneys. The more solutes to be excreted, the greater the volume of urine. In diabetes, for example, more solids are excreted than normal because of the excess glucose that "spills over" into the urine. Therefore the volume of urine secreted daily is greater than in the normal individual.

INFLUENCE OF KIDNEYS ON BLOOD PRESSURE

Clinical observation and animal experiments have established the fact that destruction of a large proportion of total kidney tissue usually results in the development of hypertension. This happens frequently, for example, in patients who have severe renal arteriosclerosis or glomerulonephritis. Many experiments have been performed and various theories devised to explain the mecha-

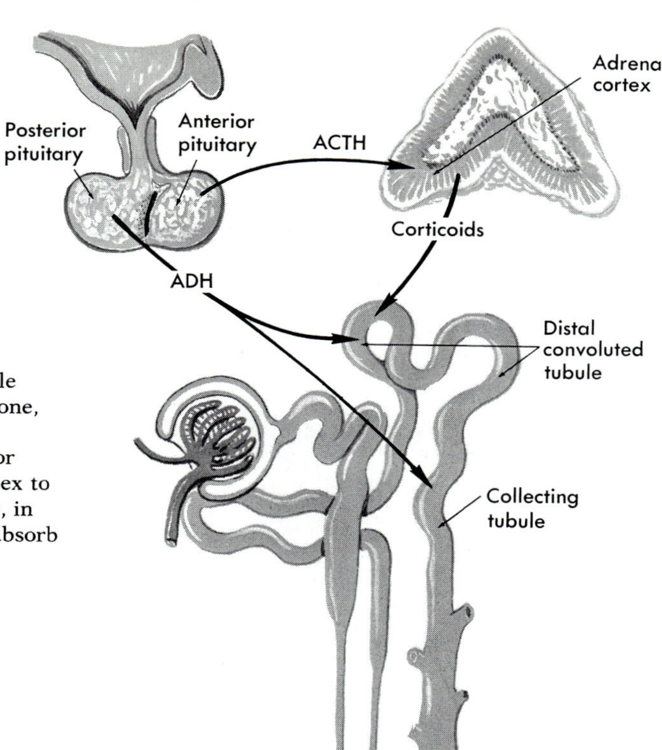

Fig. 16-7

ADH and corticoid control of distal renal tubule reabsorption. *ADH,* a posterior pituitary hormone, increases water reabsorption from distal and collecting tubules into blood. *ACTH,* an anterior pituitary hormone, stimulates the adrenal cortex to secrete chiefly glucocorticoids. Glucocorticoids, in turn, mildly stimulate the distal tubules to reabsorb sodium and excrete potassium more rapidly.

nism responsible for "renal hypertension." Ischemic kidneys are known to produce a proteolytic enzyme, *renin,* which hydrolyzes one of the blood proteins to produce *angiotensin* (Fig. 11-12). Angiotensin causes arteriolar constriction and a rise in blood pressure.

Ureters

Location and structure

The two ureters are tubes from 10 to 12 inches long. At their widest point they measure less than ½ inch in diameter. They lie behind the parietal peritoneum and extend from the kidneys to the posterior surface of the bladder. As the upper end of each ureter enters the kidney, it enlarges into a funnel-shaped basin named the *renal pelvis.* The pelvis expands into several branches called *calyces.* Each calyx contains a renal papilla. As urine is secreted, it drops out of the collecting tubules, whose openings are in the papillae, into the calyces, then into the pelvis, and down the ureters into the bladder.

The walls of the ureters are composed of three coats: a lining coat of mucous membrane, a middle coat of two layers of smooth muscle, and an outer fibrous coat.

Where the ureters empty into the bladder, there is a fold of mucous membrane that serves as a valve preventing the backflow of urine into the ureter when the bladder contracts.

Function

The ureters, together with their expanded upper portions, the pelves and calyces, collect the urine as it forms and drain it into the bladder. Peristaltic waves (about one to five per minute) force the urine down the ureters and into the bladder.

Correlations

Stones known as *renal calculi* sometimes develop within the kidney. Urine may wash them into the ureter, where they cause extreme pain if they are large enough to distend its walls.

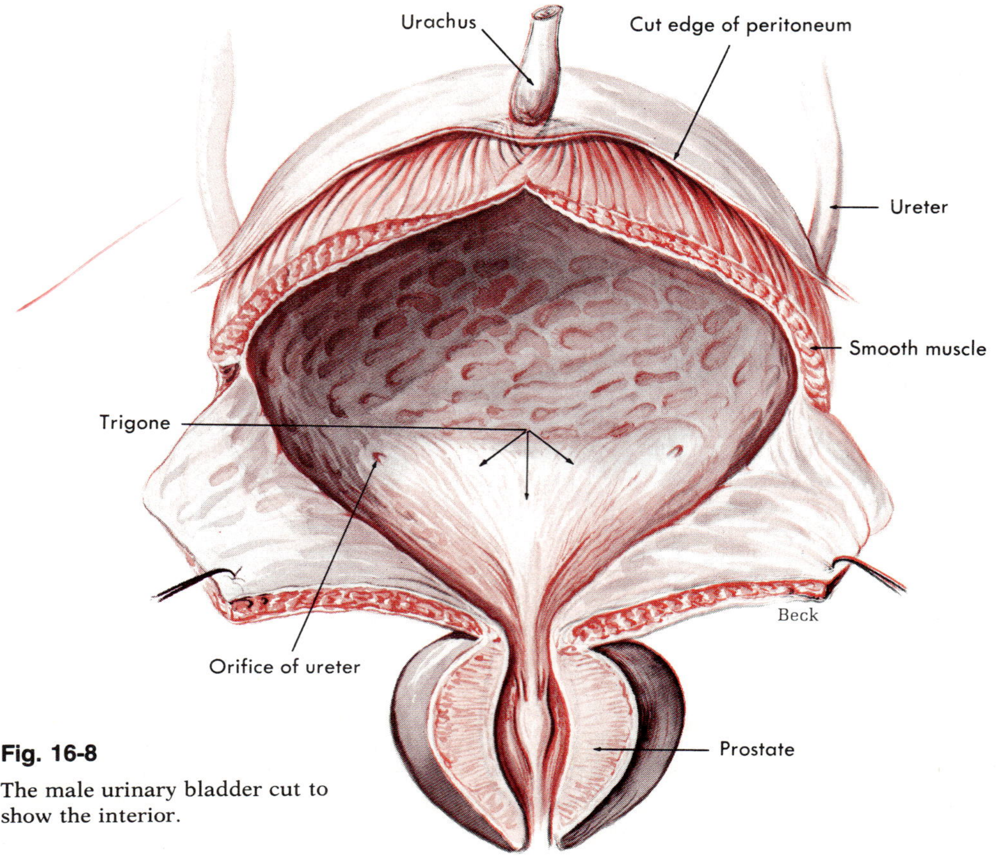

Fig. 16-8

The male urinary bladder cut to show the interior.

Bladder

Structure and location

The bladder is a collapsible bag located directly behind the symphysis pubis. It lies below the parietal peritoneum, which covers only its superior surface. Three layers of smooth muscle (known collectively as the detrusor muscle) fashion its walls, while mucous membrane, arranged in rugae, forms its lining. Because of the rugae and the elasticity of its walls, the bladder is capable of considerable distention, although its capacity varies greatly with individuals. There are three openings in the floor of the bladder—two from the ureters and one into the urethra. The ureter openings lie at the posterior corners of the triangular-shaped floor (the trigone) and the urethral opening at the anterior and lower corner (Fig. 16-8).

Functions

The bladder performs two functions.

1. It serves as a reservoir for urine before it leaves the body.
2. Aided by the urethra, it expels urine from the body.

Distention of the bladder with urine stimulates stretch receptors in the bladder wall. This initiates reflex contraction of the bladder wall but simultaneous relaxation of the internal sphincter, followed rapidly by relaxation of the external sphincter and emptying of the bladder. Parasympathetic fibers transmit the impulses that cause contractions of the bladder and relaxation of the internal sphincter. Voluntary contraction of the external sphincter, however, to prevent or terminate micturition is learned. Voluntary control of micturition is possible only if the nerves supplying the bladder and urethra, the projection tracts of the cord and brain, and the motor area of the cerebrum are all intact. Injury to any of these parts of the nervous system, by a cerebral hemorrhage or cord injury, for example, results in involuntary emptying of the bladder at intervals. Involuntary micturition is called *incontinence*. In the average bladder, 250 ml of urine will cause a moderately distended sensation and therefore the desire to void.

Occasionally an individual is unable to void even though the bladder contains an excessive amount of urine. This condition is known as *retention*. It often follows pelvic operations and childbirth. Catheterization (introduction of a rubber tube through the urethra into the bladder to remove urine) is used to relieve the discomfort accompanying retention. A more serious complication, which is also characterized by the inability to void, is called *suppression*. In this condition, the patient cannot void because the kidneys are not secreting any urine, and therefore the bladder is empty. Catheterization, of course, gives no relief for this condition.

Urethra

Structure and location

The urethra is a small tube leading from the floor of the bladder to the exterior. In the female it lies directly behind the symphysis pubis and anterior to the vagina. It extends up, in, and back for a distance of about 1 to 1½ inches. The male urethra follows a tortuous course for a distance of approximately 8 inches. Immediately below the bladder, it passes through the center of the prostate gland, then between two sheets of white fibrous tissue connecting the pubic bones, and then through the penis, the external male reproductive organ. These three parts of the urethra are known, respectively, as the prostatic portion, the membranous portion, and the cavernous portion.

The opening of the urethra to the exterior is named the *urinary meatus*.

Mucous membrane lines the urethra as well as the rest of the urinary tract.

Functions

Since it is the terminal portion of the urinary tract, the urethra serves as the passageway for eliminating urine from the body. In addition, the male urethra is the terminal portion of the reproductive tract and serves as the passageway for eliminating the reproductive fluid (semen) from the body. The female urethra serves only the urinary tract.

Urine

Physical characteristics

The physical characteristics of normal urine are listed in Table 16-5.

Chemical composition

Urine is approximately 95% water, in which are dissolved several kinds of substances, the most important of which are as follows:

1. *Nitrogenous wastes* from protein metabolism—such as urea (most abundant solute in urine), uric acid, ammonia, and creatinine
2. *Electrolytes*—mainly the following ions: sodium, potassium, ammonium, chloride, bicarbonate, phosphate, and sul-

Table 16-5

Physical characteristics of normal urine

Amount (24 hours)	Three pints (1,500 ml) but varies greatly according to fluid intake, amount of perspiration, and several other factors
Clearness	Transparent or clear; upon standing, becomes cloudy
Color	Amber or straw-colored; varies according to amount voided—less voided, darker the color, usually; diet also may change color—e.g., reddish color from beets
Odor	"Characteristic"; upon standing, develops ammonia odor from formation of ammonium carbonate
Specific gravity	1.015 to 1.020; highest in morning specimen
Reaction	Acid but may become alkaline if diet is largely vegetables; high-protein diet increases acidity; stale urine has alkaline reaction from decomposition of urea forming ammonium carbonate; normal range for urine pH 4.8-7.5; average, about 6; rarely becomes more acid than 4.5 or more alkaline than 8

fate; amounts and kinds of minerals vary with diet and other factors

3 *Toxins*—during disease, bacterial poisons leave the body in the urine—an important reason for "forcing fluids" on patients suffering with infectious diseases, so as to dilute the toxins that might damage the kidney cells if they were eliminated in a concentrated form
4 *Pigments*
5 *Hormones*
6 Various *abnormal constituents* sometimes found in urine—such as glucose, albumin, blood, casts, calculi, etc.

Definitions

1 *Glycosuria,* or *glucosuria*—sugar (glucose) in the urine
2 *Hematuria*—blood in the urine
3 *Pyuria*—pus in the urine
4 *Casts*—substances, such as mucus, that harden and form molds inside the uriniferous tubules and then are washed out into the urine; microscopic in size
5 *Dysuria*—painful urination
6 *Polyuria*—unusually large amounts of urine
7 *Oliguria*—scanty urine
8 *Anuria*—absence of urine

Outline summary

ORGANS

Kidneys

Gross anatomy

1 Size, shape, and location
  a 4½ in × 2 in × 3 in × 1 in
  b Shaped like lima beans
  c Lie against posterior abdominal wall, behind peritoneum, at level of last thoracic and first three lumbar vertebrae; right kidney slightly lower than left
2 External structure
  a Hilum, concave notch on mesial surface
  b Enveloping capsule of white fibrous tissue
3 Internal structure
  a Outer layer called cortex
  b Inner portion called medulla
  c Renal pyramids triangular wedges of medullary substance, apices of which called papillae
  d Renal columns inward extensions of cortex between pyramids

Microscopic anatomy

1 Cluster of capillaries invaginated in Bowman's capsule called *glomerulus*
2 Bowman's capsule together with glomerulus constitute renal (malpighian) corpuscle
3 Physiological unit of kidney called nephron—consists of renal corpuscle, convoluted tubules, loop of Henle, and straight tubule

Physiology

1 Functions
  a Excrete urine, by which various toxins and metabolic wastes excreted and composition and volume of blood regulated
  b Influence blood pressure
2 How kidneys excrete urine (see Table 16-3 and Fig. 16-6)
  a Filtration of substances from blood in glomeruli into Bowman's capsules

b Reabsorption of most of water and part of solutes from tubular filtrate into blood
c Secretion of K^+, H^+, NH_3, and some other substances into tubular filtrate from blood

3 Mechanisms that control volume of urine excreted are factors that change
a Amount of water reabsorbed from tubular filtrate into blood—most important determinant of amount of urine formed; posterior pituitary ADH increases water reabsorption from distal and collecting tubules; corticoids (especially aldosterone) increase sodium reabsorption and therefore also increase water reabsorption
b Rate of filtration from glomeruli—normally quite constant at about 125 ml per minute; varies directly with changes in glomerular blood pressure—i.e., increased glomerular blood pressure tends to increase glomerular filtration rate and therefore urine volume and vice versa but glomerular blood pressure normally quite constant; decreased blood colloidal osmotic pressure tends to increase glomerular filtration and urine volume and vice versa
c Total extracellular fluid volume; urine output increases following increase in total ECF and decreases following decrease in ECF
d Amount of solutes excreted in urine; urine output increases when solutes increase

Ureters

1 Location and structure
a Lie retroperitoneally
b Extend from kidneys to posterior part of bladder floor
c Ureter expands as it enters kidney, becoming renal pelvis, which is subdivided into calyces, each of which contains renal papilla
d Walls of smooth muscle with mucous lining and fibrous outer coat

2 Function—collect urine and drain it into bladder

Bladder

1 Structure and location
a Collapsible bag of smooth muscle lined with mucosa
b Lies behind symphysis pubis, below parietal peritoneum
c Three openings—one into urethra and two into ureters

2 Functions
a Reservoir for urine
b Expels urine from body by way of urethra, called micturition, urination, or voiding; retention, inability to expel urine from bladder; suppression, failure of kidneys to form urine

Urethra

1 Structure and location
a Musculomembranous tube lined with mucosa
b Lies behind symphysis, in front of vagina in female
c Extends through prostate gland, fibrous sheet, and penis in male
d Opening to exterior called urinary meatus

2 Functions
a Female—passageway for expulsion of urine from body
b Male—passageway for expulsion of urine and of reproductive fluid (semen)

URINE

1 Physical characteristics—see Table 16-5

2 Chemical composition—consists of approximately 95% water in which are dissolved
a Wastes from protein metabolism (urea, uric acid creatinine, etc.)
b Mineral salts (sodium chloride main one but various others according to diet)
c Toxins—e.g., from bacteria
d Pigments
e Sex hormones
f Abnormal constituents—e.g., glucose, albumin, blood, casts, and calculi

3 Definitions
a Glycosuria—sugar in urine
b Hematuria—blood in urine
c Pyuria—pus in urine
d Casts—microscopic bits of substances that harden and form molds inside tubules
e Dysuria—painful urination
f Polyuria—excessive amounts of urine
g Oliguria—scanty urine
h Anuria—absence of urine

Review questions

1 What four organs are excretory organs?
2 Which organs eliminate wastes of protein metabolism? Of digestion? Of carbohydrate and fat metabolism?
3 Name, locate, and give main function(s) of each organ of the urinary system.
4 How far and in which direction must a catheter be inserted to reach the bladder in the female? In the male?
5 Describe the microscopic structure of the kidney.
6 Describe the mechanism of urine formation, relating each step to part of the nephron that performs it.

Situation: An artificial kidney consists of a device in which blood flows directly from a patient's body through cellophane tubing immersed in a dialyzing fluid that contains prescribed amounts of various electrolytes and other substances. The two columns below show the composition of one patient's blood and of the dialyzing fluid used for his treatment with the artificial kidney.

	Blood plasma (in coiled tube) mEq/L	Dialyzing fluid (around coiled tube) mEq/L
Na^+	142	126
K^+	5	5
Ca^{++}	5	0
Mg^{++}	3	0
Cl^-	103	110
HCO_3^-	27	25
$HPO_4^=$	2	0
$SO_4^=$	1	0
	mg/100 ml	mg/100 ml
Glucose	100	1,750
Urea	26	0
Uric acid	4	0
Creatinine	1	0

7 What body structures do you think the cellophane tube substitutes for?
8 Which, if any, of the above substances diffuse out of the blood into the dialyzing fluid? Give your reasons.
9 Which, if any, of the above substances diffuse into the blood from the dialyzing fluid? Give your reasons.
10 Net diffusion of which, if any, of the above substances does not occur through the cellophane membrane in either direction? Give your reasons.
11 What reasons can you see for having the dialyzing fluid contain the concentration of each substance given above? What does it accomplish?

Situation: Results of a blood urea clearance test indicate that the glomerular filtration rate of a patient who has had a severe hemorrhage is less than 50% of normal.

12 Explain the mechanism responsible for the drop in the glomerular filtration rate.
13 Do you consider this a homeostatic mechanism? Does it serve a useful purpose? If so, what purpose?
14 What is the normal glomerular filtration rate?
15 What would you expect to be true of the volume of urine this patient would excrete—normal, polyuria, oliguria?
16 Define the following terms briefly: Bowman's capsule, calculi, calyces, casts, cystitis, dysuria, glomerulus, glycosuria, hematuria, incontinence, nephritis, oliguria, polyuria, ptosis, pyelitis, renal capsule, renal cortex, renal hilum, renal medulla, renal papilla, renal pelvis, renal pyramids, retention, suppression.

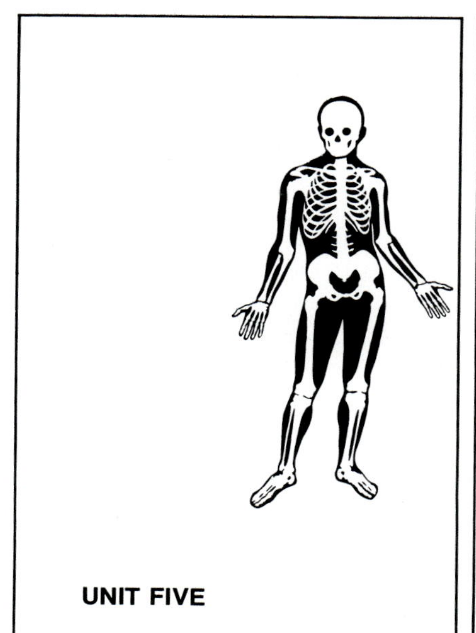

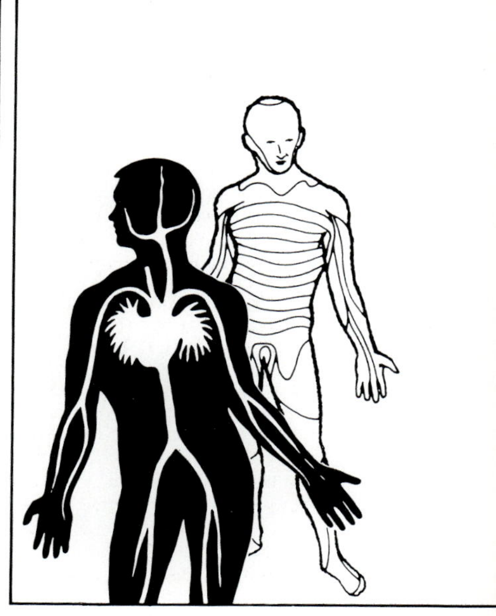

REPRODUCTION

17

Reproduction of cells

Deoxyribonucleic acid (DNA)
Mitosis
Meiosis
Spermatogenesis
Oogenesis

Cell reproduction is one of the most fundamental of all living functions. Upon it depend the creation of all individual organisms and the continued existence of all complex, multicellular organisms. The survival, therefore, of each one of us and of our human species depends upon cell reproduction. In addition, cell reproduction is one of the most fascinating stories that physiology has to tell. We shall start our version of it by telling about DNA—"the most golden molecule of them all," Watson called it.*

Deoxyribonucleic acid (DNA)

The deoxyribonucleic acid molecule is a giant among molecules. Both its size and the complexity of its shape exceed those of most molecules. The importance of its function—in a word, heredity—surpasses that of any other molecule in the world. A little more than 20 years ago, an American, James D. Watson, and two British scientists, Francis H. C. Crick and Maurice H. F. Wilkins, solved the puzzle of DNA's molecular structure. Watson and Crick announced their discovery with an understated, one-page report that appeared in the April 25, 1953, issue of the British journal *Nature*. Nine years later the coveted Nobel Prize in Medicine was awarded all three scientists for their brilliant and significant work—hailed as the greatest biological discovery of our times. Watson tells how they accomplished this in *The Double Helix*, now a classic, and a book you will probably find hard to put down once you start it.

To try to visualize the shape of the DNA molecule, picture to yourself an extremely long, narrow ladder made of a pliable material. Now see it twisting round and round on its axis and taking on the shape of a steep spiral staircase millions of turns long. This is the shape of the DNA molecule—a double helix (Greek word for spiral).

As to its structure, the DNA molecule is a polymer. This means that it is a large molecule made up of many smaller molecules joined together. DNA is a polymer of millions of pairs of nucleotides. A nucleotide is a compound formed by combining phosphoric acid with a sugar and a nitrogenous base. In the DNA molecule, each nucleotide consists of a phosphate group that attaches to the sugar deoxyribose that attaches to one of four bases—either adenine or thymine or cytosine or guanine. (Deoxyribose is a sugar that is not sweet and one whose molecules contain only five carbon atoms.) Notice what the name deoxyribonucleic acid tells you—that this compound contains deoxyribose, that it occurs in nuclei, and that it is an acid.

Fig. 17-1 reveals additional and highly significant facts about DNA's molecular structure. First, observe which compounds form the sides of the DNA spiral staircase—a long line of phosphate and deoxyribose units joined alternately one after the other. Look next at the stair steps. Notice two facts about them: that two bases join (by means of a hydrogen bond) to form each step, and that only two combinations of bases occur. The same two bases invariably pair off with each other in a DNA molecule—like teenagers "going steady." Adenine always goes with thymine (or vice versa, thymine with adenine), and guanine always goes with cytosine (or vice versa). This fact about DNA's molecular structure is called *obligatory base-pairing*. Pay particular attention to it, for it is the key to understanding how a DNA molecule is able to duplicate itself. DNA duplication, or replication as it is usually called, is one of the most important of all biological phenomena because it is an essential and crucial part of the mechanism of heredity.

Another fact about DNA's molecular struc-

*Watson, J. D.: The double helix, New York, 1969, The New American Library, p. 21.

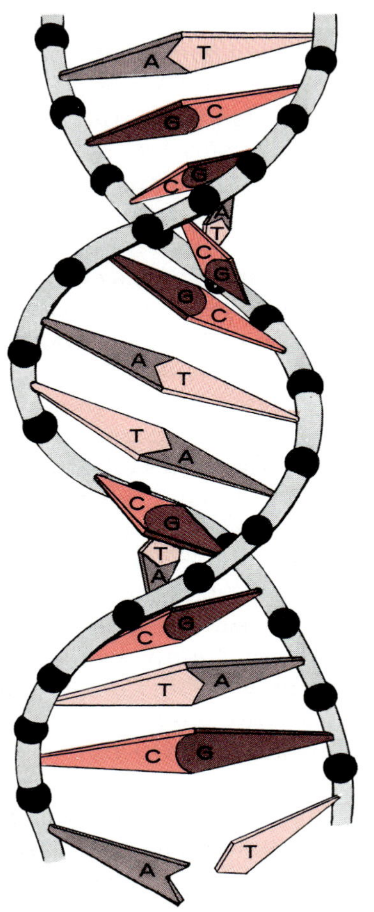

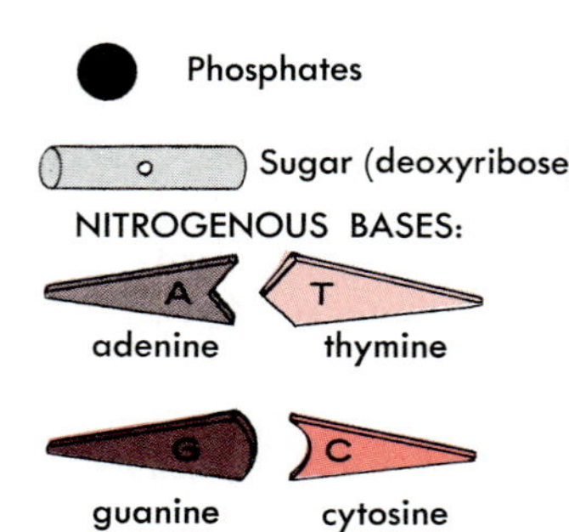

Fig. 17-1

Watson-Crick structure of DNA molecule. Note that each side of the DNA molecule consists of alternating sugar and phosphate groups. Each sugar group is united to the sugar group opposite it by a pair of nitrogenous bases, adenine-thymine or thymine-adenine and cytosine-guanine or guanine-cytosine. Differences in the sequences of base pairs establish the identity of the many different kinds of DNA.

ture that has great functional importance is the sequence of its base pairs. Although the base pairs in all DNA molecules are the same, the sequence of these base pairs is not the same in all DNA molecules. For instance, the sequence of the base pairs composing the seventh, eighth, and ninth steps of one DNA molecule might be cytosine-guanine, adenine-thymine, and thymine-adenine. In another DNA molecule, the sequence of the base pairs making up these same steps might be entirely different, perhaps thymine-adenine, guanine-cytosine, and cytosine-guanine. Perhaps these seem to be minor details. But nothing could be further from the truth, since it is the sequence of the base pairs in the nucleotides composing DNA molecules that identifies each gene. Hence, it is the sequence of base pairs that determines all hereditary traits.

A human *gene* is a segment of a DNA molecule. One gene consists of a chain of about 1,000 pairs of nucleotides joined one after the other in a precise sequence. One gene controls the production within the cell of one protein (either a structural protein or a functional protein—that is, an enzyme). Each enzyme catalyzes one chemical reaction. Therefore, as Fig. 17-2 indicates, the millions of genes constituting a cell's DNA determine both the cell's structure and its functions.

Genes control protein synthesis (anabolism). A brief synopsis of this process as now visualized follows. First, a strand of RNA (ribonucleic acid) forms along a segment of one strand of a DNA molecule. RNA differs from DNA in certain respects. Its molecules are smaller than those of DNA, and RNA contains ribose instead of deoxyribose. Also, one of the four bases in RNA is uracil instead of thymine. As a strand of RNA is forming along a strand of DNA, uracil attaches to adenine and guanine attaches to cytosine. The process is known as complementary pairing. Thus a

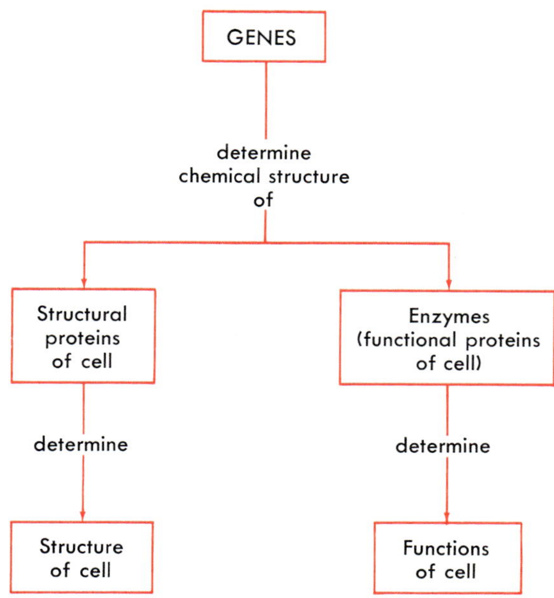

Fig. 17-2

Brief scheme to show how genes determine hereditary characteristics.

single strand molecule of messenger RNA (abbreviated mRNA) is formed. The name "messenger RNA" describes its function. As soon as it is formed, it separates from the DNA strand, diffuses out of the nucleus, and carries a "message" to a ribosome in the cell's cytoplasm, directing its synthesis of a specific protein. Ribosomes contain another type of RNA called appropriately "ribosomal RNA" (abbreviated rRNA). Next, a molecule of a third type of RNA—transfer RNA, or tRNA—present in cytoplasm attaches itself to a molecule of a specific amino acid and transfers it to a ribosome. Here the amino acid fits into the position indicated by the messenger RNA. More transfer RNA molecules, one after the other in rapid sequence, bring more amino acids to the ribosome and fit them into their proper positions. Result? A chain of amino acids joined to each other in a definite sequence—a protein, in other words—is formed. In short, protein anabolism has occurred. It is one of the major kinds of cellular work. One human cell is estimated to synthesize perhaps 2,000 different enzymes! In addition to this staggering work load, it produces many different protein compounds that help form its own structures and many cells also synthesize special proteins. Liver cells are a notable example; they synthesize prothrombin, fibrinogen, albumins, and globulins.

Mitosis

Human cells, other than sex cells, reproduce by the process of mitosis. In this process a cell divides in order to multiply. One cell divides to form two cells. Mitosis consists of a sequence of events plainly visible in suitably stained cells viewed with the light microscope. The events of mitosis occur in five phases: interphase, prophase, metaphase, anaphase, and telophase.

DNA replication occurs during the interphase of mitosis (cell reproduction). The tightly coiled DNA molecules uncoil except in small segments. Since these remaining tight little coils are denser than the thin elongated sections, they absorb more stain and appear as chromatin granules under the microscope. The thin uncoiled sections, in contrast, are invisible because they absorb so little stain. As the DNA molecule uncoils, its two strands come apart. Then, along each of the two separated strands of nucleotides, a complementary strand forms. Intracellular fluid contains many molecules of deoxyribose and nitrogenous bases and many phosphate ions. By the mechanism of obligatory base-pairing, these substances become attached at their right places along each DNA strand. Interpreted, this means that new thymine (that is, from the intracellular fluid) attaches to the "old" adenine in the original DNA strand. Conversely, new adenine attaches to old thy-

Table 17-1

Mitosis

Interphase	Prophase	Metaphase	Anaphase	Telophase
1 Period when cell prepares for division and also grows in size **2** Chromosomes elongate and become too thin to be visible as such, but chromatin granules become visible **3** DNA of each chromosome replicates itself, forming two chromatids attached only at centromere	**1** Chromosomes shorten and thicken (from coiling of DNA molecules that compose them); each chromosome consists of two *chromatids* attached at *centromere* **2** Centrioles move to opposite poles of cell; spindle fibers appear and, under control of centrioles, begin to orient between opposing poles	**1** Chromosomes align across equator of spindle fibers; each pair of chromatids attached to spindle fiber at its centromere **2** Nucleoli and nuclear membrane disappear	**1** Each centromere divides into two, thereby detaching two chromatids that compose each chromosome from each other **2** Divided centromeres start moving to opposite poles, each pulling its chromatid (now called a chromosome) along with it; there are now twice as many chromosomes as there were before mitosis started	**1** Changes occurring during telophase essentially reverse of those taking place during prophase; new chromosomes start elongating (DNA molecules start uncoiling) **2** Nucleoli and two new nuclear membranes appear, enclosing each new set of chromosomes **3** Spindle fibers disappear **4** *Cytokinesis*, or dividing of cytoplasm, usually occurs during telophase; starts as pinching in along equator of old cell and ends with division of old cell into two new cells **5** Centrioles replicate

mine. Also, new guanine joins old cytosine and, vice versa, new cytosine joins old guanine. By the end of the interphase, each of the two DNA strands of the original DNA molecules has a complete new complementary strand attached to it. Each half DNA molecule, in other words, has duplicated itself to create a whole new DNA molecule. Thus two new chromosomes now replace each original chromosome. However, at this stage (end of interphase, beginning of prophase), they are called chromatids instead of chromosomes. The two chromatids formed from each original chromosome contain duplicate copies of DNA and, therefore, the same genes as the chromosome from which they were formed. Chromatids are present as attached pairs. Centromere is the name of their point of attachment. By the time the parent cell divides to form two daughter cells, chromatids have separated and become chromosomes and one set of them has become part of the nucleus of one daughter cell and the other set has become part of the nucleus of the other daughter cell. Hence, each of the two new cells formed by mitosis contains a complete set of genes identical to those in its parent cell. Both daughter cells have the potentialities to become like their parent cell. Thus the function of mitosis is to enable cells to reproduce their own kind. By means of the mechanism of mitosis, a new generation of cells inherits both the structural and the functional characteristics of the preceding generation (Fig. 17-2). In short, mitosis is the mechanism of heredity for cells. Table 17-1 summarizes the events that characterize the five phases of mitosis. As you read the events, try to correlate them with the drawings in Fig. 17-3.

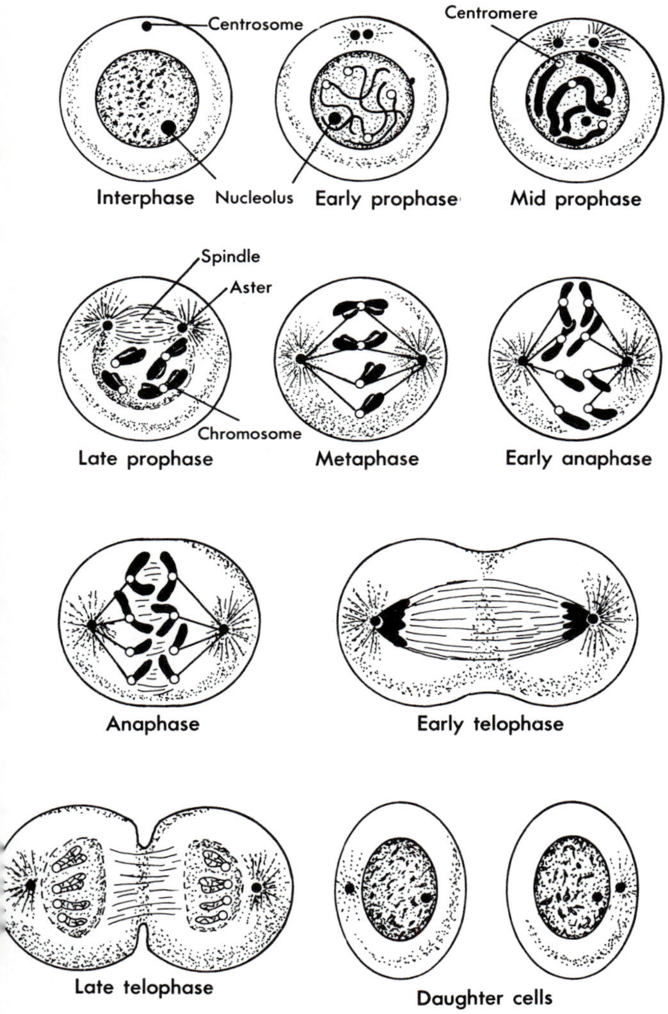

Fig. 17-3

Diagram of mitosis in an animal cell with four chromosomes. (From Levine, Louis: Biology of the gene, ed. 2, St. Louis, 1973, The C. V. Mosby Co.)

Meiosis

Meiosis is the type of cell division that occurs only in primitive sex cells during the process of their becoming mature sex cells. The scientific name of the primitive male sex cells that reproduce by meiosis is *primary spermatocytes. Primary oocytes* is the name of the primitive female sex cells. The nucleus of each primary spermatocyte and oocyte contains 23 pairs of chromosomes. This total of 46 chromosomes per cell is known as the *diploid* number of chromosomes for human cells. The scientific name for either a male or female mature sex cell is gamete. Male gametes are named spermatozoa but are usually called sperm. Female gametes are named ova (singular, ovum).

Meiosis consists of two cell divisions that take place one after the other in quick succession. They are referred to as meiotic division I and meiotic division II, and in both, an interphase, prophase, metaphase, anaphase, and telophase occur. In the interphase that precedes prophase I (of meiotic division I), the same events occur as take place in the interphase of mitosis. Specifically, the DNA of each chromosome replicates itself and thereby changes each chromosome into two chromatids, attached only at the centromere. Prophase I of meiosis consists of several stages. You can find the names of these stages in Fig. 17-4. For simplicity's sake, only two chromosomes—a black one and a white one—are shown in this figure. Look closely at the diplotene stage and you will notice that a segment of one of the chromatids of the white chromosome and a segment of one of the chromatids of the black chromosome have exchanged places. A chromatid segment of each chromosome has crossed over and become part of the other chromosome. This is a highly significant event. Since each chromatid segment consists of specific genes, the crossing-over of chromatids reshuffles the genes, so to speak

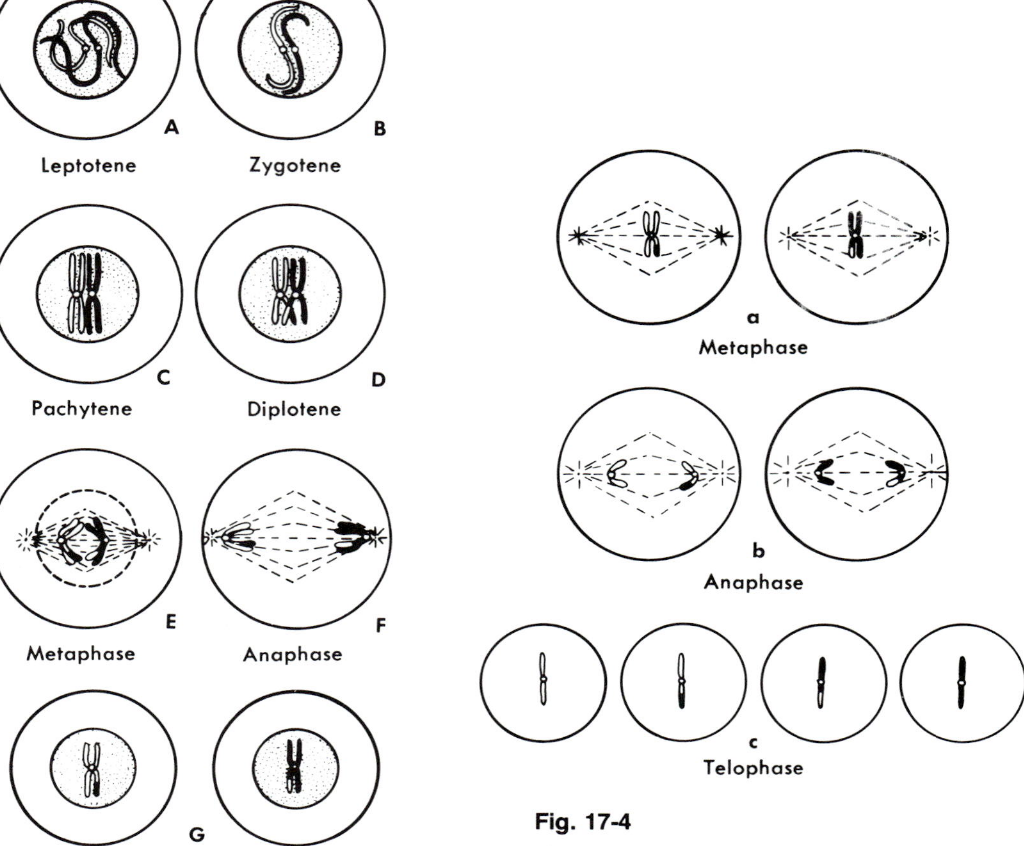

Fig. 17-4

The major stages of meiosis. *A* through *G* indicate the first meiotic division; *a* through *c* indicate the second meiotic division culminating in four daughter cells, each containing half the original number of chromosomes. (From Chinn, P. L.: Child health maintenance, St. Louis, 1974, The C. V. Mosby Co.)

—it transfers some of them from one chromosome to another.

Metaphase I follows the last stage of prophase I and, as in mitosis, the chromosomes align themselves along the equator of the spindle fibers as Fig. 17-4 shows. But in anaphase, the two chromatids that make up each chromosome do not separate from each other as they do in mitosis to form two new chromosomes out of each original one. Therefore in anaphase I of Fig. 17-4, only one chromosome moves to each pole of the parent cell. When the parent cell divides, to form two cells, each daughter cell contains only one chromosome or half as many as the parent cell had. The daughter cells formed by meiotic division I contain a *haploid* number of chromosomes, which means half as many as the diploid number present in the parent cell.

The right side of Fig. 17-4 shows meiotic

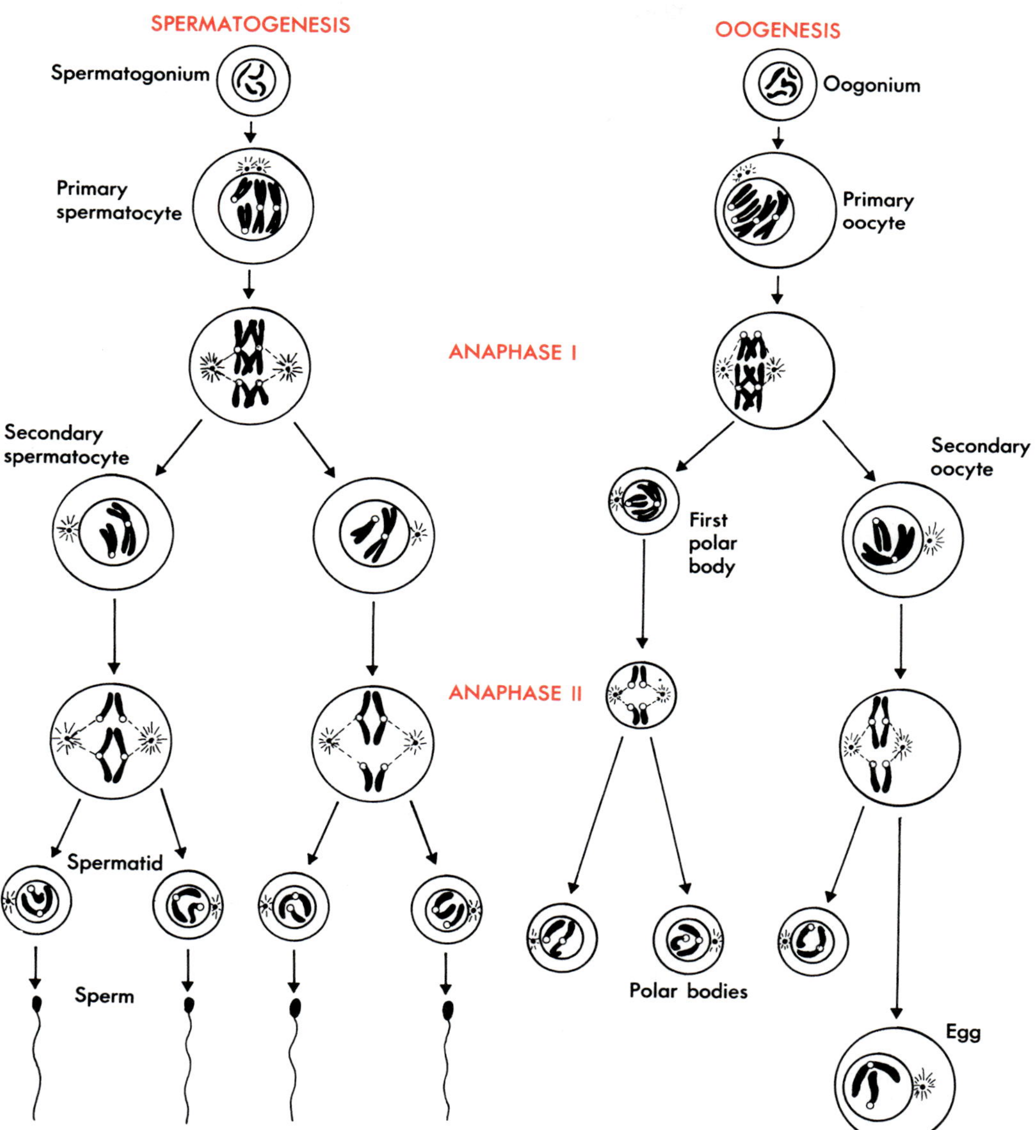

Fig. 17-5

Diagram representing the meiotic sequence in male and female animals. Left: Process of spermatogenesis, resulting in the formation of four sperm. Right: Oogenesis, resulting in the formation of one egg and three polar bodies. (From Levine, Louis: Biology of the gene, ed. 2, St. Louis, 1973, The C. V. Mosby Co.)

division II; it is essentially the same as mitosis. Notice that it reproduces each of the two cells formed by meiotic division I and so forms four cells, each with the haploid number of chromosomes.

Spermatogenesis

Spermatogenesis is the process by which the primitive sex cells (spermatogonia) present in the seminiferous tubules of a newborn baby boy become transformed into mature sperm. Spermatogenesis starts at about the time of puberty and usually continues throughout a man's life. Fig. 17-5 diagrams some of the major steps of spermatogenesis. Meiotic division I reproduces one primary spermatocyte to form two secondary spermatocytes, each with a haploid number of chromosomes (23). One of the 23 chromosomes is either an X or a Y sex chromosome. Meiotic division II then reproduces each of the two secondary spermatocytes to form a total of four spermatids. Thus spermatogenesis forms four sperm—each with only 23 chromosomes—from one primary spermatocyte that has 23 pairs or 46 chromosomes.

Oogenesis

Oogenesis is the process by which primitive female sex cells (oogonia) become mature ova. During the fetal period, oogonia reproduce in the ovaries by mitosis to form primary oocytes—about a half million of them by the time a baby girl is born. Most of the primary oocytes develop to prophase I before birth. There they stay until puberty. Then the other steps of oogenesis shown in Fig. 17-4 take place about once a month for the next 30 or 40 years, that is, until a woman reaches her menopause. Note that oogenesis forms only one mature ovum (egg) from each primitive oocyte. How many sperm are formed from each primary spermatocyte by spermatogenesis? An ovum, like a sperm, contains only 23 chromosomes, but unlike sperm, half of which have an X and half of which a Y sex chromosome, all ova have an X chromosome. This distinction between sperm and ovum makes a difference—the difference between whether the new life formed by their union will be male or female. If a sperm with an X chromosome fertilizes an ovum, the baby will be a girl. If a Y-bearing sperm fertilizes an ovum, it will be a boy. Chapter 19 will give more information about this.

Outline summary

Deoxyribonucleic acid (DNA)

1 Structure—see Fig. 17-1
2 Relation to genes
  a One gene is a segment of a DNA molecule, consisting of a chain of about 1,000 pairs of nucleotides joined in a precise sequence
  b DNA of one human chromosome made up of about 175,000 genes according to one cytologist's estimate
3 Replication—DNA of each chromosome duplicates itself during interphase of mitosis, thereby forming two chromatids out of each chromosome
4 Function—DNA replication makes possible cellular heredity

Mitosis

1 Definition—process by which somatic cells reproduce
2 Phases—interphase, prophase, metaphase, anaphase, and telophase; see Table 17-1 for summary of events during each phase
3 Function—transmits to daughter cells the same (diploid) number of chromosomes composed of same genes as present in parent cell and thereby makes possible inheritance of parent cell's traits

Meiosis

1 Definition—process by which primitive sex cells reproduce in the process of becoming mature sex cells (gametes)
2 Stages
  a Meiotic division I—consists of interphase, prophase I, metaphase I, anaphase I, and telophase I; primary sex cell reproduces to form two secondary sex cells containing haploid number of chromosomes, i.e., 23
  b Meiotic division II—consists of same phases as meiotic division I—4 sperm or 1 ovum result from this division; see Fig. 17-5
3 Function—formation of gametes with haploid number of chromosomes

Spermatogenesis

Process by which spermatogonia present at birth in seminiferous tubules undergo meiosis to become mature sperm; spermatogenesis begins at puberty and usually continues throughout a man's life

Oogenesis

Process by which oogonia become mature ova; oogonia in ovaries of fetus reproduce by mitosis to form primary oocytes at prophase I stage; beginning at puberty, and recurring about once a month for the next 30 or 40 years, remaining steps of oogenesis take place as shown in Fig. 17-4. Ovum contains 23 chromosomes, one of which is an X chromosome; sperm also contain 23 chromosomes but one of them is either an X or a Y sex chromosome

Review questions

1 Describe the size and shape of DNA molecule.
2 Where are DNA molecules located?
3 Explain the steps of DNA replication.
4 When does DNA replication occur?
5 Watson called DNA "the most golden molecule of them all," primarily because of its function. What is it?
6 Define mitosis and name its phases.
7 Define meiosis and name its stages.
8 When does the reduction of chromosomes from the diploid to haploid number take place? Meiosis is a part of what larger male process? Of what larger female process?
9 Define spermatogenesis. When and where does it occur?
10 Define oogenesis. When and where does it occur?

18

The male reproductive system

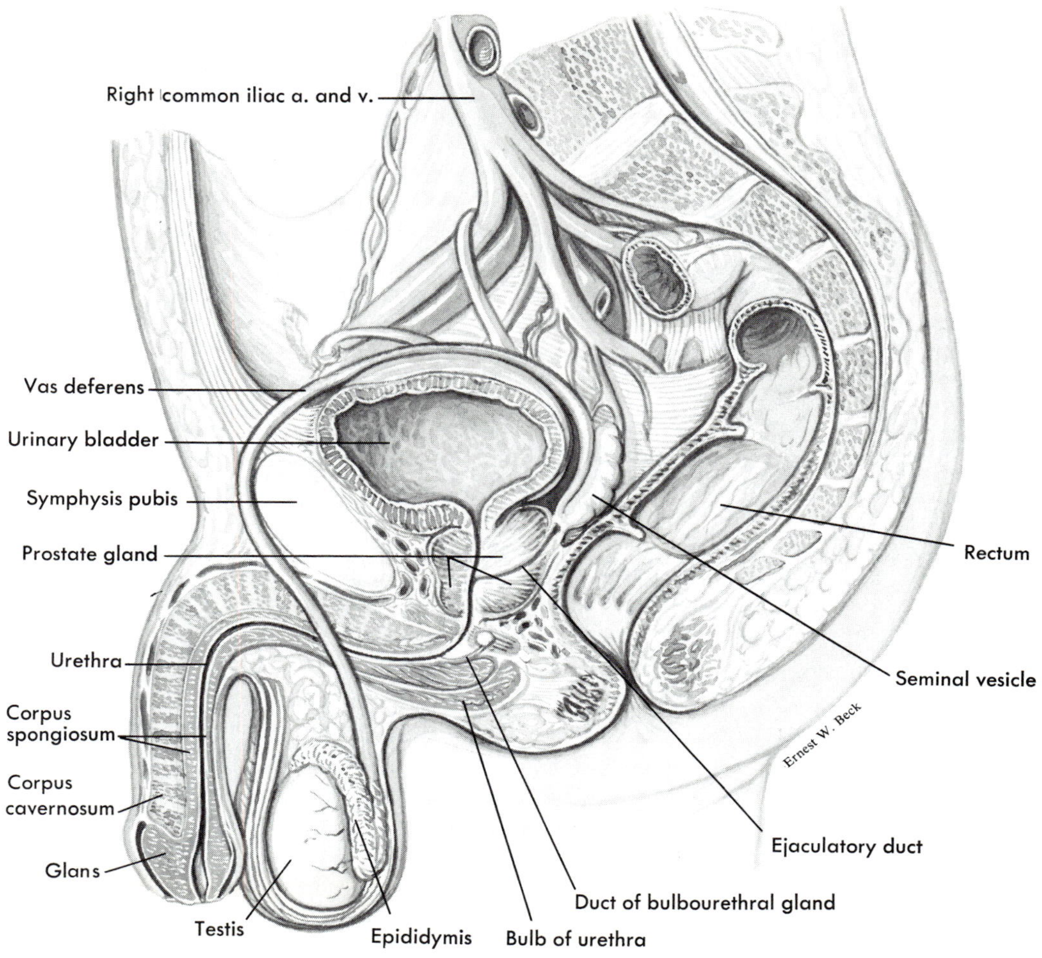

Fig. 18-1

Male pelvic organs as viewed in a median sagittal section.

Meaning and function

The reproductive system consists of those organs whose functions are to produce mature sperm and to introduce them into the female reproductive tract where they can fertilize mature ova. Also, the testes secrete the male hormones, notably, testosterone.

Male reproductive organs

Glands, ducts, and supporting structures compose the male reproductive system (Fig. 18-1). The glands are a pair of testes, a pair of seminal vesicles, one prostate, and a pair of bulbourethral glands. The ducts are a pair of epididymides (singular: epididymis), a pair of vas deferens (seminal ducts; ductus deferens), a pair of ejaculatory ducts, and one urethra. The supporting structures are the scrotum, the penis, and a pair of spermatic cords.

Testes

Structure and location

The testes are small ovoid glands that lie in a pouchlike, skin-covered structure called the *scrotum*. A white fibrous capsule encases each testis and sends partitions through its

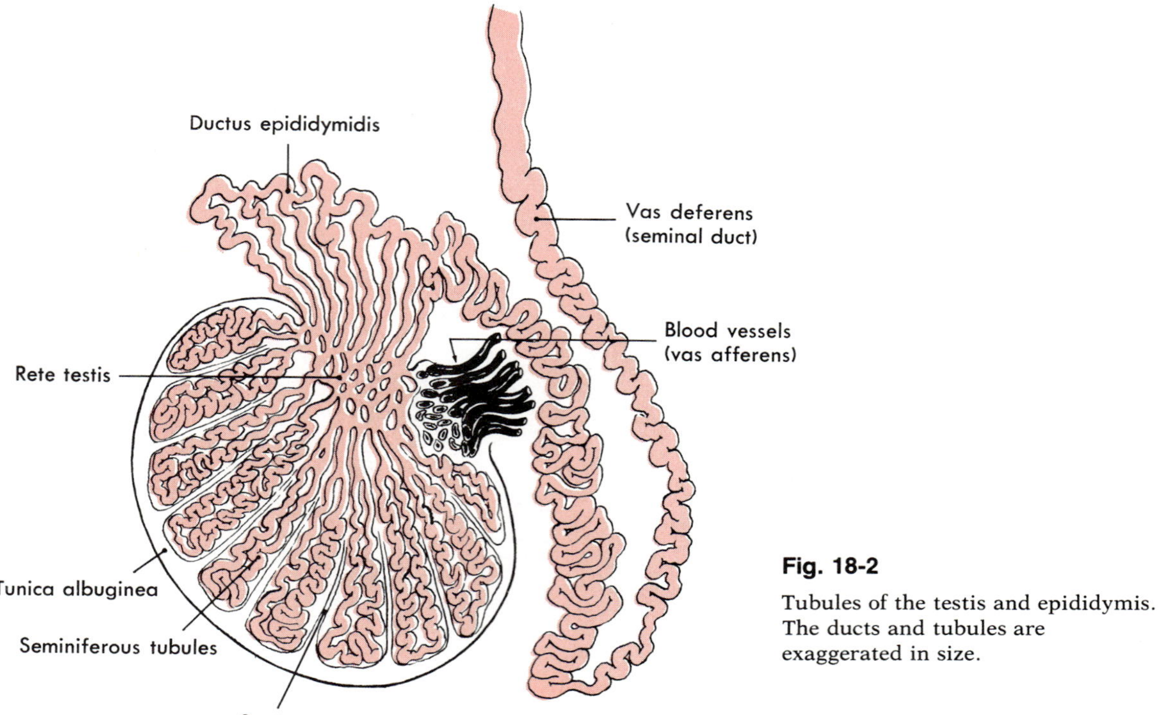

Fig. 18-2

Tubules of the testis and epididymis. The ducts and tubules are exaggerated in size.

interior, dividing it into lobules. Each lobule contains a tiny, coiled *seminiferous tubule* (Fig. 18-2) and numerous interstitial cells (of Leydig). The seminiferous tubules come together to form a plexus from which a few ducts emerge and enter the head of the epididymis.

Functions

The testes perform two primary functions: spermatogenesis and secretion of hormones.

1 Spermatogenesis, the production of spermatozoa (sperm), the male gametes or reproductive cells. The seminiferous tubules produce the sperm. (See Fig. 17-5.)

2 Secretion of hormones, chiefly testosterone (androgen or masculinizing hormone) by interstitial cells (Leydig cells). Testosterone serves the following general functions:

a It promotes "maleness"—that is, the development and maintenance of male secondary sex characteristics, of male accessory organs such as the prostate, seminal vesicles, etc., and of adult male sexual behavior.

b It helps regulate metabolism and is sometimes referred to as "the anabolic hormone" because of its marked stimulating effect on protein anabolism. By stimulating protein anabolism, testosterone promotes growth of skeletal muscles (responsible for greater muscular development and strength of male) and promotes growth of bone. However, testosterone also promotes closure of the epiphyses. Early sexual maturation, therefore, generally leads to early epiphyseal closure and shortness. The converse also holds true: late sexual maturation, delayed epiphyseal closure, and tallness tend to go together.

c It plays a part in fluid and electrolyte metabolism. Testosterone has a mild stimulating effect on kidney tubule reabsorption of sodium and therefore water; it also promotes kidney tubule excretion of potassium.

d It inhibits anterior pituitary secretion of gonadotropins—namely, FSH and ICSH (interstitial cell–stimulating hormone; called LH or luteinizing hormone in the female).

The anterior pituitary gland controls the testes by means of its gonadotropic hormones—specifically, FSH and ICSH just mentioned. FSH stimulates the seminiferous tubules to produce sperm more rapidly. ICSH stimulates interstitial cells to increase their secretion of testosterone. Soon the blood concentration of testosterone reaches a high level that inhibits anterior pituitary secretion of FSH and ICSH. Thus, a negative feedback mechanism operates between the anterior pituitary gland and the testes. A high blood concentration of gonadotropins stimulates testosterone secretion. But a high blood concentration of testosterone inhibits (has a negative effect on) gonadotropin secretion (Fig. 18-3).

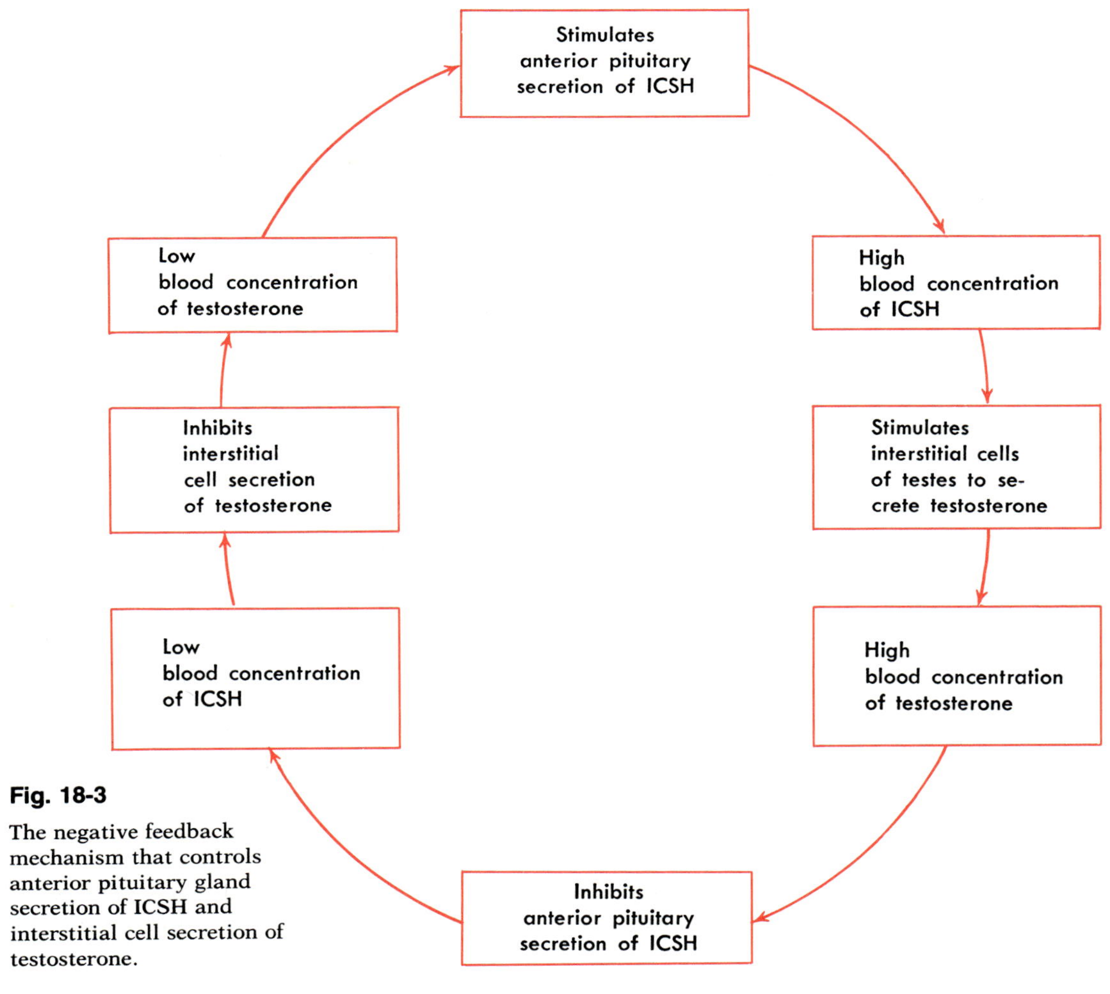

Fig. 18-3

The negative feedback mechanism that controls anterior pituitary gland secretion of ICSH and interstitial cell secretion of testosterone.

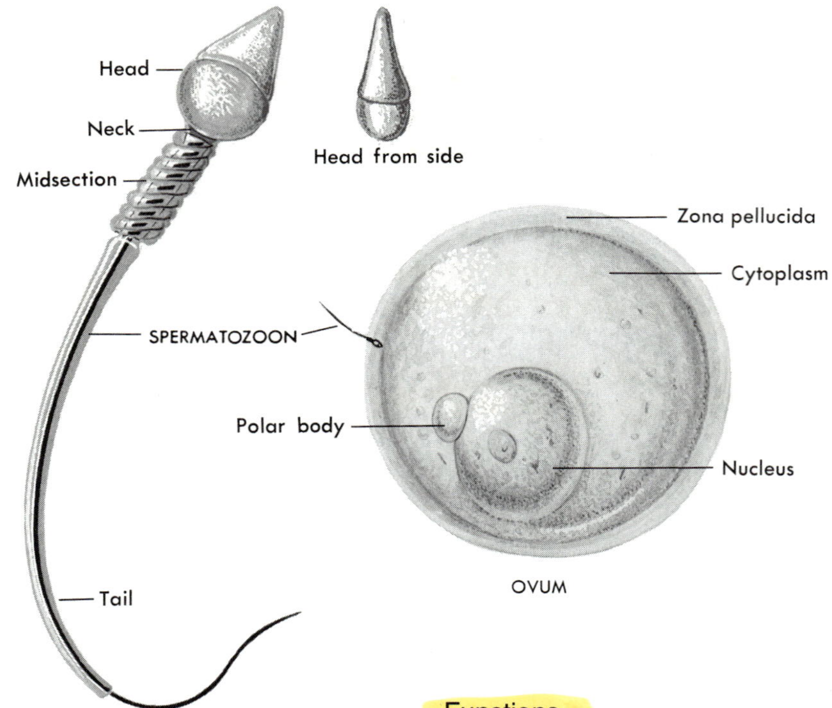

Fig. 18-4

Reproductive cells: male cell or spermatozoon and female cell or ovum.

Structure of spermatozoa

Fig. 18-4 shows the characteristic parts of a spermatozoon: head, neck, and elongated, lashlike tail.

Ducts of testes

Epididymis

Structure and location

Each epididymis consists of a single tightly coiled tube enclosed in a fibrous casing. The tube has a very small diameter (just barely macroscopic) but measures approximately 20 feet in length. It lies along the top and side of the testis. As shown in Fig. 18-2, several small ducts connect the seminiferous tubules of the testis with the epididymis. (Also see Fig. 18-7.)

Functions

The epididymis serves the following functions:

1. It serves as one of the ducts through which sperm pass in their journey from the testis to the exterior.
2. It stores a small quantity of sperm prior to ejaculation.
3. It secretes a small part of the seminal fluid (semen).

Vas deferens (seminal duct; ductus deferens)

Structure and location

The vas deferens, like the epididymis, is a tube. In fact, it is an extension of the epididymis. It passes through the inguinal canal—where it is enclosed in a fibrous cylinder, the spermatic cord—into the abdominal cavity. Here, it extends over the top and down the posterior surface of the bladder, where it joins the duct from the seminal vesicle to form the ejaculatory duct (Figs. 18-5 and 18-6).

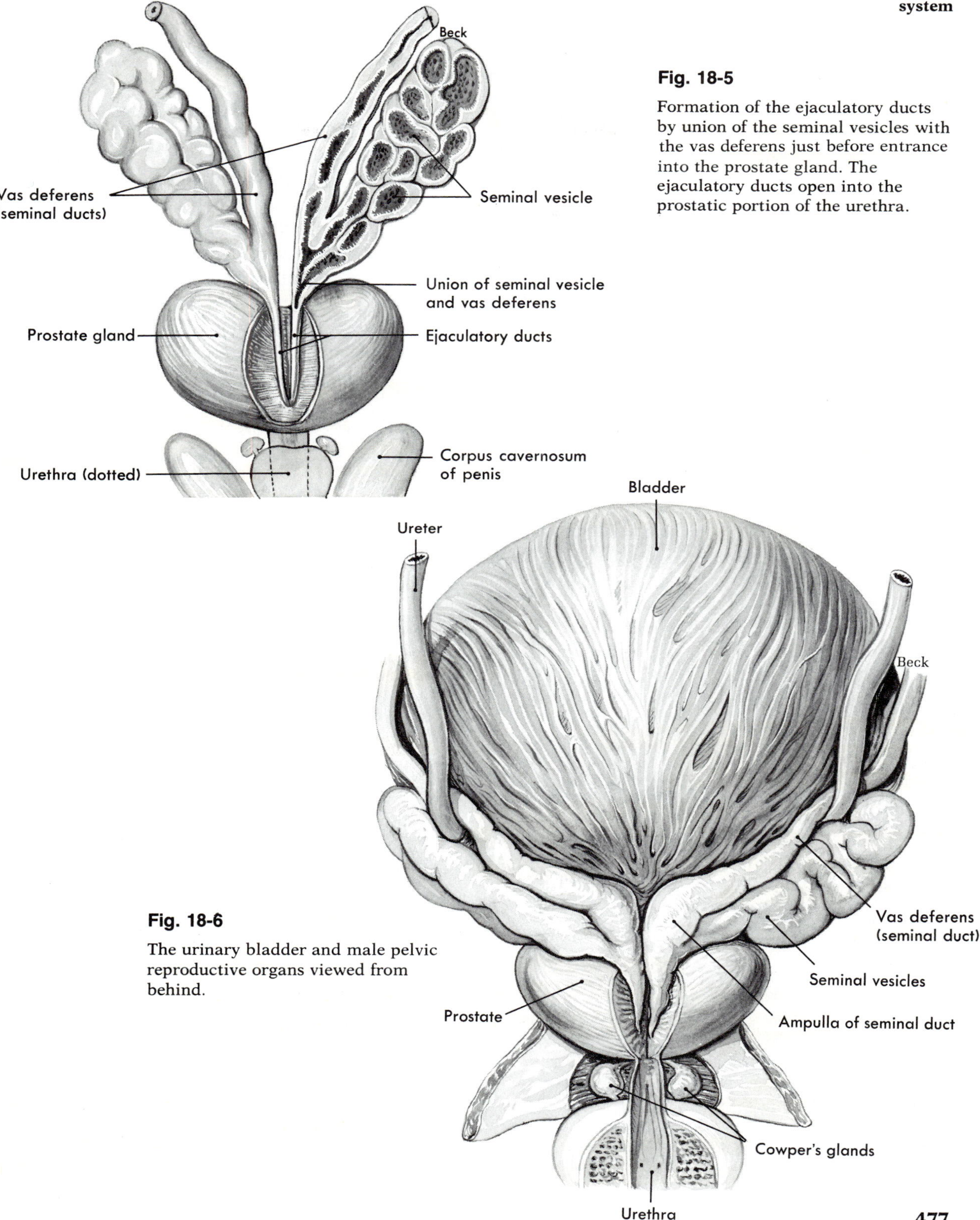

Fig. 18-5

Formation of the ejaculatory ducts by union of the seminal vesicles with the vas deferens just before entrance into the prostate gland. The ejaculatory ducts open into the prostatic portion of the urethra.

Fig. 18-6

The urinary bladder and male pelvic reproductive organs viewed from behind.

Function

The vas deferens serves as one of the ducts for the testis, connecting the epididymis with the ejaculatory duct.

Correlation

Severing of the vas deferens—that is, a vasectomy, usually done through incision in the groin—makes a man sterile. Why? Because it interrupts the route to the exterior from the epididymis. To leave the body, sperm must journey in succession through the epididymis, vas deferens, ejaculatory duct, and urethra.

Ejaculatory duct

The two ejaculatory ducts are short tubes that pass through the prostate gland to terminate in the urethra. As Fig. 18-5 shows, they are formed by the union of the vas deferens with the ducts from the seminal vesicles.

Urethra

For a discussion of the urethra, see p. 457.

Accessory reproductive glands

Seminal vesicles

Structure and location

The seminal vesicles are convoluted pouches that lie along the lower part of the posterior surface of the bladder, directly in front of the rectum.

Function

The seminal vesicles secrete the viscous liquid portion of the semen and prostaglandins, substances postulated to influence cyclic AMP formation (p. 268).

Prostate gland

Structure and location

The prostate is a compound tubuloalveolar gland that lies just below the bladder and is shaped like a doughnut. The fact that the urethra passes through the small hole in the center of the prostate is a matter of considerable clinical significance. Many older men suffer from enlargement of this gland. As it enlarges, it squeezes the urethra, frequently closing it so completely that urination becomes impossible. Urinary retention results. Surgical removal of the gland (prostatectomy) is resorted to as a cure for this condition when other less radical methods of treatment fail.

Function

The prostate secretes a thin alkaline substance that constitutes the largest part of the seminal fluid. Its alkalinity helps protect the sperm from acid present in the male urethra and female vagina and thereby increases sperm motility. (Acid depresses or, if strong enough, kills sperm. Sperm motility is greatest in neutral or slightly alkaline media.)

Bulbourethral glands

Structure and location

The two bulbourethral or Cowper's glands resemble peas in both size and shape. You can see the location of these compound tubuloalveolar glands below the prostate glands in Fig. 18-6. A duct approximately 1 inch long connects them with the membranous portion of the urethra.

Function

Like the prostate, the bulbourethral glands secrete an alkaline fluid that is important for counteracting the acid present in both the male urethra and the female vagina.

Supporting structures

Scrotum (external)

The scrotum is a skin-covered pouch suspended from the perineal region. Internally, it is divided into two sacs by a septum, each sac containing a testis, epididymis, and lower part of a spermatic cord.

Penis (external)

Structure

Three cylindrical masses of erectile or cavernous tissue, enclosed in separate fibrous coverings and held together by a covering of skin, compose the penis. The two larger and uppermost of these cylinders are named the *corpora cavernosa penis,* whereas the smaller, lower one, which contains the urethra, is called the *corpus cavernosum urethrae.*

At the distal end of the penis, there is a slightly bulging structure, the *glans penis,* over which the skin is folded doubly to form a more or less loose-fitting, retractable casing known as the *prepuce* or foreskin. If the foreskin fits too tightly about the glans, a circumcision is usually performed to prevent irritation.

Functions

The penis contains the urethra, the terminal duct for both urinary and reproductive tracts. It is the copulatory organ by means of which spermatozoa are introduced into the female vagina. The scrotum and penis together constitute the *external genitals* of the male.

Spermatic cords (internal)

The *spermatic cords* are cylindrical casings of white fibrous tissue located in the inguinal canals between the scrotum and the abdominal cavity. They enclose the vas deferens, blood vessels, lymphatics, and nerves (Fig. 18-7).

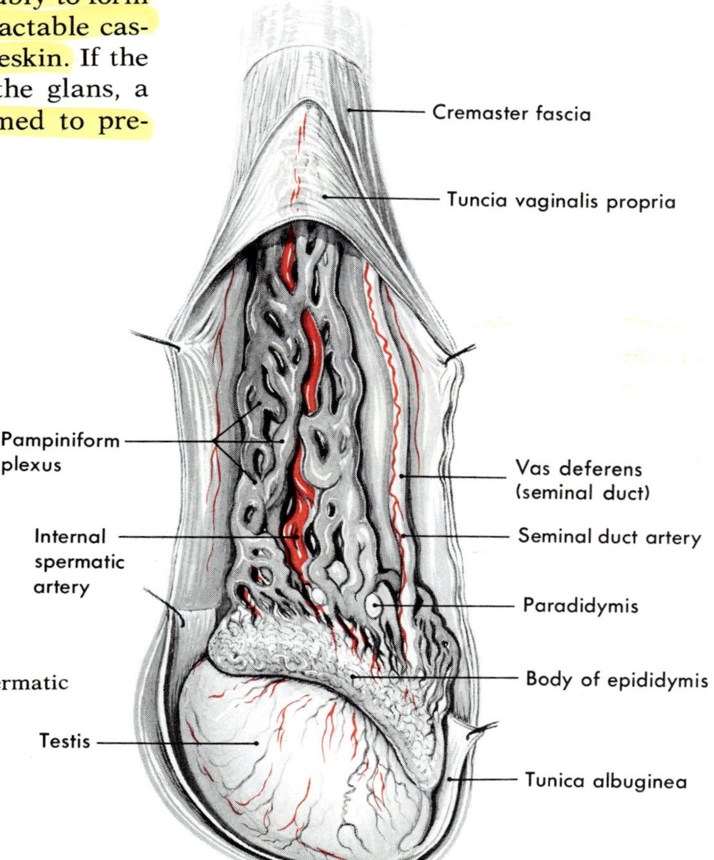

Fig. 18-7

Lateral view of the *left* spermatic cord and testis.

Composition and course of seminal fluid

The following structures secrete the substances that, together, make up the seminal fluid or semen:

1. Testes and epididymides—their secretions, according to one estimate, constitute less than 5% of the seminal fluid volume.
2. Seminal vesicles—their secretions are reported to contribute about 30% of the seminal fluid volume.
3. Prostate gland—its secretions constitute the bulk of the seminal fluid volume, reportedly about 60%.
4. Bulbourethral glands—their secretions are said to constitute less than 5% of the seminal fluid volume.

Besides contributing slightly to the fluid part of semen, the testes also add hundreds of millions of sperm. In traversing the distance from their place of origin to the exterior, the sperm must pass from the testis through the epididymis, vas deferens, ejaculatory duct, and urethra. Note that sperm originate in the testes, glands located outside the body (that is, not within a body cavity), travel inside, and finally are expelled outside.

Early in fetal life the testes are located in the abdominal cavity but normally descend in the spermatic cord (located in the inguinal canal) into the scrotum some time before birth. Occasionally a baby is born with undescended testes, a condition readily observed by palpation of the scrotum. Because the higher temperature inside the abdominal cavity makes sperm infertile, measures are taken to bring the testes down into the scrotum in order to prevent sterility.

Male fertility

Male fertility relates to many factors—most of all to the number of sperm ejaculated but also to their size and shape. Fertile sperm have a uniform size and shape. They are highly motile. Although only one sperm fertilizes an ovum, millions of sperm seem to be necessary for fertilization to occur. According to one estimate, when the sperm count falls below about 50 million per milliliter of semen, sterility results.

One hypothesis suggested to explain this puzzling fact is this: semen that contains an adequate number of sperm also contains enough hyaluronidase to liquefy the intercellular substance between the cells that encase each ovum. Without this, a single sperm cannot penetrate the layers of cells (corona radiata) around the ovum and hence cannot fertilize it.

Male functions in reproduction

All body functions but one have for their ultimate goal survival of the individual. Only the function of reproduction serves a different, a longer range, and, no doubt in nature's scheme, a more important purpose—survival of the human species. Male functions in reproduction consist of the production of male sex cells (spermatogenesis, discussed in Chapter 17) and introduction of these cells into the female body (coitus, copulation, or sexual intercourse). In order for coitus to take place, erection of the penis must first occur, and in order for sperm to enter the female body, semen must be ejaculated from the penis.

Erection is a parasympathetic reflex initiated mainly by certain tactile, visual, and mental stimuli. It consists of dilation of the arteries and arterioles of the penis, which in turn floods and distends spaces in its erectile tissue and compresses its veins. Therefore, more blood enters the penis through the dilated arteries than leaves it through the constricted veins. Hence, it becomes larger and rigid, or in other words, erection occurs.

Ejaculation of semen is also a reflex response. It is the usual outcome of the same stimuli that initiate erection. Ejaculation and various other responses—notably accelerated heart rate, increased blood pressure, hyperventilation, dilated skin blood vessels, and intense sexual excitement—characterize the climax of coitus known as an *orgasm*.

Outline summary

Meaning and function

Consists of organs that together, produce new individual

MALE REPRODUCTIVE ORGANS

1 Glands
  a Testes, gonads (paired)
  b Accessory glands
    1 Seminal vesicles (paired)
    2 Prostate gland
    3 Bulbourethral (Cowper's) glands (paired)
2 Ducts of testes
  a Epididymis (paired)
  b Vas deferens (seminal ducts; ductus deferens) (paired)
  c Ejaculatory ducts (paired)
  d Urethra
3 Supporting structures
  a External—scrotum and penis
  b Internal—spermatic cords (paired)

Testes

1 Structure and location
  a Several lobules composed of seminiferous tubules and interstitial cells (of Leydig), separated by septa, encased in fibrous capsule
  b Few ducts emerge from top of organ and enter head of epididymis
  c Located in scrotum, one testis in each of two scrotal compartments
2 Functions
  a Spermatogenesis—formation of mature male gametes (spermatozoa) by seminiferous tubules
  b Secretion of hormone (testosterone) by interstitial cells
3 Structure of spermatozoa—consists of head, neck (middle piece), and whiplike tail

Ducts of testes

Epididymis

1 Structure and location
  a Single tightly coiled tube enclosed in fibrous casing
  b Lies along top and side of each testis
2 Functions
  a Duct for seminal fluid
  b Also secretes part of seminal fluid
  c Sperm become capable of motility while they are stored in epididymis

Vas deferens (seminal duct; ductus deferens)

1 Structure and location
  a Tube, extension of epididymis
  b Extends through inguinal canal, into abdominal cavity, over top and down posterior surface of bladder to join duct from seminal vesicle
2 Function
  a One of excretory ducts for seminal fluid
  b Connects epididymis with ejaculatory duct

Ejaculatory duct

1 Formed by union of vas deferens with duct from seminal vesicle
2 Passes through prostate gland, terminating in urethra

Urethra

See p. 457

Accessory reproductive glands

Seminal vesicles

1 Structure and location—convoluted pouches on posterior surface of bladder
2 Function—secrete prostaglandins and viscous nutrient-rich part of seminal fluid

Prostate gland

1 Structure and location
  a Doughnut-shaped
  b Encircles urethra just below bladder
2 Function—adds alkaline secretion to seminal fluid

Bulbourethral glands

1 Structure and location
  a Small, pea-shaped structures with 1-inch long ducts leading into urethra
  b Lie below prostate gland
2 Function—secrete alkaline fluid that is part of semen

Supporting structures

External

1 Scrotum
  a Skin-covered pouch suspended from perineal region
  b Divided into two compartments
  c Contains testis, epididymis, and first part of vas deferens
2 Penis—composed of three cylindrical masses of erectile tissue, one of which contains urethra

Internal

1 Spermatic cords
  a Fibrous cylinders located in inguinal canals
  b Enclose seminal ducts, blood vessels, lymphatics, and nerves

Composition and course of seminal fluid

1 Consists of secretions from testes, epididymides, seminal vesicles, prostate, and bulbourethral glands
2 Each drop contains millions of sperm
3 Passes from testes through epididymis, vas deferens, ejaculatory duct, and urethra

MALE FUNCTIONS IN REPRODUCTION

1 Spermatogenesis
2 Coitus (copulation or sexual intercourse)
  a Erection—a parasympathetic reflex initiated by various kinds of stimuli and consisting of dilation of arteries and arterioles of the penis, causing its enlargement and erection
  b Ejaculation of semen—one of several responses that characterize orgasm, the climax of coitus

Review questions

1 Name the male sex glands. Where are they located?
2 Of what is the seminal fluid composed? Trace its course from its formation in the gonads to the exterior.
3 What and where are the prostate glands?
4 What and where are Cowper's glands?
5 What is the spermatic cord? From what to what does it extend, and what does it contain?
6 Removal of the testes (orchiectomy or castration) results in both sterility and various changes in the secondary sex characteristics. Why?

19

The female reproductive system

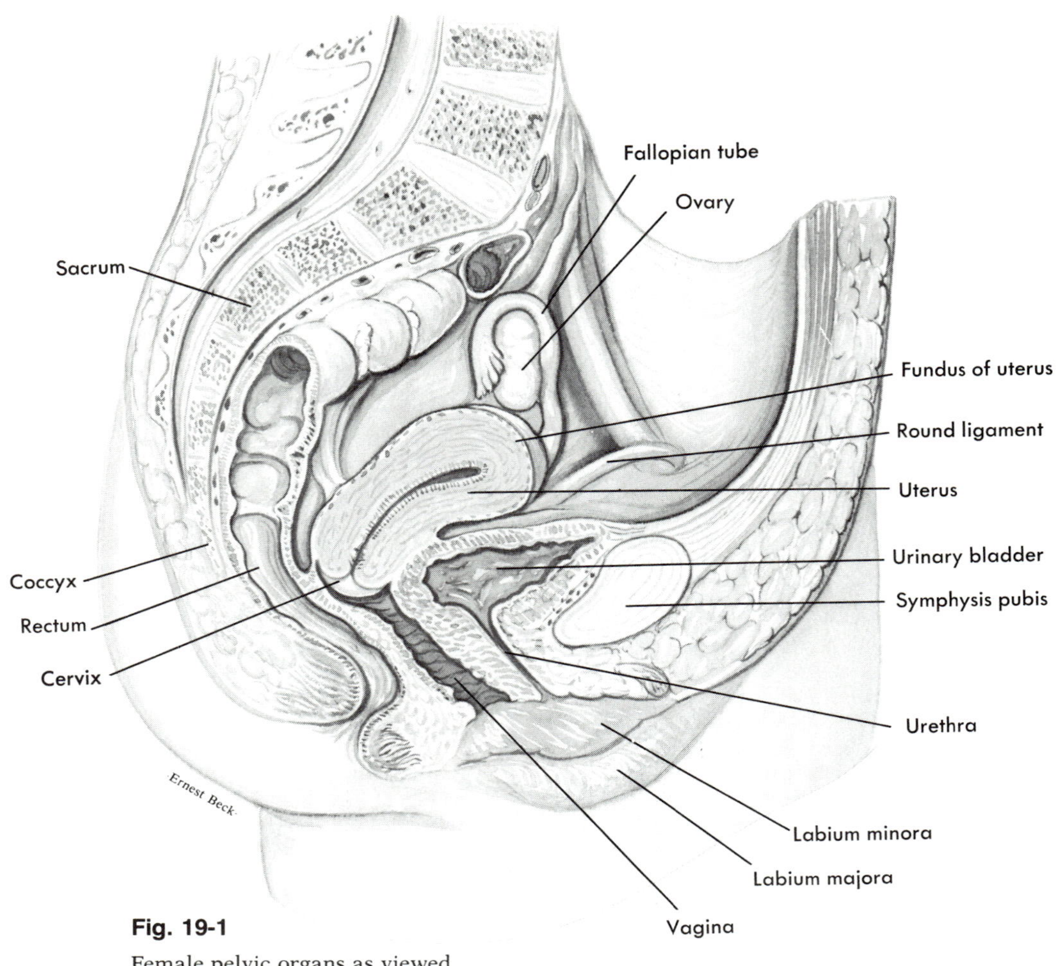

Fig. 19-1

Female pelvic organs as viewed in a median sagittal section.

Female reproductive organs

Examine Fig. 19-1 and Plate XIV of the color insert (p. 54) to identify the following organs of the female reproductive system:

1 *Primary sex organs*—the two ovaries (female gonads)
2 *Secondary sex organs*—two uterine tubes (fallopian tubes or oviducts), one uterus, one vagina, one vulva (pudendum or external genitalia), and two breasts or mammary glands

Uterus

Structure

Size, shape, and divisions. The uterus is pear-shaped in its virginal state and measures approximately 3 inches in length, 2 inches in width at its widest part, and 1 inch in thickness. Note in Fig. 19-2 that the uterus has two main parts: an upper portion, the *body*, and a lower, narrow section, the *cervix*. Did you notice that the body of the uterus rounds into a bulging prominence above the level at which the uterine tubes enter? This bulging upper surface is the *fundus*.

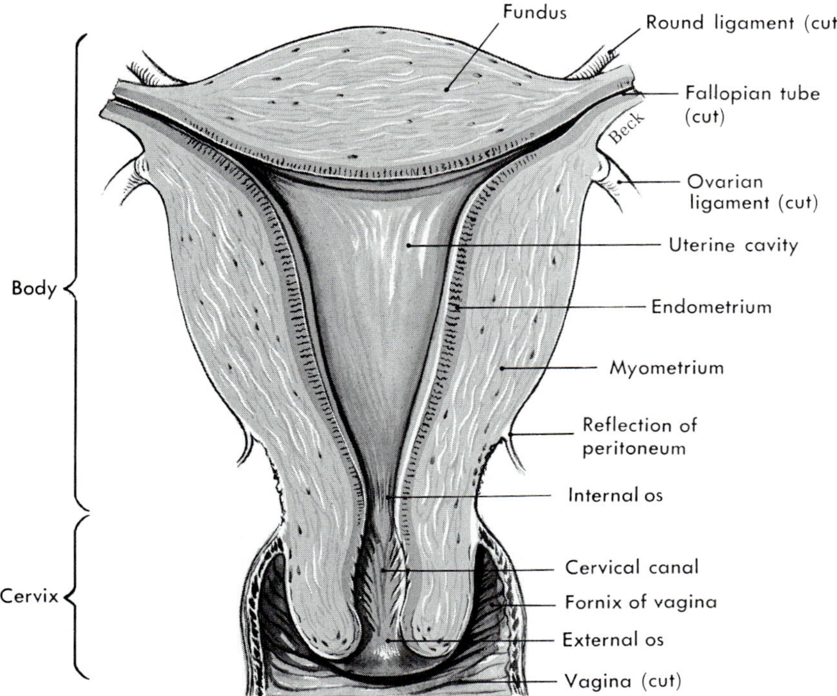

Fig. 19-2

Sectioned view of the uterus.

Wall. Three coats compose the walls of the uterus: endometrium, myometrium, and parietal peritoneum.

1 A lining of mucous membrane, called the *endometrium,* composed of three layers of tissues: a compact surface layer of columnar epithelium, a spongy middle layer of loose connective tissue, and a basal layer of dense connective tissue that attaches the endometrium to the underlying myometrium. During menstruation and following delivery of a baby, the compact and spongy layers slough off.

2 A thick, middle coat (the *myometrium*) consists of three layers of smooth muscle fibers that extend in all directions, longitudinally, transversely, and obliquely, and give the uterus great strength. The myometrium is thickest in the fundus and thinnest in the cervix—a good example of the principle of structural adaptation to function. In order to expel a fetus—that is, move it down and out of the uterus—the fundus must contract more forcibly than the lower part of the uterine wall and the cervix must be stretched or dilated.

3 An external coat of serous membrane, the *parietal peritoneum,* is incomplete since it covers none of the cervix and only part of the body (all except the lower one fourth of its anterior surface). The fact that the entire uterus is not covered with peritoneum has clinical value because it makes it possible to perform operations on this organ without the risk of infection that attends cutting into the peritoneum.

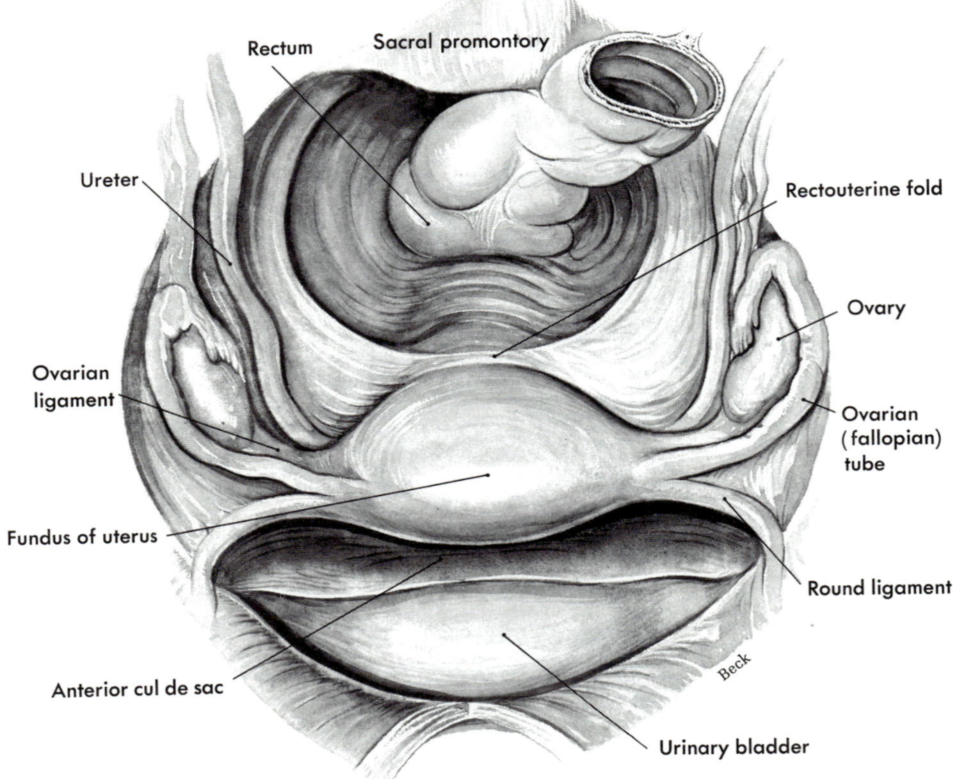

Fig. 19-3

Female pelvic contents as seen from above and in front.

Cavities. The cavities of the uterus are small because of the thickness of its walls (see Fig. 19-2). The body cavity is flat and triangular. Its apex is directed downward and constitutes the *internal os,* which opens into the *cervical canal.* The cervical canal is constricted on its lower end also, forming the *external os,* which opens into the vagina. The uterine tubes open into the body cavity at its upper, outer angles.

Blood supply. The uterus receives a generous supply of blood from uterine arteries, branches of the internal iliac arteries.

Location

Fig. 19-3 shows the location of the uterus in the pelvic cavity between the bladder and the rectum.

Position

1 *Normally* the uterus is flexed between the body and cervix, with the body lying over the superior surface of the bladder, pointing forward and slightly upward (Fig. 19-4). The cervix points downward and backward from the point of flexion, joining the vagina at approximately a right angle. Several liga-

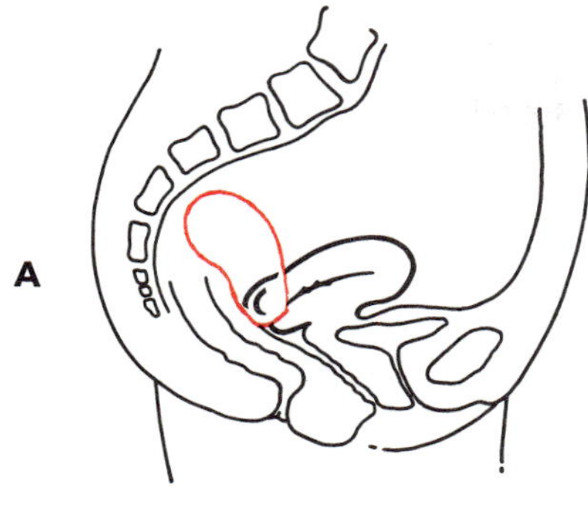

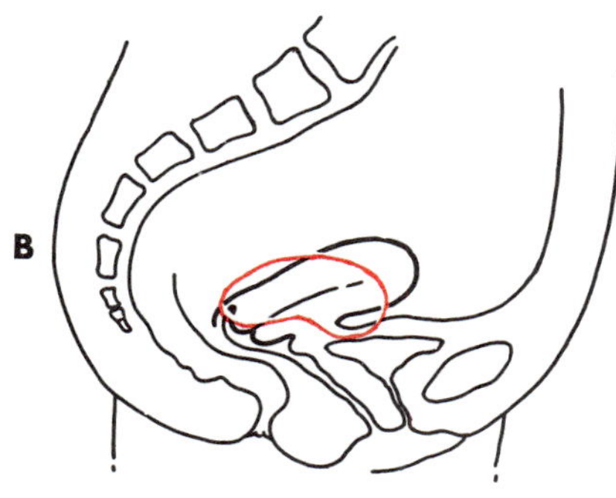

Fig. 19-4

Normal and abnormal positions of the uterus. Red lines show the abnormal positions. **A,** Retroflexion. **B,** Anteflexion.

ments hold the uterus in place but allow its body considerable movement, a characteristic that often leads to malpositions of the organ.

2 The uterus may lie in any one of several *abnormal positions.* A common one is retroversion or backward tilting of the entire organ.

3 *Eight ligaments* (three pairs, two single ones) anchor the uterus in the pelvic cavity: broad (paired), uterosacral (paired), posterior (single), anterior (single), and round (paired). Six of these so-called ligaments are actually extensions of the parietal peritoneum in different directions. The round ligaments are fibromuscular cords.

a The *two broad ligaments* are double folds of parietal peritoneum that form a kind of partition across the pelvic cavity. The uterus is suspended between these two folds.

b The *two uterosacral ligaments* are foldlike extensions of the peritoneum from the posterior surface of the uterus to the sacrum, one on each side of the rectum.

c The *posterior ligament* is a fold of peritoneum extending from the posterior surface of the uterus to the rectum. This ligament forms a deep pouch known as the *cul-de-sac of Douglas* (or rectouterine pouch) between the uterus and rectum. Since this is the lowest point in the pelvic cavity, pus collects here in pelvic inflammations. To secure drainage, an incision may be made at the top of the posterior wall of the vagina (posterior colpotomy).

d The *anterior ligament* is the fold of peritoneum formed by the extension of the peritoneum on the anterior surface of the uterus to the posterior surface of the bladder. This fold also forms a cul-de-sac but one that is less deep than the posterior pouch.

e The *two round ligaments* (Fig. 19-3) are fibromuscular cords extending from the upper, outer angles of the uterus through the inguinal canals and disappearing in the labia majora.

Functions

The uterus or womb plays a role in the accomplishment of three functions vital for survival of the human species but not for individual survival: menstruation, pregnancy, and labor.

1 *Menstruation* is a sloughing away of the compact and spongy layers of the endometrium, attended by bleeding from the torn vessels.

2 In *pregnancy,* the embryo implants itself

in the endometrium and there lives as a parasite throughout the fetal period.

3 *Labor* consists of powerful, rhythmic contractions of the muscular uterine wall that result in expulsion of the fetus or birth.

Uterine tubes—fallopian tubes or oviducts

Location

The uterine tubes are attached to the uterus at its upper outer angles (see Figs. 19-1 and 19-3). They lie between the folds of the broad ligaments and extend upward and outward toward the sides of the pelvis and then curve downward and backward.

Structure

The same three coats (mucous, smooth muscle, serous) of the uterus compose the tubes. The mucosa of the tubes, however, is ciliated. At the distal end, each tube expands into a funnel-like portion called the *infundibulum*. The open outer margin of the infundibulum resembles a fringe in its irregular outline. The fringelike projections are known as *fimbriae*. Here, the mucous lining of the tubes is directly continuous with the peritoneum—a fact of great clinical significance because the tubal mucosa is continuous with that of the uterus and vagina and, therefore, often becomes infected by gonococci or other organisms introduced into the vagina. Inflammation of the tubes (salpingitis) may readily spread to become inflammation of the peritoneum (peritonitis), a serious condition. In the male, there is no such direct route by which microorganisms can reach the peritoneum from the exterior.

Each uterine tube is approximately 4 inches long.

Function

The tubes serve as ducts for the ovaries, although they are not attached to them. Ovaries are the organs that produce ova (the female gametes). Fertilization, the union of a spermatozoon with an ovum, normally occurs in the tubes.

Ovaries—female gonads

Location and size

The ovaries are glands that resemble large almonds in size and shape and are located one on either side of the uterus, below and behind the uterine tubes. Each ovary attaches to the posterior surface of the broad ligament by the mesovarian ligament. The ovarian ligament anchors it to the uterus. The distal portion of the tube curves about the ovary in such a way that the fimbriae cup over the ovary but do not actually attach to it. Here, then, is a gland whose duct is detached from it, a fact that makes possible pregnancy in the pelvic cavity instead of in the uterus as is normal.

Microscopic structure

The surface of the ovary consists of a single layer of germinal epithelial cells, whereas its interior is made up of connective tissue in which are embedded thousands of microscopic structures known as *graafian follicles*. After puberty, the follicles are present in varying stages of development (Fig. 19-5). The primordial follicles consist of an *ovum* encased in a nest of epithelial cells. Before puberty, all the follicles are in this stage.

Functions

The ovaries perform two functions: ovulation and hormone secretion. Ova develop and mature in the ovaries and are discharged from them into the pelvic cavity between the folds of the broad ligament. The ovaries also secrete the female hormones—estrogens (chiefly estradiol and estrone) and progesterone. More details about their secretion and their functions appear on pp. 493-498.

Fig. 19-5

Mammalian ovary showing successive stages of ovarian follicle and ovum development. Begin with the first stage (egg nest near arrow at lower left) and follow the arrow counterclockwise to the final stage (corpus albicans).

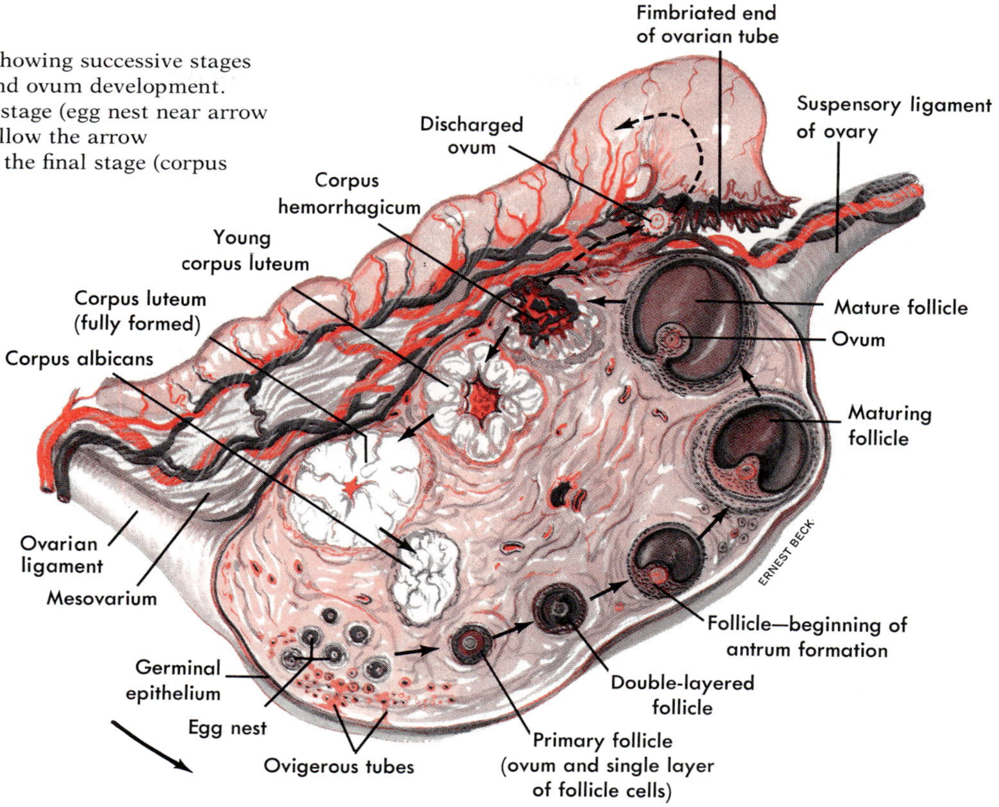

Vagina

Location

The vagina is situated between the rectum, which lies posterior to it, and the urethra and bladder, which lie anterior to it. It extends upward and backward from its external orifice.

Structure

The vagina is a collapsible tube capable of great distention, is composed mainly of smooth muscle, and is lined with mucous membrane arranged in rugae. Its anterior wall, which measures from 2½ to 3 inches in length, is about 1 inch shorter than the posterior wall because the cervix protrudes into the uppermost portion of the anterior wall. In the virginal state, a fold of mucous membrane, the *hymen* ("maidenhead"), forms a border around the external opening of the vagina, partially closing the orifice. Occasionally, this structure completely covers the vaginal outlet, a condition referred to as *imperforate hymen*. Perforation has to be done before the menstrual flow can escape.

Functions

The vagina constitutes an essential part of the reproductive tract because of the following:

1. It is the organ that receives the seminal fluid from the male.
2. It serves as the lower part of the birth canal.
3. It acts as the excretory duct for uterine secretions and the menstrual flow.

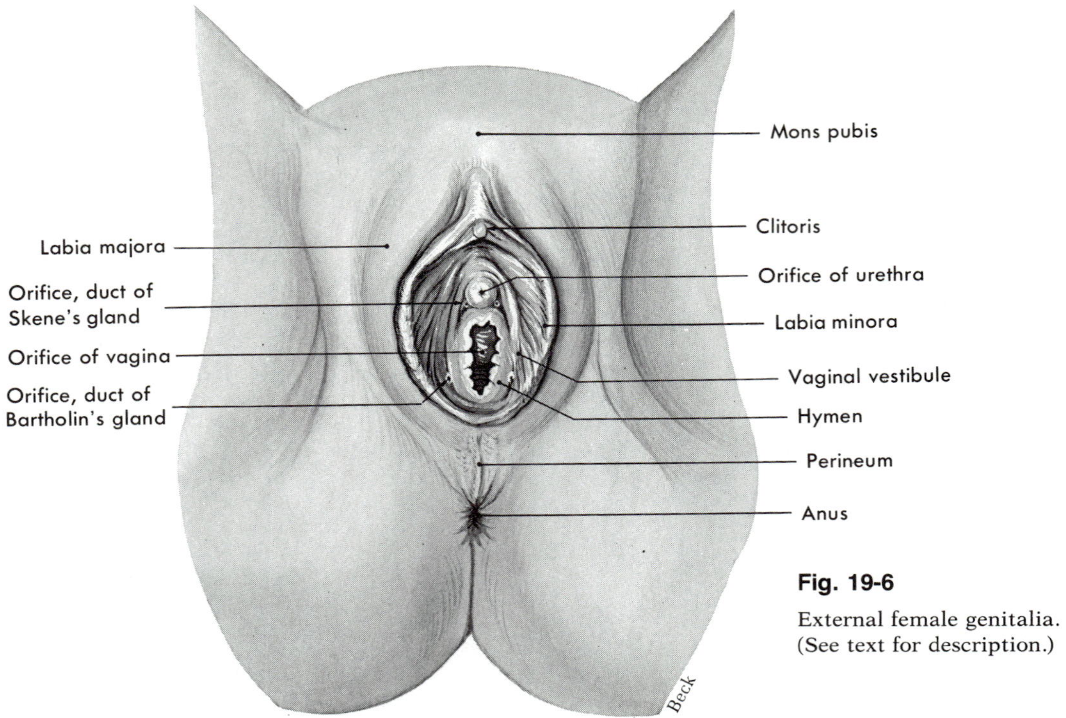

Fig. 19-6

External female genitalia. (See text for description.)

Vulva

Fig. 19-6 shows the structures that, together, constitute the female external genitals (reproductive organs) or vulva: mons veneris, labia majora, labia minora, clitoris, urinary meatus (urethral orifice), vaginal orifice, and Bartholin's or the greater vestibular glands.

1 The *mons veneris* or mons pubis is a skin-covered pad of fat over the symphysis pubis. Coarse hairs appear on this structure at puberty and persist throughout life.

2 The *labia majora* or "large lips" are covered with pigmented skin and hair on the outer surface and are smooth and free from hair on the inner surface. They are composed mainly of fat and numerous glands.

3 The *labia minora* or "small lips" are located within the labia majora and are covered with modified skin. These two lips come together anteriorly in the midline. The area between the labia minora is the *vestibule*.

4 The *clitoris* is a small organ composed of erectile tissue, located just behind the junction of the labia minora, and homologous to the corpora cavernosa and glans of the penis. The *prepuce* or foreskin covers the clitoris, as it does the glans penis in the male.

5 The *urinary meatus* (urethral orifice) is the small opening of the urethra, situated between the clitoris and the vaginal orifice.

6 The *vaginal orifice* is an opening that, in the virginal state, is usually only slightly larger than the urinary meatus because of the constricting border formed by the hymen. In the nonvirginal state, the vaginal orifice is noticeably larger than the urinary meatus. It is located posterior to the meatus.

7 *Bartholin's* or the *greater vestibular glands* are two bean-shaped glands, one on either side of the vaginal orifice. Each gland opens by means of a single, long duct into the space between the hymen and the labium minus. These glands are of clinical importance because they are frequently infected (bartholinitis or Bartholin's abscess), particularly by gonococci. They are homologous to the bulbourethral glands in the male, and they secrete a lubricating fluid. Opening into the vestibule near the urinary meatus by way of two small ducts is a group of tiny mucous glands, the *lesser vestibular* or *Skene's glands*. These have clinical interest because gonococci that lodge there are difficult to eradicate.

Perineum

The perineum is the skin-covered muscular region between the vaginal orifice and the anus. This area has great clinical importance because of the danger of its being torn during childbirth. If the tear is deep, it may extend all the way through the perineum and even through the anal sphincter, resulting in involuntary seepage from the rectum until the laceration is repaired. To avoid this possibility, an incision known as an *episiotomy* is usually made in the perineum, particularly at the birth of a first baby.

Breasts

Location and size

The breasts lie over the pectoral muscles and are attached to them by a layer of connective tissue (fascia). Estrogens and progesterone, two ovarian hormones, control their development during puberty. Estrogens stimulate growth of the ducts of the mammary glands, whereas progesterone stimulates development of the alveoli, the actual secreting cells. Breast size is determined more by the amount of fat around the glandular tissue than by the amount of glandular tissue itself. Hence, the size of the breast does not relate to its functional ability.

Structure

Each breast consists of several lobes separated by septa of connective tissue. Each lobe consists of several lobules, which, in turn, are composed of connective tissue in which are embedded the secreting cells (alveoli) of the gland, arranged in grapelike clusters around minute ducts. The ducts from the various lobules unite, forming a single excretory duct for each lobe, or between fifteen and twenty in each breast. These main ducts converge toward the nipple, like the spokes of a wheel. They enlarge slightly before reaching the nipple into ampullae or small "reservoirs" (Fig. 19-7). Each of these main ducts terminates in a tiny opening on the surface of the nipple. Adipose tissue is deposited around the surface of the gland, just under the skin, and between the lobes. The nipples are bordered by a circular pigmented area, the *areola*. It contains numerous sebaceous glands that appear as small nodules under the skin. In Caucasians (other than those with very dark complexions), the areola and nipple change color from delicate pink to brown early in pregnancy—a fact of value in diagnosing a first pregnancy. The color decreases after lactation has ceased but never entirely returns to the virginal hue. In darker-skinned women, no noticeable color change in the areola or nipple heralds the first pregnancy.

Function

The function of the mammary glands is lactation—that is, the secretion of milk for the nourishment of newborn infants.

Mechanism controlling lactation

Very briefly, lactation is controlled as follows:

1 The ovarian hormones, estrogens and progesterone, act on the breasts to make them structurally ready to secrete milk. Estrogens promote development of the ducts of the

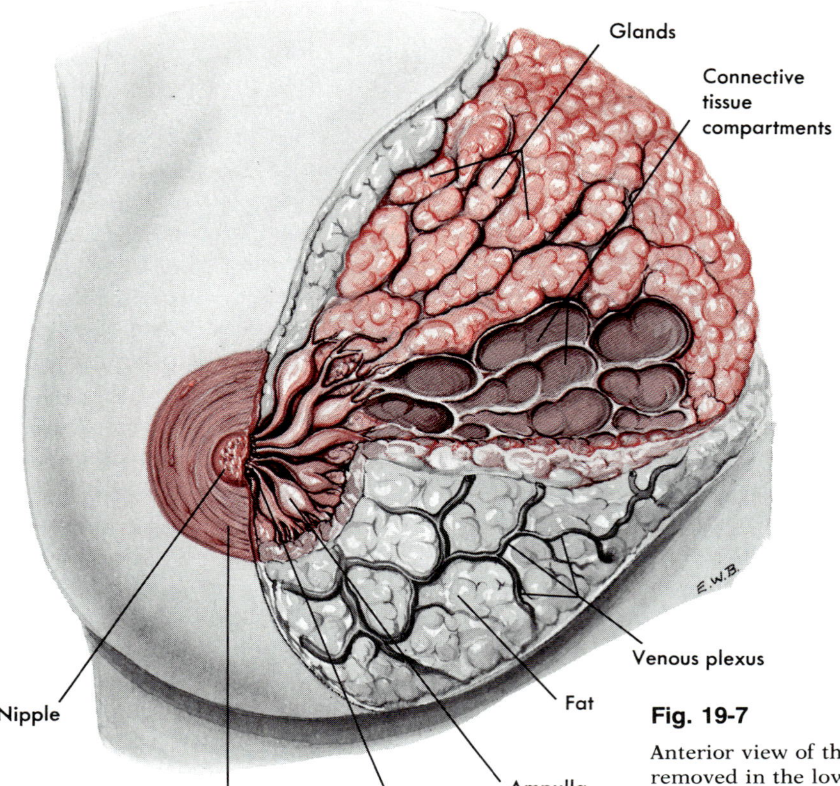

Fig. 19-7

Anterior view of the breast. The skin has been removed in the lower right quadrant to reveal the plexus overlying the adipose (fatty) tissue. In the upper right quadrant the adipose tissue has been removed to show the alveoli of the glands. The glands have been removed in a small area to reveal the connective tissue compartments that separate the lobules.

breasts. Progesterone acts on the estrogen-primed breasts to promote completion of the development of the ducts and development of the alveoli, the secreting cells of the breasts. A high blood concentration of estrogens (for example, during pregnancy) inhibits anterior pituitary secretion of lactogenic hormone.

2 Shedding of the placenta following delivery of the baby cuts off a major source of estrogens. The resulting rapid drop in the blood concentration of estrogens stimulates anterior pituitary secretion of lactogenic hormone. Also, the suckling movements of a nursing baby act in some way to stimulate both anterior pituitary secretion of lactogenic hormone and posterior pituitary secretion of oxytocin (Fig. 19-8).

3 Lactogenic hormone stimulates lactation—that is, stimulates alveoli of the mammary glands to secrete milk. Milk secretion starts about the third or fourth day after delivery of a baby, supplanting a thin, yellowish secretion called *colostrum*. With repeated stimulation by the suckling infant, plus various favorable mental and physical conditions, lactation may continue almost indefinitely.

4 Oxytocin stimulates the alveoli of the breasts to eject milk into the ducts, thereby making it accessible for the infant to remove by suckling.

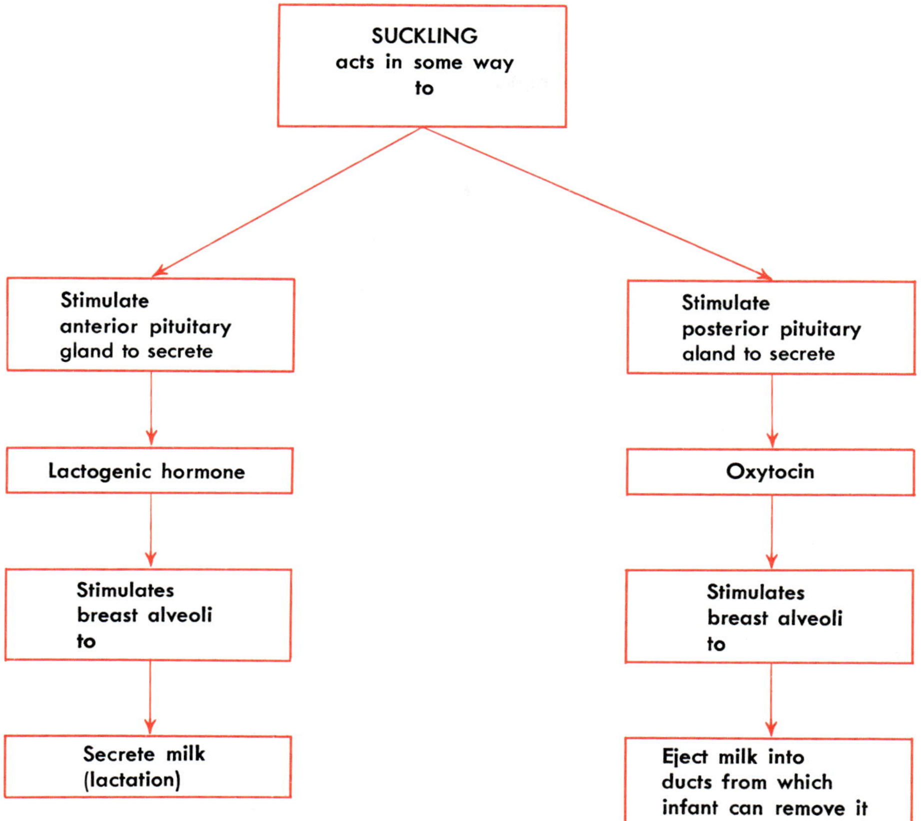

Fig. 19-8

Mechanism for controlling lactation and milk ejection.

Female sexual cycles

Recurring cycles

Many changes recur periodically in the female during the years between the onset of the menses (menarche) and their cessation (menopause or climacteric). Most obvious, of course, is menstruation—the outward sign of changes in the endometrium. Most women also note periodic changes in the breasts. But these are only two of many changes that occur over and over again at fairly uniform intervals during the approximately 30 years of female reproductive maturity. Rhythmic changes also take place in the ovaries, the myometrium, the vagina, gonadotropin secretion, body temperature, and even mood or "emotional tone." First we shall describe the major cyclical changes, and then we shall discuss the mechanisms that produce them.

Ovarian cycles

Once each month, on about the first day of menstruation, several primitive graafian follicles and their enclosed ova begin to grow and develop. The follicular cells proliferate and start to secrete estrogens (and minute amounts of progesterone). Usually, only one follicle matures and migrates to the surface of the ovary. The surface of the follicle degenerates, causing expulsion of the mature ovum into the pelvic cavity (ovulation).

When does ovulation occur? This is a question of great practical importance and one that in the past was given many answers. Today it is known that ovulation usually occurs 14 days before the next menstrual period begins. (Only in a 28-day menstrual cycle is this also 14 days after the beginning of the preceding menstrual cycle, as explained in the next column. A few women experience pain within a few hours after ovulation. This is referred to as *mittelschmerz*—German for middle pain. It has been ascribed to irritation of the peritoneum by hemorrhage from the ruptured follicle.

Shortly before ovulation, the ovum undergoes meiosis, the process by which its number of chromosomes is reduced by half. Immediately after ovulation, cells of the ruptured follicle enlarge and, because of the appearance of lipoid substances in them, become transformed into a golden-colored body, the *corpus luteum*. The corpus luteum grows for 7 or 8 days. During this time, it secretes both progesterone and estrogens in increasing amounts. Then, provided fertilization of the ovum has not taken place, the size of the corpus luteum and the amount of its secretions gradually diminish.

Endometrial or menstrual cycle

During menstruation, necrotic bits of the compact and spongy layers of the endometrium slough off, leaving denuded bleeding areas. Following menstruation, the cells of these layers proliferate, causing the endometrium to reach a thickness of 2 or 3 ml by the time of ovulation. During this period, endometrial glands and arterioles grow longer and more coiled—two factors that also contribute to the thickening of the endometrium. (See Fig. 19-12.) After ovulation, the endometrium grows still thicker (reaching a maximum of about 4 to 6 ml). Most of this increase, however, is believed to be caused by swelling produced by fluid retention rather than by further proliferation of endometrial cells. The increasingly tortuous endometrial glands start to secrete during the time between ovulation and the next menses. Then, the day before menstruation starts again, the tightly coiled arterioles constrict, producing endometrial ischemia. This leads to necrosis, sloughing, and, once again, menstrual bleeding.

The menstrual cycle is customarily divided into phases, named for major events occurring in them: menses, postmenstrual or preovulatory phase, ovulation, and postovulatory or premenstrual phase.

1 The *menses* or *menstrual period* occurs on cycle days 1 to 5. There is some individual variation, however.

2 The *postmenstrual phase* occurs between the end of the menses and ovulation. Therefore, it is the preovulatory as well as the postmenstrual phase. In a 28-day cycle, it usually includes cycle days 6 to 13 or 14. But the length of this phase varies more than do the others. It lasts longer in long cycles and ends sooner in short ones. This phase is also called the *estrogenic* or *follicular phase* because of the high blood estrogen level resulting from secretion by the developing follicle. *Proliferative phase* is still another name for it because proliferation of endometrial cells occurs at this time.

3 *Ovulation* (that is, rupture of the mature follicle with expulsion of its ovum into the pelvic cavity) occurs frequently on cycle day 15 in a 28-day cycle. However, it occurs on different days in different length cycles depending on the length of the preovulatory phase. For example, in a 32-day cycle, the preovulatory phase would probably last until cycle day 18 and ovulation would then occur on cycle day 19 instead of 15. In short, the day of ovulation cannot be predicted with certainty. One cannot know ahead of time precisely how many days the preovulatory phase will last. This physiological fact probably accounts for most of the unreliability of the rhythm method of contraception.

4 The *postovulatory* (or *premenstrual*) *phase* occurs between ovulation and the onset of the menses. This phase is also called the *luteal phase* because the corpus luteum secretes only during this time and *progesterone phase* because it secretes mainly this hormone. The length of the premenstrual phase is pretty constant, lasting usually 14 days—that is, cycle days 15 to 28 in a 28-day cycle. Differences in length of the total menstrual cycle, therefore, exist mainly because of differences in duration of the preovulatory rather than of the premenstrual phase.

Myometrial cycle

The myometrium contracts mildly but with increasing frequency during the 2 weeks preceding ovulation. Contractions decrease or disappear between ovulation and the next menses, thereby lessening the probability of expulsion of an implanted ovum.

Gonadotropic cycles

The adenohypophysis (anterior pituitary gland) secretes two hormones called gonadotropins that influence female reproductive cycles. Their names are follicle-stimulating hormone (FSH) and luteinizing hormone (LH). The amount of each gonadotropin secreted varies with a rhythmic regularity that can be related, as we shall see, to the rhythmic ovarian and uterine changes just described.

Control

Physiologists agree that hormones play the major role in producing the cyclic changes characteristic in the female during the years of reproductive maturity. The recent development of a method called radioimmunoassay has made it possible to measure blood levels of gonadotropins. By correlating these with the monthly ovarian and uterine changes, investigators have worked out the main features of the control mechanisms.

A brief description follows of the mechanisms that produce cyclical changes in the ovaries and uterus and in the amounts of gonadotropins secreted.

Control of cyclical changes in ovaries

Cyclical changes in the ovaries result from cyclical changes in the amounts of gonadotropins secreted by the anterior pituitary gland. An increasing FSH blood level has two effects: it stimulates one or more primitive graafian follicles and ova to start growing, and it stimulates the follicles to secrete estrogens. (Developing follicles also secrete very small amounts of progesterone.) Because of the influence of FSH on follicle secretion, the level of estrogens in blood increases gradually for a few days during the postmenstrual phase. Then suddenly, on about the twelfth cycle day, it leaps upward to a maximum peak. Scarcely 12 hours after this "estrogen surge," an "LH surge" occurs and presumably triggers ovulation a day or two later.* The control of cyclical ovarian changes by gonadotropins FSH and LH is summarized in Figs. 19-9 and 19-10. As Fig. 19-10 shows, LH brings about the following changes:

1 Completion of growth of the follicle and ovum with increasing secretion of estrogens before ovulation. LH and FSH act as synergists to produce these effects.

2 Rupturing of the mature follicle with expulsion of its ripe ovum (process known as *ovulation*). Because of this function, LH is sometimes called the ovulating hormone.

*Mountcastle, V. B., editor: Medical physiology, ed. 13, St. Louis, 1974, The C. V. Mosby Co., p. 1752 and Fig. 72-5.

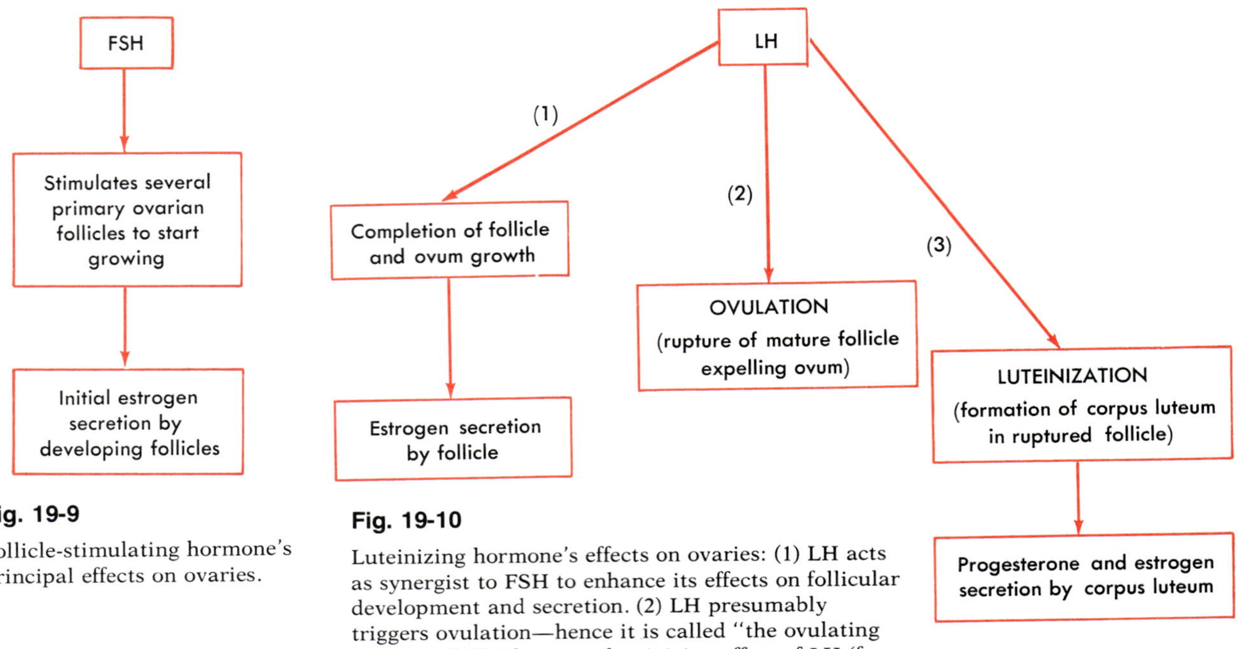

Fig. 19-9

Follicle-stimulating hormone's principal effects on ovaries.

Fig. 19-10

Luteinizing hormone's effects on ovaries: (1) LH acts as synergist to FSH to enhance its effects on follicular development and secretion. (2) LH presumably triggers ovulation—hence it is called "the ovulating hormone." (3) There is a luteinizing effect of LH (for which the hormone was named); recent evidence shows that FSH is also necessary for luteinization.

3 Formation of a golden body, the corpus luteum, in the ruptured follicle (process called *luteinization*). The name luteinizing hormone refers, obviously, to this LH function—a function to which, according to recent evidence, FSH also contributes. The corpus luteum functions as a temporary endocrine gland. It secretes only during the luteal (postovulatory or premenstrual) phase of the menstrual cycle. Its hormones are progestins (the important one of which is progesterone) and also estrogens. The blood level of progesterone rises rapidly after the "LH surge" described before. It remains at a high level for about a week, then it decreases to a very low level approximately 3 days before menstruation starts again. This low blood level of progesterone persists during both the menstrual and postmenstrual phases. Its sources? Not from the corpus luteum, which secretes only during the luteal phase, but from the developing follicles and the adrenal cortex. Blood's estrogen content increases during the luteal phase but to a lower level than develops before ovulation.

If pregnancy does not occur, the corpus luteum regresses in about 12 days. A scar tissue structure, the corpus albicans, replaces it. Fig. 19-5 shows the cyclical changes in the ovarian follicles.

Control of cyclical changes in uterus

Cyclical changes in the uterus are brought about by changing blood concentrations of estrogens and progesterone. As blood estrogens increase during the postmenstrual phase of the menstrual cycle, they produce the following main changes in the uterus.

1 Proliferation of endometrial cells, producing a thickening of the endometrium
2 Growth of endometrial glands and of its spiral arteries
3 Increase in the water content of the endometrium
4 Increased myometrial contractions

Increasing blood progesterone concentration during the premenstrual phase of the menstrual cycle produces progestational changes in the uterus—that is, changes favorable for pregnancy, specifically:

1 Secretion by endometrial glands, thereby preparing the endometrium for implantation of a fertilized ovum
2 Increase in the water content of the endometrium
3 Decreased myometrial contractions

Control of cyclical changes in amounts of gonadotropins secreted

Both negative and positive feedback mechanisms help control anterior pituitary secretion of the gonadotropins, FSH and LH. These mechanisms involve the ovaries' secretion of both estrogens and progesterone and the hypothalamus' secretion of releasing hormones (until recently, called releasing factors). Fig. 19-11 describes a negative feedback mechanism that controls gonadotropin secretion. Examine it carefully. Note par-

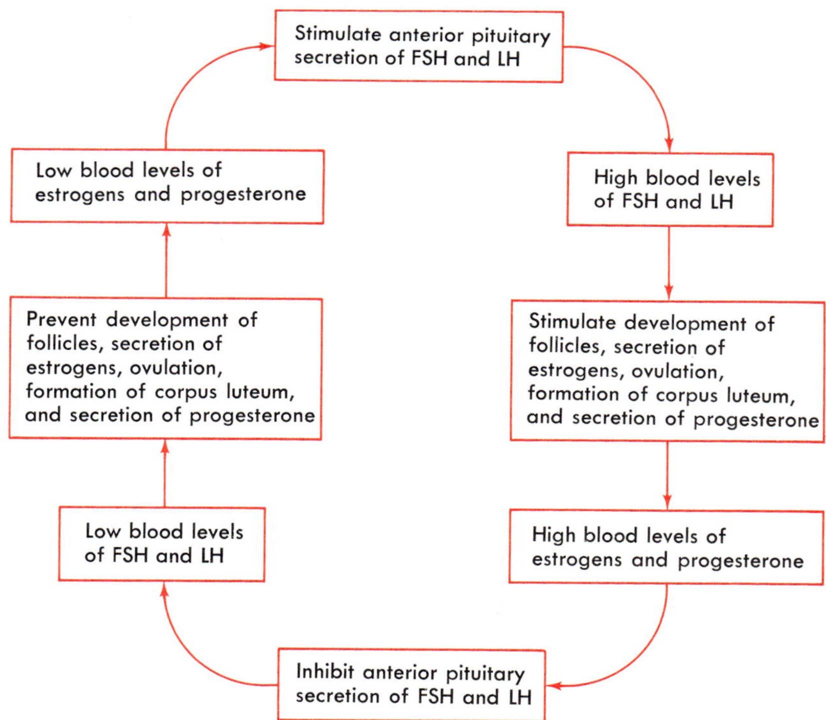

Fig. 19-11

A negative feedback mechanism that controls anterior pituitary secretion of follicle-stimulating hormone (FSH) and ovarian secretion of estrogens. A high blood level of FSH stimulates estrogen secretion, whereas the resulting high estrogen level inhibits FSH secretion. How does this compare with interstitial cell–stimulating hormone (ICSH)–testosterone feedback mechanism? (See Fig. 18-3 if you want to check your answer.) According to the above diagram, what effect does a high blood concentration of estrogens have on anterior pituitary secretion of FSH? of LH?

ticularly the effects of a high blood concentration of estrogens on anterior pituitary gland secretion and the effect of a low blood concentration of FSH on the development of a graafian follicle and ovum. Establishment of these two facts led eventually to the development of "the pill" for preventing pregnancy. Contraceptive pills contain synthetic estrogen-like or progesterone-like compounds or both. By building up a high blood concentration of these substances, they prevent the development of a follicle and its ovum that month. With no mature ovum to be expelled, ovulation does not occur, and therefore pregnancy cannot occur. The next menses, however, does take place—because the progesterone and estrogen dosage is stopped in time to allow their blood level to decrease as they normally do near the end of the cycle to bring on menstruation. Actually the effects of contraceptive pills are much more complex than our explanation indicates. They still are not completely understood.

Several observations and animal experiments strongly suggest that high blood levels of estrogens and progesterone may also act indirectly to inhibit pituitary secretion of FSH and LH. These ovarian hormones appear to inhibit certain neurons of the hypothalamus (part of the central nervous system) from secreting FSH-releasing and LH-releasing hormones into the pituitary portal vessels. Without the stimulating effects of these releasing hormones, the pituitary's secretion of FSH and LH decreases.

A positive feedback mechanism has also been postulated to control LH secretion. The rapid and marked increase in blood's estrogen content that occurs late in the follicular phase of the menstrual cycle is thought to stimulate the hypothalamus to secrete LH-releasing hormone into the pituitary portal vessels. As its name suggests, this hormone stimulates the release of LH by the anterior pituitary, which in turn accounts for the "LH surge" that triggers ovulation. The fact that a part of the brain—the hypothalamus—secretes FSH- and LH-releasing hormones has interesting implications. This may be the crucial part of the pathway by which changes in a woman's environment or in her emotional state can alter her menstrual cycle. That this occurs is a matter of common observation. For example, intense fear of either becoming or not becoming pregnant often delays menstruation.

Functions served by cycles

The major function seems to be to prepare the endometrium each month for a pregnancy. If it does not occur, the thick vascular lining, no longer needed, is shed.

If fertilization of the ovum (pregnancy) occurs, the menstrual cycle is modified as follows:

1 The corpus luteum does not disappear but persists and continues to secrete progesterone and estrogens for 6 months or more of pregnancy. If it is removed by any means during the early months of pregnancy, spontaneous abortion results.

2 The fertilized ovum, which immediately starts developing into an embryo, travels down the tube and implants itself in the endometrium, so carefully prepared for this event.

Menarche and menopause

The menstrual flow first occurs (menarche) at puberty, at about the age of 13 years, although there is wide individual variation according to race, nutrition, health, heredity, etc. Normally, it recurs about every 28 days for about 40 years, except during pregnancy, and then ceases (menopause or climacteric). The average age at which menstruation ceases is reported to have increased markedly—from about age 40 a few decades ago to between ages 45 and 50 more recently.

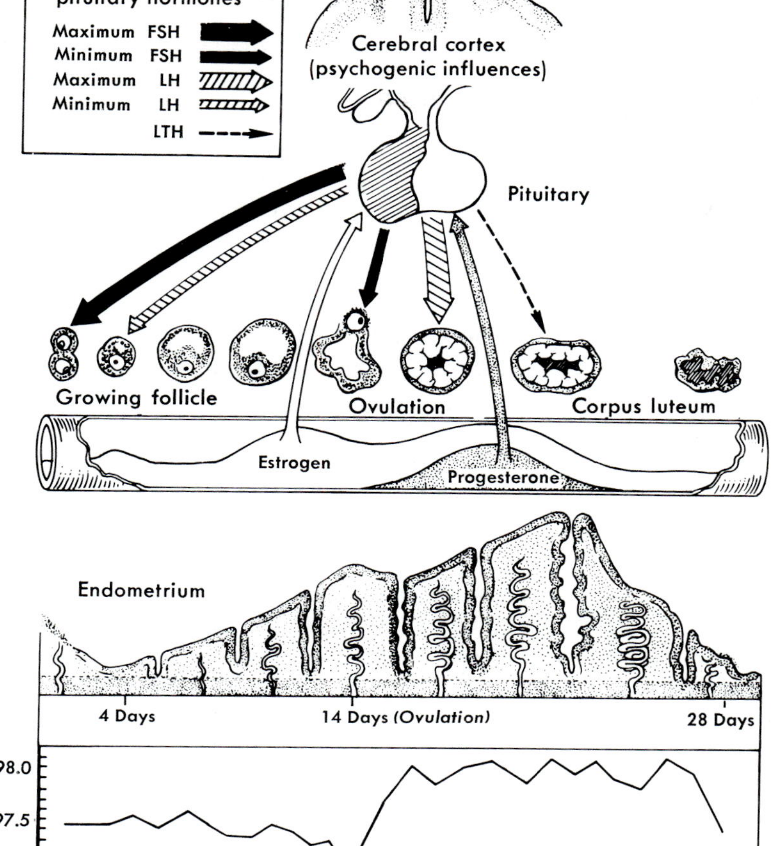

Fig. 19-12

Diagram illustrating the interrelationships among the cerebral, hypothalamic, pituitary, ovarian, and uterine functions throughout a usual 28-day menstrual cycle. The variations in basal body temperature are also illustrated, although this may not be demonstrated universally in women. (From Chinn, P. L.: Child health maintenance, St. Louis, 1974, The C. V. Mosby Co.)

Embryology

Meaning and scope

Embryology is the science of the development of the individual before birth. It is a story of miracles, describing the means by which a new human life is started and the steps by which a single microscopic cell is transformed into a complex human being. In a work of this kind, it seems feasible to include only a few of the main points in the development of the new individual since the facts amassed by the science of embryology are so many and so intricate that to tell them requires volumes, not paragraphs.

Value of knowledge of embryology

The main value to the medical profession of knowing the steps by which a new individual is evolved lies in their explanation of various congenital deformities. For example, one of the most common malformations is harelip, a condition that results from imperfect fusion of the frontal and maxillary processes during embryonic development.

Steps in development of new individual

Preliminary processes

Production of a new human being starts with the union of a spermatozoon and an ovum to form a single cell. Two preliminary steps, however, are necessary before such a union can take place: maturation of the sex cells (meiosis) and ovulation and insemination.

Maturation of sex cells (meiosis*)—reduction in number of chromosomes to half original number. Only when maturation has occurred are the ovum and spermatozoon mature and ready to unite with each other. The necessity for chromosome reduction as a preliminary to union of the sex cells is explained by the fact that the cells of each species of living organisms contain a specific number of chromosomes. Human cells, for example, contain 23 pairs or a total of 46 chromosomes. If the male and female cells united without first halving their respective chromosomes, the resulting cell would contain twice as many chromosomes as is specific for human beings. Mature ova and sperm, therefore, contain only 23 chromosomes or half as many as other human cells. Of these, one is the sex chromosome and may be either one of two types, known as X or Y. All ova contain an X chromosome. Sperm, on the other hand, have either an X or a Y chromosome. A female child results from the union of an X chromosome–bearing sperm with an ovum and a male child from the union of a Y chromosome–bearing sperm with an ovum.

Ovulation and insemination. The second preliminary step necessary for conception of a new individual consists in bringing the sperm and ovum into proximity with each other so that the union of the two can take place. Two processes are involved in the accomplishment of this step.

1 Ovulation or expulsion of the mature ovum from the graafian follicle into the pelvic cavity, from which it enters one of the uterine tubes.

2 Insemination or expulsion of the seminal fluid from the male urethra into the female vagina. Several million sperm enter the female reproductive tract with each ejaculation of semen. By lashing movements of their flagella-like tails, assisted somewhat by muscular contractions of surrounding structures, the sperm make their way into the external os of the cervix, through the cervical canal and uterine cavity, and into the tubes.

Recently *Science News* (Jan. 12, 1974) reported that an English investigator had demonstrated that X sperm swim more slowly than do Y sperm. An interesting hypothesis stems from this fact. If insemination occurs on the day of ovulation, Y sperm, being faster than X sperm, should reach the ovum first. Therefore a Y sperm would be more apt to fertilize the ovum. And since Y sperm produce males, there should be a greater probability of having a baby boy when insemination occurs on the day of ovulation. Statistical evidence supports this view.

Developmental processes

Fertilization or union of male and female gametes to produce one-celled individual called zygote. The sperm "swim" up the tube toward the ovum. Although numerous sperm surround the ovum, only one penetrates it. As soon as the head and neck of one spermatozoon enter the ovum (the tail drops off), the remaining sperm seem to be repulsed. The sperm head then forms itself into a nucleus that approaches and eventually fuses with the nucleus of the ovum, producing, at that moment, a new single-celled individual or *zygote.* Half of the 46 chromosomes in the zygote nucleus have come from the sperm and half from the ovum. Since chromosomes are composed of *genes,* the new being inherits half of its genes and the characteristics they determine from its father and half from its mother.

Normally, fertilization occurs in the uterine tube. Occasionally, however, it takes

*For details of the mechanism of meiosis, see pp. 467-470.

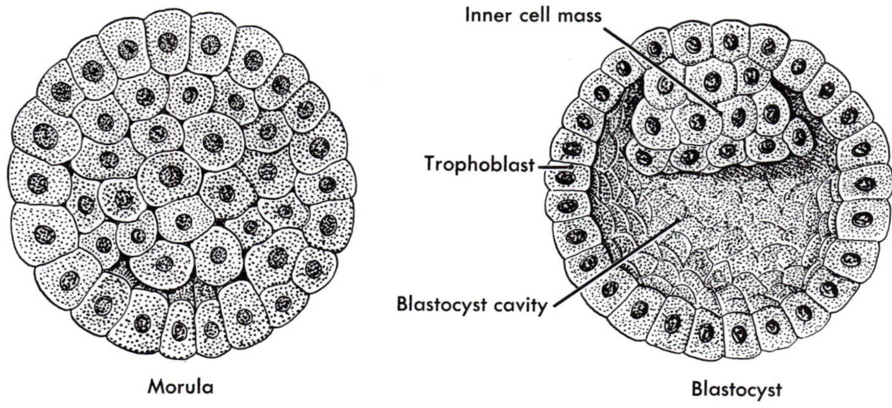

Fig. 19-13

Stages in the development of the human embryo. The morula consists of an almost solid spherical mass of cells. The embryo reaches this stage about 3 days after fertilization. The blastocyst (hollowing) stage develops later, after implantation in the uterine lining.

place in the pelvic cavity, as evidenced by pregnancies that start to develop in the pelvic cavity instead of in the uterus.

Inasmuch as the ovum lives only a short time (probably less than 48 hours) after leaving the graafian follicle, fertilization can occur only around the time of ovulation (p. 494). Sperm may live a few days after entering the female tract.

Cleavage and implantation. Cleavage consists of repeated mitotic divisions, first of the zygote to form two cells, then of those two cells to form four cells, etc., resulting in about 3 days' time in the formation of a solid, spherical mass of cells known as a *morula* (Fig. 19-13). A day or so later the embryo reaches the uterus, where it starts to implant itself in the endometrium. Occasionally, implantation occurs in the tube or pelvic cavity instead of in the uterus. The condition is known as an *ectopic pregnancy*.

As the cells of the morula continue to divide, a hollow ball of cells or *blastocyst*, consisting of an outer layer of cells and an inner cell mass, is formed. Implantation in the uterine lining is now complete (Fig. 19-13). About 10 days have elapsed since fertilization. The cells that compose the outer wall of the blastocyst are known as trophoblasts. They eventually become part of the placenta, the structure that anchors the fetus to the uterus.

Differentiation. As the cells composing the inner mass of the blastocyst continue to divide, they arrange themselves into a structure shaped like a figure-eight, containing two cavities separated by a double-layered plate of cells known as the *embryonic disk*. Young human embryos have been examined at this stage, or about 2 weeks old dated from the time of fertilization. The cells that form the cavity above the embryonic disk eventually become a fluid-filled, shock-absorbing sac (the amnion) in which the fetus floats. The cells of the lower cavity form the yolk sac, a small vesicle attached to the belly of the embryo until about the middle of the second month, when it breaks away. Only the double layer of cells that compose the embryonic disk is destined to form the new individual. The upper layer of cells is called the *ectoderm* and the lower layer the *entoderm*. A third layer of cells, known as the *mesoderm*, develops between the ectoderm and entoderm. Up to this time, all the cells have appeared alike, but now they are differentiated into three distinct types, ectodermal, mesodermal, and entodermal, known as the *primary germ layers*, each of which will give rise to definite structures. For example, the ectoderm cells will form the skin and its appendages and the nervous system; the mesoderm, the muscles, bones, and various other connective tissues; the entoderm, the epithelium of the digestive and respiratory tracts, etc.

Histogenesis and organogenesis. The story of how the primary germ layers develop into many different kinds of tissues (histogenesis) and how those tissues arrange themselves into organs (organogenesis) is long and complicated. Its telling belongs to the science of embryology. But for the beginning student of anatomy it seems sufficient to appreciate that life begins when two sex cells unite to form a single cell, that the new human body evolves by a series of processes consisting of cell multiplication, cell growth, cell differentiation, and cell rearrangements, all of which take place in definite, orderly sequence. By the end of the second month, a recognizable human figure has been formed. At the end of another month, the sex is clearly distinguishable, and from then until birth, development is mainly a matter of growth.

A Japanese physician, Dr. Motoyuki Hayaski of the Toho University School of Medicine in Tokyo, has recently completed a film for which he used time-lapse photography to show in living subjects various processes that create a new life. For many of the scenes he used human volunteers, for others he substituted animals. Ovulation, spermatogenesis, fertilization of the ovum, cleavage, differentiation, and formation of organs are all shown.*

*The Fertility Research Foundation, according to a report in *Time,* June 24, 1974, hopes to make possible showings of the film in the United States.

Outline summary

FEMALE REPRODUCTIVE ORGANS

- **1** Glands
  - **a** Ovaries (gonads)
  - **b** Accessory glands
    - 1 Bartholin's (greater vestibular)
    - 2 Skene's (lesser vestibular)
    - 3 Mammary (breasts)
- **2** Ducts
  - **a** Uterine (fallopian) tubes
  - **b** Uterus
  - **c** Vagina
- **3** External genitalia (vulva)

Uterus

- **1** Structure
  - **a** Size, shape, and divisions
    - 1 In virginal state, 3 inches × 2 inches × 1 inch
    - 2 Pear-shaped
    - 3 Consists of body and cervix; fundus bulging upper surface of body
  - **b** Wall—lining of mucosa called endometrium; thick, middle coat of muscle called myometrium; partial external coat of peritoneum
  - **c** Cavities—body cavity small and triangular in shape with three openings—two from tubes and one (internal os) into cervical canal; external os opening of cervical canal into vagina
  - **d** Blood supply—generous, from uterine arteries
- **2** Location—in pelvic cavity between bladder and rectum
- **3** Position
  - **a** Flexed between body and cervix, with body lying over bladder pointing forward and slightly upward
  - **b** Cervix joins vagina at right angles
  - **c** Capable of considerable mobility; therefore, often in abnormal positions, such as retroverted
  - **d** Eight ligaments anchor it—two broad, two uterosacral, one posterior, one anterior, and two round
- **4** Functions
  - **a** Menstruation
  - **b** Pregnancy
  - **c** Labor and expulsion of fetus

Uterine tubes—fallopian tubes or oviducts

- **1** Location—attached to uterus at upper, outer angles
- **2** Structure
  - **a** Same three coats as uterus
  - **b** Distal ends open with fimbriated margins
  - **c** Mucosa and peritoneum in direct contact here
- **3** Function
  - **a** Serve as ducts for ovaries, providing passageway by which ova can reach uterus
  - **b** Fertilization occurs here normally

Ovaries—female gonads

- **1** Location and size
  - **a** Size and shape of large almonds
  - **b** Lie behind and below uterine tubes
  - **c** Anchored to uterus and broad ligament

2 Microscopic structure
  a Consists of several thousand graafian follicles embedded in connective tissue base
  b Follicles in all stages of development; usually each month one matures, ruptures, and expels its ovum into abdominal cavity
3 Functions
  a Oogenesis—formation of mature female gametes (ova)
  b Secretion of hormones—estrogens and progesterone

Vagina

1 Location—between rectum and urethra
2 Structure
  a Collapsible, musculomembranous tube, capable of great distention
  b External outlet protected by fold of mucous membrane, hymen
3 Functions
  a Receive seminal fluid
  b Is lower part of birth canal
  c Is excretory duct for uterine secretions and menstrual flow

Vulva

1 Consists of numerous structures that, together, constitute external genitalia
2 Main structures
  a Mons veneris—skin-covered pad of fat over symphysis pubis
  b Labia majora—hairy, skin-covered lips
  c Labia minora—small lips covered with modified skin
  d Clitoris—small mound of erectile tissue just below junction of two labia minora
  e Urinary meatus—just below clitoris; opens into urethra
  f Vaginal orifice—below urethra
  g Bartholin's or greater vestibular glands
    1 Comparable to bulbourethral glands of male
    2 Open by means of long duct in space between hymen and labia minora
    3 Ducts from lesser vestibular or Skene's glands open near urinary meatus

Perineum

1 Region between vaginal orifice and anus
2 Frequently torn at childbirth

Breasts

1 Location and size
  a Just under skin, over pectoral muscles
  b Size depends on deposits of adipose tissue
2 Structure
  a Divided into lobes and lobules; latter composed of glands
  b Single excretory duct per lobe opens in nipple
  c Circular, pigmented area called areola borders nipple
3 Function—secrete milk for infant
4 Mechanism controlling lactation—see Fig. 19-8

FEMALE SEXUAL CYCLES

Recurring cycles

1 Ovarian cycles
  a Each month, follicle and ovum develop
  b Follicle secretes estrogens
  c Ovum matures and follicle ruptures (ovulation)
  d Corpus luteum forms and secretes progesterone and estrogens
  e If no pregnancy, corpus luteum gradually degenerates and is replaced by fibrous tissue; remains if there is pregnancy
2 Endometrial or menstrual cycle
  a Surface of endometrium sloughs off during menses, with bleeding from denuded area
  b Regeneration and proliferation of lining occurs
  c Endometrial glands and arterioles become longer and more tortuous
  d Glands secrete viscous mucus and cycle repeats
3 Myometrial cycle—contractility increases before ovulation and subsides following
4 Gonadotropic cycles
  a FSH—a high rate of secretion for about 10 days after menses
  b LH secretion increasing from ninth cycle day to day of ovulation (Fig. 19-10)

Control

1 Cyclical changes in ovaries controlled as follows:
  a Increasing blood level of FSH during postmenstrual phase stimulates one or more primitive graafian follicles to start growing and to start secreting estrogens (Fig. 19-9); "estrogen surge" day or two before ovulation, lowest blood level of estrogen on day of ovulation
  b Increasing blood concentration of LH (from cycle day 9 to day of ovulation) causes completion of growth of follicle and ovum, stimulates follicle to secrete estrogens, causes ovulation and luteinization, which in turn leads to progesterone secretion (Fig. 19-10)
2 Cyclical changes in uterus controlled as follows:
  a Increasing blood concentration of estrogens during preovulatory phase causes proliferation of endometrial cells with resultant thickening of endometrium, growth of endometrial glands, increased water content of endometrium, and increased contractions by myometrium
  b Increasing blood concentration of progesterone after ovulation—i.e., during premenstrual phase, causes secretion by endometrial glands and preparation of endometrium for implantation, increased water content of endometrium, and decreased contractions by myometrium
3 Cyclical changes in FSH and LH secretion—controlled by both negative and positive feedback mechanisms that include releasing hormones from hypothalamus

Functions served by cycles

1 Preparation of endometrium for pregnancy during preovulatory and premenstrual periods
2 Shedding of progestational endometrium if no pregnancy occurs

Menarche and menopause

1 Menarche—onset of menses; often between 11 and 14 years of age
2 Menopause (climacteric)—cessation of menses; usually between 45 and 50 years of age

EMBRYOLOGY

Meaning and scope

1 Science of development of individual before birth
2 Long and complex study

Value of knowledge of embryology

Interprets many abnormal formations

Steps in development of new individual

1 Preliminary processes
  a Maturation of ovum and sperm or reduction of their chromosomes to half original number—i.e., half of 46
  b Ovulation and insemination
    1 Ovulation once every 28 days, usually about 14 days before beginning of next menstrual period
    2 Insemination, indefinite occurrence; sperm introduced into vagina, swim up to meet ovum in tube
2 Developmental processes
  a Fertilization or union of ovum and sperm
    1 Normally occurs in tube
    2 Only one sperm penetrates ovum
    3 One-celled new individual called zygote formed by union
  b Cleavage or segmentation
    1 Multiplication of cells from zygote by repeated mitosis
    2 Morula or solid ball of cells formed; becomes hollow with a cluster of cells attached at one point on inner surface of sphere; called blastocyst at this stage; now implanted in endometrium
  c Differentiation of cells into three primary germ layers—ectoderm, mesoderm, and entoderm
  d Histogenesis or formation of various tissues from primary germ layers; organogenesis or formation of various organs by rearrangements of tissues, such as fusions, shiftings, foldings, etc.

Review questions

1 Name the female sex glands. Name all the internal female reproductive organs.
2 What is the perineum? Of what clinical importance is it in the female?
3 Name the three openings to the exterior from the female pelvis. How many openings to the exterior are there in the male?
4 What is a graafian follicle? What does it contain?
5 How many ova mature in a month, usually?
6 When, in the menstrual cycle, is ovulation thought to occur?
7 Discuss the mechanism thought to control menstruation.
8 Name the periods in the menstrual cycle with the approximate length in days of each and the main events.
9 What two organs are necessary for menstruation to occur?
10 Name the divisions of the uterus.
11 What and where are the fallopian tubes? Approximately how long are they? With what are they lined? Their lining is continuous on their distal ends with what? On their proximal ends? Why is an infection of the lower part of the female reproductive tract likely to develop into a very serious condition?
12 What is the cul-de-sac of Douglas?
13 Describe briefly the hormonal control of breast development and lactation.
14 Explain how contraceptive pills act to prevent pregnancy.
15 On which day of the menstrual cycle would ovulation most probably occur in a 26-day cycle? in a 32-day cycle?
16 Where does fertilization normally take place?
17 Which of the following patients will no longer menstruate? Why? (1) One who has had a bilateral salpingectomy (removal of the tubes)? (2) One who has had a panhysterectomy (removal of the entire uterus)? (3) One who has had a bilateral oophorectomy (removal of both ovaries)? (4) One who has had a cervical hysterectomy (body and fundus removed)? (5) One who has had a unilateral oophorectomy?
18 Which of the patients described in question 16 could no longer become pregnant? Explain.
19 Following menopause, which of the following hormones—estrogens, FSH, progesterone—would you expect to have a high blood concentration? Which a low blood concentration? Explain your reasoning.
20 Define or make an identifying statement about each of the following: adolescence, climacteric, colostrum, corpus luteum, ectopic pregnancy, embryology, endometrium, fertilization, fimbriae, gamete, genes, gonad, hysterectomy, maturation, meiosis, menarche, mitosis, morula, oophorectomy, ovum, ovulation, puberty, salpingitis, semen, spermatozoon, vulva, zygote.

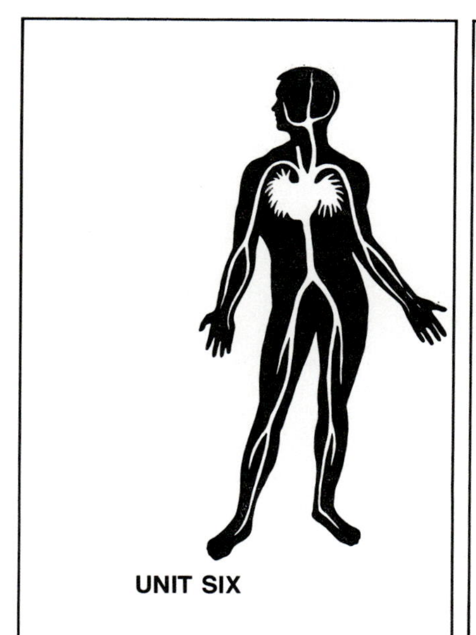

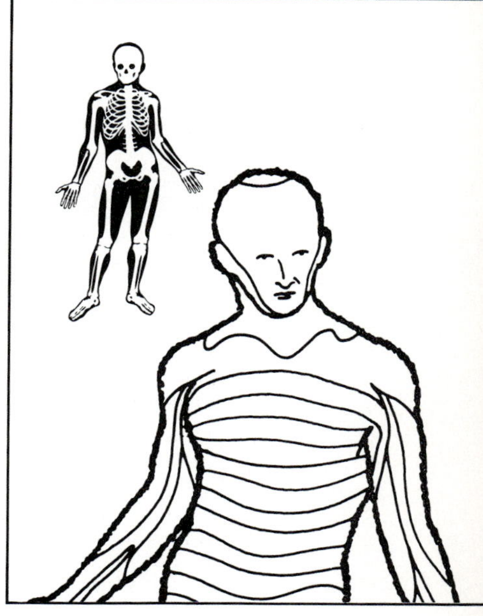

FLUID, ELECTROLYTE, AND ACID-BASE BALANCE

20

Fluid and electrolyte balance

Introduction

The term *fluid balance* means several things. It, of course, means the same thing as homeostasis of fluids. To say that the body is in a state of fluid balance is to say that the total amount of water in the body is normal and that it remains relatively constant. But fluid balance also means something more. It also means relative constancy of the distribution of water in the body's three fluid compartments. The volume of water inside the cells, in the interstitial spaces, and in the blood vessels all remains relatively constant when a condition of fluid balance exists. Fluid imbalance, then, means that both the total volume of water in the body and the amount in one or more of its fluid compartments have increased or decreased beyond normal limits.

Fluid balance and electrolyte balance are interdependent. If one deviates from normal, so does the other. A discussion of one, therefore, necessitates a discussion of the other.

Modern medicine attaches great importance to fluid and electrolyte balance. Today, a large proportion of hospital patients receive some kind of fluid and electrolyte therapy. To help you understand the rationale underlying such treatment, this chapter is included. We shall consider successively some general principles about fluid balance, the avenues by which water enters and leaves the body, the mechanisms that maintain homeostasis of total fluid volume, and the mechanisms that maintain homeostasis of fluid distribution.

Some general principles about fluid balance

1 The cardinal principle about fluid balance is this: fluid balance can be maintained only if intake equals output. Obviously, if more water or less leaves the body than enters it, imbalance will result. Total fluid volume will increase or decrease but cannot remain constant under those conditions.

2 Devices for varying output so that it equals intake constitute the most crucial mechanism for maintaining fluid balance, but mechanisms for adjusting intake to output also operate. Fig. 20-1 illustrates the aldosterone mechanism for decreasing fluid output (urine volume) to compensate for decreased intake. Fig. 20-2 diagrams a postulated mechanism for adjusting intake to compensate for excess output.

3 Mechanisms for controlling water movement between the fluid compartments of the body constitute the most rapid-acting fluid balance devices. They serve first of all to maintain normal blood volume at the expense of interstitial fluid volume.

Avenues by which water enters and leaves body

Water enters the body, as everyone knows, from the digestive tract—in the liquids one drinks and in the foods one eats. But, in addition, and less universally known, water enters the body—that is, is added to its total fluid volume—from its billions of cells. Each cell produces water by catabolizing foods, and this water enters the bloodstream. Water normally leaves the body by four exits: kidneys (urine), lungs (water in expired air), skin (by diffusion and by sweat), and intestines (feces). In accord with the cardinal principle of fluid balance, the total volume of water entering the body normally equals the total volume leaving. In short, fluid intake normally equals fluid output. Fig. 20-3 illustrates the portals of water entry and exit, and Table 20-1 gives their normal volumes. These, however, can vary considerably and still be considered normal.

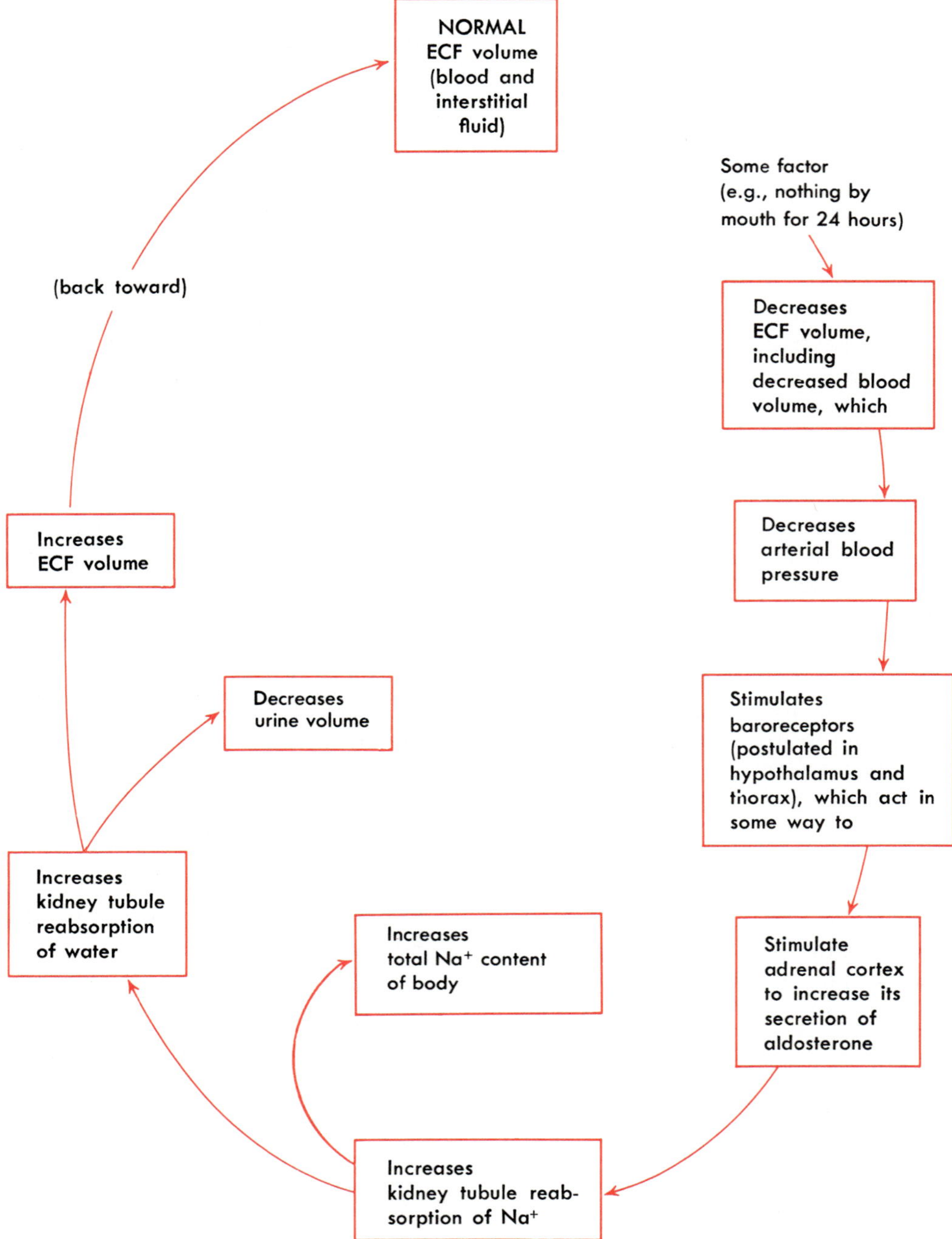

Fig. 20-1

Aldosterone mechanism that tends to restore normal extracellular fluid (ECF) volume when it decreases below normal. Excess aldosterone, however, leads to excess extracellular fluid volume—i.e., excess blood volume (hypervolemia) and excess interstitial fluid volume (edema)—and also to an excess of the total Na^+ content of the body.

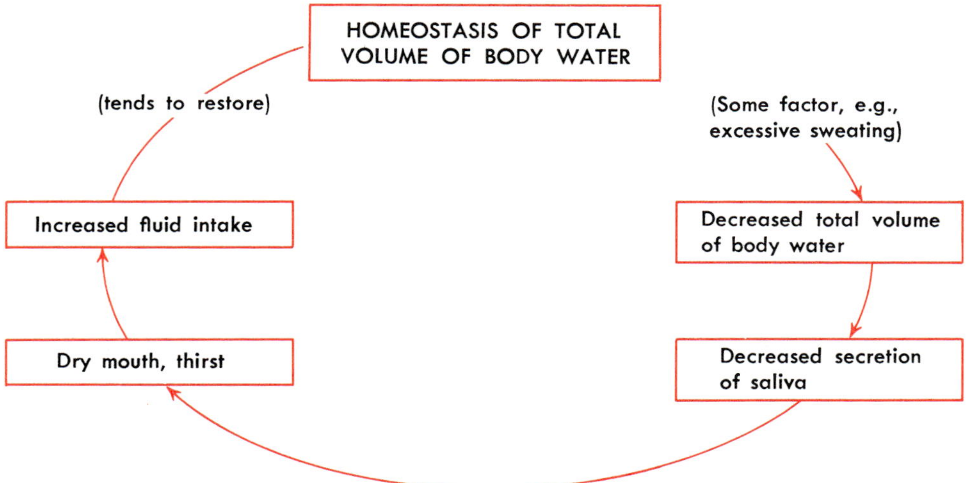

Fig. 20-2

The basic principle of a postulated homeostatic mechanism for adjusting intake to compensate for excess output.

Mechanisms that maintain homeostasis of total fluid volume

Under normal conditions, homeostasis of the total volume of water in the body is maintained or restored primarily by devices that adjust output (urine volume) to intake and secondarily by mechanisms that adjust fluid intake.

Regulation of urine volume

Two factors together determine urine volume: the glomerular filtration rate and the rate of water reabsorption by the renal tubules. The glomerular filtration rate, except under abnormal conditions, remains fairly constant—hence it does not normally cause urine volume to fluctuate. The rate of tubular reabsorption of water, on the other hand, fluctuates considerably. The rate of tubular reabsorption, therefore, rather than the glomerular filtration rate, normally adjusts urine volume to fluid intake. The amount of ADH and of aldosterone secreted regulates the amount of water reabsorbed by the kidney tubules (discussed on p. 452; also see Fig. 20-1). In other words, urine volume is regulated chiefly by hormones secreted by the posterior lobe of the pituitary gland (ADH) and by the adrenal cortex (aldosterone).

Although changes in the volume of fluid loss via the skin, the lungs, and the intestines also affect the fluid intake-output ratio, these volumes are not automatically adjusted to intake volume, as is the volume of urine.

Factors that alter fluid loss under abnormal conditions

Both the rate of respiration and the volume of sweat secreted may greatly alter fluid output under certain abnormal conditions. For example, a patient who hyperventilates for an extended time loses an excessive amount of water via the expired air. If, as frequently happens, he also takes in less water by mouth than normal, his fluid output then exceeds his intake and he develops a fluid imbalance—namely, dehydration (that is, a decrease in total body water). Other abnormal conditions such as vomiting, diarrhea, intestinal drainage, etc. also cause fluid and electrolyte output to exceed intake

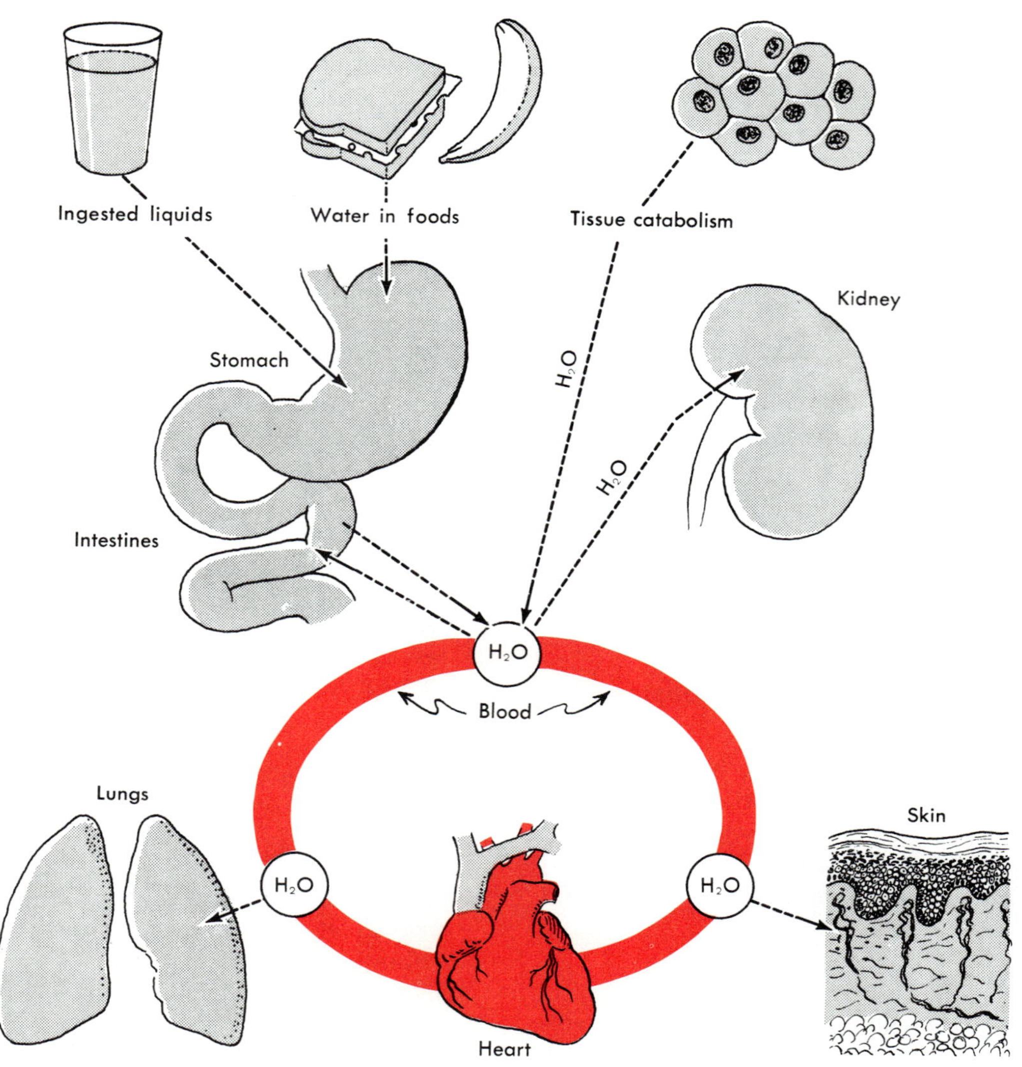

Fig. 20-3

The three sources of water entry into the body and its four avenues of exit.

Table 20-1

Typical normal values for each portal of water entry and exit

Intake		Output	
Ingested liquids	1,500 ml	Kidneys (urine)	1,400 ml
Water in foods	700 ml	Lungs (water in expired air)	350 ml
Water formed by catabolism	200 ml	Skin	
		By diffusion	350 ml
		By sweat	100 ml
		Intestines (in feces)	200 ml
Totals	2,400 ml		2,400 ml

and so produce fluid and electrolyte imbalances.

Regulation of fluid intake

Physiologists disagree about the details of the mechanism for controlling intake so that it increases when output increases and decreases when output decreases. In general, it operates in this way: when dehydration starts to develop, salivary secretion decreases, producing a "dry mouth feeling" and the sensation of thrist. The individual then drinks water, thereby increasing his fluid intake to offset his increased output, and this tends to restore fluid balance (Fig. 20-2). If, however, an individual takes nothing by mouth for several days, fluid balance cannot be maintained despite every effort of homeostatic mechanisms to compensate for the zero intake. Obviously, under this condition, the only way balance could be maintained would be for fluid output to also decrease to zero. But this cannot occur. Some output is obligatory. Why? Because as long as respirations continue, some water leaves the body by way of the expired air. Also, as long as life continues, an irreducible minimum of water diffuses through the skin.

Mechanisms that maintain homeostasis of fluid distribution

Comparison of plasma, interstitial fluid, and intracellular fluid

Structurally speaking, body fluids occupy three fluid compartments: blood vessels, tissue spaces, and cells. Thus we speak of the plasma, interstitial fluid, and intracellular fluid. Functionally, however, these three fluids may be considered as only two fluids: that which lies outside the cells and that which lies within them. Functionally, then, we speak of the extracellular fluid (ECF), meaning both the plasma and interstitial fluid, and the intracellular fluid (ICF), mean-

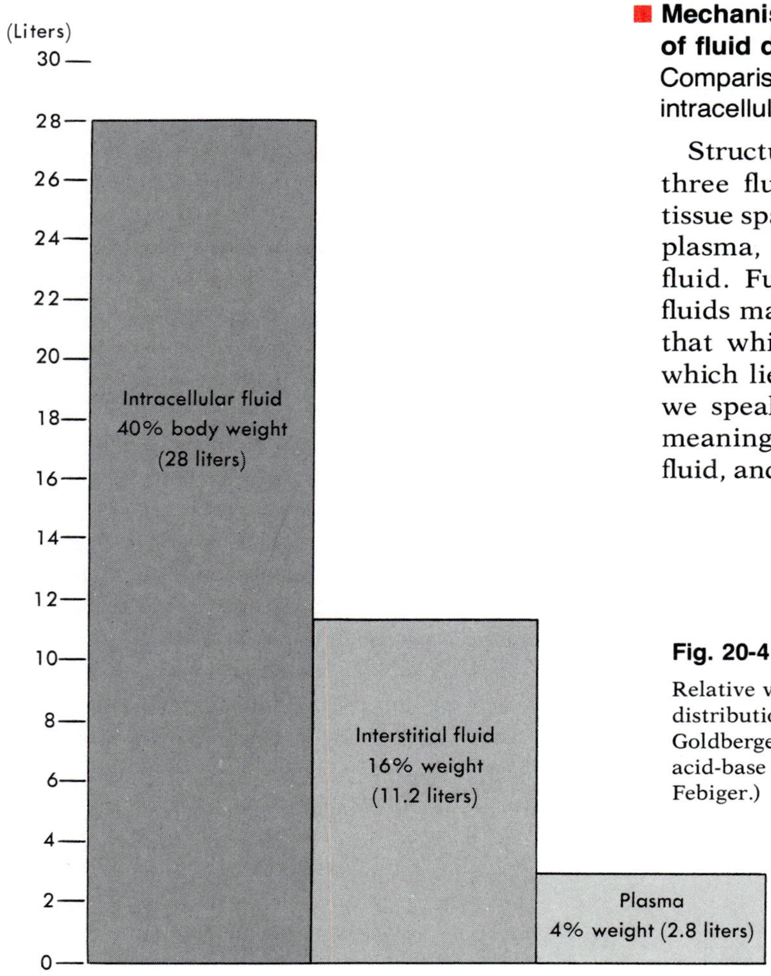

Fig. 20-4

Relative volumes of three body fluids. Normal fluid distribution in average young adult male. (Data from Goldberger, E.: A primer of water, electrolyte and acid-base syndromes, ed. 2, Philadelphia, 1962, Lea & Febiger.)

ing the water inside all the cells. Extracellular fluid, as you already know, constitutes the internal environment of the body. It therefore serves the dual vital functions of providing a relatively constant environment for cells and of transporting substances to and from them. Intracellular fluid, on the other hand, because it is a solvent, functions to facilitate intracellular chemical reactions that maintain life. Compared as to volume, intracellular fluid is the largest, plasma the smallest, and interstitial fluid in between. Fig. 20-4 gives typical normal fluid volumes in a young adult male. Note that intracellular fluid constitutes 40% of body weight, interstitial fluid 16%, and blood plasma 4%. Or, expressed differently, for every kilogram of its weight, the body contains about 400 ml of intracellular fluid, 160 ml of interstitial fluid, and 40 ml of plasma. Normal fluid volumes, however, vary considerably—mainly according to the fat content of the body. Fat people have a lower water content per kilogram of body weight than slender ones. Women have a relatively lower water content than men—but chiefly because a woman's body contains a higher percentage of fat. Total fluid volume and fluid distribution also vary with age. As a person grows older, the amount of water in his body decreases. Fluid makes up a smaller percent of his body weight. Extracellular fluid volume is proportionately larger in infants and children than in adults.

Compared chemically, plasma and interstitial fluid (the two extracellular fluids) are almost identical. Intracellular fluid, on the other hand, shows striking differences from either of the two extracellular fluids. Let us examine first the chemical structure of plasma and interstitial fluid as shown in Fig. 20-5 and Table 20-2.

Perhaps the first difference between the two extracellular fluids to catch your eye (in Fig. 20-5) is that blood contains a slightly larger total of electrolytes (ions) than does interstitial fluids. If you compare the two fluids, ion for ion, you will discover the most important difference between blood plasma and interstitial fluid. Look at the anions (negative ions) in these two extracellular fluids. Note that blood contains an appreciable amount of protein anions. Interstitial fluid, in contrast, contains hardly any protein anions. This is the only functionally important difference between blood and interstitial fluid. It exists because the normal capillary membrane is practically impermeable to proteins. Hence, almost all protein anions remain behind in the blood instead of filtering out into the interstitial fluid. Because proteins remain in the blood, certain other differences also exist between blood and interstitial fluid: notably, blood contains more sodium ions and fewer chloride ions than does interstitial fluid.*

Table 20-2

Electrolyte composition of blood plasma*

Cations	Anions
142 mEq Na^+	102 mEq Cl^-
4 mEq K^+	26 HCO_3^-
5 mEq Ca^{++}	17 protein$^-$
2 mEq Mg^{++}	6 other
	2 $HPO_4^=$
153 mEq/L plasma	153 mEq/L plasma

*Data from Ruch, T. C., and Patton, H. D.: Physiology and biophysics, ed. 19, Philadelphia, 1965, W. B. Saunders Co.

*According to the Donnan equilibrium principle, when nondiffusible anions (negative ions) are present on one side of a membrane, there are on that side of the membrane fewer diffusible anions and more diffusible cations (positive ions) than on the other side. Applying this principle to the blood and interstitial fluid: because blood contains nondiffusible protein anions, it contains fewer chloride ions (diffusible anions) and more sodium ions (diffusible cations) than does interstitial fluid.

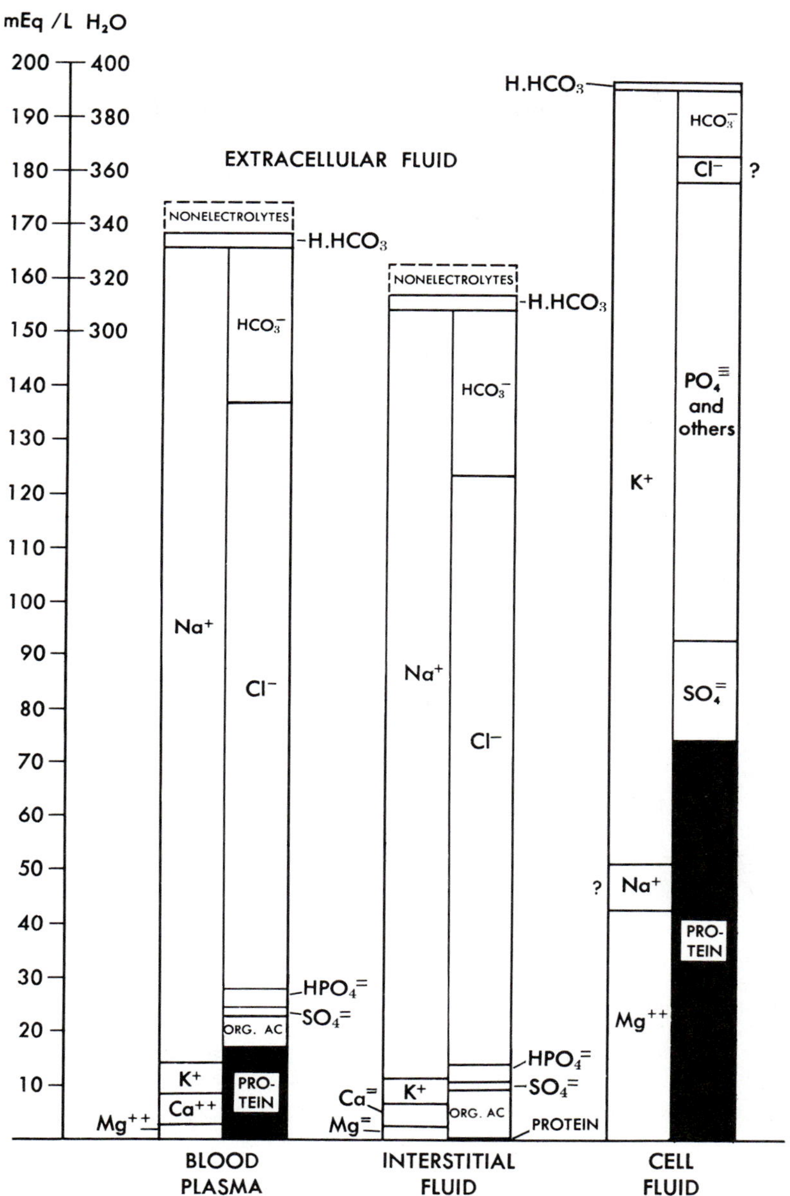

Fig. 20-5

Chief chemical constituents of three fluid compartments. The column of figures on the left (200, 190, 180, etc.) indicates amounts of cations or of anions, while the figures on the right (400, 380, 360, etc.) indicate the sum of cations and anions. Note that chloride and sodium values in cell fluid are questioned. It is probable that at least muscle intracellular fluid contains some sodium but no chloride. (Slightly modified from Mountcastle, V. B., editor: Medical physiology, St. Louis, The C. V. Mosby Co.; after Gamble, J. L.: Physiological information gained from studies on the life raft ration, Harvey Lect. **42**:247-273, 1946-1947.)

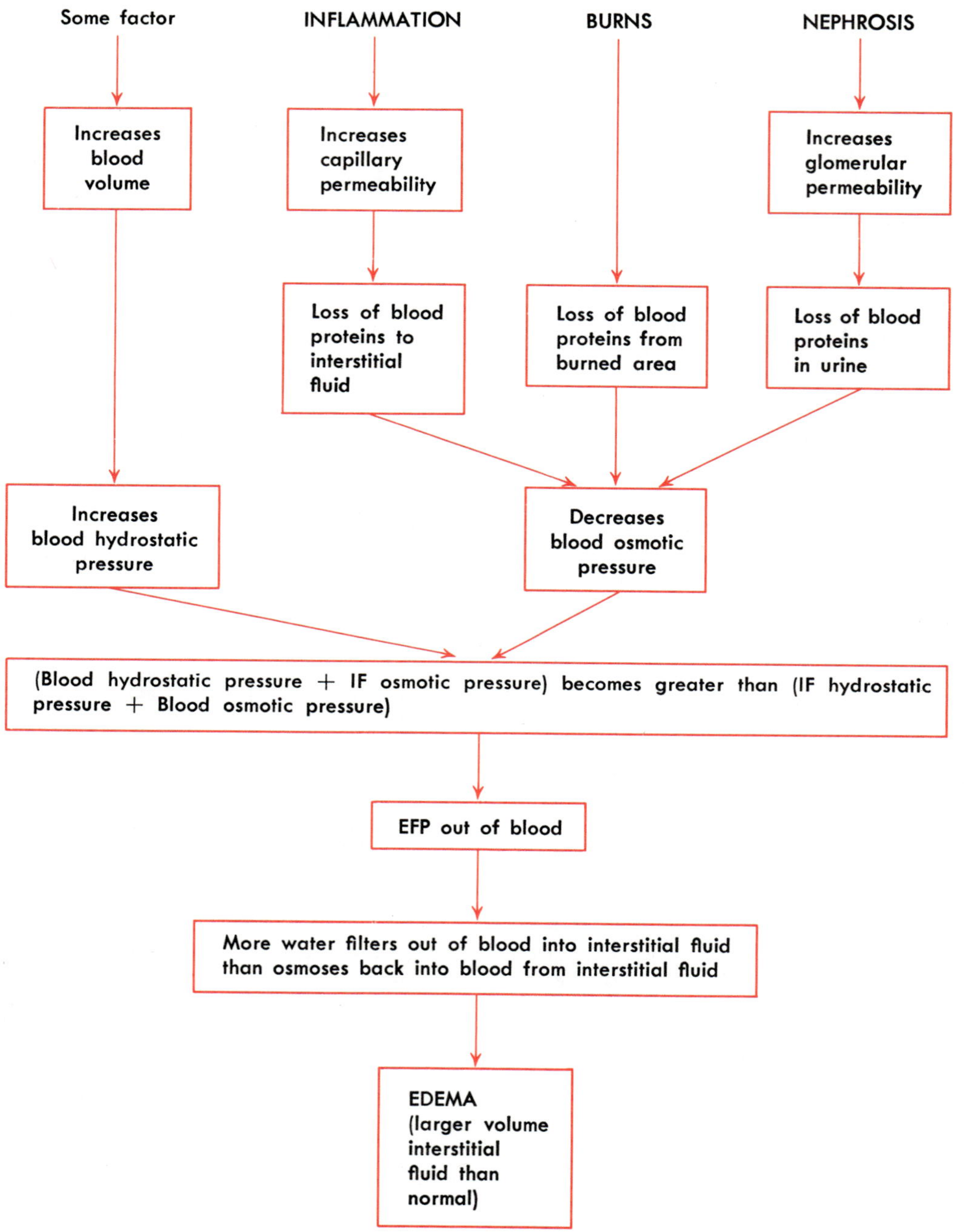

Fig. 20-6

Mechanisms of edema formation in some common conditions. *IF*, Interstitial fluid; *EFP*, effective or net filtration pressure. (Also see Fig. 20-7.)

Extracellular fluids and intracellular fluid are more unlike than alike chemically. Chemical difference predominates between the extracellular and intracellular fluids. Chemical similarity predominates between the two extracellular fluids. Study Fig. 20-5 and make some generalizations about the main chemical differences between the extracellular and intracellular fluids. For example: What is the most abundant cation in the extracellular fluids? In the intracellular fluid? What is the most abundant anion in the extracellular fluids? In the intracellular fluid? What about the relative concentrations of protein anions in extracellular fluids and intracellular fluid?

The only reason we have called attention to the chemical structure of the three body fluids is that here, as elsewhere, structure determines function. In this instance, the chemical structure of the three fluids helps control water and electrolyte movement between them. Or, phrased differently, the chemical structure of body fluids, if normal, functions to maintain homeostasis of fluid distribution and, if abnormal, results in fluid imbalance. Hypervolemia (excess blood volume) is a case in point. Edema, too, frequently stems from changes in the chemical structure of body fluids.

We are ready now to try to answer the following question: How does the chemical structure of body fluids control water movement between them and thereby control fluid distribution in the body?

Control of water movement between plasma and interstitial fluid

Over 60 years ago, Starling advanced a hypothesis about the nature of the mechanism that controls water movement between plasma and interstitial fluid—that is, across the capillary membrane. This hypothesis has since become one of the major premises of physiology and is often spoken of as Starling's "law of the capillaries." According to this law, the control mechanism for water exchange between plasma and interstitial fluid consists of four pressures: blood hydrostatic and colloid osmotic pressure* on one side of the capillary membrane and interstitial fluid hydrostatic and colloid† osmotic pressures on the other side.

According to the physical laws governing filtration and osmosis, blood hydrostatic pressure (HP) tends to force fluid out of capillaries into interstitial fluid (IF), but blood colloid osmotic pressure (OP) tends to draw it back into them. Interstitial fluid hydrostatic pressure, in contrast, tends to force fluid out of the interstitial fluid into the capillaries, and interstitial fluid colloid osmotic pressure tends to draw it back out of capillaries. In short, two of these pressures constitute vectors in one direction and two in the opposite direction. Does this remind you of another mechanism studied earlier? (To check your answer, see p. 450.)

The difference between the two sets of opposing forces obviously represents the net or effective filtration pressure—in other words, the effective force tending to produce the net fluid movement between blood and interstitial fluid. In general terms, therefore, we may state Starling's law of the capillaries this way: The rate and direction of fluid exchange between capillaries and interstitial fluid is determined by the hydrostatic and colloid osmotic pressures of the two fluids.

*Osmotic pressure from concentrations of protein in blood and interstitial fluid. Since the capillary membrane is permeable to other plasma solutes, they quickly diffuse through the membrane so cause no osmotic pressure to develop against it. Only the proteins, to which the capillary membrane is practically impermeable, cause an actual osmotic pressure against the capillary membrane (explained on pp. 26-30).

†A small amount of blood protein passes through the capillary membrane and tends to concentrate around the venous end of capillaries, hence this pressure.

Or, we may state it more specifically as a formula:

(BHP + IFOP) − (IFHP + BOP) = EFP*

Note that the factors enclosed in the first set of parentheses tend to move fluid out of capillaries and that those in the second set oppose this movement—they tend to move fluid into the capillaries.

To illustrate operation of Starling's law, let us consider how it controls water exchange at the arterial ends of tissue capillaries. Table 20-3 gives typical normal pressures. Using these figures in Starling's law of the capillaries:

(35 + 0) − (2 + 25) = 8 mm Hg net pressure (EFP), causing water to filter out of blood at arterial ends of capillaries into interstitial fluid

Table 20-3

Pressures at arterial end of tissue capillaries

Arterial end of capillary	Hydrostatic pressure	Colloid osmotic pressure
Blood	35 mm Hg	25 mm Hg
Interstitial fluid	2 mm Hg	0 mm Hg

The same law operates at the venous end of capillaries (Table 20-4). Again apply Starling's law of the capillaries. What is the net effective pressure at the venous ends of capillaries? In which direction does it cause water to move? Assuming that the figures given in Table 20-4 are normal, do you agree that theoretically "the same amount of water returns to the blood at the venous ends of capillaries as left it from the arterial ends"?

*BHP, Blood hydrostatic pressure. IFOP, Interstitial fluid osmotic pressure. IFHP, Interstitial fluid hydrostatic pressure. BOP, Blood osmotic pressure. EFP, Effective filtration pressure between blood and interstitial fluid.

Table 20-4

Pressures at venous end of tissue capillaries

Venous end of capillary	Hydrostatic pressure	Colloid osmotic pressure
Blood	15 mm Hg	25 mm Hg
Interstitial fluid	1 mm Hg	3 mm Hg*

*A small amount of blood proteins passes through the capillary membrane and tends to concentrate around the venous ends of capillaries—hence this pressure.

On the basis of our discussion thus far, we can formulate some principles about the transfer of water between blood and interstitial fluid.

1 No net transfer of water occurs between blood and interstitial fluid so long as the effective filtration pressure (EFP) equals 0—that is, when:

(Blood HP + IFOP) = (IFHP + Blood OP)

2 A net transfer of water, a "fluid shift," occurs between blood and interstitial fluid whenever the EFP does not equal 0—that is, when:

(Blood HP + IFOP)
does not equal
(IFHP + Blood OP)

3 Since (Blood HP + IFOP) is a force that tends to move water out of capillary blood, fluid shifts out of blood into interstitial fluid whenever:

(Blood HP + IFOP)
is greater than
(IFHP + Blood OP)

4 Since (IFHP + Blood OP) is a force that tends to move water out of interstitial fluid into capillary blood, fluid shifts out of interstitial fluid into blood whenever:

(IFHP + Blood OP)
is greater than
(Blood HP + IFOP)

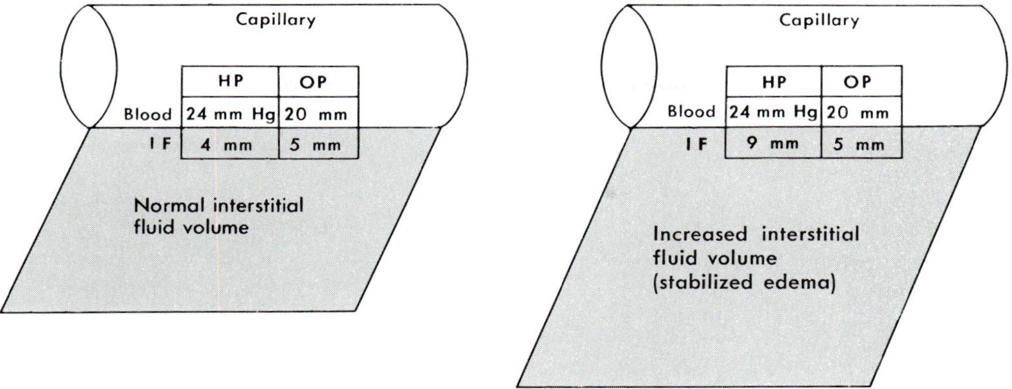

Fig. 20-7

The mechanism of edema formation initiated by a decrease in blood protein concentration and, therefore, in blood osmotic pressure. Left diagram, blood osmotic pressure has just decreased to 20 from the normal 25 mm Hg. This increases the effective filtration pressure (EFP) to 5 mm Hg from a normal of 0 (see Starling's formula, p. 516). The EFP of 5 mm Hg causes fluid to shift out of blood into interstitial fluid (IF) until the EFP again equals 0—in this case, when the interstitial fluid volume has increased enough the raise interstitial fluid hydrostatic pressure to 9 mm Hg as shown in the diagram on the right. At this point, a new equilibrium is established and equal amounts of water once more are exchanged between the blood and interstitial fluid. Thus, the increased interstitial fluid volume—that is, the edema—becomes stabilized.

Or, stated the other way around, whenever:

(Blood HP + IFOP)
is less than
(IFHP + Blood OP)

To apply these principles, answer questions 8 to 11 on p. 521 and examine Figs. 20-6 and 20-7.

Control of water movement through cell membranes between interstitial and intracellular fluids

The mechanism that regulates water movement through cell membranes is similar to the one that regulates water movement through capillary membranes. In other words, interstitial fluid and intracellular fluid hydrostatic and osmotic pressures regulate water transfer between these two fluids. But because the osmotic pressures of interstitial and intracellular fluids vary more than do their hydrostatic pressures, their osmotic pressures serve as the chief regulators of water transfer across cell membranes. Their osmotic pressures, in turn, are directly related to the electrolyte concentration gradients—notably sodium and potassium—maintained across cell membranes. As Fig. 20-5 shows, the chief electrolyte by far in interstitial fluid is sodium salt, and intracellular fluid's main electrolyte is potassium salt. Therefore, a change in the sodium or the potassium concentrations of either of these fluids causes the exchange of fluid between them to become unbalanced. For example, a decrease in interstitial fluid sodium concentration immediately decreases interstitial fluid osmotic pressure, making it hypotonic to intracellular fluid osmotic pressure. In other words, a decrease in interstitial fluid sodium concentration establishes an

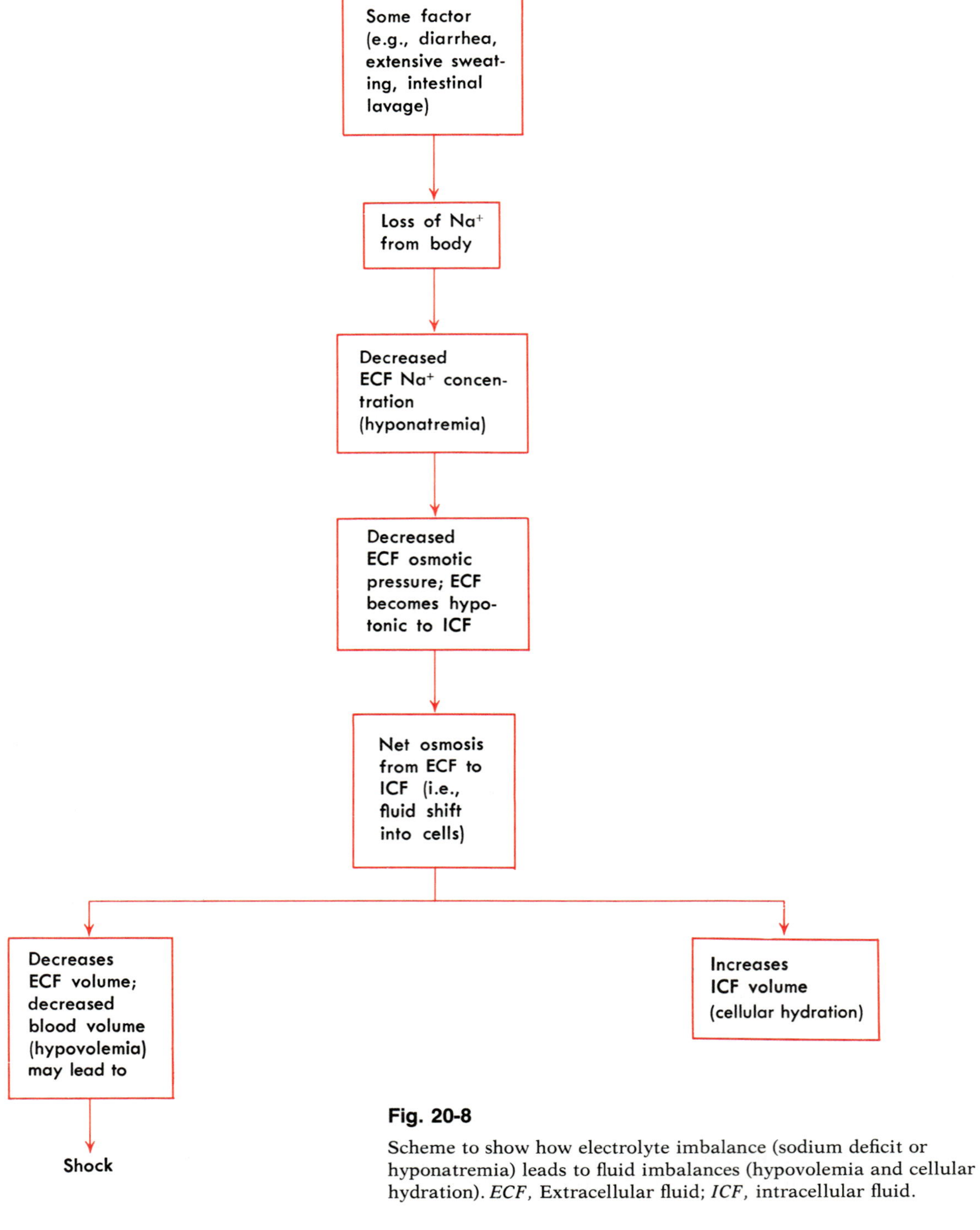

Fig. 20-8

Scheme to show how electrolyte imbalance (sodium deficit or hyponatremia) leads to fluid imbalances (hypovolemia and cellular hydration). *ECF,* Extracellular fluid; *ICF,* intracellular fluid.

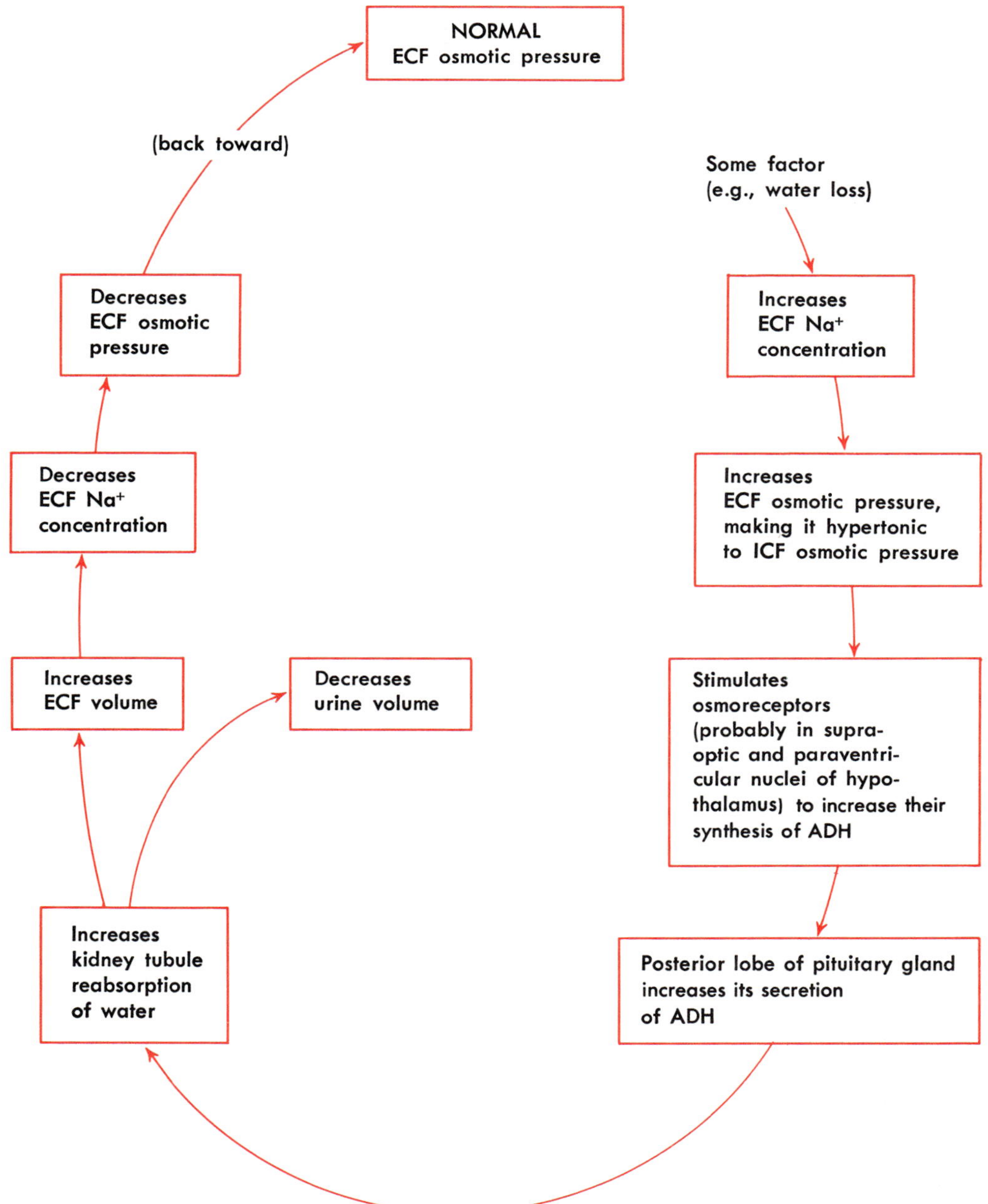

Fig. 20-9

ADH (antidiuretic hormone) mechanism that helps maintain homeostasis of extracellular fluid (ECF) osmotic pressure by regulating its volume and thereby its electrolyte concentration—that is, mainly ECF Na^+ concentration.

osmotic pressure gradient between interstitial and intracellular fluids. This causes net osmosis to occur out of interstitial fluid into cells. In short, interstitial fluid and intracellular fluid electrolyte concentrations are the main determinants of their osmotic pressures; their osmotic pressures regulate the amount and direction of water transfer between the two fluids, and this regulates their volumes. Hence, fluid balance depends upon electrolyte balance. Conversely, electrolyte balance depends upon fluid balance. An imbalance in one produces an imbalance in the other (Fig. 20-8).

Normal sodium concentration in interstitial fluid and potassium concentration in intracellular fluid depend upon many factors but especially upon the amount of ADH and aldosterone secreted. As shown in Fig. 20-9, ADH regulates extracellular fluid electrolyte concentration and osmotic pressure by regulating the amount of water reabsorbed into blood by renal tubules. Aldosterone, on the other hand, regulates extracellular fluid volume by regulating the amount of sodium reabsorbed into blood by renal tubules (Fig. 20-1).

Outline summary

Introduction

1 Meaning of fluid balance
   a Same as homeostasis of fluids—i.e., total volume of water in body normal and remains relatively constant
   b Volume of blood plasma, interstitial fluid, and intracellular fluid all remain relatively constant—i.e., homeostasis of distribution of water as well as of total volume
2 Fluid balance and electrolyte balance interdependent—see Figs. 20-1 and 20-9

Some general principles about fluid balance

1 Cardinal principle—intake must equal output
2 Fluid and electrolyte balance maintained primarily by mechanisms that adjust output to intake; secondarily by mechanisms that adjust intake to output
3 Fluid balance also maintained by mechanisms that control movement of water between fluid compartments

Avenues by which water enters and leaves body

1 Water enters body through digestive tract in
   a Liquids
   b Foods
2 Water formed in body by metabolism of foods
3 Water leaves body via kidneys, lungs, skin, and intestines

Mechanisms that maintain homeostasis of total fluid volume

1 Regulation of urine volume—under normal conditions, by factors that control reabsorption of water by distal and collecting tubules
   a ECF electrolyte concentration (crystalloid osmotic pressure) controls ADH secretion, which controls tubule H_2O reabsorption
   b ECF volume controls aldosterone secretion, which controls tubule Na^+ reabsorption and therefore water reabsorption
2 Factors that alter fluid loss under abnormal conditions—hyperventilation, hypoventilation, vomiting, diarrhea, circulatory failure, etc.
3 Regulation of fluid intake—see Fig. 20-2
   a Mechanism by which intake adjusted to output not completely known
   b One controlling factor seems to be degree of moistness of mucosa of mouth—if output exceeds intake, mouth feels dry, sensation of thirst occurs, and individual ingests liquids

Mechanisms that maintain homeostasis of fluid distribution

1 Comparison of plasma, interstitial fluid, and intracellular fluid
   a Plasma and interstitial fluid constitute extracellular fluid (ECF), internal environment of body or, in other words, environment of cells
   b Intracellular fluid (ICF) volume largest, plasma volume smallest; ICF about 40% of body weight, IF about 16%, and plasma about 4%, or ICF volume about ten times that of plasma and IF volume about four times plasma volume
   c Chemically, plasma and IF almost identical except that plasma contains slightly more electrolytes and considerably more proteins than IF; also blood contains somewhat more sodium and fewer chloride ions
   d Chemically, ECF and ICF strikingly different; sodium main cation of ECF, potassium main cation of ICF; chloride main anion of ECF; phosphate main anion of ICF; protein concentration much higher in ICF than in IF
2 Control of water movement between plasma and interstitial fluid
   a By four pressures—blood hydrostatic and colloid osmotic pressures and interstitial fluid hydrostatic and colloid osmotic pressures
   b Effect of these pressures on water movement between plasma and interstitial fluid expressed in Starling's "law of the capillaries"; only when (blood hydrostatic pressure + IF colloid osmotic pressure) − (blood colloid osmotic pressure + IF hydrostatic pressure) = 0, do equal amounts of water filter out of blood into IF and osmose back into blood from IF; in other words, water balance exists between these two fluids under these conditions

3 Control of water movement through cell membranes between interstitial and intracellular fluids—primarily by relative crystalloid osmotic pressures of ECF and ICF, which depend mainly upon sodium concentration of ECF and potassium concentration of ICF, which in turn depend upon intake and output of sodium and potassium and upon sodium-potassium transport mechanisms

Review questions

1 Explain in your own words the meaning of the term *fluid balance.*
2 How is total volume of body fluids kept relatively constant—i.e., what other factors must be controlled in order to keep the total volume of water in the body relatively constant?
3 What, if any, are functionally important differences between the chemical composition of plasma and interstitial fluid?
4 What, if any, are functionally important differences between the chemical composition of extracellular and intracellular fluids?
5 Are plasma and interstitial fluid more accurately described as "similar" or "different" as to chemical composition? Volume?
6 Support your answer to question **5** with some specific facts.
7 Are interstitial fluid and intracellular fluid more accurately described as "similar" or "different" as to chemical composition? Volume?
8 Explain Starling's law of the capillaries in your own words. Be as brief and clear as you can. This law describes the mechanism for controlling what?
9 Suppose that in one individual the normal average or mean pressures (in mm Hg) are capillary blood hydrostatic pressure 24 and osmotic pressure 25; interstitial fluid hydrostatic pressure 4 and osmotic pressure 5. Following a hemorrhage, this patient's blood hydrostatic pressure falls to 18. (a) Assuming that the other pressures momentarily stay the same, what is the EFP now? (b) The new EFP causes a fluid shift in which direction? (c) When will the fluid shift stop and an even exchange of water again go on between blood and interstitial fluids?
10 Formulate a principle by filling in the blanks in the following sentences: Anything that decreases blood hydrostatic pressure in the capillaries tends to cause a fluid shift from ________ into ________. Conversely, anything that increases blood hydrostatic pressure in the capillaries tends to cause a fluid shift out of ________ into ________.
11 Formulate a principle by completing the blanks in the following sentence: A decrease in blood protein concentration tends to decrease blood ________ pressure and therefore tends to cause a fluid shift from ________ to ________

Situation: A patient has had marked diarrhea for several days.

12 Explain the homeostatic mechanisms that would tend to compensate for this excessive fluid loss.
13 Do you think they could succeed in maintaining fluid balance or would fluid therapy probably be necessary?
14 What, if any, abnormality of fluid distribution do you think would occur in this patient without fluid therapy? Explain your reasoning.
15 Why does dehydration necessarily develop if an individual takes nothing by mouth for several days and receives no fluid therapy? Why cannot homeostatic mechanisms prevent this?

21

Acid-base balance

Acid-base balance is vitally important. Acid-base balance means maintenance of homeostasis of the hydrogen ion concentration of body fluids. Even a slight deviation from normal causes pronounced changes in the rate of cellular chemical reactions. This, in turn, threatens survival.

Mechanisms that control pH of body fluids

Meaning of term pH

The term pH is a symbol used to mean the hydrogen ion concentration of a solution. Actually, pH stands for the negative logarithm of the hydrogen ion concentration.* pH indicates the degree of acidity and alkalinity of a solution—the latter because as hydrogen ion concentration increases, OH ion concentration (alkalinity) necessarily decreases. A pH of 7 indicates neutrality (equal amounts of H^+ and OH^-), a pH of less than 7 indicates acidity (more H^+ than OH^-), and one greater than 7 indicates alkalinity (more OH^- than H^+).

Types of pH control mechanisms

Since various acids and bases continually enter the blood from absorbed foods and from the catabolism of foods, some kind of mechanism for neutralizing or eliminating these substances is necessary if blood pH is to remain constant. Actually, three different devices operate together to maintain constancy of pH. Collectively, these devices—buffers, respirations, and kidney excretion of acids and bases—might be said to constitute the pH homeostatic mechanism.

*A pH of 7, for example, means that a solution contains 10^{-7} gm hydrogen ions per liter. Or, translating this logarithm into a number, a pH of 7 means that a solution contains 0.0000001 (that is, 1/10,000,000) gm of hydrogen ions per liter. A solution of pH 6 contains 0.000001 (1/1,000,000) gm of hydrogen ions per liter and one of pH 8 contains 0.00000001 (1/100,000,000) gm of hydrogen ions per liter. Note that a solution with pH 7 contains ten times as many hydrogen ions as a solution with pH 8 and that pH decreases as hydrogen ion concentration increases.

Effectiveness of pH control mechanisms; range of pH

The most eloquent evidence of the effectiveness of the pH control mechanism is the extremely narrow range of pH, normally 7.35 to 7.45. In terms of hydrogen ion concentration, this means that normally there is a little more than 1/100,000,000 gm of hydrogen ions (pH 8) in a liter of blood but a little less than 1/10,000,000 gm (pH 7). Also, the greatest normal amount of hydrogen ions is only about 1/100,000,000 gm more than the smallest normal amount. What incredible constancy! Acids continually stream into capillary blood from cell metabolism and yet a liter of venous blood (pH 7.35) contains only about 1/100,000,000 gm more hydrogen ions than does a liter of arterial blood (pH 7.45)! The pH homeostatic mechanism does indeed control effectively—astonishingly so.

Buffer mechanism for controlling pH of body fluids

Buffers defined

In terms of action, a buffer is a substance that prevents marked changes in the pH of a solution when an acid or a base is added to it. Let us suppose that a small amount of the strong acid, hydrochloric acid, is added to a solution that contains a buffer (blood, for example) and that its pH decreases from 7.41 to 7.27. But if the same amount of hydrochloric acid were added to pure water containing no buffers, its pH would decrease much more markedly, from 7 to perhaps 3.4. In both instances, pH decreased upon addition of the acid, but much less so with buffers present than without them.

In terms of chemical composition, buffers consists of two kinds of substances and are, therefore, often referred to as "buffer pairs." Most of the body fluid buffer pairs consist of a weak acid and a salt of that acid, as shown below.

Buffer pairs present in body fluids

The main buffer pairs present in body fluids are as follows:

Bicarbonate pairs $\frac{NaHCO_3}{H_2CO_3}, \frac{KHCO_3}{H_2CO_3}$, etc.

Plasma protein pair $\frac{Na \cdot proteinate}{Proteins\ (weak\ acids)}$

Hemoglobin pairs $\frac{K \cdot Hb}{Hb}$ and $\frac{K \cdot HbO_2}{HbO_2}$

(Hb and HbO_2 are weak acids)

Phosphate buffer pair $\frac{Na_2HPO_4\ (basic\ phosphate)}{NaH_2PO_4\ (acid\ phosphate)}$

Action of buffers to prevent marked changes in pH of body fluids

Buffers react with a relatively strong acid (or base) to replace it by a relatively weak acid (or base). That is to say, an acid that highly dissociates to yield many hydrogen ions is replaced by one that dissociates less highly to yield fewer hydrogen ions. Thus, by the buffer reaction, instead of the strong acid remaining in the solution and contributing many hydrogen ions to drastically lower the pH of the solution, a weaker acid takes its place, contributes fewer additional hydrogen ions to the solution, and thereby lowers its pH only slightly. Therefore, because blood contains buffer pairs, its pH fluctuates much less widely than it would without them. In other words, blood buffers constitute one of the devices for preventing marked changes in blood pH.

Let us consider some specific examples of buffer action. More carbonic acid is formed in the body than any other acid. It continuously enters tissue capillaries, where it is buffered primarily by the potassium salt of hemoglobin inside the red blood cells, as shown in Fig. 21-1.

Note how this reaction between carbonic acid (H_2CO_3) and the potassium salt of hemoglobin (a buffer) applies the principle of buffering. As a result of the buffering action of KHb, the weaker acid, $H \cdot Hb$, replaces the stronger acid, H_2CO_3, and, therefore, the hydrogen ion concentration of blood increases much less than it would have if carbonic acid were not buffered.

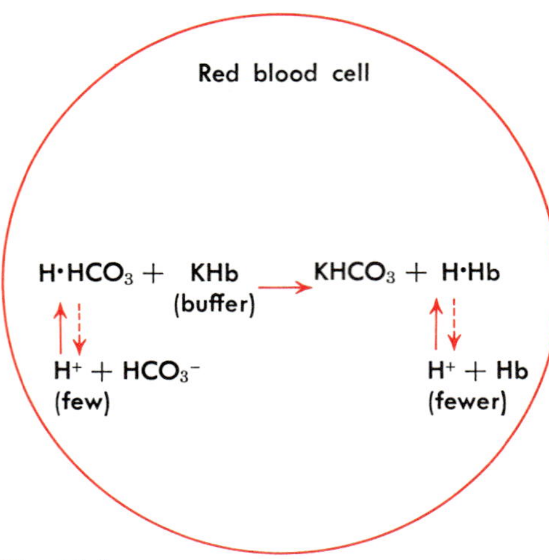

Fig. 21-1

Buffering of carbonic acid ($H \cdot HCO_3$ or H_2CO_3) inside red blood cell by potassium salt of hemoglobin. Note that each molecule of carbonic acid is replaced by a molecule of the acid hemoglobin. Since hemoglobin is a weaker acid than carbonic acid, fewer of these hemoglobin molecules dissociate to form hydrogen ions. Hence, fewer hydrogen ions are added to red blood cell intracellular fluid than would have been added by unbuffered carbonic acid. Also, since some of the carbonic acid in the red blood cells has come from plasma, fewer hydrogen ions remain in blood than would have if there were no buffering of carbonic acid.

The potassium salt of oxyhemoglobin, sodium proteinate (sodium salts of blood proteins), and basic sodium phosphate also buffer carbonic acid.

Nonvolatile or fixed acids, such as lactic acid and ketone bodies, are buffered mainly by the basic member of the bicarbonate buffer pair, mainly by sodium bicarbonate—ordinary baking soda (Fig. 21-2).

Carbonic acid buffers strong bases, as is illustrated in Fig. 21-3.

$$\begin{array}{l} H\cdot lactate + NaHCO_3 \longrightarrow Na\cdot lactate + H\cdot HCO_3 \\ \uparrow\downarrow \qquad\qquad\qquad\qquad\qquad\qquad\qquad\qquad \uparrow\downarrow \\ H^+ + lactate^- \qquad\qquad\qquad\qquad\qquad H^+ + HCO_3^- \\ \text{(few)} \qquad\qquad\qquad\qquad\qquad\qquad\qquad \text{(fewer)} \end{array}$$

Fig. 21-2

Lactic acid (H · lactate) and other nonvolatile or "fixed" acids are buffered by sodium bicarbonate, the most abundant base bicarbonate in the blood. Carbonic acid (H · HCO_3 or H_2CO_3, a weaker acid than lactic acid) replaces lactic acid. Result: fewer hydrogen ions are added to blood than would be if lactic acid were not buffered. Ketone bodies (e.g., acetoacetic acid) from fat catabolism are also buffered by base bicabonate.

$$\begin{array}{l} NaOH + H\cdot HCO_3 \longrightarrow NaHCO_3 + HOH \\ \downarrow\uparrow \qquad\qquad\qquad\qquad\qquad\qquad\qquad \uparrow\downarrow \\ Na^+ + OH^- \qquad\qquad\qquad\qquad\quad H^+ + OH^- \\ \quad\text{(many)} \qquad\qquad\qquad\qquad\qquad \text{(very few)} \end{array}$$

Fig. 21-3

Buffering of base NaOH by carbonic acid.

When blood pH is normal and a state of acid-base balance exists, components of the bicarbonate buffer pair are present in the extracellular fluid in a ratio of 20 parts of base bicarbonate (primarily $NaHCO_3$) to 1 part of carbonic acid. Actually, a liter of plasma normally contains about 27 mEq of base bicarbonate (B · HCO_3) and 1.3 mEq of carbonic acid:

$$\frac{27\ \text{mEq}\cdot B\cdot HCO_3}{1.3\ \text{mEq}\cdot H_2CO_3} = \frac{20}{1} = \text{pH } 7.4$$

An increase in this ratio causes pH to increase (uncompensated alkalosis), and a decrease in it causes pH to decrease (uncompensated acidosis).

Both carbonic acid and nonvolatile acids enter the blood in tissue capillaries and are immediately buffered. If you examine Fig. 21-2, you can discover how the buffering of nonvolatile acids changes blood's bicarbonate–carbonic acid ratio. Note what this buffering action does. It replaces some of the sodium bicarbonate in blood with carbonic acid. So blood leaving the capillaries—venous blood, that is—contains less sodium bicarbonate and more carbonic acid than does arterial blood. Venous blood's base bicarbonate–carbonic acid ratio, therefore, is somewhat lower than arterial blood's. This necessarily means that venous blood's pH is also lower than arterial blood's pH.

Evaluation of role of buffers in pH control

Buffering alone cannot maintain homeostasis of pH. As we have seen, hydrogen ions are added continually to capillary blood despite buffering. If even a few more hydrogen ions were added every time blood circulated and no way were provided for eliminating them, blood hydrogen ion concentration would necessarily increase and thereby decrease blood pH. "Acid blood," in other words, would soon develop. Respiratory and urinary devices must, therefore, function concurrently with buffers in order to remove from the blood and from the body the hydrogen ions continually being added to blood. Only then can the body maintain constancy of pH.

Respiratory mechanism of pH control

Explanation of mechanism

Respirations play a vital part in controlling pH. With every expiration, carbon dioxide and water leave the body in the expired air. The carbon dioxide has come from the venous blood—has diffused out of it as it moves through the lung capillaries. Less carbon dioxide remains, therefore, in the arterial blood leaving the lung capillaries and fewer

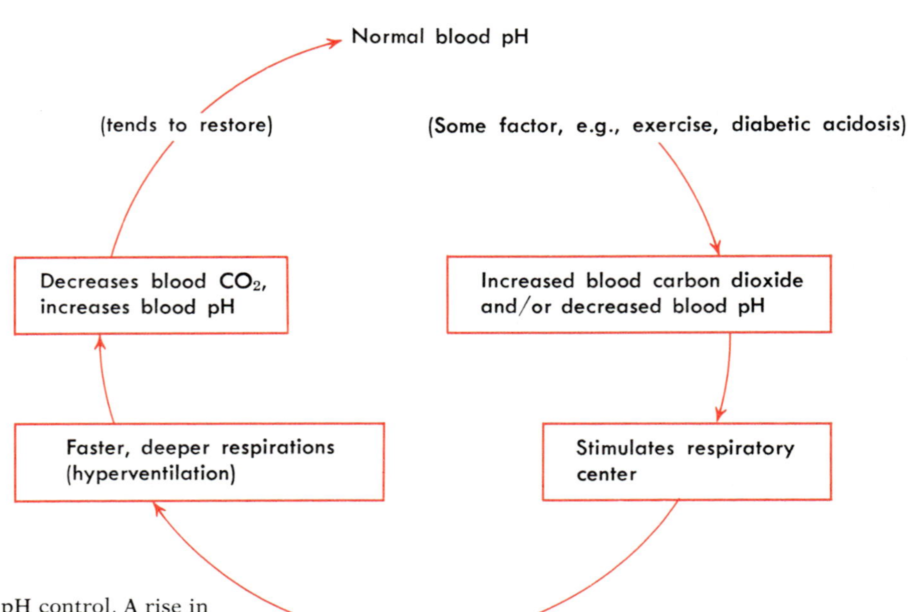

Fig. 21-4

Respiratory mechanism of pH control. A rise in arterial blood CO_2 content or a drop in its pH (below about 7.38) stimulates respiratory center neurons. Hyperventilation results. Less CO_2 and therefore less carbonic acid and fewer hydrogen ions remain in the blood so that blood pH increases, often reaching the normal level.

hydrogen ions can be formed in it by the following reactions:

$$CO_2 + H_2O \xrightarrow{\text{(carbonic anhydrase)}} H_2CO_3$$
$$H_2CO_3 \longrightarrow H^+ + HCO_3^-$$

So arterial blood has a lower hydrogen ion concentration and a higher pH than venous blood. A typical average pH for venous blood is 7.36, and 7.41 is a typical average pH for arterial blood.

Adjustment of respirations to pH of arterial blood

Obviously, in order for respirations to serve as a mechanism of pH control, there must be some mechanism for increasing or decreasing respirations as needed to maintain or restore normal pH. Suppose that blood pH has decreased (that is, hydrogen ion concentration has increased). Respirations then need to increase in order to eliminate more carbon dioxide from the body and thereby leave less carbonic acid and fewer hydrogen ions in the blood.

One mechanism for adjusting respirations to arterial blood carbon dioxide content or pH operates in this way. Neurons of the respiratory center are sensitive to changes in arterial blood carbon dioxide content and to changes in its pH. If the amount of carbon dioxide in arterial blood increases beyond a certain level, or if arterial blood pH decreases below about 7.38, the respiratory center is stimulated and respirations accordingly increase in rate and depth. This, in turn, eliminates more carbon dioxide, reduces carbonic acid and hydrogen ions, and increases pH back toward the normal level (Fig. 21-4). The carotid chemoreflexes (p. 312) are also devices by which respirations adjust to blood pH and, in turn, adjust pH.

Some principles relating respirations and pH of body fluids

1 A decrease in blood pH below normal (that is, acidosis) tends to stimulate increased respirations (hyperventilation), which tends to increase pH back toward normal. In other words, acidosis causes hyperventilation, which in turn acts as a compensating mechanism for the acidosis.

2 Prolonged hyperventilation may increase blood pH enough to produce alkalosis.

3 An increase in blood pH above normal (or alkalosis) causes hypoventilation, which serves as a compensating mechanism for the alkalosis by decreasing blood pH back toward normal.

4 Prolonged hypoventilation may decrease blood pH enough to produce acidosis.

Urinary mechanism of pH control

General principles about mechanism

Because the kidneys can excrete varying amounts of acid and base, they, like the lungs, play a vital role in pH control. Kidney tubules, by excreting many or few hydrogen ions, in exchange for reabsorbing many or few sodium ions, control urine pH and thereby help control blood pH. If, for example, blood pH decreases below normal, kidney tubules secrete more hydrogen ions from blood to urine and, in exchange for each hydrogen ion, reabsorb a sodium ion from the urine back into the blood. This, of course, decreases urine pH. But simultaneously—and of far more importance—it increases blood pH back toward normal. This urinary mechanism of pH control is a device for excreting varying amounts of hydrogen ions from the body to match the amounts entering the blood. It constitutes a much more effective device for adjusting hydrogen output to hydrogen input than does the body's only other mechanism for expelling hydrogen ions—namely, the respiratory mechanism previously described. But abnormalities of any one of the three pH control mechanisms soon throws the body into a state of acid-base imbalance. Only when all three parts of this complex mechanism—buffering, respirations, and urine secretion—function adequately can acid-base balance be maintained.

Let us turn our attention now to the mechanisms that adjust urine pH to counteract changes in blood pH.

Mechanisms that control urine pH

A decrease in blood pH accelerates the renal tubule ion-exchange mechanisms that both acidify urine and conserve blood's base and thereby tend to increase blood pH back to normal. The following paragraphs describe these mechanisms.

1 Distal and collecting tubules secrete hydrogen ions into the urine in exchange for basic ions, which they reabsorb. Refer to Fig. 21-5 as you read the rest of this paragraph. Note that carbon dioxide diffuses from tubule capillaries into distal tubule cells, where the enzyme carbonic anhydrase accelerates the combining of carbon dioxide with water to form carbonic acid. The latter dissociates into hydrogen ions and bicarbonate ions. The hydrogen ions then diffuse into the tubular urine, where they displace basic ions (most often sodium) from a basic salt of a weak acid and thereby change the basic salt to an acid salt or to a weak acid that is eliminated in the urine. While this is happening, the displaced sodium or other basic ion diffuses into a tubule cell. Here, it combines with the bicarbonate ion left over from the carbonic acid dissociation to form sodium bicarbonate. The sodium bicarbonate then diffuses—is reabsorbed, that is—into the blood. Consider the various results of this mechanism. Sodium bicarbonate (or other base bicarbonate) is conserved for the body. Instead of all the basic salts that filter out of glomerular blood leaving the body in the urine, considerable amounts are recovered into peritubular capillary blood. In addition, extra hydrogen ions are added to the urine and thereby eliminated from the body. Both the reabsorption of base bicarbonate into blood and the excretion of hydrogen ions into urine tend to increase the ratio of the bicarbonate buffer pair $B \cdot HCO_3/H \cdot HCO_3$ present in blood.

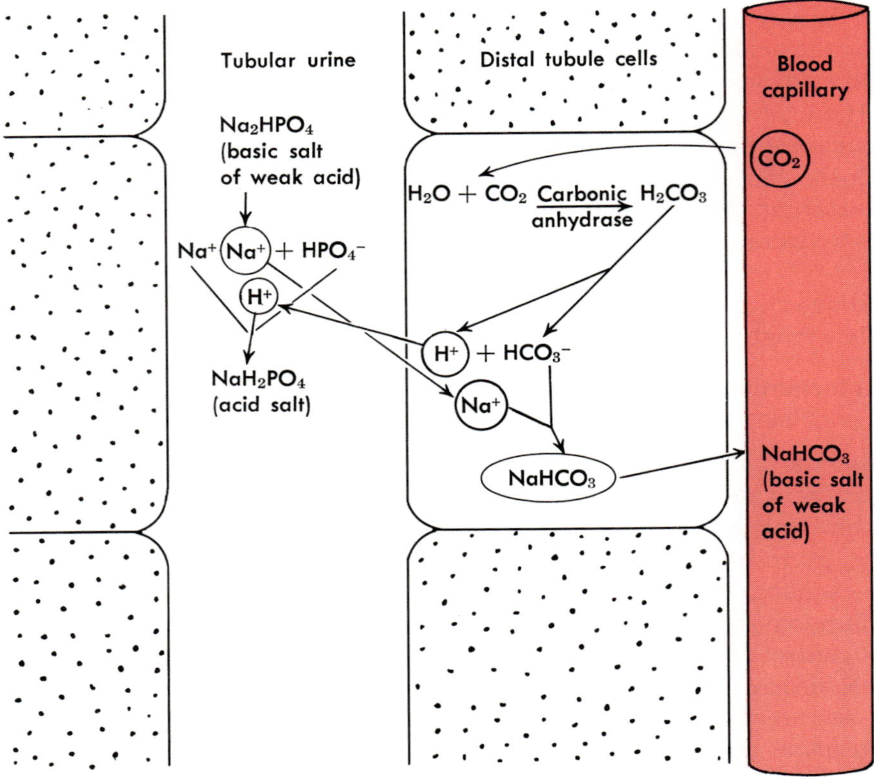

Fig. 21-5

Acidification of urine and conservation of base by distal renal tubule excretion of H ions (discussed on p. 527).

This automatically increases blood pH. In short, kidney tubule base bicarbonate reabsorption and hydrogen-ion excretion both tend to alkalinize blood by acidifying urine.

Renal tubules can excrete either hydrogen or potassium in exchange for the sodium they reabsorb. Therefore, in general, the more hydrogen ions they excrete, the fewer the potassium ions they can excrete. For example, in acidosis, tubule excretion of hydrogen ions increases markedly and potassium ion excretion decreases—an important fact because it may lead to *hyperkalemia* (excessive blood potassium), a dangerous condition because it can cause heart block and death.

2 Distal and collecting tubule cells excrete ammonia into the tubular urine. As Fig. 21-6 shows, the ammonia combines with hydrogen to form an ammonium ion. The ammonium ion displaces sodium or some other basic ion from a salt of a fixed (nonvolatile) acid to form an ammonium salt. The basic ion then diffuses back into a tubule cell and combines with bicarbonate ion to form a basic salt, which in turn diffuses into tubular blood. Thus, like the renal tubules' excretion of hydrogen ions, their excretion of ammonia and its combining with hydrogen to form ammonium ions also tends to increase the blood bicarbonate buffer pair ratio and, therefore, tends to increase blood pH. Quantitatively, however, ammonium ion excretion

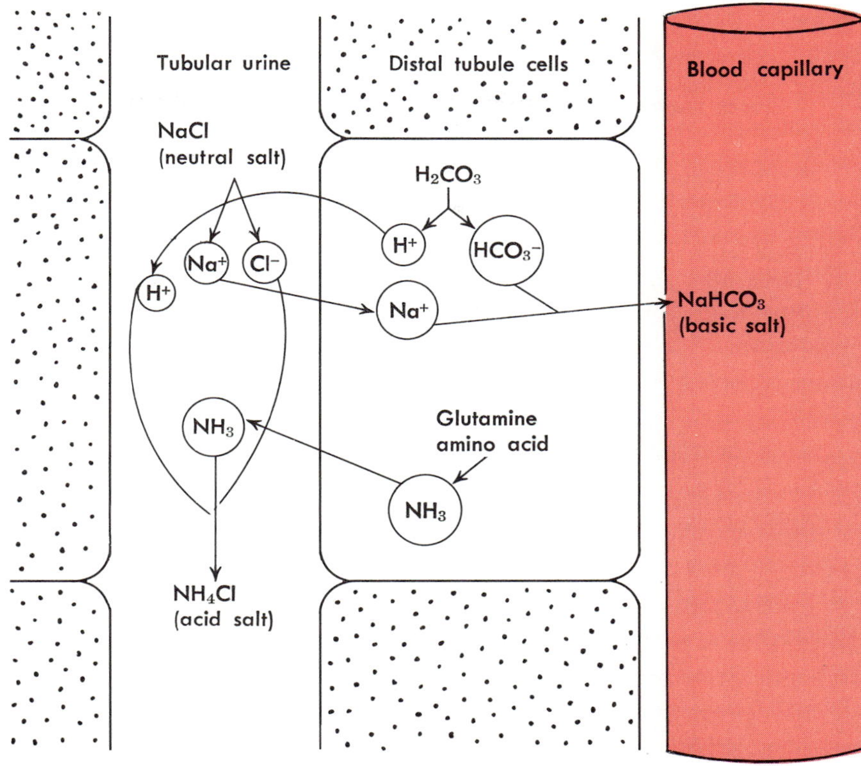

Fig. 21-6

Acidification of urine by tubule excretion of ammonia (NH_3). An amino acid (glutamine) leaves blood, enters a tubule cell, and is deaminized to form ammonia, which is excreted into urine. In exchange, the tubule cell reabsorbs a basic salt (mainly $NaHCO_3$) into blood from urine.

is more important than hydrogen ion excretion.

Renal tubule excretion of hydrogen and ammonia is controlled at least in part by the blood pH level. As indicated in Fig. 21-7, a decrease in blood pH accelerates tubule excretion of both hydrogen and ammonia. An increase in blood pH produces the opposite effects.

Acid-base imbalances

Acidosis

During the course of certain diseases, such as diabetic ketosis or starvation, for instance, abnormally large amounts of nonvolatile acids enter the blood. All three types of pH control mechanisms—buffers, respiratory, and urinary—are enlisted in an all-out effort to compensate for the excess acid and to maintain acid-base balance. Basic bicarbonate buffers immediately react with the acids (Fig. 21-2). This decreases the ratio of base bicarbonate–carbonic acid of blood and therefore decreases blood pH. The decreased blood pH stimulates the respiratory centers. Respirations become faster and deeper (hyperventilation). Hence more carbon dioxide leaves the blood and less carbonic acid remains in it. This increases the ratio $B \cdot HCO_3 / H \cdot HCO_3$ and thereby increases blood pH back up toward normal. The kidney tubules

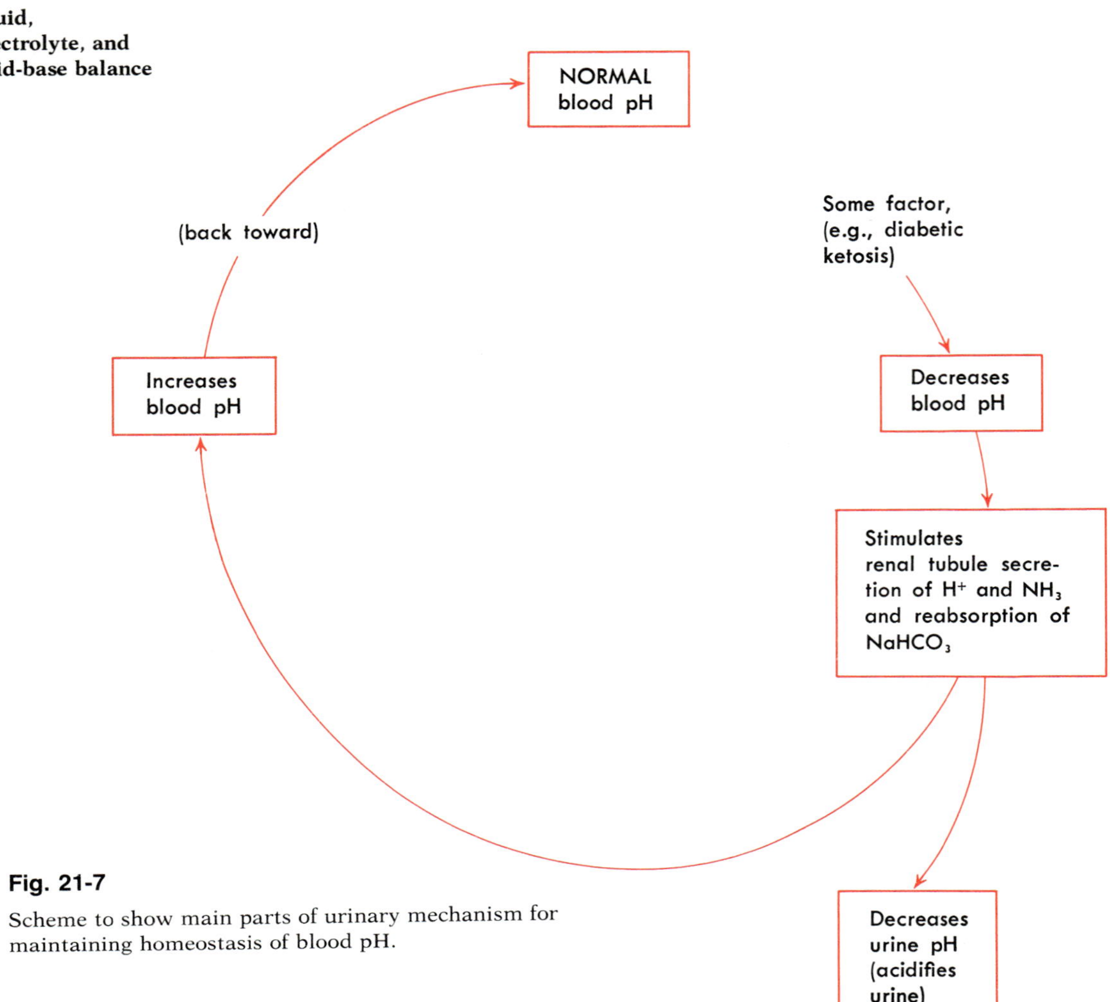

Fig. 21-7

Scheme to show main parts of urinary mechanism for maintaining homeostasis of blood pH.

increase their excretion of H^+ and NH_3 in exchange for reabsorbed Na^+ (Figs. 21-5 and 21-6). These actions also tend to increase blood's base bicarbonate–carbonic acid ratio and its pH.

If hyperventilation and urine acidification together succeed in preventing a decrease in the base bicarbonate–carbonic acid ratio, blood pH remains normal and a state of *compensated acidosis* exists. But if, despite these homeostatic devices, the ratio and pH decrease, *uncompensated acidosis* develops.

Increased blood hydrogen ion concentration (that is, decreased blood pH), as we have noted, stimulates the respiratory center. For this reason, hyperventilation is an outstanding clinical sign of acidosis. Increases in hydrogen ion concentration above a certain level depress the central nervous system and, therefore, produce such symptoms as disorientation and coma.

Alkalosis

Alkalosis develops less often than acidosis. Some circumstances, however, such as ingestion of an excessive amount of an alkaline drug, or hyperventilation, or excessive vomiting can produce alkalosis. Base bicarbonate increases above normal in alkalosis. In compensated alkalosis, carbonic acid also increases—enough to maintain a normal base bicarbonate–carbonic acid ratio and pH. In uncompensated alkalosis, the ratio and therefore the pH increase.

Outline summary

Mechanisms that control pH of body fluids

1 Meaning of term pH—negative logarithm of H ion concentration of solution
2 Types of pH control mechanisms
  a Buffers
  b Respirations
  c Secretion of urine of varying pH
3 Effectiveness of pH control mechanisms; range of pH—extremely effective, normally maintain pH within very narrow range of 7.35 to 7.45

Buffer mechanism for controlling pH of body fluids

1 Buffers defined
  a Substances that prevent marked change in pH of solution when acid or base added to it
  b Consist of weak acid (or its acid salt) and basic salt of that acid
2 Buffer pairs present in body fluids—mainly carbonic acid, proteins, hemoglobin, acid phosphate, and sodium and potassium salts of these weak acids
3 Action of buffers to prevent marked changes in pH of body fluids
  a Volatile acids, chiefly carbonic acid, buffered mainly by potassium salts of hemoglobin and oxyhemoglobin
  b Nonvolatile acids, such as lactic acid and ketone bodies, buffered mainly by sodium bicarbonate
  c Bases buffered mainly by carbonic acid

(Ratio $\frac{B \cdot HCO_3}{H_2CO_3} = \frac{20}{1}$

when homeostasis of pH at 7.4 exists)
4 Evaluation of role of buffers in pH control—cannot maintain normal pH without adequate functioning of respiratory and urinary pH control mechanisms

Respiratory mechanism of pH control

1 Explanation of mechanism
  a Amount of blood carbon dioxide directly related to amount of carbonic acid and, therefore, to concentration of H ions
  b With increased respirations, less carbon dioxide remains in blood, hence less carbonic acid and fewer H ions; with decreased respirations, more carbon dioxide remains in blood, hence more carbonic acid and more H ions
2 Adjustment of respirations to pH of arterial blood—see Fig. 21-4
3 Some principles relating respirations and pH of body fluids
  a Acidosis → hyperventilation
↓
increases elimination of CO_2
↓
decreases blood CO_2
↓
decreases blood H_2CO_3
↓
decreases blood H ions (i.e., increases blood pH)
↓
tends to correct acidosis (i.e., to restore normal pH)
  b Prolonged hyperventilation, by decreasing blood H ions excessively, may produce alkalosis
  c Alkalosis causes hypoventilation, which tends to correct alkalosis by increasing blood CO_2 and, therefore, blood H_2CO_3 and H ions
  d Prolonged hypoventilation, by eliminating too little CO_2, causes increase in blood H_2CO_3 and, consequently, in blood H ions, thereby may produce acidosis

Urinary mechanism of pH control

1 General principles about mechanism—plays vital role in acid-base balance because kidneys can eliminate more H ions from body while reabsorbing more base when pH tends toward acid side and eliminate fewer H ions while reabsorbing less base when pH tends toward alkaline side
2 Mechanisms that control urine pH
  a Secretion of H ions into urine—when blood CO_2, H_2CO_3, and H ions increase above normal, distal tubules secrete more H ions into urine to displace basic ion (mainly sodium) from a urine salt and then reabsorb sodium into blood in exchange for the H ions excreted
  b Secretion of NH_3—when blood hydrogen ion concentration increases, distal tubules secrete more NH_3, which combines with H ion of urine to form NH_4 ion, which displaces basic ion (mainly sodium) from a salt; basic ion then reabsorbed back into blood in exchange for ammonium ion excreted

Acid-base imbalances

1 Acidosis
  a Compensated metabolic acidosis—decreased alkaline reserve (mainly $NaHCO_3$), but ratio $B \cdot HCO_3/H_2CO_3$ maintained at normal 20/1 by proportionately decreasing blood carbonic acid by hyperventilation
  b Uncompensated acidosis—alkaline reserve $B \cdot HCO_3/H_2CO_3$ ratio and blood pH all decrease below normal
2 Alkalosis
  a Compensated alkalosis—opposite to compensated acidosis
  b Uncompensated alkalosis—opposite to uncompensated acidosis

Review questions

1 Explain, in your own words, what the term pH means.
2 What is the normal range for pH of body fluids?
3 Explain what a buffer is in terms of its chemical composition and in terms of its function. Cite specific equations to illustrate your explanation.
4 What is the numerical value of the ratio of $B \cdot HCO_3/H_2CO_3$ when blood pH is 7.4 and the body is in a state of acid-base balance?
5 Is the ratio $B \cdot HCO_3/H_2CO_3$ necessarily abnormal when an acid-base disturbance is present? Give reasons to support your answer.
6 Is blood pH always abnormal when an acid-base disturbance is present? Give reasons for your answer.
7 Explain how the "respiratory mechanism of pH control" operates.
8 Why is hyperventilation a characteristic clinical sign in acidosis?

Situation: A patient has suffered a severe head injury. Respirations are markedly depressed.

9 Which, if either, do you think is a potential danger—acidosis or alkalosis? State your reasons.

10 Describe the homeostatic mechanisms that would operate to try to maintain acid-base balance in this patient.

Situation: A mother brings her baby to the hospital and reports that he has seemed very sick for the past 24 hours and that he has not eaten during that time and has passed no urine.

11 Do you think acidosis or alkalosis may be present in this baby?

12 Why?

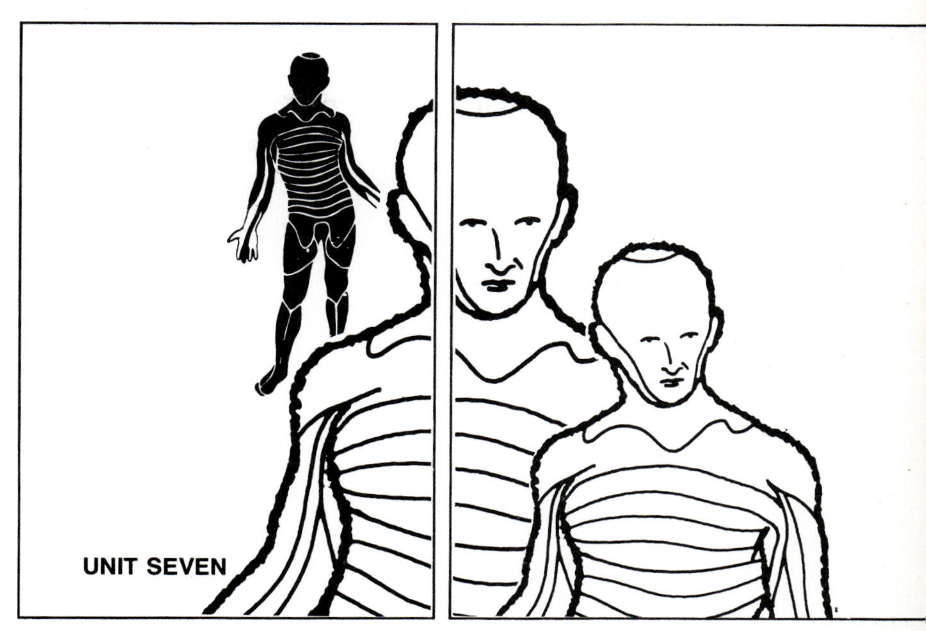

STRESS

22

Selye's concept of stress

In 1935 Hans Selye of McGill University in Montreal made an accidental discovery that launched him on a lifelong career and led him to conceive the idea of stress. This chapter will tell the story briefly of how Selye developed his stress concept and will describe the mechanism of stress that he postulated. Chapter 23 will present some current ideas about stress.

Development of Selye's concept of stress

Selye made his accidental discovery in 1935 when he was trying to learn whether there might be another sex hormone besides those already known. He had injected rats with various extracts made from ovaries and placenta, expecting to find that different changes had occurred in animals injected with different hormonal preparations. But to his surprise and puzzlement, he found the same three changes in all of the animals. The cortex of their adrenal glands were enlarged but their lymphatic organs—thymus glands, spleens, and lymph nodes—were atrophied, and bleeding ulcers of the stomach and duodenum had developed in every animal. Then he tried injecting many other substances, for example, extracts from pituitary glands, from kidneys, and from spleens, and even a poison, formaldehyde. Every time he found the same three changes: enlarged adrenals, shrunken lymphatic organs, and bleeding gastrointestinal ulcers. They seemed to be a syndrome, he thought. A syndrome, according to the classical definition, is a set of signs and symptoms that occur together and that are characteristic of one particular disease. The three changes Selye had observed occurred together but they seemed to be characteristic not of any one particular kind of injury but of any and all kinds of harmful stimuli. More experiments using a wide variety of chemicals and injurious agents confirmed for him that the three changes truly were a syndrome of injury. His first publication on the subject was a short paper entitled "A syndrome produced by diverse nocuous agents"; it appeared in the July, 1936, issue of the British journal, *Nature*. Years later, in 1956, he published his monumental technical treatise, *The Stress of Life*.

Definition of stress and stressors

Selye used the term *stress* to mean a state or condition of the body that manifests itself by a syndrome of changes. He coined the name *general adaptation syndrome* for these changes and named the stimuli that produce stress *stressors*.

Kinds of stressors

A stressor may be any kind of extreme stimuli; mild stimuli usually are not stressors. Thus extreme cold, extreme heat, and loud noise are stressors whereas coolness, warmth, and soft music are not. Stressors are frequently injurious or unpleasant—but not always. "A painful blow and a passionate kiss," Selye wrote, "can be equally stressful."* Stressors are extreme stimuli but they may not be extreme excesses. Some are extreme deficiencies of an essential kind of stimuli—for instance, social contact. Solitary confinement in a prison, a long solitary ocean voyage, travel in space, the social isolation imposed by blindness or deafness or, in some cases, by old age have all been identified as stressors. But the opposite extreme of isolation, overcrowding, also acts as a stressor.

It is impossible to say definitely that certain stimuli are stressors and that certain other stimuli are not stressors. Why? Because individuals differ in their reactions to the same stimuli. Stimuli that are stressors for you may not be stressors for your friends. Or stimuli that are stressors for you today may not have been stressors for you yesterday or

*Selye, H.: The stress syndrome, Am. J. Nurs. **65**:97-99, Mar., 1965.

may not be stressors for you tomorrow. About as definite as we can be about stressors is to say that they are generally extreme stimuli, that they are frequently unpleasant, and that many factors—especially the state of one's physical and mental health—affect individual reactions to such stimuli.

Manifestations of stress

Stress, like health or any other state or condition, is an intangible phenomenon. It cannot itself be seen, heard, tasted, smelled, felt, or measured directly. How, then, can we know that stress exists? It can be inferred to exist when certain visible, tangible, measurable responses occur. Selye, for example, inferred that the animals he experimented on were in a state of stress when he found the syndrome of the three changes before noted—hypertrophied adrenals, atrophied lymphatic organs, and bleeding gastrointestinal ulcers. Because this syndrome indicated the presence of stress and consisted of three changes, he called them the "stress triad." Eventually he found that many other changes also took place as a result of stress. The entire group of changes or responses he named the *general adaptation syndrome (GAS).* For each word in this name he gave his reasons. By the word "general" he wanted to suggest that the syndrome was "produced only by agents which have a general effect upon large portions of the body."* The word "adaptation" was meant to imply that the syndrome of changes made it possible for the body to adapt, to cope successfully with stress. Selye thought that these responses seemed to protect the animals from serious damage by extreme stimuli and to promote their healthy survival. He looked upon the general adaptation syndrome as a crucial part of the body's complex defense mechanism.

*Selye, H.: The stress of life, New York, 1956, McGraw-Hill Book Co., p. 32.

The changes that make up the general adaptation syndrome do not all take place at the same time. They occur, according to Selye, in a sequence of one, two, or three stages. He named them the alarm reaction, the stage of adaptation or resistance, and the stage of exhaustion and noted that a different syndrome of responses characterized each stage. Among the responses characteristic of the *alarm reaction,* for example, were the stress triad already described—hypertrophied adrenal cortex, atrophied lymphatic organs (thymus, spleen, lymph nodes), and bleeding gastric and duodenal ulcers. In addition, the adrenal cortex increased its secretion of glucocorticoids, the number of lymphocytes decreased markedly, and so, too, did the number of eosinophils. Also, the sympathetic nervous system and the adrenal medulla greatly increased their activity. Each of these changes, in turn, produced other widespread changes. Fig. 22-1 indicates some of the changes stemming from adrenal cortical hypertrophy. Fig. 22-2 shows responses produced by increased sympathetic activity and increased secretion by the adrenal medulla of its hormone, epinephrine (adrenaline).

Quite different responses characterize the *stage of resistance* of the general adaptation syndrome. For instance, the adrenal cortex and medulla both return to their normal rates of secreting hormones. The changes that had taken place in the alarm stage as a result of increased corticoid secretion disappear during the stage of resistance. All of us go through the first and second stages of the stress syndrome many times in our lifetimes. Stressors of one kind or another act on most of us every day. They may upset or alarm us but we soon adapt to them.

The *stage of exhaustion* develops only when stress is extremely severe—in extensive burns, for example—or when it continues over long periods of time.

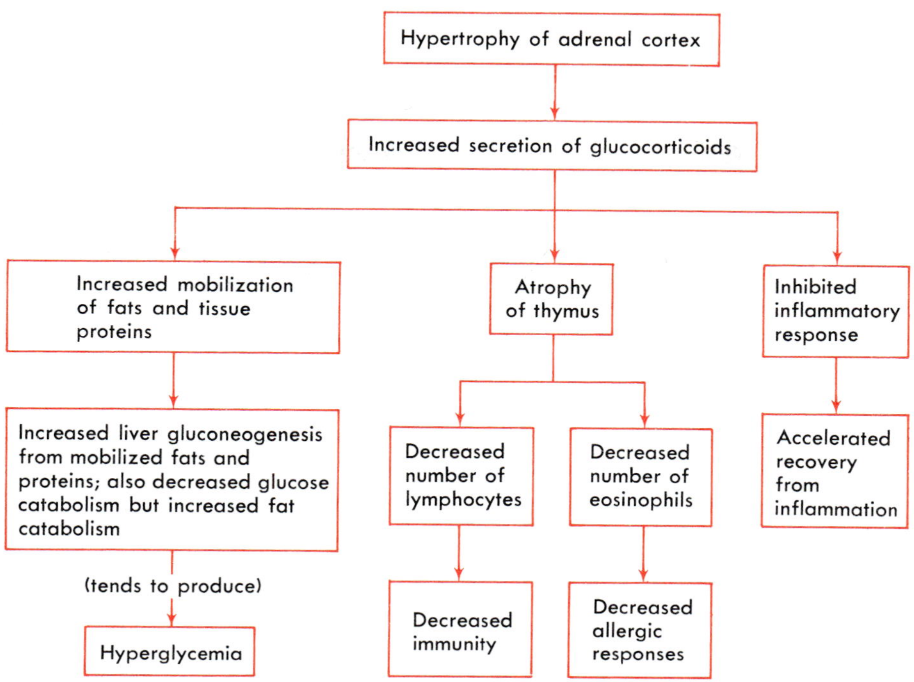

Fig. 22-1
Alarm reaction responses resulting from hypertrophy of adrenal cortex.

Some of the manifestations of stress can be measured to help determine how severe an individual's state of stress may be. Indicators of increased sympathetic activity are frequently measured to evaluate stress. Examples are the following: an increase in the rate and force of the heartbeat, a rise in systolic blood pressure, an increase in epinephrine or norepinephrine concentrations in blood and urine, sweating of the palms of the hands, and dilatation of the pupils.

The heart rate has been shown to increase in response to such varied stressors as anesthesia, annoying sounds, and the entrance of an attractive nurse into a patient's room. Even anticipation by patients in a coronary care unit of their transferral to a less closely supervised convalescent unit has been identified as a stressor that causes heart rate to speed up.

In normal persons, urinary epinephrine and norepinephrine levels have been found to increase in response to sensory deprivation (a stressor) and to decline after reestablishment of sensory stimulation.

Indicators of adrenocortical activity that are most frequently used to measure stress are eosinopenia and lymphocytopenia (a decrease in the number of circulating eosinophils and lymphocytes) and an increase in the levels of glucocorticoids (or their breakdown products) in blood and urine. Soldiers stressed by prolonged marching have been found to have fewer circulating eosinophils than normal. This same stress indicator has been observed in college oarsmen when they were anticipating performance on exhibition days and in heart patients when they were anticipating transfer from a coronary care unit to a convalescent unit.

The amount of urinary adrenocorticoids is often used as a measure of stress. It has been

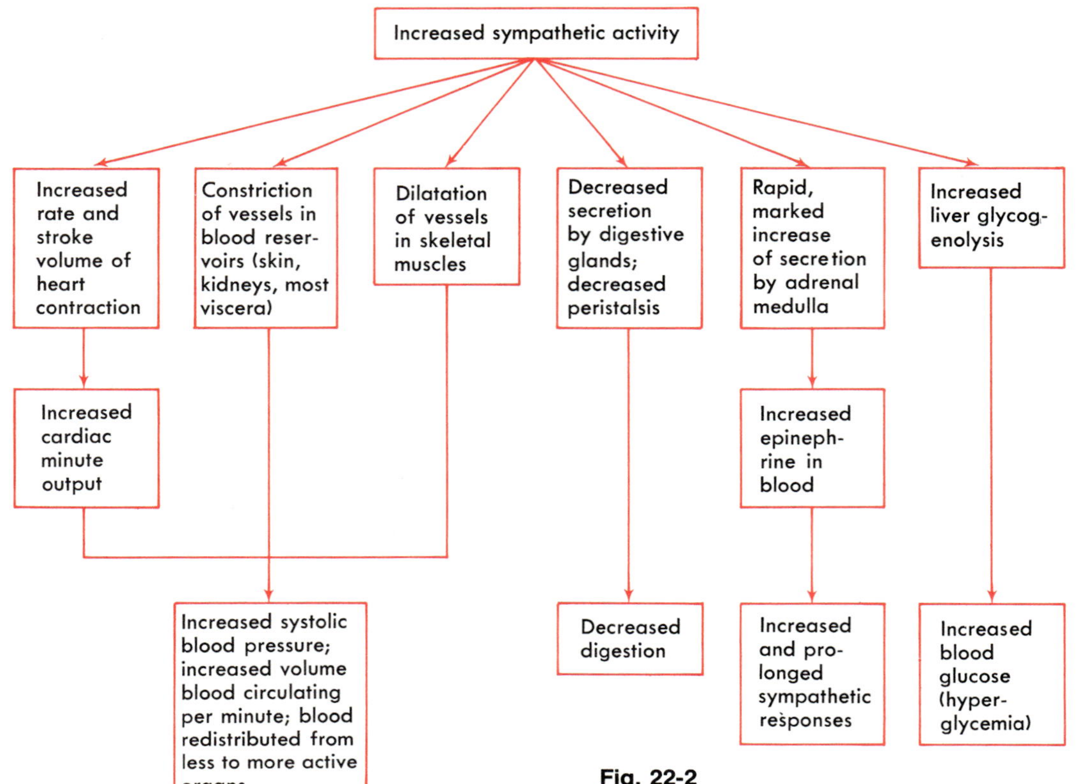

Fig. 22-2

Alarm reaction responses resulting from increased sympathetic activity. Note that these are the responses commonly referred to as the "fight or flight" syndrome.

found to increase in depressed persons who feel hopeless and doomed, in test pilots, and in college students taking examinations or attending exciting movies. In contrast, urinary corticoids were found to drop markedly in persons watching unexciting nature study films.

The level of adrenocorticoids in the blood plasma of disturbed patients having acute psychotic episodes has been found to be 70% higher than that in normal individuals or in calm patients. Another study showed that the plasma corticoid levels of chronically depressed patients were significantly lower than those of acutely anxious patients. Smoking and exposure to nicotine have also been shown to be stressors that caused a marked rise in plasma adrenocorticoids—by as much as 77% in both human beings and experimental animals.

Mechanism of stress

Stressors produce a state of stress. A state of stress, in turn, inaugurates a series of responses that Selye called the general adaptation syndrome. More simply, a state of stress turns on the stress response mechanism. It activates the organs that produce the responses that make up the general adaptation syndrome. But just how stress—a state of the body—does this is not clear. Selye could only guess about it and his terms were vague. For instance, he postulated that by some unknown "alarm signals," stress "acted

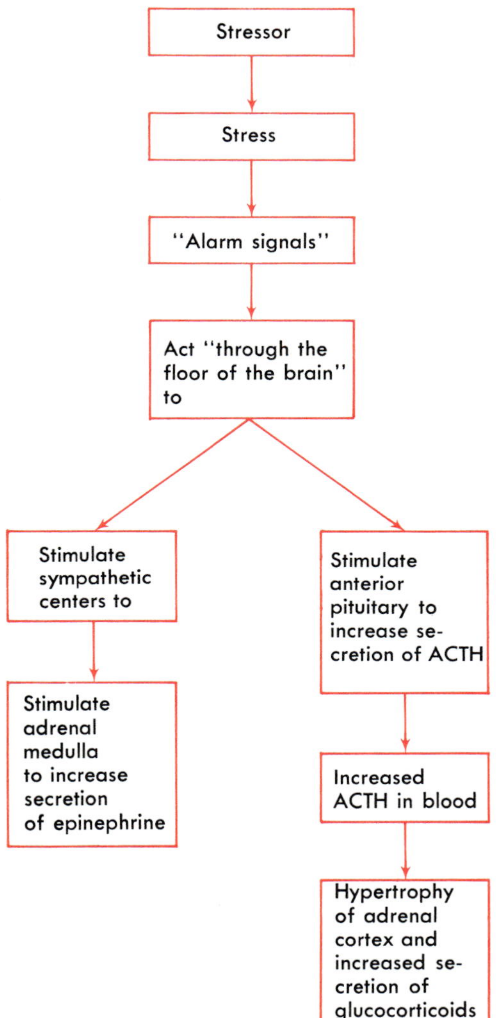

Fig. 22-3

Selye's hypothesis about activation of the stress mechanism.

through the floor of the brain"* (presumably the hypothalamus) to stimulate the sympathetic nervous system and the pituitary gland. In Fig. 22-3, you can see Selye's hypothesis in diagram form.

*Selye, H.: The stress of life, New York, 1956, McGraw-Hill Book Co., p. 114.

Stress and disease

Stress, as we have observed several times, produces different results in different individuals and in the same individual at different times. In one person, a certain amount of stress may induce responses that maintain or even enhance his health. But in another person that same amount of stress appears to "make him sick." Whether stress is "good" or "bad" for you seems to depend more upon your own body's responses to it than upon the severity of the stressors inducing it. You may recall that Selye emphasized the adaptive nature of stress responses. He coined the term *general adaptation syndrome* because he believed that stress responses usually enable the body to adapt successfully to the many stressors that assail it. He held that the state of stress alerts physiological mechanisms to meet the challenge imposed by stressors. But a challenge issued does not necessarily mean a challenge successfully met. Selye proposed that sometimes the body's adaptive mechanisms fail to meet the challenge issued by stressors and that, when they fail, disease results—diseases of adaptation, he called them. He and his co-workers succeeded in inducing a variety of pathological changes by exposing sensitized animals to intense stimuli of various kinds. Their experimental animals developed a number of different diseases, among them hypertension, arthritis, arteriosclerosis, nephrosclerosis, and gastrointestinal ulcers. As a result of these experiments, many researchers have attempted to establish stress as a cause of disease in man. Their findings, however, have not been widely accepted as convincing evidence that stress, induced in a normal individual, can produce disease.

Around the middle of this century, one of the problems studied was the relationship of blood glucocorticoid concentration to disease. If stress is adaptive, the investigators reasoned, and helps the body combat the effects of many kinds of stressors (for example, infection, injury, burns, and the like), then possibly various diseases might be treated by adding to the body's natural output of glucocorticoids. In 1949, Hench and his colleagues* at the Mayo Clinic cautiously suggested that it might be helpful to give glucocorticoids to patients afflicted with various illnesses. Subsequently, they acted on their own suggestion and reported finding that a hormone of the adrenal cortex improved some of the clinical symptoms of rheumatoid arthritis. Since that time, cortisone has been administered to thousands of patients with widely different ailments. So often has it been used and so numerous have been the articles written about cortisone that today it would be almost impossible to find an adult who has never heard of this hormone.

*Hench, P. S., Kendall, E. C., Slocumb, C. H., and Polley, H. F.: The effect of a hormone of the adrenal cortex (17-hydroxy-11-dehydrocorticosterone; compound E) and of pituitary adrenocorticotropic hormone on rheumatoid arthritis: preliminary report, Proc. Staff Meet. Mayo Clin. **24:**181-197, Apr., 1949.

Outline summary

Development of Selye's concept of stress

Developed through many experiments in which he exposed animals to wide variety of noxious agents and found that they had all responded with the same syndrome of changes

Definitions of stress and stressors

1 Stress—a state or condition of the body that manifests itself by the general adaptation syndrome
2 Stressors—stimuli that produce stress

Kinds of stressors

Any kind of extreme stimuli; frequently injurious and unpleasant, but not always; may be either extreme excesses or extreme deficiencies

Manifestations of stress

1 General adaptation syndrome—consists of as many as three successive phases, namely, alarm reaction, stage of adaptation or resistance, and stage of exhaustion
2 Measurable indicators
   a Indicators of increased sympathetic activity, e.g., increase in rate and force of heartbeat, increase in systolic blood pressure, etc.
   b Indicators of hyperactivity of adrenal cortex, e.g., decreased number of circulating eosinophils and lymphocytes

Mechanism of stress

1 Consists of group (syndrome) of responses to internal condition of stress; stress responses nonspecific in that same syndrome of responses occurs regardless of kind of extreme change that produced stress
2 Stimulus that produces stress and thereby activates stress mechanism is nonspecific in that it can be any kind of extreme change in environment
3 Stress responses, Selye thought, were adaptive; tend to enable body to adapt to and survive extreme change; Selye referred to this syndrome of stress responses as general adaptation syndrome
4 Numerous factors influence stress responses—individual's physical and mental condition, age, sex, socioeconomic status, heredity, previous experience with similar stressor, and even religious affiliation
5 Stress most often results successfully—i.e., results in adaptation, healthy survival, and increased resistance; sometimes, however, stress produces exhaustion and death
6 See Figs. 22-3, then 22-1, and 22-2 for a summary of Selye's stress mechanism

Stress and disease

1 Selye held that stress could result in disease instead of adaptation
2 To date, research evidence that stress, induced in normal individual, can produce disease has not gained wide acceptance

Review questions

1 Explain the meaning of the term *physiological stress.*
2 What causes physiological stress?
3 What responses does physiological stress cause? Construct a flow chart diagram summarizing the stress mechanism.
4 What factors affect stress responses?
5 How is physiological stress measured?
6 What is psychological stress?
7 How, if at all, does psychological stress relate to physiological stress?
8 How, if at all, does physiological stress relate to disease?

Current concepts about stress

Definitions

Physiologists today differ in their use of the words stressors and stress. Some use these terms as Selye did—stressors to indicate injurious stimuli and stress to mean the state that stressors produce in the body when they affect large portions of it. Other physiologists substitute the word stress for Selye's word stressor. They give an operational definition of stress as any stimulus that directly or indirectly stimulates neurons of the hypothalamus to release corticotropin-releasing hormone (CRH) into pituitary portal blood. Indirect stimulation of the hypothalamus can result from impulses conducted to it from the limbic lobe or other parts of the cerebral cortex. A low blood glucose concentration or certain other chemical changes in the blood can stimulate the hypothalamus directly.

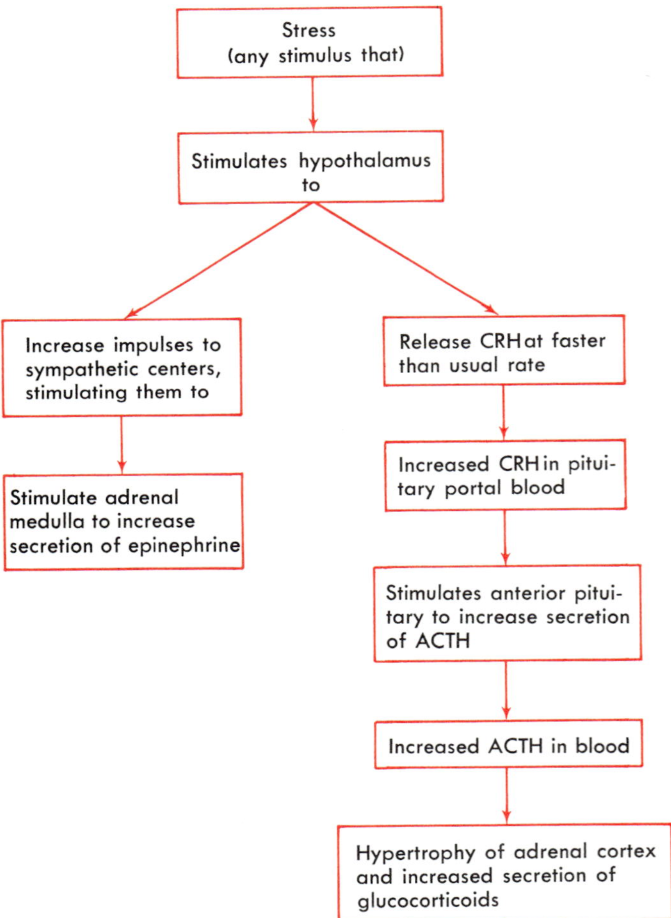

Fig. 23-1

A current concept about activation of stress mechanism.

Mechanism of stress

The current hypothesis about how the mechanism of stress is activated is somewhat more specific than was Selye's. By comparing Fig. 23-1 with Fig. 22-3, you can discover this difference.

Corticoids and resistance to stress

Selye thought that the increase in corticoids that occurred in his stressed animals enabled them to adapt to and to resist stress. Today many physiologists doubt this. No one questions that adrenocortical hormones increase during stress. That fact has been clearly established. But what many do question is how essential this increase is for resisting stress. No one has proved by an unequivocal experiment that a higher than normal blood level of corticoids increases an animal's ability to adapt to stress. Some clinical evidence, however, seems to indicate that it does increase man's coping ability. For instance, patients who have been taking cortisol for some time are known to require increased doses of this hormone in order to successfully resist stresses such as surgery or severe injury.

Psychological stress

Stress as defined by Selye is physiological stress, that is, a state of the body. Psychological stress, in contrast, might be defined as a state of the mind. It is caused by psychological stressors and manifested by a syndrome. By definition, a *psychological stressor* is anything that an individual perceives as a threat —a threat either to his survival or to his self-image. Moreover, the threat does not need to be real—it needs only to be real to the individual. He must see it as a threat although in truth it may not be. Psychological stressors produce a syndrome of subjective and objective responses. Dominant among the subjective reactions is a feeling of anxiety. Other emotional reactions such as anger, hate, depression, fear, and guilt are also common subjective responses to psychological stressors. Some characteristic objective responses are restlessness, fidgeting, criticizing, quarreling, lying, and crying. Another objective indicator of psychological stress, discovered only a few years ago, is that the concentration of lactate in the blood increases. For an interesting account of this, read "The biochemistry of anxiety" by F. N. Pitts, Jr., in the February, 1969, issue of *Scientific American.*

Does psychological stress relate to physiological stress? The answer is clearly "yes." Physiological stress usually is accompanied by some degree of psychological stress. And, conversely, in most people, psychological stress produces some physiological stress responses. Ancient peoples intuitively recognized this fact. For example, it is said that in ancient times when the Chinese suspected a person of lying, they made him chew rice powder and then spit it out. If the powder came out dry, not moistened by saliva, they judged the suspect guilty. They seemed to know that lying makes a person "nervous"—and that nervousness makes a person's mouth dry. We moderns also know these facts but we describe them with more technical language. Lying, we might say, induces psychological stress and psychological stress acts in some way to cause the sympathetic response of decreased salivation.

Within the last few years, a new scientific discipline called psychophysiology has come into being. Psychophysiologists, using accepted research methods and sophisticated instruments—including polygraphs designed especially for this type of research—have investigated a wide variety of physiological responses made by individuals being subjected to psychological stressors. Their findings amply confirm the principle that psychological stressors often produce physiological stress responses. They have found, too, that identical psychological stressors do not necessarily induce identical physiological responses in different individuals. In the words of one experimenter, "for one person the cardiovascular system may quite regularly mirror emotion most sensitively, whereas another person may be primarily a pulmonary reactor."* Another of their interesting discoveries is that some organ systems become less responsive after they have been stimulated a number of times.

Summarizing, here are some principles to remember about psychological stress.

1 Physiological stress almost always is accompanied by some degree of psychological stress.

2 In most people, psychological stress leads to some physiological stress responses. Many of these are measurable autonomic responses, for example, accelerated heart rate, and increased systolic blood pressure.

3 Identical psychological stressors do not always induce identical physiological responses in different individuals.

4 In any one individual, certain autonomic responses are better indicators of psychological stress than others.

*Smith, B. M.: The polygraph, Sci. Am. **216**:25-31, Jan., 1967.

Outline summary

Definitions

1 Stressors—some physiologists today use the word as Selye did, to mean stimuli that produce stress in the body
2 Stress—some physiologists today substitute the word stress for stressor; they define stress as any stimulus that stimulates neurons of hypothalamus to release CRH

Mechanism of stress

Current hypothesis about activation of stress mechanism more specific than Selye's—compare Fig. 23-1 with Fig. 22-3

Corticoids and resistance to stress

Still controversial; not proved that increased corticoids increase animal's or man's ability to resist stress but is proved that corticoid levels in blood increase in stress

Psychological stress

1 Psychological stressors—anything that an individual perceives as a threat to either his survival or self-image
2 Psychological stress—a mental state characterized by a syndrome of subjective and objective responses; dominant subjective response is anxiety; some characteristic objective responses are restlessness, quarrelsomeness, lying, crying
3 Relation to physiological stress—see summary of principles on p. 543

Review questions

1 What is a current operational definition of stress?
2 Describe the current hypothesis about how the stress mechanism is activated.
3 What has been proved about corticoids in relation to stress?
4 What is the controversial issue about corticoids in relation to stress?
5 Define psychological stressor, psychological stress.
6 Cite several examples of psychological stressors.
7 Are psychological stressors the same for all individuals?
8 Give some examples of subjective indicators of psychological stress.
9 Give some examples of objective responses that are part of the syndrome of psychological stress.
10 State three or four principles about the relationship between psychological and physiological stress.

Supplementary readings

Chapter 1

Anthony, C. P.: Basic concepts in anatomy and physiology—a programmed presentation, ed. 3, St. Louis, 1974, The C. V. Mosby Co.

Cannon, W. B.: The wisdom of the body, ed. 2, New York, 1963, W. W. Norton & Co., Inc.

Morris, D.: The naked ape, New York, 1967, McGraw-Hill Book Co.

Chapter 2

Anthony, C. P.: Basic concepts in anatomy and physiology—a programmed presentation, ed. 3, St. Louis, 1974, The C. V. Mosby Co.

Bevelander, G., and Ramaley, J. A.: Essentials of histology, ed. 7, St. Louis, 1974, The C. V. Mosby Co.

Brown, D. D.: The isolation of genes, Sci. Am. **229**:21-29, Aug., 1973.

Capaldi, R. A.: A dynamic model of cell membranes, Sci. Am. **230**:27-33, Mar., 1974.

Everhart, T. E., and Hayes, T. L.: The scanning electron microscope, Sci. Am. **226**:55-69, Jan. 1972.

Frieden, E.: The chemical elements of life, Sci. Am. **227**: 52-60, July, 1972.

Hayflick, L.: Human cells and aging, Sci. Am. **218:**32-37, Mar., 1968.
Howland, J. L.: Cell physiology, New York, 1973, The Macmillan Co.
Miller, O. L., Jr.: The visualization of genes in action, Sci. Am. **228:**34-41, Mar., 1973.
Neutra, M., and Lebland, C. P.: The Golgi apparatus, Sci. Am. **220:**100-107, Feb., 1969.
Nomura, M.: Ribosomes, Sci. Am. **221:**28-35, Oct., 1969.
Racker, E.: The membrane of the mitochondrion, Sci. Am. **218:**32-39, Feb., 1968.
Trotter, R. J.: Left-handedness, Sci. News **106:**220-222, Oct., 1974.

Chapters 3 and 4

Bevelander, G., and Ramaley, J. A.: Essentials of histology, ed. 7, St. Louis, 1974, The C. V. Mosby Co.

Chapter 6

Hoyle, G.: How is muscle turned on and off? Sci. Am. **222:**84-93, Apr., 1970.
Margaria, R.: The sources of muscular energy, Sci. Am. **226:**83-91, Mar., 1972.
Morehouse, L. E., and Miller, A. T., Jr.: Physiology of exercise, ed. 6, St. Louis, 1971, The C. V. Mosby Co.
Murray, J. M., and Weber, A.: The cooperative action of muscle proteins, Sci. Am. **230:**59-71, Feb., 1974.

Chapter 7

Eccles, J.: The synapse, Sci. Am. **212:**56-66, Jan., 1965.
Hyden, H.: Satellite cells in the nervous system, Sci. Am. **205:**62-70, Dec., 1961.
Mountcastle, V. B., editor: Medical physiology, ed. 13, St. Louis, 1974, The C. V. Mosby Co.

Chapter 8

Evarts, E. V.: Brain mechanisms in movement, Sci. Am. **229:**96-103, July, 1973.
Fernstrom, J. D., and Wurtman, R. J.: Nutrition and the brain, Sci. Am. **230:**84-91, Feb., 1974.
Geschwind, N.: Language and the brain, Sci. Am. **226:** 76-83, Apr., 1972.
Kimura, D.: The asymmetry of the human brain, Sci. Am. **228:**70-77, Mar., 1973.
Llinas, R. R.: The cortex of the cerebellum, Sci. Am. **232:**56-71, Jan., 1975.
Luria, A. R.: The functional organization of the brain, Sci. Am. **222:**66-79, Mar., 1970.
Ornstein, R. E.: The psychology of consciousness, San Francisco, 1972, W. H. Freeman and Co. Publishers.
Penfield, W., and Rasmussen, T.: The cerebral cortex of man, New York, 1968, Hafner Publishing Co., Inc.
Pribram, K. H.: The neurophysiology of remembering, Sci. Am. **220:**73-86, Jan., 1969.
Rosenzweig, M. R., Bennett, E. L., and Diamond, M. C.: Brain changes in response to experience, Sci. Am. **226:**22-29, Feb., 1972.
Trotter, R.: A clockwork orange in a California prison, Sci. News **101:**164-175, Mar., 1972.
Wallace, R. K., and Benson, H.: The physiology of meditation, Sci. Am. **226:**85-90, Feb., 1972.

Chapter 9

Dicara, L. V.: Learning in the autonomic nervous system, Sci. Am. **222:**30-39, Jan., 1970.
Mountcastle, V. B., editor: Medical physiology, ed. 13, St. Louis, 1974, The C. V. Mosby Co., pp. 783-834.
Trotter, R. J.: Transcendental meditation, Sci. News **104:**376-378, Dec., 1973.
Wallace R. K., and Benson, H.: The physiology of meditation, Sci. Am. **226:**85-90, Feb., 1972.

Chapter 10

Hubbard, R., and Kropf, A.: Molecular isomers in vision, Sci. Am. **216:**64-70, June, 1967.
Oster, G.: Auditory beats in the brain, Sci. Am. **227:**94-102, Oct., 1973.
Pettigrew, J. D.: The neurophysiology of binocular vision, Sci. Am. **227:**84-95, Aug., 1972.
Werblin, F. S.: The control of sensitivity in the retina, Sci. Am. **228:**71-80, Jan., 1973.
Young, R. W.: Visual cells, Sci. Am. **223:**81-91, Oct., 1970.

Chapter 11

Gardener, L. I.: Deprivation dwarfism, Sci. Am. **227:** 76-82, July, 1972.
Guillemin, R., and Burgus, R.: The hormones of the hypothalamus, Sci. Am. **227:**24-33, Nov., 1972.
McKusick, V. A., and Rimoin, D. L.: General Tom Thumb and other midgets, Sci. Am. **217:**102-106, July, 1967.
Mountcastle, V. B., editor: Medical physiology, ed. 13, St. Louis, 1974, The C. V. Mosby Co., pp. 1601-1802.
Pastan, I.: Cyclic AMP, Sci. Am. **227:**97-105, Aug., 1972.
Pike, J. E.: Prostaglandins, Sci. Am. **225:**84-91, Nov., 1971.

Chapter 12

Comroe, J. H., Jr.: The lung, Sci. Am. **214:**57-66, Feb., 1966.
Comroe, J. H., Jr.: Physiology of respiration, Chicago, 1965, Year Book Medical Publishers, Inc.
Egan, D. F.: Fundamentals of respiratory therapy, ed. 2, St. Louis, 1973, The C. V. Mosby Co.
Mountcastle, V. B., editor: Medical physiology, ed. 13, St. Louis, 1974, The C. V. Mosby Co., pp. 1361-1591.
Slonim, N. B., and Hamilton, L. H.: Respiratory physiology, ed. 2, St. Louis, 1971, The C. V. Mosby Co.

Chapter 13

Adolph, E. F.: The heart's pacemakers, Sci. Am. **213:** 32-37, Mar., 1967.

Anthony, C. P.: Basic concepts in anatomy and physiology—a programmed presentation, ed. 3, St. Louis, 1974, The C. V. Mosby Co., pp. 108-144.
Clarke, C. A.: The prevention of "Rhesus" babies, Sci. Am. **219:**46-52, Nov., 1968.
Cooper, M. D., and Lawton, A. R., III: The development of the immune system, Sci. Am. **231:**58-72, Nov., 1974.
Effler, D. B.: Surgery for coronary disease, Sci. Am. **219:**36-43, Oct., 1968.
Ingram, M., and Preston, K., Jr.: Automatic analysis of blood cells, Sci. Am. **223:**72-82, Nov., 1970.
Lerner, R. A., and Dixon, F. J.: The human lymphocyte as an experimental animal, Sci. Am. **228:**82-91, June, 1973.
Mountcastle, V. B., editor: Medical physiology, ed. 13, St. Louis, 1974, The C. V. Mosby Co., pp. 839-1043.
Porter, R. R.: The structure of antibodies, Sci. Am. **217:** 81-87, Oct., 1967.
Wood, E. J.: The venous system, Sci. Am. **218:**86-96, Jan., 1968.
Zalis, E. G., and Conover, M. H.: Understanding electrocardiography, St. Louis, 1972, The C. V. Mosby Co.

Chapter 14

Davenport, H. W.: Why the stomach does not digest itself, Sci. Am. **226:**87-93, Jan., 1972.
Dudrick, S. J., and Rhoades, J. E.: Total intravenous feeding, Sci. Am. **226:**73-79, May, 1972.
Kretchmer, N.: Lactose and lactase, Sci. Am. **227:**70-78, Oct., 1972.
Programmed instruction: potassium imbalance, Am. J. Nurs. **67:**343-366, Feb., 1967.

Chapter 15

Brady, R. O.: Hereditary fat-metabolism diseases, Sci. Am. **229:**88-97, Aug., 1973.
Morehouse, L. E., and Miller, A. T., Jr.: Physiology of exercise, ed. 6, St. Louis, 1971, The C. V. Mosby Co., p. 193.
Young, V. R., and Schrimshaw, N. S.: The physiology of starvation, Sci. Am. **225:**14-21, Oct., 1971.

Chapter 16

Anthony, C. P.: Basic concepts in anatomy and physiology—a programmed presentation, ed. **3,** St. Louis, 1974, The C. V. Mosby Co.
Cummings, J. W.: Hemodialysis—feelings, Am. J. Nurs. **70:**70-83, Jan., 1970.
Downing, S. R.: Nursing support in early renal failure, Am. J. Nurs. **69:**1212-1216, June, 1969.
Schumann, B.: The renal donor, Am. J. Nurs. **74:**105-110, Jan., 1974.

Chapter 17

Chinn, P. L.: Child health maintenance, St. Louis, 1974, The C. V. Mosby Co., pp. 63-67.
Kornberg, A.: The synthesis of DNA, Sci. Am. **219:**64-78, Oct., 1968.
Levine, L.: Biology of the gene, ed. 2, St. Louis, 1973, The C. V. Mosby Co., pp. 46-57.
Mazia, D.: The cell cycle, Sci. Am. **231:**55-63, Jan., 1974.

Chapter 18

Mountcastle, V. B., editor: Medical physiology, ed. 13, St. Louis, 1974, The C. V. Mosby Co., pp. 1763-1770.

Chapter 19

Edwards, R. G., and Fowler, R. E.: Human embryos in the laboratory, Sci. Am. **222:**45-54, Dec., 1970.
Mountcastle, V. B., editor: Medical physiology, ed. 13, St. Louis, 1974, The C. V. Mosby Co., pp. 1741-1762.

Chapter 20

Anthony, C. P.: Basic concepts in anatomy and physiology—a programmed presentation, ed. 3, St. Louis, 1974, The C. V. Mosby Co.
Gamble, J. L.: Chemical anatomy, physiology and pathology of extracellular fluid, ed. 6, Cambridge, Mass., 1958, Harvard University Press.
Mountcastle, V. B., editor: Medical physiology, ed. 13, St. Louis, 1974, The C. V. Mosby Co., pp. 1049-1064.
Voda, A. M.: Body water dynamics, Am. J. Nurs. **70:** 2594-2601, Dec., 1970.

Chapter 21

Anthony, C. P.: Basic concepts in anatomy and physiology—a programmed presentation, ed. 3, St. Louis, 1974, The C. V. Mosby Co.
Burke, S. R.: The composition and function of body fluids, St. Louis, 1972, The C. V. Mosby Co., pp. 57-84.
Weldy, N. J.: Body fluids and electrolytes, St. Louis, 1972, The C. V. Mosby Co., pp. 35-60, 79-100.

Chapter 22

Selye, H.: The stress of life, New York, 1956, McGraw-Hill Book Co.
Selye, H.: The stress syndrome, Am. J. Nurs. **65:**97-99, Mar., 1965.

Chapter 23

Levine, S.: Stress and behavior, Sci. Am. **224:**26-31, Jan., 1971.
Mountcastle, V. B., editor: Medical physiology, ed. 13, St. Louis, 1974, The C. V. Mosby Co., pp. 832-834, 1696-1698, 1700-1703.
Pitts, F. N., Jr.: The biochemistry of anxiety, Sci. Am. **220:**69-75, Feb., 1969.
Smith, B. M.: The polygraph, Sci. Am. **216:**25-31, Jan., 1967.
Trotter, R. J.: The biological depths of loneliness, Sci. News **103:**140-142, Mar., 1973.
Weiss, J. M.: The psychological factors in stress and disease, Sci. Am. **226:**104-113, June, 1972.

Additional references

Biochemistry

Orten, J. M., and Neuhaus, O. W.: Biochemistry, ed. 8, St. Louis, 1970, The C. V. Mosby Co.

Developmental anatomy

Arey, L. B.: Developmental anatomy—a textbook and laboratory manual of embryology, ed. 7, Philadelphia, 1965, W. B. Saunders Co.

Gross anatomy

Goss, C. M., editor: Gray's anatomy of the human body, ed. 29, Philadelphia, 1973, Lea & Febiger.
Hamilton, W. J., editor: Textbook of human anatomy, New York, 1957, The Macmillan Co.
Sobotta, J., and Figge, F. H. J.: Atlas of human anatomy, ed. 8, New York, 1963, Hafner Publishing Co., Inc. (3 vols.).
Spalteholz, W.: Atlas of human anatomy, ed. 16 (revised and re-edited by R. Spanner), New York, 1967, F. A. Davis Co.

Microscopic anatomy

Bloom, W., and Fawcett, D. W.: A textbook of histology, ed. 9, Philadelphia, 1968, W. B. Saunders Co.
Ham, A. W., and Leeson, T. S.: Histology, ed. 6, Philadelphia, 1969, J. B. Lippincott Co.

Physiology

Cannon, W. B.: The wisdom of the body, rev. ed., New York, 1963, W. W. Norton & Co., Inc.
Guyton, A. D.: Textbook of medical physiology, ed. 4, Philadelphia, 1971, W. B. Saunders Co.
Mountcastle, V. B., editor: Medical physiology, ed. 13, St. Louis, 1974, The C. V. Mosby Co.

Periodicals

American Journal of Nursing
American Journal of Physiology
Annual Review of Physiology
Journal of the American Medical Association
Science Newsletter
Scientific American

Abbreviations and prefixes

Abbreviations

Å Angstrom unit
ACh acetylcholine
AChE acetylcholinesterase
ACTH adrenocorticotropic hormone
ADP adenosine diphosphate
ASC altered state of consciousness
ATP adenosine triphosphate
BMR basal metabolic rate
BNA Basle Nomina Anatomica (see Glossary)
C centigrade
CA catecholamines
Cal large calorie
cm centimeter
CNS central nervous system
COMT catechol-O-methyl transferase
CP creatine phosphate
CRF corticotropin-releasing factor
CVA cardiovascular accident
DA dopamine
DNA deoxyribonucleic acid
DPN diphosphopyridine nucleotide
ECF extracellular fluid

EFP effective filtration pressure
EPSP excitatory postsynaptic potential
ER endoplasmic reticulum
ERV expiratory reserve volume
F Fahrenheit
FRC functional respiratory capacity
FSH follicle-stimulating hormone
GABA gamma aminobutyrate
GH growth hormone
Hb hemoglobin
HbO_2 oxyhemoglobin
HP hydrostatic pressure
5-HT 5-hydroxytryptamine (serotonin)
IC inspiratory capacity
ICF intracellular fluid
ICSH interstitial cell–stimulating hormone
IF interstitial or intercellular fluid
IPSP inhibitory postsynaptic potential
IRV inspiratory reserve volume
kcal kilocalorie
kg kilogram
LH luteinizing hormone
LSD lysergic acid diethylamide
MAO monoamine oxidase
mEq milliequivalents
mg milligram
μ micron
ml milliliter
mm millimeter
mm Hg pres millimeter mercury pressure
MSH melanocyte-stimulating hormone
mv millivolt
MW molecular weight
NE norepinephrine
OP osmotic pressure
PAH para-aminohippuric acid
PBI protein-bound iodine
Pco_2 partial pressure of carbon dioxide
pH hydrogen-ion concentration; negative logarithm of hydrogen-ion concentration
Po_2 partial pressure of oxygen
PNS peripheral nervous system
RBC, rbc red blood cells
REM rapid eye movements
RNA ribonucleic acid
RV residual volume
SD systolic discharge
SDA specific dynamic action
STH somatotropic hormone
SV stroke volume
SWS slow-wave sleep
TH thyrotropic hormone
TMR total metabolic rate
TSH thyroid-stimulating hormone
TV tidal volume
VC vital capacity
WBC, wbc white blood cells

Prefixes

ab- away from
ad- to, toward
adeno- glandular
amphi- on both sides
ante- before, forward
anti- against
bi- two, double, twice
circum- around, about
contra- opposite, against
de- away from, from
dys- difficult
ecto- outside
endo- in, within
ento- inside, within
epi- on, upon
eu- well
ex- from out of, from
extra- outside, beyond, in addition
hemi- half
hyper- over, excessive, above
hypo- under, deficient
infra- underneath, below
inter- between, among
intra- within, on the side
para- beside, to side of
peri- round about, beyond
post- after, behind
pre- before, in front of
pro- before, in front of
retro- backward, back
semi- half
sub- under, beneath
super- above, over
supra- above, on upper side
syn- with, together
trans- across, beyond

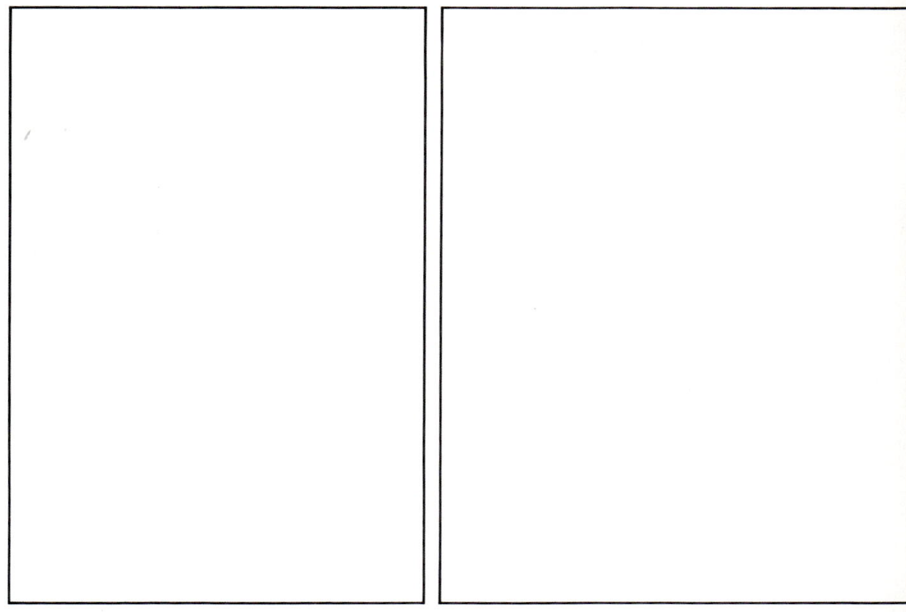

Glossary

abdomen body area between the diaphragm and pelvis.

abduct to move away from the midline; opposite of adduct.

absorption passage of a substance through a membrane (e.g., skin or mucosa) into blood.

acapnia marked decrease in blood carbon dioxide content.

acetabulum socket in the hip bone (os coxae or innominate bone) into which the head of the femur fits.

acetone bodies ketone bodies, acids formed during the first part of fat catabolism—viz., acetoacetic acid, betahydroxybutyric acid, and acetone.

Achilles tendon tendon inserted on calcaneus; so-called because of the Greek myth that Achilles' mother held him by the heels when she dipped him in the river Styx, thereby making him invulnerable except in this area.

acidosis condition in which there is an excessive proportion of acid in the blood.

acromion bony projection of the scapula; forms point of the shoulder.

adduct to move toward the midline; opposite of abduct.

adenohypophysis anterior pituitary gland.

adenoids glandlike; adenoids or pharyngeal tonsils are paired lymphoid structures in the nasopharynx.
adolescence period between puberty and adulthood.
adrenergic fibers axons whose terminals release norepinephrine and epinephrine.
adventitia, externa outer coat of a tube-shaped structure such as blood vessels.
aerobic requiring free oxygen; opposite of anaerobic.
afferent neuron transmitting impulses to the central nervous system.
albuminuria albumin in the urine.
aldosterone a hormone secreted by the adrenal cortex.
alkali reserve bicarbonate salts present in body fluids; mainly sodium bicarbonate.
alkalosis condition in which there is an excessive proportion of alkali in the blood; opposite of acidosis.
alveolus literally a small cavity; alveoli of lungs are microscopic saclike dilations of terminal bronchioles.
ameboid movement movement characteristic of amebae —i.e., by projections of protoplasm (pseudopodia) toward which the rest of the cell's protoplasm flows.
amenorrhea absence of the menses.
amino acid organic compound having an NH_3 and a COOH group in its molecule; has both acid and basic properties; amino acids are the structural units from which proteins are built.
amphiarthrosis slightly movable joint.
ampulla saclike dilation of a tube or duct.
anabolism synthesis by cells of complex compounds (e.g., protoplasm, hormones) from simpler compounds (amino acids, simple sugars, fats, minerals); opposite of catabolism, the other phase of metabolism.
anaerobic not requiring free oxygen; opposite of aerobic.
anastomosis connection between vessels—e.g., the circle of Willis is an anastomosis of certain cerebral arteries.
anemia deficient number of red blood cells or deficient hemoglobin.
anesthesia loss of sensation.
aneurysm blood-filled saclike dilation of the wall of an artery.
angina any disease characterized by spasmodic suffocative attacks—e.g., angina pectoris, paroxysmal thoracic pain with feeling of suffocation.
Angstrom unit 0.1 mμ or $^1/_{10}$ millionth of a meter or about $^1/_{250}$ millionth of an inch.
anorexia loss of appetite.
anoxemia deficient blood oxygen content.
anoxia deficient oxygen supply to tissues.
antagonistic muscles those having opposite action—e.g., muscles that flex the upper arm are antagonistic to muscles that extend it.
anterior front or ventral; opposite of posterior or dorsal.
antibody, immune body substance produced by the body that destroys or inactivates a specific substance (antigen) that has entered the body—e.g., diphtheria antitoxin is the antibody against diphtheria toxin.
antigen substance that, when introduced into the body, causes formation of antibodies against it.
antrum cavity—e.g., the antrum of Highmore, the space in each maxillary bone, or the maxillary sinus.
anus distal end or outlet of the rectum.
apex pointed end of a conical structure.
aphasia loss of a language faculty such as the ability to use words or to understand them.
apnea temporary cessation of breathing.
aponeurosis flat sheet of white fibrous tissue that serves as a muscle attachment.
aqueduct tube for conduction of liquid—e.g., the cerebral aqueduct conducts cerebrospinal fluid from the third to the fourth ventricle.
arachnoid delicate, weblike middle membrane of the meninges.
areola small space; the pigmented ring around the nipple.
arteriole small branch of an artery.
artery vessel carrying blood away from the heart.
arthrosis joint or articulation.
articular referring to a joint.
articulation joint.
arytenoid ladle-shaped; two small cartilages of the larynx.
ascites accumulation of serous fluid in the abdominal cavity.
asphyxia loss of consciousness from deficient oxygen supply.
aspirate to remove by suction.
asthenia bodily weakness.
astrocytes star-shaped neuroglia, connective tissue cells in brain and cord.
ataxia loss of power of muscle coordination.
atherosclerosis arteriosclerosis or hardening of artery walls characterized by lipid deposits in tunica intima.
atrium chamber or cavity—e.g., atrium of each side of the heart.
atrophy wasting away of tissue; decrease in size of a part.
auricle part of the ear attached to the side of the head; earlike appendage of each atrium of heart.
autonomic self-governing, independent.
axilla armpit.
axon nerve cell process that transmits impulses away from the cell body.

baroreceptor receptor stimulated by change in pressure.
Bartholin seventeenth century Danish anatomist.
basophil white blood cell that stains readily with basic dyes.
biceps two headed.
bilirubin red pigment in the bile.

biliverdin green pigment in the bile.
BNA (Basle Nomina Anatomica) anatomic terminology accepted at Basle by the Anatomical Society in 1895.
Bowman nineteenth century English physician.
brachial pertaining to the arm.
bronchiectasis dilation of the bronchi.
bronchiole small branch of a bronchus.
bronchus one of the two branches of the trachea.
buccal pertaining to the cheek.
buffer compound that combines with an acid or with a base to form a weaker acid or base, thereby lessening the change in hydrogen ion concentration that would occur without the buffer.
bursa fluid-containing sac or pouch lined with synovial membrane.
buttock prominence over the gluteal muscles.

calcitonin a hormone secreted by the parathyroid glands.
calculus stone formed in various parts of the body; may consist of different substances.
calorie heat unit; a large Calorie is the amount of heat needed to raise the temperature of 1 kg of water 1° C.
calyx cup-shaped division of the renal pelvis.
canaliculus little canal.
capillary microscopic blood vessel; capillaries connect arterioles with venules; also, microscopic lymphatic vessels.
carbhemoglobin, carbaminohemoglobin compound formed by union of carbon dioxide with hemoglobin.
carbohydrate organic compounds containing carbon, hydrogen, and oxygen in certain specific proportions —e.g., sugars, starches, cellulose.
carboxyhemoglobin compound formed by union of carbon monoxide with hemoglobin.
carcinoma cancer, a malignant tumor.
caries decay of teeth or of bone.
carotid from Greek word meaning to plunge into deep sleep; carotid arteries of the neck so called because pressure on them may produce unconsciousness.
carpal pertaining to the wrist.
casein protein in milk.
cast mold—e.g., formed in renal tubules.
castration removal of testes or ovaries.
catabolism breakdown of food compounds or of protoplasm into simpler compounds; opposite of anabolism, the other phase of metabolism.
catalyst substance that alters the speed of a chemical reaction.
cataract opacity of the lens of the eye.
catecholamines norepinephrine and epinephrine.
caudal pertaining to the tail of an animal; opposite of cephalic.
cecum blind pouch; the pouch at the proximal end of the large intestine.
celiac pertaining to the abdomen.
cellulose polysaccharide, the main plant carbohydrate.
centimeter $^1/_{100}$ of a meter, about $^2/_5$ of an inch.
centrioles two dots seen (with light microscope) in centrosphere; active during mitosis.
centromere structure that joins each pair of chromatids produced by chromosome duplication.
centrosphere, centrosome spherical area or body near center of cell.
cephalic pertaining to the head; opposite of caudal.
cerumen earwax.
cervix neck; any necklike structure.
chemoreceptor distal end of sensory dendrites especially adapted for chemical stimulation.
chiasm crossing; specifically, a crossing of the optic nerves.
cholecystectomy removal of the gallbladder.
cholesterol organic alcohol present in bile, blood, and various tissues.
cholinergic fibers axons whose terminals release acetylcholine.
cholinesterase enzyme; catalyzes breakdown of acetylcholine.
choroid, chorioid skinlike.
chromatids newly formed chromosomes.
chromatin deep-staining substance in the nucleus of cells; divides into chromosomes during mitosis.
chromosomes deep-staining, rod-shaped bodies in cell nucleus; composed of genes.
chyle milky fluid; the fat-containing lymph in the lymphatics of the intestine.
chyme partially digested food mixture leaving the stomach.
cilia hairlike projections of protoplasm.
circadian daily.
cochlea snail shell or structure of similar shape.
coenzyme nonprotein substance that activates an enzyme.
collagen principal organic constituent of connective tissue.
colloid solute particles with diameters of 1 to 100 $m\mu$.
colostrum first milk secreted after childbirth.
commissure bundle of nerve fibers passing from one side to the other of the brain or cord.
concha shell-shaped structure—e.g., bony projections into the nasal cavity.
condyle rounded projection at the end of a bone.
congenital present at birth.
contralateral on the opposite side.
coracoid like a raven's beak in form.
corium true skin or derma.
coronal of or like a crown.
coronary encircling; in the form of a crown.
corpus body.
corpuscles very small body or particle.
cortex outer part of an internal organ—e.g., of the cerebrum and of the kidneys.

cortisol (hydrocortisone, compound F) a glucocorticoid secreted by adrenal cortex.
costal pertaining to the ribs.
crenation, plasmolysis shriveling of a cell because of water withdrawal.
cretinism dwarfism caused by hypofunction of the thyroid gland.
cribriform sievelike.
cricoid ring shaped; a cartilage of this shape in the larynx.
cruciate cross shaped.
crystalloid solute particle less than 1 $m\mu$ in diameter.
cubital pertaining to the forearm.
cutaneous pertaining to the skin.
cyanosis bluish appearance of the skin from deficient oxygenation of blood.
cytokinesis dividing of cytoplasm to form two cells
cytology study of cells.
cytoplasm the protoplasm of a cell exclusive of the nucleus.

deamination chemical reaction by which the amino group NH_3 is split from an amino acid.
deciduous temporary; shedding at a certain stage of growth—e.g., deciduous teeth.
decussation crossing over like an X.
defecation elimination of waste matter from the intestines.
deferens carrying away.
deglutition swallowing.
deltoid triangular—e.g., deltoid muscle.
dendrite, dendron branching or treelike; a nerve cell process that transmits impulses toward the cell body.
dens tooth.
dentate having toothlike projections.
dentine main part of a tooth, under the enamel.
dentition teething; also, number, shape, and arrangement of the teeth.
dermatome area of skin supplied by sensory fibers of a single dorsal root.
dermis, corium true skin.
dextrose glucose, a monosaccharide, the principal blood sugar.
dialysis separation; the separation of crystalloids from colloids by the faster diffusion of the former through a membrane.
diapedesis passage of blood cells through intact blood vessel walls.
diaphragm membrane or partition that separates one thing from another; the muscular partition between the thorax and abdomen; the midriff.
diaphysis shaft of a long bone.
diarthrosis freely movable joint.
diastole relaxation of the heart interposed between its contractions; opposite of systole.
diencephalon "tween" brain; parts of the brain between the cerebral hemispheres and the mesencephalon or midbrain.
diffusion spreading—e.g., scattering of solute particles.
digestion conversion of food into assimilable compounds.
diplopia double vision; seeing one object as two.
disaccharide sugar formed by the union of two monosaccharides; contains twelve carbon atoms.
distal toward the end of a structure; opposite of proximal.
diverticulum outpocketing from a tubular organ such as the intestine.
dorsal, posterior pertaining to the back; opposite of ventral.
Douglas Scottish anatomist of the late seventeenth and early eighteenth centuries.
dropsy accumulation of serous fluid in a body cavity, in tissues; edema.
duct canal or passage.
dura mater literally strong or hard mother; outermost layer of the meninges.
dyspnea difficult or labored breathing.
dystrophy faulty nutrition.

ectopic displaced; not in the normal place—e.g., extrauterine pregnancy.
edema excessive fluid in tissues; dropsy.
effector responding organ—e.g., voluntary and involuntary muscle, the heart, and glands.
efferent carrying from, as neurons that transmit impulses from the central nervous system to the periphery; opposite of afferent.
electrocardiogram graphic record of heart's action potentials.
electroencephalogram graphic record of brain's action potentials.
electrolyte substance that ionizes in solution rendering the solution capable of conducting an electric current.
electron minute, negatively charged particle.
elimination expulsion of wastes from the body.
embolism obstruction of a blood vessel by foreign matter carried in the bloodstream.
embryo animal in early stages of intrauterine development; the human fetus the first 3 months after conception.
emesis vomiting.
emphysema dilation of pulmonary alveoli.
empyema pus in a cavity—e.g., in the chest cavity.
encephalon brain.
endocrine secreting into the blood or tissue fluid rather than into a duct; opposite of exocrine.
endoplasm cytoplasm located toward center of cell, as distinguished from ectoplasm located nearer periphery of cell.

endoplasmic reticulum network of tubules and vesicles in cytoplasm.
energy power or ability to do work; *kinetic* or *active,* caused by moving particles (e.g., mechanical energy, heat); *potential* or *stored,* caused by attraction between particles (e.g., chemical energy).
enteron intestine.
enzyme catalytic agent formed in living cells.
eosinophil, acidophil white blood cell readily stained by eosin.
epidermis "false" skin; outermost layer of the skin.
epinephrine secretion of the adrenal medulla.
epiphyses ends of a long bone.
erythrocyte red blood cell.
ethmoid sievelike.
eupnea normal respiration.
Eustachio Italian anatomist of the sixteenth century.
exocrine secreting into a duct; opposite of endocrine.
exophthalmos abnormal protrusion of the eyes.
extrinsic coming from the outside; opposite of intrinsic.

facilitation decrease in a neuron's resting potential to a point above its threshold of stimulation.
facilitated diffusion carrier-mediated diffusion.
Fallopius sixteenth century Italian anatomist.
fascia sheet of connective tissue.
fasciculus little bundle.
fetus unborn young, especially in the later stages; in human beings, from third month of intrauterine period until birth.
fiber threadlike structure.
fibrin insoluble protein in clotted blood.
fibrinogen soluble blood protein that is converted to insoluble fibrin during clotting.
fibroblasts connective tissue cells that synthesize interstitial fibers and gels.
fibrocytes old fibroblasts.
filtration passage of water and solutes through a membrane from hydrostatic pressure gradient.
fimbria fringe.
fissure groove.
flaccid soft, limp.
follicle small sac or gland.
fontanel "soft spots" of the infant's head; unossified areas in the infant's skull.
foramen small opening.
fossa cavity or hollow.
fovea small pit or depression.
fundus base of a hollow organ—e.g., the part farthest from its outlet.

ganglion cluster of nerve cell bodies outside the central nervous sytem.
gasserian named for Gasser, a sixteenth century Austrian surgeon.
gastric pertaining to the stomach.
gene part of the chromosome that transmits a given hereditary trait.
genitalia reproductive organs.
gestation pregnancy.
gland secreting structure.
glomerulus compact cluster—e.g., of capillaries in the kidneys.
glossal of the tongue.
glucagon hormone secreted by alpha cells of the islands of Langerhans.
glucocorticoids hormones that influence food metabolism; secreted by adrenal cortex.
glucokinase enzyme; catalyzes conversion of glucose to glucose-6-phosphate.
gluconeogenesis formation of glucose from protein or fat compounds.
glucose monosaccharide or simple sugar; the principal blood sugar.
gluteal of or near the buttocks.
glycerin, glycerol product of fat digestion.
glycogen "animal starch"; main polysaccharide stored in animal cells.
glycogenesis formation of glycogen from glucose or from other monosaccharides, fructose or galactose.
glycogenolysis hydrolysis of glycogen to glucose-6-phosphate or to glucose.
glyconeogenesis the formation of glycogen from protein or fat compounds.
gonad sex gland in which reproductive cells are formed.
graafian named for Graaf, a seventeenth century Dutch anatomist.
gradient a slope or difference between two levels—e.g., concentration gradient; a difference between the concentrations of two substances.
gustatory pertaining to taste.
gyrus convoluted ridge.

haversian named for Havers, English anatomist of the late seventeenth century.
helix spiral; coil.
hemiplegia paralysis of one side of the body.
hemoglobin iron-containing protein in red blood cell.
hemolysis destruction of red blood cells with escape of hemoglobin from them into surrounding medium.
hemopoiesis blood cell formation.
hemorrhage bleeding.
hepar liver.
heparin substance obtained from the liver that inhibits blood clotting.
heredity transmission of characteristics from a parent to a child.
hernia, "rupture" protrusion of a loop of an organ through an abnormal opening.
hilus, hilum depression where vessels enter an organ.
His German anatomist of the late nineteenth century.
histology science of minute structure of tissues.

homeostasis relative uniformity of the normal body's internal environment.
hormone substance secreted by an endocrine gland.
hyaline glasslike.
hydrocortisone (cortisol, compound F) a hormone secreted by the adrenal cortex.
hydrolysis literally "split by water"; chemical reaction in which a compound reacts with water.
hymen Greek for skin; mucous membrane that may partially or entirely occlude the vaginal outlet.
hyoid shaped like the letter U; bone of this shape at the base of the tongue.
hypercapnia abnormally high blood CO_2 concentration.
hyperemia increased blood in a part.
hyperkalemia higher than normal concentration of potassium in the blood.
hypernatremia higher than normal concentration of sodium in the blood.
hyperopia farsightedness.
hyperplasia increase in the size of a part from an increase in the number of its cells.
hyperpnea abnormally rapid breathing; panting.
hypertension abnormally high blood pressure.
hyperthermia fever; body temperature above 37° C.
hypertrophy increased size of a part from an increase in the size of its cells.
hypervolemia larger volume of blood than normal.
hypokalemia lower than normal concentration of potassium in the blood.
hyponatremia lower than normal concentration of sodium in the blood.
hypophysis Greek for undergrowth; hence the pituitary gland, which grows out from the undersurface of the brain.
hypothalamus part of the diencephalon; gray matter in the floor and walls of the third ventricle.
hypothermia subnormal body temperature; below 37° C.
hypovolemia subnormal volume of blood.
hypoxia oxygen deficiency.

inclusions any foreign or heterogenous substance contained in a cell or in any tissue or organ that was not introduced as a result of trauma.
incus anvil; the middle ear bone that is shaped like an anvil.
inferior lower; opposite of superior.
inguinal of the groin.
inhalation inspiration or breathing in; opposite of exhalation or expiration.
inhibition an increase in a neuron's resting potential above its usual level.
innominate not named, anonymous—e.g., ossa coxae (hip bones) formerly known as innominate bones.
insulin hormone secreted by beta cells of the islands of Langerhans in the pancreas.
intercellular between cells; interstitial.
interneurons (internuncial or *intercalated neurons)* conduct impulses from sensory to motor neurons; lie entirely within the central nervous system.
interstitial of or forming small spaces between things; intercellular.
intima innermost.
intrinsic not dependent upon externals; located within something; opposite of extrinsic.
involuntary not willed; opposite of voluntary.
involution return of an organ to its normal size after enlargement; also retrograde or degenerative change.
ion electrically charged atom or group of atoms.
ipsilateral on the same side; opposite of contralateral.
irritability excitability; ability to react to a stimulus.
ischemia local anemia; temporary lack of blood supply to an area.
isotonic of the same tension or pressure.

keratin protein compound present in the human body, chiefly in hair and nails.
ketones acids (acetoacetic, beta-hydroxybutyric, and acetone) produced during fat catabolism.
ketosis excess amount of ketone bodies in the blood.
kilogram 1,000 gm, approximately 2.2 pounds.
kinesthesia "muscle sense"—i.e., sense of position and movement of body parts.

labia lips.
lacrimal pertaining to tears.
lactation secretion of milk.
lactose milk sugar, a disaccharide.
lacuna space of cavity—e.g., lacunae in bone contain bone cells.
lamella thin layer, as of bone.
lateral of or toward the side; opposite of medial.
lemniscus, medial a flat band of sensory fibers extending up from the medulla, through the pons and midbrain to the thalamus.
leukocyte white blood cell.
ligament bond or band connecting two objects; in anatomy, a band of white fibrous tissue connecting bones.
limbic lobe or *system* cerebral cortex on the medial surface of the brain that forms a border around the corpus callosum; older name rhinencephalon.
lipid fats and fatlike compounds.
loin part of the back between the ribs and hip bones.
lumbar of or near the loins.
lumen passageway or space within a tubular structure.
luteum golden yellow.
lymph watery fluid in the lymphatic vessels.
lymphocyte one type of white blood cells.
lysis the destruction of cells and other antigens by a specific lysin antibody.
lysosomes membranous organelles containing various enzymes that can dissolve most cellular compounds;

hence called "digestive bags" or "suicide bags" of cells.

malleolus small hammer; projections at the distal ends of the tibia and fibula.
malleus hammer; the tiny middle ear bone that is shaped like a hammer.
Malpighii seventeenth century Italian anatomist.
maltose disaccharide or "double" sugar.
mamillary like a nipple.
mammary pertaining to the breast.
manometer instrument used for measuring the pressure of fluids.
manubrium handle; upper part of the sternum.
mastication chewing.
matrix ground substance in which cells are embedded.
meatus passageway.
medial of or toward the middle; opposite of lateral.
mediastinum middle section of the thorax—i.e., between the two lungs.
medulla Latin for marrow; hence, the inner portion of an organ in contrast to the outer portion or cortex.
meiosis nuclear division in which the number of chromosomes are reduced to half their original number before the cell divides in two.
membrane thin layer or sheet.
menstruation monthly discharge of blood from the uterus.
mesencephalon midbrain.
mesentery fold of peritoneum that attaches the intestine to the posterior abdominal wall.
mesial situated in the middle; median.
metabolism complex process by which food is utilized by a living organism.
metabolite any substance produced by metabolism.
metacarpus "after" the wrist; hence, the part of the hand between the wrist and fingers.
metatarsus "after" the instep; hence, the part of the foot between the tarsal bones and toes.
meter about 39.5 inch.
microglia one type connective tissue cell found in the brain and cord.
micron 0.001 ml; about 1/25,000 of an inch.
micturition urination, voiding.
milliequivalent a unit of chemical equivalence computed by the following formula:

$$\frac{\text{mg}}{\text{atomic weight}} \times \text{Valence} = \text{mEq}$$

$$\text{e.g., } \frac{3{,}266 \text{ mg Na}^+}{23 \text{ (At wt Na)}} \times 1 = 142 \text{ mEq Na}^+$$

millimeter 0.001 meter; about 1/25 of an inch.
mineralocorticoids hormones that influence mineral salt metabolism; secreted by adrenal cortex.
mitochondria threadlike structures.
mitosis indirect cell division involving complex changes in the nucleus.
mitral shaped like a miter.
molar concentration number of grams solute per liter of solution divided by the solute's molecular weight.
mole molecular weight of a compound in grams; also called gram molecular weight.
monosaccharide simple sugar.
Monro eighteenth century English surgeon.
morphology study of shape and structure of living organisms.
motoneurons (motor or *efferent neurons)* transmit nerve impulses away from the brain or spinal cord.
myelin lipoid substance found in the myelin sheath around some nerve fibers.
myocardium muscle of the heart.
myopia nearsightedness.

nares nostrils.
neurilemma nerve sheath.
neurohypophysis posterior pituitary gland.
neuron nerve cell, including its processes.
neurovesicles microscopic sacs in axon terminals; contain transmitter substance.
neutrophil white blood cell that stains readily with neutral dyes.
nuchal pertaining to the nape of the neck.
nucleotide component of DNA and RNA; a nucleotide is composed of a sugar (deoxyribose or ribose), a nitrogenous base (adenine, thymine, cytosine, or guanine), and a phosphate group.
nucleus spherical structure within a cell; a group of neuron cell bodies in the brain or cord.

occiput back of the head.
olecranon elbow.
olfactory pertaining to the sense of smell.
oligodendroglia a type of connective tissue cell found in the brain and cord.
ophthalmic pertaining to the eyes.
organelle cell organ; one of the specialized parts of a single-celled organism (protozoon), serving for the performance of some individual function.
os Latin for mouth and for bone.
osmosis movement of a fluid through a semipermeable membrane.
ossicle little bone.
oxidation loss of hydrogen or electrons from a compound or element.
oxyhemoglobin a compound formed by union of oxygen with hemoglobin.

palate roof of the mouth.
palpebrae eyelids.
papilla small nipple-shaped elevation.
paralysis loss of the power of motion or sensation, especially voluntary motion.
parenchyma the distinguishing, functional cells of an organ.

parietal of the walls of an organ or cavity.
parotid located near the ear.
parturition act of giving birth to an infant.
patella small, shallow pan; the kneecap.
Pavlov Russian physiologist of the late nineteenth and early twentieth centuries.
pectineal pertaining to the pubic bone.
pectoral pertaining to the chest or breast.
pelvis basin or funnel-shaped structure.
peripheral pertaining to an outside surface.
peroneus, peroneal of or near the fibula.
petrous rocklike.
Peyer Swiss anatomist of the late seventeenth and early eighteenth centuries.
pH hydrogen ion concentration; the negative logarithm of hydrogen ion concentration.
phagocytosis process by which a segment of cell membrane forms a small pocket around a bit of solid outside the cell, breaks off from rest of the membrane, and moves into the cell; briefly, ingestion and digestion of particles by a cell.
phalanges finger or toe bones.
phrenic pertaining to the diaphragm.
pia mater gentle mother; the vascular innermost covering (meninges) of the brain and cord.
pilomotor mover of a hair.
pineal shaped like a pine cone.
pinocytosis a process by which a segment of cell membrane forms a small pocket around a bit of fluid outside the cell, breaks off from rest of membrane, and moves into the cell.
piriformis pear shaped.
pisiform pea shaped.
plantar pertaining to the sole of the foot.
plasma liquid part of the blood.
plasmolysis shrinking of a cell from water loss by osmosis.
plexus network.
plica fold.
polymorphonuclear having many-shaped nuclei.
polyribosomes a group of ribosomes working together to synthesize proteins.
polysaccharide a carbohydrate containing a large number of saccharide groups ($C_6H_{10}O_5$)—e.g., starch, glycogen.
pons bridge.
popliteal behind the knee.
posterior following after; hence, located behind; opposite of anterior.
potential or *potential difference* difference in electrical charges—e.g., on outer and inner surfaces of cell membrane.
potential, action difference in electrical charges on inner and outer surfaces of the cell membrane during impulse conduction; outer surface negative to inner.
potential, resting difference in electrical charges on inner and outer surfaces of the cell membrane when it is not conducting impulses; outer surface positive to inner.
Poupart seventeenth century French anatomist.
presbyopia "oldsightedness"; farsightedness of old age.
pressoreceptors receptors stimulated by a change in pressure; baroreceptors.
pronate to turn palm downward.
proprioreceptors receptors located in the muscles, tendons, and joints.
protoplasm living substance.
proximal next or nearest; located nearest the center of the body or the point of attachment of a structure.
psoas pertaining to the loin, the part of the back between the ribs and hip bones.
psychosomatic pertaining to the influence of the mind (notably the emotions) on body functions.
pterygoid wing-shaped.
puberty age at which the reproductive organs become functional.

racemose like a cluster of grapes.
ramus branch.
Ranvier French pathologist of the late nineteenth and early twentieth centuries.
receptor peripheral ending of a sensory neuron.
reflex involuntary action.
refraction bending of a ray of light as it passes from a medium of one density to one of a different density.
refractory resisting stimulation.
renal pertaining to the kidney.
reticular netlike.
reticulum a network.
rhinencephalon see *limbic lobe.*
ribosomes organelles in cytoplasm of cells; synthesize proteins so nicknamed "protein factories."
rugae wrinkles or folds.

sagittal like an arrow; longitudinal.
salpinx tube; oviduct.
sartorius tailor; hence, the thigh muscle used to sit cross-legged like a tailor.
sciatic pertaining to the ischium.
sclera from Greek for hard.
scrotum bag.
sebum Latin for tallow; secretion of sebaceous glands.
sella turcica Turkish saddle; saddle-shaped depression in the sphenoid bone.
semen Latin for seed; male reproductive fluid.
semilunar half-moon shaped.
senescence old age.
serratus saw toothed.
serum any watery animal fluid; clear, yellowish liquid that separates from a clot of blood.
sesamoid shaped like a sesame seed.
sigmoid S shaped.

sinus cavity.
soleus pertaining to a sole; a muscle in the leg shaped like the sole of a shoe.
somatic of the body framework or walls, as distinguished from the viscera or internal organs.
sphenoid wedge shaped.
sphincter ring-shaped muscle.
splanchnic visceral.
squamous scalelike.
stapes stirrup; tiny stirrup-shaped bone in the middle ear.
Starling English physiologist of the late nineteenth and early twentieth centuries.
stereognosis awareness of the shape of an object by means of touch.
stimulus agent that causes a change in the activity of a structure.
stratum layer.
stress, physiological according to Selye, a condition in the body produced by all kinds of injurious factors that he calls "stressors" and manifested by a syndrome.
stressor any injurious factor that produces biological stress—e.g., emotional trauma, infections, severe exercise, etc.
striated marked with parallel lines.
stroma Greek for mattress or bed; hence, the framework or matrix of a structure.
sudoriferous secreting sweat.
sulcus furrow or groove.
superior higher; opposite of inferior.
supinate to turn the palm of the hand upward; opposite of pronate.
Sylvius seventeenth century anatomist.
symphysis Greek for a growing together.
synapse joining; point of contact between adjacent neurons.
synovia literally "with egg"; secretion of the synovial membrane resembles egg white.
synthesis putting together of parts to form a more complex whole.
systole contraction of the heart muscle.

talus ankle; one of the bones of the ankle.
tarsus instep.
tegmentum part of the brainstem that covers the posterior surface of the cerebral peduncles and the pons.
tendon band or cord of fibrous connective tissue that attaches a muscle to a bone or other structure.
thorax chest.
thrombosis formation of a clot in a blood vessel.
tibia Latin for shin bone.
tonus continued, partial contraction of muscle.
tract bundle of axons located within the central nervous system.
trauma injury.
trochlear pertaining to a pulley.
trophic having to do with nutrition.
tropic having to do with a turning or a change.
tunica covering.
turbinate shaped like a cone or like a scroll or spiral.
tympanum drum.

umbilicus navel.
utricle little sac.
uvula Latin for a little grape; a projection hanging from the soft palate.

vagina sheath.
vagus Latin for wandering.
valve structure that permits flow of a fluid in one direction only.
vas vessel or duct.
vastus wide, of great size.
Vater German anatomist of the late seventeenth and early eighteenth centuries.
vein vessel carrying blood to the heart.
ventral of or near the belly; in man, front or anterior; opposite of dorsal or posterior.
ventricle small cavity.
vermiform worm shaped.
villus hairlike projection.
viscera internal organs.
volar pertaining to the palm of the hand or the sole of the foot; palmar; plantar.
vomer ploughshare.

Willis seventeenth century English anatomist.

xiphoid sword shaped.

yoga meditation, an altered state of consciousness.

zygoma yoke.

INDEX

Index

H

O

P

Q